AF326521

The Unified SuperStandard Model
In Our Universe and The Megaverse

Quarks, Enhanced Standard Model, Faster-Than-Light Tachyons, Higgs Particles, Dark Matter, Gravitation, Cosmology, and Megaverse Features, Matter, Starships, and Life

Stephen Blaha Ph. D.
Blaha Research

MMXVII

Pingree-Hill Publishing

To Margaret

Some Other Books by Stephen Blaha

All the Megaverse! Starships Exploring the Endless Universes of the Cosmos using the Baryonic Force (Blaha Research, Auburn, NH, 2014)

SuperCivilizations: Civilizations as Superorganisms (McMann-Fisher Publishing, Auburn, NH, 2010)

PHYSICS IS LOGIC PAINTED ON THE VOID: Origin of Bare Masses and The Standard Model in Logic, U(4) Origin of the Generations, Normal and Dark Baryonic Forces, Dark Matter, Dark Energy, The Big Bang, Complex General Relativity, A Megaverse of Universe Particles (Blaha Research, Auburn, NH, 2015).

The Origin of Higgs ("God") Particles and the Higgs Mechanism: Physics is Logic III, Beyond Higgs – A Revamped Theory With a Local Arrow of Time, The Theory of Everything Enhanced, Why Inertial Frames are Special, Universes of the Mind (Blaha Research, Auburn, NH, 2015).

New Types of Dark Matter, Big Bang Equipartition, and A New U(4) Symmetry in the Theory of Everything: Equipartition Principle for Fermions, Matter is 83.33% Dark, Penetrating the Veil of the Big Bang, Explicit QFT Quark Confinement and Charmonium, Physics is Logic V (Blaha Research, Auburn, NH, 2015).

New Boson Quantum Field Theory, Dark Matter Dynamics, Dark Matter Fermion Layer Mixing, Genesis of Higgs Particles, New Layer Higgs Masses, Higgs Coupling Constants, Non-Abelian Higgs Gauge Fields, Physics is Logic VII (Blaha Research, Auburn, NH, 2015)

CQMechanics: A Unification of Quantum & Classical Mechanics, Quantum/Semi-Classical Entanglement, Quantum/Classical Path Integrals, Quantum/Classical Chaos (Blaha Research, Auburn, NH, 2016).

All the Universe! Faster Than Light Tachyon Quark Starships & Particle Accelerators with the LHC as a Prototype Starship Drive Scientific Edition (Pingree-Hill Publishing, Auburn, NH, 2011).

From Asynchronous Logic to The Standard Model to Superflight to the Stars; Volume 2: Superluminal CP and CPT, U(4) Complex General Relativity and The Standard Model, Complex Vierbein General Relativity, Kinetic Theory, Thermodynamics (Blaha Research, Auburn, NH, 2012)

New Boson Quantum Field Theory, Dark Matter Dynamics, Dark Matter Fermion Layer Mixing, Genesis of Higgs Particles, New Layer Higgs Masses, Higgs Coupling Constants, Non-Abelian Higgs Gauge Fields, Physics is Logic VII (Blaha Research, Auburn, NH, 2015)

The Origin of Fermions and Bosons,and Their Unification (Pingree-Hill Publishing, Auburn, NH, 2017).

Megaverse: The Universe of Universes (Pingree Hill Publishing, Auburn, NH, 2017).

SuperSymmetry and the Unified SuperStandard Model (Pingree Hill Publishing, Auburn, NH, 2017).

Available on Amazon.com, bn.com Amazon.co.uk and other international web sites as well as at better bookstores (through Ingram Distributors).

CONTENTS

FIGURES and TABLES

INTRODUCTION

This book clarifies, extends and revises parts of The Unified SuperStandard Model and its extension from our universe to the Megaverse of universes. The naturalness and simplicity of the derivation from first principles (primarily Relativity) leads us to suggest that the derivation is correct. More importantly, the directness of the derivation to a unified theory encompassing the known features of The Standard Model of elementary particles and Gravitation also enables us to exclude possibilities that might have been conjectured for elementary particle theory and gravity based on their unnaturalness within the framework of the derivation.

The book begins by providing an explanation of the number of space-time dimensions in our universe (4) based on the number of qubit interactions and also based on the requirement that parts of physical processes must be able to run in parallel. Parallelism necessitates Asynchronous Logic which also requires a minimum dimension of 4 space-time dimensions.

From this start we show Complex Special and General Relativity lead to the known spectrum of fermions and vector boson interactions – The Standard Model – plus much more.

Assuming a lagrangian formalism we then show that additional groups emerge – the Generation group that gives fermion generations, and Layer groups that give four layers of fermions and vector bosons. We know only of one layer of fermions at present.

Having developed the backbone of a Quantum Field Theory we eliminate all the infinities that crop up in perturbation using the fuzzy, string-like, coordinates of Two-Tier Quantum Field Theory. We also eliminate issues with the particle interpretation of states by using our PseudoQuantum Field Theory. Having removed the diseases of Quantum Field Theory we turn to the Megaverse. First we describe evidence for something big beyond our universe. Then we create a Megaverse composed of universes and describe its size, age, contents and other attributes including its likely atomic nuclei, Chemistry, and life forms.

We show that a universe has surface tension analogous to drops of liquid. From this concept we suggest starship designs that could lead to an exit from our universe via sling shot trajectories around neutron stars as well as other approaches. These possibilities are not likely to be feasible for tens of thousands of years yet they merit consideration.

The book also shows that our Unified SuperStandard Model can be viewed as a distinct variant of SuperString theory due to the presence of strings within dressed free particles and the presence of SuperSymmetry. In brief Two-Tier theory dresses particles with strings unlike SuperString theories which make strings into particles.

The easy flow of the derivation is perhaps the best argument in its favor. In a sense our derivation yields a culmination by derivation. Other theories do not start on the bedrock of Relativity. As the derivation progresses one sees an example of a Principle of Maximal Serendipity – processes (such as derivations) tend to proceed with maximal simplicity to better-than-expected ends. The concept of this principle was first expressed by Horace Walpole (1754) who defined serendipity as "[people] making discoveries, by accidents and sagacity, which they were not in quest of."

In conclusion, here, and there, within this fundamental work there arise hints of a yet deeper level of Physics. The author believes that this work – although an attempt at a final theory – is not the end.

0.Unified SuperStandard Model Derivation

In this chapter we will outline the process deriving our Unified SuperStandard Model from fundamental axioms.[1] Our theory has the somewhat unique feature of being derivable from first principles. Most other fundamental theories are constructed in an ad hoc fashion, often based on symmetry considerations, with features being added rather than derived.

The goal of this book is to derive the Unified SuperStandard Model in the manner of Euclid with a clear connection between the steps of the derivation just as Euclid developed geometry from a progression of theorems.

0.1 Primitive Terms and Axioms

Primitive terms can be as simple as those of Euclid or they can be more complex. The level of simplicity depends on the nature of the theory and the Physical Laws that emerge from it. In the case at hand, a fundamental unified theory, the constructs that emerge in the construction of the theory are mathematically complex. Consequently, the choice of primitive terms and axioms may be expected to be mathematically complex as well unless one wishes to expand the primitive terms into an extensive term by term description in simpler, more basic primitives. We will not pursue that alternative here since the terms that we use are 'self-explanatory' to the Elementary Particle Physics theorist knowledgable about quantum field theory and particle symmetries.

0.1.1 Primitive Terms for the Unified SuperStandard Model

The primitive terms of the theory are:

> Qubits
> Speed of Light
> Space and Time Dimensions
> Space and Time Coordinates
> Covariance
> Asynchronous
> Parallel Processes
> Reference Frame
> Complex Lorentz Group

---

[1] Appendix C outlines the general process of deriving a unified theory of everything from a set of axioms. It is described in detail in Blaha (2017c) and (2017d) as well as in versions in earlier books.

General Coordinate Transformations
Particle Masses
Fermions
Bosons
Particle State
Particle Rest State
Particle Momenta
Spin
Parity
Canonical Quantization
Quantum Field Theory
Asymptotic Particle State
Internal Symmetries
Coupling Constant
Discrete Symmetries
Yang-Mills Local Gauge Theory
Functionals
Gravity
Universe
Surface Tension
Megaverse

In choosing these primitives, we understand that they each embody a significant theoretic description or body of knowledge. However their meaning is clear to the experienced theorist. We do not include names used in the mapping to reality (such as quark) in the list of primitives since the mapping to reality is a separate issue in our view.

0.1.2 Axioms for the Unified SuperStandard Model

The axioms that we will discuss below include the Decision Axioms of section 3.1.3. The physical axioms are

1. The dimensions of space-time are determined by the number of fundamental interactions.
2. The Asynchronicity Principle supporting parallel processes is required.
3. The Constancy of the speed of light in all reference frames.
4. Covariance for flat space-times of symmetry under the Complex Lorentz group of transformations.
5. Covariance for curved space-times under Complex General Coordinate transformations.
6. Physically acceptable reference frames have real-valued coordinates. These coordinates can be obtained by transformations from complex-value coordinate systems.

7. Fermion and Boson vacua can be defined that are valid in all coordinate systems.
8. The number of particles in an asymptotic state of any given type is invariant in all reference frames.
9. Free fundamental fermions must have a real-valued energy.
10. All fields must be canonically quantized.
11. All interactions have a local Yang-Mills gauge theory formulation.
12. Gravity may cause space-time to be curved.
13. The complete theory has a lagrangian formulation.
14. The vector bosons, and the interactions among them, are determined by terms in the Riemann-Christoffel Curvature Tensor.

0.2 The Derivation of the Unified SuperStandard Model

The derivation of the Unified SuperStandard Model has been a multi-year process undertaken by the author. Much of the derivation appears in Blaha (2015a), (2016f), (2017b), (2017c) and (2017d). Earlier work, upon which these books are based, is referenced in these books and listed in the References in this book.

In the following chapters we describe the derivations presented in detail in earlier books with some changes. The goal is to show a clear logical development of the Unified SuperStandard Model from first principles in a manner reminiscent of the derivation of Euclidean geometry. This derivation will be seen to be primarily based on a 'simple' concept – the origin of space-time in our universe. The derivation explains the construction process of the physical theory. The manner of the derivation embodies the mapping of the theory to physical reality. The question of the 'Unmoved Mover' (Appendix C) is necessarily beyond the scope of Physics.

0.3 The 'Time' Order of the Derivation

An examination of the derivation presented here will show that the derivation has an implicit ordering which is analogous to the ordering of Euclidean Geometry. It begins with the most 'primitive' aspects of the theory and progressively develops and adds features. Thus it differs completely from the dynamical progression of the universe from the Big Bang, and the dynamical evolution of physical processes in time. To some extent the ordering is: the derivation of the basic nature of the fundamental fermion and boson spectrum, and the Standard Model interactions, from Complex Lorentz group considerations, the derivation of the number of fermion generations, and the Generation group interaction, from the form of the free lagrangian fermion terms and their associated conservation laws, the derivation of the fermion layers, and the Layer groups interactions, from conservation laws implicit in the free lagrangian plus ElectroWeak terms, and the derivation of the boson layers from Complex Lorentz group considerations.

0.4 Emergence vs. Derivation

Emergence is an interesting approach that is being successfully used in Physics and Biology.The use of emergent concepts to 'derive' the features of complex phenomena from simpler constructs is conceptually different from derivation along Euclidean lines. However the procedure is quite similar. The most tangible difference is that emergence suggests that complex phenomena somehow mask the underlying simpler structure at their base. However, an examination of the derivation of the Unified SuperStandard Model presented here shows that the primitives of a derivation – terms and axioms – may be sufficiently masked so as to justify describing the derived model as quasi-emergent, as it is based primarily on the Complex Lorentz group and Complex General Relativity.

1. Space-Time Dimensions

SUMMARY

The number of space-time dimensions (4) is determined by the number of interactions and the need for asynchronous physical processes, parts of which can execute in parallel. The dimension of our universe is jointly set by the U(2) rotation group of qubits and Asynchronous Logic considerations. In the case of the Megaverse it is set to 192 by the number of internal symmetry interactions in the Unified SuperStandard Model.

1.1 Determination of the Number of Space-Time Dimensions

The *a priori* determination of the dimensions of a space-time is guesswork in the absence of a fundamental principle(s).

This chapter derives the dimensions of our 4-dimensional space-time using two principles that yield the same space-time dimension of four. The first principle (axiom)uses the most fundamental and general logic construct, qubits, to determine the dimension of space-time. We assume the axiom (principle) that the dimension of space-time[2] equals the number of fundamental interactions in a theoretically totally empty space-time devoid of particles and energy.[3] Particles, energy, and particle interactions appear at a later stage in the theoretical development. The qubit concept alone occupies the stage at the beginning of the theoretical development.

The second principle (axiom) specifies the requirement that space-time must allow physical processes to run in parallel following the principles of Asynchronous Logic.[4] This requirement is physically necessary. This principle is related to the qubit principle through quantum entangled parallel processes evolving at a distance.

Consequently it is not surprising that both principles lead to a space-time dimension of four for our space-time (and a spatial dimension of 192 for the Megaverse.[5])

1.2 The Determination of the Dimension from the Number of Interactions

We will assume that interactions have a dual role in fundamental physics: they determine the dynamics of particles, and they act to determine space-time dimensions.[6]

---

[2] This approach was used in Blaha (2017c) and (2017d) to determine the dimensions of the Megaverse.
[3] A flat empty space-time is allowed in General Relativity.
[4] This approach is developed in (2012a) and (2015a).
[5] Blaha (2017d).
[6] Blaha (2017c).

The motivation for the second role can be discerned from considering a 2-dimensional space, and introducing a simple 1/r potential such as:

$$V = g^2/(x^2 + y^2)^{-½}$$

where g is a coupling constant.

One can view V as the potential of a force. However the values of V suitably extended to the range $[-\infty, +\infty]$ can be viewed as a third dimension. In addition, the basic feature of General Relativity is the role of gravity in determining the curvature of the universe. One can simply say that the universe is curved. One could also say that the universe is 'curved' into a surface in an implicit higher dimensional space-time.

Perhaps the most compelling reason for associating dimensions and interactions is the simple observation that, in the absence of interactions, all physical reality exists at a mathematical point and there is no dynamical evolution. Only interactions cause extension in space-time.

These considerations lead us to propose the axiom:

The number of interactions in a fundamental set of interactions equals the dimension of the space-time.

In section 1.3 we will determine the number of dimensions of our universe from the most fundamental construct in Logic and Reality.

1.3 The Determination of the Dimensions of Our Universe from the Only Essential Interactions

If we consider the universe as a thing in itself without reference to its matter content or their interactions there is only one construct that presents itself as an absolute necessity for a physical theory: Logic. In its most fundamental and most general form Logic is based on *qubits*. A qubit is a unit of quantum information that represents a state of a quantum system. Qubits are fundamental to Quantum Logic, and, in a restricted form, furnish the truth values of true and false in 'classical' Logic.

Being a two-state system with the capability of having complex values, and with values being able to be superposed, qubits can be viewed as having an associated U(2) unitary group whose operators serve to 'rotate' qubit values. Thus the U(2) group can be viewed as providing the 'interaction' that transforms qubits from one value to another.

The U(2) qubit group has four generators. Taking the qubit as the fundamental construct of our theory, since all Physics is based on Logic and ultimately Quantum Logic, we identify the source of our 4-dimensional space-time in the four U(2) interaction generators. In Appendix C we show that the correct fundamental theory of Physics is a logical (mental) construct. With the choice made here we can base Physics on Quantum Logic and qubit interactions in particular.

In the next section we will show that this analysis is consistent with a second axiom based on the requirement that parts of physical processes must be able to execute in parallel. We thus achieve a unified framework for understanding the dimensionality of our space-time. From the dimension of space-time, and the minimal requirement that the dynamical evolution of physical systems needs a time coordinate, we are led to the Complex Lorentz group in flat space-time and Complex General Relativity in curved space-time.[7]

1.4 Asynchronous Logic and the Dimension of Space-Time

In this section we show that a consideration of the physical requirement (axiom) that Nature must allow physical processes to evolve dynamically in parallel leads to the principle of Asynchronous Logic that, in turn, necessitates a space-time dimendion of four.

Thus we have two approaches based in Logic that inexorably lead to 4-dimensional space-time.

1.4.1 Synchronization of Non-Local Physical Processes

In earlier books we discussed the central role of Logic and the need for synchronization of non-local physical processes. The need for synchronized non-local physical processes requires the introduction of a new principle: the Principle of Asynchronicity.[8] When processes take place in parallel whether it is Quantum Mechanical entangled processes at small/large distances, or in high order Feynman diagrams (or their old fashioned time ordered perturbation theory predecessor) the synchronicity of a process is a physical requirement. It is implicitly resolved by physical laws which prevent asynchronicities (situations when parallel processes get "out of sync" resulting in the failure of an entire physical process to complete properly.) The Principle of Asynchronicity is described in the following pages. Asynchronicity can be briefly described as:

> In computation asynchronicity issues can arise. For example parallel computations or computer processes on a chip or set of chips have to be carefully managed for a parallel computer process to complete properly. In the case of computer chip design (VLSI chips and so on) techniques have been developed for the design of chips based on multi-valued logic. One conceptual approach uses 4-valued logic to define clockless computer logic circuits. The 4-valued logic developed by Fant (2005) has the four logic values TRUE, FALSE, NULL, and INTERMEDIATE. It is an extension of Boolean Logic that can accommodate time asynchronicities in asynchronous computer circuits. It enables circuits to avoid the use of system clocks to implement synchronization.[9] Thus the synchronization is explicit in 4-valued logic and

---

[7] We use similar arguments in the Megaverse (the universe of universes) discussion in Blaha (2017d) to determine the Megaverse dimension is 192.

[8] Much of this chapter is covered in Blaha (2011c) and printed in smaller type. Some might argue that it should be called the principle of synchronicity since the goal is synchronization of the parts of an evolving process. We chose to follow the terminology in the field of Asynchronous Logic as exemplified by Fant (2005) – a classic in that field.

[9] Remarkably Bjorken (1965) pp 220-226 presents an analogy of Feynman diagrams with electrical circuits where momenta map to currents, coordinates to voltages, Feynman parameters to resistance, and free particle equations of

non-logical constructs are not needed.[10] Concurrent transitions are coordinated solely by logical relationships with no need for any time constraints or relationships.

Now, realizing that The Standard Model, and physical theories that are ultimately derived from it such as Quantum Mechanics, potentially contain asynchronicities, we suggest that a Principle of Asynchronicity must be embodied in the fundamental theory of Physics.

1.4.2 Four-Valued Asynchronous Logic

The basic defining features of asynchronous circuits and Asynchronous Logic are:

1. An *asynchronous circuit* is a circuit in which the component parts are autonomous and can act in parallel at various rates of time evolution. They are not controled by a clock mechanism but proceed or wait for signals indicating that they can proceed.
2. *Asynchronous logic* is the logic used in the design of asynchronous circuits. The logic embodies the asynchronicity, and so the circuits built using the logic do not use a clock to control the execution speed of the various parts of an asynchronous circuit. Consequently logic elements do not necessarily have a distinct true or false state at any given point in time. 2-valued Boolean Logic is not sufficient and so asynchronous logic is multi-valued. The logic embodies states that allow for "stop and go" states within an executing asynchronous circuit.

In Fant's asynchronous 4-valued logic the four possible truth values of a state are:

True – status is true and all data is current
False – status is false and all data is current
Intermediate – status is indefinite with some data current
NULL – status is indefinite with no data present – results in a suspension of processing of the circuit part in a NULL state until current data becomes present

"Data" is the information flowing through all or part of a circuit. Using these truth values the evolution in time of the parts of an asynchronous circuit are effectively synchronized by the logic without the use of a clock mechanism. (A clock mechanism effectively is a subsidiary time constraint or set of time constraints.) See Fant (2005) for further details.

An implicit aspect of asynchronous logic is the coordination of spatially separated parts of a circuit. Since spatial separations in a circuit can be mapped to time delays using the speed of data propagation between parts, spatial asynchronicites are subsumed under time asynchronicities. This is particularly true for computer chips which are kept small to minimize delays.

1.4.3 Principle of Asynchronicity

An obvious feature of elementary particle phenomena is the coordination of the parts of a physical process in time and space. Complex Feynman diagrams embody the coordination of the parts of interacting particles. Quantum entanglement phenomena embody the coordination

---

motion to Ohm's Law plus the equivalent of Kirchhoff's Laws. Thus Feynman diagrams and computer circuits are completely analogous.
[10] A two-valued asynchronous logic is also possible – just as the Dirac equation can be expressed as two 2-dimensional equarions. See Fant (2005) and Bjorken (1965).

of the parts of a physical phenomena separated by large distances and perhaps times. Examples of these types, which could be multiplied indefinitely, lead to a Principle of Asynchronicity.

Principle (Axiom): Nature requires asynchronicity. Aasynchronicity is coordinated by 4-valued physico-logical structures for matter.

Elaboration: Elementary particle physical phenomena must support extended coordinated physical phenomena in space and time. The fundamental laws of particle physics must be such as to permit coordinated physical phenomena with coordination between the parts of a physical phenomenon at small/large distances and small/large time intervals. The coordination must be embodied within physical laws. We make it an axiom of our theory.

This principle will be applied below to justify Dirac-like equations for particle dynamics.

Coordination is an obvious feature of physical phenomena. This principle goes beyond that by asserting that extended coordinated physical phenomena must exist. If particles exist, then their antiparticles must also exist to provide asynchronous behavior in interaction regions. If only particles existed then all interactions would proceed forward in time and the state of the interaction at any point in time would be known. With the addition of antiparticles, asynchronicity issues are introduced and at various 'time slices' (if one thinks in terms of old-fashioned time ordered perturbation theory) of the progress of an interaction, the state can be ambiguous since antiparticles are negative energy particles moving backward in time.

Asynchronicities are common in the many subcircuits of a computer chip. Asynchronicities are also common in the many interaction subregions of a set of particles in interaction. Page 7 of Fant (2005) has a diagram of a circuit with a set of subcircuits with five time slices of the interacting subcircuits showing five states of the "'data' wavefront" at five points in time. This diagram is similar to the time-sliced diagram of an interacting system of particles in "old fashioned" time-ordered perturbation theory. Page 29 of Blaha (2005b) displays a similar diagram (Fig. 5.1.4) in a description of a Standard Model Quantum Langauge Grammar – a language representation of particle physics. Blaha's diagram[11] is remarkably similar to Fant's diagram in overall features as one might expect since both address time asynchronicity.

The asynchronicity that appears in perturbation theory diagrams is intimately related to the appearance of antiparticles in diagrams. As noted earlier antiparticles are interpretable as negative energy particles traveling backwards in time. The time orderings, which are implicit in the Feynman diagram approach and explicit in old fashioned perturbation theory, evidence time asynchronicity and the effects of the dynamics. They coordinate asynchronicities such that correct results follow from perturbative calculations.

---

[11] Created without knowledge of Fant's work.

1.4.4 Why add Space to Logic?

Space is necessarily a part of physical theory because propositions often depend on a spatial location. Thus we must add spatial (and time) dimensions to our specification of the physical Reality as well as Theory. Later we will connect the spin ½ matrix formulation of Operator Logic[12] augmented with space-time dimensions to the Dirac equation for spin ½ particles. Thus we map Operator Logic spinors and the physical spinors of physical Reality. Fermion particles (spin ½ particles) in physical Reality emerge from a map from Operator Logic spinors.

Now we address the issue of the number of space-time dimensions. Clearly if spinors exist in Reality, as we know they do, then they must be "spinning" in spatial dimensions. The number of components of a spinor is related to the total number of time and space dimensions.[13] For the case of an even number of dimensions d a spinor has $2^{d/2}$ components. For the case of an odd number of dimensions d a spinor has $2^{(d-1)/2}$ components. Based on these formulas we find the results in the following Table 1.1.

The case of d = 1 is immediately ruled out because Operator Logic supports, at minimum, 2 component spinors or 4 component spinors. The case d = 2 is also ruled out because spinor particles in a one-dimensional space reduce to scalar particles, and Reality has true spinors. The case d = 3 is ruled out because in two spatial dimensions there is no difference between left-handedness and right-handedness. *Thus the minimal number of spatial dimensions that yield true physical spinors and support "handedness" is three spatial dimensions.* This case meets Leibniz's Decision Axiom (Appendix C). The simplest features associated with space are spin (represented by spinors) and handedness. They yield a rich spectrum of particle types and interaction types (maximal complexity).

Total Number of Space-Time Dimensions d	Number of Spinor Components
1	1
2	2
3	2
4	**4**

Table 1.1. The number of spinor components for various numbers of space-time dimensions.

---

[12] See Blaha (2010a).
[13] Weinberg (1995) p. 216.

Thus we have a rationale for the extension of Operator Logic to include one time and three spatial dimensions.

The Principle of Asynchronous Logic and the parallel nature of Physical processes in general imply our space-time dimension is 4.

1.4.5 Truth is Generally Local – Space-Time Dependent

The extension of Operator Logic to include space-time is further buttressed by the dependence of the truth of statements on location and time in general. For example: "Today it rained in Concord, New Hampshire." is a space and time dependent statement.

Thus we find that statements are *local* in general – they depend on the time and spatial location.

The locality that we find in Logic naturally leads to locality in physical theories – particularly the locality of the Yang-Mills rotations in quantum field theories such as The Standard Model where the values of quantum numbers can vary from space-time point to space-time point but in such a way that their variation is compensated by local rotations of quantum fields. The locality of logical statements thus supports the connection of Logic to fundamental Physics.

1.4.6 Matrix Representation of Asynchronous Logic

The four possible logic states of Asynchronous Logic can be mapped to a matrix representation with four component columns and 4×4 matrices that transform between logic values.[14] The basic four pure logic states can be labeled using a notation that connects with physics:

$$u(+\tfrac{1}{2}) = \begin{bmatrix} 1 \\ 0 \\ 0 \\ 0 \end{bmatrix} \qquad u(-\tfrac{1}{2}) = \begin{bmatrix} 0 \\ 1 \\ 0 \\ 0 \end{bmatrix}$$

and

$$v(-\tfrac{1}{2}) = \begin{bmatrix} 0 \\ 0 \\ 1 \\ 0 \end{bmatrix} \qquad v(+\tfrac{1}{2}) = \begin{bmatrix} 0 \\ 0 \\ 0 \\ 1 \end{bmatrix} \tag{9.2}$$

---

[14] See Blaha (2010a) for more details.

The arguments of u and v will become physically values of particle spin: $+\frac{1}{2}$ represents an "up" spin state and $-\frac{1}{2}$ represents a "down" spin state. Linear combinations of these four states obtained using linear combinations of the sixteen 4×4 matrices with complex coefficients form *qubits*.[15] Any bit or qubit transformation can be constructed from a sum of the sixteen Dirac matrices[16] multiplied by complex coefficients. The states listed above are constants. They will become variable through the use of Lorentz transformations considered later.

A set of sixteen independent 4×4 matrices can be constructed from the four basic Dirac matrices, usually denoted γ^μ, by multiplications and summations. They can be found in most books on quantum field theory.

Applying these basic facts about the 4×4 matrix representation of Asynchronous Logic to particles we can develop the Dirac equation formulation of spin $\frac{1}{2}$ particles after introducing space-time coordinates.

We can view fermion particles as parametrized by coordinates and interacting via interactions (forces) that primarily result from the need to impose symmetry requirements on coordinate transformations that lead to parallel internal symmetries.[17]

1.5 Matter is Insubstantial Neglecting Particle Interactions

Philosophers and physicists have debated the nature of matter for milleniums. Differing opinions have been the norm. Perhaps one of the most interesting expressions of opinion was that of Dr. Johnson, the 19th century author of the Dictionary. Upon hearing of Bishop Berkeley's philosophic view that matter was insubstantial and "not real", Dr. Johnson proceeded to kick a rock while exclaiming "I refute it thus" according to his biographer Boswell. While Dr. Johnson's riposte cannot be denied in its succinctness, our theoretical development of fermion particles, and the further development of our construction, later, leads to a refinement of the Berkeley-Johnson conflict as well as that of many other philosophers and physicists.

For we propose that matter is truly insubstantial with Bishop Berkeley, and yet gains substantiality through interactions (forces) without which all matter would be interpenetrable and could reside at a single point.[18] Thus Dr. Johnson does not truly refute Bishop Berkeley but does show that forces exist amongst matter that gives it "substantiality" (and dimension.)

---

[15] S. Weisner, "Conjugate coding". Association for Computing Machinery, Special Interest Group in Algorithms and Computation Theory **15**, 78–88 (1983).

[16] See Bjorken (1964).

[17] We note that Asynchronous Logic can also be formulated with two dimensional vectors and 2×2 matrices just as the Dirac matrix formalism for spin $\frac{1}{2}$ particles can be expressed in a 2×2 matrix formalism.

[18] To some extent we see an approximation of this proposal in the approximate interpenetrability of normal matter and Dark Matter which would be exact if there were not a very weak force between matter and Dark Matter as well as the gravitational interaction.

Since all matter might have been concentrated[19] at an 'essentially' mathematical point in the absence of interactions we can call that point the *void* and view it as the 'location' of the Big Bang that presumably existed at the Beginning. Coordinates and interactions may well have originated at that point as the result of quantum fluctuations.

---

[19] This comment is based on our version of quantum field thory in which all interactions (including gravity) go to zero at zero distance. See Blaha (2005a) for a detailed discussion of our divergence-free form of quantum field theories. With zero forces, all particles can concentrate at a single point.

2. Iotas, Four Fermion Species and Their Second Quantization

Summary

Under transformations of the Complex Lorentz group[20] from a state of rest, a spin ½ fermion can have one of four forms which we call *fermion species*.[21] These forms (species) are Dirac fermions and tachyon fermions with real-valued momenta; and complexon (Dirac-like) fermions and tachyon fermions with real energies and complex-valued spatial momenta (subject to the condition that the spin, real part of the spatial momentum, and the imaginary part of the spatial momentum are orthogonal to each other.) The energy of each boosted fundamental particle must be real-valued in the free field case since free fundamental particles do not decay.

The number of distinct species is not changed by the extension of the set of coordinate transformations to Complex General Coordinate Transformations.

We map the four fermion species to charged leptons, neutrinos, up-type quarks, and down-type quarks respectively.

This derivation of four types (species) of fermions *directly* from Complex Lorentz group boosts from a rest state to a state with real-valued energy (thus giving stable fundamental fermions in the absence of interactions) is the only derivation of four fermion species that does not make any assumptions about internal symmetries.

2.0 The Logic Building Block of Fermions – The Iota

If we consider all possible 'things' that might constitute a fundamental building block for a fundamental theory they are all, at best, *ad hoc* and raise questions of their necessity and whether they are composed of yet a more fundamental substructure.

There is only one choice of building block that avoids these issues – a logic unit or qubit. A qubit is a fundamental entity that is a complex form of bit. A bit (and thus a qubit) is known to have an energy or equivalently a mass, and has no constituents of a more primitive form.[22] We will call a unit of logic that will form the core of a particle an *iota*. It has a conceptual value. But, in itself, it has no physical form or material existence in the sense of lumps of matter unless features, such as coordinates, supporting interactions are introduced by

---

[20] Complex Lorentz group boosts are require to obtain four species. It is one of our axioms and a requirement of proofs in axiomatic quantum field theory. See Streater (2000).

[21] Most of the material in this chapter was presented in -Blaha (2012a), (2015a) and (2017b) as well as earlier books.

[22] Ab iota is a physical manifestation of a logical value. The relation of an iota to a logical value is analogous to the relation of a penciled point placed on paper to the concept of a point as a primitive in geometry.

construction. We view iotas as fermion field theory functionals. (See chapter 8.) Later in this chapter and following chapters we will introduce physical features that will cloak iotas with properties and interactions making them into fermions.

2.0.1 Mass of an Iota

Recent experiments have shown that a logical value (a qubit or iota).has an energy.associated with it. One bit of information has about 3×10^{-21} joules of energy[23] or a rest mass, m_0, or about 0.02 eV using $m_0 = E/c^2$. This result was confirmed by E. Lutz et al.[24] who showed that there is a minimum amount of heat produced per bit of erased data. This minimal heat is called the *Landauer*[25] *limit*. The equivalent mass we will call the *Landauer mass* and denote it as m_0. We will assume that a fundamental Landauer mass exists in our discussions although the precise value of the mass will not be used since we may expect all physical particle masses to be renormalized to different values when interactions are taken into account.

However, it is intriguing that the mass of the electron neutrino has been measured in a variety of experiments and found to be within an order of magnitude or so larger than our estimate of the Landauer mass (as we would expect since particles acquire a 'cloud of virtual particles' due to interactions.) This 'cloud' can be expected to increase its mass above the Landauer mass. Since neutrinos only have the weak interaction it is not surprising that the increase due to interactions should not be large. The Mainz Neutrino Mass Experiment, for example, estimates the electron neutrino mass to be less than 2 eV.

A number of astronomical studies have also generated estimates of neutrino masses. In July 2010 the 3-D MegaZ DR7 galaxy survey found a limit for the combined mass of the three neutrino varieties to be less than 0.28 eV.[26] A smaller upper bound for the sum of neutrino masses, 0.23 eV, was found in March 2013 by the Planck collaboration,[27] In February 2014 a new estimate of the sum was found to be 0.320 ± 0.081 eV due to discrepancies between the Planck's measurements of the Cosmic Microwave Background, and other predictions, combined with the assumption that neutrinos are the cause of weaker gravitational lensing than implied by massless neutrinos.[28]

Thus the experimentally measured values of neutrino masses are consistent with the iota Landauer mass estimate of 0.02 eV given above. We can thus assume that *a fermion*

---

[23] E. Muneyuki et al, *Nature Physics*, DOI: 10.1038/NPHYS1821.

[24] E. Lutz et al, Nature **483** (7388): 187–190,10.1038/nature10872, (2012).

[25] R. Landauer, "Irreversibility and heat generation in the computing process", IBM Journal of Research and Development **5** (3): 183–191, (1961).

[26] S. Thomas et al, "Upper Bound of 0.28 eV on Neutrino Masses from the Largest Photometric Redshift Survey", Physical Review Letters **105**: 031301 (2010).

[27] Planck Collaboration, arXiv:1303.5076 (2013).

[28] R. A. Battye et al, "Evidence for Massive Neutrinos from Cosmic Microwave Background and Lensing Observations", Phys. Rev. Lett. **112,** 051303 (2014).

particle consists of an iota with a certain mass,[29] that is renormalized, together with other features. These features will emerge later in the derivation of the complete theory.[30] We view reality as ultimately a representation (or painting) of logic values evolving through interactions in time and space.[31]

2.0.2 Iotas as Fermion Field Functionals

At this point iotas have an insubstantial appearance with only the attribute of mass. Later in chapter 8 we will suggest that they can be mathematically represented as fermion field functionals[32] and used to develop the internal symmetry structure of the fermion spectrum.

2.0.3 Iota Dirac-like Equations for Particles and Antiparticles

In developing our theory of fermions we will assume all fermions are quantum fields and conform to the rules of canonical quantum field theory. We begin by defining energy and momentum as Fourier transform variables for functions of coordinates. For any field f(x) where x is a real or complex 4-vector we define its Fourier transform using an inner product of the coordinates with a 4-vector p that we call the momentum 4-vector, which consists of an energy component and a spatial 3-momentum:

$$h(p) = \int d^4x \exp(ip{\cdot}x)\, g(x) \tag{2.1}$$

where $p{\cdot}x = g_{\mu\nu}p^{\mu}x^{\nu}$.

2.1 Matrix Representation of Complex Lorentz Group L_C Boosts

We begin with Complex Lorentz Group (L_C) boosts because they will be crucial in the determination of the equations of motion of various types of spin ½ particles. An L_C boost can be expressed in the form

$$\Lambda_C(\mathbf{v_c}) = \exp[i\omega\hat{\mathbf{w}}{\cdot}\mathbf{K}] \tag{2.2}$$

where

$$\omega = (\omega_r^2 - \omega_i^2 + 2i\omega_r\omega_i\,\hat{\mathbf{u}}_r{\cdot}\hat{\mathbf{u}}_i)^{\frac{1}{2}} \tag{2.3}$$

and

$$\hat{\mathbf{w}} = (\omega_r\hat{\mathbf{u}}_r + i\omega_i\hat{\mathbf{u}}_i)/\omega \tag{2.4}$$

---

[29] Leibniz first proposed the idea of logic 'particles' which he called monads. Our definition of a logic 'particle' does not include (or exclude) the presence of a spiritual part which was part of the definition of Leibniz's monads.

[30] A recent experiment claims to separate the spin part (which we identify as a logical value later) of a molecule from the rest of the molecule.

[31] Those who might suggest matter is substantial, and logic values are not, should remember that matter would be completely insubstantial if there were no forces in nature. Neutrinos which are close to insubstantial would be completely insubstantial if there were no weak interactions.

[32] Functionals are a mathematical primitive of our theory. They have been used extensively by Feynman and others in quantum theories.

Since $\hat{\mathbf{u}}_r\cdot\hat{\mathbf{u}}_r = 1 = \hat{\mathbf{u}}_i\cdot\hat{\mathbf{u}}_i$

$$\hat{\mathbf{w}}\cdot\hat{\mathbf{w}} = 1 \tag{2.5}$$

and the complex relative velocity is

$$\mathbf{v_c} = \hat{\mathbf{w}}\,\tanh(\omega) \tag{2.6}$$

We now analytically continue to complex ω and complex unit vectors $\hat{\mathbf{w}}$. The resulting complex generalization will be the matrix form of proper L_C boosts:

$$\Lambda_C(\mathbf{v_c}) = \exp[i\omega\hat{\mathbf{w}}\cdot\mathbf{K}] \equiv \Lambda_C(\omega,\,\hat{\mathbf{w}})$$

$$= \begin{bmatrix} \cosh(\omega) & -\sinh(\omega)\hat{w}_x & -\sinh(\omega)\hat{w}_y & -\sinh(\omega)\hat{w}_z \\[6pt] -\sinh(\omega)\hat{w}_x & 1+(\cosh(\omega)-1)\hat{w}_x^{\,2} & (\cosh(\omega)-1)\hat{w}_x\hat{w}_y & (\cosh(\omega)-1)\hat{w}_x\hat{w}_z \\[6pt] -\sinh(\omega)\hat{w}_y & (\cosh(\omega)-1)\hat{w}_x\hat{w}_y & 1+(\cosh(\omega)-1)\hat{w}_y^{\,2} & (\cosh(\omega)-1)\hat{w}_y\hat{w}_z \\[6pt] -\sinh(\omega)\hat{w}_z & (\cosh(\omega)-1)\hat{w}_x\hat{w}_z & (\cosh(\omega)-1)\hat{w}_y\hat{w}_z & 1+(\cosh(\omega)-1)\hat{w}_z^{\,2} \end{bmatrix}$$

$$\tag{2.7}$$

Since analytic continuations are unique, the above form for $\Lambda_C(\mathbf{v_c})$ is well-defined and unique. It spans the complete set of proper L_C boosts.

2.2 Left-handed and Right-handed Parts of L_C

We now describe the Left-handed and Right-handed parts[33] of L_C boosts.

2.2.1 Left-handed Part of L_C

If we let

$$\hat{\mathbf{u}}_i = \hat{\mathbf{u}}_r \equiv \hat{\mathbf{u}} \tag{2.8}$$

so that the vector $\hat{\mathbf{u}}_i$ is parallel to $\hat{\mathbf{u}}_r$, and

$$\omega_i = \pi/2 \tag{2.9}$$

then $\Lambda_C(\mathbf{v_c})$ becomes a Left-handed L_C boost:

---

[33] The designations Left-handed and Right-handed are chosen to reflect the Left-handed and Right-handed fermion fields that will be used to construct The Standard Model later. See Blaha (2007b) for more detail.

$$\Lambda_C(\mathbf{v_c}) = \Lambda_L(\omega_r\,,\,\mathbf{u}) \qquad (2.10)$$

2.2.2 Right-handed part of L_C

If we let

$$\hat{\mathbf{u}}_i = -\hat{\mathbf{u}}_r \equiv -\hat{\mathbf{u}} \qquad (2.11)$$

so that the vector $\hat{\mathbf{u}}_i$ is anti-parallel to $\hat{\mathbf{u}}_r$, and

$$\omega_i = -\pi/2 \qquad (2.12)$$

then $\Lambda_C(\mathbf{v_c})$ becomes a Right-handed L_C boost:

$$\Lambda_C(\mathbf{v_c}) = \Lambda_R(\omega_r\,,\,\mathbf{u}) \qquad (2.13)$$

as described in Blaha (2007b).

2.3 Difference between the Parts of L_C Reduced to Parallelism of $\hat{\mathbf{u}}_r$ and $\hat{\mathbf{u}}_i$

Since the Left-handed L_C part leads to the Standard Model's left-handed features, it seems that the parallel case $\hat{\mathbf{u}}_i = \hat{\mathbf{u}}_r \equiv \hat{\mathbf{u}}$ is more favored by Nature.[34] To some extent this concept of parallel vectors $\hat{\mathbf{u}}_i$ and $\hat{\mathbf{u}}_r$, which leads to the Left-handed L_C, is more intuitively satisfying then the anti-parallel case that leads to the Right-handed L_C part. However, a deeper reason for Nature's choice remains to be found.

2.4 Free Spin ½ Particles – Leptons & Quarks

In this section we begin by developing dynamical equations for spin ½ particles based on the L_C parts. These spin ½ particles are conventional Dirac particles (Majorana particles are also allowed but not discussed), spin ½ tachyons, and "color" versions of both types totalling four species. We will identify leptons and quarks with these fields.

2.4.1 Introduction

Tachyons are particles that move faster than the speed of light. As we saw in earlier books tachyons exist inside Black Holes, and within current theories – particularly SuperString theories. There are also experimental indications that neutrinos are tachyons.

Attempts to create canonical tachyon quantum field theories began in the 1960's. These attempts were made within the framework of the Lorentz group and, consequently, were limited

---

[34] It is possible that parity violation might disappear at ultra-high energies. Then we would view the parity symmetric theory as broken to the left-handed Standard Model currently established by experiment with right-handed parts at higher energy.

to spin 0 theories since there are no finite dimensional representations of the Lorentz group for negative m^2 except for the one-dimensional representation. None of these attempts, or attempts since then, succeeded in creating a canonically quantized spin 0 tachyon quantum field theory.[35]

In this section we will formulate a free spin ½ tachyon[36] Quantum Field Theory. We choose to develop a normal spin ½ theory first. Then we develop a free spin ½ tachyon theory because, as we will see, spin ½ tachyon particles (quarks and leptons) play an extraordinary role in the Standard Model.

We will develop our spin ½ tachyon theory from the "ground up" by applying a Left-Handed L_C boost to the Dirac equation, and its Dirac spinor wave function, for a particle at rest. This procedure will give a tachyon spinor wave function, and the momentum space tachyon equation equivalent of the Dirac equation. Then we will obtain the coordinate space tachyon Dirac equation, define a lagrangian, and proceed to create a canonical quantum field theory for spin ½ tachyons.

The need for dynamical equations arises when we clothe each iota with coordinates to "make" a fermion. Having coordinates leads to describing the motion of particles. Dynamical equations specify the motion of particles. If they have a finite number of terms they can be derived from lagrangians. A lagrangian formalism yields a Hamiltonian (the energy) and the momentum.

2.4.2 First Step - Deriving the Conventional Dirac Equation

In this section we will review a method of obtaining the equation of motion of a particle using a free Dirac equation that is obtained by a Lorentz boost of a spinor wave function[37] of a particle at rest.

In the case of a Lorentz transformation the 4×4 matrix form of a Lorentz transformation of Dirac matrices is

$$S^{-1}(\Lambda(v))\gamma^{\nu}S(\Lambda(v)) = \Lambda^{\nu}{}_{\mu}(v)\gamma^{\mu} \qquad (2.14)$$

where $S(\Lambda(v))$ is

$$S(\Lambda(v)) = \exp(-i\omega\sigma_{0i}v_i/(2|\mathbf{v}|)) = \exp(-\omega\gamma^0\boldsymbol{\gamma}\cdot\mathbf{v}/(2|\mathbf{v}|))$$
$$= \cosh(\omega/2)I + \sinh(\omega/2)\gamma^0\boldsymbol{\gamma}\cdot\mathbf{p}/|\mathbf{p}| \qquad (2.15)$$

with $\omega = \text{arctanh}(|\mathbf{v}|)$, $\cosh(\omega/2) = [(E+m)/(2m)]^{\frac{1}{2}}$ and $\sinh(\omega/2) = |\mathbf{p}|[2m(E+m)]^{-\frac{1}{2}}$. Also

$$S^{-1}(\Lambda(v)) = \gamma^0 S^{\dagger}(\Lambda(v))\gamma^0 = \exp(\omega\gamma^0\boldsymbol{\gamma}\cdot\mathbf{v}/(2|\mathbf{v}|))$$
$$= \cosh(\omega/2)I - \sinh(\omega/2)\gamma^0\boldsymbol{\gamma}\cdot\mathbf{p}/|\mathbf{p}| \qquad (2.16)$$

---

[35] Except Blaha (2006).

[36] It fiffers significantly from tachyon theories such as those of G. Feinberg and E. C. Sudarshan.

[37] The spinor wave function of a particle at rest is a 4-vector of the 4×4 matrix representation of 4-valued Asynchronous Logic.

In constructing fermion dynamical equations *we shall assume that they are linear in derivatives* (although a quadratic form is possible.) We will use the sixteen 4×4 matrices that span the set of transformations of the four values of Asynchronous Logic. Since by theorem[38] all 4×4 γ matrices are equivalent up to a unitary transformation we can rotate any constant matrix into a multiple of γ^0 without loss of generality.

We begin by defining a generic positive energy plane wave solution of the Dirac equation for a normal fermion particle at rest with rest mass m, *which we take to be the iota bare mass in the absence of interactions,*[39] as

$$\psi(x) = e^{-imt}w(0) \tag{2.17}$$

with w(0) a four component logic spinor column vector. *For a free particle at rest, the rest energy $m = m_0$, the iota mass.* The wave function satisfies the momentum space Dirac equation for a fermion at rest:

$$(m\gamma^0 - m)e^{-imt}w(0) = 0 \tag{2.18}$$

Subsequently we will use a similar procedure to construct the free tachyonic Dirac equation. If we now apply $S(\Lambda(v))$ we find

$$0 = S(\Lambda(v))(m\gamma^0 - m)e^{-imt}w(0) = [mS(\Lambda(v))\gamma^0 S^{-1}(\Lambda(v)) - m]S(\Lambda(v))w(0)$$

A straightforward evaluation shows

$$mS(\Lambda(v))\gamma^0 S^{-1}(\Lambda(v)) = g_{\mu\nu}p^\mu\gamma^\nu = \not{p} \tag{2.19}$$

where $p^0 = (p^2 + m^2)^{\frac{1}{2}}$, $\mathbf{p} = \gamma m\mathbf{v}$, and $p = |\mathbf{p}|$. In addition

$$S(\Lambda(v))w(0) = w(p) \tag{2.20}$$

is a positive energy Dirac spinor. Therefore the Dirac equation for a fermion in motion in momentum space has the form:

$$(\not{p} - m)e^{-ip\cdot x}w(p) = 0 \tag{2.21}$$

where the exponential factor, mt, is also boosted to p·x. Eq. 2.21 implies the well-known free, coordinate space Dirac equation:

---

[38] R. H. Good, Rev. Mod. Phys., **27**, 187 (1955).
[39] As state earlier the derivation proceeds in steps from the Complex Lorentz group to free fermions and thence to interacting fermions. **We use the bare iota mass throughout the free fermion discussions in this chapter. m = m$_0$.**

$$(i\gamma^{\mu}\partial/\partial x^{\mu} - m)\psi(x) = 0 \qquad (2.22)$$

2.4.3 Derivation of the Tachyon Dirac Equation

The Left-handed boost has the form:

$$\Lambda_L(\omega, \mathbf{u}) = \Lambda(\omega + i\pi/2, \mathbf{u}) = \exp[i\omega_L\hat{\mathbf{u}}\cdot\mathbf{K}] \qquad (2.23)$$

where $\omega_L = \omega + i\pi/2$ and

$$\cosh(\omega_L) = i\sinh(\omega) = -\gamma = i\gamma_s$$

$$(2.24)$$

$$\sinh(\omega_L) = i\cosh(\omega) = -\beta\gamma = i\beta\gamma_s$$

with, $\beta = v > 1$, $\gamma_s = (\beta^2 - 1)^{-\frac{1}{2}}$, and $\boldsymbol{\omega \geq 0}$. Thus

$$\sinh(\omega) = \gamma_s$$

$$(2.25)$$

$$\cosh(\omega) = \beta\gamma_s$$

The corresponding spinor transformation is:

$$S_L(\Lambda_L(\omega, \mathbf{u})) = \exp(-i\omega_L\sigma_{0i}v_i/(2|\mathbf{v}|)) = \exp(-\omega_L\gamma^0\boldsymbol{\gamma}\cdot\mathbf{v}/(2|\mathbf{v}|))$$
$$= \cosh(\omega_L/2)I + \sinh(\omega_L/2)\gamma^0\boldsymbol{\gamma}\cdot\mathbf{p}/|\mathbf{p}| \qquad (2.26)$$

The inverse transformation is

$$S_L^{-1}(\Lambda_L(\omega, \mathbf{u})) = \gamma^2\gamma^0 K^{-1}S_L^{\dagger}K\gamma^0\gamma^2 = \gamma^2\gamma^0 S_L^{T}\gamma^0\gamma^2 = \exp(\omega_L\gamma^0\boldsymbol{\gamma}\cdot\mathbf{v}/(2|\mathbf{v}|))$$
$$= \cosh(\omega_L/2)I - \sinh(\omega_L/2)\gamma^0\boldsymbol{\gamma}\cdot\mathbf{p}/|\mathbf{p}| \qquad (2.27)$$

where the superscript T denotes the transpose and K is the complex conjugation operator (that also appears in the time-reversal operator). Note that S_L is not unitary just as the equivalent spinor Lorentz transformation $S(\Lambda(v))$ is not unitary.

We can now apply a left-handed superluminal transformation to the generic positive energy plane wave solution of the Dirac equation for a particle of mass m at rest. The result is

$$0 = S_L(\Lambda_L(\omega, \mathbf{u}))(m\gamma^0 - m)e^{-imt}w(0)$$
$$= [mS_L\gamma^0 S_L^{-1} - m]e^{-imt}S_L w(0)$$

where $S_L = S_L(\Lambda_L(\omega, \mathbf{u}))$. After some algebra

$$mS_L\gamma^0 S_L^{-1} = m[\cosh(\omega_L)\gamma^0 - \sinh(\omega_L)\boldsymbol{\gamma}\cdot\mathbf{p}/|\mathbf{p}|]$$

$$= i\gamma^0 E - i\boldsymbol{\gamma}\cdot\mathbf{p} = i\not{p} \tag{2.28}$$

using the tachyon energy and momentum expressions

$$\mathbf{p} = m\mathbf{v}\gamma_s \qquad\qquad E = m\gamma_s \tag{2.29}$$

Also

$$S_L w(0) = w_T(p) \tag{2.30}$$

is a tachyon spinor. See Appendix 2-A (at the end of this section) for a discussion of tachyon spinors.

The momentum space tachyonic Dirac equation is

$$(i\not{p} - m)e^{ip\cdot x}w_T(p) = 0 \tag{2.31}$$

where p·x = Et − $\mathbf{p}\cdot\mathbf{x}$ after performing a corresponding left-handed superluminal coordinate transformation in the exponential factor. Thus a positive energy wave is transformed into a negative energy wave by the superluminal transformation.

If we apply $i\not{p}$ to we find the tachyon mass condition is satisfied

$$-E^2 + \mathbf{p}^2 = m^2 \tag{2.32}$$

Transforming back to coordinate space we obtain the *tachyon Dirac equation*:

$$(\gamma^\mu \partial/\partial x^\mu - m)\psi_T(x) = 0 \tag{2.33}$$

The "missing" factor of i in the first term of eq. 2.33 requires the lagrangian to be different from the conventional Dirac lagrangian in order for the lagrangian to be real. The simplest, physically acceptable, free spin ½ tachyon lagrangian density is:

$$\mathcal{L}_T = \psi_T^{\;S}(\gamma^\mu \partial/\partial x^\mu - m)\psi_T(x) \tag{2.34}$$

where

$$\psi_T^{\;S} = \psi_T^{\;\dagger} i\gamma^0\gamma^5 \tag{2.35}$$

The corresponding action is

$$I = \int d^4x\, \mathcal{L}_T \tag{2.36}$$

Appendix 3-B of Blaha (2007b) proves I is real. The Hamiltonian density is

$$\mathcal{H} = \pi_T\dot{\psi}_T - \mathcal{L} = i\psi_T^{\;\dagger}\gamma^5(\boldsymbol{\alpha}\cdot\nabla + \beta m)\psi_T = -i\psi_T^{\;\dagger}\gamma^5\dot{\psi}_T \tag{2.37}$$

using the tachyon Dirac equation to obtain the last equality. The reader will note that the tachyon hamiltonian is hermitean by explicit calculation up to an irrelevant total spatial divergence.

2.4.3.1 Probability Conservation Law

The tachyon Dirac equation implies a probability conservation law:

$$\partial\rho_5/\partial t = \nabla\cdot\mathbf{j}_5 \qquad (2.38)$$

where

$$\rho_5 = \psi_T^{\,\dagger}\gamma^5\psi_T \qquad\qquad \mathbf{j}_5 = \psi_T^{\,\dagger}\gamma^5\boldsymbol{\alpha}\psi_T \qquad (2.39)$$

We are thus led to define the conserved axial charge Q_5

$$Q_5 = \int d^3x\, \psi_T^{\,\dagger}\gamma^5\psi_T \qquad (2.40)$$

2.4.3.2 Energy-Momentum Tensor

The tachyon energy-momentum tensor is

$$\mathscr{T}_{T\mu\nu} = - g_{\mu\nu}\,\mathscr{L}_T + \partial\mathscr{L}_T/\partial(\partial\psi_T/\partial x_\mu)\, \partial\psi_T/\partial x^\nu \qquad (2.41)$$
$$= i\psi_T^{\,\dagger}\gamma^0\gamma^5\gamma_\mu\partial\psi_T/\partial x^\nu \qquad (2.42)$$

and thus the conserved energy and momentum are

$$P^0 = H = \int d^3x\, \mathscr{T}_T^{\,00} = i\int d^3x\,\psi_T^{\,\dagger}\gamma^5(\boldsymbol{\alpha}\cdot\nabla + \beta m)\psi_T \qquad (2.43)$$

and

$$P^i = \int d^3x\, \mathscr{T}_T^{\,0i} = - i\int d^3x\, \psi_T^{\,\dagger}\gamma^5\partial\psi_T/\partial x_i \qquad (2.44)$$

Both the energy and momentum differ significantly from the corresponding quantities for conventional Dirac fields.

2.4.4 Tachyon Canonical Quantization

Having defined a suitable tachyon lagrangian we can now proceed to its canonical quantization. The conjugate momentum can be calculated from the above lagrangian density:

$$\pi_{Ta} = \partial\mathscr{L}_T/\partial\dot{\psi}_{Ta} \equiv \partial\mathscr{L}_T/\partial(\partial\psi_{Ta}/\partial t) = -i(\psi_T^{\,\dagger}\gamma^5)_a \qquad (2.45)$$

The resulting non-zero, canonical anti-commutation relations are

$$\{\pi_{Ta}(x), \psi_{Tb}(x')\} = i\,\delta_{ab}\,\delta^3(x - x')$$

or

$$\{\psi_{T\,a}^{\dagger}(x), \psi_{Tb}(x')\} = -\,[\gamma^5]_{ab}\,\delta^3(x - x') \qquad (2.46)$$

At this point we might attempt to complete the canonical quantization procedure in the conventional manner by fourier expanding the quantum field and specifying anti-commutation relations for the fourier component amplitudes. However the incompleteness of the set of plane waves, which are limited by the restriction $|p| \geq m$, causes the anti-commutator of the fields not to yield a $\delta^3(x - x')$. Thus the conventional approach fails to yield the required anti-commutation relations.[40]

Other approaches: 1) decompose the tachyon field into left-handed and right-handed parts and then second quantize each part; and 2) second quantize in light-front coordinates ($x^{\pm} = (x^0 \pm x^3)/\sqrt{2}$). These approaches also both fail.[41]

The only approach that does succeed[42] *is to decompose the tachyon field into left-handed and right-handed parts and then second quantize in light-front coordinates. We follow that procedure in the following subsections.*

2.4.4.1 Separation into Left-Handed and Right-Handed Fields

We will use a transformed set of Dirac matrices to develop our left handed and right-handed tachyon formulations:

$$\gamma^0 = \begin{bmatrix} 0 & -I \\ -I & 0 \end{bmatrix} \qquad \gamma^i = \begin{bmatrix} 0 & \sigma_i \\ -\sigma_i & 0 \end{bmatrix} \qquad \gamma^5 = \begin{bmatrix} I & 0 \\ 0 & -I \end{bmatrix}$$

$$(2.47)$$

which are obtained from the usual Dirac matrices by applying the unitary transformation $U = 2^{\frac{1}{2}}(I + \gamma^5\gamma^0)$. *I is the 4x4 identity matrix in eq. 2.47.* The γ^5 chirality operator's eigenvalues define handedness: +1 corresponds to right-handed; and −1 corresponds to left-handed:

$$\gamma^5\psi_L = -\,\psi_L \qquad\qquad \gamma^5\psi_R = \psi_R \qquad (2.48)$$

Consequently, we can define left-handed and right-handed tachyon fields with the projection operators:

$$C^{\pm} = \tfrac{1}{2}(I \pm \gamma^5)$$

---

[40] See G. Feinberg, Phys. Rev. **159**, 1089 (1967) for example.
[41] See the first edition Blaha (2006) where these possibilities were considered and found to fail.
[42] Blaha (2006) discusses this case in detail.

$$C^+ + C^- = I \tag{2.49}$$
$$C^{\pm\,2} = C^\pm$$
$$C^+ C^- = 0$$

with the result

$$\psi_{TL} = C^- \psi_T \tag{2.50}$$
$$\psi_{TR} = C^+ \psi_T$$

We can calculate the commutation relations of the left-handed and right-handed tachyon fields from eq. 2.46 by pre-multiplying and post-multiplying by $\tfrac{1}{2}(1 - \gamma^5)$ and $\tfrac{1}{2}(1 + \gamma^5)$. The results are:

$$\{\psi_{TLa}^{\dagger}(x),\ \psi_{TLb}(x')\} = \tfrac{1}{2}(1 - \gamma^5)_{ab}\,\delta^3(x - x') \tag{2.51}$$

$$\{\psi_{TRa}^{\dagger}(x),\ \psi_{TRb}(x')\} = -\tfrac{1}{2}(1 + \gamma^5)_{ab}\,\delta^3(x - x') \tag{2.52}$$

$$\{\psi_{TLa}^{\dagger}(x),\ \psi_{TRb}(x')\} = \{\psi_{TRa}^{\dagger}(x),\ \psi_{TLb}(x')\} = 0 \tag{2.53}$$

The lagrangian density above decomposes into left-handed and right-handed parts:

$$\mathcal{L}_T = \psi_{TL}^{\dagger}\gamma^0 i\gamma^\mu \partial_\mu \psi_{TL} - \psi_{TR}^{\dagger}\gamma^0 i\gamma^\mu \partial_\mu \psi_{TR} - im[\psi_{TR}^{\dagger}\gamma^0 \psi_{TL} - \psi_{TL}^{\dagger}\gamma^0 \psi_{TR}] \tag{2.54}$$

2.4.4.2 Further Separation into + and – Light-Front Fields

There have been many studies of light-front (infinite momentum frame) physics in the past forty years.[43] Light-front coordinates *cannot* be obtained by a Lorentz transformation, or by a superluminal transformation, from a standard set of coordinate system variables even in a limiting sense. Instead they are a defined set of variables that have been used to develop quantum field theories that have been shown to be equivalent to quantum field theories based on conventional coordinates. In particular, light-front quantum field theories have been shown to yield fully Lorentz covariant S matrix elements that are the same as S matrix elements calculated in the conventional way.

Light-front variables can be defined by:

$$x^\pm = (x^0 \pm x^3)/\sqrt{2} \tag{2.55}$$
$$\partial/\partial x^\pm \equiv \partial^\mp \equiv (\partial/\partial x^0 \pm \partial/\partial x^3)/\sqrt{2}$$

---

[43] L. Susskind, Phys. Rev. **165**, 1535 (1968); K. Bardakci and M. B. Halpern Phys. Rev. **176**, 1686 (1968), S. Weinberg, Phys. Rev. **150**, 1313 (1966); J. Kogut and D. Soper, Phys. Rev. **D1**, 2901 (1970); J. D. Bjorken, J. Kogut, and D. Soper, Phys. Rev. **D3**, 1382 (1971); R. A. Neville and F. Rohrlich, Nuov. Cim. **A1**, 625 (1971); F. Rohrlich, Acta Phys Austr. Suppl. **8**, 277 (1971); S-J Chang, R. Root, and T-M Yan, Phys. Rev. **D7**, 1133 (1973); S-J Chang, and T-M Yan, Phys. Rev. **D7**, 1147 (1973); T-M Yan, Phys. Rev. **D7**, 1761 (1973); T-M Yan, Phys. Rev. **D7**, 1780 (1973); C. Thorn, Phys. Rev. **D19**, 639 (1979); and references therein.

with the "transverse" coordinate variables, x^1 and x^2, unchanged.

The inner product of two 4-vectors has the form

$$x{\cdot}y = x^+y^- + y^+x^- - x^1y^1 - x^2y^2 \qquad (2.56)$$

and the light-front definition of Dirac matrices is:

$$\gamma^\pm = (\gamma^0 \pm \gamma^3)/\sqrt{2} \qquad (2.57)$$

with transverse matrices γ^1 and γ^2 defined as usual. Note the useful identity:

$$\gamma^{\pm\,2} = 0$$

We define "+" and "–" tachyon fields with the projection operators:

$$R^\pm = \tfrac{1}{2}(I \pm \gamma^0\gamma^3) \qquad (2.58)$$

They are:

Left-handed, $\pm$ light-front fields: $\qquad \psi_{TL}{}^\pm = R^\pm C^-\psi_T$

$$(2.59)$$

Right-handed, $\pm$ light-front fields: $\qquad \psi_{TR}{}^\pm = R^\pm C^+\psi_T$

Now if we transform to light-front variables and fields as above we obtain the light-front free tachyon lagrangian:

$$
\begin{aligned}
\mathcal{L}_T = {}& 2^{\frac12}\psi_{TL}{}^{++}i\partial^-\psi_{TL}{}^+ + 2^{\frac12}\psi_{TL}{}^{-+}i\partial^+\psi_{TL}{}^- - \psi_{TL}{}^{++}\gamma^0 i\gamma^j\partial^j\psi_{TL}{}^- - \psi_{TL}{}^{-+}\gamma^0 i\gamma^j\partial^j\psi_{TL}{}^+ - \\
& 2^{\frac12}\psi_{TR}{}^{++}i\partial^-\psi_{TR}{}^+ - 2^{\frac12}\psi_{TR}{}^{-+}i\partial^+\psi_{TR}{}^- + \psi_{TR}{}^{++}\gamma^0 i\gamma^j\partial^j\psi_{TR}{}^- + \psi_{TR}{}^{-+}\gamma^0 i\gamma^j\partial^j\psi_{TR}{}^+ - \\
& im[\psi_{TR}{}^{++}\gamma^0\psi_{TL}{}^- - \psi_{TL}{}^{++}\gamma^0\psi_{TR}{}^- + \psi_{TR}{}^{-+}\gamma^0\psi_{TL}{}^+ - \psi_{TL}{}^{-+}\gamma^0\psi_{TR}{}^+]
\end{aligned}
\qquad (2.60)
$$

with implied sums over $j = 1,2$. In contrast to the light-front tachyon lagrangian we note the corresponding light-front "normal" Dirac fermion lagrangian is

$$
\begin{aligned}
\mathcal{L}_{Dirac} = {}& 2^{\frac12}\psi_L{}^{++}i\partial^-\psi_L{}^+ + 2^{\frac12}\psi_L{}^{-+}i\partial^+\psi_L{}^- - \psi_L{}^{++}\gamma^0 i\gamma^j\partial^j\psi_L{}^- - \psi_L{}^{-+}\gamma^0 i\gamma^j\partial^j\psi_L{}^+ - \\
& 2^{\frac12}\psi_R{}^{++}i\partial^-\psi_R{}^+ + 2^{\frac12}\psi_R{}^{-+}i\partial^+\psi_R{}^- - \psi_R{}^{++}\gamma^0 i\gamma^j\partial^j\psi_R{}^- - \psi_R{}^{-+}\gamma^0 i\gamma^j\partial^j\psi_R{}^+ - \\
& im[\psi_R{}^{++}\gamma^0\psi_L{}^- + \psi_L{}^{++}\gamma^0\psi_R{}^- + \psi_R{}^{-+}\gamma^0\psi_L{}^+ + \psi_L{}^{-+}\gamma^0\psi_R{}^+]
\end{aligned}
\qquad (2.61)
$$

The difference in signs between these lagrangians will turn out to be a crucial factor in the derivation of features of the Standard Model later.

Returning to the tachyon lagrangian eq. 2.60 we obtain equations of motion through the standard variational techniques:

$$2^{\frac{1}{2}}i\partial^-\psi_{TL}{}^+ - \gamma^0 i\gamma^j\partial^j\psi_{TL}{}^- + im\gamma^0\psi_{TR}{}^- = 0$$
$$2^{\frac{1}{2}}i\partial^-\psi_{TR}{}^+ - \gamma^0 i\gamma^j\partial^j\psi_{TR}{}^- + im\gamma^0\psi_{TL}{}^- = 0$$
$$2^{\frac{1}{2}}i\partial^+\psi_{TL}{}^- - \gamma^0 i\gamma^j\partial^j\psi_{TL}{}^+ + im\gamma^0\psi_{TR}{}^+ = 0$$
$$2^{\frac{1}{2}}i\partial^+\psi_{TR}{}^- - \gamma^0 i\gamma^j\partial^j\psi_{TR}{}^+ + im\gamma^0\psi_{TL}{}^+ = 0$$

(2.62)

Eqs. 2.62 show that $\psi_{TL}{}^-$ and $\psi_{TR}{}^-$ are dependent fields that are functions of $\psi_{TL}{}^+$ and $\psi_{TR}{}^+$ on the light-front where x^+ equals a constant. They can be expressed in an integral form as well. (The independent fields $\psi_{TL}{}^+$ and $\psi_{TR}{}^+$ play a fundamental role in tachyon theory and are used to define "in" and "out" tachyon states in perturbation theory.)

The conjugate momenta are

$$\pi_{TL}{}^+ = \partial\mathcal{L}/\partial(\partial^-\psi_{TL}{}^+) = 2^{\frac{1}{2}}i\psi_{TL}{}^{+\dagger}$$
$$\pi_{TL}{}^- = \partial\mathcal{L}/\partial(\partial^-\psi_{TL}{}^-) = 0$$

(2.63)

$$\pi_{TR}{}^+ = \partial\mathcal{L}/\partial(\partial^-\psi_{TR}{}^+) = -2^{\frac{1}{2}}i\psi_{TR}{}^{+\dagger}$$
$$\pi_{TR}{}^- = \partial\mathcal{L}/\partial(\partial^-\psi_{TR}{}^-) = 0$$

(2.64)

Quantization on surfaces of constant x^+ (light-front surfaces) has been shown to support satisfactory formulations of Quantum Electrodynamics and other quantum field theories. Thus x^+ plays the role of the "time" variable in light-front quantized theories. So we will define canonical equal x^+ anti-commutation relations for spin ½ tachyons.

The resulting canonical equal-light-front ($x^+ = y^+$) anti-commutation relations of the independent fields are:

$$\{\psi_{TL}{}^{+\dagger}{}_a(x),\, \psi_{TL}{}^+{}_b(y)\} = 2^{-1}[C^-R^+]_{ab}\,\delta(x^- - y^-)\delta^2(x - y)$$

(2.65)

$$\{\psi_{TR}{}^{+\dagger}{}_a(x),\, \psi_{TR}{}^+{}_b(y)\} = -2^{-1}[C^+R^-]_{ab}\,\delta(x^- - y^-)\delta^2(x - y)$$

(2.66)

$$\{\psi_{TL}{}^+{}_a{}^\dagger(x),\, \psi_{TR}{}^+{}_b(y)\} = \{\psi_{TR}{}^+{}_a{}^\dagger(x),\, \psi_{TL}{}^+{}_b(y)\} = 0$$

(2.67)

$$\{\psi_{TL}{}^+{}_a(x),\, \psi_{TR}{}^+{}_b(y)\} = \{\psi_{TR}{}^+{}_a{}^\dagger(x),\, \psi_{TL}{}^{+\dagger}{}_b(y)\} = 0$$

(2.68)

where the factors of 2^{-1} are the result of the $2^{\frac{1}{2}}$ factor in eqs. 2.63 and 2.64, and the factor of $2^{-\frac{1}{2}}$ in the definition of x^- above.

If we compare eqs. 2.65 and 2.66 with the corresponding anti-commutation relations of *conventional <u>Dirac</u>* quantum fields:

$$\{\psi_L{}^{+\dagger}{}_a(x), \psi_L{}^{+}{}_b(y)\} = 2^{-1}[C^-R^+]_{ab}\,\delta(x^- - y^-)\delta^2(x - y) \quad (2.69)$$

$$\{\psi_R{}^{+\dagger}{}_a(x), \psi_R{}^{+}{}_b(y)\} = 2^{-1}[C^+R^+]_{ab}\,\delta(x^- - y^-)\delta^2(x - y) \quad (2.70)$$

we see that the right-handed tachyon anti-commutation relation has a minus sign relative to the corresponding right-handed conventional anti-commutation relation. The right-handed tachyon anti-commutation relation with its minus sign will require compensating minus signs in its creation and annihilation Fourier component operators' anti-commutation relations.

The sign differences between the lagrangian terms in eqs. 2.63 and 2.64 ultimately lead to parity violating features in the Standard Model lagrangian and thus resolve the long-standing question:

Why parity violation? Answer: Nature preferentially chooses the Left-handed part of the complex Lorentz group.. This choice is not a consequence of Ockham's Razor. But it does conform to Leibniz's Minimax Principle – a minor differentiation based on parity results in "maximal" physical consequences.

2.4.4.3 Left-Handed Tachyons

The free, "+" light-front, left-handed tachyon wave function Fourier expansion is:

$$\psi_{TL}{}^{+}(x) = \sum_{\pm s}\int d^2pdp^+N_{TL}{}^{+}(p)\theta(p^+)[b_{TL}{}^{+}(p, s)u_{TL}{}^{+}(p, s)e^{-ip\cdot x} + d_{TL}{}^{+\dagger}(p, s)v_{TL}{}^{+}(p, s)e^{+ip\cdot x}] \quad (2.71)$$

and its hermitean conjugate is

$$\psi_{TL}{}^{+\dagger}(x) = \sum_{\pm s}\int d^2pdp^+N_{TL}{}^{+}(p)\theta(p^+)\,[b_{TL}{}^{+\dagger}(p, s)u_{TL}{}^{+\dagger}(p,s)e^{+ip\cdot x} + d_{TL}{}^{+}(p, s)v_{TL}{}^{+\dagger}(p, s)e^{-ip\cdot x}] \quad (2.72)$$

where † indicates hermitean conjugate, where

$$N_{TL}{}^{+}(p) = [2m|\mathbf{p}|/((2\pi)^3(p^+(p^+ - p^-) + p_\perp{}^2))]^{\frac{1}{2}} \quad (2.73)$$

where the anti-commutation relations of the Fourier coefficient operators are

$$
\begin{aligned}
\{b_{TL}{}^{+}(q,s), b_{TL}{}^{+\dagger}(p,s')\} &= \delta_{ss'}\delta^2(\mathbf{q} - \mathbf{p})\delta(q^+ - p^+) \\
\{d_{TL}{}^{+}(q,s), d_{TL}{}^{+\dagger}(p,s')\} &= \delta_{ss'}\delta^2(\mathbf{q} - \mathbf{p})\delta(q^+ - p^+) \\
\{b_{TL}{}^{+}(q,s), b_{TL}{}^{+}(p,s')\} &= \{d_{TL}{}^{+}(q,s), d_{TL}{}^{+}(p,s')\} = 0 \\
\{b_{TL}{}^{+\dagger}(q,s), b_{TL}{}^{+\dagger}(p,s')\} &= \{d_{TL}{}^{+\dagger}(q,s), d_{TL}{}^{+\dagger}(p,s')\} = 0 \\
\{b_{TL}{}^{+}(q,s), d_{TL}{}^{+\dagger}(p,s')\} &= \{d_{TL}{}^{+}(q,s), b_{TL}{}^{+\dagger}(p,s')\} = 0 \\
\{b_{TL}{}^{+\dagger}(q,s), d_{TL}{}^{+\dagger}(p,s')\} &= \{d_{TL}{}^{+}(q,s), b_{TL}{}^{+}(p,s')\} = 0
\end{aligned}
\quad (2.74)
$$

and where the spinors are

$$u_{TL}^{\ +}(p, s) = C^- R^+ S_L(\Lambda_L(\mathbf{p}))w^1(0)$$
$$u_{TL}^{\ +}(p, -s) = C^- R^+ S_L(\Lambda_L(\mathbf{p}))w^2(0)$$
$$v_{TL}^{\ +}(p, s) = C^- R^+ S_L(\Lambda_L(\mathbf{p}))w^3(0)$$
$$v_{TL}^{\ +}(p, -s) = C^- R^+ S_L(\Lambda_L(\mathbf{p}))w^4(0) \qquad (2.75)$$
$$u_{TL}^{\ +\dagger}(p, s) = w^{1T}(0)S_L^{\ \dagger}(\Lambda_L(\mathbf{p}))R^+C^-$$
$$u_{TL}^{\ +\dagger}(p, -s) = w^{2T}(0)S_L^{\ \dagger}(\Lambda_L(\mathbf{p}))R^+C^-$$
$$v_{TL}^{\ +\dagger}(p, s) = w^{3T}(0)S_L^{\ \dagger}(\Lambda_L(\mathbf{p}))R^+C^-$$
$$v_{TL}^{\ +\dagger}(p, -s) = w^{4T}(0)S_L^{\ \dagger}(\Lambda_L(\mathbf{p}))R^+C^-$$

where the superscript "T" indicates the transpose. (These spinors are described in Appendix 2-A.)

The canonical left-handed, light-front anti-commutation relation results in:

$$\{\psi_{TL}^{\ +}{}_a(x), \psi_{TL}^{\ +\dagger}{}_b(y)\} = \sum_{\pm s,s'} \int d^2pdp^+ \int d^2p'dp'^+ N_{TL}^{\ +}(p)N_{TL}^{\ +}(p')\theta(p^+)\theta(p'^+)\cdot$$

$$\cdot[\{b_{TL}^{\ +\dagger}(p',s'),b_{TL}^{\ +}(p,s)\}u_{TL}^{\ +}{}_a(p,s)u_{TL}^{\ +\dagger}{}_b(p',s')e^{+ip'\cdot y - ip\cdot x} +$$

$$+ \{d_{TL}^{\ +}(p',s'),d_{TL}^{\ +\dagger}(p,s)\}v_{TL}^{\ +}{}_a(p,s)v_{TL}^{\ +\dagger}{}_b(p',s')e^{-ip'\cdot y + ip\cdot x}]$$

$$= \sum_{\pm s} \int d^2pdp^+ N_{TL}^{\ +2}(p)\theta(p^+)[u_{TL}^{\ +}{}_a(p,s)u_{TL}^{\ +\dagger}{}_b(p,s)e^{+ip\cdot(y-x)} +$$

$$+ v_{TL}^{\ +}{}_a(p,s)v_{TL}^{\ +\dagger}{}_b(p,s)e^{-ip\cdot(y-x)}]$$

$$= -i\int d^2pdp^+ \theta(p^+)N_{TL}^{\ +2}(p)(2m|\mathbf{p}|)^{-1}\{[\ C^-R^+(i\not{p} - m)\gamma\cdot\mathbf{p}R^+C^-]_{ab}e^{+ip\cdot(y-x)} +$$

$$+ [C^-R^+(i\not{p} + m)\gamma\cdot\mathbf{p}R^+C^-]_{ab}e^{-ip\cdot(y-x)}\}$$

$$= -i\int d^2p_\perp \int_0^\infty dp^+ N_{TL}^{\ +2}(p)\{[C^-R^+(ip^+(p^+ - p^-) + ip_\perp^2 - mp_\perp\cdot\gamma_\perp)C^-]_{ab}e^{+ip^+(y^- - x^-) - ip_\perp\cdot(y_\perp - x_\perp)} -$$

$$- [C^-R^+(-ip^+(p^+ - p^-) - ip_\perp^2 - mp_\perp\cdot\gamma_\perp)C^-]_{ab}e^{-ip^+(y^- - x^-) + ip_\perp\cdot(y_\perp - x_\perp)}\}/(2m|\mathbf{p}|)$$

$$= \int d^2p_\perp \int_{-\infty}^\infty dp^+ N_{TL}^{\ +2}(p)[C^-R^+(p^+(p^+ - p^-) + p_\perp^2)]_{ab}\, e^{+ip^+(y^- - x^-) - ip_\perp\cdot(y_\perp - x_\perp)}/(2m|\mathbf{p}|)$$

upon letting $p^+ \rightarrow -p^+$ and $\mathbf{p}_\perp \rightarrow -\mathbf{p}_\perp$ in the second term after using $N_{TL}^{+2}(p)(p^+(p^+ - p^-) + p_\perp^2) = 1$. The result

$$= \tfrac{1}{2}\int d^2 p_\perp \int_{-\infty}^{\infty} dp^+ (2\pi)^{-3}[C^-R^+]_{ab}e^{+ip^+(y^- - x^-) - ip_\perp\cdot(y_\perp - x_\perp)}$$

$$= 2^{-1}[C^-R^+]_{ab}\delta(y^- - x^-)\delta^2(\mathbf{y} - \mathbf{x}) \tag{2.76}$$

Therefore we have left-handed, light-front quantized tachyons with canonical commutation relations and localized tachyons. As a result we have a canonical Tachyon Quantum Field Theory unlike previous efforts.

2.4.4.4 Right-Handed Tachyons

The case of right-handed tachyons is similar to the left-handed case with only two differences: a minus sign in the creation and annihilation operator anti-commutation relations, and the use of right-handed projection operators. The right-handed tachyon wave function light-front Fourier expansion is:

$$\psi_{TR}^{+}(x) = \sum_{\pm s}\int d^2 p\, dp^+ N_{TR}^{+}(p)\theta(p^+)[b_{TR}^{+}(p, s)u_{TR}^{+}(p, s)e^{-ip\cdot x} + d_{TR}^{++}(p, s)v_{TR}^{+}(p, s)e^{+ip\cdot x}]$$

$$\tag{2.77}$$

and its hermitean conjugate is

$$\psi_{TR}^{++}(x) = \sum_{\pm s}\int d^2 p\, dp^+ N_{TR}^{+}(p)\theta(p^+)\,[b_{TR}^{++}(p, s)u_{TR}^{++}(p, s)e^{+ip\cdot x} + d_{TR}^{+}(p, s)v_{TR}^{++}(p, s)e^{-ip\cdot x}]$$

$$\tag{2.78}$$

where $N_{TR}^{+}(p) = N_{TL}^{+}(p)$, where the anti-commutation relations of the Fourier coefficient operators are

$$\{b_{TR}^{+}(q,s),\, b_{TR}^{++}(p,s')\} = -\delta_{ss'}\delta^2(\mathbf{q} - \mathbf{p})\delta(q^+ - p^+) \tag{2.79}$$

$$\{d_{TR}^{+}(q,s),\, d_{TR}^{++}(p,s')\} = -\delta_{ss'}\delta^2(\mathbf{q} - \mathbf{p})\delta(q^+ - p^+)$$

$$\{b_{TR}^{+}(q,s),\, b_{TR}^{+}(p,s')\} = \{d_{TR}^{+}(q,s),\, d_{TR}^{+}(p,s')\} = 0$$

$$\{b_{TR}^{++}(q,s),\, b_{TR}^{++}(p,s')\} = \{d_{TR}^{++}(q,s),\, d_{TR}^{++}(p,s')\} = 0$$

$$\{b_{TR}^{+}(q,s),\, d_{TR}^{++}(p,s')\} = \{d_{TR}^{+}(q,s),\, b_{TR}^{++}(p,s')\} = 0$$

$$\{b_{TR}^{++}(q,s),\, d_{TR}^{++}(p,s')\} = \{d_{TR}^{+}(q,s),\, b_{TR}^{+}(p,s')\} = 0$$

and where the spinors are

$$u_{TR}^{\;\;+}(p,\, s) = C^+ R^+ u_T(p,s) \qquad\qquad (2.80)$$

$$v_{TR}^{\;\;+}(p,\, s) = C^+ R^+ v_T(p,s) \qquad\qquad (2.81)$$

by Appendix 2-A (eq. 2-A.7).

The right-handed anti-commutation relation with the minus sign follows in particular because of the minus signs found earlier.

2.4.5 Interpretation of Tachyon Creation and Annihilation Operators

To properly discuss the physical interpretation of tachyon creation and annihilation operators we must first determine the Hamiltonian and momentum operators in terms of creation and annihilation operators.

The energy-momentum tensor density is the symmetrized version of

$$\mathcal{T}^{\mu\nu} = \sum_i \partial\mathcal{L}/\partial(\partial\chi_i/\partial x_\mu)\; \partial\chi_i/\partial x_\nu - g^{\mu\nu}\mathcal{L} \qquad\qquad (2.82)$$

where the sum over i is over the fields. The light-front hamiltonian is

$$H \equiv P^- = T^{+-} = \int dx^- d^2x\, \mathcal{T}^{+-} \qquad\qquad (2.83)$$

and the "momenta" are

$$P^+ = T^{++} = \int dx^- d^2x\, \mathcal{T}^{++} \qquad\qquad (2.84)$$

$$P^i = T^{+i} = \int dx^- d^2x\, \mathcal{T}^{+i} \qquad\qquad (2.85)$$

for i = 1,2.

The light-front, left-handed and right-handed tachyon lagrangian $\mathcal{L}_T$ and its equations of motion imply

$$H = i2^{-\frac{1}{2}}\int dx^- d^2x\; [\psi_{TL}^{\;+\dagger}\partial^-\psi_{TL}^{\;+} - \partial^-\psi_{TL}^{\;+\dagger}\psi_{TL}^{\;+} + \psi_{TL}^{\;-\dagger}\partial^+\psi_{TL}^{\;-} - \partial^+\psi_{TL}^{\;-\dagger}\psi_{TL}^{\;-} -$$
$$- \psi_{TR}^{\;+\dagger}\partial^-\psi_{TR}^{\;+} + \partial^-\psi_{TR}^{\;+\dagger}\psi_{TR}^{\;+} - \psi_{TR}^{\;-\dagger}\partial^+\psi_{TR}^{\;-} + \partial^+\psi_{TR}^{\;-\dagger}\psi_{TR}^{\;-} + \text{mass terms}]$$

$$(2.86)$$

After substituting for the various fields we find the *independent fields* (which create the in and out particle states) have the hamiltonian terms:

$$H = \sum_{\pm s}\int d^2pdp^+\; p^-[b_{TL}^{\;+\dagger}(p,s)b_{TL}^{\;+}(p,s) - d_{TL}^{\;+}(p,s)d_{TL}^{\;+\dagger}(p,s) - b_{TR}^{\;+\dagger}(p,s)b_{TR}^{\;+}(p,s) +$$
$$+ d_{TR}^{\;+}(p,s)d_{TR}^{\;+\dagger}(p,s)] \qquad\qquad (2.87)$$

$$= \sum_{\pm s} \int d^2 p \, dp^+ \, p^- [b_{TL}^{\dagger\dagger}(p,s) b_{TL}^{+}(p,s) + d_{TL}^{\dagger\dagger}(p,s) d_{TL}^{+}(p,s) - b_{TR}^{\dagger\dagger}(p,s) b_{TR}^{+}(p,s) -$$

$$- d_{TR}^{\dagger\dagger}(p,s) d_{TR}^{+}(p,s)] \qquad (2.88)$$

up to the usual infinite constants due to left-handed operator rearrangement and right-handed operator rearrangement that are discarded. Eq. 2.88 is the basis for our particle interpretation of tachyon creation and annihilation operators based on Dirac's hole theory. Dirac hole theory as applied in light-front coordinates assumes all negative p^- ("energy") states are filled.

2.4.5.1 Left-Handed Tachyon Creation and Annihilation Operators

1.　We identify $b_{TL}^{\dagger\dagger}(p,s)$ and $d_{TL}^{+}(p,s)$ as creation operators for left-handed tachyons. $b_{TL}^{\dagger\dagger}(p,s)$ creates a positive p^- ("energy") state and $d_{TL}^{+}(p,s)$ creates a negative p^- ("energy") state.

2.　$b_{TL}^{+}(p,s)$ and $d_{TL}^{\dagger\dagger}(p,s)$ are the corresponding annihilation operators for left-handed tachyons. $b_{TL}^{+}(p,s)$ annihilates a positive p^- ("energy") state and $d_{TL}^{\dagger\dagger}(p,s)$ annihilates a negative p^- ("energy") state.

3.　We assume Dirac hole theory holds for the left-handed tachyon vacuum with all negative energy states filled. There is no tachyon energy gap as there is for Dirac fermions. There is also the problem that the left-handed tachyon vacuum is not invariant under ordinary Lorentz transformations or Superluminal transformations. *However if we confine ourselves to light-front coordinates for computations no ambiguity can result and the Lorentz covariant quantities that we calculate, such as the S matrix, are well-defined.*

4.　Using tachyon hole theory we identify $b_{TL}^{+}(p,s)$ and $d_{TL}^{\dagger\dagger}(p,s)$ as annihilation operators for left-handed tachyons. $b_{TL}^{+}(p,s)$ annihilates a positive p^- ("energy") state and $d_{TL}^{\dagger\dagger}(p,s)$ annihilates a negative p^- ("energy") state – thus creating a hole in the tachyon sea that we view as the creation of a positive p^- ("energy"), left-handed antitachyon. $d_{TL}^{+}(p,s)$ annihilates a positive p^- ("energy"), left-handed antitachyon.

2.4.5.2 Right-Handed Tachyon Creation and Annihilation Operators

　　The anti-commutation relations of right-handed tachyon creation and annhilation operators and the right-handed Hamiltonian terms have the "wrong" sign compared to corresponding Dirac operators and left-handed tachyon operators. This situation is completely analogous to the situation of time-like photons in the covariant formulation of quantum Electrodynamics.[44] In the case of time-like photons it was possible to introduce an indefinite

---

[44] Bogoliubov (1959) pp. 130-136.

metric (Gupta-Bleuler formulation), and then to use the subsidiary condition $\partial A^v/\partial x^v = 0$ to reduce the dynamics of QED to the transverse components. Thus the time-like photons were intermediate artifacts needed to have a manifestly covariant formulation while QED observables depended solely on the transverse components of the electromagnetic field.

In the present case of free tachyons, and in leptonic ElectroWeak Theory there is no evident "subsidiary condition" to eliminate the right-handed tachyon fields. But since the only manner in which the right-handed leptonic tachyon fields[45] interact is through mass terms, which can be easily 'integrated out", right-handed leptonic tachyon fields are removed from the observable part of the leptonic ElectroWeak Theory by their "lack of interaction" with left-handed fields.

In the case of quark ElectroWeak Theory right-handed tachyon quark fields have charge $(-1/3)$ and thus experience an electromagnetic interaction as well as a Z interaction. However, since quarks are totally confined, right-handed tachyon quarks will not be able to continuously emit photons or Z's due to energy conservation and their confinement to bound states of fixed positive energy. Earlier, when we consider complex Lorentz group boosts, we will suggest that quarks may not consist of Dirac particles or tachyons of the type considered up to this point in this chapter. Rather they may be variants on Dirac particles and tachyons satisfying different dynamical equations. However, the preceding comments on quarks would still apply.

Thus right-handed tachyons are analogous to time-like photons – necessary theoretically but prevented from causing a negative energy disaster by the forms of their interactions. We discuss this subject in more detail in the following chapters.

2.4.6 Tachyon Feynman Propagator

In this section we develop the light-front propagator for tachyons. We begin with a subsection describing the light-front propagators of Dirac fields.

2.4.6.1 Dirac Field Light-Front Propagators

The light-front Feynman propagator for the ψ^+ field of a Dirac fermion is

$$iS^+_F(x,y)\gamma^0 = \theta(x^+ - y^+)\langle 0|\psi^+(x)\psi^{+\dagger}(y)|0\rangle - \theta(y^+ - x^+)\langle 0|\psi^{+\dagger}(y)\psi^+(x)|0\rangle \qquad (2.89)$$

and does not contain a non-covariant piece due to the projection operators:

$$iS^+_F(x,y) = \int d^2p\,dp^+\theta(p^+)[1/(2(2\pi)^3 p^+)]\{\theta(x^+ - y^+)[R^+(\not{p} +m)R^-]\,e^{-ip\cdot(x-y)} +$$
$$+ \theta(y^+ - x^+)[R^+(-\not{p}+m)R^-]e^{+ip\cdot(x-y)}\}$$
$$= R^+iS_F(x,y)R^- \qquad (2.90)$$

---

[45] The tachyon fields are provisionally assumed to be neutrino fields in the leptonic sector, and d, s and b quarks in the quark sector.

where $S_F(x,y)$ is the usual Feynman propagator.

The light-front Feynman propagator for a *left-handed* <u>Dirac</u> field ψ^+ is

$$iS^+_{LF}(x,y) = \int d^2p\,dp^+\theta(p^+)[1/(2(2\pi)^3p^+)]\{\theta(x^+-y^+)[C^-R^+(\not{p}+m)R^-C^-]e^{-ip\cdot(x-y)} +$$
$$+ \theta(y^+-x^+)[C^-R^+(-\not{p}+m)R^-C^-]e^{+ip\cdot(x-y)}\}$$

$$= C^-R^+iS_F(x,y)R^-C^- \tag{2.91}$$

2.4.6.2 Tachyon Field Light Front Propagators

Turning now to tachyons, the light-front Feynman propagator for the left-handed ψ_{TL}^+ *tachyon* field is (using the previous Fourier expansion of the left-handed tachyon field):

$$iS^+_{TLF}(x,y) = \theta(x^+-y^+)<0|\psi_{TL}^+(x)\psi_{TL}^{++}(y)\gamma^0|0> - \theta(y^+-x^+)<0|\psi_{TL}^{++}(y)\gamma^0\psi_{TL}^+(x)|0>$$
$$= -i\int d^2p\,dp^+\theta(p^+)N_{TL}^{+2}(2m|\mathbf{p}|)^{-1}C^-R^+\{\theta(x^+-y^+)[(i\not{p}-m)\boldsymbol{\gamma}\cdot\mathbf{p}]e^{-ip\cdot(x-y)} +$$
$$+ \theta(y^+-x^+)[(i\not{p}+m)\boldsymbol{\gamma}\cdot\mathbf{p}]e^{+ip\cdot(x-y)}\}R^+C^-\gamma^0$$

If we define the on-shell momentum variable

$$p_0^- = (p_0^1p_0^1 + p_0^2p_0^2 - m^2)/(2p_0^+),\ p_0^+ = p^+,\ p_0^j = p^j\ (\text{for } j = 1, 2),\ p_{\perp 0}^2 = p_0^jp_0^j$$

and

$$\not{p}_0 = p_0\cdot\gamma$$

then the above equation can be rewritten as

$$S^+_{TLF}(x,y) = -C^-R^+\int d^4p[32\pi^4(p_0^+(p_0^+ - p_0^-) + p_{0\perp}^2)]^{-1}e^{-ip\cdot(x-y)}\{\theta(p^+)(i\not{p}_0 - m)\boldsymbol{\gamma}\cdot\mathbf{p}_0]/[p^- - p_0^- + i\varepsilon] +$$

$$+ \theta(-p^+)(i\not{p}_0 + m)\boldsymbol{\gamma}\cdot\mathbf{p}_0]/[p^- + p_0^- - i\varepsilon]\}R^+C^-\gamma^0$$

$$= -\tfrac{1}{2}\, i\int d^4p(2\pi)^{-4}[C^-R^+(i\not{p} - m)\boldsymbol{\gamma}\cdot\mathbf{p}R^+C^-\gamma^0]e^{-ip\cdot(x-y)}[(p^2 + m^2 +i\varepsilon)(p^+(p^+ - p^-) + p_\perp^2))]^{-1}$$

and using $C^-R^+(i\not{p} - m)\boldsymbol{\gamma}\cdot\mathbf{p}R^+C^- = i\,C^-R^+(p^+(p^+ - p^-) + p_\perp^2)$ we find

$$iS^+_{TLF}(x,y) = \tfrac{1}{2}C^-R^+\gamma^0\int d^4p(2\pi)^{-4}\,p^+e^{-ip\cdot(x-y)}/(p^2 + m^2 +i\varepsilon) \tag{2.92}$$

Similarly the light-front Feynman propagator for the right-handed ψ_{TR}^+ tachyon field is

$$iS^+_{TRF}(x,y) = \theta(x^+-y^+)<0|\psi_{TR}^+(x)\psi_{TR}^{++}(y)\gamma^0|0> - \theta(y^+-x^+)<0|\psi_{TR}^{++}(y)\gamma^0\psi_{TR}^+(x)|0>$$

$$= -\tfrac{1}{2}C^+R^+\gamma^0\!\int d^4p(2\pi)^{-4}\,p^+e^{-ip\cdot(x-y)}/(p^2 + m^2 + i\varepsilon) \qquad (2.93)$$

where the relative minus sign between eqs. 2.92 and 2.93 is due to the relative minus signs of the Fouier component operator anti-commutation relations.

Thus we find *tachyon* pole terms in the tachyon propagators as one would expect.

2.5 Complex Space and 3-Momentum & Real-Valued Energy Fermions (Quarks)

In this section we will use L_C boosts to develop a wider set of dynamical equations for free spin ½ fermions with real-valued energy and complex-valued 3-momentum.[46] Earlier we defined L_C boosts with

$$\Lambda_C(\mathbf{v_c}) = \exp[i\omega\hat{\mathbf{w}}\cdot\mathbf{K}] \qquad (2.94)$$

$$\omega = (\omega_r^{\,2} - \omega_i^{\,2} + 2i\omega_r\omega_i\,\hat{\mathbf{u}}_r\cdot\hat{\mathbf{u}}_i)^{\frac{1}{2}} \qquad (2.95)$$

$$\hat{\mathbf{w}} = (\omega_r\hat{\mathbf{u}}_r + i\omega_i\hat{\mathbf{u}}_i)/\omega \qquad (2.96)$$

$$\hat{\mathbf{w}}\cdot\hat{\mathbf{w}} = \hat{\mathbf{u}}_r\cdot\hat{\mathbf{u}}_r = \hat{\mathbf{u}}_i\cdot\hat{\mathbf{u}}_i = 1 \qquad (2.97)$$

$$\mathbf{v_c} = \hat{\mathbf{w}}\,\tanh(\omega) \qquad (2.98)$$

2.5.1 L_C Spinor "Normal" Lorentz Boosts & More Spin ½ Particle Types

Spinor boost transformations were used in previous sections to develop the dynamical equations for Dirac fields and tachyon fields. In this section we will use L_C spinor boosts to generate additional fermion field dynamical equations.

The form of the L_C spinor boost transformation corresponding to the coordinate transformation is:

$$S_C(\omega, \mathbf{v_c}) = \exp(-i\omega\sigma_{0k}\hat{w}_k/2) = \exp(-\omega\gamma^0\gamma\cdot\hat{\mathbf{w}}/2)$$
$$= \cosh(\omega/2)I + \sinh(\omega/2)\gamma^0\gamma\cdot\hat{\mathbf{w}} \qquad (2.99)$$

The inverse transformation is

$$S_C^{-1}(\omega, \mathbf{v_c}) = \gamma^2\gamma^0K^{-1}S_C^\dagger K\gamma^0\gamma^2 = \gamma^2\gamma^0S_C^{\,T}\gamma^0\gamma^2 = \exp(\omega\gamma^0\gamma\cdot\hat{\mathbf{w}}/2)$$
$$= \cosh(\omega/2)I - \sinh(\omega/2)\gamma^0\gamma\cdot\hat{\mathbf{w}} \qquad (2.100)$$

where the superscript T denotes the transpose and K is the complex conjugation operator (that also appears in the time-reversal operator). Note that S_C is not unitary just as in previous cases considered in this chapter.

---

[46] The complexon theory that we develop and use for quark dynamics in the Standard Model is <u>not</u> required. Our Standard Model could use Dirac fermion dynamics for the up-type quarks and tachyon dynamics for down-type quarks. We choose to use complexon dynamics for all quark types because they have an internal SU(3)-like structure suggestive of color SU(3). More importantly, their spin dynamics is different and thus may resolve the differences between theory and experiment – particularly for the deep inelastic parton spin-dependent structure functions.

We now redo the development of spin ½ dynamical equations of motion of earlier sections for this more general case of complex ω and $\hat{\mathbf{w}}$. Again we apply a boost to a Dirac equation for a positive energy plane wave particle of mass m at rest:

$$0 = S_C(\omega, \mathbf{v}_c))(m\gamma^0 - m)e^{-imt}w(0)$$
$$= [mS_C\gamma^0 S_C^{-1} - m]e^{-imt}S_C w(0) \qquad (2.101)$$

where $S_C = S_C(\omega, \hat{\mathbf{w}})$. After some algebra

$$mS_C\gamma^0 S_C^{-1} = m[\cosh(\omega)\gamma^0 - \sinh(\omega)\gamma\cdot\hat{\mathbf{w}}] \qquad (2.102)$$

2.5.1.1 Case 1: Parallel Real and Imaginary Relative Vectors

If the real and imaginary relative vectors parts of $\hat{\mathbf{w}}$, namely $\hat{\mathbf{u}}_r$ and $\hat{\mathbf{u}}_i$, are parallel, then $\hat{\mathbf{u}}_r\cdot\hat{\mathbf{u}}_i = 1$ and

$$\omega = \omega_r + i\omega_i \qquad (2.103)$$

Eq. 2.102 can be re-expressed as

$$mS_C\gamma^0 S_C^{-1} = m[\cosh(\omega_r)\cos(\omega_i) + i\sinh(\omega_r)\sin(\omega_i)]\gamma^0 - m[\sinh(\omega_r)\cos(\omega_i) +$$
$$+ i\cosh(\omega_r)\sin(\omega_i)]\gamma\cdot\hat{\mathbf{u}}_r \qquad (2.104)$$

or equivalently

$$mS_C\gamma^0 S_C^{-1} = \cos(\omega_i)\gamma\cdot p_r + i\sin(\omega_i)\gamma\cdot p_i \qquad (2.105)$$

where

$$p_r{}^0 = m\cosh(\omega_r) \qquad\qquad p_i{}^0 = m\sinh(\omega_r) \qquad (2.106)$$

and

$$\mathbf{p}_r = m\hat{\mathbf{u}}_r\sinh(\omega_r) \qquad\qquad \mathbf{p}_i = m\hat{\mathbf{u}}_r\cosh(\omega_r) \qquad (2.107)$$

If $\omega_i = 0$, then we recover the momentum space Dirac equation. If $\omega_i = \pi/2$, then we obtain the left-handed momentum space tachyon equation. Since the range of ω_i is $[0, \infty>$ (due to the cut along the real ω-plane axis) eq. 2.105 corresponds to the results of the Left-Handed Lorentz boost part discussed earlier.

2.5.1.2 Case 2: Anti-Parallel Real and Imaginary Relative Vectors

If the real and imaginary relative vectors parts of $\hat{\mathbf{w}}$, $\hat{\mathbf{u}}_r$ and $\hat{\mathbf{u}}_i$, are anti-parallel $\hat{\mathbf{u}}_r = -\hat{\mathbf{u}}_i$, then $\hat{\mathbf{u}}_r\cdot\hat{\mathbf{u}}_i = -1$ and

$$\omega = \omega_r - i\omega_i \qquad (2.108)$$

We can then express eq. 2.105 as

$$mS_C\gamma^0 S_C^{-1} = m[\cosh(\omega_r)\cos(\omega_i) - i\sinh(\omega_r)\sin(\omega_i)]\gamma^0 - m[\sinh(\omega_r)\cos(\omega_i) -$$
$$- i\cosh(\omega_r)\sin(\omega_i)]\gamma\cdot\hat{\mathbf{u}}_r \qquad (2.109)$$

or

$$mS_C\gamma^0S_C^{-1} = \cos(\omega_i)\gamma\cdot p_r - i\sin(\omega_i)\gamma\cdot p_i \qquad (2.110)$$

where

$$p_r^0 = m\cosh(\omega_r) \qquad\qquad p_i^0 = m\sinh(\omega_r) \qquad (2.111)$$

and

$$\mathbf{p_r} = m\hat{\mathbf{u}}_r\sinh(\omega_r) \qquad\qquad \mathbf{p_i} = m\hat{\mathbf{u}}_r\cosh(\omega_r) \qquad (2.112)$$

If $\omega_i = 0$, then we again recover the momentum space Dirac equation, If $\omega_i = \pi/2$, then we obtain the right-handed momentum space tachyon equation. (The range of ω_i is again $[0, \infty>$.)

Note: Since the matrix elements in the boost depend on $\gamma = (1 - \beta^2)^{-\frac{1}{2}}$ with a singularities at $\beta = \pm 1$, which in turn corresponds to $\omega = \pm\infty$, there is a branch cut along the ω axis in the complex ω-plane. Therefore we point out again the product of three Left-handed transformations is not equivalent to a Right-handed transformation.

2.5.1.3 Case 3: Complexons: A New Type of Particle with Perpendicular Real and Imaginary 3-Momenta

If the real and imaginary relative vectors parts of $\hat{\mathbf{w}}$, namely $\hat{\mathbf{u}}_r$ and $\hat{\mathbf{u}}_i$, are perpendicular, $\hat{\mathbf{u}}_r\cdot\hat{\mathbf{u}}_i = 0$, then

$$\omega = (\omega_r^2 - \omega_i^2)^{\frac{1}{2}} \qquad (2.113)$$

Thus ω is either pure real ($\omega_r \geq \omega_i$) or pure imaginary ($\omega_r < \omega_i$).

The momentum space equation generated by the corresponding L_C spinor boost is

$$\{m\cosh(\omega)\gamma^0 - m\sinh(\omega)\gamma\cdot(\omega_r\hat{\mathbf{u}}_r + i\omega_i\hat{\mathbf{u}}_i)/\omega - m\}e^{-ip\cdot x}w_c(p) = 0 \qquad (2.114)$$

Defining the momentum 4-vector

$$p = (p^0, \mathbf{p}) \qquad (2.115)$$

where

$$p^0 = m\cosh(\omega) \qquad\qquad \mathbf{p} = \mathbf{p_r} + i\mathbf{p_i} \qquad (2.116)$$
$$\mathbf{p_r} = m\omega_r\hat{\mathbf{u}}_r\sinh(\omega)/\omega \qquad \mathbf{p_i} = m\omega_i\hat{\mathbf{u}}_i\sinh(\omega)/\omega \qquad (2.117)$$

and

$$\mathbf{p_r}\cdot\mathbf{p_i} = 0 \qquad (2.118)$$

then we obtain a positive energy Dirac-like equation with complex 3-momentum

$$[p\cdot\gamma - m]e^{-ip\cdot x}w_c(p) = 0$$

or, explicitly,

$$\qquad (2.119)$$

$$[p^0\gamma^0 - (\mathbf{p_r} + i\mathbf{p_i})\cdot\gamma - m]e^{-ip\cdot x}w_c(p) = 0$$

with a complex 3-momentum $\mathbf{p}$ and the 4-momentum mass shell condition:

$$p^2 = p^{0\,2} - \mathbf{p_r}\cdot\mathbf{p_r} + \mathbf{p_i}\cdot\mathbf{p_i} = m^2 \tag{2.120}$$

Note

$$|\mathbf{v}| = |\mathbf{p}|/p^0 = [(\mathbf{p_r} + i\mathbf{p_i})\cdot(\mathbf{p_r} + i\mathbf{p_i})]^{\frac{1}{2}}/p^0 = \tanh(\omega) \tag{2.121}$$

and thus the Lorentz factor

$$\gamma = \cosh(\omega) \tag{2.122}$$

Eq. 2.119 is the momentum space equivalent of the wave equation

$$[i\gamma^0\partial/\partial t + i\gamma\cdot(\nabla_r + i\nabla_i) - m]\psi_C(t, \mathbf{x_r}, \mathbf{x_i}) = 0 \tag{2.123}$$

where

$$x_c = (t, \mathbf{x_r} - i\mathbf{x_i}) \tag{2.123a}$$

and where the grad operators ∇_r and ∇_i are with respect to $\mathbf{x_r}$ and $\mathbf{x_i}$ respectively. Since $\hat{\mathbf{u}}_r\cdot\hat{\mathbf{u}}_i = 0$, we see that there is a subsidiary condition on the wave function

$$\nabla_r\cdot\nabla_i\,\psi_C(t, \mathbf{x_r}, \mathbf{x_i}) = 0 \tag{2.124}$$

We will call the particles satisfying eqs.2.123 and 2.124 *complexons*. In addition eq. 2.118 implies the anti-commutation relation

$$\{\gamma\cdot\mathbf{p_r}, \gamma\cdot\mathbf{p_i}\} = 0 \tag{2.125}$$

which in turn implies

$$\gamma\cdot\nabla_r\gamma\cdot\nabla_i\psi_C(t, \mathbf{x_r}, \mathbf{x_i}) = \gamma\cdot\nabla_i\gamma\cdot\nabla_r\psi_C(t, \mathbf{x_r}, \mathbf{x_i}) = 0 \tag{2.126}$$

We note that eq. 2.125 is covariant under the real Lorentz group and eq. 2.126 can be easily put into covariant form since the difference of these 4-vectors squared is a real Lorentz group invariant: $[\gamma^0\partial/\partial t + \gamma\cdot(\nabla_r + i\nabla_i)]^2 - [\gamma^0\partial/\partial t + i\gamma\cdot(\nabla_r - i\nabla_i)]^2 = 4\nabla_r\cdot\nabla_i$.

Before considering a lagrangian formulation and the Fourier operator representation of $\psi_C(t, \mathbf{x_r}, \mathbf{x_i})$ we will define the spinors and associated real and imaginary spin operators.

The spinor generated from a spin up Dirac spinor at rest by a complex boost is

$$w_c(p) = S_C(p)w(0) = [\cosh(\omega/2)I + \sinh(\omega/2)\gamma^0\gamma\cdot\hat{\mathbf{w}}]w(0) \tag{2.127}$$

Following a procedure similar to Appendix 2-A (which the reader may wish to examine first) we define four spinors for Dirac particles at rest:

$$w^k(0) = \begin{bmatrix} \delta_{1k} \\ \delta_{2k} \\ \delta_{3k} \\ \delta_{4k} \end{bmatrix} \qquad (2\text{-}A.2)$$

where Kronecker deltas appear in the brackets. Then by applying eq. 2.127 to the spinors defined by eq. 2-A.2 we find the L_C spinors

$$S_C w^k(0) = w_{Cr}^{\ k}(p) + i w_{Ci}^{\ k}(p) \qquad (2.128)$$

where

$$S_{Cr} = \cosh(\omega/2)I + (\omega_r/\omega)\sinh(\omega/2)\gamma^0\boldsymbol{\gamma}\cdot\hat{\mathbf{u}}_r$$
$$= [(m + E)/(2m)]^{\frac{1}{2}}I + [m(m + E)]^{-\frac{1}{2}}\gamma^0\boldsymbol{\gamma}\cdot\mathbf{p}_r = aI + b\gamma^0\boldsymbol{\gamma}\cdot\mathbf{p}_r \qquad (2.129)$$

Thus the "real" spinors $w_{Cr}^{\ k}(p)$ are the columns of

$$\underline{w_{Cr}^{\ 1}(p)} \qquad \underline{w_{Cr}^{\ 2}(p)} \qquad \underline{w_{Cr}^{\ 3}(p)} \qquad \underline{w_{Cr}^{\ 4}(p)}$$

$$S_{Cr} = \begin{bmatrix} a & 0 & bp_{rz} & bp_{r-} \\ 0 & a & bp_{r+} & -bp_{rz} \\ bp_{rz} & bp_{r-} & a & 0 \\ bp_{r+} & -bp_{rz} & 0 & a \end{bmatrix}$$

$$(2.130)$$

where $p_{r\pm} = p_{rx} \pm ip_{ry}$. The "imaginary" spinors are the columns of

$$S_{Ci} = (\omega_i/\omega)\sinh(\omega/2)\gamma^0\boldsymbol{\gamma}\cdot\hat{\mathbf{u}}_i = [m(m + E)]^{-\frac{1}{2}}\gamma^0\boldsymbol{\gamma}\cdot\mathbf{p}_i = b\gamma^0\boldsymbol{\gamma}\cdot\mathbf{p}_i \qquad (2.131)$$

$$\underline{w_{Ci}^{\ 1}(p)} \qquad \underline{w_{Ci}^{\ 2}(p)} \qquad \underline{w_{Ci}^{\ 3}(p)} \qquad \underline{w_{Ci}^{\ 4}(p)}$$

$$S_{Ci} = \begin{bmatrix} 0 & 0 & bp_{iz} & bp_{i-} \\ 0 & 0 & bp_{i+} & -bp_{iz} \\ bp_{iz} & bp_{i-} & 0 & 0 \\ bp_{i+} & -bp_{iz} & 0 & 0 \end{bmatrix}$$

$$(2.132)$$

where $p_{i\pm} = p_{ix} \pm ip_{iy}$.

Eqs. 2.127 through 2.132 imply that the wave function solution of eq. 2.123, subject to the subsidiary condition eq. 2.124, is[47, 48]

$$\psi_C(x_r, x_i) = \sum_{\pm s} \int d^3p_r d^3p_i \, N_C(p)\delta(\mathbf{p_r \cdot p_i}/m^2)[b_C(p,s)u_C(p, s)e^{-i(p \cdot x + p^* \cdot x^*)/2} +$$
$$+ d_C^\dagger(p,s)v_C(p, s)e^{+i(p \cdot x + p^* \cdot x^*)/2}] \qquad (2.133)$$

where $\mathbf{p} = \mathbf{p_r} + i\mathbf{p_i}$ (eq. 3.95), $\mathbf{x} = \mathbf{x_r} - i\mathbf{x_i}$, $p \cdot x = p^0 x^0 - \mathbf{p \cdot x}$, and where we use

$$(p \cdot x + p^* \cdot x^*)/2 = p^0 x^0 - \mathbf{p_r \cdot x_r} - \mathbf{p_i \cdot x_i} \qquad (2.134)$$

in the exponentials in order to avoid divergences that would appear in the calculation of the equal-time commutator, the Feynman propagator and other quantities of interest after second quantization. Note that

$$(\nabla_r + i\nabla_i)e^{-i(p \cdot x + p^* \cdot x^*)/2} = i(\mathbf{p_r} + i\mathbf{p_i})e^{-i(p \cdot x + p^* \cdot x^*)/2} \qquad (2.135)$$

and

$$(\nabla_r + i\nabla_i)e^{-ip^* \cdot x^*} = 0 \qquad (2.136)$$

for all p.

The wave function's conjugate (the hermitean conjugate modified by letting $\mathbf{x_i} \rightarrow -\mathbf{x_i}$ in addition to hermitean conjugation) is

$$\psi_C^\dagger(x) = \psi_C^\dagger(x_r, -x_i) = \sum_{\pm s} \int d^3p_r d^3p_i \, \delta(\mathbf{p_r \cdot p_i}/m^2)N_C(p^*) \cdot$$
$$\cdot [b_C^\dagger(p^*,s)u_C^\dagger(p^*,s)e^{+i(p \cdot x^* + p^* \cdot x)/2} + d_C(p^*,s)v_C^\dagger(p^*,s)e^{-i(p \cdot x^* + p^* \cdot x)/2}]$$
$$(2.137)$$

where $\mathbf{p} = \mathbf{p_r} + i\mathbf{p_i}$, $\mathbf{x} = \mathbf{x_r} - i\mathbf{x_i}$, $p \cdot x = p^0 x^0 - \mathbf{p \cdot x}$, and $\dagger$ indicates hermitean hermitean conjugation.

The spinors are

$$u_C(p, s) = S_C(p)w^1(0)$$
$$u_C(p, -s) = S_C(p)w^2(0)$$

---

[47] Note that when $|\mathbf{p_i}| \geq |\mathbf{p_r}|$ (for imaginary $\omega = (\omega_r^2 - \omega_i^2)^{\frac{1}{2}}$) the 3-momentum becomes imaginary $\mathbf{p \cdot p} < 0$. However, since we will be identifying confined quarks with this type of particle – much modified by a confining color quark interaction – the issue of an imaginary 3-momentum in the hypothetical free quark case becomes moot. We note the energy gap between positive and negative energy states disappears so E = 0 is possible. Thus real Lorentz transformations can mix positive and negative energy states. The solution is to do all calculations in the light-front frame as we do for tachyons. Then the mixing issue is resolved. In the present case we second quantize on the "time-front" for illustrative purposes.

[48] We scale $\mathbf{p_r \cdot p_i}$ with m^2 in the delta function for convenience. All fermions have at least a minimal mass – the mass of the iota.

$$v_C(p, s) = S_C(p)w^3(0)$$
$$v_C(p, -s) = S_C(p)w^4(0)$$
$$u_C^\dagger(p^*, s) = w^{1T}(0)S_C^\dagger(p^*) = w^{1T}(0)S_C(p)$$
$$u_C^\dagger(p^*, -s) = w^{2T}(0)S_C^\dagger(p^*) = w^{2T}(0)S_C(p)$$
$$v_C^\dagger(p^*, s) = w^{3T}(0)S_C^\dagger(p^*) = w^{3T}(0)S_C(p)$$
$$v_C^\dagger(p^*, -s) = w^{4T}(0)S_C^\dagger(p^*) = w^{4T}(0)S_C(p)$$

$$(2.138)$$

with the superscript "T" indicating the transpose. Note that

$$S_C^\dagger(p^*) = [S_C(p^*)]^\dagger = S_C(p) \qquad (2.139)$$

The normalization factor $N_C(p)$ is

$$N_C(p) = [2m/((2\pi)^6 p^0)]^{\frac{1}{2}} \qquad (2.140)$$

Since $\mathbf{p_r} = \mathbf{p_i} = 0$ in the particle rest frame prior to the complex group boost, the boosted particle spin 4-vector s^μ satisfies

$$s^\mu p_r{}^\mu = s^\mu p_i{}^\mu = 0 \qquad (2.141)$$

Note that s^μ is itself complex[49] and, if the spin points in the z-direction prior to the complex boost, then the boosted s^μ has the form

$$s^\mu = (-\sinh(\omega)\hat{w}_z, (0,0,1) + (\cosh(\omega) - 1)\hat{w}_z\hat{\mathbf{w}}) \qquad (2.142)$$

with $\hat{\mathbf{w}}$ defined earlier: $\hat{\mathbf{w}} = (\omega_r\hat{\mathbf{u}}_r + i\omega_i\hat{\mathbf{u}}_i)/\omega = \mathbf{p}/(m\sinh(\omega))$.

2.5.1.4 A Global SU(3) Symmetry Revealed

Before proceeding to consider the second quantization of this case, we will consider a global SU(3) symmetry implicit in the previous equations. The defining property of the group SU(3) is that it preserves the invariance of inner products of complex 3-vectors of the form:

$$u^*\cdot v = u^1{}^*v^1 + u^2{}^*v^2 + u^3{}^*v^3 \qquad (2.143)$$

If we examine the dynamical equation eq. 2.123 we see that the differential operator is invariant under an SU(3) transformation U (using $\nabla_c = (\nabla_c{}^*)^* = \mathbf{D}_c{}^*$)

$$[i\gamma^0\partial/\partial t + i\mathbf{D}_c{}^*\cdot\gamma - m] = [i\gamma^0\partial/\partial t + i\mathbf{D}_c'{}^*\cdot\gamma' - m] \qquad (2.144)$$

---

[49] This feature of partons, which is not present in ordinary Dirac particles, might be the source of the discrepancies between theory and experiment in deep inelstic parton spin physics which is based on conventional real parton spins.

where

$$\mathbf{D_c}^* = \nabla_c = \nabla_r + i\nabla_i$$

and

$$\gamma'^a = U^{ab}\gamma'^b$$
$$\mathbf{D_c}'^{*a} = \mathbf{D_c}'^{*b}U^{\dagger ab}$$

where U is a global SU(3) transformation and $U^\dagger = U^{-1}$. By theorem[50] all 4×4 γ matrices such as γ' are equivalent up to a unitary transformation V. Thus $V^\dagger\gamma'V = \gamma$ and eq. 2.144 is equivalent to

$$[i\gamma^0\partial/\partial t + i\mathbf{D_c}^*\cdot\boldsymbol{\gamma} - m] = [i\gamma^0\partial/\partial t + i\mathbf{D_c}'^*\cdot\boldsymbol{\gamma} - m] \qquad (2.145)$$

$$= [i\gamma^0\partial/\partial t + i\nabla_c'\cdot\boldsymbol{\gamma} - m]$$

where $\nabla_c'_a = U^{ab}\nabla_{cb}$, This demonstrates that eq. 2.123 is invariant under an SU(3) transformation if

$$\psi_C(t, \mathbf{x_c}) = \psi_C(t, U\mathbf{x_c}) = \psi_C'(t, \mathbf{x_c}') \qquad (2.146)$$

where $\psi_C(t, \mathbf{x_c}) \equiv \psi_C(t, \mathbf{x_r}, \mathbf{x_i})$.

The subsidiary condition eq. 2.124 can be seen to transform as

$$\nabla_r\cdot\nabla_i\,\psi_C(t, \mathbf{x_c}) = \nabla_r^*\cdot\nabla_i\,\psi_C(t, \mathbf{x_c}) = \nabla_r'^*\cdot\nabla_i'\psi_C'(t, \mathbf{x_c}') = 0 \quad (2.147)$$

under an SU(3) rotation. The invariance of the orthogonality condition is preserved.

The wave function (eq. 2.123) transforms in the following way under the SU(3) transformation U. If we define

$$q^{*\mu} = (q^0, \mathbf{q}^*) = (p^0, \mathbf{p_r} + i\mathbf{p_i}) = (p^0, \mathbf{p}) = p^\mu \qquad (2.148)$$

then eq. 2.133 can be rewritten in an invariant form under a SU(3) transformation:

$$\psi_C(x) = \sum_{\pm s}\int d^3q_r d^3q_i\, N_C(p^0)\delta(\mathbf{q_r}^*\cdot\mathbf{q_i}/m^2)[b_C(q^*,s)u_C(q^*,s)e^{-i(q^*\cdot x + q\cdot x^*)/2} +$$
$$+ d_C^\dagger(q^*,s)v_C(q^*,s)e^{+i(q^*\cdot x + q\cdot x^*)/2}] \qquad (2.149)$$

where $x = x_c$ subject to an examination of the transformation properties of the fourier coefficients and spinors. Note both terms in each exponential are separately invariant under global SU(3). (Note also $\mathbf{q_r}^* = \mathbf{q_r}$ since $\mathbf{q_r}$ is real.)

---

[50] R. H. Good, Rev. Mod. Phys., **27**, 187 (1955).

From the form of S_C above it is clear that an argument similar to that for the dynamical equations shows S_C is invariant under an SU(3) transformation and thus their spinors are also invariant under SU(3) transformations. The fourier coefficients, if second quantized in a direct generalization of the usual manner, have covariant anti-commutation relations under an SU(3) transformation. For example

$$\{b_C(q,s), b_C^\dagger(q'^*,s')\} = \delta_{ss'}\delta^3(q_r - q'_r)\delta^3(q_i - q'_i) \qquad (2.150)$$

Under an SU(3) transformation, $z = Uq$ and $z' = Uq'$, the right side of eq. 2.150 transforms to

$$\delta^3(q_r - q'_r)\delta^3(q_i - q'_i) \rightarrow \delta^3(z_r - z'_r)\delta^3(z_i - z'_i)/|\partial(q)/\partial(z)| = \delta^3(z_r - z'_r)\delta^3(z_i - z'_i) \quad (2.151)$$

where

$$|\partial(q)/\partial(z)| = |\partial(q_r^1,q_r^2,q_r^3,q_i^1, q_i^2, q_i^3)/\partial(z_r^1,z_r^2,z_r^3,z_i^1, z_i^2, z_i^3)| = 1 \quad (2.152)$$

is the Jacobian of the transformation U. Thus the fourier coefficients transform trivially under SU(3). For example,

$$b_C(q^*,s) \rightarrow b_C(z^*,s) \qquad (2.153)$$

Since the integrand transforms as

$$\int d^3q_r d^3q_i \rightarrow \int d^3z_r d^3z_i \, |\partial(q)/\partial(z)| = \int d^3z_r d^3z_i \qquad (2.154)$$

the wave function $\psi_C(t, \mathbf{x})$ transforms as an SU(3) scalar up to an inessential unitary transformation V of γ matrices: $\psi_C(t, \mathbf{x}) \rightarrow V\psi_C(t, \mathbf{x})$.[51]

2.5.1.5 Global SU(3) Spin ½ Complexon Fields

Having uncovered an SU(3) symmetry in the scalar field equations of Case 3A the generalization of the scalar field equations to the <u>3</u> representation of SU(3) is direct:

$$\psi_C^a(x) = \sum_{\pm s}\int d^3p_r d^3p_i \, N_C(p)\delta(\mathbf{p}_r\cdot\mathbf{p}_i/m^2)[b_C(p,a,s)u_C^a(p, s)e^{-i(p\cdot x + p^*\cdot x^*)/2} +$$
$$+ d_C^\dagger(p,a,s)v_C^a(p, s)e^{+i(p\cdot x + p^*\cdot x^*)/2}] \qquad (2.155)$$

where $x = x_c$ for $a = 1,2, 3$ with $u_C^a(p, s)$ and $v_C^a(p, s)$ being the product a spinor of type eq. 2.138 and a 3 element column vector c^a with b^{th} element

---

[51] The spinors $u_C(q^*,s)$ and $v_C(q^*,s)$ are unchanged up to a unitary transformation of the γ matrices $(V^\dagger\gamma'V = \gamma)$. Thus the term $(U\mathbf{w})^*\cdot\gamma' = \mathbf{w}^*\cdot V\gamma V^\dagger \equiv \mathbf{w}^*\cdot\gamma$ in the expressions for the $u_C(q^*,s)$ and $v_C(q^*,s)$ spinors.

$$b^a(b) = \delta^{ab} \tag{2.156}$$

Under a global SU(3) transformation U the $\underline{3}$ complexon wave functions transform as

$$\psi_C'^a(x) = U^{ab}\psi_C^b(x) \tag{2.157}$$

In a subsequent discussion we will extend the global SU(3) symmetry described in these subsections to be color local SU(3) upon the introduction of the Yang-Mills color gluon interaction.

2.5.1.6 Lagrangian Formulation and Second Quantization of Complexons

In this subsection we will outline the canonical quantization of SU(3) singlet complexons with the quantum field equation

$$[i\gamma^0\partial/\partial t + i\gamma\cdot(\nabla_r + i\nabla_i) - m]\psi_C(t, \mathbf{x_r}, \mathbf{x_i}) = 0 \tag{2.158}$$

and subsidiary condition

$$\nabla_r\cdot\nabla_i \ \psi_C(t, \mathbf{x_r}, \mathbf{x_i}) = 0 \tag{2.159}$$

We begin with the Lagrangian density

$$\mathcal{L} = \bar{\psi}_C(i\gamma^\mu D_\mu - m)\psi_C(x) \tag{2.160}$$

where $\bar{\psi}_C = \psi_C^\dagger\gamma^0$:

$$\psi_C^\dagger = [\psi_C(\mathbf{x_r}, \mathbf{x_i})]^\dagger\big|_{\mathbf{x_i} = -\mathbf{x_i}} \tag{2.161}$$

$$D_0 = \partial/\partial x^0$$
$$D_k = \partial/\partial x^k + i\,\partial/\partial x_i^k \tag{2.162}$$

with $x^k = x_r^k$ for $k = 1, 2, 3$. The invariant action (under real Lorentz transformations) is

$$I = \int d^7x\,\mathcal{L} \tag{2.163}$$

It is easy to show that the action is real

$$I^* = I \tag{2.164}$$

in a manner similar to the case considered in Appendix 2-A due to the form of $\psi_C^\dagger$ in eq. 2.161. (One has to change the integration over $\mathbf{x_i}$ to $-\mathbf{x_i}$ after taking the complex conjugate of I and performing manipulations similar to those in Appendix 2-A.)

The conjugate momentum is

$$\pi_{Ca} = \partial\mathcal{L}/\partial\dot{\psi}_{Ca} \equiv \partial\mathcal{L}/\partial(\partial\psi_{Ca}/\partial x^0) = i\psi_{C\,a}^\dagger \tag{2.165}$$

where a is a spinor index. It yields the non-zero anti-commutation relation

$$\{\psi_{C\,a}^{\dagger}(x),\ \psi_{Cb}(y)\} = \delta_{ab}\,\delta^3(x_r - y_r)\delta^3(x_i - y_i) \tag{2.166}$$

where x and y are complex. However we will see that the constraint eq. 2.159 is required. So the correct anti-commutator turns out to be

$$\{\psi_{C\,a}^{\dagger}(x),\ \psi_{Cb}(y)\} = -\delta_{ab}\delta'(\nabla_r\cdot\nabla_i/m^2)[\delta^3(x_r - y_r)\delta^3(x_i - y_i)] \tag{2.167}$$

where all ∇_r and ∇_i are ∇ derivatives with respect to x, and where $\delta'(\nabla_r\cdot\nabla_i)$ is the derivative of a delta function with the argument being differential operators such as those in eq. 2.159. The minus sign is due to the presence of a *derivative* of a delta-function and is not an issue.

The hamiltonian density is

$$\mathcal{H} = \pi_C\dot{\psi}_C - \mathcal{L} = \psi_C^{\dagger}(-i\boldsymbol{\alpha}\cdot\mathbf{D} + \beta m)\psi_C \tag{2.168}$$

and the (unsymmetrized) energy-momentum tensor is

$$\mathcal{T}_{\mu\nu} = -\,g_{\mu\nu}\mathcal{L} + \partial\mathcal{L}/\partial(D^{\mu}\psi_C)D_{\nu}\psi_C \tag{2.169}$$

The conserved energy and momentum are

$$P^0 = H = \int d^3x_r d^3x_i\ \mathcal{T}^{00} = \int d^3x_r d^3x_i\ \mathcal{H} \tag{2.170}$$

and

$$P^i = \int d^3x_r d^3x_i\ \mathcal{T}^{0i} \tag{2.171}$$

We now proceed to establish the canonical anti-commutation relations. First, the second quantization of the complexon field uses the above fourier coefficient anti-commutation relations (suitably rewritten):

$$\begin{aligned}
\{b_C(p,s),\ b_C^{\dagger}(p'^*,s')\} &= \delta_{ss'}\delta^3(\mathbf{p}_r - \mathbf{p}'_r)\delta^3(\mathbf{p}_i + \mathbf{p}'_i)\\
\{d_C(p,s),\ d_C^{\dagger}(p'^*,s')\} &= \delta_{ss'}\,\delta^3(\mathbf{p}_r - \mathbf{p}'_r)\delta^3(\mathbf{p}_i + \mathbf{p}'_i)\\
\{b_C(p,s),\ b_C(p'^*,s')\} &= \{d_C(p,s),\ d_C(p'^*,s')\} = 0\\
\{b_C^{\dagger}(p,s),\ b_C^{\dagger}(p'^*,s')\} &= \{d_C^{\dagger}(p,s),\ d_C^{\dagger}(p'^*,s')\} = 0\\
\{b_C(p,s),\ d_C^{\dagger}(p'^*,s')\} &= \{d_C(p,s),\ b_C^{\dagger}(p'^*,s')\} = 0\\
\{b_C^{\dagger}(p,s),\ d_C^{\dagger}(p'^*,s')\} &= \{d_C(p,s),\ b_C(p'^*,s')\} = 0
\end{aligned} \tag{2.172}$$

The delta-function arguments $\delta^3(\mathbf{p}_i + \mathbf{p}'_i)$ above have a positive sign in order to obtain $\delta^3(\mathbf{x}_i - \mathbf{y}_i)$ in the field anti-commutator eq. 2.167.

The spinors, eq. 2.138, satisfy

$$\sum_{\pm s} u_\alpha(p, s)\bar{u}_\beta(p^*, s) = (2m)^{-1}(\not{p} + m)_{\alpha\beta} \tag{2.173}$$

$$\sum_{\pm s} v_\alpha(p, s)\bar{v}_\beta(p^*, s) = (2m)^{-1}(\not{p} - m)_{\alpha\beta}$$

remembering

$$\bar{u}_C(p^*,s) = w^{1T}(0)S_C(p)\gamma^0 = w^{1T}(0)[\cosh(\omega/2)I + \sinh(\omega/2)\gamma^0\gamma\cdot\hat{w}]\gamma^0 \tag{2.174}$$

by eqs. 2.137 since $\hat{w}^{**} = \hat{w}$.

We will now evaluate the equal-time anti-commutation relation using eqs. 2.136 and 2.137:

$$\{\psi_{C\,a}^{\dagger}(x), \psi_{Cb}(y)\} = \sum_{\pm s,\,s'} \int d^3p_r d^3p_i\, d^3p'_r d^3p'_i\, \delta(\mathbf{p}_r\cdot\mathbf{p}_i/m^2)\delta(\mathbf{p}'_r\cdot\mathbf{p}'_i/m^2)\, N_C(p')N_C(p)\cdot$$

$$\cdot[\{b_C^{\dagger}(p^*,s)u_{Ca}^{\dagger}(p^*,s)e^{+i(p\cdot x^* + p^*\cdot x)/2}, b_C(p',s')u_{Cb}(p', s')e^{-i(p'\cdot y + p'^*\cdot y^*)/2}\}+$$

$$+ \{d_C(p^*,s)v_{Ca}^{\dagger}(p^*,s)e^{-i(p\cdot x^* + p^*\cdot x)/2}, d_C^{\dagger}(p',s')v_{Cb}(p', s')e^{+i(p'\cdot y + p'^*\cdot y^*)/2}\}]$$

$$= \int d^3p_r d^3p_i\, N_C^2(p)[\delta(\mathbf{p}_r\cdot\mathbf{p}_i/m^2)]^2[((\not{p} + m)\gamma^0)_{ba}\, e^{+i(p\cdot x^* + p^*\cdot x)/2 - i(p^*\cdot y + p\cdot y^*)/2} +$$

$$+((\not{p} - m)\gamma^0)_{ba}\, e^{-i(p\cdot x^* + p^*\cdot x)/2 + i(p^*\cdot y + p\cdot y^*)/2}]/(2m)$$

Next we use eq. 2.140 and the identity

$$[\delta(x - y)]^2 = -\tfrac{1}{2}\,\delta'(x - y) \equiv -\tfrac{1}{2}\, d\delta(x - y)/dx \tag{2.175}$$

which can be derived from the step function identity $\theta(x - y) = [\theta(x - y)]^2$ to obtain

$$\{\psi_{C\,a}^{\dagger}(x),\psi_{Cb}(y)\} = -\tfrac{1}{2}\int d^3p_r d^3p_i N_C^2(p)\delta'(\mathbf{p}_r\cdot\mathbf{p}_i/m^2)[((\not{p}+m)\gamma^0)_{ba}\, e^{-ipr\cdot(xr - yr) + ipi\cdot(xi - yi)} +$$

$$+ ((\not{p} - m)\gamma^0)_{ba}\, e^{+ipr\cdot(xr - yr) - ipi\cdot(xi - yi)}]/(2m)$$

$$= -\tfrac{1}{2}\delta_{ba}\int d^3p_r d^3p_i N_C^2(p)\delta'(\mathbf{p}_r\cdot\mathbf{p}_i/m^2)p^0 e^{-ipr\cdot(xr - yr) + ipi\cdot(xi - yi)}/m$$

$$= -\,\delta_{ab}\,\delta'(\nabla_r\cdot\nabla_i/m^2)[\delta^3(x_r - y_r)\delta^3(x_i - y_i)] \tag{2.176}$$

The grad operators, ∇_r and ∇_i, are derivatives are with respect to x in the Dirac delta functions. The factor[52] $\delta'(\nabla_r\cdot\nabla_i)$ expresses the orthogonality constraint in coordinate space on the momenta. It is analogous to the transversality constraint on the electromagnetic vector potential commutator:

---

[52] A derivative of a delta function containing grad operators.

$$[\pi_A{}^j(x), A_k(y)] = -i\,\delta^{tr}{}_{jk}(x - y) \qquad (2.177)$$

$$\delta^{tr}{}_{jk}(x - y) = (\delta_{jk} - \partial_j\partial_k/\nabla^2)\,\delta^3(x - y) \qquad (2.178)$$

where $\partial_k = \partial/\partial x_k$.

2.5.1.7 Complexon Feynman Propagator

The complexon Feynman propagator for ψ_C is[53]

$$iS_C(x, y) = \theta(x^0 - y^0)\langle 0|\psi_C(x)\psi_C{}^\dagger(y)\gamma^0|0\rangle - \theta(y^0 - x^0)\langle 0|\psi_C{}^\dagger(y)\gamma^0\psi_C(x)|0\rangle \qquad (2.179)$$

$$= \int d^3p_r d^3p_i N_C{}^2(p)[\delta(\mathbf{p_r}\cdot\mathbf{p_i}/m^2)]^2 \{\theta(x^0 - y^0)(\not{p} + m)e^{-i(p^*\cdot(x-y)\,+\,p\cdot(x^*-y^*))/2} -$$
$$- \theta(y^0 - x^0)(\not{p} - m)e^{+i(p^*\cdot(x - y)\,+\,p\cdot(x^*-y^*))/2}\}/(2m)$$

$$= -(4\pi)^{-1}\int dp^0 d^3p_r d^3p_i (2\pi)^{-6}\delta'(\mathbf{p_r}\cdot\mathbf{p_i}/m^2)(\not{p}+m)e^{-i(p^*\cdot(x-y)\,+\,p\cdot(x^*-y^*))/2}/(p^2 - m^2 + i\varepsilon)$$

$$= -\tfrac{1}{2}\int dp^0 d^3p_r d^3p_i\,\delta'(\mathbf{p_r}\cdot\mathbf{p_i}/m^2)(\not{p} + m)(2\pi)^{-7}\exp[-ip^0(x^0 - y^0) +$$
$$+ i\mathbf{p_r}\cdot(\mathbf{x_r} - \mathbf{y_r}) - i\mathbf{p_i}\cdot(\mathbf{x_i} - \mathbf{y_i})]/(p^2 - m^2 + i\varepsilon) \qquad (2.180)$$

The integral can be written in the form:

$$I = \int dp^0 d^3p_r d^3p_i \delta'(\mathbf{p_r}\cdot\mathbf{p_i}/m^2)(\not{p}+m)\exp[-ip^0(x^0-y^0)+i\mathbf{p_r}\cdot(\mathbf{x_r}-\mathbf{y_r})-i\mathbf{p_i}\cdot(\mathbf{x_i} - \mathbf{y_i})]/(p^2 - m^2 + i\varepsilon)$$
$$= \int d^4p_r dM^2 \delta'(\nabla_r\cdot\nabla_i/m^2)(p^0\gamma^0-(\mathbf{p_r}-\nabla_i)\cdot\gamma+m)\exp[-ip^0(x^0-y^0)+i\mathbf{p_r}\cdot(\mathbf{x_r}-\mathbf{y_r})]\cdot$$
$$\cdot J(\mathbf{x_i} - \mathbf{y_i}, M^2)/(p_r{}^2 - M^2 + i\varepsilon) \qquad (2.181)$$

where $p_r{}^2 = p^{0\,2} - \mathbf{p_r}\cdot\mathbf{p_r}$ and

$$J(\mathbf{x_i} - \mathbf{y_i}, M^2) = (2\pi)^{-3}\int d^3p_i\,\delta(M^2 + \mathbf{p_i}{}^2 - m^2)\,\exp[-i\mathbf{p_i}\cdot(\mathbf{x_i} - \mathbf{y_i})] \qquad (2.182)$$
$$= (2\pi)^{-2}|\mathbf{x_i} - \mathbf{y_i}|^{-1}\theta(m^2 - M^2)\sin((m^2 - M^2)^{1/2}|\mathbf{x_i} - \mathbf{y_i}|)$$

The complexon Feynman propagator can be rearranged into the form of a spectral integral:

---

[53] The reader, upon seeing the additional integrations $\int d^3p_i$ might suspect that they would ultimately lead to divergence issues in perturbation theory calculations. However the $\delta'(\mathbf{p_r}\cdot\mathbf{p_i}/m^2)$ term compensates in part for the additional integrations by four powers of momentum since $\delta'(\mathbf{p_r}\cdot\mathbf{p_i}/m^2) = (|\mathbf{p_r}||\mathbf{p_i}|/m^2)^{-2}\delta'(\cos\theta_{ri})$ where θ_{ri} is the angle between the momenta. As a result only 2 fermion and 3 fermion loop integrations would potentially have difficulties if one uses the conventional approach to perturbation theory. If one uses the approach of Blaha (2003) and (2005a) then there are no divergences.

$$iS_C(x, y) = -\int dM \, (i\gamma^0\partial/\partial x^0 - i(\nabla_r - i\nabla_i)\cdot\gamma + m)\delta'(\nabla_r\cdot\nabla_i/m^2)J(\mathbf{x_i} - \mathbf{y_i}, M^2)\triangle_F(x - y, M) \qquad (2.183)$$

where

$$\triangle_F(x - y, M) = (2\pi)^{-4}\int d^4p_r \, \exp[-ip^0(x^0 - y^0) + i\mathbf{p_r}\cdot(\mathbf{x_r} - \mathbf{y_r})]/(p_r^2 - M^2 + i\varepsilon) \qquad (2.184)$$

2.5.1.8 Case 4: Left-handed Tachyon Complexons

In this case $\hat{\mathbf{u}}_r\cdot\hat{\mathbf{u}}_i = 0$ again. However we add an imaginary term to ω to obtain a manifest Left-handed L_C boost[54]

$$\Lambda_{CL}(\mathbf{v_c}) = \exp[i(\omega + i\pi/2)\hat{\mathbf{w}}\cdot\mathbf{K}] \qquad (2.185)$$

where ω remains

$$\omega = (\omega_r^2 - \omega_i^2)^{\frac{1}{2}} \qquad (2.186)$$

and

$$\hat{\mathbf{w}} = (\omega_r\hat{\mathbf{u}}_r + i\omega_i\hat{\mathbf{u}}_i)/\omega \qquad (2.187)$$
$$\hat{\mathbf{w}}\cdot\hat{\mathbf{w}} = \hat{\mathbf{u}}_r\cdot\hat{\mathbf{u}}_r = \hat{\mathbf{u}}_i\cdot\hat{\mathbf{u}}_i = 1 \qquad (2.188)$$
$$\mathbf{v_c} = \hat{\mathbf{w}} \tanh(\omega + i\pi/2) = \hat{\mathbf{w}} \cotanh(\omega) \qquad (2.189)$$

Letting $\omega_L = \omega + i\pi/2$ we find, as before,

$$\cosh(\omega_L) = i \sinh(\omega) = -\gamma = i\,\gamma_s \qquad (2.190)$$
$$\sinh(\omega_L) = i \cosh(\omega) = -\beta\gamma = i\beta\gamma_s$$

with, $\beta = v_c = |\mathbf{v_c}| > 1$, $\gamma_s = (\beta^2 - 1)^{-\frac{1}{2}}$, and

$$\sinh(\omega) = \gamma_s \qquad (2.191)$$
$$\cosh(\omega) = \beta\gamma_s$$

Thus we denote $\Lambda_{CL}(\mathbf{v_c})$ by

$$\Lambda_{CL}(\mathbf{v_c}) \equiv \Lambda_{CL}(\omega, \hat{\mathbf{w}}) \qquad (2.192)$$

The corresponding spinor boost transformation is:

$$S_{CL}(\Lambda_{CL}(\omega, \hat{\mathbf{w}})) = \exp(-i\omega_L\sigma_{0i}\hat{w}_i/2) = \exp(-\omega_L\gamma^0\gamma\cdot\hat{\mathbf{w}}/2)$$
$$= \cosh(\omega_L/2)I + \sinh(\omega_L/2)\gamma^0\gamma\cdot\hat{\mathbf{w}} \qquad (2.193)$$

The momentum space equation generated by $S_{CL}(\Lambda_{CL}(\omega, \hat{\mathbf{w}}))$ is

$$\{m \cosh(\omega_L)\gamma^0 - m \sinh(\omega_L)\gamma\cdot(\omega_r\hat{\mathbf{u}}_r + i\omega_i\hat{\mathbf{u}}_i)/\omega - m\}e^{+ip\cdot x}w_{cL}(p) = 0 \qquad (2.194)$$

or

$$\{im \sinh(\omega)\gamma^0 - im \cosh(\omega)\gamma\cdot(\omega_r\hat{\mathbf{u}}_r + i\omega_i\hat{\mathbf{u}}_i)/\omega - m\}e^{+ip\cdot x}w_{cL}(p) = 0 \qquad (2.195)$$

---

[54] The reader can readily verify the form is consistent that generated by an L_C boost transformation.

where p·x = Et − **p·x** after performing a corresponding left-handed superluminal coordinate transformation in the exponential factor. Thus the positive energy wave is transformed into a negative energy wave by the transformation.

The momentum 4-vector is defined by

$$p = (p^0, \mathbf{p})$$
(2.196)

where

$$p^0 = m\,\sinh(\omega) \qquad\qquad \mathbf{p} = \mathbf{p_r} + i\mathbf{p_i}$$
(2.197)

with

$$\mathbf{p_r} = m\omega_r \hat{\mathbf{u}}_r \cosh(\omega)/\omega \qquad \mathbf{p_i} = m\omega_i \hat{\mathbf{u}}_i \cosh(\omega)/\omega$$
(2.198)

and

$$\mathbf{p_r}\cdot\mathbf{p_i} = 0$$
(2.199)

then eq. 2.195 becomes the complexon tachyon equation

$$[i p\cdot\gamma - m]e^{+ip\cdot x} w_{cL}(p) = 0$$
(2.200)

with a complex 3-momentum **p** and the tachyon 4-momentum mass shell condition:[55]

$$p^2 = p^{0\,2} - \mathbf{p_r}^2 + \mathbf{p_i}^2 = -m^2$$
(2.201)

Eq. 2.200 is the momentum space equivalent of the wave equation

$$[\gamma^0 \partial/\partial t + \gamma\cdot(\nabla_r + i\nabla_i) - m]\psi_{CL}(t, \mathbf{x_r}, \mathbf{x_i}) = 0$$
(2.202)

or

$$[\gamma\cdot\nabla - m]\psi_{CL}(t, \mathbf{x_r}, \mathbf{x_i}) = 0$$
(2.203)

with the subsidiary condition on the wave function

$$\nabla_r\cdot\nabla_i\ \psi_{CL}(t, \mathbf{x_r}, \mathbf{x_i}) = 0$$
(2.204)

also holds. We note that eq. 2.202 is covariant under the real Lorentz group and eq. 2.204 can be easily put into (real Lorentz group) covariant form.

Before considering a lagrangian formulation and the Fourier operator representation of $\psi_{CL}(t, \mathbf{x_r}, \mathbf{x_i})$ we will define the tachyon spinors, and its associated real and imaginary spin operators.

The spinor generated from a spin up Dirac spinor at rest by the L_C spinor boost eq. 2.193 is

---

[55] Note that the presence of the $\mathbf{p_i}^2$ term does not change the tachyon requirement that $\mathbf{p_r}^2 \geq m^2$ as seen in the previous cases.

$$w_{cL}(p) = S_{CL}w(0) = [\cosh(\omega_L/2)I + \sinh(\omega_L/2)\gamma^0\gamma\cdot\hat{\mathbf{w}}]w(0) \qquad (2.205)$$

Following a procedure similar to Appendix 2-A (which the reader may wish to examine first) we define four spinors for Dirac particles at rest with eq. 2-A.2. Then by applying a boost to these rest spinors we find the L_C tachyon spinors:

$$S_{CL}w^k(0) = w_{cL}{}^k(p) \qquad (2.206)$$

and from these tachyon spinors we generalize to tachyon spinors $u_{CL}(p, s)$ and $v_{CL}(p, s)$ in a manner similar tothat of the previous case.

Eqs. 2.200 through 2.204 imply that the wave function solution of eq. 2.200, subject to the subsidiary condition eq. 2.204, has the form

$$\psi_{CL}(x) = \sum_{\substack{\pm s \\ p_r^2 \geq m^2}} \int d^3p_r d^3p_i \, N_{CL}(p)\delta(\mathbf{p_r}\cdot\mathbf{p_i}/m^2)[b_{CL}(p,s)u_{CL}(p, s)e^{-i(p\cdot x + p^*\cdot x^*)/2} +$$
$$+ d_{CL}{}^\dagger(p,s)v_{CL}(p, s)e^{+i(p\cdot x + p^*\cdot x^*)/2}] \qquad (2.207)$$

where $\mathbf{p} = \mathbf{p_r} + i\mathbf{p_i}$, $\mathbf{x} = \mathbf{x_r} - i\mathbf{x_i}$, $p\cdot x = p^0x^0 - \mathbf{p}\cdot\mathbf{x}$, and $b_{CL}(p, s)$ and $d_{CL}(p,s)$ are tachyon fourier coefficients.

2.5.1.9 Global SU(3) Symmetry

We can show that there is also a global SU(3) symmetry present here as shown in the previous case. The demonstration is similar to that of eqs. 2.143 – 2.156.

2.5.1.10 Light-Front Quantization of Tachyonic Complexons

Because of the momentum constraint $\mathbf{p_r}^2 \geq m^2$ the set of solutions of the form of eq. 2.207 is incomplete and the result of second quantization would not be an equal time anti-commutator expression consisting of derivatives of delta functions (eq. 2.176) but rather an analogue to previous unsuccessful attempts to create a second quantized tachyon theory.[56]

Therefore we will use light-front coordinates, and left and right handed field operators (as previously) to obtain a successful second quantization of this new type of tachyon.

The "missing" factor of i in the first term of eq. 2.203 requires the lagrangian to be different from the conventional Dirac lagrangian in order for the lagrangian to be real. The simplest, physically acceptable, free spin ½ tachyon lagrangian density for ψ_{CL} is:

$$\mathcal{L}_{CL} = \psi_{CL}{}^C(x)(\gamma\cdot\nabla - m)\psi_{CL}(x) \qquad (2.208)$$

where

---

[56] Such as G. Feinberg, Phys. Rev. **159**, 1089 (1967).

$$\psi_{CL}{}^{C}(x) = [\psi_{CL}(x)]^{\dagger}\big|_{\mathbf{x_i} = -\mathbf{x_i}}\ i\gamma^0\gamma^5 \qquad (2.209)$$

is similar to eq. 2.161. In words, eq. 2.209 states: take the hermitean conjugate of $\psi_{CL}(x)$; change $\mathbf{x_i}$ to $-\mathbf{x_i}$; and then post-multiply by the indicated factors.

The free complexon invariant action (under real Lorentz transformations) is

$$I = \int d^7x\, \mathcal{L}_{CL} \qquad (2.210)$$

The action can be shown to be real

$$I^* = I \qquad (2.211)$$

in a manner similar to the case considered in Appendix 2-A. The tachyonic complexon's energy-momentum tensor is

$$\mathcal{T}_{CL\mu\nu} = - g_{\mu\nu}\,\mathcal{L}_{CL} + \partial\mathcal{L}_{CL}/\partial(D^\mu\psi_{CL})\, D_\nu\psi_{CL}$$
$$= i\psi_{CL}{}^{C}\gamma^0\gamma^5\gamma_\mu D_\nu\psi_{CL} \qquad (2.212)$$

where

$$D_0 = \partial/\partial x^0$$
$$D_k = \partial/\partial x_r{}^k + i\,\partial/\partial x_i{}^k \qquad (2.213)$$

and thus the conserved energy and momentum are

$$P^0 = H = \int d^3x_r d^3x_i\,\mathcal{T}_{CL}{}^{00} = i\int d^3x_r d^3x_i\psi_{CL}{}^{C}\gamma^5(\boldsymbol{\alpha}\cdot\mathbf{D} + \beta m)\psi_{CL} \qquad (2.214)$$
$$P^k = \int d^3x_r d^3x_i\,\mathcal{T}_{CL}{}^{0k} = - i\int d^3x_r d^3x_i\,\psi_{CL}{}^{C}\gamma^5 D^k\psi_{CL} \qquad (2.215)$$

Having defined a suitable tachyon lagrangian we can now proceed to its canonical quantization. The conjugate momentum can be calculated from the lagrangian density eq. 2.212:

$$\pi_{CLa} = \partial\mathcal{L}_{CL}/\partial\dot{\psi}_{CLa} \equiv \partial\mathcal{L}_{CL}/\partial(\partial\psi_{CLa}/\partial t) = -i([\psi_{CL}(x)]^{\dagger}\big|_{\mathbf{x_i} = -\mathbf{x_i}}\gamma^5)_a \qquad (2.216)$$

The resulting non-zero, canonical anti-commutation relations are

$$\{\pi_{CLa}(x),\, \psi_{CLb}(y)\} = i\,\delta_{ab}\,\delta^3(x_r - y_r)\delta^3(x_i - y_i)$$

based on locality in both real and imaginary coordinates:

$$\{\psi_{CL}{}^{\dagger}{}_a(x)\big|_{\mathbf{x_i} = -\mathbf{x_i}},\, \psi_{Tb}(y)\} = - [\gamma^5]_{ab}\,\delta^3(x_r - y_r)\delta^3(x_i - y_i) \qquad (2.217)$$

At this point we might attempt to complete the canonical quantization procedure in the conventional manner by Fourier expanding the field and specifying anti-commutation relations for the fourier component amplitudes. However the incompleteness of the set of plane waves, which are limited by the restriction $\mathbf{p_r}^2 \geq m^2$, causes the equal time anti-commutator of the fields *not* to yield a δ-functions.

Therefore we turn to the previous successful approach to tachyon quantization[57] and decompose the tachyonic complexon field into left-handed and right-handed parts and then second quantize in light-front coordinates.

2.5.2 Separation into Left-Handed and Right-Handed Fields

As before we will use a transformed set of Dirac matrices to develop our left-handed and right-handed tachyon formulations. The γ^5 chirality operator's eigenvalues define handedness: +1 corresponds to right-handed; and −1 corresponds to left-handed:

$$\gamma^5 \psi_{CLL} = -\psi_{CLL} \qquad\qquad \gamma^5 \psi_{CLR} = \psi_{CLR} \qquad (2.218)$$

We define left-handed and right-handed tachyon fields with the projection operators:

$$\begin{aligned}
C^\pm &= \tfrac{1}{2}(I \pm \gamma^5) \\
C^+ + C^- &= I \\
C^{\pm\,2} &= C^\pm \\
C^+ C^- &= 0
\end{aligned} \qquad (2.219)$$

with the result

$$\begin{aligned}
\psi_{CLL} &= C^- \psi_{CL} \\
\psi_{CLR} &= C^+ \psi_{CL}
\end{aligned} \qquad (2.220)$$

We can calculate the commutation relations of the left-handed and right-handed tachyonic complexon fields from eq. 2.217 by pre-multiplying and post-multiplying by $\tfrac{1}{2}(1 - \gamma^5)$ and $\tfrac{1}{2}(1 + \gamma^5)$. The results are:

$$\{\psi_{CLLa}^{\dagger}(x)|_{\mathbf{x_i}\,=\,-\mathbf{x_i}},\ \psi_{CLLb}(y)\} = C^-_{ab}\,\delta^6(x-y) \qquad (2.221)$$

$$\{\psi_{CLRa}^{\dagger}(x)|_{\mathbf{x_i}\,=\,-\mathbf{x_i}},\ \psi_{CLRb}(y)\} = -C^+_{ab}\,\delta^6(x-y) \qquad (2.222)$$

$$\{\psi_{CLLa}^{\dagger}(x)|_{\mathbf{x_i}\,=\,-\mathbf{x_i}},\ \psi_{CLRb}(y)\} = \{\psi_{CLRa}^{\dagger}(x)|_{\mathbf{x_i}\,=\,-\mathbf{x_i}},\ \psi_{CLLb}(x')\} = 0 \qquad (2.223)$$

where

$$\delta^6(x-y) = \delta^3(x_r - y_r)\delta^3(x_i - y_i) \qquad (2.224)$$

---

[57] Blaha (2006) discusses this case in detail.

The lagrangian density of eq. 2.208 decomposes into left-handed and right-handed parts: (The change x_i to $-x_i$ will be understood in $\psi_{CLL}^{\dagger}(x)$ and $\psi_{CLR}^{\dagger}(x)$ in the following.)

$$\mathcal{L}_{CL} = \psi_{CLL}^{\dagger}\gamma^0 i\gamma^\mu \partial_\mu \psi_{CLL} - \psi_{CLR}^{\dagger}\gamma^0 i\gamma^\mu \partial_\mu \psi_{CLR} - im[\psi_{CLR}^{\dagger}\gamma^0\psi_{CLL} - \psi_{CLL}^{\dagger}\gamma^0\psi_{CLR}] \qquad (2.225)$$

2.5.3 Further Separation into + and – Light-Front Complexon Fields

As previously, we now use light-front coordinates and quantization to obtain a successful second quantization of this form of tachyon field. Light-front variables, in the present case where we have to contend with complex 3-vectors, are defined by real coordinates and derivatives:

$$x^{\pm} = (x^0 \pm x_r^3)/\sqrt{2}$$
$$\partial/\partial x^{\pm} \equiv \partial^{\mp} \equiv (\partial/\partial x^0 \pm \partial/\partial x_r^3)/\sqrt{2} \qquad (2.226)$$

with the "transverse" real coordinate variables, x_r^1 and x_r^2, and imaginary coordinate variables x_i^1, x_i^2, and x_i^3.

The inner product of two 4-vectors has the form

$$x \cdot y = x^+ y^- + y^+ x^- + i[y_i^3(x^+ - x^-) + x_i^3(y^+ - y^-)]/\sqrt{2} + x_i^3 y_i^3 - (\mathbf{x}_{r\perp} - i\mathbf{x}_{i\perp})\cdot(\mathbf{y}_{r\perp} - i\mathbf{y}_{i\perp}) \qquad (2.227)$$

with

$$\mathbf{x}_{r\perp} = (x_r^1, x_r^2) \qquad \mathbf{x}_{i\perp} = (x_i^1, x_i^2)$$
$$\mathbf{y}_{r\perp} = (y_r^1, y_r^2) \qquad \mathbf{y}_{i\perp} = (y_i^1, y_i^2) \qquad (2.228)$$

where $x = (x^0, \mathbf{x} = \mathbf{x}_r - i\mathbf{x}_i)$ and $y = (y^0, \mathbf{y} = \mathbf{y}_r - i\mathbf{y}_i)$. Momenta are always defined as $p = (p^0, \mathbf{p} = \mathbf{p}_r + i\mathbf{p}_i)$.

The light-front definition of Dirac matrices is:

$$\gamma^{\pm} = (\gamma^0 \pm \gamma^3)/\sqrt{2} \qquad (2.229)$$

with transverse matrices γ^1 and γ^2 defined as usual. Note:

$$\gamma^{\pm 2} = 0$$

We define "+" and "–" tachyon fields with the projection operators:

$$R^{\pm} = \tfrac{1}{2}(I \pm \gamma^0\gamma^3) \qquad (2.230)$$

Left-handed, $\pm$ light-front fields:
$$\psi_{CLL}{}^{\pm} = R^{\pm}C^{-}\psi_{CL} \qquad (2.231)$$

Right-handed, $\pm$ light-front fields:
$$\psi_{CLR}{}^{\pm} = R^{\pm}C^{+}\psi_{CL}$$

Transforming to light-front variables and fields as above we obtain the light-front free tachyon lagrangian:

$$
\begin{aligned}
\mathcal{L}_{CL} = {} & 2^{\frac{1}{2}}\psi_{CLL}{}^{+\dagger}i\partial^{-}\psi_{CLL}{}^{+} + 2^{\frac{1}{2}}\psi_{CLL}{}^{-\dagger}i\partial^{+}\psi_{CLL}{}^{-} - \psi_{CLL}{}^{+\dagger}\gamma^{0}[i\boldsymbol{\gamma}_{\perp}\cdot\nabla_{r\perp} - \boldsymbol{\gamma}\cdot\nabla_{i}]\psi_{CLL}{}^{-} - \\
& - \psi_{CLL}{}^{-\dagger}\gamma^{0}[i\boldsymbol{\gamma}_{\perp}\cdot\nabla_{r\perp} - \boldsymbol{\gamma}\cdot\nabla_{i}]\psi_{CLL}{}^{+} - 2^{\frac{1}{2}}\psi_{CLR}{}^{+\dagger}i\partial^{-}\psi_{CLR}{}^{+} - 2^{\frac{1}{2}}\psi_{CLR}{}^{-\dagger}i\partial^{+}\psi_{CLR}{}^{-} + \\
& + \psi_{CLR}{}^{+\dagger}\gamma^{0}[i\boldsymbol{\gamma}_{\perp}\cdot\nabla_{r\perp} - \boldsymbol{\gamma}\cdot\nabla_{i}]\psi_{CLR}{}^{-} + \psi_{CLR}{}^{-\dagger}\gamma^{0}[i\boldsymbol{\gamma}_{\perp}\cdot\nabla_{r\perp} - \boldsymbol{\gamma}\cdot\nabla_{i}]\psi_{CLR}{}^{+} - \\
& - im[\psi_{CLR}{}^{+\dagger}\gamma^{0}\psi_{CLL}{}^{-} - \psi_{CLL}{}^{+\dagger}\gamma^{0}\psi_{CLR}{}^{-} + \psi_{CLR}{}^{-\dagger}\gamma^{0}\psi_{CLL}{}^{+} - \psi_{CLL}{}^{-\dagger}\gamma^{0}\psi_{CLR}{}^{+}] \quad (2.232)
\end{aligned}
$$

(Note the similarity to the previous tachyon case.) Again the difference in signs between the left-handed and right-handed terms will be a crucial factor in the derivation of the left-handed features of the Standard Model.

Eq. 2.232 generates the equations of motion:

$$
\begin{aligned}
2^{\frac{1}{2}}i\partial^{-}\psi_{CLL}{}^{+} - \gamma^{0}[i\boldsymbol{\gamma}_{\perp}\cdot\nabla_{r\perp} - \boldsymbol{\gamma}\cdot\nabla_{i}]\psi_{CLL}{}^{-} + im\gamma^{0}\psi_{CLR}{}^{-} = 0 \qquad (2.233) \\
2^{\frac{1}{2}}i\partial^{-}\psi_{CLR}{}^{+} - \gamma^{0}[i\boldsymbol{\gamma}_{\perp}\cdot\nabla_{r\perp} - \boldsymbol{\gamma}\cdot\nabla_{i}]\psi_{CLR}{}^{-} + im\gamma^{0}\psi_{CLL}{}^{-} = 0 \\
2^{\frac{1}{2}}i\partial^{+}\psi_{CLL}{}^{-} - \gamma^{0}[i\boldsymbol{\gamma}_{\perp}\cdot\nabla_{r\perp} - \boldsymbol{\gamma}\cdot\nabla_{i}]\psi_{CLL}{}^{+} + im\gamma^{0}\psi_{CLR}{}^{+} = 0 \\
2^{\frac{1}{2}}i\partial^{+}\psi_{CLR}{}^{-} - \gamma^{0}[i\boldsymbol{\gamma}_{\perp}\cdot\nabla_{r\perp} - \boldsymbol{\gamma}\cdot\nabla_{i}]\psi_{CLR}{}^{+} + im\gamma^{0}\psi_{CLL}{}^{+} = 0
\end{aligned}
$$

Eqs. 2.233 show that $\psi_{CLL}{}^{-}$ and $\psi_{CLR}{}^{-}$ are dependent fields that are functions of $\psi_{CLL}{}^{+}$ and $\psi_{CLR}{}^{+}$ on the light-front where x^{+} equals a constant. They can be expressed in an integral form as well. (The independent fields $\psi_{CLL}{}^{+}$ and $\psi_{CLR}{}^{+}$ play a fundamental role in tachyonic complexon theory and are used to define "in" and "out" tachyon states in perturbation theory.)

The conjugate momenta implied by eq. 2.232 are

$$
\begin{aligned}
\pi_{CLL}{}^{+} &= \partial\mathcal{L}/\partial(\partial^{-}\psi_{CLL}{}^{+}) = 2^{\frac{1}{2}}i\psi_{CLL}{}^{+\dagger} \qquad (2.234) \\
\pi_{CLL}{}^{-} &= \partial\mathcal{L}/\partial(\partial^{-}\psi_{CLL}{}^{-}) = 0 \\
\pi_{CLR}{}^{+} &= \partial\mathcal{L}/\partial(\partial^{-}\psi_{CLR}{}^{+}) = -2^{\frac{1}{2}}i\psi_{CLR}{}^{+\dagger} \qquad (2.235) \\
\pi_{CLR}{}^{-} &= \partial\mathcal{L}/\partial(\partial^{-}\psi_{CLR}{}^{-}) = 0
\end{aligned}
$$

x^{+} plays the role of the "time" variable in light-front quantized theories. So we define canonical equal x^{+} anti-commutation relations for spin $\frac{1}{2}$ tachyonic complexons also.

The canonical equal-light-front $(x^{+} = y^{+})$ anti-commutation relations of the independent fields would normally be:

$$\{\psi_{CLL}^{+\dagger}{}_a(x), \psi_{CLL}^{+}{}_b(y)\} = 2^{-1}[C^-R^+]_{ab}\delta(x^- - y^-)\delta^2(x_r - y_r)\delta^3(x_I - y_i) \tag{2.236}$$

$$\{\psi_{CLR}^{+\dagger}{}_a(x), \psi_{CLR}^{+}{}_b(y)\} = -2^{-1}[C^+R^+]_{ab}\,\delta(x^- - y^-)\delta^2(x_r - y_r)\delta^3(x_I - y_i) \tag{2.237}$$

$$\{\psi_{CLL}^{+}{}_a^{\dagger}(x), \psi_{CLR}^{+}{}_b(y)\} = \{\psi_{CLR}^{+}{}_a^{\dagger}(x), \psi_{CLL}^{+}{}_b(y)\} = 0 \tag{2.238}$$

$$\{\psi_{CLL}^{+}{}_a(x), \psi_{CLR}^{+}{}_b(y)\} = \{\psi_{CLR}^{+}{}_a^{\dagger}(x), \psi_{CLL}^{+\dagger}{}_b(y)\} = 0 \tag{2.239}$$

But as in the previous case they will be modified.

Again we see that the right-handed tachyon anti-commutation relation (eq. 2.237) has a minus sign relative to the corresponding conventional right-handed anti-commutation relation.

The sign differences between the left-handed and right-handed lagrangian terms ultimately lead to parity violating features in the Standard Model lagrangian.

2.5.3.1 Left-Handed Tachyonic Complexons

The free, "+" light-front, left-handed tachyonic complexon Fourier expansion is:

$$\psi_{CLL}^{+}(x_r, x_i) = \sum_{\pm s} \int d^2p_r dp^+ d^3p_i\, N_{CLL}^{+}(p)\theta(p^+)\delta((p_i^3(p^+ - p^-)/\sqrt{2} + \mathbf{p}_{r\perp}\cdot\mathbf{p}_{i\perp})/m^2)\cdot$$

$$\cdot[b_{CLL}^{+}(p, s)u_{CLL}^{+}(p, s)e^{-i(p\cdot x + p^*\cdot x^*)/2} + d_{CLL}^{+\dagger}(p, s)v_{CLL}^{+}(p, s)e^{+i(p\cdot x + p^*\cdot x^*)/2}] \tag{2.240}$$

Its hermitean conjugate is

$$\psi_{CLL}^{+\dagger}(x_r, x_i) = \sum_{\pm s} \int d^2p_r dp^+ d^3p_i\, N_{CLL}^{+}(p)\theta(p^+)\delta((p_i^3(p^+ - p^-)/\sqrt{2} + \mathbf{p}_{r\perp}\cdot\mathbf{p}_{i\perp})/m^2)\cdot$$

$$\cdot[b_{CLL}^{\dagger}(p^*,s)u_{CLL}^{\dagger}(p^*,s)e^{+i(p^*\cdot x + p\cdot x^*)/2} + d_{CLL}(p^*,s)v_{CLL}^{\dagger}(p^*,s)e^{-i(p^*\cdot x + p\cdot x^*)/2}] \tag{2.241}$$

where $\mathbf{p} = \mathbf{p}_r + i\mathbf{p}_i$, $x = x_r - ix_i$, $p\cdot x = p^0x^0 - \mathbf{p}\cdot\mathbf{x}$, and † indicates hermitean conjugate. The spinors are

$$u_{CLL}^{+}(p, s) = C^- R^+ S_{CL}w^1(0)$$
$$u_{CLL}^{+}(p, -s) = C^- R^+ S_{CL}w^2(0)$$
$$v_{CLL}^{+}(p, s) = C^- R^+ S_{CL}w^3(0)$$
$$v_{CLL}^{+}(p, -s) = C^- R^+ S_{CL}w^4(0)$$
$$u_{CLL}^{+\dagger}(p^*, s) = w^{1T}(0)S_{CL}R^+C^- \tag{2.242}$$

$$u_{CLL}^{++}(p^*, -s) = w^{2T}(0)S_{CL}R^+C^-$$
$$v_{CLL}^{++}(p^*, s) = w^{3T}(0)S_{CL}R^+C^-$$
$$v_{CLL}^{++}(p^*, -s) = w^{4T}(0)S_{CL}R^+C^-$$

where the superscript "T" indicates the transpose (These spinors are described in Appendix 2-A.) and

$$N_{CLL}^+(p) = (2\pi)^{-3}(2m/p^+)^{1/2} \tag{2.243}$$

The anti-commutation relations of the Fourier coefficient operators are

$$\{b_{CLL}(p,s), b_{CLL}^\dagger(p'^*,s')\} = 2^{-1/2}\delta_{ss'}\delta(p^+ - p'^+)\delta^2(\mathbf{p}_r - \mathbf{p}'_{r'})\delta^3(\mathbf{p}_i + \mathbf{p}'_{i'})$$
$$\{d_{CLL}(p,s), d_{CLL}^\dagger(p'^*,s')\} = 2^{-1/2}\delta_{ss'}\delta(p^+ - p'^+)\delta^2(\mathbf{p}_r - \mathbf{p}'_{r'})\delta^3(\mathbf{p}_i + \mathbf{p}'_{i'})$$
$$\{b_{CLL}(p,s), b_{CLL}(p'^*,s')\} = \{d_{CLL}(p,s), d_{CLL}(p'^*,s')\} = 0$$
$$\{b_{CLL}^\dagger(p,s), b_{CLL}^\dagger(p'^*,s')\} = \{d_{CLL}^\dagger(p,s), d_{CLL}^\dagger(p'^*,s')\} = 0 \tag{2.244}$$
$$\{b_{CLL}(p,s), d_{CLL}^\dagger(p'^*,s')\} = \{d_{CLL}(p,s), b_{CLL}^\dagger(p'^*,s')\} = 0$$
$$\{b_{CLL}^\dagger(p,s), d_{CLL}^\dagger(p'^*,s')\} = \{d_{CLL}(p,s), b_{CLL}(p'^*,s')\} = 0$$

The delta-function arguments $\delta^3(\mathbf{p}_i + \mathbf{p}'_{i'})$ above have a positive sign in order to obtain $\delta^3(\mathbf{x}_i - \mathbf{y}_i)$ in the field anti-commutators.

The spinors, eq. 2.242, satisfy

$$\sum_{\pm s} u_{CLL}^{+}{}_\alpha(p, s)\bar{u}_{CLL}^{+}{}_\beta(p^*, s) = (2m)^{-1}[C^-R^+(i\not{p} + m)R^-C^+]_{\alpha\beta} \tag{2.245}$$

$$\sum_{\pm s} v_{CLL}^{+}{}_\alpha(p, s)\bar{v}_{CLL}^{+}{}_\beta(p^*, s) = (2m)^{-1}[C^-R^+(i\not{p} - m)R^-C^+]_{\alpha\beta}$$

where $\bar{u}_{CLL}^{+} = u_{CLL}^{++\dagger}\gamma^0$ and $\bar{v}_{CLL}^{+} = v_{CLL}^{++\dagger}\gamma^0$.

We now evaluate the canonical left-handed, light-front anti-commutation relation:

$$\{\psi_{CLL}^{+}{}_a(x), \psi_{CLL}^{++}{}_b(y)\} = \sum_{\pm s,s'} \int d^3p_i d^2p dp^+ \int d^3p_i' d^2p' dp'^+ N_{CLL}^+(p) N_{CLL}^+(p')\cdot$$

$$\cdot\theta(p^+)\theta(p'^+)\delta((p_i^3(p^+ - p^-)/\sqrt{2} + \mathbf{p}_{r\perp}\cdot\mathbf{p}_{i\perp})/m^2)\,\delta((p_i'^3(p'^+ - p^-)/\sqrt{2} + \mathbf{p}'_{r\perp}\cdot\mathbf{p}'_{i\perp})/m^2)\cdot$$

$$\cdot[\{b_{CLL}^{++}(p'^*,s'),b_{CLL}^{+}(p,s)\}u_{CLL}^{+}{}_a(p,s)u_{CLL}^{++}{}_b(p'^*,s')e^{+i(p'^*\cdot y+p'\cdot y^*)2 - i(p\cdot x+p^*\cdot x^*)/2} +$$

$$+\{d_{CLL}^{+}(p'^*,s'),d_{CLL}^{++}(p,s)\}v_{CLL}^{+}{}_a(p,s)v_{CLL}^{++}{}_b(p'^*,s')e^{-i(p'^*\cdot y+p'\cdot y^*)/2 + i(p\cdot x + p^*\cdot x^*)/2}]$$

$$= 2^{-1/2}\sum_{\pm s} \int d^3p_i d^2p_r dp^+ [N_{CLL}^+(p)]^2\theta(p^+)[\delta((p_i^3(p^+ - p^-)/\sqrt{2} + \mathbf{p}_{r\perp}\cdot\mathbf{p}_{i\perp})/m^2)]^2 \cdot$$

$$\cdot [u_{CLL\ a}{}^{+}(p,s)u_{CLL\ b}{}^{+\dagger}(p^*,s)e^{+i(p^*\cdot(y-x)+p\cdot(y^*-x^*))/2} + v_{CLL\ a}{}^{+}(p,s)v_{CLL\ b}{}^{++}(p^*,s)e^{-i(p^*\cdot(y-x)+p\cdot(y^*-x^*))/2}]$$

$$= -2^{-3/2}\int d^3p_i d^2p\,dp^+\theta(p^+)[N_{CLL}{}^{+}(p)]^2\delta'((p_i^3(p^+-p^-)/\sqrt{2} + \mathbf{p}_{r\perp}\cdot\mathbf{p}_{i\perp})/m^2)(2m)^{-1}\cdot$$
$$\cdot\{[\,C^-R^+(i\not{p} + m)\gamma^0R^+C^-]_{ab}e^{+i(p^*\cdot(y-x)+p\cdot(y^*-x^*))/2} + [C^-R^+(i\not{p} - m)\gamma^0R^+C^-]_{ab}e^{-i(p^*\cdot(y-x)+p\cdot(y^*-x^*))/2}\}$$

$$= -(1/2)C^-R^+\delta_{ab}\int d^3p_i\,d^2p_\perp\int_0^\infty dp^+\,\delta'((p_i^3(p^+-p^-)/\sqrt{2} + \mathbf{p}_\perp\cdot\mathbf{p}_{i\perp})/m^2)(2\pi)^{-6}\cdot$$

$$\cdot\{e^{+i\{p^+(y^--x^-)-\mathbf{p}_{r\perp}\cdot(\mathbf{y}_{r\perp}-\mathbf{x}_{r\perp}) + \mathbf{p}_i\cdot(\mathbf{y}_i-\mathbf{x}_i)\}} + e^{-i\{p^+(y^--x^-)-\mathbf{p}_{r\perp}\cdot(\mathbf{y}_{r\perp}-\mathbf{x}_{r\perp}) + \mathbf{p}_i\cdot(\mathbf{y}_i-\mathbf{x}_i)\}}\}$$

$$= -C^-R^+\delta_{ab}(4\pi)^{-1}\int_0^\infty dp^+\,\delta'(\nabla_r\cdot\nabla_i/m^2)\delta^3(\mathbf{y}_i-\mathbf{x}_i)\,\delta^2(\mathbf{y}_r-\mathbf{x}_r)\{e^{+ip^+(y^--x^-)}+e^{-ip^+(y^--x^-)}\}$$

whereupon we revert back to the original form of the constraint: $\delta(\nabla_r\cdot\nabla_i/m^2)$

$$\{\psi_{CLL\ a}{}^{+}(x),\ \psi_{CLL\ b}{}^{++}(y)\} = -(1/2)C^-R^+\delta_{ab}\,\delta'(\nabla_r\cdot\nabla_i/m^2)\delta(y^--x^-)\delta^2(\mathbf{y}_r-\mathbf{x}_r)\delta^3(\mathbf{y}_i-\mathbf{x}_i) \quad (2.246)$$

The result is the left-handed, light-front equivalent of the earlier non-tachyon result. Again the constraint is apparent in the anti-commutator. (The factor of 2 difference is due to light-front coordinate definitions.)

 Therefore we have left-handed, light-front quantized tachyonic complexons with the equivalent of canonical anti-commutation relations, and with localized tachyonic complexons. As a result we have a canonical tachyonic complexon Quantum Field Theory.

2.5.3.2 Left-handed Case 4: Tachyonic Complexon Feynman Propagator

 The light-front Feynman propagator for the left-handed $\psi_{CLL}{}^{+}$ *tachyonic* complexon field is

$$iS^+{}_{CLLF}(x,y) = \theta(x^+ - y^+)\langle0|\psi_{CLL}{}^{+}(x)\psi_{CLL}{}^{++}(y)\gamma^0|0\rangle - \theta(y^+ - x^+)\langle0|\psi_{CLL}{}^{++}(y)\gamma^0\psi_{CLL}{}^{+}(x)|0\rangle \quad (2.247)$$
$$= -\tfrac{1}{2}\int d^3p_i d^2p_r dp^+\theta(p^+)N_{CLL}{}^{+2}\delta'((p_i^3(p^+-p^-)/\sqrt{2} + \mathbf{p}_{r\perp}\cdot\mathbf{p}_{i\perp})/m^2)(2m)^{-1}C^-R^+\cdot$$
$$\cdot\{\theta(x^+- y^+)[(i\not{p} + m)\gamma^0]e^{+i(p^*\cdot(y-x)+p\cdot(y^*-x^*))/2} +$$
$$+ \theta(y^+- x^+)[(i\not{p} - m)\gamma^0]e^{-i(p^*\cdot(y-x)+p\cdot(y^*-x^*))/2}\}R^+C^-\gamma^0$$

If we define the on-shell momentum variables

$$p_0{}^- = (p_{r0}{}^1 p_{r0}{}^1 + p_{r0}{}^2 p_{r0}{}^2 - \mathbf{p}_{i0}\cdot\mathbf{p}_{i0} - m^2)/(2p_0{}^+)$$
$$p_0{}^+ = p^+,\ p_{r0}{}^j = p_r{}^j \quad (\text{for } j = 1, 2),$$
$$\mathbf{p}_{i0} = \mathbf{p}_i,\ p_{r\perp0}{}^2 = p_{r0}{}^j p_{r0}{}^j$$

$$\not{p}_0 = p_0 \cdot \gamma$$

with $p_0 = (p^0, \mathbf{p}_{r0} + i\mathbf{p}_{r0})$ then the above equation can be rewritten as

$$= -\tfrac{1}{2} C^- R^+ \!\int\! d^4p\, d^3p_i N_{CLL}^{+2} \delta'((p_{i0}^{\;3}(p_0^+ - p_0^-)/\sqrt{2} + \mathbf{p}_{r\perp 0} \cdot \mathbf{p}_{i\perp 0})/m^2)(4\pi m)^{-1} e^{+i(p^* \cdot (y - x) + p \cdot (y^* - x^*))/2} \; .$$

$$\cdot \{\theta(p^+)(i\not{p} + m)\gamma^0]/[p^- - p_0^- + i\varepsilon] + \theta(-p^+)(i\not{p} - m)\gamma^0]/[p^- + p_0^- - i\varepsilon]\} R^+ C \gamma^0$$

$$= -\tfrac{1}{2} \int\! d^4p_r d^3p_i \, N_{CLL}^{+2} \delta'((p_{i0}^{\;3}(p^+ - p^-)/\sqrt{2} + \mathbf{p}_{r\perp} \cdot \mathbf{p}_{i\perp})/m^2)(p^+/4\pi m)\, e^{+i(p^* \cdot (y - x) + p \cdot (y^* - x^*))/2} \; .$$
$$\cdot [C^- R^+ (i\not{p} + m)\gamma^0 R^+ C^- \gamma^0][(p^2 + m^2 + i\varepsilon)]^{-1}$$

with $p_r = (p^0, \mathbf{p}_r)$ and $p = (p^0, \mathbf{p}_r + i\mathbf{p}_r)$. Substituting for N_{CLL} and using $x\delta'(x) = -\delta(x)$ we obtain

$$= -\tfrac{1}{2} \int\! d^4p_r d^3p_i (2\pi)^{-7} \delta'(\mathbf{p}_r \cdot \mathbf{p}_i/m^2) \exp[ip^0(y^0 - x^0) - i\mathbf{p}_r \cdot (\mathbf{y}_r - \mathbf{x}_r) + i\mathbf{p}_i \cdot (\mathbf{y}_i - \mathbf{x}_i)] \cdot$$
$$\cdot [C^- R^+ (i\not{p} + m)R^- C^+]/(p^2 + m^2 + i\varepsilon)$$

since $C^- R^+ (i\not{p} + m)\gamma^0 R^+ C^- \gamma^0 = C^- R^+ (i\not{p} + m)R^- C^+$. The integral can be written:

$$= \int\! d^4p_r d^3p_i \delta'(\mathbf{p}_r \cdot \mathbf{p}_i/m^2) C^- R^+ (i\not{p} + m)R^- C^+ \cdot$$
$$\cdot \exp[-ip^0(x^0 - y^0) + i\mathbf{p}_r \cdot (\mathbf{x}_r - \mathbf{y}_r) - i\mathbf{p}_i \cdot (\mathbf{x}_i - \mathbf{y}_i)]/(p^2 + m^2 + i\varepsilon)$$
$$= \int\! d^4p_r dM^2 \delta'(\nabla_r \cdot \nabla_i/m^2) C^- R^+ (ip^0\gamma^0 - (\nabla_r - i\nabla_i) \cdot \gamma + m)R^- C^+ \cdot$$
$$\cdot \exp[-ip^0(x^0 - y^0) + i\mathbf{p}_r \cdot (\mathbf{x}_r - \mathbf{y}_r)] J_2(\mathbf{x}_i - \mathbf{y}_i, M^2)/(p_r^2 + M^2 + i\varepsilon)$$

where

$$J_2(\mathbf{x}_i - \mathbf{y}_i, M^2) = (2\pi)^{-3} \!\int\! d^3p_i \, \delta(M^2 - \mathbf{p}_i^2 - m^2) \exp[-i\mathbf{p}_i \cdot (\mathbf{x}_i - \mathbf{y}_i)] \qquad (2.248)$$
$$= (2\pi)^{-2} |\mathbf{x}_i - \mathbf{y}_i|^{-1} \theta(M^2 - m^2) \sin((M^2 - m^2)^{\frac{1}{2}} |\mathbf{x}_i - \mathbf{y}_i|)$$

This tachyonic complexon Feynman propagator can be rearranged into the form of a spectral integral:

$$iS^+_{CLLF}(x, y) = -\!\int\! dM \, C^- R^+ (\gamma^0 \partial/\partial x^0 + (\nabla_r - i\nabla_i) \cdot \gamma - m) R^- C^+ \delta'(\nabla_r \cdot \nabla_i/m^2) J_2(\mathbf{x}_i - \mathbf{y}_i, M^2) \triangle_{FT}(x - y, M)$$
$$(2.249)$$

with ∇_r and ∇_i derivatives with respect to $\mathbf{x}_r$ and $\mathbf{x}_i$ and where

$$\triangle_{FT}(x - y, M) = (2\pi)^{-4} \!\int\! d^4p_r \exp[-ip^0(x^0 - y^0) + i\mathbf{p}_r \cdot (\mathbf{x}_r - \mathbf{y}_r)]/(p_r^2 + M^2 + i\varepsilon) \qquad (2.250)$$

2.5.3.3 Case 5: Right-Handed Tachyonic Complexons

The case of right-handed tachyonic complexons is similar to left-handed complexons with only one difference: a minus sign in the canonical right-handed equal-time commutation relations resulting in a minus sign in the creation and annihilation operator anti-commutation relations. The right-handed tachyonic complexon wave function light-front Fourier expansion is:

$$\psi_{CLR}^{+}(x_r, x_i) = \sum_{\pm s} \int d^2 p_r dp^+ d^3 p_i\, N_{CLR}^{+}(p)\theta(p^+)\delta((p_i^3(p^+ - p^-)/\sqrt{2} + \mathbf{p}_{r\perp}\cdot\mathbf{p}_{i\perp})/m^2)\cdot$$
$$\cdot[b_{CLR}^{+}(p, s)u_{CLR}^{+}(p, s)e^{-i(p\cdot x + p^*\cdot x^*)/2} + d_{CLR}^{+\dagger}(p, s)v_{CLR}^{+}(p, s)e^{+i(p\cdot x + p^*\cdot x^*)/2}]$$

$$(2.251)$$

where

$$N_{CLR}^{+}(p) = (2\pi)^{-3}(2m/p^+)^{\frac{1}{2}} \qquad (2.252)$$

Its hermitean conjugate is

$$\psi_{CLR}^{+\dagger}(x_r, x_i) = \sum_{\pm s} \int d^2 p_r dp^+ d^3 p_i\, N_{CLR}^{+}(p)\theta(p^+)\delta((p_i^3(p^+ - p^-)/\sqrt{2} + \mathbf{p}_{r\perp}\cdot\mathbf{p}_{i\perp})/m^2)\cdot$$
$$\cdot[b_{CLR}^{\dagger}(p^*,s)u_{CLR}^{\dagger}(p^*,s)e^{+i(p^*\cdot x + p\cdot x^*)/2} + d_{CLR}(p^*,s)v_{CLR}^{\dagger}(p^*,s)e^{-i(p^*\cdot x + p\cdot x^*)/2}]$$

$$(2.253)$$

where $\mathbf{p} = \mathbf{p}_r + i\mathbf{p}_i$, $\mathbf{x} = \mathbf{x}_r - i\mathbf{x}_i$, $p\cdot x = p^0 x^0 - \mathbf{p}\cdot\mathbf{x}$, and † indicates hermitean conjugate. The right-handed spinors are

$$u_{CLR}^{+}(p, s) = C^+ R^+ S_{CR}w^1(0)$$
$$u_{CLR}^{+}(p, -s) = C^+ R^+ S_{CR}w^2(0)$$
$$v_{CLR}^{+}(p, s) = C^+ R^+ S_{CR}w^3(0)$$
$$v_{CLR}^{+}(p, -s) = C^+ R^+ S_{CR}w^4(0)$$
$$u_{CLR}^{+\dagger}(p^*, s) = w^{1T}(0)S_{CR}R^+C^+$$
$$u_{CLR}^{+\dagger}(p^*, -s) = w^{2T}(0)S_{CR}R^+C^+$$
$$v_{CLR}^{+\dagger}(p^*, s) = w^{3T}(0)S_{CR}R^+C^+$$
$$v_{CLR}^{+\dagger}(p^*, -s) = w^{4T}(0)S_{CR}R^+C^+$$

$$(2.254)$$

where the superscript "T" indicates the transpose. The anti-commutation relations of the Fourier coefficient operators are

$$\{b_{CLR}(p,s), b_{CLR}^{\dagger}(p'^*,s')\} = -2^{-\frac{1}{2}}\delta_{ss'}\delta(p^+ - p'^+)\delta^2(\mathbf{p}_r - \mathbf{p}'_{r'})\delta^3(\mathbf{p}_i + \mathbf{p}'_{i'})$$
$$\{d_{CLR}(p,s), d_{CLR}^{\dagger}(p'^*,s')\} = -2^{-\frac{1}{2}}\delta_{ss'}\delta(p^+ - p'^+)\delta^2(\mathbf{p}_r - \mathbf{p}'_{r'})\delta^3(\mathbf{p}_i + \mathbf{p}'_{i'})$$
$$\{b_{CLR}(p,s), b_{CLR}(p'^*,s')\} = \{d_{CLR}(p,s), d_{CLR}(p'^*,s')\} = 0$$
$$\{b_{CLR}^{\dagger}(p,s), b_{CLR}^{\dagger}(p'^*,s')\} = \{d_{CLR}^{\dagger}(p,s), d_{CLR}^{\dagger}(p'^*,s')\} = 0$$
$$\{b_{CLR}(p,s), d_{CLR}^{\dagger}(p'^*,s')\} = \{d_{CLR}(p,s), b_{CLR}^{\dagger}(p'^*,s')\} = 0$$

$$(2.255)$$

$$\{b_{CLR}^{\dagger}(p,s), d_{CLR}^{\dagger}(p'^{*},s')\} = \{d_{CLR}(p,s), b_{CRR}(p'^{*},s')\} = 0$$

The spinors satisfy

$$\sum_{\pm s} u_{CLR}^{+}{}_{\alpha}(p, s)\bar{u}_{CLR}^{+}{}_{\beta}(p^{*}, s) = (2m)^{-1}[C^{+}R^{+}(-i\not{p} + m)R^{-}C^{-}]_{\alpha\beta} \tag{2.256}$$

$$\sum_{\pm s} v_{CLR}^{+}{}_{\alpha}(p, s)\bar{v}_{CLR}^{+}{}_{\beta}(p^{*}, s) = (2m)^{-1}[C^{+}R^{+}(-i\not{p} - m)R^{-}C^{-}]_{\alpha\beta}$$

where $\bar{u}_{CLR}^{+} = u_{CLR}^{+\dagger}\gamma^{0}$ and $\bar{v}_{CLR}^{+} = v_{CLR}^{+\dagger}\gamma^{0}$.

The right-handed anti-commutation relation with a minus sign follows in particular because of the minus signs in eqs. 2.255.

2.5.3.4 Right-handed Case 5: Tachyonic Complexon Feynman Propagator

The Feynman propagator for right-handed tachyonic complexons can be obtained from eqs. 2.249 and 2.250 by changing the parity projection operator and some numerator signs in the integral (basically $p \rightarrow -p$) resulting in

$$iS^{+}{}_{CLRF}(x, y) = \int dM\, C^{+}R^{+}(\gamma^{0}\partial/\partial x^{0} + (\nabla_{r} - i\nabla_{i})\cdot\gamma - m)R^{-}C^{-}\,\delta'(\nabla_{r}\cdot\nabla_{i}/m^{2})J_{2}(x_{i} - y_{i}, M^{2})\triangle_{FT}(x - y,M) \tag{2.257}$$

with $\nabla_{r} + i\nabla_{i}$ derivatives with respect to x_{r} and x_{i} and where

$$\triangle_{FT}(x - y, M) = (2\pi)^{-4}\int d^{4}p_{r}\, \exp[-ip^{0}(x^{0} - y^{0}) + ip_{r}\cdot(x_{r} - y_{r})]/(p_{r}^{2} + M^{2} + i\varepsilon) \tag{2.258}$$

2.5.3.5 Other Cases? No

The four cases considered above are the only cases having symmetry under the real Lorentz group L and a single real energy (with a corresponding single real time parameter) that is independent of the direction of the boost thus preserving (real) spatial rotation invariance. The reality of the time variable survives the breakdown to conventional Lorentz invariance.

One might think that using the other type of spinor boost operator.

$$S_{CR}(\Lambda_{CR}(\omega, \hat{w})) = \exp(-i\omega_{R}\sigma_{0i}w_{i}/2) = \exp(-\omega_{R}\gamma^{0}\gamma\cdot\hat{w}/2) \tag{2.259}$$
$$= \cosh(\omega_{R}/2)I + \sinh(\omega_{R}/2)\gamma^{0}\gamma\cdot\hat{w}$$

where $\omega_{R} = \omega - i\pi/2$ might lead to more possible forms of spin ½ wave equations and particles. In fact it merely leads to the same particle types but with the role of the left-handed and right-handed fields reversed. The result would be a "right-handed" Standard Model contrary to experiment.

2.6 Spinor Boosts Generate 4 Species of Particles: Leptons and Quarks

In this chapter we have found four types of fermions using complex Lorentz boosts that correspond in a natural way with the four general *species* (types) of known fermions: charged leptons, neutrinos, up-type color quarks and down-type color quarks.[58]

2.6.1 Charged lepton fermions

The conventional Dirac equation and solutions.

2.6.2 Neutrinos

Simple tachyons with real energy and 3-momentum. Their free field equation is:

$$(\gamma^\mu \partial/\partial x^\mu - m)\psi_T(x) = 0 \tag{2.260}$$

and their left-handed ψ_{TL}^{+} Feynman propagator is:

$$iS^{+}{}_{TLF}(x, y) = \tfrac{1}{2}C^{-}R^{+}\gamma^0 \!\int\! d^4p(2\pi)^{-4}\, p^{+}e^{-ip\cdot(x-y)}/(p^2 + m^2 + i\varepsilon) \tag{2.261}$$

Similarly the light-front Feynman propagator for the right-handed ψ_{TR}^{+} tachyon field is

$$iS^{+}{}_{TRF}(x,y) = -\tfrac{1}{2}C^{+}R^{+}\gamma^0 \!\int\! d^4p(2\pi)^{-4}\, p^{+}e^{-ip\cdot(x-y)}/(p^2 + m^2 + i\varepsilon) \tag{2.262}$$

2.6.3 Up-type Color Quarks

Up-type quarks are assumed[59] to be fermions with complex 3-momenta - complexons, and an internal color SU(3) symmetry, that satisfy $p^2 = m^2$. Their field equation with a color SU(3) index, denoted a, inserted is

$$[i\gamma^0 \partial/\partial t + i\boldsymbol{\gamma}\cdot(\nabla_r + i\nabla_i) - m]\psi_C{}^a(t, \mathbf{x}_r, \mathbf{x}_i) = 0 \tag{2.263}$$

with the subsidiary condition

$$\nabla_r\cdot\nabla_i\, \psi_C{}^a(t, \mathbf{x}_r, \mathbf{x}_i) = 0 \tag{2.264}$$

The free field solution is:

---

[58] We call each type of fermion a *species*. Each species has three known generations.

[59] The complexon theory that we develop and use for quark dynamics in the Standard Model is <u>not</u> required. Our Standard Model could use Dirac fermion dynamics for the up-type quarks and tachyon dynamics for down-type quarks. Then the (broken) Left-handed complex Lorentz boosts would have the basic space-time group rather than L_C. We choose to use complexon dynamics for quarks because they have an internal SU(3)-like structure suggestive of color SU(3). More importantly, their spin dynamics is different and thus may resolve the differences between theory and experiment for the deep inelastic parton spin-dependent structure functions.

$$\psi_C^{\ a}(x) = \sum_{\pm s} \int d^3 p_r d^3 p_i \ N_C(p)\delta(\mathbf{p_r}\cdot\mathbf{p_i}/m^2)[b_C(p,a,s)u_C^{\ a}(p, s)e^{-i(p\cdot x + p^*\cdot x^*)/2} +$$
$$+ d_C^{\ \dagger}(p,a,s)v_C^{\ a}(p, s)e^{+i(p\cdot x + p^*\cdot x^*)/2}] \qquad (2.265)$$

The free Feynman propagator arranged into the form of a spectral integral is

$$iS_C^{\ ab}(x,y)= -\delta^{ab}\int dM \ (i\gamma^0\partial/\partial x^0 - i(\nabla_r - i\nabla_i)\cdot\gamma + m)\delta'(\nabla_r\cdot\nabla_i/m^2)J(\mathbf{x_i} - \mathbf{y_i}, M^2)\triangle_F(x - y, M) \qquad (2.266)$$

where

$$\triangle_F(x - y, M) = (2\pi)^{-4}\int d^4 p_r \ \exp[-ip^0(x^0 - y^0) + i\mathbf{p_r}\cdot(\mathbf{x_r} - \mathbf{y_r})]/(p_r^{\ 2} - M^2 + i\varepsilon) \qquad (2.267)$$

and

$$J(\mathbf{x_i}, M^2) = (2\pi)^{-3}\int d^3 p_i \delta(M^2 + \mathbf{p_i}^2 - m^2) \ \exp[-i\mathbf{p_i}\cdot(\mathbf{x_i} - \mathbf{y_i})] \qquad (2.268)$$
$$= (2\pi)^{-2}|\mathbf{x_i} - \mathbf{y_i}|^{-1}\theta(m^2 - M^2)\sin((m^2 - M^2)^{\frac{1}{2}}|\mathbf{x_i} - \mathbf{y_i}|)$$

2.6.4 Down-type Color Quarks

Tachyonic complexons with complex 3-momenta, and an internal global SU(3) symmetry, that have mass shell condition $p^2 = -m^2$. Their field equation with a color SU(3) index, denoted a, inserted is

$$[\gamma^0\partial/\partial t + \gamma\cdot(\nabla_r + i\nabla_i) - m]\psi_{CL}^{\ a}(t, \mathbf{x_r}, \mathbf{x_i}) = 0 \qquad (2.269)$$

with the subsidiary condition on the wave function

$$\nabla_r\cdot\nabla_i \ \psi_{CL}^{\ a}(t, \mathbf{x_r}, \mathbf{x_i}) = 0 \qquad (2.270)$$

Its free field left-handed solution is:

$$\psi_{CLL}^{\ +a}(\mathbf{x_r}, \mathbf{x_i}) = \sum_{\pm s} \int d^2 p_r dp^+ d^3 p_i \ N_{CLL}^{\ +}(p)\theta(p^+)\delta((p_i^{\ 3}(p^+ - p^-)/\sqrt{2} + \mathbf{p_{r\perp}}\cdot\mathbf{p_{i\perp}})/m^2)\cdot$$
$$\cdot[b_{CLL}^{\ +}(p,a,s)u_{CLL}^{\ a}(p,a,s)e^{-i(p\cdot x + p^*\cdot x^*)/2} + d_{CLL}^{\ +\dagger}(p,a,s)v_{CLL}^{\ +a}(p,a,s)e^{+i(p\cdot x + p^*\cdot x^*)/2}]$$
$$(2.271)$$

and its right-handed solution is

$$\psi_{CLR}^{\ +a}(\mathbf{x_r}, \mathbf{x_i}) = \sum_{\pm s} \int d^2 p_r dp^+ d^3 p_i \ N_{CLR}^{\ +}(p)\theta(p^+)\delta((p_i^{\ 3}(p^+ - p^-)/\sqrt{2} + \mathbf{p_{r\perp}}\cdot\mathbf{p_{i\perp}})/m^2)\cdot$$

$$\cdot[b_{CLR}^{\ +}(p,a,s)u_{CLR}^{\ +a}(p,a,s)e^{-i(p\cdot x + p^*\cdot x^*)/2} + d_{CLR}^{\ +\dagger}(p,a,s)v_{CLR}^{\ +a}(p,a,s)e^{+i(p\cdot x + p^*\cdot x^*)/2}]$$
$$(2.272)$$

The free left-handed Feynman propagator arranged into the form of a spectral integral is

$$iS^+_{CLLF}{}^{ab}(x,y) = -\delta^{ab}\int dM\ C^-R^+(\gamma^0\partial/\partial x^0 + (\nabla_r - i\nabla_i)\cdot\gamma - m)R^-C^+\delta'(\nabla_r\cdot\nabla_i/m^2)\cdot$$
$$\cdot J_2(\mathbf{x_i} - \mathbf{y_i}, M^2)\triangle_{FT}(x - y, M) \qquad (2.273)$$

with ∇_r and ∇_i derivatives with respect to $\mathbf{x_r}$ and $\mathbf{x_i}$ and where

$$\triangle_{FT}(x - y, M) = (2\pi)^{-4}\int d^4p_r\ \exp[-ip^0(x^0 - y^0) + i\mathbf{p_r}\cdot(\mathbf{x_r} - \mathbf{y_r})]/(p_r^2 + M^2 + i\varepsilon) \qquad (2.274)$$

and

$$J_2(\mathbf{x_i}, M^2) = (2\pi)^{-3}\int d^3p_i\ \delta(M^2 - \mathbf{p_i}^2 - m^2)\ \exp[-i\mathbf{p_i}\cdot(\mathbf{x_i} - \mathbf{y_i})] \qquad (2.275)$$

$$= (2\pi)^{-2}|\mathbf{x_i} - \mathbf{y_i}|^{-1}\theta(M^2 - m^2)\sin((M^2 - m^2)^{1/2}|\mathbf{x_i} - \mathbf{y_i}|)$$

The free right-handed Feynman propagator arranged into the form of a spectral integral is

$$iS^+_{CLRF}{}^{ab}(x, y) = \delta^{ab}\int dM\ C^+R^+(\gamma^0\partial/\partial x^0 + (\nabla_r - i\nabla_i)\cdot\gamma - m)R^-C^-\delta'(\nabla_r\cdot\nabla_i/m^2)\cdot$$
$$\cdot J_2(\mathbf{x_i} - \mathbf{y_i}, M^2)\triangle_{FT}(x - y, M) \qquad (2.276)$$

with ∇_r and ∇_i derivatives with respect to $\mathbf{x_r}$ and $\mathbf{x_i}$, and where

$$\triangle_{FT}(x - y, M) = (2\pi)^{-4}\int d^4p_r\ \exp[-ip^0(x^0 - y^0) + i\mathbf{p_r}\cdot(\mathbf{x_r} - \mathbf{y_r})]/(p_r^2 + M^2 + i\varepsilon) \qquad (2.277)$$

2.7 First Step Towards The SuperStandard Model

Thus we have derived a set of four fermion species that corresponds to the known fermions of one fermion generation from the Complex Lorentz Group.[60] In subsequent chapters we will derive the one generation form of the model in detail from Complex Lorentz group features. Then we will derive the four fermion generation form of the model based on a U(4) group that we call the Generation group. We derive the Generation group from conservation laws for baryon and lepton number.

The overall pattern that begins to emerge from the developments in this chapter divides particles and interactions into two categories (as seen in Nature):

Particles with real 4-Momenta	Complexons (Complex 3-Momenta)
Leptons	Color quarks
SU(2)⊗U(1) Vector Bosons	Color SU(3) gluons
Higgs Particles	Possibly Higgs Particles

---

[60] Complex Lorentz group boosts lead to tachyons. Appendix 2-B presents evidence for the existence of tachyons.

Basically the leptons, SU(2)⊗U(1) Vector Bosons and a set of Higgs particles appear to be primarily based on the Left-handed boosts. These particles have real energies and momenta although some are "normal" and some are tachyons.

Another category of particles, complexons, emerges from our study of L_C. These particles have real energies and complex 3-momenta. In perturbation theory the loop integrations of loops of these particles would consist of a 7-fold integration over energy and complex 3-momenta with corresponding 7-fold delta functions to enforce energy-momentum conservation. As pointed out earlier the complex 3-momenta of these types of fermions has an SU(3) symmetry that it is natural to generalize to local color SU(3). (The other category of fermions, leptons. lack global SU(3) symmetry.) Thus we see the beginnings of the structure of the SuperStandard Model in this chapter on spin ½ particles.

2.8 Dirac-like Equations of Matter from 4-Valued Logic

In our derivation every truly fundamental particle of matter, whether quark or lepton, has spin ½. We have seen in chapter 10 of Blaha (2011c) that the basic algebra of Operator Logic eigenvalue operators, and that of its raising and lowering operators, is the same as the algebra of creation and annihilarion operators for free spin ½ particles. Our goal is to build our theory on the scaffolding of Operator Logic. We view a fermion particle as an iota core which is dressed in spatial coordinates (and internal symmetries):

$$\text{Iota core} + \text{coordinates} \rightarrow \text{fermion particle} \qquad (2.278)$$

The creation and annihilation operators $b(p,s)$ and $d^\dagger(p,s)$ (and their hermitean conjugates $b^\dagger(p,s)$ and $d(p,s)$) are mathematically similar to the raising and lowering operators of Operator (Matrix) Logic. They satisfy the anticommutation relations

$$\{b(q,s), b^\dagger(p,s')\} = \delta_{ss'}\delta^3(\mathbf{q} - \mathbf{p}) \qquad (2.279)$$
$$\{d(q,s), d^\dagger(p,s')\} = \delta_{ss'}\delta^3(\mathbf{q} - \mathbf{p})$$

Thus we see spin ½ particle wave functions originating from the Dirac-like spinors, and raising and lowering operators of the spinor formulation of Operator Logic.

When particles interact the quantum field theory interaction terms use fermion creation operators, $b(q,s)$ *and* $d^\dagger(q,s)$, *and annihilation operators,* $b^\dagger(p,s')$ *and* $d(q,s)$, *to implement the transformations between the Iotas of the interacting particles. Thus the mathematics of the embedded Iotas' logic values is automatically implemented within quantum field theoretic calculations.*

An interesting point that emerges from this discussion is the nature of spin ½ particle states such as

$$|p, s\rangle = b^\dagger(p, s)|0\rangle \qquad (2.280)$$

This state is interpreted as a one particle state. It also has an analogous interpretation in Operator Logic as creating a one term universe of discourse – a construct which is in part linguistic and in part logic. Thus particles are embodiments of Logic values and particle interactions change the logic values of the initial particles to those of the emergent particles. All in all, our universe can be viewed as an extraordinarily intricate logic machine. Serendipitously we are now seeing the use of particles to create quantum computers, which, in a sense, is bringing us full circle. Particles are Logic; Logic machines emerge from particle interactions.

2.9 Why Second Quantization of Fields?

One might have argued that the fermion field types that we have found could be treated as ordinary c-number fields and not be second quantized. However, particles are discrete entities that can be enumerated with integers. Second quantization implements the discrete particle concept in the most direct way and thus by Leibiz's Principle as well as Ockham's Razor second quantization is the best solution to obtain particle discreteness.

Quantum Theory arises in the discreteness of particles.

Appendix 2-A. Leptonic Tachyon Spinors

The general form of the solutions of the free tachyon Dirac equation can be written

$$\psi_T^{\ r}(x) = e^{-i\chi_r p \cdot x} w^r(p) \tag{2-A.1}$$

where $\chi_r = +1$ for $r = 1, 2$ and $\chi_r = -1$ for $r = 3, 4$. Denoting the spinors $w^r(p) = w^r(0)$ for a particle is at rest in a frame ($E = m$) we see they can take the form

$$w^r(0) = \begin{bmatrix} \delta_{1r} \\ \delta_{2r} \\ \delta_{3r} \\ \delta_{4r} \end{bmatrix} \tag{2-A.2}$$

where Kronecker deltas appear in the brackets. From eq. 2.30 we find

$$S_L(\Lambda_L(\omega, \mathbf{u}))w^r(0) = w_T^{\ r}(p) \tag{2-A.3}$$

Using eq. 2.66 for $S_L(\Lambda_L(\omega, \mathbf{u}))$ and

$$\mathbf{p} = m\mathbf{v}\gamma_s \qquad\qquad E = m\gamma_s \tag{2-A.4}$$

we see that eq. 2-A.3 implies the columns of the resulting $S_L(\Lambda_L(\omega, \mathbf{u}))$ matrix are

$$
\begin{array}{cccc}
\underline{w_T^{\ 3}(p)} & \underline{w_T^{\ 4}(p)} & \underline{w_T^{\ 1}(p)} & \underline{w_T^{\ 2}(p)}
\end{array}
$$

$$
S_L(\Lambda_L(\omega, \mathbf{u})) = \begin{bmatrix}
\cosh(\omega_L/2) & 0 & \sinh(\omega_L/2)p_z/p & \sinh(\omega_L/2)p_-/p \\
0 & \cosh(\omega_L/2) & \sinh(\omega_L/2)p_+/p & -\sinh(\omega_L/2)p_z/p \\
\sinh(\omega_L/2)p_z/p & \sinh(\omega_L/2)p_-/p & \cosh(\omega_L/2) & 0 \\
\sinh(\omega_L/2)p_+/p & -\sinh(\omega_L/2)p_z/p & 0 & \cosh(\omega_L/2)
\end{bmatrix}
\tag{2-A.5}
$$

based on the superluminal transformation of positive energy states to negative energy states with $p_\pm = p_x \pm i p_y$ and where $p = |\mathbf{p}|$. It is easy to verify

$$(i\not{p} - \chi_r m)w_T^{\,r}(p) = 0 \qquad\qquad (2\text{-A}.6)$$

where $\chi_r = -1$ for $r = 1, 2$ and $\chi_r = +1$ for $r = 3, 4$.

The spinors that we defined earlier can be generalized in a manner similar to Dirac spinors. We will use a similar notation to the Dirac spinor notation:

$$\begin{aligned}
u_T(p, s) &= w_T^{\,1}(p) \\
u_T(p, -s) &= w_T^{\,2}(p) \\
v_T(p, s) &= w_T^{\,3}(p) \\
v_T(p, -s) &= w_T^{\,4}(p)
\end{aligned} \qquad\qquad (2\text{-A}.7)$$

We define "double dagger" spinors:

$$\begin{aligned}
u_T^{\ddagger}(p, s) &= u_T^{\dagger}(p, s)\,i\boldsymbol{\gamma}\cdot\mathbf{p}/|\mathbf{p}| \\
u_T^{\ddagger}(p, -s) &= u_T^{\dagger}(p, -s)\,i\boldsymbol{\gamma}\cdot\mathbf{p}/|\mathbf{p}| \\
v_T^{\ddagger}(p, s) &= v_T^{\dagger}(p, s)\,i\boldsymbol{\gamma}\cdot\mathbf{p}/|\mathbf{p}| \\
v_T^{\ddagger}(p, -s) &= v_T^{\dagger}(p, -s)\,i\boldsymbol{\gamma}\cdot\mathbf{p}/|\mathbf{p}|
\end{aligned} \qquad\qquad (2\text{-A}.8)$$

where $\dagger$ indicates hermitean conjugate, which appear in important spinor "completeness" sums:

$$\sum_{\pm s} u_{T\alpha}(p, s)u_{T\,\beta}^{\ddagger}(p, s) = (2m)^{-1}(i\not{p} - m)_{\alpha\beta} \qquad\qquad (2\text{-A}.9)$$

$$\sum_{\pm s} v_{T\alpha}(p, s)v_{T\,\beta}^{\ddagger}(p, s) = (2m)^{-1}(i\not{p} + m)_{\alpha\beta} \qquad\qquad (2\text{-A}.10)$$

or

$$\sum_{\pm s} u_{T\alpha}(p, s)u_{T\,\beta}^{\dagger}(p, s) = -i(2m)^{-1}[(i\not{p} - m)\boldsymbol{\gamma}\cdot\mathbf{p}/|\mathbf{p}|]_{\alpha\beta} \qquad\qquad (2\text{-A}.11)$$

$$\sum_{\pm s} v_{T\alpha}(p, s)v_{T\,\beta}^{\dagger}(p, s) = -i(2m)^{-1}[(i\not{p} + m)\boldsymbol{\gamma}\cdot\mathbf{p}/|\mathbf{p}|]_{\alpha\beta} \qquad\qquad (2\text{-A}.12)$$

Lastly we define light-front, left-handed tachyon spinors by

$$\begin{aligned}
u_{TL}^{\,+}(p, s) &= C^- R^+ S_L(\Lambda_L(\omega, \mathbf{u}))w^1(0) \\
u_{TL}^{\,+}(p, -s) &= C^- R^+ S_L(\Lambda_L(\omega, \mathbf{u}))w^2(0) \\
v_{TL}^{\,+}(p, s) &= C^- R^+ S_L(\Lambda_L(\omega, \mathbf{u}))w^3(0) \\
v_{TL}^{\,+}(p, -s) &= C^- R^+ S_L(\Lambda_L(\omega, \mathbf{u}))w^4(0)
\end{aligned} \qquad (2\text{-A}.13)$$

$$\begin{aligned}
u_{TL}^{\,+\dagger}(p, s) &= w^{1T}(0)\, S_L^{\dagger}(\Lambda_L(\omega, \mathbf{u}))\, R^+C^- \\
u_{TL}^{\,+\dagger}(p, -s) &= w^{2T}(0)\, S_L^{\dagger}(\Lambda_L(\omega, \mathbf{u}))R^+C^- \\
v_{TL}^{\,+\dagger}(p, s) &= w^{3T}(0)\, S_L^{\dagger}(\Lambda_L(\omega, \mathbf{u}))R^+C^- \\
v_{TL}^{\,+\dagger}(p, -s) &= w^{4T}(0)\, S_L^{\dagger}(\Lambda_L(\omega, \mathbf{u}))R^+C^-
\end{aligned} \qquad (2\text{-A}.14)$$

where the superscript "T" indicates the transpose and $\dagger$ indicates hermitean conjugate.

Appendix 2-B. Experimental Evidence for Faster-Than-Light Particles & Physics

Among the key assumptions of our Unified SuperStandard Models are 1) that the speed of light is the same in all inertial reference frames and 2) that some fundamental particles (neutrinos and down-type quarks) travel faster than the speed of light.

In this appendix we describe convincing evidence for faster than light physics. Until 1907 physicists thought that there was no limit on the speed of a particle or lump of matter. In 1907 Einstein and Poincaré showed that there was an inherent limit on the speed of a massive object – the speed of light. For the past 100 odd years physicists have generally accepted the speed of light as the limiting speed for particles with mass. Several theoretical physicists in the 1960's (E. C. Sudarshan and Gerald Feinberg) investigated the possibility of faster than light particles. They found that faster than light particles were theoretically possible but their theories – particularly their quantum field theories – had numerous discrepancies from canonical quantum field theory. These differences were taken by many to indicate that faster than light particles (called tachyons) were not present in nature. This belief was further supported by the happenings at particle accelerators where it was impossible to accelerate normal charged particles such as protons faster than the speed of light.

In the past fifteen years this author[61] developed a satisfactory quantum field theory of faster than light particles and found that if neutrinos and down-type quarks were faster than light particles he could derive the form of The Standard Model of Elementary Particles in detail. This theoretical development seems to have stimulated experimental groups at the new Linear Hadron Collider (LHC) at the CERN laboratory in Switzerland and the Gran Sasso Laboratory in Italy to measure the speed of neutrinos emitted in LHC particle collisions. The results, described below, were mixed and one can fairly say they neither proved nor disproved that neutrinos were tachyons.

However there is other experimental data that strongly indicate that neutrinos are tachyons, and that quantum mechanics requires – not just faster than light behavior – but in some circumstances instantaneous effects at a distance – infinite speed of transmission!

In this appendix we will look at experimentally proven instantaneous Quantum Mechanical effects, at tritium decay experiments over the past 20 years that imply faster than light neutrinos, at neutrino speed measurements at the CERN LHC and Gran Sasso, at tachyonic particle behavior inside of Black Holes, and at the tachyonic behavior of Higgs particles, the

---

[61] See Blaha (2012b) and earlier books extending back nine years.

"so-called God particle." *The cumulative result of these considerations is that faster than light particles, and physics, are a part of nature.*

2-B.1 Instantaneous Quantum Mechanical Effects

Quantum entanglement is a quantum phenomenon wherein parts of a physical system are in a certain quantum state but are separated by a space-like distance. If a change is made in part of a quantum entangled system then it is known theoretically, and experimentally, that other parts of the system change instantaneously.[62] Many experiments have shown that the change in other parts of a system is instantaneous and thus can be viewed as taking place at infinite speed – obviously beyond the speed of light.[63] The most recent experiment by Juan Yin et al[64] has shown directly that quantum mechanical effects travel faster than 10,000 times the speed of light. These experimental results are consistent with the instantaneous speed predicted by quantum mechanics. Thus faster than light behavior is implicit in quantum theory and is experimentally verified.

2-B.2 Tritium Decay Experiments Yielding Neutrinos

Fact: Particles with negative values for the square of their mass are tachyons – particles moving faster than light.

A series of experiments by various groups over recent years imply that electron neutrinos produced in tritium decay have negative mass squared despite the best efforts of experimenters to obtain positive values for the neutrino mass squared.

Experiment	measured mass squared	Year
Mainz	$-1.6 \pm 2.5 \pm 2.1$	2000
Troitsk	$-1.0 \pm 3.0 \pm 2.1$	2000
Zürich	$-24 \pm 48 \pm 61$	1992
Tokyo INS	$-65 \pm 85 \pm 65$	1991
Los Alamos	$-147 \pm 68 \pm 41$	1991
Livermore	$-130 \pm 20 \pm 15$	1995
China	$-31 \pm 75 \pm 48$	1995
1998 Average	-27 ± 20	1998

Table 2-B.1 Electron neutrino mass squared values found in various tritium decay experiments. (Masses are in units of eV.) The average mass squared is negative suggesting electron neutrinos are tachyons.

---

[62] Matson, John, "Quantum Teleportation Achieved Over Record Distances" *Nature* **13**, August 2012.

[63] Francis, Matthew, "Quantum Entanglement Shows that Reality Can't be Local", *Ars Technica*, 30 October 2012.

[64] Juan Yin et al, arXiv[quant-ph]: 1303.0614V1 (March 4, 2013).

Table 2-B.1 summarizes the measured electron mass squared in these experiments. These experiments strongly suggest that neutrinos have negative mass squared and are thus faster-than-light particles - tachyons. However their small masses indicate that they only exceed the speed of light by a small amount.

2-B.3 LHC/Gran Sasso Direct Measurements of Neutrino Speeds

Two groups performed experiments at Gran Sasso Laboratory in Italy. They detected neutrinos emitted in interactions at the CERN LHC in Switzerland. The LVD collaboration in an exhaustive study of neutrino velocities found that the question was still open according to their data. Their refereed Physical Review Letter Abstract stated:

> We report the measurement of the time of flight of v_μ on the CNGS baseline (732 km) with the Large Volume Detector (LVD) at the Gran Sasso Laboratory. The CERN-SPS accelerator has been operated from May 10th to May 24th 2012, with a tightly bunched-beam structure to allow the velocity of neutrinos to be accurately measured on an event-by-event basis. LVD has detected 48 neutrino events, associated with the beam, with a high absolute time accuracy. These events allow us to establish the following limit on the difference between the neutrino speed and the light velocity: $-3.8\times10^{-6} < (v_v -c)/c < 3.1\times10^{-6}$ (at 99% C.L.). This value is an order of magnitude lower than previous direct measurements.[65]

These results (involving at least 35 neutrino detections) slightly favor, and do not rule out, faster-than-light neutrinos. Another experiment at the same locations by the ATLAS group stated that they found neutrino velocities (Five neutrinos were measured.) were below c. This group has not published their results as yet. We conclude that the published data appears to support faster than light neutrinos – consistent with our theory of The Standard Model.

A new project is in the planning stages to measure neutrino beams at larger distances. The hope is that the masses of the various neutrinos will be determined by the experiment. If the neutrino mass squared values turn out to be negative then it will constitute additional proof that neutrinos are tachyons (confirming tritium decay data), and thus support this author's formulation of The Standard Model of Elementary Particles.

2-B.4 Tachyonic Behavior Within Black Holes

Inside a black hole (such as the Schwarzschild solution of General Relativity) the time coordinate effectively becomes a spatial coordinate and the radius coordinate effectively becomes a time coordinate. An in-falling particle has a constantly decreasing radial distance from the center of the black hole just as time always increases outside a black hole.

---

[65] N. Yu. Agafonova et al. (LVD Collaboration), "Measurement of the Velocity of Neutrinos from the CNGS Beam with the Large Volume Detector" Phys. Rev. Lett. **109**, 070801 (15 August 2012).

As a result of the interchange of the roles of time and radius the velocity of a particle descending radially inside a Black Hole has a speed faster than light and is tachyonic.

2-B.5 Higgs Fields are Tachyons

Recently groups at the LHC CERN laboratory have announced the discovery of Higgs particles. The dynamic equations for Higgs bosons in The Standard Model have a negative mass squared. The mass squared must be negative or the Higgs Mechanism could not generate particle masses. Having negative mass terms implies that Higgs fields are tachyonic – faster than light particles. Their tachyonic nature is masked by a quartic self-interaction that generates a condensate and thereby the masses of other particles.

2-B.6 Conclusion: Faster-Than-Light Particles – Tachyons Exist in Nature

The bulk of the experimental and theoretical evidence presented in previous sections strongly favors the existence of faster-than-light particles such as neutrinos. Tachyonic neutrinos are an important part of our form of The Standard Model. This form of the theory also strongly suggests that quarks are tachyonic in parallel with tachyonic neutrinos in order to obtain the symmetries of The Standard Model.

3. Complex Lorentz Transformations, the Coordinate Reality Group, and the Internal Symmetry Reality Group

Chapters 3 through 7 derive the Coordinate Reality Group in detail based on the Complex Lorentz Group. *The elements of the Coordinate Reality Group are local transformations that map complex-valued coordinate systems generated by complex Lorentz transformations to real-valued coordinate systems.* We will see that the Coordinate Reality Group consists of the subgroups (SU(3), SU(2)⊗U(1), and SU(2)⊗U(1)) with 16 generators. The 16 generators are linear combinations of U(4) group generators. Local U(4) transformations can map any complex 4-dimensional space-time to real values.

The reader will notice that SU(3) and SU(2)⊗U(1) are the known symmetry groups of The Standard Model. However the Coordinate Reality Group cannot be an internal symmetry group without running into the difficulties of 'No Go' theorems that caused the demise of SU(6) in the 1960's.

So we are led to consider the possibility that the Standard Model internal symmetries are the result of an Internal Symmetry Reality Group[66] that is the analogue of the Coordinate Reality Group. We describe this in detail in chapter 8.

We therefore postulate that the Standard Model internal symmetries are the consequence of an Internal Symmetry Reality Group. We will also postulate that the additional SU(2)⊗U(1) Coordinate Reality subgroup has a corresponding Internal Symmetry Group analogue for a Dark Matter SU(2)⊗U(1) ElectroWeak symmetry.[67] Thus the full Internal Symmetry Reality Group of the Unified SuperStandard Model is

$$SU(3) \otimes SU(2) \otimes U(1) \otimes SU(2) \otimes U(1)$$

---

[66] In previous books we implicitly proceeded from the Coordinate Reality Group to the Internal Symmetry Reality Group. In this volume it becomes necessary to explicitly discuss the Internal Symmetry Reality Group. We will see in chapter 8 that it follows from an Internal Symmetry Complex Lorentz Group, which, in turn, will be seen to be related to the 'Rotation of Interactions' Θ-Symmetry group described in previous books by the author and discussed later in this book.

[67] This postulate is based on the Decision Axiom Replication Principle, which, in brief, states that physical phenomena/models tend to be repeated in Nature. A good example is the appearance of harmonic oscillators in many areas of Physics. (See Appendix C.)

The origin of the Internal Symmetry Reality Group and, in particular, why we insert the word 'Reality' in its name, we will address in chapter 8. In the remainder of this chapter and chapters 4 through 7 we will explore the Coordinate Reality Group in some detail since it will map directly to the structure of the Internal Symmetry Reality Group.

3.1 Coordinate Reality Group

The Real Lorentz group transforms between coordinate systems that are in relative motion at a velocity below the speed of light. Consequently it transforms real-valued coordinate systems to real-valued coordinate systems. The Complex Lorentz group includes the Real Lorentz group as a subgroup. It also includes transformations that transform real-valued coordinate systems into complex-valued coordinate systems. A subset of these transformations transform between coordinate systems at a velocity whose magnitude is below the speed of light. Another subset of transformations transform between coordinate systems at a relative velocity whose magnitude is above the speed of light.[68] These transformations correspond to faster than light motion and provide boosts discussed in chapter 2 that generate tachyonic fermions.

We showed in earlier books such as Blaha (2017b) that faster than light physics is consistent and physically acceptable – even to the point of deriving faster-than-light Thermodynamics from the Maxwell-Boltzmann distribution including the law of increasing Entropy. (See Appendix H.)

3.2 Internal Symmetry Reality Group

In this section[69] we will outline the physical basis of the known Standard Model symmetry group SU(3)⊗SU(2)⊗U(1) and show that it should generalize to

$$SU(3)\otimes SU(2)\otimes U(1)\otimes SU(2)\otimes U(1)$$

with the extra SU(2)⊗U(1) factor, which we *postulate* describes Dark Matter ElectroWeak interactions.[70] Normal matter SU(2)⊗U(1) symmetry emerges from the consideration of boosts at real-valued velocities greater than the speed of light. The Dark Matter sector SU(2)⊗U(1) symmetry emerges from the consideration of boosts at complex-valued velocities less than or greater than the speed of light.

Before beginning the discussion of these cases we note that a local U(4) transformation can change any complex-valued coordinate system to a real-valued coordinate system in 4-

---

[68] This was not noted by the rigorously mathematical derivation of Axiomatic Quantum Field Theory by Streater (2000) and others.

[69] This chapter outlines the detailed discussion presented in chapters 5 – 7 of Blaha (2017b) as well as earlier books such as Blaha (2015a).

[70] The Dark Matter ElectroWeak interactions must be distinct from the normal matter ElectroWeak interactions or Dark Matter would have been found experimentally by now.

dimensional flat space-time. However because of the peculiar nature of the Lorentz group, and the defining relation of the Real and Complex Lorentz groups:

$$\Lambda(\mathbf{v}, \mathbf{\theta})^{\mathrm{T}} G \Lambda(\mathbf{v}, \mathbf{\theta}) = G$$

where G is the flat space-time metric G = diag(1, −1, −1, −1), and where the superscript ^{T} specifies the transpose of the matrix; it is possible to *physically* specify the origin of each of the subgroups of U(4): SU(2)⊗U(1), SU(3), and 'Dark' SU(2)⊗U(1). (See below.) These subgroups do not commute with each other. The total number of their generators is 16.

Now following Decision Axiom Replication Principle: 'Nature Tends to Repeat Successful Strategies' we assume that the fundamental internal symmetry group of elementary particles is *analogous* to the set of subgroups of the subgroups of the Complex Lorentz group that transform complex-valued coordinate systems into real-valued coordinate systems.

We construct it by creating a product of the subgroup types with the assumption that the subgroups, which are now internal symmetry subgroups, commute with each other. Thus the choice of the internal symmetry Reality group is the tensor product: SU(3)⊗SU(2)⊗U(1)⊗SU(2)⊗U(1). We denote it as R and call it the Internal Symmetry Reality group:

$$R = SU(3)⊗SU(2)⊗U(1)⊗SU(2)⊗U(1)$$

Each of the factors has an associated set of local Yang-Mills vector boson fields that become particle interactions.

3.3 Boosts at Real-Valued Velocities Greater Than the Speed of Light

In chapter 5 of Blaha (2017b) we showed that Complex Lorentz boosts for real-valued velocities with magnitude greater than the speed of light transform a rest frame coordinate system to a new complex-valued coordinate system. In general an SU(2)⊗U(1) transformation is needed to transform the target complex-valued coordinate system to a real-valued coordinate system for this type of boosts. We can symbolize these transformations with

Real-valued Coordinate System $\rightarrow$ $R_{SU(2)⊗U(1)}\Lambda(|\mathbf{v}| > \mathbf{c}) \rightarrow$ Real-valued Coordinate System

where $\Lambda(|\mathbf{v}| > \mathbf{c})$ is a Complex Lorentz transformation with real-valued relative velocity $|v| > c$, and $R_{SU(2)⊗U(1)}$ is an SU(2)⊗U(1) transformation from complex-valued coordinates to real-valued coordinates.

The Complex Lorentz transformation $\Lambda(|\mathbf{v}| > \mathbf{c})$ when applied to a fermion at rest transforms it into a tachyon with real-valued energy. (See chapter 3 of Blaha (2017b).) The tachyons of this type are those of the neutral lepton species and of the Dark neutral lepton species.

3.4 Boosts at Complex-Valued Velocities

There are two types of boosts of this kind: those whose magnitude of velocity exceeds the speed of light yielding tachyons, and those whose magnitude of velocity is below the speed of light yielding a non-tachyon fermion. (See chapter 2, or chapters 6 and 7 of Blaha (2017b) for details.) The forms of the tachyons of these varieties are described chapter 2. The fermions generated by these boosts are identified as the up and down type normal and Dark quarks.

3.5 Mapping of the Boosted Fermions to Normal and Dark Fermion Species

Four types of fermion species, and the internal symmetries and interactions of particles, emerge from the Complex Lorentz group. We can summarize the map from our theory to the real world in the following way:

<u>Normal Matter Fermions</u>
Dirac fermions – Charged Leptons – Fields generated by Real Lorentz boosts – real $v < c$
Tachyon fermions – Neutrinos – Fields generated by Real Lorentz boosts – real $v > c$
Complexon fermions – Up-Type Quark Triplets – Fields generated by Lorentz boosts – complex $v < c$
Tachyon Complexon fermions – Down-Type Quark Triplets – From Lorentz boosts – complex $v > c$

<u>Dark Matter Fermions</u>
Dirac fermions – Dark Charged Leptons – Fields generated by Real Lorentz boosts – real $v < c$
Tachyon fermions – Dark Neutrinos – Fields generated by Real Lorentz boosts – real $v > c$
Complexon fermions – Dark Up-Type Quark Singlets – Fields generated by Lorentz boosts – complex $v < c$
Tachyon Complexon fermions – Dark Down-Type Quark Singlets – From Lorentz boosts – complex $v > c$

<u>Normal Matter Gauge Bosons</u>
$SU(2) \otimes U(1)$ - real space-time coordinates – not complexon coordinates
$SU(3)$ - complex space-time coordinates – complexon coordinates

<u>Dark Matter Gauge Bosons</u>
$SU(2) \otimes U(1)$ - real space-time coordinates – not complexon coordinates

4. Coordinate Reality Group Analogue to ElectroWeak Doublets

4.0 Introduction

In chapter 2 we established the four species of fermions based on Complex Lorentz group (L_C) boosts. In this chapter[71] we will introduce the coordinate space analogue of the ElectroWeak interactions based on Complex Lorentz group considerations. We begin by generalizing the free Dirac equation to a 2×2 matrix of Dirac-like equations that have a larger group covariance. This matrix equation is applied to a doublet consisting of a normal Dirac particle wave function and a tachyon wave function. We will identify these doublets as analogues of ElectroWeak lepton doublets.

Then starting in section 4.5 we consider a generalized 2×2 equation matrix (covariant under the L_C group) for doublets of complexon particles with complex 3-momenta consisting of an up-type complexon and a down-type tachyonic complexon. Because of an inherent SU(3) symmetry we will identify these doublets as analogues of quark ElectroWeak doublets. SU(3) symmetry leads us to identify each complexon quark in a doublet as a color SU(3) triplet, and leads thence to SU(3) color quark confinement.

4.1 Transformations of Dirac and Tachyon Equations

A Left-handed boost of the Dirac equation transforms the Dirac equation into the spin ½ tachyon equation, and vice versa:

$$S_L(\Lambda_L(\omega, \mathbf{u}))\psi(x) \rightarrow \psi_T'(x') \tag{4.1a}$$
$$S_L(\Lambda_L(\omega, \mathbf{u}))\psi_T(x) \rightarrow \psi'(x')$$

Also, noting the appearance of a γ^5,

$$S_L(\Lambda_L(\omega, \mathbf{u}))(\gamma^\mu \partial/\partial x^\mu - m)S_L^{-1}(\Lambda_L(\omega, \mathbf{u})) = (i\gamma^\mu \partial/\partial x'^\mu - m) \tag{4.1b}$$
$$S_L(\Lambda_L(\omega, \mathbf{u}))\gamma^5(i\gamma^\mu \partial/\partial x^\mu - m)\gamma^5 S_L^{-1}(\Lambda_L(\omega, \mathbf{u})) = (\gamma^\mu \partial/\partial x'^\mu - m)$$

where

$$x'^\mu = i\Lambda_L{}^\mu{}_\nu(\omega, \mathbf{u})x^\nu \tag{4.1c}$$
$$\partial/\partial x'^\mu = -i\Lambda_L{}^\nu{}_\mu(\omega, \mathbf{u})\partial/\partial x^\nu$$

with

---

[71] This chapter is extracted from Blaha (2007b).

$$x' = E(\mathbf{v})x = i\Lambda_L(\mathbf{v})x$$

Eqs. 4.1a – 4.1c imply

$$S_L(\Lambda_L(\omega, \mathbf{u}))(\gamma^\mu \partial/\partial x^\mu - m)\psi_T(x) = (i\gamma^\mu \partial/\partial x'^\mu - m)S_L(\Lambda_L(\omega, \mathbf{u}))\psi_T(x)$$
$$= (i\gamma^\mu \partial/\partial x'^\mu - m)\psi'(x') \qquad (4.1d)$$

and

$$S_L(\Lambda_L(\omega, \mathbf{u}))\gamma^5(i\gamma^\mu \partial/\partial x^\mu - m)\psi(x) = (\gamma^\mu \partial/\partial x'^\mu - m)S_L(\Lambda_L(\omega, \mathbf{u}))\gamma^5\psi(x)$$
$$= (\gamma^\mu \partial/\partial x'^\mu - m)\psi_T'(x') \qquad (4.1e)$$

where

$$\psi'(x') = S_L(\Lambda_L(\omega, \mathbf{u}))\psi_T(x) \qquad (4.1f)$$

and

$$\psi_T'(x') = S_L(\Lambda_L(\omega, \mathbf{u}))\gamma^5\psi(x) \qquad (4.1g)$$

Note the Dirac equation is not left-handed Complex Lorentz covariant.

4.2 Doublet Extended Dirac Equations

We will now consider the issue of generalizing the Dirac equation so that the extended equation is covariant under both Lorentz transformations and Left-handed Complex Lorentz transformations.

The only obvious method to obtain an extended Dirac equation that is covariant under Complex Lorentz transformations is to define an 8×8 matrix generalization. Let

$$\text{đ}(x) = \begin{bmatrix} (\gamma^\mu \partial/\partial x^\mu - m) & 0 \\ 0 & (i\gamma^\mu \partial/\partial x^\mu - m) \end{bmatrix} \qquad (4.2)$$

be an 8×8 matrix operator with the 4×4 matrix elements shown, and let

$$\Psi(x) = \begin{bmatrix} \psi_T(x) \\ \psi(x) \end{bmatrix} \qquad (4.3)$$

be an 8 component column vector composed of a Dirac field and a tachyon field. Then the extended free Dirac equation is

$$\text{đ}(x)\Psi(x) = 0 \qquad (4.4)$$

We now define the 8×8 Left-handed Complex Lorentz transformation

$$S_{L8}(\Lambda_L(v)) = \begin{bmatrix} 0 & S_L(\Lambda_L(v))\gamma^5 \\ S_L(\Lambda_L(v)) & 0 \end{bmatrix} \qquad (4.5)$$

with inverse transformation

$$S_{L8}^{-1}(\Lambda_L(v)) = \begin{bmatrix} 0 & S_L^{-1}(\Lambda_L(v)) \\ \gamma^5 S_L^{-1}(\Lambda_L(v)) & 0 \end{bmatrix} \qquad (4.6)$$

Note: we use the notations $S_L(\Lambda_L(v))$ and $S_L(\Lambda_L(\omega, \mathbf{u}))$ interchangeably. Applying S_{L8} to eq. 4.4 yields

$$0 = S_{L8}(\Lambda_L(v))đ(x)\Psi(x) = đ(x')\Psi'(x') \qquad (4.7)$$

where

$$\Psi'(x') = \begin{bmatrix} S_L\gamma^5\psi(x) \\ S_L\psi_T(x) \end{bmatrix} = \begin{bmatrix} \psi_T'(x') \\ \psi'(x') \end{bmatrix} \qquad (4.8)$$

Thus the extended Dirac equation is covariant under generalized Left-handed Complex Lorentz transformations such as eqs. 4.5-4.6. Covariance requires the tachyon and the Dirac particles must have the same absolute value for the mass which is the iotal mass in the free fermion case.

It is easy to show that the extended Dirac equation eq. 4.4 is also covariant under conventional Lorentz transformations in the 8×8 representation:

$$S_8(\Lambda(v)) = \begin{bmatrix} S(\Lambda(v)) & 0 \\ 0 & S(\Lambda(v)) \end{bmatrix} \qquad (4.9)$$

with inverse

$$S_8^{-1}(\Lambda(v)) = \begin{bmatrix} S^{-1}(\Lambda(v)) & 0 \\ 0 & S^{-1}(\Lambda(v)) \end{bmatrix} \qquad (4.10)$$

and non-diagonal Lorentz transformations:

$$S_{8A}(\Lambda(v)) = \begin{bmatrix} 0 & S(\Lambda(v)) \\ S(\Lambda(v)) & 0 \end{bmatrix} \tag{4.11}$$

with inverse transformation

$$S_{8A}^{-1}(\Lambda(v)) = \begin{bmatrix} 0 & S^{-1}(\Lambda(v)) \\ S^{-1}(\Lambda(v)) & 0 \end{bmatrix} \tag{4.12}$$

Under a conventional Lorentz transformation we find

$$0 = S_8(\Lambda(v))đ(x)\Psi(x) = đ(x')\Psi'(x') \tag{4.13}$$

$$0 = S_{8A}(\Lambda(v))đ(x)\Psi(x) = đ(x')\Psi'(x')$$

The lagrangian density that corresponds to our 8-dimensional construction is

$$\mathcal{L}_8 = \overline{\Psi}(x)đ(x)\Psi(x) \tag{4.14}$$

where

$$\overline{\Psi}(x) = \Psi^\dagger\Gamma^0 \tag{4.15}$$

and

$$\Gamma^0 = \begin{bmatrix} i\gamma^0\gamma^5 & 0 \\ 0 & \gamma^0 \end{bmatrix} \tag{4.16}$$

The action

$$I = \int d^4x \mathcal{L}_8 \tag{4.17}$$

is invariant under Lorentz transformations S_8 and S_{8A}.   The Hamiltonian density for the 8-dimensional theory is

$$\mathcal{H}_8(x) = \begin{bmatrix} i\psi_T^\dagger\gamma^5(\boldsymbol{\alpha}\cdot\nabla + \beta m)\psi_T & 0 \\ 0 & \psi^\dagger(-i\boldsymbol{\alpha}\cdot\nabla + \beta m)\psi \end{bmatrix} \tag{4.18}$$

4.3 Non-Invariance of the Extended Free Action under a Left-handed Extended Lorentz Transformation

The action eq. 4.17 is not invariant under Left-handed Complex Lorentz transformations. The fundamental cause of this non-invariance is the three dimensional nature of space. In the case of Dirac particles one can define a Lorentz invariant action because time is one- dimensional. Thus one can use $\psi^\dagger\gamma^0 = \bar\psi$ to form the Dirac field lagrangian and action. A key factor in Lorentz invariance is the relation between the inverse and hermitean conjugate of the spinor boost operator

$$\gamma^0 S^{-1}\gamma^0 = S^\dagger \tag{4.19}$$

In the case of the tachyon lagrangian and action, Left-handed Complex Lorentz invariance is not possible because the tachyonic equivalent to eq. 4.19 is

$$S_L^{-1}(\Lambda(\mathbf{v}))\gamma\cdot\mathbf{p}/|\mathbf{p}| = i\gamma^0 S_L^{\dagger}(\Lambda(\mathbf{v})) \tag{4.20}$$

where $\mathbf{p} = m\gamma_s\mathbf{v}$. The appearance of $\gamma\cdot\mathbf{p}/|\mathbf{p}|$ in eq. 4.20 precludes the invariance of the free tachyon action.

We will now show the effect of a Left-handed Complex Lorentz transformation (eqs. 4.5 and 4.6) on the lagrangian density eq. 4.14. The two non-zero parts of the lagrangian density $\mathscr{L}_8$ (eq. 4.14) are

$$\mathscr{L}_1 = \psi_T^{\dagger} i\gamma^0\gamma^5(\gamma^\mu\partial/\partial x^\mu - m)\psi_T(x) \tag{4.21}$$

and

$$\mathscr{L}_2 = \psi^\dagger\gamma^0(i\gamma^\mu\partial/\partial x^\mu - m)\psi(x) \tag{4.22}$$

The effect of the transformation, eqs. 4.5-4.6, on these terms is

$$
\begin{aligned}
\mathscr{L}_1' &= \psi_T^{\dagger} i\gamma^0\gamma^5 S_L^{-1} S_L(\gamma^\mu\partial/\partial x^\mu - m)\ S_L^{-1} S_L\psi_T(x)\\
&= \psi_T^{\dagger} i\gamma^0\gamma^5 S_L^{-1}(i\gamma^\mu\partial/\partial x'^\mu - m)S_L\psi_T(x)\\
&= -\psi_T^{\dagger} S_L^{\dagger}\gamma^5(\gamma\cdot\mathbf{p}/|\mathbf{p}|)(i\gamma^\mu\partial/\partial x'^\mu - m)S_L\psi_T(x)\\
&= \psi'^{\dagger}(x')(\gamma\cdot\mathbf{p}/|\mathbf{p}|)\gamma^5(i\gamma^\mu\partial/\partial x'^\mu - m)\psi'(x')
\end{aligned} \tag{4.23}
$$

and

$$
\begin{aligned}
\mathscr{L}_2' &= \psi^\dagger\gamma^0\gamma^5 S_L^{-1} S_L\gamma^5(i\gamma^\mu\partial/\partial x^\mu - m)\gamma^5 S_L^{-1} S_L\gamma^5\psi(x)\\
&= \psi^\dagger\gamma^0\gamma^5 S_L^{-1}(\gamma^\mu\partial/\partial x'^\mu - m)S_L\gamma^5\psi(x)\\
&= i\psi^\dagger\gamma^5 S_L^{\dagger}(\gamma\cdot\mathbf{p}/|\mathbf{p}|)(\gamma^\mu\partial/\partial x'^\mu - m)S_L\gamma^5\psi(x)\\
&= i\psi_T'^{\dagger}(x')(\gamma\cdot\mathbf{p}/|\mathbf{p}|)(\gamma^\mu\partial/\partial x'^\mu - m)\psi_T'(x')
\end{aligned} \tag{4.24}
$$

using eqs. 4.20, 4.1f and 4.1g, where $\psi'(x')$ is a solution of the Dirac equation obtained by Left-handed Complex Lorentz boosting (by $\mathbf{v} = \mathbf{p}/(\gamma m)$) of a tachyon field and where $\psi_T'(x')$ is a solution of the tachyon equation obtained by Left-handed Complex Lorentz boosting (by $\mathbf{v} = \mathbf{p}/(\gamma m)$) of a Dirac field. Eqs. 4.23-4.24 clearly show that $\mathcal{L}_8$ is *not* invariant under Left-handed Complex Lorentz transformations.

Consequently the action of eq. 4.17 is only invariant under inhomogeneous Lorentz transformations. *This state of affairs is actually an advantage when we derive features of the SuperStandard Model because it will be seen to prevent any interplay between unbroken internal symmetry ElectroWeak SU(2) rotations and Left-handed Complex Lorentz transformations.*

4.4 The Diracian Dilemma – To what do Left-handed Extended Lorentz Boost Particles Correspond? Answer: Left-handed Leptons

The development of this 8-dimensional formalism, and in particular, the "bi-spinor" wave function consisting of a Dirac spinor and and a tachyon spinor, raises the question, "Is there a particle interpretation for the "bi-spinor" wave function?" Dirac faced a similar issue in 1928-1930 with the negative energy states of the Dirac equation. He developed "hole theory" which eventually led to the interpretation of holes in the sea of filled negative energy states as *positrons*. We now face the same problem: with what pairs of particles do we identify the doublets consisting of a Dirac particle and a tachyon?

The obvious natural interpretation of these 8-spinors is ElectroWeak isodoublets such as:

$$\Psi_\ell(x) = \begin{bmatrix} \psi_{\ell T} \\ \psi_\ell \end{bmatrix} \sim \begin{bmatrix} \nu \\ e \end{bmatrix} \tag{4.25}$$

is for leptons to have "e" represent a charged lepton and ν represent a neutrino. With this interpretation we can introduce SU(2) gauge interactions and develop one-generation, leptonic Weak theory naturally.

4.5 To what do Complexons Correspond? Quarks

We have identified two of the four types of spin ½ fermions as leptons. The remaining two types of spin ½ fermions – complexons – ψ_C and ψ_{CT} seem to naturally correspond to quarks since their equations of motion and wave functions have a natural SU(3) symmetry as we pointed out earlier. We therefore associate a color SU(3) symmetry with these two types of spin ½ complexons. The Electroweak doublet of quarks then is

$$\Psi_q{}^a(x) = \begin{bmatrix} \psi_C{}^a \\ \\ \psi_{CT}{}^a \end{bmatrix} \sim \begin{bmatrix} u^a \\ \\ d^a \end{bmatrix} \tag{4.26}$$

where u is an "up" type quark and d is a "down" type quark.[72]

The rationale for constructing quark doublets is the same as in the leptonic case: We wish to define a generalization of the "Dirac-like" equations of motion that is covariant under L_C boosts.

4.6 Quark Doublets

We assume that quark doublets consist of a complexon[73] and a tachyonic complexon[74] and to this extent they mirror lepton doublets. In this section we will develop a generalized free complexon equation and describe its features.

4.6.1 Summary of L_C Boosts to Generate Spin ½ Equations

We begin by recapitulating L_C boost features for coordinates and spinors:[75]

$$\Lambda_C(\mathbf{v_c}) = \exp[i\omega\hat{\mathbf{w}}\cdot\mathbf{K}] \tag{2.61}$$

---

[72] While the lepton situation is clear in the sense that charged leptons cannot be tachyons since their masses are known (Thus only tachyonic neutrinos are the only currently allowed possibility.), the quark situation is somewhat unclear. We have provisionally chosen the "down" type of quark (d, s, and b) as tachyonic. The association of bound states of these quarks such as the K^0 and B^0 systems which are known to have CP violation, and the CP violation engendered by tachyons, encourages this interpretation.

In addition, $W^\pm$ charge asymmetry in $p\bar{p}$ collisions indicate the d sea in a proton is greater than the u sea (K. Abe et al, PRL **74**, 850 (1995)) as does the asymmetry of Drell-Yan production in deep inelastic scattering on p and n targets (A. Baldit et al, Phys. Lett. **B332**, 244 (1994)). These results are to be expected since there is no mass gap for a d tachyon sea while there is a mass gap for a u Dirac particle sea. Complexon quarks may explicate the discrepancies between theory and experiment in the spin structure functions of the parton model for nucleons.

[73] **An "ordinary" complexon can "exceed the speed of light" just like a tachyonic complexon because a complexon has a complex valued velocity enabling it to evade the real-valued singularity at v = c.**

[74] The global SU(3) symmetry of complexons makes their identification with quarks reasonable. However, the complexon theory that we develop and use for quark dynamics in the Standard Model is not required. Our SuperStandard Model could use Dirac fermion dynamics for the up-type quarks and tachyon dynamics for down-type quarks. Then the (broken) Left-handed Extended Lorentz group would be the basic space-time group rather than L_C. We choose to use complexon dynamics for quarks because they have an internal SU(3)-like structure suggestive of color SU(3). More importantly, their spin dynamics is different and thus may resolve the differences between theory and experiment for the deep inelastic parton spin-dependent structure functions. Nevertheless, quarks could be similar to leptons in this regard and form a doublet of a Dirac fermion and an ordinary tachyon. Whether quarks are complexons or not is an experimental question!

[75] **The equation numbering of this subsection 4.6.1 follows that of Blaha (2007b).**

$$\omega = (\omega_r^2 - \omega_i^2 + 2i\omega_r\omega_i\,\hat{\mathbf{u}}_r\cdot\hat{\mathbf{u}}_i)^{\frac{1}{2}} \qquad (2.62)$$

$$\hat{\mathbf{w}} = (\omega_r\hat{\mathbf{u}}_r + i\omega_i\hat{\mathbf{u}}_i)/\omega \qquad (2.63)$$

$$\hat{\mathbf{w}}\cdot\hat{\mathbf{w}} = \hat{\mathbf{u}}_r\cdot\hat{\mathbf{u}}_r = \hat{\mathbf{u}}_i\cdot\hat{\mathbf{u}}_i = 1 \qquad (2.64a)$$

$$\mathbf{v}_c = \hat{\mathbf{w}}\,\tanh(\omega) \qquad (2.64b)$$

The corresponding L_C spinor boost for $m^2 > 0$ particles with complex 3-momenta is

$$S_C(\omega, \mathbf{v}_c) = \exp(-i\omega\sigma_{0k}\hat{w}_k/2) = \exp(-\omega\gamma^0\boldsymbol{\gamma}\cdot\hat{\mathbf{w}}/2)$$
$$= \cosh(\omega/2)I + \sinh(\omega/2)\gamma^0\boldsymbol{\gamma}\cdot\hat{\mathbf{w}} \qquad (3.78)$$

with inverse transformation

$$S_C^{-1}(\omega, \mathbf{v}_c) = \gamma^2\gamma^0 K^{-1}S_C^{\dagger}K\gamma^0\gamma^2 = \gamma^2\gamma^0 S_C^{\mathrm{T}}\gamma^0\gamma^2 = \exp(\omega\gamma^0\boldsymbol{\gamma}\cdot\hat{\mathbf{w}}/2)$$
$$= \cosh(\omega/2)I - \sinh(\omega/2)\gamma^0\boldsymbol{\gamma}\cdot\hat{\mathbf{w}} \qquad (3.79)$$

The Dirac-like complexon equation resulting from the boost is

$$[i\gamma^0\partial/\partial t + i\boldsymbol{\gamma}\cdot(\nabla_r + i\nabla_i) - m]\psi_C(t, \mathbf{x}_r, \mathbf{x}_i) = 0 \qquad (3.101)$$

where $\mathbf{x} = \mathbf{x}_r - i\mathbf{x}_i$. The subsidiary condition is

$$\nabla_r\cdot\nabla_i\,\psi_C(t, \mathbf{x}_r, \mathbf{x}_i) = 0 \qquad (3.102a)$$

The L_C coordinate boost that leads to $m^2 < 0$ tachyonic complexons with complex 3-momenta is

$$\Lambda_{CL}(\mathbf{v}_c) \equiv \Lambda_{CL}(\omega, \hat{\mathbf{w}}) = \exp[i(\omega + i\pi/2)\hat{\mathbf{w}}\cdot\mathbf{K}] \qquad (3.151)$$

where

$$\omega = (\omega_r^2 - \omega_i^2)^{\frac{1}{2}} \qquad (2.62)$$

$$\hat{\mathbf{w}} = (\omega_r\hat{\mathbf{u}}_r + i\omega_i\hat{\mathbf{u}}_i)/\omega \qquad (2.63)$$

$$\hat{\mathbf{w}}\cdot\hat{\mathbf{w}} = \hat{\mathbf{u}}_r\cdot\hat{\mathbf{u}}_r = \hat{\mathbf{u}}_i\cdot\hat{\mathbf{u}}_i = 1 \qquad (2.64a)$$

$$\mathbf{v}_c = \hat{\mathbf{w}}\,\tanh(\omega + i\pi/2) = \hat{\mathbf{w}}\,\cotanh(\omega) \qquad (3.152)$$

$$\omega_L = \omega + i\pi/2$$

The L_C spinor boost for tachyonic complexons is

$$S_{CL}(\Lambda_{CL}(\omega, \hat{\mathbf{w}})) = \exp(-i\omega_L\sigma_{0i}\hat{w}_i/2) = \exp(-\omega_L\gamma^0\boldsymbol{\gamma}\cdot\hat{\mathbf{w}}/2)$$
$$= \cosh(\omega_L/2)I + \sinh(\omega_L/2)\gamma^0\boldsymbol{\gamma}\cdot\hat{\mathbf{w}} \qquad (3.154)$$

The resulting Dirac-like tachyonic complexon equation is

$$[\gamma^0 \partial/\partial t + \boldsymbol{\gamma} \cdot (\nabla_{\mathbf{r}} + i\nabla_{\mathbf{i}}) - m]\psi_{CL}(t, \mathbf{x_r}, \mathbf{x_i}) = 0 \qquad (3.163a)$$

with the subsidiary condition

$$\nabla_{\mathbf{r}} \cdot \nabla_{\mathbf{i}}\, \psi_{CL}(t, \mathbf{x_r}, \mathbf{x_i}) = 0 \qquad (3.164)$$

4.6.2 L_C Boosts between Complexons and Tachyonic Complexons

An L_C spinor boost of a complexon can change it into a tachyonic complexon and vice versa:

$$S_{CL}(\Lambda_{CL}(\omega, \hat{\mathbf{w}}))\psi_C(x) \rightarrow \psi_{CT}'(x') \qquad (4.27)$$
$$S_{CL}(\Lambda_{CL}(\omega, \hat{\mathbf{w}}))\psi_{CT}(x) \rightarrow \psi_C'(x')$$

Similarly the differential operator used in the equations of motion can also be transformed.

$$S_{CL}(\Lambda_{CL}(\omega, \hat{\mathbf{w}}))(\gamma^\mu D_\mu - m)S_{CL}^{-1}(\Lambda_{CL}(\omega, \hat{\mathbf{w}})) = (i\gamma^\mu D'_\mu - m) \qquad (4.28)$$
$$S_{CL}(\Lambda_{CL}(\omega, \hat{\mathbf{w}}))\gamma^5(i\gamma^\mu D_\mu - m)\gamma^5 S_{CL}^{-1}(\Lambda_{CL}(\omega, \hat{\mathbf{w}})) = (\gamma^\mu D'_\mu - m)$$

where

$$x'^\mu = i\Lambda_{CL}{}^\mu{}_\nu(\omega, \mathbf{u})x^\nu \qquad (4.29)$$
$$D'_\mu = -i\Lambda_{CL}{}^\nu{}_\mu(\omega, \mathbf{u})D_\nu$$

or in matrix form

$$X' = E_{CL}(\omega, \hat{\mathbf{w}})X \equiv i\Lambda_{CL}(\omega, \hat{\mathbf{w}})X \qquad (4.30)$$

Eqs. 4.27 – 4.29 imply

$$S_{CL}(\Lambda_{CL}(\omega, \hat{\mathbf{w}}))(\gamma^\mu D_\mu - m)\psi_{CT}(x) = (i\gamma^\mu D'_\mu - m)S_{CL}(\Lambda_{CL}(\omega, \hat{\mathbf{w}}))\psi_{CT}(x)$$
$$= (i\gamma^\mu D'_\mu - m)\psi_C'(x') \qquad (4.31)$$

and

$$S_{CL}(\Lambda_{CL}(\omega,\hat{\mathbf{w}}))\gamma^5(i\gamma^\mu D_\mu - m)\psi_C(x) = (\gamma^\mu D'_\mu - m)S_{CL}(\Lambda_{CL}(\omega,\hat{\mathbf{w}}))\gamma^5\psi_C(x)$$
$$= (\gamma^\mu D'_\mu - m)\psi_{CT}'(x') \qquad (4.32)$$

where

$$\psi_C'(x') = S_{CL}(\Lambda_{CL}(\omega, \hat{\mathbf{w}}))\psi_{CT}(x) \qquad (4.33)$$

and

$$\psi_{CT}'(x') = S_{CL}(\Lambda_{CL}(\omega, \hat{\mathbf{w}}))\gamma^5\psi_C(x) \qquad (4.34)$$

Thus neither complexon dynamical equation is L_C covariant.

4.6.3 Doublet Dynamical Equation for Complexons

We will now consider the issue of generalizing the complexon dynamical equations so that the generalized equation is covariant under both Lorentz transformations and L_C boosts.

The only obvious method to obtain a generalized equation that is covariant under L_C boosts is to define an 8×8 matrix generalization. Let

$$\mathrm{d}_C(x) \;=\; \begin{bmatrix} (i\gamma^\mu D_\mu - m) & 0 \\[2ex] 0 & (\gamma^\mu D_\mu - m) \end{bmatrix} \tag{4.35}$$

be an 8×8 matrix operator with the 4×4 matrix elements shown, and let

$$\Psi_C(x) \;=\; \begin{bmatrix} \psi_C(x) \\[2ex] \psi_{CT}(x) \end{bmatrix} \tag{4.36}$$

be an 8 component column vector composed of a complexon field and a tachyonic complexon field. Then the generalized complexon equation is

$$\mathrm{d}_C(x)\Psi_C(x) = 0 \tag{4.37}$$

We now define the 8×8 Left-handed L_C boost transformation

$$S_{CL8} \equiv S_{CL8}(\Lambda_{CL}(\omega, \hat{\mathbf{w}})) = \begin{bmatrix} 0 & S_{CL}(\Lambda_{CL}(\omega, \hat{\mathbf{w}})) \\[2ex] S_{CL}(\Lambda_{CL}(\omega, \hat{\mathbf{w}}))\gamma^5 & 0 \end{bmatrix} \tag{4.38}$$

with inverse transformation

$$S_{CL8}^{-1} \equiv S_{CL8}^{-1}(\Lambda_{CL}(\omega, \hat{\mathbf{w}})) = \begin{bmatrix} 0 & \gamma^5 S_{CL}^{-1}(\Lambda_{CL}(\omega, \hat{\mathbf{w}})) \\[2ex] S_{CL}^{-1}(\Lambda_{CL}(\omega, \hat{\mathbf{w}})) & 0 \end{bmatrix} \tag{4.39}$$

Applying S_{CL8} to eq. 4.37 yields

$$0 = S_{CL8}\mathrm{d}_C(x)\Psi_C(x) = \mathrm{d}_C(x')\Psi_C'(x') \tag{4.40}$$

where

$$\Psi_C{}'(x') = \begin{bmatrix} S_{CL8}\psi_{CT}(x) \\[2mm] S_{CL8}\gamma^5\psi_C(x) \end{bmatrix} = \begin{bmatrix} \psi_C{}'(x') \\[2mm] \psi_{CT}{}'(x') \end{bmatrix} \tag{4.41}$$

Thus the generalized complexon equation is covariant under L_C boosts. Covariance requires the complexon, and the tachyonic complexon, must have the same absolute value for the mass.

It is easy to show that the generalized complexon equation is also covariant under conventional Lorentz transformations represented as 4×4 diagonal blocks in an 8×8 matrix representation. (The demonstration is analogous to eqs. $4.9 - 4.13$.)

The lagrangian density that corresponds to our 8-dimensional construction is

$$\mathcal{L}_{C8} = \overline{\Psi}_C(x)\mathit{d}_C(x)\Psi_C(x) \tag{4.42}$$

where

$$\overline{\Psi}_C(x) = \Psi_C{}^\dagger\big|_{\mathbf{x_i} = -\mathbf{x_i}}\Gamma_C{}^0 \tag{4.43}$$

and

$$\Gamma_C{}^0 = \begin{bmatrix} \gamma^0 & 0 \\[2mm] 0 & i\gamma^0\gamma^5 \end{bmatrix} \tag{4.44}$$

The action

$$I = \int d^4x\,\mathcal{L}_{C8} \tag{4.45}$$

is invariant under Lorentz transformations S_8 and S_{8A} (eqs. $4.9 - 4.12$).

The Hamiltonian density for the 8-dimensional theory is

$$\mathcal{H}_{C8}(x) = \begin{bmatrix} \psi_C{}^\dagger(-i\boldsymbol{\alpha}\cdot\nabla_C + \beta m)\psi_C & 0 \\[2mm] 0 & i\psi_{CT}{}^\dagger\gamma^5(\boldsymbol{\alpha}\cdot\nabla_C + \beta m)\psi_{CT} \end{bmatrix} \tag{4.46}$$

where the spatial vector part of D^μ is

$$\nabla_C = \mathbf{D} \tag{4.47}$$

4.6.4 Non-Invariance of the Generalized Free Complexon Action under an L_C Boost

The action 4.45 is not invariant under L_C boosts. The reason is similar to that of section 4.3 for the "leptonic" type of particle: there is no simple relation between the hermitean conjugate of an L_C spinor boost and its inverse (a situation similar to eq. 4.19 for the Dirac boost case).

Consequently the action of eq. 4.45 is only invariant under inhomogeneous Lorentz transformations. *This state of affairs is again an advantage when we derive features of the SuperStandard Model because it prevents any interplay between unbroken internal symmetry ElectroWeak SU(2) rotations and L_C transformations in the complexon (quark) sector.*

5. Coordinate Reality Group Analogue for ElectroWeak SU(2)⊗U(1) due to Real Superluminal Velocities

In the preceding chapter we developed a fermion doublet framework for the ElectroWeak SU(2) interactions. We now turn to develop the ElectroWeak interactions with an SU(2)⊗U(1) group structure, from the geometry of Complex Lorentz transformations.

In the discussions up to this point we have not considered imaginary (and more generally complex) coordinates resulting from a superluminal Lorentz transformation.[76] In this chapter we show that the coordinates generated from real-valued coordinates are complex-valued in general and require us to introduce another transformation that maps complex coordinates to real coordinates.[77] This transformation, which we will call a *Coordinate Reality Group* transformation, will be of significance because it has an SU(2)⊗U(1) group symmetry. It emerges when we consider superluminal transformations but is not required for ordinary sublight Lorentz transformations. This new SU(2)⊗U(1) symmetry is the analogue of the SU(2)⊗U(1) symmetry of the ElectroWeak sector of The SuperStandard Model.

We introduce this new transformation by reconsidering the previous simple example wherein one coordinate system is traveling at a speed v in the x direction with respect to the "laboratory" system. See Fig. 5.1.

The (left-handed[78]) Lorentz transformation is given by eq. 5.1, and the coordinates in the two reference frames are related by eq. 5.2.

$$\Lambda_L(\omega, \mathbf{u} = (1,0,0)) = \begin{bmatrix} i\gamma_s & -i\beta\gamma_s & 0 & 0 \\ -i\beta\gamma_s & i\gamma_s & 0 & 0 \\ 0 & 0 & 1 & 0 \\ 0 & 0 & 0 & 1 \end{bmatrix} \tag{5.1}$$

implementing the coordinate transformation:

$$X' = \Lambda_L(\omega, \mathbf{u} = (1,0,0))X$$

---

[76] Superluminal transformations are a subset of Complex Lorentz transformations.

[77] The complex coordinates resulting from a superluminal transformation are physically viewed as real-valued by an observer in the new coordinate system. The apparent complexity of the coordinates resulting from a superluminal transformation are an artifact of the transformation.

[78] The right-handed Lorentz transformation case is analogous.

or

$$t' = i\gamma_s(t - \beta x)$$
$$x' = i\gamma_s(x - \beta t)$$
$$y' = y \qquad\qquad (5.2)$$
$$z' = z$$

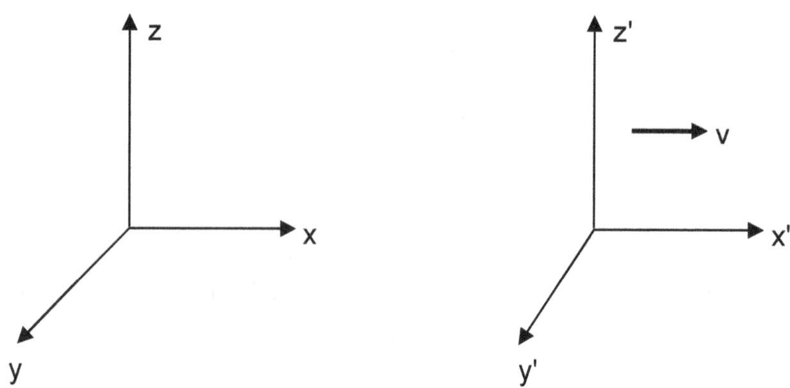

Figure 5.1. Depiction of two coordinate systems. The "primed" coordinate system is moving with velocity v in the positive x direction with respect to the "unprimed" coordinate system. We choose parallel axes for convenience.

We now define a Coordinate Reality group transformation $\Pi_L(\mathbf{u})$ that maps the real coordinates of the unprimed reference frame to real coordinates in the primed reference frame.

$$\Pi_L(\mathbf{u}) \;=\; \begin{bmatrix} -i & 0 & 0 & 0 \\ 0 & -i & 0 & 0 \\ 0 & 0 & 1 & 0 \\ 0 & 0 & 0 & 1 \end{bmatrix} \qquad (5.3)$$

where $\mathbf{u}$ is the unit vector corresponding to the direction of $\mathbf{v}$ (the positive x direction in this example). Using $\Pi_L(\mathbf{u})$ we obtain an overall transformation from real coordinates to real coordinates in the case considered:

$$X'' = \Pi_L(\mathbf{u})\Lambda_L(\omega, \mathbf{u} = (1,0,0))X$$

or

$$t'' = \gamma_s(t - \beta x)$$
$$x'' = \gamma_s(x - \beta t)$$
$$y'' = y \qquad\qquad (5.4)$$
$$z'' = z$$

where $\gamma_s = (\beta^2 - 1)^{-\frac{1}{2}}$. An observer in the primed reference frame would consider his/her time to be real when measured on a clock, and distances along the x axis to be real when measured with a ruler. Thus eq. 5.4 makes good sense physically because in any reference frame, observers measure real distances and real times. For this reason we will call combined transformations of the type of eq. 5.4 – from real coordinates to real coordinates – *physical* superluminal transformations for real-valued velocities.

It is important to note that $\Pi_L(\mathbf{u})$ is position dependent in general for more complicated Λ_L transformations and so the Coordinate Reality group is a local group of the Yang-Mills type. This is clear from eqs. 5.1 and 5.2. We will see the Coordinate Reality group is the analogue of local (Yang-Mills) SU(2)⊗U(1) ElectroWeak symmetry.

This simple example generalizes to arbitrary relative real velocities $\mathbf{v}$. First we note that the Lorentz transformation for a velocity $\mathbf{v}$ that is a rotation of the velocity in the x-direction ($\mathbf{v} = |\mathbf{v}|\mathbf{R}\mathbf{u}$ where R is the relevant rotation matrix) has the form

$$\Lambda_L(\omega, \mathbf{v}) = \mathcal{R}(\mathbf{v}/v, \mathbf{u})\Lambda_L(\omega, \mathbf{u} = (1,0,0))\mathcal{R}^{-1}(\mathbf{v}/v, \mathbf{u}) \tag{5.5}$$

where $\mathcal{R}(\mathbf{v}/v, \mathbf{u})$ is a rotation from the velocity direction $\mathbf{u}$ to direction $\mathbf{v}/v$.

The original transformation (eq. 5.2) can be written as

$$\Pi_L(\mathbf{u})\Lambda_L(\omega, \mathbf{u} = (1,0,0)) = \Pi_L(\mathbf{u})\mathcal{R}^{-1}(\mathbf{v}/v, \mathbf{u})\Lambda_L(\omega, \mathbf{v})\mathcal{R}(\mathbf{v}/v, \mathbf{u}) \tag{5.6}$$

Consequently the combined transformation for velocity $\mathbf{v}$ is

$$\begin{aligned}
\mathcal{R}(\mathbf{v}/v, \mathbf{u})\Pi_L(\mathbf{u})&\Lambda_L(\omega, \mathbf{u} = (1,0,0))\mathcal{R}^{-1}(\mathbf{v}/v, \mathbf{u}) \\
&= \mathcal{R}(\mathbf{v}/v, \mathbf{u})\Pi_L(\mathbf{u})\mathcal{R}^{-1}(\mathbf{v}/v, \mathbf{u})\Lambda_L(\omega, \mathbf{v}) \\
&= \Pi_L(\mathbf{v}/v)\Lambda_L(\omega, \mathbf{v})
\end{aligned} \tag{5.7}$$

Thus for a Lorentz transformation $\Lambda_L(\omega, \mathbf{v})$ for velocity $\mathbf{v}$ we see that we can define a subsidiary transformation $\Pi_L(\mathbf{v}/v)$ of the form

$$\Pi_L(\mathbf{v}/v) = \mathcal{R}(\mathbf{v}/v, \mathbf{u})\Pi_L(\mathbf{u})\mathcal{R}^{-1}(\mathbf{v}/v, \mathbf{u}) \tag{5.8}$$

The general form of $\mathcal{R}(\mathbf{v}/v, \mathbf{u})$, is

$$\mathcal{R}(\mathbf{v}/v, \mathbf{u}) = \begin{bmatrix} 1 & 0 & 0 & 0 \\ 0 & & & \\ 0 & & \mathcal{R}_3(\mathbf{v}/v, \mathbf{u}) & \\ 0 & & & \end{bmatrix} \tag{5.9}$$

where $\mathcal{R}_3(\mathbf{v}/v, \mathbf{u})$ is a 3×3 rotation matrix that can be expressed in terms of the generators of the 3-dimensional rotation group as

$$\mathcal{R}_3(\mathbf{v}/v, \mathbf{u}) = \exp(i\boldsymbol{\theta}\cdot\mathbf{J}) \tag{5.10}$$

The rotation angles $\boldsymbol{\theta}$ are real numbers since we are rotating the real vector $\mathbf{u}$ to the real vector $\mathbf{v}/v$. Given the form of eq. 5.10 then we see that the form of $\Pi_L(\mathbf{v}//v)$ is

$$\Pi_L(\mathbf{v}/v) = \begin{bmatrix} -i & 0 & 0 & 0 \\ 0 & & & \\ 0 & & \mathcal{R}_3(\mathbf{v}/v, \mathbf{u})\Pi_{L3}(\mathbf{u})\mathcal{R}_3^{-1}(\mathbf{v}/v, \mathbf{u}) & \\ 0 & & & \end{bmatrix} \tag{5.11}$$

where

$$\Pi_{L3}(\mathbf{u}) = \begin{bmatrix} -i & 0 & 0 \\ 0 & 1 & 0 \\ 0 & 0 & 1 \end{bmatrix} \tag{5.12}$$

If we consider the case of an infinitesimal rotation $\boldsymbol{\theta}$ to first order in $\boldsymbol{\theta}$

$$\mathcal{R}_3(\mathbf{v}/v, \mathbf{u}) \simeq I + i\boldsymbol{\theta}\cdot\mathbf{J} \tag{5.13}$$

then

$$\begin{aligned}\Pi_{L3}(\mathbf{v}/v) = \mathcal{R}_3(\mathbf{v}/v, \mathbf{u})\Pi_{L3}(\mathbf{u})\mathcal{R}_3^{-1}(\mathbf{v}/v, \mathbf{u}) &\simeq \Pi_{L3}(\mathbf{u}) + i\boldsymbol{\theta}\cdot\mathbf{J}\Pi_{L3}(\mathbf{u}) - i\Pi_{L3}(\mathbf{u})\boldsymbol{\theta}\cdot\mathbf{J} \\ &\simeq \Pi_{L3}(\mathbf{u})[I + i\Pi_{L3}^{-1}(\mathbf{u})[\boldsymbol{\theta}\cdot\mathbf{J}, \Pi_{L3}(\mathbf{u})]\end{aligned} \tag{5.14}$$

where $\Pi_{L3}^{-1}(\mathbf{u})$ is the inverse of $\Pi_{L3}(\mathbf{u})$ and $[...]$ represents the commutator. Thus for arbitrary rotations eq. 5.14 implies

$$\Pi_{L3}(\mathbf{v}/v) = \mathcal{R}_3(\mathbf{v}/v, \mathbf{u})\Pi_{L3}(\mathbf{u})\mathcal{R}_3^{-1}(\mathbf{v}/v, \mathbf{u}) = \Pi_{L3}(\mathbf{u})\exp\{i\Pi_{L3}^{-1}(\mathbf{u})[\boldsymbol{\theta}\cdot\mathbf{J}, \Pi_{L3}(\mathbf{u})]\} \tag{5.15}$$

We can find the general form of $\Pi_{L3}(\mathbf{v}/v)$ by considering the case of eq. 5.6 in more detail. The exponentiated matrix expression in 5.15 can be written

$$\Pi_{L3}^{-1}(\mathbf{u})[\boldsymbol{\theta}\cdot\mathbf{J}, \Pi_{L3}(\mathbf{u})] = \Pi_{L3}^{-1}(\mathbf{u})\boldsymbol{\theta}\cdot\mathbf{J}\Pi_{L3}(\mathbf{u}) - \boldsymbol{\theta}\cdot\mathbf{J} = \boldsymbol{\theta}\cdot\mathbf{Q} \tag{5.16}$$

where

$$\mathbf{Q} = \Pi_{L3}^{-1}(\mathbf{u})\mathbf{J}\Pi_{L3}(\mathbf{u}) - \mathbf{J} = \mathbf{Q}' - \mathbf{J} \tag{5.17}$$

The matrices Q_i can be evaluated using eq. 5.12 and the matrix representations of rotation generators J_i: which are equivalent in form to the SU(2) generators T_i:

$$J_1 = \begin{bmatrix} 0 & 0 & 0 \\ 0 & 0 & -i \\ 0 & i & 0 \end{bmatrix} = T_1 \tag{5.18}$$

$$J_2 = \begin{bmatrix} 0 & 0 & i \\ 0 & 0 & 0 \\ -i & 0 & 0 \end{bmatrix} = T_2 \tag{5.19}$$

$$J_3 = \begin{bmatrix} 0 & -i & 0 \\ i & 0 & 0 \\ 0 & 0 & 0 \end{bmatrix} = T_3 \tag{5.20}$$

The rotation generators satisfy the commutation relations

$$[J_i, J_j] = i\epsilon_{ijk}J_k \tag{5.21}$$

as do the SU(2) generators:

$$[T_i, T_j] = i\epsilon_{ijk}T_k \tag{5.22}$$

We can calculate Q' and obtain

$$Q'_1 = \begin{bmatrix} 0 & 0 & 0 \\ 0 & 0 & -i \\ 0 & i & 0 \end{bmatrix} \tag{5.23}$$

$$Q'_2 = \begin{bmatrix} 0 & 0 & -1 \\ 0 & 0 & 0 \\ -1 & 0 & 0 \end{bmatrix} \tag{5.24}$$

$$Q'_3 = \begin{bmatrix} 0 & 1 & 0 \\ 1 & 0 & 0 \\ 0 & 0 & 0 \end{bmatrix} \tag{5.25}$$

We note that each Q'_i is hermitean and the Q'_i satisfy the commutation relations:

$$[Q'_i, Q'_j] = i\epsilon_{ijk}Q'_k \tag{5.26}$$

Consequently the set of Q'_i are also equivalent to SU(2) generators. As a result the exponential factor

$$\Pi_{L3}(\mathbf{v}/v) = \Pi_{L3}(\mathbf{u})\exp\{i\boldsymbol{\theta}\cdot(\mathbf{Q}' - \mathbf{J})\} \tag{5.27}$$

is equivalent to a combination of SU(2) rotations not only in this case but in general for superluminal transformations. The factor $\Pi_{L3}(\mathbf{u})$ is not an SU(2) matrix since its determinant is not 1. However

$$\Pi'_{L3}(\mathbf{u}) = -i\Pi_{L3}(\mathbf{u}) \tag{5.28}$$

is an SU(2) matrix since

$$\Pi'_{L3}{}^{-1}(\mathbf{u}) = \Pi'_{L3}{}^{\dagger}(\mathbf{u}) \tag{5.29}$$
$$\det \Pi'_{L3}(\mathbf{u}) = 1 \tag{5.30}$$

and

$$\Pi'_{L3}(\mathbf{v}/v) = \Pi'_{L3}(\mathbf{u})\exp\{i\boldsymbol{\theta}\cdot(\mathbf{Q'} - \mathbf{J})\} \tag{5.31}$$

is similarly an SU(2) rotation.

Thus the general form of superluminal, *real* velocity, transformation from a real set of coordinates to a real set of coordinates is[79]

$$\Pi_L(\mathbf{v}/v)\Lambda_L(\omega, \mathbf{v}) \tag{5.32}$$

where

$$\Pi_L(\mathbf{v}/v) = \begin{bmatrix} -i & 0 & 0 & 0 \\ 0 & & & \\ 0 & & \Pi_{L3}(\mathbf{u})\exp\{i\,\boldsymbol{\theta}\cdot(\mathbf{Q'} - \mathbf{J})\} & \\ 0 & & & \end{bmatrix} \tag{5.33}$$

The Lorentz condition for real to real physical transformations generalizes to

$$\Lambda(\mathbf{v})^T\Pi_L(\mathbf{v}/v)^{\dagger}G\,\Pi_L(\mathbf{v})\Lambda(\mathbf{v}/v) = G \tag{5.34}$$

Since superluminal transformations $\Lambda_L(\omega, \mathbf{v})$ transform real coordinates to complex coordinates in general, we can generalize the form of a real-to-real superluminal transformation to

$$e^{i\varphi}\Pi_L(\mathbf{v'}/v')\Lambda_L(\omega, \mathbf{v}) \tag{5.35}$$

where φ is a constant phase and $\mathbf{v'}$ is an arbitrary velocity. This generalization will satisfy the generalized Lorentz condition

$$\Lambda(\mathbf{v})^T\Pi_L(\mathbf{v'}/v')^{\dagger}e^{-i\varphi}G\,e^{i\varphi}\Pi_L(\mathbf{v'}/v')\Lambda(\mathbf{v}) = G \tag{5.36}$$

---

[79] The choice of the unit vector $\mathbf{u}$ and the angle vector $\boldsymbol{\theta}$ must be such that applying the transformation to a real set of coordinates yields a real set of coordinates.

but the transformation will, in general, yield a complex set of coordinates when applied to a set of real coordinates.

These considerations imply:

1. Any observer in a coordinate system will treat a complex 4-dimensional coordinate system as if it were a real 4-dimensional coordinate system with complex-valued straight lines along each dimension (assuming rectangular coordinates).

2. The transformation $e^{i\varphi}\Pi'_{L3}(\mathbf{v}/v)$ is a $SU(2)\otimes U(1)$ transformation that takes complex 3-dimensional spatial coordinates to complex 3-dimensional spatial coordinates. In particular straight lines map to straight lines.

3. Physical observations in the observer's coordinate system are invariant under $SU(2)\otimes U(1)$ rotations of the spatial coordinates and the multiplication of the time component by an arbitrary phase.

4. The matrix

$$\Pi'_{L}(\mathbf{v}/v, \chi, \varphi) = \begin{bmatrix} e^{i\chi} & 0 & 0 & 0 \\ 0 & & & \\ 0 & e^{i\varphi}\Pi'_{L3}(\mathbf{u})\exp\{i\,\boldsymbol{\theta}{\cdot}(\mathbf{Q'} - \mathbf{J})\} & & \\ 0 & & & \end{bmatrix} \qquad (5.37)$$

(where χ and φ are real numbers and $\mathbf{u}$ is a unit vector along any convenient coordinate axis) is a $SU(2)\otimes U(1)$ transformation that transforms complex 4-dimensional coordinates to complex 4-dimensional coordinates. Note, $\Pi_{L}(\mathbf{v}/v) = \Pi'_{L}(\mathbf{v}/v, 3\pi/2, \pi/2)$ is a special case of $\Pi'_{L}(\mathbf{v}/v, \chi, \varphi)$. Due to the manifest form of 5.37 we see

$$\Pi'_{L}{}^{\mu}{}_{\alpha}{}^{*}\Pi'_{L}{}^{\mu}{}_{\beta} = [\Pi'_{L}{}^{\dagger}\Pi'_{L}]_{\alpha\beta} = I_{\alpha\beta} \qquad (5.38)$$

(with an implied sum over μ) or, in matrix form,

$$\Pi'_{L}{}^{\dagger}\,\Pi'_{L} = I \qquad (5.39)$$

and also[80]

$$\Pi'_{L}{}^{\dagger}G\Pi'_{L} = G \qquad (5.40)$$

---

[80] Eq.5.40 is close to the defining condition for a Lorentz group element but the presence of complex conjugation rather than a transpose means Π'_{L} is outside the real and complex Lorentz groups.

5.   Complex coordinate values of the type generated by superluminal transformations with real-valued velocities are transformable to real coordinates. The complex coordinates are thus physically equivalent to corresponding real coordinate values in the sense that an observer in that frame would automatically use the real coordinates so obtained since rulers and clocks always measure real spatial coordinates and times. *Therefore physical theory is invariant under global SU(2)⊗U(1) coordinate transformations since complex coordinates, so generated, can be rotated back to real coordinates.*

6.   The complex coordinates of any point obtained through a superluminal transformation can be transformed to a real set of coordinates by the above SU(2)⊗U(1) transformation. This SU(2)⊗U(1) invariance is the analogue of the SU(2)⊗U(1) symmetry of the ElectroWeak interactions.

6. Coordinate Reality Group Analogue for Dark ElectroWeak SU(2)⊗U(1)

In this chapter[81] we will consider superluminal transformations based on *complex-valued relative velocities*. The previous chapter considered the case of real-valued velocities. That case led to the Coordinate Reality Group analogue of ElectroWeak SU(2)⊗U(1).

We now consider Complex Lorentz transformations for complex-valued relative velocities. These transformations will require us to introduce another Coordinate Reality group with transformations that map complex coordinates to real coordinates. These transformations will be of significance because they lead to a hitherto unstated SU(2)⊗U(1) symmetry. We identify this SU(2)⊗U(1) symmetry as the analogue of the symmetry of the *Dark* ElectroWeak interactions of the SuperStandard Model. Dark matter and interactions remain to be found experimentally but there may be some preliminary suggestive data from the CERN LHC.

We introduce these new Dark transformations by extending the previous simple example to Fig. 6.1 wherein one coordinate system is traveling at a complex-valued velocity u in the x direction with respect to the "laboratory" system. In these new transformations the relative velocity is complex-valued and has two components: a real-valued component in the x direction and an imaginary-valued component in the y direction. $\mathbf{u} = u_x\mathbf{i} + iu_y\mathbf{j}$. In the complex case $\beta = \tanh(\omega_L)$ is real-valued by eqs. 2.20 – 2.22 where $\omega_L = \omega + i\pi/2$ and ω is real.

The (left-handed[82]) Lorentz transformation is given by eq. 2.23:

$$\Lambda_L(\omega, \mathbf{u}) = \begin{bmatrix} i\gamma_s & -i\beta\gamma_s u_x & \beta\gamma_s u_y & 0 \\ -i\beta\gamma_s u_x & 1 + (i\gamma_s - 1)u_x^2 & i(i\gamma_s - 1)u_x u_y & 0 \\ \beta\gamma_s u_y & i(i\gamma_s - 1)u_x u_y & 1 - (i\gamma_s - 1)u_y^2 & 0 \\ 0 & 0 & 0 & 1 \end{bmatrix} \tag{6.1}$$

$$= \Lambda(\omega + i\pi/2, \mathbf{u})$$

implementing the coordinate transformation:

$$X' = \Lambda_L(\omega, \mathbf{u} = (u_x, iu_y, 0))X$$

or

$$t' = i\gamma_s(t - \beta u_x - i\beta u_y)$$

---

[81] Most of the material in this chapter appeared in Blaha (2011c) originally.
[82] The right-handed Lorentz transformation case is analogous.

$$x' = -i\gamma_s\beta u_x t + i\gamma_s x + u_x x - u_x^2 x + i(i\gamma_s - 1)u_x u_y y$$
$$y' = \gamma_s\beta u_y t + i(i\gamma_s - 1)u_x u_y x + [1 - (i\gamma_s - 1)u_y^2]y$$
$$z' = z$$

(6.2)

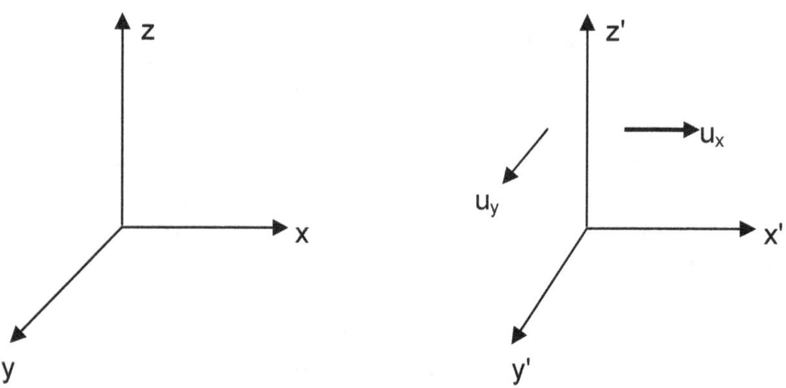

Figure 6.1. Depiction of two coordinate systems. The "primed" coordinate system is moving with velocity u = u_xi + iu_yj with respect to the "unprimed" coordinate system. We choose parallel axes for convenience.

We now define a transformation that maps the real coordinates of the unprimed reference frame to real coordinates in the primed reference frame.

$$\Pi_L(\mathbf{u}, \mathbf{X}) = \begin{bmatrix} e^{ia} & e^{ib} & e^{ic} & 0 \\ e^{id} & e^{ie} & e^{if} & 0 \\ e^{ig} & e^{ih} & e^{ij} & 0 \\ 0 & 0 & 0 & 1 \end{bmatrix}$$

(6.3)

where **u** is the unit vector corresponding to the direction of the relative velocity. It is important to note that $\Pi_L(\mathbf{u}, \mathbf{X})$ is position dependent and so the Coordinate Reality group is a local group. This is clear from eqs. 6.1 and 6.2. The Coordinate Reality Group in this case is SU(2)⊗U(1). It appears similar to the Reality group of chapter 5 except for the crucial difference that the Coordinate Reality group here mixes time and spatial rows while the Coordinate Reality group of chapter 5 only mixed spatial rows (See eqs. 5.35 and 5.37.).Again we have a local theory. Thus the combined Coordinate Reality group from this chapter and chapter 5 is SU(2)⊗U(1)⊗DSU(2)⊗DU(1) where we prepend 'D' to the Dark Reality group parts.

Earlier we defined complex boosts. We summarize the definition below:

$$\Lambda_L(\omega, \mathbf{u}) = \Lambda_L(\mathbf{v_c}) = \exp[i(\omega + i\pi/2)\mathbf{u}\cdot\mathbf{K}]$$

(6.4)

where ω remains

$$\omega = (\omega_r^2 - \omega_i^2)^{\frac{1}{2}}$$

and

$$\mathbf{u} = (\omega_r\mathbf{u}_r + i\omega_i\mathbf{u}_i)/\omega$$
$$\mathbf{u}\cdot\mathbf{u} = \mathbf{u}_r\cdot\mathbf{u}_r = \mathbf{u}_i\cdot\mathbf{u}_i = 1$$
$$\mathbf{v}_c = \mathbf{u}\tanh(\omega + i\pi/2) = \mathbf{u}\cotanh(\omega)$$

In the example, that we are considering, we set $\mathbf{u}_x = \omega_r\mathbf{u}_r$ and $\mathbf{u}_y = \omega_i\mathbf{u}_i$.

Using $\Pi_L(\mathbf{u}, \mathbf{X})$ we obtain an overall transformation from real coordinates to real coordinates:

$$X'' = \Pi_L(\mathbf{u}, \mathbf{X})\Lambda_L(\omega, \mathbf{u} = (u_x, u_y, 0))X \tag{6.5}$$

with the coordinates of X and X'' real-valued. An observer in the double primed reference frame would consider his/her time to be real when measured on a clock, and distances along the x and y axes to be real when measured with a ruler.

The velocity vectors: $u_x\mathbf{i}$ and $iu_y\mathbf{j}$ in our example define a plane in space. There are two types of rotations that are possible. 1) An angular rotation in the plane defined by the vectors. This is a U(1) transformation. 2) a spatial rotation of the plane that is an SU(2) rotation. Thus the joint rotations of $\mathbf{u}$ have an SU(2)⊗U(1) symmetry group. The R group – the Coordinate Reality group for 4-dimensions – has two SU(2)⊗U(1) factors that we denote SU(2)⊗U(1)⊗DSU(2)⊗DU(1). We see that the "newly found" group can be *assumed* to be the analogue of the Dark ElectroWeak symmetry group. We note that a further Coordinate Reality group part SU(3) will be introduced in the next chapter.

We will consider Dark Matter and its interactions in greater detail later.

7. Coordinate Reality Group Analogue for Color SU(3)

7.1 Two Possible Approaches to Color SU(3)

There are two approaches to obtaining the Strong interaction and Color SU(3) symmetry:

1. Assume up-type and down-type quarks are in $\underline{3}$ representations of Color SU(3). This assumption sheds no light on a deeper origin of the Strong interaction and Color SU(3). It simply assumes the color SU(3) of the Strong interaction sector of the Standard Model. Thus our understanding is not deepened. A postulate corresponding to this assumption is:

Possible Postulate: Quarks are in the $\underline{3}$ representation of Color SU(3). The SU(3) symmetry is gauged with local Yang-Mills SU(3) fields called gluons that constitute the Strong interaction of the quark sector. Quarks are minimally coupled to the gluons in a gauge covariant fashion.

Comment: Not in this derivation.

2. In the preceding chapters the Internal Symmetry ElectroWeak interactions of the Unified SuperStandard Model (modulo generations and their mixing) were shown to follow as an analogue to the Coordinate Reality Group associated with the Complex Lorentz group. The Reality group included SU(2)⊗U(1)⊗SU(2)⊗U(1). Thus we found a significant geometrical analogue for the form of the ElectroWeak interactions of normal and Dark matter.

3. We now establish a similar geometrical analogue of the Strong interaction and Color SU(3). *If we extend the parameters to be real functions of the space-time coordinates (i.e. local SU(3) transformations), then we obtain an analogue for color SU(3). A key factor in this interpretation is the global covariance of complexon equations of motion under global SU(3).*[83]

Comment: We follow this approach followed in this derivation.

---

[83] This chapter was extracted from chapter 17 of Blaha (2011c) with some changes.

7.2 A Global SU(3) Symmetry of Complexon Quarks

We will now consider a global SU(3) covariance implicit in eqs.2.123 – 2.127. The defining property of the SU(3) group is that it preserves the invariance of inner products of complex 3-vectors of the form:

$$u^* \cdot v = u^1 {}^* v^1 + u^2 {}^* v^2 + u^3 {}^* v^3 \qquad (7.1)$$

If we examine the dynamical equation eq. 2.123 we see that the differential operator is covariant under a global SU(3) transformation U of the complex spatial 3-coordinates:

$$[i\gamma^0 \partial/\partial t + i\mathbf{D_c}^* \cdot \mathbf{\gamma} - m] = [i\gamma^0 \partial/\partial t + i\mathbf{D_c}'^* \cdot \mathbf{\gamma}' - m] \qquad (7.2)$$

where

$$\mathbf{D_c}^* = \nabla_c = \nabla_r + i\nabla_i$$

and

$$\gamma^a = U^{ab}\gamma'^b \qquad (7.3a)$$
$$D_c^{*a} = D_c'^{*b}U^{ab*} \qquad (7.3b)$$

where $U^\dagger = U^{-1}$. We now exhibit the covariance of eq. 2.123. Since we can view the three spatial γ-matrices as SU(3) 3-vectors, we can express eq. 7.3 as the result of a SU(3) rotation V of the γ-matrices (on the spinor indices)

$$V\gamma^a V^{-1} = U^{ab}\gamma'^b \qquad (7.4)$$

where V is a 4×4 reducible representation of SU(3), namely, $\underline{3} \oplus \underline{1}$. Since V commutes with γ^0 in the Pauli matrix representation of the γ matrices we see that V can have the form

$$V = \begin{bmatrix} A\exp(i\alpha_i\sigma_i) & 0 \\ 0 & B\exp(i\beta_i\sigma_i) \end{bmatrix}$$

where A, B, α_i and β_i are constants, and the zeroes represent 2×2 zero matrices. The inverse of V is $V^\dagger$. Thus eq. 7.4 becomes

$$V\gamma^a V^{-1} = \begin{bmatrix} 0 & AB^*\exp(i\alpha_i\sigma_i)\sigma_a\exp(-i\beta_i\sigma_i) \\ -A^*B\exp(i\beta_i\sigma_i)\sigma_a\exp(-i\alpha_i\sigma_i) & 0(-i\beta_i\sigma_i) \end{bmatrix}$$

We now note the generators of the global SU(3) symmetry under discussion have a 4×4 matrix reducible representation $(\underline{3} \oplus \underline{1})$. The generators of this reducible representation are F_i and F_0 (a diagonal matrix diag(0,0,0,0,0,0,0,0,0,1) with F_i being the Gell-Mann SU(3) generators for i = 1, 2, …, 8.

Projection operators can be defined to project out the $\underline{3}$ representation piece P_3 and the $\underline{1}$ representations piece P_1 of the complexon spinor fields:

Thus the $\underline{3}$ complexon field is

$$\psi_{C3}(t, \mathbf{x_r}, \mathbf{x_i}) = P_3\psi_C(t, \mathbf{x_r}, \mathbf{x_i}) \tag{7.5}$$

while the $\underline{1}$ complexon field is

$$\psi_{C1}(t, \mathbf{x_r}, \mathbf{x_i}) = P_1\psi_C(t, \mathbf{x_r}, \mathbf{x_i}) \tag{7.6}$$

Since P_1 and P_3 do not commute with Lorentz transformations, a Lorentz transformation mixes ψ_{C1} and ψ_{C3}.[84] Since P_1 and P_3 do not commute with γ_5, left-handed and right-handed complexons would also be mixed by these projection operators. The matrix V has a $\underline{3}\oplus\underline{1}$ reducible representation.

In a manner similar to the covariance proof of the Dirac equation[85] we see that eq. 7.2 is covariant under SU(3) transformations:

$$V[i\gamma^0\partial/\partial t + i\gamma\cdot\mathbf{D_c}^* - m]V^{-1}V\psi_C(t, \mathbf{x_r}, \mathbf{x_i}) = 0$$

or

$$[i\gamma^{0'}\partial/\partial t' + i\mathbf{D_c'}^*\cdot\gamma' - m]V\psi_C(t, \mathbf{x_r}, \mathbf{x_i}) = 0 \tag{7.7}$$

(Note $\gamma^{0'} = V\gamma^0 V^{-1}$ and t' = t.) The SU(3) transformed wave function $\psi_C'(t, \mathbf{x'})$ is

$$\psi_C'(t', \mathbf{x'}) = V\psi_C(t, \mathbf{x}) = V\psi_C(t', U\mathbf{x'}) \tag{7.8}$$

Thus the complexon Dirac equation is covariant under global coordinate SU(3).

The subsidiary condition,

$$\nabla_r\cdot\nabla_i\,\psi_{Cu}(t, \mathbf{x_r}, \mathbf{x_i}) = 0 \tag{7.9}$$

is also covariant under an SU(3) rotation:

---

[84] At this point it is worth noting that the construction of complexon fields, based on a boost from a particle rest state, guarantees that a reference frame exists in which any complexon particle has a single real time variable. Similarly a reference frame exists for a set of complexon particles (that is within a Lorentz of the center of momentum frame) with a single real time variable. The time variables of the individual complexon particles in the set are complex in general but are functions of the center of momentum real time variable. So there is only one real time variable for each complexon in the set although the time variable of an individual particle may be a complex function of the real-valued center of momentum time variable.

[85] For example see Bjorken (1964) pp. 18 – 20.

$$\nabla_r'^* \cdot \nabla_i' \psi_C'(t, \mathbf{x}') = \nabla_r \cdot \nabla_i \, V\psi_C(t, \mathbf{x}) = V\nabla_r^* \cdot \nabla_i \, \psi_C(t, \mathbf{x}) = 0 \qquad (7.10)$$

We now examine the transformation of the wave function eq. 7.8 under the SU(3) transformation U. If we define

$$q^{*\mu} = (q^0, \mathbf{q}^*) = (p^0, \mathbf{p}_r + i\mathbf{p}_i) = (p^0, \mathbf{p}) = p^\mu \qquad (7.11)$$

then $\psi_C(t, \mathbf{x})$ will be seen to be covariant form under an SU(3) transformation:

$$\psi_C(t, x) = \sum_{\pm s} \int d^3 q_r d^3 q_i \, N_C(p^0) \delta(\mathbf{q}_r^* \cdot \mathbf{q}_i/m^2)[b_C(q^*,s)u_C(q^*,s)e^{-i(q^* \cdot x + q \cdot x^*)/2} +$$
$$+ \, d_C^\dagger(q^*,s)v_C(q^*,s)e^{+i(q^* \cdot x + q \cdot x^*)/2}] \qquad (7.12)$$

Note both terms in each exponential are separately invariant under global SU(3). ($\mathbf{q}_r^* = \mathbf{q}_r$ since $\mathbf{q}_r$ is real.)

Eq. 7.8 implies that the spinors appearing in eq. 7.12 are covariant under SU(3) transformations

$$u_C'(q'^*,s') = Vu_C(q^*,s) \qquad (7.13)$$
$$v_C'(q'^*,s') = Vv_C(q^*,s) \qquad (7.14)$$

The fourier coefficients, if second quantized in a complex spatial coordinate generalization of the usual manner, also have covariant anti-commutation relations under an SU(3) transformation:

$$\{b_C(q,s), b_C^\dagger(q'^*,s')\} = \delta_{ss'}\delta^3(q_r - q'_r)\delta^3(q_i - q'_i) \qquad (7.15)$$

Under an SU(3) transformation, $z = Uq$ and $z' = Uq'$, the right side of eq. 7.15 transforms to

$$\delta^3(q_r - q'_r)\delta^3(q_i - q'_i) \rightarrow \delta^3(z_r - z'_r)\delta^3(z_i - z'_i)/|\partial(q)/\partial(z)| = \delta^3(z_r - z'_r)\delta^3(z_i - z'_i) \qquad (7.16)$$

where

$$|\partial(q)/\partial(z)| = |\partial(q_r^1,q_r^2,q_r^3,q_i^1, q_i^2, q_i^3)/\partial(z_r^1,z_r^2,z_r^3,z_i^1, z_i^2, z_i^3)| = 1 \qquad (7.17)$$

is the Jacobian of the transformation U. The fourier coefficients transform trivially under SU(3):

$$b_C(q^*,s) \rightarrow b_C(z^*,s) \qquad (7.18)$$

Since the integrand transforms as

$$\int d^3q_r d^3q_i \;\rightarrow\; \int d^3z_r d^3z_i \,|\partial(q)/\partial(z)| = \int d^3z_r d^3z_i \qquad (7.19)$$

we see that the wave function $\psi_C(t, \mathbf{x})$ transforms covariantly.

7.3 Local Color SU(3) and the Internal Symmetry Strong Interactions

In the previous section we showed that the equations of motion of free Dirac-like, complexon, up-type quarks are covariant under global SU(3) coordinate rotations.. The free, tachyon, complexon, down-type quark equations of motion are also easily seen to be covariant under this SU(3) subgroup. In this section we will show this covariance is the analogue of local Color SU(3) symmetry of quarks, and then we will introduce the Internal Symmetry Strong interaction analogue via minimal coupling to SU(3) Yang-Mills gluons in gauge covariant derivatives.

We now introduce a complexon field with a global SU(3) index a which takes values from 1 to 3 making the field a member of the $\underline{3}$ representation of global SU(3):

$$\psi_C{}^a(t, \mathbf{x}) \qquad (7.20)$$

Due to the SU(3) index the transformation property of $\psi_C{}^a(t, \mathbf{x})$ changes from eq. 7.8 to

$$\psi_C{}''^a(t, \mathbf{x}') = U^{ab}V\psi_C{}^b(t, \mathbf{x}) = U^{ab}V\psi_C{}^b(t, U\mathbf{x}') \qquad (7.21)$$

where U^{ab} is an SU(3) rotation of $\underline{3}$ representation "vectors" such as $\psi_C{}^b$ and $\mathbf{x}$. V is the corresponding rotation of the spinor indices of $\psi_C{}^b(t, \mathbf{x})$.

Note that the coordinate SU(3) rotation of the field factorizes into an SU(3) rotation of the three fields $\psi_C{}^b$ by U^{ab} and an SU(3) rotation of the four spinor components of each individual field $\psi_C{}^b$ by V.

This factorization enables us to consider a global SU(3) rotation of the $\psi_C{}^b$ fields while holding the coordinates fixed:

$$\psi_C{}'^a(t, \mathbf{x}) = U^{ab}\psi_C{}^b(t, \mathbf{x}) \qquad (7.22)$$

The equations of motion are covariant under this global transformation

$$0 = U^{ab}[i\gamma^0\partial/\partial t + i\boldsymbol{\gamma}\cdot\mathbf{D_c}^* - m]\psi_C{}^b(t, \mathbf{x_r}, \mathbf{x_i})$$

$$= [i\gamma^0\partial/\partial t + i\boldsymbol{\gamma}\cdot\mathbf{D_c}^* - m]\psi'_C{}^a(t, \mathbf{x_r}, \mathbf{x_i}) \qquad (7.23)$$

We now note the form of eq. 7.22 is the same as that of a *local* Yang-Mills rotation:

$$\psi_C'^a(t,\,\mathbf{x}) = \Theta^{ab}(t,\,\mathbf{x})\psi_C{}^b(t,\,\mathbf{x}) \tag{7.24}$$

where $\mathbf{x} = \mathbf{x_r} + i\mathbf{x_i}$. Therefore if we introduce a local SU(3) Yang-Mills field $A_{Cv}(t,\,\mathbf{x_r},\,\mathbf{x_i})$ and define a covariant derivative we can convert eq. 7.21 to the analogue case of Internal Symmetry local, color SU(3) if we do <u>not</u> perform the spinor rotation V.[86] The covariant derivative is

$$\mathcal{D}_v = D_v - igA_{Cv} \tag{7.25}$$

where

$$A_{Cv} = A_C{}^a{}_v t^a \tag{7.26}$$

and where $D_v = D_{qv}$ is given by

$$\begin{aligned} D_0 &= \partial/\partial x^0 \\ D_k &= \partial/\partial x_r{}^k + i\,\partial/\partial x_i{}^k \end{aligned} \tag{7.27}$$

The SU(3) 3×3 matrix generators satisfy

$$[t^a,\,t^b] = if^{abc}t^c \tag{7.28}$$

We can represent $\Theta_{ab}(x)$ in the form:

$$\Theta_{ab}(x) = [\exp(-i\varphi_c(x)t^c)]_{ab} \tag{7.29}$$

where $\varphi_c(x)$ is a local parameter dependent on $x = (x^0,\,\mathbf{x} = \mathbf{x_r} + i\mathbf{x_i})$, and t^c is an SU(3) generator.

Applying a gauge transformation to the gauge covariant derivative of a complexon fermion field $\mathcal{D}_v\psi_C(x)$:

$$\begin{aligned} \Theta\mathcal{D}_v\psi_C(x) &= \Theta D_v\psi_C(x) - ig\Theta A_{Cv}\Theta^{-1}\Theta\psi_C(x) \\ &= D_v\psi_C'(x) - igA_C'_v\psi_C'(x) = (\mathcal{D}_v\psi_C(x))' \end{aligned} \tag{7.30}$$

where

$$\psi_C'(x) = \Theta(x)\psi_C(x) \tag{7.31}$$

we find

$$A_C'_v = (-i/g)(D_v\Theta(x))\Theta^{-1}(x) + \Theta(x)A_{Cv}(x)\Theta^{-1}(x) \tag{7.32}$$

The reader will note that the form of eqs. 7.25 – 7.31 is identical to those associated with a conventional non-abelian gauge interaction with the replacement:

---

[86] This approach, by analogy, enables us to avoid the dilemmas associated with mixing coordinate and internal symmetries as described by Coleman, S., Phys. Rev. **138** B1262 (1965) and others in the case of SU(6) in the 1960's. Note that the spinor rotation V is expressed in terms of numerical matrices while, in the second quantized formulation, the U^{ab} rotation is expressed in terms of second quantized fields as well as numeric matrices. Thus the factorization is reflected in the form of the transformation.

$$\partial/\partial x^{\nu} \rightarrow D_{\nu} \qquad (7.33)$$

with D_{ν} given by eq. 7.27. Note that $\varphi_c(x)$, the local parameter in eq. 7.29 is dependent in general, on time, and the real and imaginary parts of the complex spatial 3-vector.

Introducing the SU(3) gauge covariant derivative transforms eq. 7.23 to

$$0 = [i\gamma^{\nu}\mathcal{D}_{\nu} - m]\psi_C^{\,a}(t, \mathbf{x_r}, \mathbf{x_i}) \qquad (7.34)$$

The preceding argumentation supports the following postulate:

Postulate: Quarks are in a $\underline{3}$ representation of an internal symmetry global SU(3) group..

We note the case of tachyon complexon quarks differs only in small details from the above discussion of Dirac-type complexon quarks.

7.4 Internal Symmetry Interactions Resulting by Analogy from Complex Space-Time Projected to Real Physical Space-Time

This chapter, and the preceding chapters, have shown that the Complex Lorentz group and the Coordinate Reality Group generate analogues of the familiar interactions of The Standard Model: SU(3)⊗SU(2)⊗U(1) plus an additional set of SU(2)⊗U(1) interactions that we take to be the interactions of Dark Matter.

8. An Internal Symmetry Reality Group based on an Internal Symmetry Complex Lorentz Group

8.1 Particle – Coordinate Duality

Elementary particles seem to be distinctly different from coordinates. However, there is a connection. One cannot measure a location's coordinates without the use of particles. This fact is evidenced by the number-phase uncertainty relation in Quantum Electrodynamics:

$$\Delta\varphi\Delta N \geq \hbar$$

where φ is the phase of a wave, N is the number of photons and $\hbar$ is Heisenberg's constant.

In this chapter we will suggest that elementary particles and coordinates have some analogous features. We will see that fermions, which have an iota at their 'core,' have a Lorentz symmetry group that can be viewed as generating the SU(3)⊗SU(2)⊗U(1)⊗SU(2)⊗U(1) part of the Unified SuperStandard Model symmetry based on considerations that parallel the coordinate space Lorentz symmetry group. As a result the Internal Symmetry Reality group discussed in earlier chapters emerges.

8.2 Coordinates' Lorentz Group

The Lorenz group that governs coordinate transformations has a number of non-commuting subroups: SU(3), SU(2), U(1), SU(2), and U(1). These groups are familiar from The Standard Model with the addition of SU(2), and U(1) subgroups that we add to The Stadard Model for Dark Matter.

We defined a Coordinate Reality group SU(3)⊗SU(2)⊗U(1)⊗SU(2)⊗U(1) that transformed complex-valued coordinate systems to real-valued coordinate systems. This group acts on coordinates, and to avoid issues with NOGo Theorems, we distinguished this group from an Internal Symmetry Reality group that acted on particles within the Unified SuperStandard Model. The question that presents itself is the source of the Internal Symmetry Reality group? It appears that there is a 'Particles Lorentz group' that acts to generate the Unified SuperStandard Model Internal Symmetry Reality group.

8.3 Iota Field Functionals

In chapter 2 and earlier books we developed the concept of the iota – the germ – if you will – of each fermion giving it a bare mas – the Landauer mass – but devoid of other features. We now extend the role of the iota to be a fermion field functional that acts as a 'template' for the addition of the wave function to create a fermion field. Thus when we create a fermion particle by applying a fermion field to the vacuum, the particle that is generated automatically contains an iota giving it a bare mass that is subsequently modified by quantum effects such as the Higgs Mechanism.

We can thus represent a fermion field functional as

$$f_\alpha$$

where α is a four component index. A simple fermion field then is denoted

$$\psi_\alpha(x) = f_\alpha(x)$$

where $\psi_\alpha(x)$ is a fourier representation of a quantum field with coordinates and q-number creation and annihilation operators.

We will not generate boson fields in this manner since they are initially massless and acquire masses, if they do, via the Higgs Mechanism. Bosons do not contain iotas.

8.4 A Particles' Lorentz Group

We now consider an iota functional, treat it as a type of coordinate, and apply Particles' Lorentz group transformations to it. First we apply all transformations of the SU(3) subgroup to f_α creating a set of coordinate systems denoted $f_{SU(3)}$.[87] We then define an independent SU(3) group whose sole purpose is to rotate the coordinate systems of $f_{SU(3)}$. Next we follow the same procedure independently for each of SU(2), U(1), SU(2), and U(1) generating sets of coordinate systems for each. Then we again define SU(2), U(1), SU(2), and U(1) groups to transform amongst each set.

We thus end up with five independent groups (not the Lorentz subgroups) that commute with each other and thus give us the Internal Symmetry group for fields SU(3)⊗SU(2)⊗U(1)⊗SU(2)⊗U(1).

---

[87] Each coordinate system in the set is one of the members of the set of items upon which the newly defined SU(3) group operates. The individual coordinates of any coordinate system are not relevant to the newly defined SU(3) group and the set upon which it operates.

8.5 Internal Symmetry Reality Group

We map each member of each set of coordinate systems for each of the groups to the fundamental representation of the respective group. Thus each coordinate system within the set $f_{SU(3)}$ is mapped to an element in the set of elements in the $\underline{3}$ representation of SU(3).

Following this procedure leads to the Standard Model fermion multiplets with the group structure SU(3)⊗SU(2)⊗U(1)⊗SU(2)⊗U(1).

8.6 A Space of Fermion Functionals?

The procedure we have developed puts particles on a footing similar to coordinates. Having developed the beginnings of coordinate spaces of fermion functionals it is natural to consider additional extensions of this concept. We leave that topic for future investigation being content with establishing the basis of the Internal Symmetry Reality group.However we do see the possibility of a connection of these fermion functional with the 'rotation of interactions' Θ-Symmetry group.

9. Two-Tier Coordinates

Originally Two-Tier coordinates were developed by this author to remove infinities that appear in perturbation theory calculations. We showed that the quantum smeared coordinates of Two-Tier Quantum Field Theory succeeded in removing all ultra-violet infinities in perturbation theory including the fermion triangle infinities. Remarkably the high precision, low energy[88] predictions of QED remained true in Two-Tier QED and thus remained consistent with experiment to a hitherto unsurpassed level of accuracy. 'Low' energy predictions in other quantum field theories also remained unchanged. At high energies, Two-Tier perturbation theory results are finite and consequently all ultra-violet infinities, to any order in perturbation theory, in *any number of space-time dimensions* were eliminated.

In addition to removing perturbation theory infinities Two-Tier coordinates enable us to define finite theories of Quantum Gravity and 'non-renormalizable' quantum field theories based on polynomial lagrangians, to tame vacuum fluctuations, to eliminate infinities associated with the Big Bang, and possibly to generate the explosive growth of the universe in its role as Dark Energy.[89]

Two-Tier Quantum Field Theory is established on the most fundamental level.

9.1 Two-Tier Features in 4-Dimensional Space-Time

Two-Tier Quantum Field Theory,[90] which was based on a new method[91] in the Calculus of Variations, uses two 'layers' of fields to introduce quantum coordinates. We shall consider this technique for the specific case of a massless vector field $V^i(y)$ analogous to the electromagnetic field.

In 4-dimensional space-time the massless vector field has the form $Y^\mu(y)$ where the index μ ranges from 0 through 3. The X^μ coordinate system, where it appears, has a c-number real part and a q-number imaginary part. Thus particle fields which are normally defined on four-dimensional real space-time will now be defined on a complex four-dimensional space-time where four imaginary dimensions will appear as *Quantum Dimensions* embodied in a vector quantum field $Y^\mu(y)$:

$$X^\mu(y) = y^\mu + \; i \; Y^\mu(y)/M_c^{\;2}$$

---

[88] Relative to a mass scale that was perhaps of the order of the Planck mass.

[89] See Blaha (2017b) and earlier books for details. This section is basically a summary of some features.

[90] See Blaha (2005a), and Blaha (2002), for discussions of this new method to eliminate infinities in quantum field theory calculations.

[91] Appendix D describes our method for the composition of extrema in some detail.

where M_c is an extremely large mass of the order of the Planck mass or perhaps much larger.

The $Y''(y)$ field is a function of the subspace y coordinates. The real part of the space-time dimensions will be taken to be the space of real-valued y coordinates.[92]

The imaginary part of space-time coordinates is the a massless $Y''(y)$ vector quantum field that is suppressed further by a very large mass scale – perhaps of the order of the Planck mass – that reduces the imaginary Quantum Dimensions to the infinitesimal except at large momenta. The effects of Quantum Dimensions only become appreciable in quantum field theory at energies of the order of M_c. At these energies exponential Gaussian factors in each particle (and ghost) propagator are generated by the Quantum Dimensions and serve to make perturbation theory calculations ultra-violet finite – including calculations in Quantum Gravity.

The formalism introduces a new form of interaction that does not have the form of the simple polynomial interactions that have hitherto dominated quantum field theories. This form of interaction takes place via the composition of quantum fields and can be called a *Dimensional Interaction* or an *Interdimensional Interaction* since it affects particle behavior through Quantum Dimensions.

The basic ansatz of the Two-Tier formalism is to replace every appearance of a coordinate x in a quantum field with the variable

$$x^\mu \to X^\mu = (y^0, \mathbf{y} + \mathbf{Y}(y^0, \mathbf{y})/M_c^2)$$

where $\mathbf{Y}(y^0, \mathbf{y})$ is the spatial part of a free massless vector field with features that are identical to the free QED field in the Radiation gauge.

Then one finds that the momentum space free field Feynman propagators G(k) of all particles acquires a Gaussian factor exp(h(k)):

$$G(k) \to G(k) \, \exp(h(k))$$

so that all perturbation theory diagrams are finite. The result is finite perturbative results for all calculations to any order in perturbation theory. Blaha (2005a) shows that Two-Tier theories are finite, Poincare covariant, and unitary. (See Blaha (2005a), chapter 5, for a complete discussion.)

9.2 Simple Two-Tier X^μ Formalism

In this subsection we will describe the basic Two-Tier formalism. Taking the lagrangian described in Blaha (2005a):[93]

---

[92] In a deeper theory the real part might also be a quantum field that undergoes a condensation to generate c-number coordinates. We will not consider this possibility in this book.
[93] Eq. 7.1. See Appendix D for more detail.

$$\mathcal{L}(y) = \mathcal{L}_F(X^\mu(y))J + \mathcal{L}_C(X^\mu(y), \partial X^\mu(y)/\partial y^\nu, y) \qquad (9.1)$$

where

$$X^\mu(y) = y^\mu + i\, Y^\mu(y)/M_c^2 \qquad (9.2)$$

with M_c being a large mass scale, $Y_\mu(y)$ a vector quantum field, and where J is the absolute value of the Jacobian of the transformation from X to y coordinates:

$$J = |\partial(X)/\partial(y)|$$

The lagrangian term $\mathcal{L}_C$ is

$$\mathcal{L}_C = +\tfrac{1}{4}\, M_c^{\,4} F^{\mu\nu} F_{\mu\nu}$$

with

$$F_{\mu\nu} = \partial X_\mu/\partial y^\nu - \partial X_\nu/\partial y^\mu \qquad (9.3)$$
$$\equiv i\,(\partial Y_\mu/\partial y^\nu - \partial Y_\nu/\partial y^\mu)/M_c^2$$

The lagrangian term $\mathcal{L}_F(X^\mu(y))$ contains the terms for scalar, fermion and other gauge terms in general. The sign in $\mathcal{L}_C$ is not negative – contrary to the conventional electromagnetic Lagrangian. The reason for this difference is that the quantum field part of X^μ is imaginary. Thus $\mathcal{L}_C$ ends up having the correct sign after taking account of the factor of i in the field strength $F_{\mu\nu}$.

Defining

$$F_{Y\mu\nu} = (\partial Y_\mu/\partial y^\nu - \partial Y_\nu/\partial y^\mu)$$

we see the Lagrangian assumes the form of the conventional electromagnetic Lagrangian:

$$\mathcal{L}_C = -\tfrac{1}{4}\, F_Y^{\mu\nu} F_{Y\mu\nu}$$

The action of this theory has the form

$$I = \int d^4y\, \mathcal{L}(y)$$

9.3 Y^μ Gauge

The gauge invariance of the Lagrangian allows us to choose a convenient gauge. The gauge invariance of the full Lagrangian

$$\mathcal{L}_s = L_F(\phi(X), \partial\phi/\partial X^\mu)\, J + \mathcal{L}_C(X^\mu(y), \partial X^\mu(y)/\partial y^\nu)$$

is based on the standard gauge invariance of $\mathscr{L}_C$, and the gauge invariance of $J\mathscr{L}_F$ in the form of translational invariance

$$X^\mu(y) \rightarrow X^\mu(y) + \delta X^\mu(y)$$

for the special case of a translation of X with the form of a gauge transformation:

$$\delta X^\mu(y) = \partial\Lambda(y)/\partial y_\mu$$

In this case we find

$$\int d^4y \; \Lambda(y) \; \partial \left[J \, \partial/\partial X^\mu \; \mathscr{T}_{F\mu\nu} \right]/\partial y_\nu = 0 \qquad (9.4)$$

after a partial integration and so we have the differential conservation law:

$$\partial \left[J \, \partial\mathscr{T}_{F\mu\nu}/\partial X^\mu \right]/\partial y_\nu = 0$$

since $\Lambda(y)$ is arbitrary. This conservation law is trivially obeyed:

$$\partial\mathscr{T}_{F\mu\nu}/\partial X^\mu = 0 \qquad (9.5)$$

Thus translational invariance in the $\mathscr{L}_F$ sector together with standard gauge invariance in the $\mathscr{L}_C$ sector automatically guarantees Y field gauge invariance of the total Lagrangian. We use the separate invariance of each term of

$$L = \int d^4y \left[\mathscr{L}_F J + \mathscr{L}_C \right] = \int d^4X \; \mathscr{L}_F + \int d^4y \; \mathscr{L}_C = L_F + L_C$$

under a constant translation $X^\mu \rightarrow X^\mu + \delta X^\mu$ where δX^μ is constant. Then we consider a position dependent translation/gauge transformation, which taken together with the above equation, establishes the invariance under the position dependent translation/gauge transformation.

An alternate approach that leads to the same result is to start with the particle part of the Lagrangian $\mathscr{L}_F$ rewritten to be invariant under general coordinate transformations, as it must, when we generalize to include General Relativity. Since position dependent translations are a form of general coordinate transformation the full theory must be invariant under position dependent translations due to invariance under general coordinate transformations.

Having established invariance under gauge transformations we now choose to use the most convenient gauge – the radiation gauge[94]:

$$\partial Y^i / \partial y^i = 0 \qquad (9.6)$$

where i = 1, 2, 3, which, in the absence of external sources, allows us to set

$$Y^0 = 0$$

since Y^0 does not have a canonically conjugate momentum. A conventional treatment leads to the equal time commutation relations:

$$[Y^\mu(\mathbf{y}, y^0), Y^\nu(\mathbf{y'}, y^0)] = [\pi^\mu(\mathbf{y}, y^0), \pi^\nu(\mathbf{y'}, y^0)] = 0 \qquad (9.7)$$

$$[\pi^j(\mathbf{y}, y^0), Y_k(\mathbf{y'}, y^0)] = -i\, \delta^{tr}_{jk}(\mathbf{y} - \mathbf{y'})$$

(Note the locations of the j indexes above introduce a minus sign.) where

$$\pi^k = \partial \mathcal{L}_C / \partial Y_k{}'$$
$$\pi^0 = 0$$

$$\delta^{tr}_{jk}(\mathbf{y} - \mathbf{y'}) = \int d^3k \; e^{i\, \mathbf{k} \cdot (\mathbf{y} - \mathbf{y'})}(\delta_{jk} - k_j k_k / \mathbf{k}^2)/(2\pi)^3$$

$$Y_k{}' = \partial Y_k / \partial y^0$$

The Radiation gauge reveals the two degrees of freedom that are present in the vector potential. The Fourier expansion of the vector potential is:

$$Y^i(y) = \int d^3k \; N_0(k) \sum_{\lambda=1}^{2} \varepsilon^i(k, \lambda)[a(k,\lambda)\, e^{-ik\cdot y} + a^\dagger(k,\lambda)\, e^{ik\cdot y}] \qquad (9.8)$$

where

$$N_0(k) = [(2\pi)^3 2\omega_k]^{-\frac{1}{2}}$$

and (since m = 0)

$$\omega_k = (\mathbf{k}^2)^{1/2} = k^0$$

with $\vec{\epsilon}(k, \lambda)$ being the polarization unit vectors for $\lambda = 1,2$ and $k^\mu k_\mu = 0$.

The further development of this theory is described in Part 3 of Blaha (2005a).

9.4 Scalar Field Quantization Using X^μ

We will begin by considering the case of a scalar quantum field theory. We assume a real underlying y subspace. Since X^μ is a set of coordinates, we choose to define a scalar field ϕ as a function of X^μ, which, in turn, is a function of the y^ν coordinates. We will provisionally second quantize ϕ treating X^μ as c-number coordinates using a conventional approach.[95]

We assume a Lagrangian, with the momentum conjugate to ϕ:

$$\pi_\phi = \partial L_F / \partial \phi' \equiv \partial L_F / \partial(\partial\phi/\partial X^0) \tag{9.9}$$

Following the canonical quantization procedure, π and ϕ become hermitian operators with equal time ($X^0 = X^{0\prime}$) commutation rules:

$$[\phi(X), \phi(X')] = [\pi_\phi(X), \pi_\phi(X')] = 0 \tag{9.10}$$

$$[\pi_\phi(X), \phi(X')] = -i\,\delta^3(\mathbf{X} - \mathbf{X}')$$

The standard Fourier expansion of the solution to the Klein-Gordon equation is:

$$\phi(X) = \int d^3p\, N_m(p)\, [a(p)\, e^{-ip\cdot X} + a^\dagger(p)\, e^{ip\cdot X}]$$

where

$$N_m(p) = [(2\pi)^3 2\omega_p]^{-1/2}$$

and

$$\omega_p = (\mathbf{p}^2 + m^2)^{1/2}$$

The commutation relations of the Fourier coefficient operators are:

$$[a(p), a^\dagger(p')] = \delta^3(\mathbf{p} - \mathbf{p}')$$

---

[95] Some texts are: Bogoliubov, N. N., Shirkov, D. V., *Introduction to the Theory of Quantized Fields* (Wiley-Interscience Publishers Inc., New York, 1959); Bjorken, J. D., Drell, S. D., *Relativistic Quantum Fields* (McGraw-Hill, New York, 1965); Huang, K., *Quarks, Leptons & Gauge Fields Second Edition* (World Scientific, River Edge, NJ, 1992); Kaku, M., *Quantum Field Theory* (Oxford University Press, New York, 1993); Weinberg, S., *The Quantum Theory of Fields* (Cambridge University Press, New York, 1995).

$$[a^\dagger(p), a^\dagger(p')] = [a(p), a(p')] = 0$$

The reader will recognize the quantization procedure is formally identical to the standard canonical quantization procedure of a free scalar quantum field.

In the case of spin ½, spin 1 and spin 2 fields the standard quantization procedure *in terms of the X coordinate system* can also be followed in a way similar to the procedure in standard texts.

9.5 Scalar Feynman Propagators

The momentum space free field Feynman propagators $G...(k)$ of all particles and ghosts in all Two-Tier Quantum Field Theories acquires a Gaussian factor $\exp(h(k))$:

$$G...(k) \rightarrow G...(k)\, \exp(h(k))$$

so that all perturbation theory diagrams are finite. The result is a finite perturbative result in all calculations to any order in perturbation theory. Blaha (2005a) shows that Two-Tier theories are finite, Poincare covariant, and unitary.

An example of the Two-Tier effect on propagators is the case of the Two-Tier photon propagator. The Two-Tier photon propagator[96] is:

$$iD_F^{TT}(y_1 - y_2)_{\mu\nu} = -i \int \frac{d^4p\, e^{-ip\cdot z}\, g_{\mu\nu}\, R(\mathbf{p}, z)}{(2\pi)^4\,(p^2 + i\varepsilon)} \tag{9.11}$$

(since the imaginary parts can be taken to be zero: $y_{1i}{}^\mu - y_{2i}{}^\mu = 0$) where

$$z^\mu = y_{1r}{}^\mu - y_{2r}{}^\mu$$

$$R(\mathbf{p}, z) = \exp[-p^i p^j \Delta_{Tij}(z)/M_c{}^4]$$

$$= \exp\{-\mathbf{p}^2[A(v) + B(v)\cos^2\theta]\,/\,[4\pi^2 M_c{}^4|\mathbf{z}|^2]$$

with $i, j = 1, 2, 3$, and with $\Delta_{Tij}(z)$ being the commutator of the positive frequency part $Y^+{}_k(y)$ and the negative frequency part $Y^-{}_k(y)$ of $Y_k(y)$:

$$\Delta_{Tij}(z) = [Y^+{}_j(y_{1r}), Y^-{}_k(y_{2r})] = \int d^3k\, e^{ik\cdot(y_{1r} - y_{2r})}\,(\delta_{jk} - k_j k_k/\mathbf{k}^2)/[(2\pi)^3 2\omega_k] \tag{9.12}$$

and

---

[96] Blaha (2005a).

$$v = |z^0|/|\mathbf{z}|$$
$$A(v) = (1 - v^2)^{-1} + .5v \ln[(v - 1)/(v + 1)]$$
$$B(v) = v^2(1 - v^2)^{-1} - 1.5v \ln[(v - 1)/(v + 1)]$$
$$\mathbf{p}\cdot\mathbf{z} = |\mathbf{p}| \, |\mathbf{z}| \cos\theta$$

with $|\mathbf{p}|$ denoting the length of a spatial vector $\mathbf{p}$, $|\mathbf{z}|$ denoting the length of a spatial vector $\mathbf{z}$, and with $|z^0|$ being the absolute value of z^0.

The gaussian factors $R(\mathbf{p}, z)$ which appear in all Two-Tier propagators damp the large momentum behavior of all perturbation theory integrals producing a completely finite perturbation theory and yet give the usual results of perturbation theory at energies that are small compared to the mass scale M_c.

9.6 String-like Substructure of the Theory

Two-tier Quantum field Theory endows each particle with an extended structure that resembles the extended structure seen in bosonic string and Superstring theories.[97] For example, Bailin (1994) use the operator[98]

$$V_\Lambda(k) = \int d^2\sigma \, \sqrt{-h} \, W_\Lambda(\tau, \sigma) \, e^{-ik\cdot X}$$

where X^μ is a quantized fourier expansion of the string fields (see eq. 7.22 of Bailin (1994)).

We note our X^μ coordinate-field has two transverse degrees of freedom due to gauge invariance, which also invites comparison to the bosonic string. A point of difference is that we have a well-defined quantum field theoretic formulation in conventional space-time that has the Standard Model as its "large distance" behavior thus introducing a note of reality that is not apparent in Superstring theories. We see that the interacting quantum field theories based on this approach also have good, finite, short distance behavior just as string theories.

The scalar, and other particles', Feynman propagators can be viewed as describing the propagation of a particle cloaked (accompanied) by a cloud of Y particles (which generates the $R(\mathbf{p}, y_1 - y_2)$ factor in the above propagator). If we examine the fourier transform of $R(p, z)$ we see:

$$(2\pi)^4 R(\mathbf{p}, q) = \int d^4z \, e^{iq\cdot z} \, R(\mathbf{p}, z) = \int d^4z \, e^{iq\cdot z} \exp[-p^i p^j \Delta_{Tij}(z)/M_c^4]$$

$$(9.13)$$

and we find

$$R(\mathbf{p},q) = \sum_{n=0}^{\infty} [i(2\pi M_c)^4]^{-n} (n!)^{-1} \prod_{j=1}^{n} [\int d^4k_j \, \theta(k_j^0)(p^2 - (\mathbf{p}\cdot\mathbf{k}_j)^2/\mathbf{k}_j^2)/(k_j^2 + i\varepsilon)] \, \delta^4(q - \Sigma \, k_r)$$

---

[97] Chapter 28 discussed the string interpretation of Two-Tier Quantum Field Theory in more detail.
[98] D. Bailin and A. Love, *Supersymmetric Gauge Field Theory and String Theory* (Institute of Physics Publishing, Philadelphia, PA, 1994) page 272.

which can be interpreted as a "cloud" of Y particles dressing the "bare" particle propagator. (The apparent divergences for R(p, q) are an artifact of the expansion and the subsequent fourier transformation. They are not present in the $R(\mathbf{p}, y_1 - y_2)$ factor in the propagator. See Fig. 9.1 for the Feynman diagram of the Two-Tier 'cloaked' propagator as compared to the normal scalar particle Feynman propagator. The Two-Tier Feynman propagator is basically a conventional scalar propagator that is modified by coherent Y particle emission.[99]

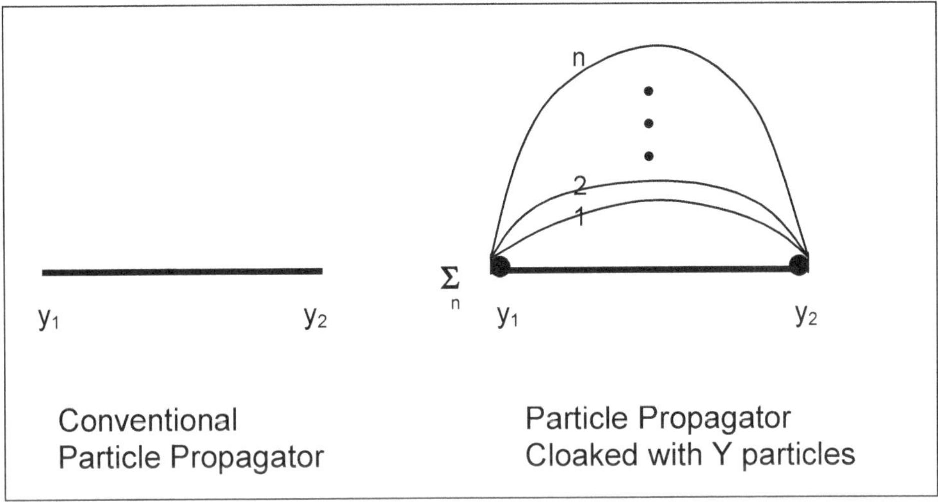

Figure 9.1. Feynman diagram for conventional and the nth diagram of a cloaked Two-Tier propagator.

We note that R(p, q) satisfies the convolution theorem:

$$\int d^4k\, R(\mathbf{p}, k)\, R(\mathbf{p}, q - k) = [R(\mathbf{p}, q)]^2$$

or

$$(2\pi)^4 \int d^4z\, e^{iq \cdot z}\, R(\mathbf{p}, z)\, R(\mathbf{p}, z) = [\, \int d^4z\, e^{iq \cdot z}\, R(\mathbf{p}, z)\,]^2 \tag{9.14}$$

The proof follows from the Binomial theorem.

---

[99] T. W. B. Kibble, Phys. Rev. **173**, 1527 (1968) and references therein. In particular see p. 1532 of Kibble's paper.

9.7 Two-Tier Complexon Quantum Fields

In the case of the Complexon Standard Model we will need two variables X_r^{μ} and X_i^{μ} since we have complex spatial 3-coordinates. We define them similarly to the previous case:

$$X_r^{\mu}(y_r) = y_r^{\mu} + i\, Y_r^{\mu}(y_r)/M_c^2$$

$$X_i^{\mu}(y_i) = y_i^{\mu} + i\, Y_i^{\mu}(y_i)/M_c^2$$

where we choose the same mass scale for both the "real" and "imaginary" variables. The Two-Tier, single generation, version of the Complexon Standard Model then has an action of the form

$$I_{CSMtt} = \int dy^0 d^3y_r d^3y_i \left(\mathscr{L}_{CSM}(X_r^{\mu}(y_r),\, \mathbf{X}_i^{k}(y_i))J_2\right)\Big|_{y_i{}^0 = 0,\ Y_r{}^0 = Y_i{}^0 = 0} + \tag{9.15}$$

$$+ \int dy_r{}^0 d^3y_r\, \mathscr{L}_C(X_r^{\mu}(y_r),\, \partial X_r^{\mu}(y_r)/\partial y_r{}^{\nu},\, y_r) +$$

$$+ \int dy_i{}^0 d^3y_i\, \mathscr{L}_C(X_i^{\mu}(y_i),\, \partial X_i^{\mu}(y_i)/\partial y_i{}^{\nu},\, y_i)$$

where the replacements

$$x^{\mu} \equiv x_r^{\mu} \;\rightarrow\; X_r^{\mu}(y_r)$$

$$x_i{}^k \;\rightarrow\; X_i^{k}(y_i)$$

for $\mu = 0, 1, 2, 3$ and $k = 1, 2, 3$ are made, followed by defining $y_r{}^0 = y^0$ and making a Complex Lorentz transformation to a frame where $y_i{}^0 = 0$. J_2 is the absolute value of the Jacobian of the transformation from (X_r, X_i) to (y_r, y_i) coordinates:

$$J_2 = |\partial(X_r, X_i)/\partial(y_r, y_i)|$$

We also choose gauges where $Y_r{}^0 = Y_i{}^0 = 0$. These types of transformations and gauge choices are discussed in detail in Blaha (2005a). The lagrangian terms $\mathscr{L}_C(X_r^{\mu}(y_r),\, \partial X_r^{\mu}(y_r)/\partial y_r{}^{\nu},\, y_r)$ and $\mathscr{L}_C(X_i^{\mu}(y_i),\, \partial X_i^{\mu}(y_i)/\partial y_i{}^{\nu},\, y_i)$ have the same form:

$$\mathscr{L}_C = +\tfrac{1}{4}\, M_c^4 F^{\mu\nu} F_{\mu\nu} \tag{9.16}$$

with

$$F_{\mu\nu} = \partial X_\mu/\partial y^\nu - \partial X_\nu/\partial y^\mu$$
$$\equiv i\,(\partial Y_\mu/\partial y^\nu - \partial Y_\nu/\partial y^\mu)/M_c^{\,2}$$

or defining

$$F_{Y\mu\nu} = (\partial Y_\mu/\partial y^\nu - \partial Y_\nu/\partial y^\mu)$$

we see each lagrangian assumes the form of the conventional electromagnetic Lagrangian:

$$\mathcal{L}_C = -\tfrac{1}{4}\,F_Y^{\,\mu\nu}F_{Y\mu\nu}$$

The lagrangian is supplemented with the following condition on all complexon fields $\Phi_{...}$:

$$(\partial/\partial X_r^{\,k}(y_r))\,(\partial/\partial X_i^{\,k}(y_i))\Phi... = 0 \qquad\qquad (9.17)$$

summed over $k = 1, 2, 3$. Non-complexon fields $\Omega...$ in our left-handed formulation satisfy the subsidiary condition:

$$\{(\partial/\partial X_r^{\,k}(y_r))(\partial/\partial X_i^{\,k}(y_i)) - [(\partial/\partial X_r^{\,k}(y_r))^2(\partial/\partial X_i^{\,m}(y_i))^2]^{1/2}\}\Omega... = 0 \qquad (9.18)$$

summed over $k = 1, 2, 3$ and over $m = 1, 2, 3$ separately in each of the two terms.

9.8 Complexon Feynman Propagator

In the case of complexons, the Two-Tier Feynman propagator differs from the non-complexon case by having an integration over imaginary spatial 3-momenta, a derivative of a delta function embodying the orthogonality of the real and imaginary 3-momenta, and two factors of $R(\mathbf{p}, z)$: one factor being $R(\mathbf{p}_r, z_r)$ and the other factor being $R(\mathbf{p}_i, z_i)$ (where the time components $z_r^{\,0} = z^0$ and $z_i^{\,0} = 0$ since there is only one real time coordinate[100]) thus providing large momentum convergence for both real and imaginary 3-momentum integrations.

For a normal scalar particle the Feynman propagator is:

$$i\Delta_{CTF}(x - y) = \theta(x^+ - y^+)\langle0|\phi_{CT}(x)\,\phi_{CT}(y)|0\rangle + \theta(y^+ - x^+)\langle0|\phi_{CT}(y)\phi_{CT}(x)|0\rangle$$
$$= i\!\int d^4p_r d^3p_i (2\pi)^{-7}\delta'(\mathbf{p}_r\!\cdot\!\mathbf{p}_i/m^2)e^{-ip^+(x^- - y^-)-ip^-(x^+ - y^+)+i\mathbf{p}_\perp\cdot(\mathbf{x}_\perp-\mathbf{y}_\perp)-i\mathbf{p}_i\cdot(\mathbf{x}_i-\mathbf{y}_i)}/(p^2 + m^2+i\varepsilon)$$
$$(9.19)$$

in conventional quantum field theory.

In the case of Two-Tier quantum field a scalar *complexon* particle has the the Feynman propagator

---

[100] We can arrange for $z_i^{\,0} = 0$ by making a Complex Lorentz transformation to an inertial frame where z is real.

$$i\Delta_{CTFtt}(x - y) = i\int d^4p_r d^3p_i (2\pi)^{-7}\delta'(\mathbf{p_r \cdot p_i}/m^2)\, R(\mathbf{p_r}, z_r)R(\mathbf{p_i}, z_i)\cdot \tag{9.20}$$

$$\cdot e^{-ip^+(x^- - y^-) - ip^-(x^+ - y^+) + ip_\perp \cdot (x_\perp - y_\perp) - ip_i \cdot (x_i - y_i)}/(p^2 - m^2 + i\varepsilon)$$

where the time components $z_r^0 = z^0$ and $z_i^0 = 0$ since there is only one time coordinate, where $R(\mathbf{p}, z)$ is given in the previous subsection, and where $p^2 = p^{0\,2} - p_r^2 + p_i^2$.

Propagators for other types of particles are similarly modified in the Two-Tier formalism (See Blaha 2005a).

9.9 Vacuum Fluctuations

While the expectation value of a *conventional* free scalar field $\phi_{conv}(x)$ is zero in a conventional quantum field theory:

$$<0|\phi_{conv}(x)|0> = 0 \tag{9.21}$$

the vacuum fluctuations of *conventional* scalar quantum field theory are quadratically divergent:

$$<0|\phi_{conv}(x)\phi_{conv}(x)|0> = \int d^3p/[(2\pi)^3 2\omega_p] \tag{9.22}$$

In "Two-Tier" quantum field theory we find the vacuum expectation value of a free field is zero and the expectation value of the square of the field is also zero:

$$<0|\phi(X)\phi(X)|0> = \int d^3p\, e^{-p^i p^j \Delta_{Tij}(0)/Mc^4}/[(2\pi)^3 2\omega_p] = 0$$

since the exponential factor in the integral is $-\infty$. The exponent contains

$$\Delta_{Tij}(z) = \int d^3k\, e^{-ik \cdot z}\, (\delta_{ij} - k_i k_j/\mathbf{k}^2)/[(2\pi)^3 2\omega_k] \tag{9.23}$$

where "T" is for "Two-Tier". Thus *vacuum fluctuations are zero in Two-Tier quantum field theory*. Correspondingly, we will see that renormalization constants are finite in the Two-Tier versions of QED, Electroweak Theory, the Standard Model and Quantum Gravity. See Blaha (2017b) and references therein for more details.

9.10 Time Intervals in General Relativity

Wigner[101] has studied the measurement of time intervals in General Relativity and sees a problem in the measurement of extremely short intervals. According to Wigner, the measurement of a time inteval in a region of space requires the measurement of the length of

---

[101] E. P. Wigner, Rev. Mod. Phys. **29**, 255 (1957); J. Math. Phys. **2**, 207 (1961).

time required for an event to happen. The measurement requires an accurate clock. But the accuracy of the clock is limited by the energy-time uncertainty relation:

$$\Delta E \Delta t \geq \hbar \qquad (9.24)$$

Thus the uncertainty in the clock's time measurement is related to the uncertainty in the clock's energy which is, in turn, related to the uncertainty in the clock's mass:

$$\Delta E = (\Delta m)c^2$$

To obtain "infinite" accuracy the uncertainty (fluctuations) in the clock's mass must be infinite and thus the clock's mass must be infinite. Infinite fluctuations in the clock's mass will produce corresponding infinite fluctuations in the gravitational field.

$$\Delta h \propto \Delta E \qquad \text{(in conventional General Relativity)}$$

As a result the notion of space-time and time intervals (which depend on the geometry through General Relativity) become uncertain. Thus, according to Wigner, and others, the concept of time intervals and space-time points becomes questionable.

The Two-Tier version of Quantum Gravity offers a way out of this dilemma. The gravitational force becomes stronger as one goes to shorter distances (higher energies) down to a distance (up to an energy) whose scale is set by M_c. At shorter distances (higher energies) the gravitational force becomes weaker and declines to zero at zero distance. Thus at very high energy the gravitational field fluctuations (Δh) are at worst inversely proportional to the energy (and probably decline by a higher power of inverse energy.) (The same considerations would apply if one chooses to consider fluctuations in the Riemann-Christoffel symbols.)

$$\Delta h < c_1/E < c_1/(\Delta E) \qquad \text{(in Two-Tier Quantum Gravity)} \qquad (9.25)$$

where c_1 is a constant. Thus Wigner's conclusion does not hold in the Two-Tier version of Quantum Gravity as gravitational fluctuations actually become smaller at energies above a critical energy whose scale is set by M_c.

In fact, combining the above equations we see

$$c_1 \Delta t / \Delta h \geq \hbar$$

at sufficiently high energy. Therefore the time uncertainty Δt, and the gravitational field fluctuations Δh, can both decrease while maintaining the energy-time uncertainty relation. *Thus the notion of a space-time point "is saved" in Two-Tier quantum gravity.*

9.11 Vacuum Fluctuations in the Gravitation Fields

While the expectation value of the free graviton field $h_{\mu\nu conv}(x)$ (weak field approximation) is zero in a conventional quantum field theoric approach:

$$<0|h_{\mu\nu conv}(x)|0> = 0 \qquad (9.26)$$

the vacuum fluctuations of the *conventional* quantum graviton field is quadratically divergent since

$$<0|h_{\mu\nu conv}(x)h_{\alpha\beta conv}(x)|0> = \int d^3p \; b'_{\mu\nu\alpha\beta}(p)/[(2\pi)^3 \, 2\omega_p] = \infty \qquad (9.27)$$

where $b'_{\mu\nu\alpha\beta}(p)$ is a rational function of the momentum p.

In "Two-Tier" quantum field theory we find

$$<0|h_{\mu\nu}(X)h_{\alpha\beta}(X)|0> = \int d^3p \; b'_{\mu\nu\alpha\beta}(p) \; e^{-p^ip^j\Delta_{Tij}(0)}/[(2\pi)^3 2\omega_p] = 0 \qquad (9.28)$$

since the exponential factor in the integrand is $-\infty$. The exponent contains

$$\Delta_{Tij}(z) = \int d^3k \; e^{-ik\cdot z}(\delta_{ij} - k_ik_j/\mathbf{k}^2)/[(2\pi)^3 2\omega_k]$$

Thus the vacuum fluctuations of $h_{\mu\nu}$ are zero in "Two-Tier" quantum field theory and, correspondingly, the weak field Two-Tier quantization of Quantum Gravity is consistently finite (and weak in perturbation theory calculations.)

9.12 Two-Tier Features in D-Dimensional Space-Time (such as the Megaverse)

Since a field, quantized in D-dimensional conventional coordinates (D > 4), would lead to divergences in perturbation theory calculations, we can use D-dimensional Two-Tier coordinates to avoid divergences in perturbation theory:

$$Y^i(y) = y^i + i \, Y_u^{\;i}(y)/M_u^{D/2} \qquad (9.29)$$

where $Y_u^{\;i}(y)$ for $i = 1, ..., D$ is a D-dimensional free gauge field and M_u is a mass of the order of the Planck mass or greater. The $Y_u^{\;i}(y)$ term adds a quantum field to the D coordinates making them a set of quantum coordinates. Quantum coordinate derivatives are defined by

$$\partial_i = \partial/\partial Y^i(y) = \partial/\partial(y^i - Y_u^{\;i}(y)/M_u^{D/2}) \qquad (9.30)$$

The use of these coordinates to quantize particle fields leads to a completely finite perturbation theory. We applied them in Blaha (2017b) to create a finite fundamental theory of mater. We applied them to fields in the Megaverse[102] to achieve a finite theory of Megaverse dynamics for elementary particles and universe particles.

The second quantization of a vector gauge field $V^i(y)$ is analogous to the second quantization of the electromagnetic field. The lagrangian density terms for the free $V^i(Y(y))$ fields is

$$\mathscr{L}_{Vu} = -\tfrac{1}{4}\, F_{Vu}{}^{ij}(Y(y))F_{Vuij}(Y(y)) \tag{9.31}$$

The lagrangian is

$$L_{Vu} = \int d^D y\, \mathscr{L}_{Vu}(Y(y))$$

with

$$F_{Vuij} = \partial V_i(Y(y))/\partial Y^j(y) - \partial V_j(Y(y))/\partial Y^i(y)$$

where the values of i and j range from 1 to D in this section.

The equal time commutation relations, using the Dth coordinate as the time coordinate, are specified in the usual way:

$$[V^i(Y(\mathbf{y}, y^0)), V^j(Y(\mathbf{y}', y^0))] = [\pi^i(Y(\mathbf{y}, y^0)), \pi^j(Y(\mathbf{y}', y^0))] = 0$$
$$[\pi_j(Y(\mathbf{y}, y^0)), V_k(Y(\mathbf{y}', y^0))] = -i\, \delta^{(D-1)tr}{}_{jk}(Y(\mathbf{y},0) - Y(\mathbf{y}',0))$$

where

$$\pi_u{}^k = \partial \mathscr{L}_{Vu}(V(Y(y)))/\partial V_k'(Y(y))$$
$$\pi_u{}^D = 0$$

for $k = 1, \ldots, (D-1)$, and

$$\delta^{(D-1)tr}{}_{jk}(\mathbf{y} - \mathbf{y}') = \int d^{(D-1)}k\, e^{i\,\mathbf{k}\cdot(Y(\mathbf{y},0) - Y(\mathbf{y}',0))}\, (\delta_{jk} - k_j k_k/\mathbf{k}^2)/(2\pi)^{D-1} \tag{9.32}$$
$$V_k'(Y(y)) = \partial V_k(Y(y))/\partial y^{1D}$$

for $j, k = 1, 2, \ldots, (D-1)$.

If we choose the Radiation gauge for $V_k(Y(y))$:

$$V^D(Y(y)) = 0$$
$$\partial V^j(Y(y))/\partial Y^j(y) = 0 \tag{9.33}$$

for $j = 1, 2, \ldots, (D-1)$ then $(D-2)$ degrees of freedom (polarizations) are present in the vector potential.[103] The Fourier expansion of the vector potential $V^i(Y(y))$ is:

$$V^i(Y(y)) = \int d^{(D-1)}k\, N_{0V}(k) \sum_{\lambda=1}^{D-2} \varepsilon^i(k, \lambda)[a_V(k,\lambda) :e^{-ik\cdot Y(y)}: + a_V^\dagger(k,\lambda) :e^{ik\cdot Y(y)}:] \tag{9.34}$$

---

[102] Blaha (2017c).
[103] Note we use the Radiation gauge for $Y^\mu(y)$ also.

for $i = 1, \ldots, (D-2)$ where

$$N_{0V}(k) = [(2\pi)^{(D-1)}2\omega_k]^{-\frac{1}{2}}$$

and (since the field is massless)

$$k^D = \omega_k = (\mathbf{k}^2)^{\frac{1}{2}}$$

where k^D is the energy, and where the $\varepsilon^i(k, \lambda)$ are the polarization unit vectors for $\lambda = 1, \ldots, (D-2)$ and $k^\mu k_\mu = k^{D\,2} - \mathbf{k}^2 = 0$.

The commutation relations of the Fourier coefficient operators are:

$$[a_V(k,\lambda), a_V^\dagger(k',\lambda')] = \delta_{\lambda\lambda'}\delta^{D-1}(\mathbf{k} - \mathbf{k}')$$
$$[a_V^\dagger(k,\lambda), a_V^\dagger(k',\lambda')] = [a_V(k,\lambda), a_V(k',\lambda')] = 0$$

and the polarization vectors satisfy

$$\sum_{\lambda=1}^{D-2} \varepsilon_i(k, \lambda)\varepsilon_j(k, \lambda) = (\delta_{ij} - k_i k_j/\mathbf{k}^2)$$

The V^μ Feynman propagator is

$$iD_F^{\text{trTT}}(y_1 - y_2)_{jk} = <0|T(V_j(Y(y_1))V_k(Y(y_2)))|0> \tag{9.35}$$

$$= -\,ig_{jk} \int \frac{d^D k\, e^{-ik\cdot(y_1 - y_2)}\, R(\mathbf{k}, y_1 - y_2)}{(2\pi)^{16}\,(k^2 + i\varepsilon)}$$

where g_{jk} is the D-dimensional Lorentz metric and where $R(\mathbf{k}, y_1 - y_2)$ is given by

$$R(\mathbf{k}, y_1 - y_2) = \exp[-k^i k^j \Delta_{Tij}(y_1 - y_2)/M_u^D]$$
$$= \exp\{-k^2[A(v) + B(v)\cos^2\theta]\,/\,[(2\pi)^{D-2}M_u^4 z^2]\}$$

where k^2 is *the sum of the squares of the D – 1 spatial components* with

$$z^\mu = y_1{}^\mu - y_2{}^\mu$$
$$z = |\mathbf{z}| = |\mathbf{y_1} - \mathbf{y_2}|$$
$$k = |\mathbf{k}|$$
$$v = |z^0|/z$$
$$A(v) = (1 - v^2)^{-1} + .5v\,\ln[(v-1)/(v+1)]$$
$$B(v) = v^2(1 - v^2)^{-1} - 1.5v\,\ln[(v-1)/(v+1)]$$

$$\mathbf{k \cdot z} = kz \cos\theta$$

and $|\mathbf{k}|$ denoting the length of a spatial $(D - 1)$-vector $\mathbf{k}$ while $|z^0|$ is the absolute value of $z^0 \equiv z^D$.

As the above equations indicate, the Gaussian damping factor $R(k, z)$ for *all* large spatial momentum k^j is the same for both the positive and negative frequency parts of the (Two Tier) V Feynman propagator. We are assuming the spatial momentum is real-valued in this discussion. It is also important to note that $R(k, z)$ does not depend on $k^0 = k^D$ (in the V and Y_u Radiation gauges) and thus the integration over k^0 proceeds in the usual way to produce time-ordered positive and negative frequency parts.

The Gaussian exponential factor in *all* spatial coordinates causes the Feynman propagator to be finite and, together with the Gaussian factor in universe particle propagators, causes all perturbation theory calculations when interactions are introduced to be finite as we have seen in Blaha (2017b).

For small momentum much less than M_u then $R(\mathbf{k}, y_1 - y_2) \rightarrow 1$ and the Feynman propagator is the "normal" propagator of conventional D-dimensional quantum field theory. For large momentum the corresponding potential approaches r^{D-3} in contrast to the electromagnetic Coulomb potential r^{-1}. The V potential is highly non-singular at large energies.

Thus using Two-Tier Quantum Field Theory we can perform perturbation theory caluculations that always yield a finite result.[104] This is not true if conventional Quantum Field is used.[105]

---

[104] In particular, the fermion triangle divergence (anomaly) does not occur in our Two Tier Quantum Field Theory of the fermion sector. Thus there is no requirement for axion-like particles in the Megaverse (or in universes) although the possible existence of this type of particle is not ruled out.

[105] Blaha (2005a) provides a complete discussion of Two-Tier Quantum Field Theory.

10. PseudoQuantum Field Theory

PseudoQuantum Field Theory (and its Quantum Mechanics analogue CQ Mechanics[106]) originate in the need to second quantize in unusual coordinate systems, and in curved space-time coordinate systems. The papers in Appendices I and J provide a detailed introduction to PseudoQuantum Field Theory to which the reader is referred.

In this subsection we point out its advantages in a variety of field theory contexts that are relevant for the Unified SuperStandard Model. The advantages of PseudoQuantum Field Theory are:

1. Quantization in any coordinate system in flat or curved space-times with an invariant definition of asymptotic particle states. An n particle asymptotic state in one coordinate system is a unitarily equivalent n particle asymptotic state in any other coordinate system. Therefore particle number is invariant under change of coordinate system. This is important for the Unified SuperStandard Model in curved space-times. It is also important for quantization in higher dimensional Euclidean spaces such as the Megaverse. The method was developed in the late 1970's by the author to provide a quantization procedure which supports a unique particle interpretation of states in arbitrary non-static space-times where no global timelike coordinate (Killing vector) exists. PseudoQuantum Field Theory which we developed in a series of books[107] also can be formulated in the Megaverse. Thus we can use it in the Megaverse to implement the Higgs Mechanism to generate particle masses and symmetry breaking.

2. PseudoQuantum Field Theory enables one to define Higgs particle dynamics in such a way that a non-zero vacuum expectation value cleanly separates from the quantum field part of the Higgs fields. This technique can be used in symmetry breaking mechanisms, mass generation, and possible generation of coupling constants as vacuum expectation values.

3. It supports the canonical definition of higher derivative field theories through the use of the Ostrogradski bootstrap. See Appendix B where a fourth order theory of the Strong

---

[106] See Appendix E for details, which contains Blaha (2016f). CQ Mechanics encompasses both classical mechanics and quantum mechanics, and provides a method of rotating between them. It has applications to transitions between Quantum/Semi-Classical Entanglement, and Quantum/Classical Path Integrals, and Quantum/Classical Chaos.

[107] See Blaha (2017b) for the discussion of the PseudoQuantum field theory formalism for Higgs particles in our Extended Standard Model. See chapter 20 of Blaha (2017b), and earlier books, for a more detailed view than that presented here.

interaction is defined that has color confinement and a linear r potential. The potential part of this theory was used by the Cornell group to calculate the Charmonium spectrum. (See Blaha (2017b) for details.)

An associated advantage of using PseudoQuantum Field Theory is that it provides for retarded propagators and an Arrow of Time.

10.1 General Case of PseudoQuantization in Differing Coordinate Systems

Appendix A describes the PseudoQuantization procedure that relates seond quantizations in differing coordinate systems. We can epitomize the general concept in the following short example.

Consider the case of a scalar particle in D space-time dimensions that we second quantize in coordinate system denoted 1 with coordinates x based on a timelike Killing vector

$$\varphi(x) = \sum_{\alpha} [\chi_\alpha(x)A_\alpha + \chi_\alpha^*(x)A_\alpha^\dagger] \qquad (10.1)$$

where the $\chi_\alpha(x)$ are positive frequency with respect to a definition of positive frequency within a universe – following the notation of Appendix A.

Consider now the second quantization of the particle field in a second coordinate system denoted 2 with coordinates y based on a different timelike Megaverse Killing vector

$$\varphi(y) = \sum_{\beta} [\psi_\beta(y)b_\beta + \psi_\beta^*(y)b_\beta^\dagger] \qquad (10.2)$$

where the $\psi_\beta(y)$ are positive frequency with respect to 2's definition of positive frequency.

Comparing above definitions we see the difference in the definition of the coordinates used in the field expansions as well as the implicit difference in the definitions of positive frequency. To relate the quantizations to each other, we must use the relation between the x and y coordinates:

$$y_i = f_i(x)$$

or, in vector form,

$$y = f(x)$$

for i = 1, 2, ... , D. Thus

$$\varphi(f(x)) = \sum_{\beta} [\psi_\beta(f(x))b_\beta + \psi_\beta^*(f(x))b_\beta^\dagger] \qquad (10.3)$$

Inverting the above equations to obtain the relation of the fourier coefficient operators we see:

$$A_\alpha = \sum_{\beta} [C_{\alpha\beta} b_\beta + C'_{\alpha\beta} b_\beta^\dagger]$$

where $C_{\alpha\beta}$ and $C'_{\alpha\beta}$ are c-number functions of α and β:

$$C_{\alpha\beta} = (\chi_\alpha(x), \varphi(f(x))) \qquad\qquad (10.4)$$
$$C'_{\alpha\beta} = (\chi_\alpha{}^*(x), \varphi(f(x)))$$

The above equations imply an N particle state in one coordinate system will appear as a superposition of states of various numbers of particles in the other coordinate system IF the standard quantum field theory formulation is used.

TO REMEDY this situation – which we take to be unphysical – we must reformulate quantum field theory using the PseudoQuantum formulation presented Appendix A. The scalar particle case is discussed in Appendix A between eqs. 6 – 31, to which the reader is referred.

The conclusions of that section, and the sections following it, in Appendix A are:

1. One can define corresponding unitarily equivalent particle states in two quantizations with invariant particle numbers.
2. The fourier coefficient operators of the two quantizations are related by Bogoliubov transformations and are unitarily equivalent.
3. The group of the local Bogoliubov transformations is an infinite tensor product of $SU_{1,1}$ groups.
4. The vacua of the particle are invariant under Bogoliubov transformations that relate the the Megaverse and the universe quantizations.
5. Unitarily equivalent perturbation theories of both quantizations can be defined.

We now consider the case of Two-Tier PseudoQuantization, and then turn to various applications of PseudoQuantization.

10.2 Two-Tier PseudoQuantum Field Theory

The combination of the Two-Tier procedure with the PseudoQuantiztion procedure leads to a somewhat more complicated situation. In principle, both are required for a Unified SuperStandard Model in any coordinate system in flat or curved space-times in any number of dimensions. However their direct combination is both complicated and unphysical.

The main purpose of PseudoQuantization is to have particle number invariance under a change of coordinate system. Two-Tier Field Theory 'cloaks' each particle in infinite 'clouds' of Y^μ quanta as Fig. 9.1 illustrates. We define PseudoQuantization as implementing particle number invariance for 'bare' particles without their clouds of Y^μ quanta. Thus an asymptotic particle state of n particles (neglecting its Y^μ quanta cloud) remains a unitarily equivalent n particle state (neglecting its Y^μ quanta cloud) under a change of coordinate system.

To implement this concept we first define quantizations of a particle in coordinate systems without Two-tier quanta. We then 'dress' the quantizations by replacing the coordinates y^μ in each coordinate system with the corresponding Two-Tier coordinates:

$$y^\mu \rightarrow X^\mu(y) = y^\mu + \text{ i } Y^\mu(y)/M_c^{\,2} \qquad (10.5)$$

It appears the most convenient gauge in each coordinate system is the Lorentz gauge:

$$\partial Y^\mu/\partial y^\mu = 0 \qquad (10.6)$$

We now briefly consider the case of a scalar particle PseudoQuantization. This case is considered in more detail in Appendix A. Following Appendix A we must introduce two fields $\varphi_1(y)$ and $\varphi_2(y)$ with the free fields' lagrangian

$$\mathcal{L}(y) = \partial^\mu\varphi_1\partial_\mu\varphi_2 - \tfrac{1}{2}\,\partial^\mu\varphi_1\partial_\mu\varphi_1 - m^2\varphi_1\varphi_2 + \tfrac{1}{2}\,m^2\,\varphi_1^{\,2} \qquad (10.7)$$

in a coordinate system with coordinates y. Then following the steps indicated in Appendix A from eq. 7 onward we arrive at a PseudoQuantum formulation in the coordinate system with coordinates y that is unitarily equivalent to that of a different coordinate system defined a similar manner.

From eq. 43 onwards we can replace the c-number coordinates x and y with Two-Tier coordinates of the form

$$X^\mu(y) = y^\mu + \text{i } Y^\mu(y)/M_c^{\,2}$$

and proceed to calculate propagators and perturbation theory diagrams.

Thus we have a straight-forward procedure to unite the PseudoQuantum formalism with Two-Tier coordinates to obtain finite perturbation theory results with unitary equivalence to quantization in other coordinate systems in both flat and curved space-times.

The use of two fields per particle of PseudoQuantum field theory will be seen to part of the applications consider in the remainder of this subsection. We will put aside the consideration of quantizations in other coordinate systems in what follows to keep the presentation as simple as possible.

10.3 PseudoQuantum Higgs Scalar Particle Field Theory in D-dimensional Space-Time

10.3.1 The Enigma of Higgs Particles and the Higgs Mechanism

In our previous work on the Standard Model, and its generalization to The Unified SuperStandard Model described in a series of books entitled *Physics is Logic ...*, we showed that the fermion spectrum results from Complex Special Relativity, the gauge interactions result from the Reality group, the fermion generations result from the Generation group, and the Theory of Everything results from a combination with Complex General Relativity. The Higgs particles and the Higgs Mechanism were inserted to generate particle masses and symmetry breaking effects.

Whence comes Higgs particles? A more fundamental cause has not been suggested until our analysis, which is presented here in chapter 11. So the Higgs sector appeared to be an expedient mechanism to insert much needed symmetry breaking and masses into the theory.

There are a number of peculiarities in the implementation of the Higgs Mechanism:

1. First, it is selective in the sense that some gauge fields have associated Higgs particles and utilize the Higgs Mechanism, and some gauge fields do not have associated Higgs particles. In particular, the ElectroWeak gauge fields, the Generation group gauge fields, the Layer group fields, and the complex gravity Species gauge fields have associated Higgs particles. The strong interaction (gluon) gauge fields do not.[108]

2. The Higgs potentials have a quadratic mass term of the "wrong" sign plus a quartic interaction term, which together, generate non-zero vacuum expectation values. They obviously accomplish their goal. But the source of these potentials, and why they have the same form, is unknown. One expects a fundamental principle should be operative here.

3. One can imagine creating a Higgs microscope at some super-accelerator. Using this microscope in the presence of a (classical) condensate could enable the Uncertainty Principle to be violated. This possibility, in the case of a microscope using electromagnetic fields, was the source of a heuristic argument for the need to quantize the electromagnetic field.[109]

4. The formulation of the Higgs Mechanism uses classical fields under the assumption that a path integral formulation justifies their use. While this may be true, the path integral formulation relies on implicit, unstated boundary conditions that obscure the physics of the quantum field theoretic nature of the mechanism. A direct quantum field theoretic study of the Higgs Mechanism is needed and would further elucidate its character.

5. Scalar fields have a cloud hanging over them that spin ½ fields do not. A spin ½ particle cannot transition to negative energy because there is a filled sea of negative energy particles. No additional particles can fall into the sea due to the Pauli Exclusion Principle that forbids two fermions with the same 4-momentum and quantum numbers. In the case of scalar particles the Pauli Exclusion Principle does

---

[108] See section 4.2.8 for an explanation.
[109] Heitler (1954) p. 86 provides a good discussion of the need to quantize the electromagnetic field.

not apply and so a *filled* negative energy sea of scalar particles is not possible and positive energy scalar particles can transition to negative energy without hindrance. This problem has been "resolved" by an appropriate definition of the scalar particle vacuum to exclude transitions to negative energy. But the rationale for the definition is lacking. Dirac was asked about this issue many years ago. He said he had a solution to the problem. However he did not present it – in keeping with his well-known taciturn nature. So the issue remains an open question.

For the above reasons we will show that a more satisfactory method of achieving the goals of mass generation and symmetry breaking exists.[110] This method relies on a larger Fock space that enables the appearance of a vacuum expectation value for Higgs particles to be understood within a truly quantum framework. More importantly, this method is a consequence the PseudoQuantization procedure described above that enables unitarily equivalent quantizations in different coordinate systems. So a profound fundamental justification for our Higgs boson formulation exists. One major consequence of this approach is the appearance of a local Arrow of Time – a concept that has been a subject of interest for over one hundred years. Another consequence is a rationale for ElectroWeak Higgs bosons and for their absence for the strong (gluon) interaction.

10.3.2 PseudoQuantization of Scalar Particles

We now consider the PseudoQuantization[111] of a scalar particle field that will become a Higgs particle with a non-zero vacuum expectation value.[112] We begin by defining two fields that correspond to the scalar particle: $\varphi_1(x)$ and $\varphi_2(x)$.[113] These fields will be assumed to have the equal time commutators

$$[\varphi_i(x), \pi_j(y)] = i(1 - \delta_{ij})\delta^3(\mathbf{x} - \mathbf{y}) \tag{10.8}$$
$$[\varphi_i(x), \varphi_j(y)] = 0$$
$$[\pi_i(x), \pi_j(y)] = 0$$

---

[110] In the Extended Standard Model of Blaha (2015a) we have shown that the basic particles have a mass, the Landauer mass, so that the theory is symmetry violating from the very start. We have also shown that our Two-Tier formalism for quantum field theories always yields finite results in perturbation theory calculations – making the renormalization approach of t'Hooft and others, which relied on initially massless gauge fields, unnecessary.

[111] PseudoQuantization in a D-dimensional space-time is described in Blaha (2017c). This discussion is relevant to PseudoQuantization in the Megaverse, or in other universes.

[112] Much of this section appears in Blaha (2016c), and earlier books, as well as in S. Blaha, Phys. Rev. **D17**, 994 (1978). The case of fermion PseudoQuantization is also discussed in Appendix A – S. Blaha, Il Nuovo Cimento **49A**, 35 (1979).

[113] The subscripts on the fields are not gauge symmetry indices but simply identifiers distinguishing the fields from each other.

where δ_{ij} is the Kronecker δ and where $\pi_i(x)$ is the canonically conjugate momentum to $\varphi_i(x)$. The fields $\varphi_1(x)$ and $\pi_1(y)$ will be observable classical fields. The fields $\varphi_2(x)$ and $\pi_2(y)$ will not be observables so that $\varphi_1(x)$ and $\pi_1(y)$ can both be sharp on the set of physical states.

We now specify the lagrangian density for a scalar Klein-Gordon particle:

$$\mathcal{L} = \partial\varphi_1/\partial x_\mu \, \partial\varphi_2/\partial x^\mu \tag{10.9}$$

with hamiltonian density

$$\mathcal{H} = \pi_1 \, \pi_2 \; + \partial\varphi_1/\partial x_i \, \partial\varphi_2/\partial x^i$$

where i labels spatial coordinates, and $\pi_1 = \partial\varphi_2/\partial t$ and $\pi_2 = \partial\varphi_1/\partial t$. The lagrangian $\mathcal{L}$ is without a potential or mass term.

The lagrangian and hamiltonian for a massive scalar particle in this formalism are

$$\mathcal{L} = \partial\varphi_1/\partial x_\mu \, \partial\varphi_2/\partial x^\mu - m^2 \, \varphi_1\varphi_2 \tag{10.10}$$

with hamiltonian density

$$\mathcal{H} = \pi_1 \, \pi_2 \; + \partial\varphi_1/\partial x_i \, \partial\varphi_2/\partial x^i + m^2 \, \varphi_1\varphi_2$$

The fields can be fourier expanded in terms of creation and annihilation operators:

$$\varphi_i(\mathbf{x}, t) = \int d^3k \, [a_i(k)f_k(x) + \; a_i^{\dagger}(k)f_k{}^*(x)] \tag{10.11}$$

for i = 1, 2 where

$$f_k(x) = e^{-ik\cdot x} \, /(2\omega_k(2\pi)^3)^{\frac{1}{2}}$$

with $\omega_k = |\mathbf{k}|$.

The creation and annihilation operators satisfy the commutation relations:

$$[a_i(k), a_j^{\dagger}(k')] = (1 - \delta_{ij})\delta^3(\mathbf{k} - \mathbf{k}')$$
$$[a_i(k), a_j(k')] = 0$$
$$[a_i^{\dagger}(k), a_j^{\dagger}(k')] = 0$$

for i, j = 1, 2.

In this formulation the defining properties of a physical state are:

$$\varphi_1(x)|\Phi, \Pi\rangle = \Phi(x)|\Phi, \Pi\rangle \tag{10.12}$$
$$\pi_1(x)|\Phi, \Pi\rangle = \Pi(x)|\Phi, \Pi\rangle$$

where $\Phi(x)$ and $\Pi(x)$ are sharp on the states and thus classical fields with

$$\Phi(\mathbf{x}, t) = \int d^3k \, [\alpha(k)f_k(x) + \; \alpha^*(k)f_k{}^*(x)] \tag{10.13}$$

and correspondingly for $\Pi(x)$.

10.3.3 Vacuum States for Scalar (Higgs) Particles with Non-Zero Vacuum Expectation Values

When we implement the mass mechanism, Φ is constant. We can define a set of states

$$a_1(k)|\alpha> = \alpha(k)|\alpha>$$
$$a_1^\dagger(k)|\alpha> = \alpha^*(k)|\alpha>$$

and correspondingly a set of coherent states

$$|\alpha> = C\exp\left\{\int d^3k\,[\alpha(k)a_2^\dagger(k) + \alpha^*(k)a_2(k)]\right\}|0> \qquad (10.14)$$

where C is a normalization constant and where the vacuum state $|0>$ satisfies

$$a_1(k)|0> = a_1^\dagger(k)|0> = 0 \qquad (10.15)$$

$$a_2(k)|0> \neq 0 \qquad\qquad a_2^\dagger(k)|0> \neq 0$$

The dual vacuum state satisfies

$$<0|a_2(k) = <0|a_2^\dagger(k) = 0$$
$$<0|a_1(k) \neq 0 \qquad\qquad <0|a_1^\dagger(k) \neq 0$$

With this coherent state formalism, which gives purely classical fields and yet also has quantum fields through the use of φ_2 and its creation and annihilation operators, we now have the machinery to define a mass mechanism without the introduction of a potential whose origin can only be described as dubious.

For we can define a coherent state for some k as

$$|\Phi, \Pi> = C\exp\{[(2\pi)^3\omega_k/2]^{\frac{1}{2}}\Phi[a_2^\dagger(k) + a_2(k)]\}|0> \qquad (10.16)$$

where C is a normalization constant, that yields a non-zero vacuum expectation value:

$$\varphi_1(x)|\Phi, \Pi> = \Phi|\,\Phi, \Pi> \qquad (10.17)$$

where Φ is a constant. Evaluating a fermion interaction term we find a mass term emerges[114]

$$\overline{\psi}(\varphi_1 + \varphi_2)\psi \;\; \rightarrow \;\; \overline{\psi}(\Phi + \varphi_2)\psi \qquad (10.18)$$

---

[114] When matrix elements with a "vacuum state" are taken.

It generates a mass for an interaction with a gauge field of the form

$$A^{\mu}(\varphi_1 + \varphi_2)^2 A_{\mu} \;\rightarrow\; A^{\mu}(\Phi + \varphi_2)^2 A_{\mu} \qquad (10.19)$$

It also yields a quantum field theoretic interaction that would result in the production of ElectroWeak particles from these scalar fields. The production of Higgs particles that decay into ElectroWeak gauge particles has recently been found at CERN.

The present formalism provides a clean way to separate the vacuum expectation value of a scalar particle from its quantum field part in contrast to the Higgs Mechanism where one has to separate a Higgs field into parts manually.

10.3.4 Interpretation of Negative Energy Scalar Particle States

As we noted earlier, scalar particle physics has the problem of no barrier to the decay of positive energy states to negative energy states due to the absence of a Pauli Exclusion Principle for bosons. The PseudoQuantization procedure that we developed in 1978 and describe here allows negative energy states as one would physically expect and raises the possibility of disastrous particle decays to negative energy. The above equations show that negative energy states are possible in this theory.

However they also show that combined positive and negative energy boson states can be interpreted as classical field states. In addition, the ability of any number of boson particles to have the same 4-momentum and quantum numbers shows that a *macroscopic* classical scalar field state can be constructed.

Thus we can view states containing negative energy particles as classical field states and thus solve[115] the issue of interpreting negative energy particle states – a more satisfactory approach than the standard quantization procedure does – with due respect to Professor Dirac.

We note that macroscopic many particle fermion states can only have one particle in any mode unlike bosons. Therefore we cannot use this formalism to create macroscopic classical fermion field states.[116] And the filled Dirac sea of negative energy fermions precludes the transition of a positive energy Dirac fermion to a negative energy state. *Thus there is a certain complementarity between fermions that cannot become classical fields but have a filled sea precluding decays to negative energy states, and bosons that can become classical fields but support decays to negative energy states.*

---

[115] Also a boson that has no interactions cannot transition from to a positive energy state to a negative energy state due to conservation of energy.

[116] However we can create PseudoQuantum fermion states. See S. Blaha, Phys. Rev. **D17**, 994 (1978) (reproduced in Appendix I) and references therein to earlier papers by the author.

10.3.5 Contrast with Conventional Second Quantization of Scalar Particles

The PseudoQuantization procedure followed here uses different boundary conditions than the usual scalar particle quantization procedure. The essence of the difference is embodied in a comparison of the definition of the vacuum above and the definition of the conventional second quantized field vacuum:

$$a|0> = 0 \qquad \text{Conventional Approach}$$
$$a^{\dagger}|0> \neq 0$$

In the conventional approach the creation of negative energy boson states is eliminated *ab initio* whereas in our approach it is allowed in order to support classical field states with non-zero vacuum expectation values that are a form of classical field. While one cannot discredit the conventional choice for conventional scalar fields, one can see that our approach yields a physically more important result – particularly for Higgs fields – because it leads to an Arrow of Time *locally* – an important feature of physical phenomena that has been a subject of much discussion and dispute. One can say that the conventional approach sweeps the issue "under the rug" rather than seeking a deeper justification – differing from Dirac's implied notion that the issue merited attention. We will discuss the "Arrow of Time" within the framework of our PseudoQuantization approach later.

10.3.6 Why Inertial Reference Frames are Special

The great physicists of the early 20th century raised numerous questions about Special Relativity after Einstein and Poincarè's discovery. Prominent among them was the question of why inertial reference frames are of especial importance in Special Relativity, and afterwards in General Relativity.

It appears that our formulation of the mass generation mechanism sheds significant light on the reason for the special prominence of inertial frames. Earlier we considered the case of a massless PseudoQuantized scalar. We now consider massive scalars since experiments at CERN have apparently discovered a Higgs particle with a 125 GeV/c mass. The above equations describe a massive scalar particle. If the scalar is massive, then the "vacuum" state that yields a non-zero expectation value must change to

$$|\Phi, \Pi> = C\exp\{(2\pi)^{3}m/2]^{\frac{1}{2}}[a_2^{\dagger}(\mathbf{0},m) + a_2(\mathbf{0},m)]\}|0> \qquad (10.20)$$

to have operators for a particle of mass m in its rest frame. Then, having established this preferred frame for a Higgs particle, in The Unified SuperStandard Model, and requiring that invariant intervals

$$ds^2 = dt^2 - d\mathbf{x}^2 \quad \text{(in rectangular coordinates)}$$

are unchanged by a (complex or real) Lorentz transformation, we find that inertial reference frames are singled out as "special" in the sense that they are the only accessible reference frames that can be generated by a Lorentz boost/transformation from the Higgs particle rest frame. *The Higgs particle vacuum state singles out the class of inertial reference frames.*

Thus Higgs particles play a central role in establishing the basis of physical reality.

10.3.7 PseudoQuantization Reveals More Physical Consequences than the Higgs Mechanism of Scalar Particles

Earlier we pointed out that our PseudoQuantization theory of Higgs particles reveals more physical consequences than the conventional approach, which implements the Higgs Mechanism by simply using a potential term that has a minimum at a non-zero vacuum expectation value. This section shows the major results of a properly implemented mechanism. We find a better explanation of the negative energy state problem of boson field theories. We find a local arrow of time that explains the direction of time that we, and all of nature, experiences. We find the reason why inertial reference frames have a special physical significance – a result long sought by physicists.

In addition we will see in chapter 11 that real gauge fields should have an associated Higgs particle, while necessarily complex gauge fields (the Strong interaction gauge field in The Unified SuperStandard Model) do not have an associated gauge field. These results correspond to experimental reality.

10.3.8 The T Invariance Issues of Our PseudoQuantized Scalar Particle Theory

The PseudoQuantized scalar particle hamiltonian equations are invariant under time reversal $t \rightarrow t' = -t$. The 'new' vacuum states defined above break the time reversal invariance of the theory resulting in retarded particle propagators.

The hamiltonian equations

$$[H, \varphi_1(\mathbf{x}, t)] = -i\partial\varphi_1/\partial t \qquad (10.21)$$
$$[H, \varphi_2(\mathbf{x}, t)] = -i\partial\varphi_2/\partial t$$

are invariant under time reversal. If we define a time reversal operator transformation U then the time reversed equations are

$$[UHU^{-1}, \varphi_1(\mathbf{x}, -t)] = +i\partial\varphi_1(\mathbf{x}, -t)/\partial(-t)$$
$$[UHU^{-1}, \varphi_2(\mathbf{x}, -t)] = +i\partial\varphi_2(\mathbf{x}, -t)/\partial (-t)$$

The operator U, which is unitary, transforms H into –H. This operation is legal because the hamiltonian – in this case – is not positive definite and admits negative energy states.[117] Thus

---

[117] Unlike the usual case of second quantized Klein-Gordon quantum field theory.

$$[H, \varphi_1(\mathbf{x}, -t)] = -i\partial\varphi_1(\mathbf{x}, -t)/\partial\,(-t)$$
$$[H, \varphi_2(\mathbf{x}, -t)] = -i\partial\varphi_2(\mathbf{x}, -t)/\partial\,(-t)$$

and the time reversal invariance of the equations of motion is established for this case.

Time reversal invariance is broken by our choice of vacuum states. This choice is necessary to obtain classical field states as we showed earlier. A demonstration of the time reversal symmetry breaking is presented later where we show theory has retarded propagators for particle propagation to and from asymptotic states.

Within the interaction region the particle propagators are the sum of retarded and advanced parts that combine to yield principle value propagators – not Feynman propagators. Many years ago Feynman and Wheeler championed principle value propagators for electrodynamics to obtain an action-at-a distance theory of Quantum Electrodynamics. While their theory, and ours, differ from the standard quantum field theory approach there is no reason to view them as faulty, or having serious physical defects. The only question is whether nature chooses conventional quantum field theory or PseudoQuantized quantum field theory. In our case the need for a classical scalar particle non-zero vacuum expectation value strongly motivates our choice of psedoquantized Higgs particles.

10.3.9 Retarded Propagators for Our Quantized Higgs Particles

In the previous section we pointed out that our PseudoQuantization Higgs theory has an arrow of time due to is boundary conditions as expressed by its definition of the vacuum state and its dual. In this section we will show that the theory uses retarded propagators for propagation to and from the interaction region to asymptotic in-states and out-states. Within an interaction region the theory uses half-retarded – half-advanced propagators. We discuss aspects of the perturbation theory and propagators of our scalar particles in this chapter.

First we note that in-states at $t = -\infty$ are composed of superpositions of $a_2(k)$ and $a_2^{\dagger}(k)$ creation and annihilation operators:

$$a_2(k)|0> \neq 0 \qquad\qquad a_2^{\dagger}(k)|0> \neq 0$$

while the out-states composed of superpositions of $a_1(k)$ and $a_1^{\dagger}(k)$ creation and annihilation operators:

$$<0|a_1(k) \neq 0 \qquad\qquad <0|a_1^{\dagger}(k) \neq 0$$

Consequently when in-state particles (x_1) propagate into the interaction region (x_2) the relevant propagators are retarded propagators with the form

$$G_{in}(x_2, x_1) = <0|T(\varphi_{1\,in}(x_2), \varphi_{2\,in}(x_1))|0> \qquad (10.22)$$
$$= \theta(x_{20} - x_{10})<0|[\varphi_{1\,in}(x_2), \varphi_{2\,in}(x_1)]\,|0>$$

This is a manifestly retarded propagator. The choice of vacuums clearly results in a time asymmetry giving a retarded propagation reflecting the familiar Arrow of Time.

A similar situation prevails for propagation to out-states (x_3) from the interaction (x_2) region:

$$G_{out}(x_3, x_2) = <0|T(\varphi_{1\,out}(x_3), \varphi_{2\,out}(x_2))|0> \qquad (10.23)$$
$$= \theta(x_{30} - x_{20})<0|[\varphi_{1\,out}(x_3), \varphi_{2\,out}(x_2)]\,|0>$$

Within the interaction region the Higgs particles have principle value propagators.

Thus we find PseudoQuantized Higgs particles embody a local Arrow of Time. The locality of the Arrow of Time is embodied in all the particles that interact with the Higgs particle. Since the mass of *every* particle – bosons and fermions – has a Higgs contribution, and thus *every* particle interacts with the Higgs particles, the Arrow of Time permeates The Unified SuperStandard Model as well as the more familiar Standard Model known from experiment.

10.3.10 The Local Arrow of Time

In the *Physics is Logic* series of monographs we saw that complex coordinates led to the form of the fermion spectrum, that the mapping of complex coordinates to real-valued coordinates yielded the Reality group and The Unified SuperStandard Model gauge interactions, that Complex General Relativity led to Higgs particles that were directly united with elementary particle masses and gave us the equality of inertial mass and gravitational mass. Later we will see the reduction of complex gauge fields to real gauge fields explains the appearance of Higgs fields in The Unified SuperStandard Model.

The PseudoQuantization procedure leads to retarded Higgs field propagators and thence to a *local* arrow of time. Many arguments have been put forward over the past hundred plus years for the Arrow of Time. Many arguments based on Statistical Mechanics, Entropy, and Boltzmann's statistical atomic theory have suggested the Arrow of Time is a global statistical consequence. This view seems to contradict the results of elementary particle experiments where a *local* Arrow of Time is evident.

Our rationale for the Arrow of Time begins with retarded Higgs fields. Then we note that Higgs field quantum interactions appear for all fermions and gauge particles. Thus all particle interactions are imbued with an Arrow of Time. Particles united to form macroscopic matter inherit their combined Arrows of Time producing the global Arrow of Time we experience.

Thus our PseudoQuantization approach offers a more satisfactory solution of the origin of the Arrow of Time.

It is remarkable that complex quantities – coordinates and fields – through the Higgs phenomena that we have considered, lead to the equality of inertial mass and gravitational mass, and an Arrow of Time. This unity of mass and time phenomena may reflect the deeper fact that we can have no practical Arrow of Time if all particles were massless, for particle dynamics at light speed would then be pointless. This view has been expressed by DeWitt,

Unruh, and others who have pointed out that, physically, time is meaningful and measurable only if masses exist; the larger the mass, the more accurate the time measurement in principle.[118]

10.3.11 Space-Time Dependent Particle Masses

It is possible that the ultimate Unified SuperStandard Model has masses that evolve with time and may also be spatially varying – different values in different parts of the universe. Presently there is no decisive evidence for this possibility although astrophysical studies continue. In this section we will describe the mechanism for space-time dependent masses.

Consider a classical field (time and spatially varying):

$$\Phi(\mathbf{x}, t) = \int d^3k \, [\alpha(k)f_k(x) + \alpha^*(k)f_k^*(x)] \qquad (10.24)$$

If we define the coherent vacuum state

$$|\alpha\rangle = C \exp\left\{\int d^3k \, [\alpha(k)a_2^\dagger(k) + \alpha^*(k)a_2(k)]\right\} |0\rangle \qquad (10.25)$$

then

$$\varphi_1(x)|\Phi, \Pi\rangle = \Phi(x)|\Phi, \Pi\rangle$$
$$\pi_1(x)|\Phi, \Pi\rangle = \Pi(x)|\Phi, \Pi\rangle$$

where

$$\varphi_i(\mathbf{x}, t) = \int d^3k \, [a_i(k)f_k(x) + a_i^\dagger(k)f_k^*(x)] \qquad (10.26)$$

for i = 1, 2 and where

$$f_k(x) = e^{-ik\cdot x} /(2\omega_k(2\pi)^3)^{1/2}$$

with ω_k equal to the energy.

10.3.12 Inertial Mass Equals Gravitational Mass

From the days of Newton through Einstein[119] to the present the equality of gravitational mass and inertial mass has been a topic of interest. Mach, who played an important role, in this ongoing discussion, thought distant masses in the universe were the source of the equality. However the origin of the equality, which has been shown experimentally to very high accuracy, remained uncertain until the *Physics is Logic* series of books, in which we showed the interconnection of the Unified SuperStandard Model and Complex Gravitation via Higgs generated masses that united gravitational and inertial mass.

---

[118] No mass, no clock; no clock, no physical time. See Blaha (2015a) pp. 368-371 for a discussion including comments by DeWitt and Unruh.

[119] For example, Einstein and Grossman in 1913 stated, "The theory herein described originates in the conviction that the proportionality between the inertial and gravitational mass of a body is an exact law of nature that must be expressed as a foundation principle of theoretical physics."

In Blaha (2016h) we showed that a Complex General Relativity transformation can be factored into the product of a complex-valued transformation and a real-valued General Coordinate transformation. The set of complex valued transformations form a U(4) group that we called the General Coordinate Reality group. Later we will define the Internal Symmetry Species Group as the corresponding analogue. The Species Group has gauge fields that undergo spomtaneous symmetry breaking and generate contributions to all fermion masses.

Since fermion field masses are now sums of ElectroWeak Higgs contributions, Generation group Higgs contributions, Layer group Higgs contributions, and Species group contributions, and since the gravitational Higgs fields appear in all fermion masses, the equality of inertial and gravitational mass is proven. The gravitational Higgs particles' equations depend, in part, on the gravitational field by Blaha (2016h) and so set the mass scale of gravitational mass, and thereby of all Higgs mass contributions. They set the scale of inertial masses equal to the scale of gravitational masses. **Since an expression cannot mix mass scales, the gravitational mass scale must be the same as the inertial mass scale. Inertial Mass equals gravitational mass.**

We have established the equality of inertial and gravitational mass at the short distance quantum level. In our view, this explanation is far more satisfying than basing the equality on a combination of large distance phenomena and quantum phenomena. As Einstein and Weyl have pointed out, all fundamental physics phenomena should be based on a local theory. Complex Gravity as we have constructed it, combined with the Unified SuperStandard Model, furnishes a completely local basic Theory of Everything.

The equation above contains a coherent state $|\alpha>$ for a time and spatially varying mass. The above equations can be generalized to the case of multiple space-time varying masses.[120]

$$|\Phi_1,\Phi_2, \ldots ,\Phi_n;\Pi_1,\Pi_2, \ldots ,\Pi_n> = C \prod_{i=1}^{n} \exp\left\{\int d^3k \,[\alpha_i(k)a_{2i}^\dagger(k) + \alpha_i^*(k)a_{2i}(k)]\right\}|0> \quad (10.27)$$

Then all n mass vacuum expectation values are space-time dependent:

$$\varphi_{1i}(x)\,|\,\Phi_1, \Phi_2, \ldots , \Phi_n; \Pi_1, \Pi_2, \ldots , \Pi_n> = \Phi_i(x)\,|\,\Phi_1, \Phi_2, \ldots , \Phi_n; \Pi_1, \Pi_2, \ldots , \Pi_n>$$
$$(10.28)$$

Thus our formalism can accommodate space-time varying masses should they be found in the Cosmos.

---

[120] The "vacuum" state $|0>$ also implicitly has factors for the vacuum expectation values used for fields that give masses to fermions and vector bosons as described in Blaha (2016h).

10.3.13 Benefits of the PseudoQuantization Method

In this book and earlier work we showed that a more physically satisfactory method for avoiding the negative energy state problem exists. This method relies on the use of a larger Fock space in which negative energy states (or partially negative energy states) are interpreted as states containing classical fields or a mix of classical fields and individual boson particles. This approach resolves the negative energy boson issue and provides a common framework for boson particles and classical boson fields.

One consequence of the PseudoQuantization method is that it enables the appearance of a vacuum expectation value for Higgs particles (a constant classical field) to be understood within a truly quantum framework. Another major consequence of this approach is the appearance of a *local* Arrow of Time due to the Higgs mass generation mechanism – a concept that has been a subject of interest for over one hundred years. A macroscopic arrow of time is often described as a statistical result. But our approach yields an arrow of time at the single particle level.

The conventional approach to boson field quantization sweeps these issues "under the rug" rather than seeking a deeper justification. It differs from Dirac's implied notion that the issue merited attention.

Another important consequence of the PseudoQuantization method is that it singles out inertial reference frames when applied to the case of Higgs particles.

Yet another more subtle consequence of boson PseudoQuantization is that it provides a rationale/explanation for the presence of ElectroWeak Higgs bosons, *and for their absence for the strong (gluon) interactions. The question of why there are no strong interaction Higgs bosons has not been previously considered to the best of this author's knowledge.*

11. Higgs Particles, Gauge Fields, and Higgs Mechanism Generated Coupling Constants

Higgs particles appear in many contexts in the Unified SuperStandard Model. In this subsection we consider a possible origin of Higgs particles in complex-valued gauge fields that explains why there are no Strong Interaction gauge field Higgs particles. We also show it is possible to define Higgs particles that generate gauge field coupling constants. Using this mechanism we show that the known coupling constants appear to correspond to Higgs values of the same order of magnitude suggesting that we are close to a form of unification. We also show the Higgs Mechanism for fermion particle masses may explain the equality of inertial and gravitational mass – a topic of continuing interest for many years.

11.1 The Genesis of Scalar (Higgs?) Particle Fields from Complex Gauge Fields

In the past[121] we showed that scalar particles can be 'extracted' from all spin 1 gauge fields except color SU(3).

Since our Unified SuperStandard Model is ultimately based on the Complex General Coordinate Transformations and the Complex Lorentz group (thus complex-valued coordinate systems), it appears reasonable to assume all spin 1 gauge fields to initially be similarly complex-valued. Most of these gauge fields can be rotated to real values. However we shall see that color SU(3) gauge fields are *necessarily* complex-valued. All other gauge fields can be rotated to real values. The price of rotation is the introduction of scalar fields. Some of these fields may be Higgs particle fields and generate gauge boson masses (symmetry breaking) and fermion masses.

Thus we can view scalar particles including Higgs particles as inherently associated with most gauge fields. From the viewpoint of our derivation from basic axioms, the origin of Higgs bosons in complex-valued gauge fields gives a 'tighter' derivation of the overall theory. Higgs fields are an inherently a part of the theory.

---

[121] Blaha (2015c) and (2016c).

11.2 The Difference between the Strong Gauge Field and the Other Gauge Fields in the SuperStandard Model

In our Unified SuperStandard Model the only gauge field without an associated Higgs particle is the strong interaction gluon gauge field. *We view this exception as a particularly important clue as to the nature of the relation between gauge fields and Higgs particles.*

How does the strong interaction gauge field differ from all other gauge fields in the Unified SuperStandard Model? An examination of the gauge fields dynamic equations (and other lagrangian terms) of our Unified SuperStandard Model reveals that all gauge field dynamic equation kinetic terms *except those of the strong interaction gauge field* have the form:

$$\partial/\partial x_\mu \, F^a_{\mu\nu} + gf^{abc}A^{b\mu} F^c_{\mu\nu} = j^a_\nu \qquad (11.1)$$

where

$$F^a_{\mu\nu} = \partial/\partial x^\nu A^a_\mu - \partial/\partial x^\mu A^a_\nu + gf^{abc}A^b_\mu A^c_\nu \qquad (11.2)$$

where the coordinates x^ν *are real-valued,*[122] where a, b, c are structure constant indices, where g is a coupling constant, and where j^a_ν is the corresponding current. The gauge field A^a_μ is real for all normal and Dark ElectroWeak gauge fields, Generation group gauge fields, and Layer group gauge fields. Thus the above equations are real-valued.

The strong interaction gauge field[123] in our Unified SuperStandard Model differs from the other gauge fields by being *necessarily* complex[124] due to the complex 3-space complexon derivatives that appear in the corresponding dynamic equations:

$$D^\mu F_{C\,\mu\nu}^{\,a} + gf^{abc}A_C^{\,b\mu} F_{C\,\mu\nu}^{\,c} = j^a_\nu \qquad (11.3)$$

with

$$F_{C\,\mu\nu}^{\,a} = D_\nu A_{C\,\mu}^{\,a} - D_\mu A_{C\,\nu}^{\,a} + gf^{abc}A_{C\,\mu}^{\,b} A_{C\,\nu}^{\,c} \qquad (11.4)$$

where

$$D_k = \partial/\partial x_r^{\,k} + i\,\partial/\partial x_i^{\,k}$$
$$D_0 = \partial/\partial x^0$$

for k = 1, 2, 3 where $A_{C\,\mu}^{\,a}$ is the complexon color Strong interaction gauge field. The complexon spatial coordinates have the form $x_r^{\,k} + i\,x_i$. The time coordinate is real-valued. These equations are eqs. 10.16 and 5.162 of Blaha (2015a) for complexon gauge fields,[125] the carriers of the strong interaction in the Unified SuperStandard Model.

---

[122] Before the introduction of the Two-Tier formalism.

[123] This field is called a complexon gauge field in Blaha (2017b), (2015a) and earlier books.

[124] One cannot cleanly separate the real and imaginary parts of its dynamic equations.

[125] In The Unified SuperStandard Model we also identify quark species particles as having complex 3-momentum. We call them complexon fermions.

This difference enables us to differentiate the strong gauge field from all other gauge fields in The Unified SuperStandard Model. Thereby we can develop a unified formalism for the non-strong gauge fields and their corresponding Higgs particles.

The necessarily complex nature of the color SU(3) field is the reason that the Strong Interaction gauge fields do not acquire a mass via the Higgs Mechanism. As shown below, the necessary complexity of Strong Interaction gauge fields precludes the generation of Higgs fields from Strong Yang-Mills gauge fields.

11.3 Generation of Higgs Fields from Non-Abelian Gauge Fields

In the prior section we considered the difference between the strong gauge field and the other gauge fields of The Unified SuperStandard Model. Unlike strong gauge fields the other gauge fields (ElectroWeak and so on) could be real or complex. In a manner similar to what we did in the preceding *Physics is Logic* books (and earlier books) we can assume gauge fields are initially complex, and then transform them to real-valued fields using a phase transformation that introduces scalar fields, some of which we will take to be Higgs fields.

We define a complex phase transformation for a gauge field $A^{b\mu}$ with

$$A'^{a\mu}(x) = \Phi(x)^a{}_b A^{b\mu}(x) \tag{11.5}$$

where $\Phi(x) = \text{diag}(\exp[i\varphi_1(x)], \exp[i\varphi_2(x)], \ldots , \exp[i\varphi_n(x)])$, and n is the number of symmetry components of $A^{b\mu}$. Inserting $A'^{a\mu}(x)$ above we find:

$$\partial/\partial x_\mu\, F'^{a}{}_{\mu\nu} + gf^{abc} A'^{b\mu}\, F'^{c}{}_{\mu\nu} = \dot{j}^{a}{}_{\nu} \tag{11.6}$$

where

$$F'^{a}{}_{\mu\nu} = \partial/\partial x^\nu\{\exp[i\varphi_a(x)]A^{a}{}_\mu\} - \partial/\partial x^\mu\{\exp[i\varphi_a(x)]A^{a}{}_\nu\} + gf^{abc}\exp[i\varphi_b(x)]\, A^{b}{}_\mu \exp[i\varphi_c(x)]A^{c}{}_\nu$$

$$\tag{11.7}$$

If we now assume that $\varphi_a(x)$ is small for all a then

$$\exp[i\varphi_a(x)] \simeq 1 + i\varphi_a(x)$$

to first order. Substituting above, and keeping terms to leading order yields the real part:

$$\partial/\partial x_\mu\, F^{a}{}_{\mu\nu} + gf^{abc} A^{b\mu}\, F^{c}{}_{\mu\nu} = \dot{j}^{a}{}_{\nu} \tag{11.8}$$

where $F^a_{\mu\nu}$ is given above, and the imaginary part is:

$$\partial/\partial x_\mu \, F^a_{i\,\mu\nu} + gf^{abc}A^{b\mu} \, F^c_{i\,\mu\nu} = 0 \qquad (11.9)$$

to leading order where

$$F^a_{i\,\mu\nu} = \partial/\partial x^\nu \, \varphi_a(x)A^a_\mu - \partial/\partial x^\mu \, \varphi_a(x)A^a_\nu$$

Then we find

$$A^a_\nu\Box\varphi_a(x) - A^a_\mu \, \partial/\partial x_\mu\partial/\partial x^\nu \, \varphi_a(x) - gf^{abc}A^{b\mu}[A^c_\mu \, \partial/\partial x^\nu \, \varphi_a(x) - A^c_\nu \, \partial/\partial x^\mu \, \varphi_a(x)] = 0$$

$$(11.10)$$

in the Lorentz gauge, with no sum over a. This equation is a form of Klein-Gordon equation having interaction terms with the gauge field. If the gauge field is weak then only the first two terms are important.

Note that only derivatives of $\varphi_a(x)$ appear above. Consequently shifts of the $\varphi_a(x)$ field by a constant still yield solutions. This feature makes $\varphi_a(x)$ a candidate to be a Higgs particle field.

Note also that complexon gauge fields cannot have such a phase change, with a subdivision into real and imaginary dynamic equations, due to the complexity of the spatial coordinates. This difference appears to be the reason why the strong interaction gauge field does not have an associated Higgs particle.

The $\varphi_a(x)$ particles can be made into Higgs particles by adding an appropriate potential:

$$V = A\,\varphi_a^2(x) + B\,\varphi_a^4(x) \qquad (11.11)$$

where A and B are constants. Approximating with its first two terms and inserting the potential term (with an A^a_ν factor) we find the Higgs-like equation:

$$A^a_\nu\Box\varphi_a(x) - A^a_\mu \, \partial/\partial x_\mu\partial/\partial x^\nu \, \varphi_a(x) + A^a_\nu\partial V/\partial\varphi_a = 0 \qquad (11.12)$$

$\varphi_a(x)$ has a minimum at the minimum of the potential in the corresponding lagrangian. The second and third terms constitute the interaction. Neglecting these terms we see that we obtain the free, massless, field Klein-Gordon equation

$$\Box\varphi_a(x) = 0 \qquad (11.13)$$

The pairing of Higgs particles with real-valued gauge fields is thus established.[126] The non-existence of a matching Higgs field for the strong interaction is due to the inherently complex nature of the strong interaction (complexon) gauge field in the Unified SuperStandard Model also follows.

The derivation presented here is analogous to the derivation of Higgs fields in Complex General Relativity – also a gauge theory – in *Physics is Logic Part II*.

One of the remarkable aspects of The Unified SuperStandard Model is its ability to directly prove qualitative properties of elementary particles: four fermion species, Parity violation, the distinction between leptons and quarks, the match of the SuperStandard Model's (broken) symmetries with the internal symmetry Reality group, and now the existence of Higgs gauge fields in all interaction sectors except for the strong interactions. We take these successes to be indicators of the correctness of The Unified SuperStandard Model.

11.4 General Higgs Formulation of Gauge and Fermion Particle Masses

We have seen seven of the interactions present in our Unified SuperStandard Model. Four more interactions will be presented later. One of them is the Species gauge field A_S^{μ} generated from the Reality group of complex General Relativistic transformations. There are two more interactions associated with General Relativistic transformations and an all-encompassing interaction $A_\Theta^{\mu}(x)$ considered later. We shall discuss the Higgs Mechanism associated with seven of the spin 1 gauge field interactions that appear in fermion covariant derivatives. They can be put in a vector form:[127]

$$\mathbf{A_I}^{\mu} = (g_1\mathbf{A}_{SU(3)}^{\mu}(x_C),\ g_2\mathbf{W}^{\mu}(x)\ ,\ g_3\mathbf{A}_E^{\mu}(x),\ g_4\mathbf{W}_D^{\mu}(x),\ g_5\mathbf{A}_{DE}^{\mu}(x),\ g_6\mathbf{U}^{\mu}(x),\ g_7\mathbf{V}^{\mu}(x))$$

$$(11.14)$$

where each element is a vector of the gauge fields in the group of the gauge field and the respective coupling constants are labeled g_1, g_2, ... , g_7. The subscript 'D' labels Dark matter interactions. 'W' labels Weak fields, 'E' labels Electromagnetic fields. 'U' labels U(4) Generation group fields., and 'V' labels one of the U(4) Layer groups fields.

The interactions' symmetry[128] is $[SU(3)\otimes SU(2)\otimes U(1)\otimes SU(2)\otimes U(1)\otimes U(4)\otimes U(4)]^4$ since, as we shall see, each of the four fermion layers due to the Layer groups discussed later, has a separate set of the seven interactions.[129] In each layer the number of fields for the seven interactions is 8. 3, 1, 3, 1, 16, and 16 – totaling 48 fields.

---

[126] Some of the Higgs fields so generated may not have vacuum expectation values and so may only play a role in interactions.

[127] Later we will reformulate this discussion in terms of PseudoQuantum field theory.

[128] Excepting the Species group and the Θ-Symmetry group.

[129] Otherwise the various layers would have interactions between them which would have appeared in experiments. Then the upper three layers would not be Dark.

Similarly we define a 7-vector of 48 generators

$$\mathbf{T_I} = (\mathbf{T}_{SU(3)}, \tau_{SU(2)}, \mathbf{I}_{U(1)}, \tau_{DSU(2)}, \mathbf{I}_{DU(1)}, \mathbf{G}_{U(4)}, \mathbf{G}_{LU(4)}) \qquad (11.15)$$

Then the total gauge fields interaction term within a covariant derivative corresponding to the seven interactions, the General Relativistic U(4) Species interaction $A_S{}^\mu$, the spinor interaction B^μ, and the Θ-interaction $A_\Theta{}^\mu$ can be expressed as

$$\mathbf{A_I}{}^\mu{}_k\mathbf{T}_{Ik} + A_S{}^\mu{}_k\mathbf{G}_{Sk} + g_B B^\mu + g_\Theta A_\Theta{}^\mu(x) \qquad (11.16)$$

summed separately over k for each interaction. The remaining additional interactions are real-valued gravitational connections that we will describe later. The covariant derivative of a fermion field (neglecting General Relativistic terms) for each layer is

$$\{\partial^\mu + i\,[\mathbf{A_{Itot}}{}^\mu + A_S{}^\mu + g_B B^\mu + g_\Theta A_\Theta{}^\mu(x)]\}\gamma_\mu\psi = 0 \qquad (11.17)$$

where

$$\mathbf{A_{Itot}}{}^\mu = \mathbf{A_I}{}^\mu{}_k\mathbf{T}_{Ik}$$

Note the complexon nature of the SU(3) gauge field makes us use the covariant derivative

$$\{\partial\,/\partial x_{C\mu} + i\,[\mathbf{A_{Itot}}{}^\mu + A_S{}^\mu + g_B B^\mu + g_\Theta A_\Theta{}^\mu(x)]\}\gamma^\mu\psi = 0 \qquad (11.18)$$

for the SU(3) quark dynamic equations where the other gauge fields are functions of $x_r = \mathrm{Re}\; x_C$.

We now consider the combined effects of the seven interactions, $\mathbf{A_I}{}^\mu$, on generating gauge boson masses (symmetry breaking) and fermion masses.[130] We begin by defining a composite Higgs field for all 7 interactions:

$$\eta = \prod_{k=1}^{7} \eta_{kTSLg}$$

where k labels the group, T labels the type of matter, S labels the species, L labels the layer and g labels the generation. We now consider

$$D^\mu\eta = \{\partial^\mu + i\,[\mathbf{A_{Itot}}{}^\mu + g_B B^\mu + g_\Theta A_\Theta{}^\mu(x)]\}\eta \qquad (11.19)$$

Letting η be a product of real fields $\rho..$, whose elements are composed of zeroes and non-zero real fields

---

[130] We developed the Higgs Mechanism in detail in Blaha (2017b) and earlier books for fermions and bosons for all seven interactions above plus the Reality group $A_S{}^\mu$ of Complex General Relativistic transformations, the Generation group, the Layer groups, and the Θ-group gauge field $A_\Theta{}^\mu(x)$.

$$\eta = \prod_{k=1}^{7} \rho_{kTSLg} \tag{11.20}$$

Then we find that

$$(D_\mu\eta)^\dagger D^\mu\eta = \sum_{kTSLg} \partial_\mu\rho_{kTSLg}\, \partial^\mu\rho_{kTSLg} + \sum_{kTSL} g_k^2\beta_{kTSL}U_{kTSL}^2 \tag{11.21}$$

where β_{kTSL} is a sum of terms, each of which is quadratic in the vacuum expectation value of a Higgs field. The second term above yields the masses of the gauge fields of non-zero mass.

The lagrangian terms that generate fermion masses have the form

$$\mathscr{L}_{FermionMasses} = \sum_{kTSLgh} \bar\Psi_{LkTSLg}\, \rho_{0kTSL}m_{kTSLgh}\Psi_{RkTSLh} \tag{11.22}$$

where the sums over g and h are over the generations of a specific T, S, and L of a group labeled k, and ρ_{0kTSL} is the vacuum expectation value of a Higgs particle. The initial 'L' and 'R' subscripts represent Left and Right.

In addition to the mass contributions of the seven interactions the mass contributions of the Θ-Symmetry group, denoted $m_{\Theta...}$, and the Species group, denoted $m_{G...}$, must eventually also be taken into account (later).

We note that the total fermion mass matrix can be diagonalized using a matrix A_{TSL}:

$$m_{TSLphys} = A_{LTSL}\sum_{k} \rho_{0kTSL}m_{kTSL}A_{RTSL}^{-1} \tag{11.23}$$

where $m_{TSLphys}$ is the diagonal mass matrix for the generations specified by T, S, and L.

Thus the lagrangian fermion mass terms for physical fermions become

$$\mathscr{L}_{FermionMasses} = \sum_{TSL} \bar\Psi_{LphysTSL}\, m_{TSLphys}\Psi_{RphysTSL} + c.c. \tag{11.24}$$

with diagonal mass matrices $m_{TSLphys}$.

11.5 The Mixing Pattern in the Fermion Periodic Table

The preceding discussions describe the pattern of mixing resulting from the ElectroWeak, Generation group, and Layer group. Fig. 11.1 pictorially presents an example of the mixing pattern within the Periodic Table of Fermions.

11.6 The Full Unified SuperStandard Model Fermion Mass Matrices

The above equations lead to the total mass matrices *for the four layers* listed below. The masses for each particle in the various layers are different although we use the same symbol for each type of mass. We include the mass contributions, m_{Gi}, from the Species group with gauge fields $A_S{}^\mu$.

The below list is for the 'lowest' generation of the lowest layer. The other generations and the other three layers have a similar pattern. The masses listed in the list are symbolic and are not of the same value for each particle type. We denote them generally as: m_{Wi} for the Weak group contribution, m_{Li} for the Layer group contribution, m_{Geni} for the Generation group contribution, m_{Gi} for the (Gravitational) Species group contribution with gauge fields $A_S{}^\mu$, and $m_{\Theta i}$ for the Θ-Symmetry group contribution.[131]

Charged Lepton Species Total Mass Matrix
$$m_{etot} = m_{We} + m_{Le} + m_{Ge} + m_{\Theta e}$$
Neutral Lepton Species Mass Matrix
$$m_{\upsilon tot} = m_{W\upsilon} + m_{L\upsilon} + m_{G\upsilon} + m_{\Theta \upsilon}$$
Up-Type Quark Species Mass Matrix (for each color)
$$m_{utot} = m_{Wu} + m_{Lu} + m_{Gen\text{-}u} + m_{Gu} + m_{\Theta u}$$
Down-Type Quark Species Mass Matrix (for each color)
$$m_{dtot} = m_{Wd} + m_{Ld} + m_{Gen\text{-}d} + m_{Gd} + m_{\Theta d}$$
Dark Charged Lepton Species Total Mass Matrix
$$m_{Detot} = m_{DWe} + m_{DLe} + m_{Ge} + m_{D\Theta e}$$
Dark Neutral Lepton Species Mass Matrix
$$m_{D\upsilon tot} = m_{DW\upsilon} + m_{DL\upsilon} + m_{G\upsilon} + m_{D\Theta \upsilon}$$
Dark Up-Type Quark Species Mass Matrix
$$m_{Dutot} = m_{DWu} + m_{DLu} + m_{DGen\text{-}u} + m_{Gu} + m_{D\Theta u}$$
Dark Down-Type Quark Species Mass Matrix
$$m_{Ddtot} = m_{DWd} + m_{DLd} + m_{DGen\text{-}d} + m_{Gd} + m_{D\Theta d}$$

*The (gravitational) Species group contribution to each fermion mass, m_{Gi} for each fermion type i, sets the scale for all fermion masses (and secondarily of massive gauge bosons' masses) yielding the "principle" of Newton, Einstein and others that **inertial mass equals gravitational mass**.*

The generation group contributions, in the spontaneous breakdown that we described, appear only in quark and Dark quark mass matrices possibly providing a reason why quark masses are so much larger than lepton masses. See Blaha (2017b).

11.7 Higgs Mechanism for Coupling Constants

Particle masses are attributed to the operation of the Higgs Mechanism. Coupling constants are also constants that appear in the Unified SuperStandard Model. In this section[132]

---

[131] The Species group and the Θ-Symmetry groups are discussed in subsequent chapters.
[132] This chapter is largely extracted from Blaha (2015d).

we define a Higgs Mechanism that yields the values of coupling constants as vacuum expectation values of Higgs particles.

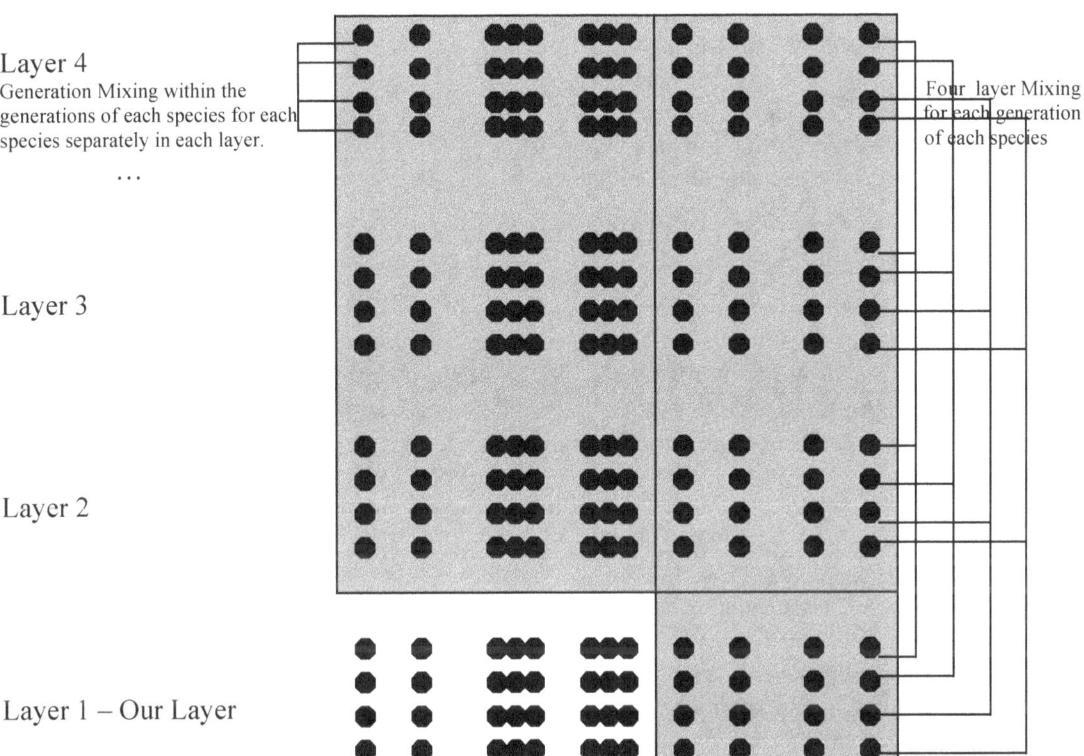

Figure 11.1. Partial example of pattern of mass mixing of the Generation group and of the Layer group. Dark parts of the periodic table are gray. Light parts are the known fermions with an additional, as yet not found, 4th generation shown. The lines on the left side show an example of the Generation mixing within one species. The Generation mixing applies to each species in each layer. The lines on the right side show an example of Layer mixing within one species with the mixing amongst all four layers of the species for each generation individually.

11.7.1 The Interaction Coupling Constants and their PseudoQuantum Field Vacuum Expectation Values

Ten of our Unified SuperStandard Model coupling constants are:[133]

- The Strong interaction coupling constant field g_S.
- The Weak SU(2) coupling constant g_W.
- The Electromagnetic U(1) coupling constant g_E.
- The Dark Weak SU(2) coupling constant g_{DW}.
- The Dark Electromagnetic U(1) coupling constant g_{DE}.
- The U(4) Generation group coupling constant g_G.
- The U(4) Layer group coupling constant g_V.
- The U(4) Species group coupling constant g_S.
- The U(192) Θ-interaction group coupling constant g_Θ.
- The complex gravitational coupling constant $g_{GR} = \kappa^{-1} = (4\pi G)^{-\frac{1}{2}}$.

Based on the discussions of Blaha (2015d) we can define Higgs vacuum expectation values for these coupling constants using a mass factor to obtain the correct coupling constant dimensions. We use the PseudoQuantum formalism of section 10.3.3 where we set the constants as follows:

- The Strong interaction coupling constant field $\Phi_1 = m_1 g_S$.
- The Weak SU(2) coupling constant $\Phi_2 = m_2 g_W$.
- The Electromagnetic U(1) coupling constant $\Phi_3 = m_3 g_E$.
- The Dark Weak SU(2) coupling constant $\Phi_4 = m_4 g_{DW}$.
- The Dark Electromagnetic U(1) coupling constant $\Phi_5 = m_5 g_{DE}$.
- The U(4) Generation group coupling constant $\Phi_6 = m_6 g_G$.
- The U(4) Layer group coupling constant $\Phi_7 = m_7 g_V$.
- The U(4) Species group coupling constant $\Phi_8 = m_8 g_S$.
- The U(192) Θ-interaction group coupling constant $\Phi_9 = m_9 g_\Theta$.
- The complex gravitational coupling constant $\Phi_{10} = m_{10} g_{GR} = \kappa^{-1} = (4\pi G)^{-\frac{1}{2}}$.

The ten masses, m_1 , m_2 , ... , m_{10} may be equal or they may have different values. It is also possible they all may be equal to κ^{-1}, which would yield

---

[133] The following groups and their coupling constants are duplicated four-fold for the four layers. The Strong interaction coupling constant field g_S, The Weak SU(2) coupling constant g_W, The Electromagnetic U(1) coupling constant g_E, The Dark Weak SU(2) coupling constant g_{DW}, The Dark Electromagnetic U(1) coupling constant g_{DE}, The U(4) Generation group coupling constant g_G, and The U(4) Layer group coupling constant g_V. We will only consider the known first layer here The following group constants appear once – the same for all four layers – the U(4) Species group, Gravitational coupling constant, and the Θ-symmetry group. The interactions of these groups will be discussed later.

- The Strong interaction coupling constant field $\Phi_1 = \kappa^{-1}g_S$.
- The Weak SU(2) coupling constant $\Phi_2 = \kappa^{-1}g_W$.
- The Electromagnetic U(1) coupling constant $\Phi_3 = \kappa^{-1}g_E$.
- The Dark Weak SU(2) coupling constant $\Phi_4 = \kappa^{-1}g_{DW}$.
- The Dark Electromagnetic U(1) coupling constant $\Phi_5 = \kappa^{-1}g_{DE}$.
- The U(4) Generation group coupling constant $\Phi_6 = \kappa^{-1}g_G$.
- The U(4) Layer group coupling constant $\Phi_7 = \kappa^{-1}g_V$.
- The U(4) Species group coupling constant $\Phi_8 = \kappa^{-1}g_S$.
- The U(192) Θ-interaction group coupling constant $\Phi_9 = \kappa^{-1}g_\Theta$.
- The complex gravitational coupling constant $\Phi_{10} = \kappa^{-1}g_{GR} = \kappa^{-1} = (4\pi G)^{-\frac{1}{2}}$.

Then scaling the above vacuum expectation values by κ^{-1} would give:[134]

- The strong interaction coupling constant[135] vacuum expectation value $\Phi_1' = g_S = 1.22$
- The Weak SU(2) coupling constant vacuum expectation value $\Phi_2' = g_W = 0.619$.
- The Electromagnetic U(1) coupling constant vacuum expectation value $\Phi_3' = e = g_E = 0.303$.
- The Dark Weak SU(2) coupling constant vacuum expectation value $\Phi_4' = \Phi_2'$. (?)
- The Dark Electromagnetic U(1) coupling constant vacuum expectation value $\Phi_5' = = \Phi_3'$. (?)
- The U(4) Generation group coupling constant $\Phi_6' = g_G$. (?)
- The U(4) Layer group coupling constant $\Phi_7' = g_V$. (?)
- The U(4) Species group coupling constant $\Phi_8' = g_S$. (?)
- The U(192) Θ-interaction group coupling constant $\Phi_9' = g_\Theta$. (?)
- The complex gravitational coupling constant $\Phi_{10}' = 1$.

The known *scaled* vacuum expectation values,[136] which are in fact the scaled Higgsian coupling constants, have a comparable range of values[137] as opposed to the range of values for the unscaled constants which range from the ultra-small gravitational vacuum expectation value to values, perhaps, within a few orders of magnitude of unity.

Given the range of known values above, it appears reasonable to conjecture that the unknown values would also be of the order of unity.

The known coupling constant values above are of comparable value, which suggests that our Unified SuperStandard Model, at current energies, may be close to the GUT level at which properly scaled coupling constants are equal.

---

[134] All coupling constant values are based on data extracted from K. A. Olive et al (Particle Data Group), Chinese Physics **C38**, 090001 (2014).

[135] Based on the running coupling constant value $\alpha_s(M_Z^2) = 0.1193 \pm 0.0016$.

[136] The closeness of all the values to one is suggestive: The value $\alpha = 1$ (or $e = (4\pi)^{\frac{1}{2}} = 3.54$) was the value found in our calculation in the Johnson, Baker, Willey model of QED. Perhaps a larger calculation along the lines of our paper in massless ElectroWeak theory might yield scaled coupling constant values near unity.

[137] The weakness of the Weak interactions is primarily due to the large masses of the Z and W vector bosons – not the values of their coupling constants g and g'.

11.7.2 Unified SuperStandard Model Lagrangian Coupling Constants

We begin with the Unified SuperStandard Model lagrangian density $\mathcal{L}_{TE}$ with coupling constants explicitly displayed[138]

$$\mathcal{L}_{TE} = \mathcal{L}_{TE}(g_S, g_W, g_E, g_{DW}, g_{DE}, g_G, g_V, g_S, g_{GR}, g_\Omega) \tag{11.25}$$

and fields and space-time coordinates not displayed.

In terms of vacuum expectation values as discussed earlier we see we can write[139]

$$\mathcal{L}_{TE} = \mathcal{L}_{TE}(\Phi_1/m_1, \Phi_2/m_2, \ldots, \Phi_{11}/m_{11}) \tag{11.26}$$

where

$$| \Phi_1, \Phi_2, \ldots, \Phi_{11}; \Pi_1, \Pi_2, \ldots, \Pi_{11}> = C \prod_{i=1}^{11} \left\{ \exp[[(2\pi)^3 m_i/2]^{1/2} \Phi_i[a_{i2}^\dagger(0,m_i) + a_{i2}(0,m_i)]] \right\} |0>$$

Assuming all $m_i = \kappa^{-1}$ we obtain

$$| \Phi_1, \Phi_2, \ldots, \Phi_{11}; \Pi_1, \Pi_2, \ldots, \Pi_{11}> = C \prod_{i=1}^{11} \left\{ \exp[[(2\pi/\kappa)^3/2]^{1/2} \Phi_i'[a_{i2}^\dagger(0, \kappa^{-1}) + a_{i2}(0, \kappa^{-1})]] \right\} |0>$$

$$\tag{11.27}$$

Then we can write

$$\mathcal{L}_{TE} = \mathcal{L}_{TE}(\Phi_1', \Phi_2', \ldots, \Phi_{11}') \tag{11.28}$$

Setting $m_i = g_{CG} = \kappa^{-1} = $ the Planck mass, simplifies the above expressions and *supports the belief that we are close to the unification of all interactions.* However having particles of such large mass makes them undetectable by accelerators. It also seems too large from the viewpoint of physical intuition. Consequently the above may be the correct expressions with masses perhaps in the TeV range.

We finally note

$$\varphi_{1i}| \Phi_1, \Phi_2, \ldots, \Phi_{11}; \Pi_1, \Pi_2, \ldots, \Pi_{11}> = \Phi_{1i}| \Phi_1, \Phi_2, \ldots, \Phi_{11}; \Pi_1, \Pi_2, \ldots, \Pi_{11}> \tag{11.29}$$

11.7.3 Big Bang Vacuum

At the origin of the universe – the Big Bang – there was a vacuum state in principle. In our earlier books[140] we showed that the universe existed in an ultra-small, but finite, region for

---

[138] The 11th coupling constant g_Θ and the Θ symmetry group are discussed later.

[139] The "vacuum" state $|0>$ above also has factors for the vacuum expectation values used for fields that give masses to fermions and vector bosons as described in Blaha (2015b).

[140] Blaha (2015a) and Blaha (2004).

an infinitesimal time before it began an explosive inflationary expansion to become the familiar universe. In this time period there were no infinities – a finite temperature and so on.

Thus it is reasonable to assume one of two possibilities for the above ten coupling constants: 1) they have remained unchanged since the beginning, or 2) they have changed with time.

In this section we note, that if our scaling with the Planck mass κ^{-1} in preceding discussions is correct, then it is reasonable to assume that the vacuum state in the beginning is that defined above with $|0>$ including factors setting fermion and vector boson masses as described in Blaha (2015b).

11.7.4 Evolving/Space-Time Dependent Coupling Constants

It is possible that the Unified SuperStandard Model coupling constants evolve with time and may also be spatially varying – different constants in different parts of the universe. Presently there is no decisive evidence for either possibility. In this section we will describe the mechanism to support either or both possibilities.

Consider a classical field (time and spatially varying):

$$\Phi(\mathbf{x}, t) = \int d^3k \, [\alpha(k)f_k(x) + \alpha^*(k)f_k^*(x)] \qquad (11.30)$$

If we define the coherent vacuum state

$$|\alpha> = C \exp\left\{\int d^3k \, [\alpha(k)a_2^\dagger(k) + \alpha^*(k)a_2(k)]\right\}|0> \qquad (11.31)$$

then

$$\varphi_1(x)|\Phi, \Pi> = \Phi(x)|\Phi, \Pi>$$
$$\pi_1(x)|\Phi, \Pi> = \Pi(x)|\Phi, \Pi>$$

where

$$\varphi_i(\mathbf{x}, t) = \int d^3k \, [a_i(k)f_k(x) + a_i^\dagger(k)f_k^*(x)] \qquad (11.32)$$

for $i = 1, 2$ with

$$f_k(x) = e^{-ik\cdot x} /(2\omega_k(2\pi)^3)^{\frac{1}{2}}$$

where $\omega_k = |\mathbf{k}|$.

The coherent vacuu state $|\alpha>$ has a time and spatially varying vacuum expectation value (classical) field. The above equations can be generalized to the case of the eleven coupling constant vacuum expectation values:[141]

---

[141] The "vacuum" state $|0>$ also has factors for the vacuum expectation values used for fields that give masses to fermions and vector bosons as described in Blaha (2015b).

$$| \Phi_1, \Phi_2, \dots, \Phi_{11}; \Pi_1, \Pi_2, \dots, \Pi_{11}> = C \prod_{i=1}^{11} \exp \left\{ \int d^3k \, [\alpha_i(k)a_{2i}^{\dagger}(k) + \alpha_i^{*}(k)a_{2i}(k)] \right\} |0>$$

$$(11.33)$$

Then all eleven coupling constant vacuum expectation values are space-time dependent:

$$\varphi_{1i}(x) \, | \, \Phi_1, \Phi_2, \dots, \Phi_{11}; \Pi_1, \Pi_2, \dots, \Pi_{11}> = \Phi_i(x) \, | \, \Phi_1, \Phi_2, \dots, \Phi_{11}; \Pi_1, \Pi_2, \dots, \Pi_{11}>$$

$$(11.34)$$

and the Unified SuperStandard Model lagrangian becomes

$$\mathcal{L}_{TE} = \mathcal{L}_{TE}(\Phi_1(x), \Phi_2(x), \dots, \Phi_{11}(x)) \qquad (11.35)$$

for matrix elements between the vacuum defined by $|\alpha>$ and its conjugate,

Thus our formalism can accommodate space-time varying coupling constants should they be found in the Cosmos.

11.7.5 A Unified SuperStandard Model Lagrangian Without Any Constants

The preceding sections put coupling constants within the same framework as particle masses completing the process of eliminating constants from The Unified SuperStandard Model lagrangian. Instead the vacuum contains the values of all coupling constants and particle masses. In one sense this new formulation is a tradeoff. The values of all constants are shifted to the vacuum. However the shift has some advantages technically. One advantage is the ability to have space-time dependent coupling constants as shown above. It would also be straightforward to make masses, and mixing angles, space-time dependent. The possible space-time dependence of coupling constants and particle masses has been an active area of experimental interest for many years although cosmological data seems to indicate these quantities have not changed significantly since the universe began.

The question of changes in lagrangian coupling constant physical values is of great philosophical importance since it appears that the existence of life, as we know it, depends sensitively on their values. This dependence has been embodied in the Anthropomorphic Principle and studied by a number of physicists and philosophers.

Since our formulation allows space-time varying physical constants the question of the Anthropomorphic Principle attains new importance. As we saw in the case of the theory of Black Holes, which was a theory without evidence for over forty years before Black Holes were discovered, Nature seems to provide phenomena that have been shown to be theoretically possible. Many other cases of this sort have also occurred – the most recent example at the time of this writing is Weyl fermions.

Lastly, our approach opens the possibility of a study of all the many constants in The Unified SuperStandard Model lagrangian *on the same footing* rather than in the piecemeal fashion used up to the present. It replaces the scattered hodge-podge of constants in the lagrangian with a centralized location for all constants in the vacuum state permitting a direct study of their interconnection. *The study of the vacuum now becomes of central importance.*

11.7.6 The Form is Determined But Not the Constants

The derivation of The Unified SuperStandard Model here, and in the Blaha (2015) books and earlier work, was based on Asynchronous Logic (to support parallel physical processes spread in space and time); on complex space-time coordinates, the Complex Lorentz group and complex General Coordinate transformations; the Reality group to map complex coordinates to the real-valued coordinates that we observe; the Generations group that yields the four fermion generations and particle number interactions such as the baryon number and lepton number interactions, and the Species group for complex general coordinate transformations.

This firm basis in fundamental considerations enables us to forge a path to The Unified SuperStandard Model, which included the known features of The Standard Model. We thus were able to avoid the many possible variants and extensions of The Standard Model that have been considered in the Physics literature over the past thirty years.

Two remarkable features of The Unified SuperStandard Model derivation were:

1.   A precise fixing of the form of the Unified SuperStandard Model.
2.   The absence of any constraints on the values of its coupling constants or masses.

Particle masses were fixed by either the original Higgs Mechanism or by our new mechanism that was based on an extension of Quantum Field Theory to include classical fields such as the vacuum expectation values that cropped up in the original Higgs Mechanism and were handled "by hand." (See Blaha (2015c).)

Thus, up to this point, we have a Unified SuperStandard Model (known) except for a basis for the values of the coupling constants that appear in the theory. The coupling constants have a wide range of values. A fundamental basis for their values has been wanting.

11.7.7 How Can Coupling Constants be Determined?

The renormalized coupling constants of The Unified SuperStandard Model can be determined experimentally. However a theoretical determination is lacking. This gap in our understanding suggests that there is a major aspect of fundamental physics that is not understood. The fact that we can determine the form – but not the values of coupling constants – so directly from basic principles suggests that a new basic principle(s) is needed to complete The Unified SuperStandard Model. A similar comment applies to fermion and boson masses – both our mechanism,[142] and the vanilla Higgs Mechanism, arbitrarily fixes particle masses. (Attempts to relate particle masses using various symmetries beg the question. As Isidore Rabi

---

[142] Blaha (2015c).

(Columbia) once said in a different context, "Who ordered them?" Proposed symmetries are typically "pulled out of a hat.")

The *one* meaningful attempt to determine a coupling constant in a non-trivial 4-dimensional quantum field theory was that of Johnson, Baker and Willey[143] in a 4-dimensional model – massless Quantum Electrodynamics. They developed the theory to the point where if one function, that they called the eigenvalue function, had a zero at the value of the fine structure constant $\alpha \approx 1/137$ then the theory would have no infinities. This author then developed an approximate solution for the eigenvalue function in perhaps the most comprehensive 4-dimensional quantum field theory calculation to all orders in α. The approximate calculation agreed with known exact results to 6th order in e.[144] This author found a zero at $\alpha = 1$. Thus the eigenvalue function zero made the Johnson, Baker, Willey model QED finite removing all divergences.

While the Johnson, Baker, Willey model QED was not successful in finding $\alpha = 1/137$, its method illustrates one possible approach to determining the coupling constants of The Unified SuperStandard Model. It might be possible to use a consistency condition(s) to fix coupling constant values. However, since The Unified SuperStandard Model does not have infinities, the motivation of Johnson, Baker and Willey is absent. A fundamental set of consistency conditions is not apparent and so this approach is not currently viable.

What other approaches are possible? There is an anthropomorphic approach which posits the necessity of certain ranges of some coupling constants for human life, and life in general, to exist. We are not comfortable with this approach since it seems to "beg the question." The input is equivalent to the output mitigating is character as fundamental.

One could also study the set of coupling constants in a 10-dimensional space looking for the set of values.

Given these considerations we have chosen to pursue a less ambitious approach: to specify the coupling constants as vacuum expectation values of a set of new Higgs-like scalar fields. This approach conceptually parallels the determination of particle masses as vacuum expectation values of scalar Higgs fields.

After reducing coupling constants to vacuum expectation values we considered the possibility that the vacuum state at the Big Bang point determined the coupling constants. We also considered the possibility that coupling constants evolve slowly with time and/or may vary in differing spatial locations.

11.8 Serendipitous Determination of Physical Constants?

Our Principle of Serendipity (Appendix C) suggests that Nature is most helpful in guiding us to Physical Reality. Perhaps the best examples of this principle are the Curies' discovery of radioactivity, the Periodic Table of Chemistry with missing elements filled in and then found, and t'Hooft and others discovery of the renormalizability of the ElectroWeak

---

[143] M. Baker and K. Johnson, Phys. Rev. **D8**, 1110 (1973) and references therein.
[144] Equation 1 in our paper S. Blaha, Phys. Rev. **D9**, 2246 (1974).

Model, which we view as a precursor to our Two-Tier formulation eliminating infinities and automatically yielding renormalizability for all lagrangian interactions with any order of derivatives in any number of dimensions.

Given the above observations on coupling constants, we would hope that the calculation of the values of coupling constants might eventually happen through the consideration of consistency conditions in the spirit of the Johnson-Baker-Willey model QED.

Status of the Derivation Now

Up to this point in our derivation of the Unified SuperStandard Model we have relied directly on the Complex Lorentz group for the derivation of the form of the fermion spectrum with four fermion species (three up-type and down-type quark subspecies) in one generation,[145] and for the derivation of the form of the Internal Symmetry Reality group as the analogue of the subgroups of the Coordinate Reality group of the Complex Lorentz group. The Internal Symmetry Reality group is identical to that of the Standard Model with the addition of a Dark $SU(2)\otimes U(1)$ symmetry and Dark fermions (and Higgs fields to generate particle masses.) We also developed the formalisms for Two-Tier Field Theory and PseudoQuantum Field Theory to remedy shortcomings in the conventional formulation of Quantum Field Theory.

Now we turn to derive two additional symmetry groups based on the form of the one generation Standard Model lagrangian with $SU(3)\otimes SU(2)\otimes U(1)\otimes SU(2)\otimes U(1)$ symmetry group. An examination of this lagrangian with its dynamic terms and interaction terms corresponding to this symmetry group reveals that there is a set of conserved particle numbers. (The most prominent of these conserved particle numbers is Baryon number. Baryon number conservation violation has been fruitlessly sought in many high energy experiments. To high accuracy Baryon Number is conserved.) Since these particle numbers are conserved in the one generation Standard Model we find that they imply additional group symmetries that we call the U(4) Generation group, and the Layer groups consisting of a set of four U(4) groups.

Thus we use the Complex Lorentz group to derive the interactions and species of fermions: neutrino, charged lepton, up-type quarks and down-type quarks (with four Dark species equivalents) in the one generation case. We then use the structure of the corresponding lagrangian to derive the additional Generation and Layers groups[146] with their additional interactions and levels of fermions. Since these groups are all U(4) groups we find four generations of fermions in four layers. (The fourth fermion generation remains to be found. We expect the fourth generation fermions – excepting neutrinos – to be very massive.)

---

[145] Generations and layers will be described next.

[146] These group symmetries are badly broken.

12. Generation Group

We now turn to the derivation of the Generation Group and, in the following chapter, the Layer Groups, which are based on conserved (and almost conserved) particle numbers such as Baryon Number Conservation.

In this chapter we consider the extension of the one generation SuperStandard theory to four generations based on a new U(4) symmetry group that we call the Generation Group.

It is based on four conservation laws for Baryon, Lepton, Dark Baryon, and Dark Lepton Numbers. The conservation laws are manifest in the fermion terms of the one generation version of The Standard Model which all have the form:

$$\mathcal{L} \sim \overline{\psi}_\alpha \gamma^\mu D_{\mu\alpha\beta} \, \psi_\beta$$

where the covariant derivative terms preserve the four conservation laws. Symmetry breaking is brought in at a later point through Higgs fields terms. There is an evident $SU(3) \otimes SU(2) \otimes U(1) \otimes SU(2) \otimes U(1)$ symmetry in the sum of these lagrangian terms. The next relevant question is whether the particle number conservation laws have associated gauge fields that provide interactions. This question can be partially answered in our current state of experimental knowledge.

In Blaha (2017b), and earlier books, we showed that the existence of a long range baryonic force[147] supports baryon number conservation in a manner similar to electric charge conservation due to the electromagnetic force. Lepton number conservation suggests a very weak long range force as well. By analogy we postulate two similar Dark particle number conservation laws and forces.

Having four conserved (or almost conserved) particle numbers with their attendant forces (interactions) leads naturaly to a U(4) symmetry with four 'diagonal' generators. We call the U(4) group the Generation Group.

If Baryon Number and Lepton Number are both conserved quantities then any linear combination of them is also conserved. Therefore

$$B' = aB + bL$$

is also conserved.

---

[147] See Blaha (2017b) and earlier books for a discussion of evidence for a baryonic force and conservation law. We found that gravity experiments suggested a possible baryonic potential with the (order of magnitude) coupling constant $\alpha_B = \beta^2/4\pi \simeq .118\, Gm_H^2$ where m_H is the mass of the hydrogen atom. Section 12.1 provides details.

If we consider the Dark Matter sector of the Unified SuperStandard Model it is reasonable to assume that *Dark Baryon Number B_D and Dark Lepton Number L_D are conserved* also (although there is no experimental evidence available as yet to confirm (or deny) these assumptions.)

Thus we have four conserved particle Numbers. Linear combinations of these numbers are also conserved:

$$B' = aB + bL + cB_D + dL_D$$
$$L' = eB + fL + gB_D + hL_D$$
$$B_D' = iB + jL + kB_D + lL_D$$
$$L_D' = mB + nL + oB_D + pL_D$$

The set of 4×4 matrices form an U(4) group if we wish to perform these transformations within lagrangians of the type of the Unified SuperStandard Model. The choice of U(4) rather than SU(4) is required since there are four independent particle numbers. U(4) has four diagonal matrices in its algebra while SU(4) only has three diagonal matrices. U(4) preserves the independence of the four independent particle Numbers. It also allows complex rotations.

The new U(4) Generation Group symmetry leads immediately to four fermion generations. We add an index to each fermion field ranging from 1 through four, add U(4) gauge fields to the covariant derivative,[148] and thus Yang-Mills local gauge field terms to the lagrangian.

Then we perform the following tasks:[149]

- Define the Two-Tier Lepton and Quark Sectors

- Introduce Symmetry Breaking via the Higgs Mechanism for Fermions and U(4) Gauge Fields

Thus we now have four fermion generations, a broken Generation Group symmetry, and masses for fermions and Generation Group gauge fields.

12.1 Baryon Number Conservation and a Possible Baryonic Force

We have considered baryon number conservation and a possible baryonic force in Blaha (2014a) and (2014b). The primary forces involved in the interactions and collisions of baryons include the forces of The Standard Model, the force of gravity, and possibly a fifth force which we take to be the baryonic force, a much discussed force that depends on the baryon numbers of objects experiencing it. The Gravitation and baryonic forces (neglecting other SuperStandard Model interactions) between two clumps of baryonic matter containing

---

[148] See subsection 4.2.8.4.

[149] These tasks are described in detail in chapters 12 and 13 in Blaha (2017b) as well as earlier books.

baryons and other particles: clump1 being of mass m_1 and baryon number n_1, and clump2 being of mass m_2 and baryon number n_2 is

$$F = -Gm_1m_2/r^2 + (\beta^2/4\pi)n_1n_2/r^2 \qquad (12.1)$$

where G is the gravitational constant, β is the baryonic coupling constant and r is the distance between the 'widely' separated clumps. Experimentally a baryonic force between baryons has not been detected with any degree of certainty. Sakurai (1964) discusses early efforts to determine the force of gravity in detail. Eőtvős experiments on the ratio of the observed gravitational mass to the inertial mass showed that that the gravity force is constant to within one part in 100,000,000 as far back as 1922 indicating the baryonic force, if it exists, as we believe it does, is extremely weak compared to the gravitational force. Eőtvős et al[150] found

$$(\beta^2/4\pi)/(Gm_p^2) < 10^{-5}$$

where m_p is the proton mass.

Since then, the experiment has been redone with improved accuracy by Dicke and collaborators.[151] They have improved the accuracy to one part in 100 billion. A further analysis showed a very small discrepancy that suggested the ratio, while small, was non-zero, implying the equivalence principle might not be exact and that the discrepancy changed with the material used in the experiment – just what one might expect if a very small baryonic force was present – often called the "fifth force." At present the existence and amount of the discrepancy is unclear. *Nevertheless, we will assume a fifth force – Baryonic force. We will also assume there are three additional forces corresponding to the three other conserved particle numbers. These forces are not relevant to these considerations.*

The conservation of baryon number has been repeatedly investigated by experimenters and found to be true to extremely high accuracy. For decades theorists have suggested that a baryon conservation law[152] follows from the existence of a gauge field in a manner much like electric charge conservation follows from the properties of the electromagnetic gauge field.

12.1.1 Estimate of the Baryonic Coupling Constant

The baryonic force, and coupling constant, if it exists, is known to be very small in comparison to gravity and the other known forces. However, measurements of the gravitational constant G are significantly different.[153,154] The reason(s) for these discrepancies is not known. We will assume that both the 2010 and 2013 measurements of G are experimentally correct but

---

[150] Eőtvős, R. V., Pekár, D., Fekete, E., Ann. d. Physik **68**, 11 (1922).

[151] P. G. Roll, R. Krotkov, R. H. Dicke, Annals of Physics, 26, 442, 1964.

[152] See Gell-Mann, M. and Levy, M. *Nuovo Cimento* 16, 705 (1960) for a proof and Sakurai (1964) for a discussion of the relation of the baryonic gauge field to gravity experimentally.

[153] T. Quinn et al, Phys. Rev. Lett. **111**, 101102 (2013).

[154] P. J. Mohr, B.N. Taylor, and D. B. Newell, Rev. Mod. Phys. 84, 1527 (2012).

disagree because of the baryonic force term in eq. 12.1 that would create a difference in effective G values if the experiments used different masses and thus baryon numbers. Quinn et al found a value for the gravitational constant of $G_1 = 6.67545 \times 10^{-11}$ $m^3 kg^{-1} s^{-2}$. The combined 2010 CODATA value for the gravitational constant was $G_2 = 6.67384 \times 10^{-11}$ $m^3 kg^{-1} s^{-2}$. Both values are subject to estimated uncertainties.

Suppose these values are correct and due to a difference in the chemical composition (metals) of the test masses used in the experiment. Quinn et all use 1.2 kg test masses composed of Cu-0.7% Te free machining alloy. The CODATA value being a composite of many experiments does not have an effective equivalent test mass value or composition specified.[155] Suppose the test mass value is $N_1^2 m_1^2 + N_{1e}^2 m_e^2$ for the G_1 result giving

$$-(N_1^2 m_1^2 + N_{1e}^2 m_e^2)G_1 = [-G(m_1^2 N_1^2 + N_{1e}^2 m_e^2) + (\beta^2/4\pi)N_1^2] \qquad (12.2)$$

where G is the *real* value of the gravitational constant. The total test mass is $(m_1^2 N_1^2 + N_{1e}^2 m_e^2)$ with N_1 baryons of average mass m in each test mass and N_{1e} leptons of average mass m_e.

Suppose further the test mass value is $N_2^2 m_2^2 + N_{2e}^2 m_e^2$ for the G_2 result giving

$$-(N_2^2 m_2^2 + N_{2e}^2 m_e^2)G_2 = [-G(m_2^2 N_2^2 + N_{2e}^2 m_e^2) + (\beta^2/4\pi)N_2^2] \qquad (12.3)$$

where G is again the *real* value of the gravitational constant. The total test mass is $(m_2^2 N_2^2 + N_{2e}^2 m_e^2)$ with N_2 baryons of average mass m_2 in each test mass and N_{2e} leptons of average mass m_e. Since the test masses are electrically neutral and there are approximately equal numbers of protons and neutrons in a test mass it follows approximately that

$$N_{1e} = \tfrac{1}{2}N_1 \quad \text{and} \quad N_{2e} = \tfrac{1}{2}N_2 \qquad (12.4)$$

Subtracting eq. 12.2 from eq. 12.3 after some algebra[156] we find

$$\Delta G = -G_2 + G_1 = (\beta^2/4\pi)/(m_2^2 + m_e^2/2) - (\beta^2/4\pi)/(m_1^2 + m_e^2/2)$$
$$\simeq (\beta^2/4\pi)(1/m_2^2 - 1/m_1^2) \qquad (12.5)$$

The masses m_1 and m_2 can differ. For example, if m_H is mass of the hydrogen atom, then $m^{-1} = 1.0 m_H^{-1}$ for hydrogen, for carbon $m^{-1} = 1.00782 m_H^{-1}$, for copper $m^{-1} = 1.00895 m_H^{-1}$, and for

---

[155] The Eötvös' experiment used a 0.1 gm test mass of $RaBr_2$. R. v. Eötvös, D. Pekár, E. Fekete, Annalen der Physik (Leipzig) 68, 11, 1922.

[156] The reduction of the calculation to algebra reminds the author of Nobelist Hans Bethe's remark that he only felt he understood a physical phenomenon when he could reduce it to algebra. This was quite evident when the author collaborated with Professor Bethe on a study of pion condensation in neutron stars some years ago.

lead $m^{-1} = 1.00794 m_H^{-1}$.[157] Thus using the Quinn et al results, and CODATA results, and assuming copper and lead test masses, we find the order of magnitude *estimate*:

$$\alpha_B = \beta^2/4\pi \simeq \Delta G/[(1.00895^2 - 1.00794^2)\, m_H^2]$$
$$\simeq \Delta G/G\; G\; m_H^2/.002037$$
$$\simeq (0.000241/0.002037) G m_H^2$$
$$\simeq 0.118\; G m_H^2 \qquad\qquad (12.6)$$

indicating a very weak baryonic force consistent with our general view of the universe. The baryon fine structure constant is minute in comparison to the electromagnetic fine structure constant $\alpha \simeq 1/137$.

Due to our assumptions in the calculation of α_B, which makes it merely an order of magnitude estimate at best, we suggest that an experimental group measure G with differing test masses in the same apparatus to obtain a better value for α_B.

12.2 Four Generation Unified SuperStandard Model

Previously we derived the form of a one generation Unified SuperStandard Model that included the known parts of the Standard Model (excepting the Higgs sector) and an $SU(2) \otimes U(1)$ part for Dark Matter.

In this section we generalize to the four generation SuperStandard Model that results.[158] Covariant derivatives acquire another interaction term with 16 U(4) fields U_i^μ. In addition we add another index to each fermion field specifying its generation. Lastly a set of initially massless gauge field dynamics terms is added to the Unified SuperStandard Model lagrangian to specify U(4) gauge field evolution and interactions.

12.2.1 Two-Tier Lepton Sector

We begin with the definition of a quadruplet of leptons – a pair of doublets, one normal and one Dark, instead of a single doublet. We define left and right lepton quadruplets with[159]

$$\Psi_{L,Ra}(X) = \begin{bmatrix} \psi_{DL,Ra}(X) \\ \psi_{NL,Ra}(X) \end{bmatrix} \qquad\qquad (12.7)$$

---

[157] "One Hundred Years of the Eötvös Experiment", l. Bod, E. Fischbach, G. Marx and Maria Náray-Ziegler, August, 1990.

[158] It is based on the three principles based on Ockham's Razor ("The simplest choice is often the best."): 1) The only connecting interaction is a weak interaction, 2) The form of ElectroWeak theory remains unchanged, and 3) Dark Matter parallels normal matter in its general characteristics: four generations, SU(3) singlets, an $SU(2) \otimes U(1)$ symmetry analogous to ElectroWeak symmetry, $SU(2) \otimes U(1)$ dark lepton and dark quark doublets.

[159] The X's are Two-Tier coordinates.

where a is a generation index ranging from 1 to 4, $\psi_{NL,R}(X)$ is a "normal" ElectroWeak-like lepton doublet, and where $\psi_{DL,R}(X)$ is a similar Dark ElectroWeak-like lepton doublet consisting of a Dark electron-like fermion and a Dark neutrino-like fermion.

We define covariant derivative terms which we express in matrix form are

$$D_{L,R}(X) \;=\; \begin{bmatrix} \gamma^{\mu}D_{DL,R\mu} & 0 \\[2ex] 0 & \gamma^{\mu}D_{NL,R\mu} \end{bmatrix} \tag{12.8}$$

where the normal matter left-handed covariant derivative is

$$D_{NL\mu} = \partial/\partial X^{\mu} - \tfrac{1}{2}ig\boldsymbol{\sigma}\cdot\mathbf{W}_{\mu} + \tfrac{1}{2}ig'\mathbf{B}_{\mu} - \tfrac{1}{2}ig_G\mathbf{G}\cdot\mathbf{U}_{\mu} \tag{12.9}$$

where g_G is an ultra-weak generational coupling constant, $\mathbf{G}\cdot\mathbf{U}_{\mu}$ is the sum of the inner product of 16 U(4) generators G_i and gauge fields $U_i(X)$, and where the Dark matter left-handed covariant derivative is

$$D_{DL\mu} = \partial/\partial X^{\mu} - \tfrac{1}{2}ig_D\boldsymbol{\sigma}\cdot\mathbf{W'}_{\mu} + \tfrac{1}{2}ig_D'\mathbf{B'}_{\mu} - \tfrac{1}{2}ig_G\mathbf{G}\cdot\mathbf{U}_{\mu} \tag{12.10}$$

with $\boldsymbol{\sigma}$ a vector composed of the Pauli matrices. The right-handed covariant derivatives have a simpler form. The normal matter right-handed covariant derivative is

$$D_{NR\mu} = \partial/\partial X^{\mu} + \tfrac{1}{2}ig'\mathbf{B}_{\mu} - \tfrac{1}{2}ig_G\mathbf{G}\cdot\mathbf{U}_{\mu} \tag{12.11}$$

and the Dark matter right-handed covariant derivative is

$$D_{DR\mu} = \partial/\partial X^{\mu} + \tfrac{1}{2}ig_D'\mathbf{B'}_{\mu} - \tfrac{1}{2}ig_G\mathbf{G}\cdot\mathbf{U}_{\mu} \tag{12.12}$$

The normal and Dark electroweak fields above are functions of a Two-Tier X. The Faddeev-Popov mechanism operative for these types of fields is described in appendix 19-A of Blaha (2011c).

12.2.2 Quark Sector

In the *quark* sector we define left and right quark quadruplets with

$$\Psi_{qL,Ra}(X_c) \;=\; \begin{bmatrix} \psi_{DqL,Ra}(X_c) \\ \psi_{NqL,Ra}(X_c) \end{bmatrix} \tag{12.13}$$

where $\psi_{NqL,Ra}(X_c)$ is a "normal" ElectroWeak-like quark doublet, and where $\psi_{DqL,Ra}(X_c)$ is a Dark ElectroWeak-like quark doublet consisting of a SU(3) singlet Dark up-quark of unit Dark charge and a SU(3) singlet Dark down-quark of zero Dark charge in the ath generation.

The covariant derivative terms are contained in $D_q(X_c)$ which we express in matrix form as

$$D_{qL,R}(X_c) \; = \; \begin{bmatrix} \gamma^\mu D_{qDL,R\mu}(X_c) & 0 \\ \\ 0 & \gamma^\mu D_{qNL,R\mu}(X_c) \end{bmatrix} \qquad (12.14)$$

where the normal quark matter left-handed covariant derivative is

$$D_{qNL\mu} = \partial/\partial X_c{}^\mu - \tfrac{1}{2}ig\boldsymbol{\sigma}{\cdot}\mathbf{W}_\mu - ig'B_\mu/6 - \tfrac{1}{2}ig_G\mathbf{G}{\cdot}\mathbf{U}_\mu + ig_C\boldsymbol{\tau}{\cdot}\mathbf{A}_{C\mu} \qquad (12.15)$$

and where the Dark quark left-handed covariant derivative is

$$D_{qDL\mu} = \partial/\partial X_c{}^\mu - \tfrac{1}{2}ig_D\boldsymbol{\sigma}{\cdot}\mathbf{W'}_\mu + \tfrac{1}{2}ig_D'B'_\mu - \tfrac{1}{2}ig_G\mathbf{G}{\cdot}\mathbf{U}_\mu \qquad (12.16)$$

since Dark quarks are SU(3) singlets with unit or zero Dark charge. The right-handed quark covariant derivatives have a simpler form. The normal quark right-handed covariant derivative is

$$D_{qNR\mu} = \partial/\partial X_c{}^\mu + \tfrac{1}{2}ig'B_\mu/3 - \tfrac{1}{2}ig_G\mathbf{G}{\cdot}\mathbf{U}_\mu + ig_C\boldsymbol{\tau}{\cdot}\mathbf{A}_{C\mu} \qquad (12.17)$$

and the Dark quark right-handed covariant derivative is

$$D_{qDR\mu} = \partial/\partial X_c{}^\mu + \tfrac{1}{2}ig_D'B'_\mu - \tfrac{1}{2}ig_G\mathbf{G}{\cdot}\mathbf{U}_\mu \qquad (12.18)$$

The normal and Dark gauge boson fields are functions of $X_c. = (X_{r\mu}(y_r), X_{i\mu}(y_i))$. The Faddeev-Popov mechanism is operative for gauge boson fields and is described later in this book and in Appendix 19-A of Blaha (2011c).[160] The *complexon* quark SuperStandard Model ElectroWeak Sector covariant derivatives in quadruplet matrix form are

---

[160] Those who might be concerned about the propagator term $\langle W_i(X), W_j(X_c)\rangle$ and similar propagators where one field is a function of X and the other field is a function of X_c should note that such terms are to very good approximation equal to $\langle W_i(X), W_j(X)\rangle$ for energies much less than M_c (which could be as large as the Planck energy.)

$$D_{qL,R}(X_c) = \begin{bmatrix} \gamma^\mu D_{qDL,R\mu} & 0 \\ 0 & \gamma^\mu D_{qNL,R\mu} \end{bmatrix} \qquad (12.19)$$

The remaining parts of the complexon Standard Model are described in chapter 23 of Blaha (2011) and summarized below. The addition of singlet Dark quark Higgs terms is also required. The lagrangian density and action is

$$\mathcal{L}_{CSM} = \Psi_{La}{}^\dagger\gamma^0 i\gamma^\mu D_{L\mu}\Psi_{La} - \Psi_{Ra}{}^\dagger\gamma^0 i\gamma^\mu D_{R\mu}\Psi_{3Ra} + \Psi_{CLa}{}^\dagger\gamma^0 i\gamma^\mu \mathcal{D}_{qL\mu}\Psi_{CLa} + \Psi_{CRa}{}^\dagger\gamma^0 i\gamma^\mu \mathcal{D}_{qR\mu}\Psi_{CRa} - $$
$$ - \mathcal{L}_{BareMasses} + \mathcal{L}_{Gauge} + \mathcal{L}_{Mass} + \mathcal{L}_{Ufields} \qquad (12.20)$$

where a is the generation index. $\mathcal{L}_{BareMasses}$ contains the fermion bare mass terms. Also,

$$\mathcal{L}_{Gauge} = \mathcal{L}_{GaugeEW} + \mathcal{L}_{GaugeC} + \mathcal{L}_{GaugeEWD} \qquad (12.21)$$

with

$$\mathcal{L}_{GaugeEW} = -\tfrac{1}{4} F_W{}^{a\mu\nu}F_W{}^a{}_{\mu\nu} - \tfrac{1}{4} F_B{}^{\mu\nu}F_{B\mu\nu} + \mathcal{L}_{EW}{}^{ghost} \qquad (12.22)$$

$$\mathcal{L}_{GaugeEWD} = -\tfrac{1}{4} F'_W{}^{a\mu\nu}F'_W{}^a{}_{\mu\nu} - \tfrac{1}{4} F_{B'}{}^{\mu\nu}F_{B'\mu\nu} + \mathcal{L}_{W'}{}^{ghost} \qquad (12.23)$$

and

$$\mathcal{L}_{GaugeC} = \mathcal{L}_{CCG} + \mathcal{L}_C{}^{ghost} + \mathcal{L}_{CC}{}^{ghost} \qquad (12.24)$$

$$\mathcal{L}_{Ufields} = -\tfrac{1}{4} F_U{}^{a\mu\nu}F_{U\mu\nu\nu} + \mathcal{L}_U{}^{ghost} + \mathcal{L}_U{}^{UHiggs} \qquad (12.25)$$

where $\mathcal{L}_U{}^{UHiggs}$ is discussed later. The ElectroWeak gauge bosons $W_\mu{}^a$, B_μ and B'_μ field tensors are:

$$F_W{}^a{}_{\mu\nu} = \partial W^a{}_\mu/\partial X^\nu - \partial W^a{}_\nu/\partial X^\mu + g_2 f^{abc}W^b{}_\mu W^c{}_\nu \qquad (12.26)$$

$$F_{B\mu\nu} = \partial B_\mu/\partial X^\nu - \partial B_\nu/\partial X^\mu \qquad (12.27)$$

and the Dark ElectroWeak gauge bosons $W'_\mu{}^a$ and B'_μ field tensors are:

$$F_{B'\mu\nu} = \partial B'_\mu/\partial X^\nu - \partial B'_\nu/\partial X^\mu$$

$$F'_W{}^a{}_{\mu\nu} = \partial W'^a{}_\mu/\partial X^\nu - \partial W'^a{}_\nu/\partial X^\mu + g_2 f^{abc}W'^b{}_\mu W'^c{}_\nu \qquad (12.28)$$

The U fields' tensor is:

$$F_U{}^a{}_{\mu\nu} = \partial U^a{}_\mu/\partial X^\nu - \partial U^a{}_\nu/\partial X^\mu + g_G f_4{}^{abc} U^b{}_\mu U^c{}_\nu \qquad (12.29)$$

where $f_4{}^{abc}$ are the U(4) algebra commutator constants.

$\mathcal{L}_{EW}{}^{ghost}$ contains the Faddeev-Popov ghost terms for the ElectroWeak $W_\mu{}^a$ gauge bosons. The complexon color gluon lagrangian $\mathcal{L}_{CCG}$ is defined by

$$\mathcal{L}_{CCG} = -\tfrac{1}{4}\, F_{CC}{}^{a\mu\nu}(X) F_{CC}{}^a{}_{\mu\nu}(X) \qquad (12.30)$$

where

$$F_{CC}{}^a{}_{\mu\nu} = \partial/\partial X_c{}^\nu\, A_C{}^a{}_\mu - \partial/\partial X_c{}^\mu\, A_C{}^a{}_\nu + g f_{su(3)}{}^{abc} A_C{}^b{}_\mu A_C{}^c{}_\nu \qquad (12.31)$$

where $A_C{}^a{}_\nu$ is the color gluon gauge field, g is the color coupling constant, and the $f_{su(3)}{}^{abc}$ are the SU(3) structure constants.

In addition $\mathcal{L}_C{}^{ghost}$ is the color SU(3) Faddeev-Popov ghost terms which we discuss later.[161] The mass sector $\mathcal{L}_{Mass}$ is presumably based on the Higgs Mechanism.which creates the fermion and ElectroWeak vector boson masses, and generation mixing.

The lagrangian is supplemented with the following condition on all complexon fields $\Phi_{...}$:[162]

$$\nabla_r\cdot\nabla_i\Phi... = 0 \qquad (12.32)$$

The subsidiary condition:

$$[\nabla_r\cdot\nabla_i - (\nabla_r{}^2\nabla_i{}^2)^{1/2}]\Omega... = 0 \qquad (12.33)$$

would guarantee a particle's real momentum is parallel to its imaginary momentum. The Faddeev-Popov Method can be used to implement the eq. 12.32 constraint as we show later.

12.3 U(4) Gauge Symmetry Breaking and Long Range Forces

Above we showed that there was good experimental evidence for a conserved Baryon Number B and we proceeded to develop a simple U(1) gauge theory that would imply Baryon Number conservation in a manner analogous to QED's implying electric charge conservation. In section 12.2 we used a new symmetry group local U(4) to generalize the one generation

---

[161] Faddeev-Popov gauge fixing is described below, and in appendix 19-A of Blaha (2011c), for the complexon Lorentz gauge. $\mathcal{L}_{CC}{}^{ghost}$ is the complexon color SU(3) constraint ghost terms defined through the Faddeev-Popov mechanism.

[162] These conditions implement the orthogonality of the real and imaginary parts of complexon 3-momentum.

Unified SuperStandard Model to a four generation Unified SuperStandard Model based on four conserved particle numbers: B, L, B_D, and L_D.[163]

We now assume in our construction that the four generation Unified SuperStandard Model has a local U(4) symmetry that is broken by mass terms gewnerated by the Higgs Mechanism.

Further, we will assume that the Higgs breakdown yields two massless (long range) fields which we associate with Baryon Number B and Dark Baryon Number B_D. The remaining fields acquire masses and generate short range forces.

We use the following U(4) diagonal matrices:

$$G_1 = \text{diag}(1, 1, 1, 1)$$
$$G_2 = \text{diag}(0, 1, 0, 0)$$
$$G_3 = \text{diag}(0, 0, 1, 0)$$
$$G_4 = \text{diag}(0, 0, 0, 1)$$

$$(12.34)$$

The U(4) algebra has 16 hermitean matrices that satisfy

$$G_i^\dagger = G_i \tag{12.35}$$

The particle numbers can be expressed in terms of the diagonal generators as

$$B \ = G_1 - G_2 - G_3 - G_4$$
$$B_D = G_2$$
$$L \ = G_3$$
$$L_D = G_4$$

$$(12.36)$$

The covariant derivatives have the general form:

$$D_{...\mu} = \partial / \partial X^\mu + ... - \tfrac{1}{2} i g_G \mathbf{G} \cdot \mathbf{U}_\mu \tag{12.37}$$

where the ellipses indicate the other details of the particular covariant derivative. We now wish to express the four gauge fields $U_i(X)$ for i = 1, 2, 3, 4 corresponding to the diagonal generators in terms of the fields of the four particle number gauge fields: B_μ, L_μ, $B_{D\mu}$, and $L_{D\mu}$.

$$U_{i\mu} = A_{ik} N_{k\mu} \tag{12.38}$$

where A_{ik} are the elements of a matrix of constants and

------

[163] Charge, although a conserved number, is a part of the ElectroWeak sector, account of which has already been taken.

$$N_\mu = \begin{bmatrix} B_\mu(X) \\ L_\mu(X) \\ B_{D\mu}(X) \\ L_{D\mu}(X) \end{bmatrix} \tag{12.39}$$

is a column vector consisting of the gauge fields corresponding to each of the conserved particle numbers.

The matrix A must have non-zero determinant so that eq. 12.38 can be inverted to express the particle number fields in terms of the four $U_i(X)$ gauge fields:

$$N_\mu = A^{-1}U_\mu \tag{12.40}$$

resulting in

$$B_\mu(X) = U_{1\mu} \tag{12.41}$$
$$L_\mu(X) = U_{1\mu} + U_{2\mu}$$
$$B_{D\mu}(X) = U_{1\mu} + U_{3\mu}$$
$$L_{D\mu}(X) = U_{1\mu} + U_{4\mu}$$

Then

$$D_{\dots\mu} = \partial/\partial X^\mu + \dots - \tfrac{1}{2}ig_G[\sum_{i=5}^{16} G_i U_{i\mu} + BB_\mu(X) + LL_\mu(X) + B_D B_{D\mu}(X) + L_D L_{D\mu}(X)] \tag{12.42}$$

where the particle numbers, which are analogous to the charges Q and Q' in ElectroWeak theory, are B, L, B_D, and L_D. They are expressed in terms of U(4) generators by eqs. 12.36.

In section 15.2 we describe the Higgs mass mechanism for Generation group gauge fields using the above notation.

13. Layer Group

This chapter[164] describes the four Layer Groups. The transformations of the Complex Lorentz group transformations lead to the internal symmetry Reality group: R = SU(3)⊗SU(2)⊗U(1)⊗SU(2)⊗U(1), which is the symmetry group of the 'original' Standard Model plus Dark SU(2)⊗U(1). The U(4) Generation group was shown to follow from the four particle number conservation laws,[165] and increased the Standard Model group to SU(3)⊗SU(2)⊗U(1)⊗SU(2)⊗U(1)⊗U(4).

13.1 New Conserved Particle Numbers

The four generations of enhanced form of the model lead to a new set of particle conservation laws. If we examine the form of the SU(3)⊗SU(2)⊗U(1)⊗SU(2)⊗U(1)⊗U(4) free fermion lagrangian terms of the Unified SuperStandard Model

$$\mathcal{L} \sim \overline{\psi}_\alpha \gamma^\mu D_{\mu\alpha\beta} \psi_\beta$$

we find that the SU(3)⊗SU(2)⊗U(1)⊗SU(2)⊗U(1)⊗U(4) symmetry of the terms not only yields the B, L, B_D, and L_D particle conservation laws but it also yields four quartets of conservation laws – one quartet for each of the four generations – totaling 16 conserved particle numbers.

The four 'conserved' layer numbers per generation are:

L_{iB} – The Baryon layer particle number for the ith generation
L_{iDB} – The Dark Baryon particle layer number for the ith generation
L_{iL} – The Lepton layer particle number for the ith generation
L_{iDL} – The Dark Lepton particle layer number for the ith generation

for each generation i = 1, 2, 3, 4. Fermions have positive L_{ia} = +1 values and anti-fermions have negative L_{ia} = –1 values for species a = 1, 2, 3, 4 (with the three color subspecies treated as part of one species.)

13.2 Four Layers of Fermions

These four sets of particle numbers lead to four U(4) Layer groups. The rationale for the new Layer groups is similar to that of the Generation group which was based on the four

---

[164] Much of this chapter appears in Blaha (2017c).
[165] Baryon number, Lepton number, and their Dark analogues.

particle numbers B, L, B_D, and L_D. The Generation group was based on U(4) rotations related to the four number operators B, L, B_D, and L_D in the one generation SuperStandard Model.

Similarly we base four Layer groups on the four quartets of conserved layer numbers. Each generation has a Layer group. Consequently, for each fermion species, *we assume the fermion in each generation acquires a Layer index and 'becomes' a set of four fermions in a Layer U(4) 4 representation.* Thus the set of four generations becomes four layers of four generations as shown in Figs. 13.1 and 13.2.

13.3 Layer Group Rotations

A U(4) rotation of the ith generation transforms the four i^{th} generation fermions of the four layers amongst each other (Fig. 13.3). The vertical rotations for generation k can be symbolized by:

$$
\begin{array}{ll}
G_k \uparrow & \text{Layer 4} \\
G_k & \text{Layer 3} \\
G_k & \text{Layer 2} \\
G_k \downarrow & \text{Layer 1}
\end{array}
$$

We assume layer 1 is the known layer of fermions.

Figure 13.1. The four layers of fermions. Each layer (oval) has four generations of fermions. Layer 1 is the layer that we have found experimentally. The 4th generation of layer 1, the Dark part of layer 1, and the remaining three more massive layers constitute Dark matter.

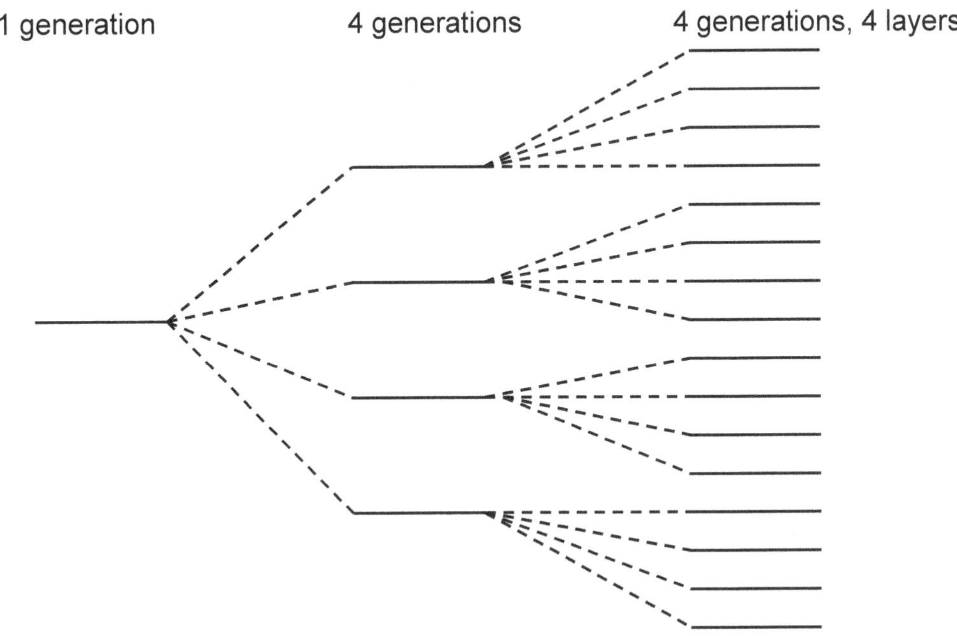

Figure 13.2. The 'splitting of a single generation fermion into four generations and then into four layers.

13.4 Layer Groups

We assume that each of the four Layer groups has associated gauge fields and particle interactions. Each generation in the four layers has a U(4) Layer symmetry group mixing the fermions in its generation across all four layers.(Fig. 13.3) Since the coefficients in Layer group transformations can be local functions, the new Layer groups are implemented with Yang-Mills fields.

It is important to note that the Layer particle numbers are independent of the baryon and lepton particle numbers that form the basis of the Generation group, and so the physics embodied in the Generation group is not the same as the physics of the Layer groups

Layer numbers are conserved under strong and electromagnetic interactions but broken by the Electromagnetic and Weak interactions as well as their Dark counterparts.

Further the gauge fields for SU(3)⊗SU(2)⊗U(1)⊗SU(2)⊗U(1)⊗U(4) must now be different for each layer.Thus interactions of these types between fermions in different layers is prevented. As a result the SuperStandard Model symmetry now is

$$[SU(3)\otimes SU(2)\otimes U(1)\otimes SU(2)\otimes U(1)\otimes U(4)\otimes U(4)]^4$$

where, for compactness, we place the four Layer groups within the quartic expression.

THE FERMION PERIODIC TABLE

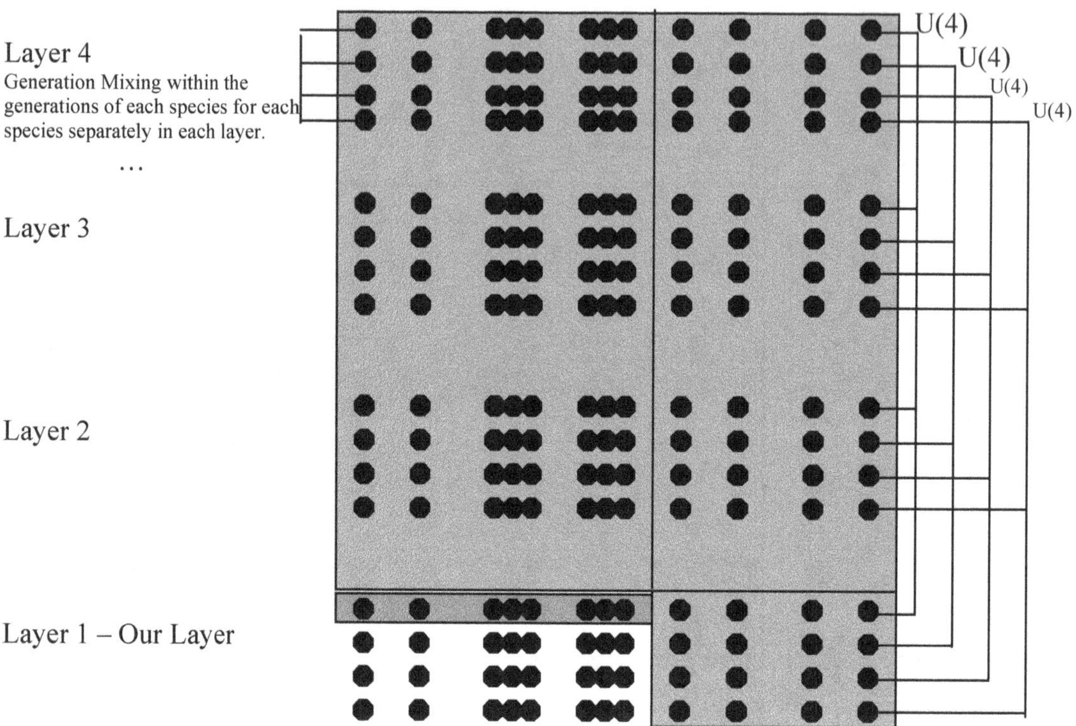

Figure 13.3. Partial example of pattern of particle transformations of the Generation group and of the Layer groups.. Dark parts of the periodic table are gray. Light parts are the known fermions with an additional, as yet not found, 4th generation shown. The lines on the left side show an example of the Generation mixing within one species. The Generation mixing applies to each species in each layer. The lines on the right side show the Layer mixing generation by generation.among all four layers for each generation individually.

13.5 Steps to Introduce the Layer Groups in the SuperStandard Model

To implement this new Layer symmetry we expand the SuperStandard Model lagrangian with the following steps:

1, All covariant derivatives must expand to incude four U(4) Layer gauge fields terms $V^{\mu}{}_{i}$. for i = 1, 2, 3, 4. The new terms are for four layers of

SU(3)⊗SU(2)⊗U(1)⊗SU(2)⊗U(1)⊗U(4) fields. Each gauge field then has an additionl index specifying its layer. The rationale for the choice of these sets of groups to be duplicated is that they, and only they, all play a necessary role in determining the structure of the Periodic Table of Fermions.

2. The new Layer groups are the same as in the Unified SuperStandard Model of Blaha (2017b). Each fermion particle field has an index labeling the layer of the particle making four layers of four generations of fermions.

3. Each layer should have its own set of Higgs particles (modulo mixing) contributing to fermion masses. A layer index number must be added to each Higgs field. One expects that the masses of fermions should be substantially larger for the three 'upper' layers beyond our layer.[166] Otherwise we would have found particles from these upper layers.

THE VECTOR BOSON PERIODIC TABLE

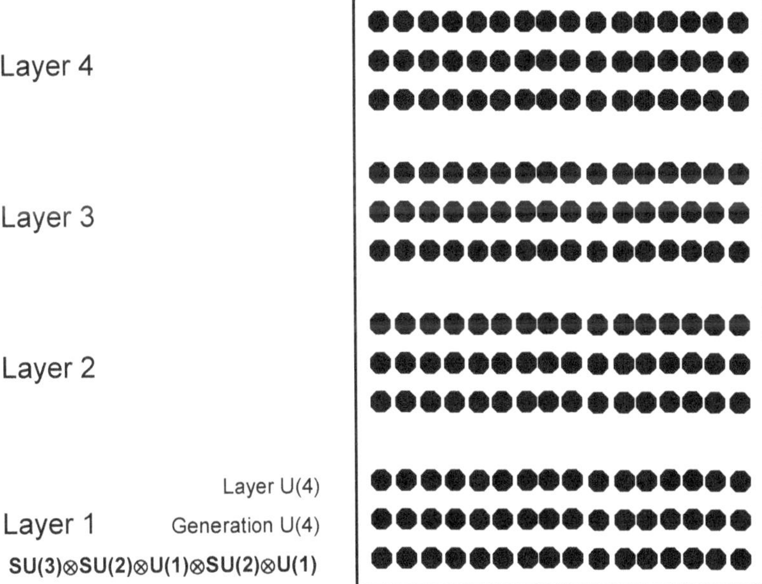

Figure 13.4. The known vector bosons are in the lowest row. The Layer groups are distributed by layer symbolically although they each straddle all four layers.

---

[166] The interplayed mixing of the particles in each generation between different layers may partly explain the vast increases seen in fermion masses as one goes from generation to generation in each species. The Higgs particles in different layers are different and a possible source of the growth of fermion masses.

The form of the "periodic tables" of fermion and vector bosons that results appears in Figs. 13.4 and 13.5. The anticipated[167] large breakdown of Layer groups symmetries causes the fermion masses of each generation of each species to be significantly different. *This difference may explain the large increase in masses in each species as one goes up from the lowest mass fermion in the species (such as the e mass in the charged lepton species) to the highest (known) mass in the species (such as the τ mass in the charged lepton species.)*

13.6 New Labeling of Fermion Periodic Table Particles

Fig. 13.5 has the Periodic Table of Fermions with rows and columns labeled with quadruplets of numbers. In the case of normal quark species, which each consist of a color triplet, a fifth integer would be needed to identify each color quark within a triplet: (T, S, L, G, C) where C = 1, 2, 3 identifies the color ("red, white or blue"). For example, the second generation, fourth layer, normal up-type red quark is (1, 3, 4, 2, 1) where we treat 'red' as having the value 1.

The quadruplet numbering patern is as follows. Species are numbered from 1 through 4 – separately for normal and Dark species: charged lepton, neutral lepton, three up-type quarks, three down-type quarks, Dark charged lepton, Dark neutral lepton, one Dark up-type quark species, and one Dark down-type quark species. Layers are numbered from 1 through 4 with our layer being layer 1. Generations are numbered from 1 through 4 from lightest to the heaviest.[168] (e, v_e, u and d constitute the known part of generation 1 of our layer.) We will call the quadruplet (or quintet) of a fermion its *ID number*. See Fig. 13.5.

Since the Periodic table of Fermions is 2-dimensional – like the Chemical Periodic Table of Elements one might ask: Why not use two numbers to identify fermions in a manner similar to the Chemistry table? The answer is related to the group structure of the Extended Standatd Model interactions. Except for T (the type of matter), the other three digits in a quadruplet are each related to a group. Thus G specifies the generation and thus the position of a fermion in the 4 of the U(4) Generation group. The integer L specifies the layer and thus the position of the fermion in the 4 of the form of the U(4) Layer group.

The integer S specifies the position of a fermion amongst the species. In Blaha (2017b) we saw that the four species of each type of matter follow from a U(4) group, that we called the Species Group which was derived from a consideration of the structure of Complex General Coordinate transformations. The U(4) Species gauge fields are denoted $A_S{}^\mu(x)$ in discussions below.

Thus the fermion quadruplet labeling is physically motivated by the group structure of the SuperStandard Model is now: $[SU(3) \otimes SU(2) \otimes U(1) \otimes SU(2) \otimes U(1) \otimes U(4) \otimes U(4)]^4$. The additional interactions will be explained later.

---

[167] Experimentally the top three layers of fermions have not been found. Therefore we expect that they have extremely large masses.
[168] The ordering by mass may not hold in the currently Dark part of the fermion spectrum.

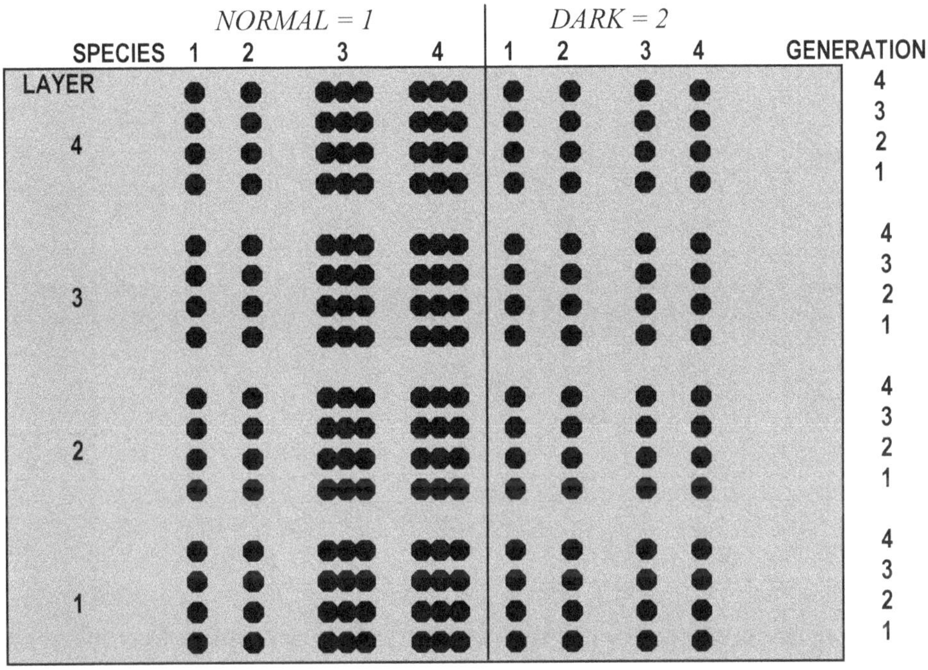

Figure 13.5. The Periodic Table of Fermions with each fermion identified by a quadruplet of integers: (T, S, L, G) where T identifies Normal or Dark, S identifies the species, G identifies the generation, and L identifies the layer. For example the electron has the triplet (1, 1, 1, 1), and the second generation, fourth layer, Dark up-type quark is (2, 3, 4, 2).

13.7 Number of Fermions and Bosons

The number of fermions and vector bosons now is 192 for both as Figs. 13.4 and 13.5 show. The equality of these numbers leads us to suggest a SuperSymmetry aspect to the Unified SuperStandard Model that we will consider later.

14. Equipartition Principle for Fermions and Gauge Fields

We now[169] will suggest a rationale for the dominant abundance of Dark mass-energy:

14.1 Equipartition Principle for Particle Degrees of Freedom

In a closed system at equilibrium the thermal energy of a system is equally partitioned (distributed) among its degrees of freedom. This Equipartition Principle is well known. The application of this principle to the beginning of the universe *when all particles were massless* and all symmetries were unbroken suggests that the distribution of mass-energy should be the same for all degrees of freedom at that time. Thus there should be approximately equal numbers and energies of 192 fermions and 192 vector bosons with the same fraction of the total thermal energy.

We now estimate the relative proportion of Normal and Dark matter in the universe at its beginning based on this Equipartition Principle.

14.2 Proportion of Dark Mass-Energy in the Universe

First we note that 8 of the 12 fermion species (counting quarks of each color as a separate species) in layer 1 – the layer with which we are familiar are Normal matter fermions. (Our discussion is based on our SuperStandard Model.) Four of the 12 first layer species are Dark.

The other three fermion layers are all Dark from our point of view since they have not been detected. Thus we find that 40 of the 48 species are Dark yielding a *percentage of Dark Matter equal to 40/48 = 83.33%.* (The same counting could have been done by counting individual fermions with the same results.)

Recent studies of the proportion of Dark Matter in the universe have yielded two estimates: 84.5% by Aghanim et al in Astronomy and Astrophysics 1303;5062 and 81.5% from a NASA fit to various models.

Thus our estimate based on our fermion Equipartition Principle is midway between these experimental estimates.

Two possibilities emerge with respect to the present proportion of Dark Matter:

1. The percentage has not changed from the Beginning and the approximate estimates are

---

[169] Some of the material in this section appeared in Blaha (2016a).

slightly off. The lack of change could be due to the extremely small decay rates of the fermions in the higher layers.

2. The percentage of matter in the upper layers has decreased due to decay and so the current proportion may be somewhat below 83.33%.

14.3 Proportion of Dark Mass-Energy in the Universe

We know of 12 of the 192 vector bosons in the SuperStandard Model and 24 fermions. Thus we find a total of 348 out of 384 particles are Dark yielding a Dark mass-energy of 91% of the universes mass-energy at the beginning of the universe.

The sum of Dark energy in the universe currently has been estimated to be 68% of the total energy and the energy of Dark Matter is estimated to be 26.8%. The total is 95% - a value that compares favorably with our above approximate estimate of 91%. The dfference in these values can be attributed to various factors. One distinct possibility is the decay of Dark mass-energy from higher layers to the known layer in the 13.8 billion years since the Big Bang.

15. Higgs Symmetry Breaking

There are five broken symmetries in the Unified SuperStandard Model: ElectroWeak symmetry breaking, Generation group symmetry breaking, Layer groups symmetry breaking, 'Rotation of Interactions' Θ-group symmetry breaking, and Species group symmetry breaking. In this chapter we will consider ElectroWeak symmetry breaking, Generation group symmetry breaking, and Layer groups symmetry breaking. Each group has a different set of gauge fields[170] in each of the four layers of fermions and vector bosons. While the parameters of these groups differ from layer to layer *the overall form of the symmetry breakng for each layer can be assumed to be the same.*

When we consider the 'Rotation of Interactions' Θ-group and the Species group later we will examine symmetry breaking for those groups. *These groups have the same gauge fields in all layers – a necessary assumption given their nature.*

15.1 ElectroWeak Symmetry Breaking using the Higgs Mechanism

In this section, and the following sections, we consider the case of symmetry breaking in one layer. Symmetry breaking in other layers are assumed to have the same pattern.

15.1.1 Higgs Mechanism for ElectroWeak Gauge Field Masses

We require that there is one massless field, the electromagnetic field coupled to electric charge and three massive fields that receive masses via the Higgs Mechanism which breaks $SU(2) \otimes U(1)$ symmetry[171]. The Dark sector is assumed to be analogous to the normal particle sector in this respect since it has a $SU(2) \otimes U(1)$ symmetry in the Unified SuperStandard Model. The massive fields become the short range Weak interactions. These assumptions in the construction of the Unified SuperStandard Model conform to the Leibniz Minimax Principle.

We assume that a doublet Higgs field exists with two components:

$$\eta = \begin{bmatrix} \varphi_+ \\ \varphi_0 \end{bmatrix} \tag{15.1}$$

---

[170] This would appear to be a necessary assumption since the higher layers have not been found experimentally although ultra-high masses and/or ultra-low coupling constants might explain the lack of experimental evidence for higher layers at present.

[171] We will use the standard Higgs Mechanism formulation because of its familiarity rather than the PseudoQuantum formulation that we developed in an earlier chapter.

with conjugate Higgs doublet

$$\eta' = \begin{bmatrix} \varphi_0 \\ -\varphi_- \end{bmatrix} \qquad (15.2)$$

The Higgs sector lagrangian has the form:

$$\mathcal{L}_{EW}^{Higgs} = (\partial\eta^\dagger/\partial X^\mu)(\partial\eta/\partial X^\mu) - \lambda(\eta^\dagger\eta - \rho^2)^2 + \mathcal{L}_{EW}^{Higgs}{}_{EWMasses} \qquad (15.3)$$

where the symmetry breaking follows from the choice of unitary gauge

$$\eta = \begin{bmatrix} 0 \\ \rho \end{bmatrix} \qquad (15.4)$$

where ρ is a real field. Then the covariant derivative of η is

$$D_{...\mu}\,\eta = \{\partial/\partial X^\mu + ... + ig t\cdot W_\mu + + ig' t_0 W_{0\mu}\} \begin{bmatrix} 0 \\ \rho \end{bmatrix} \qquad (15.5)$$

where g and g' are coupling constants, t_0.is a ½ the identity matrix, and the **t** matrices are ½ the vector of Pauli matrices.The ellipses indicate additional indices and additional terms respectively.

Then

$$D_{...\mu}\eta = \begin{bmatrix} \tfrac{1}{2}ig\rho(W_{1\mu} - iW_{2\mu}) \\ \partial\rho/\partial X^\mu - \tfrac{1}{2}ig\rho(\cos\theta_W)^{-1}Z_\mu \end{bmatrix} \qquad (15.6)$$

where θ_W is the Weinberg angle and

$$W_3^\mu = Z^\mu\cos\theta_W + A^\mu\sin\theta_W$$
$$W_0{}^\mu = -Z^\mu\sin\theta_W + A^\mu\cos\theta_W \qquad (15.7)$$

From eq. 15.6 we find the corresponding Higgs field kinetic terms in the lagrangian are

$$(D_{...\mu}\eta)^\dagger D_{...}{}^\mu\eta = \partial\rho/\partial X^\mu\partial\rho/\partial X_\mu + g^2\rho^2[W_1{}^\mu W_{1\mu} + W_2{}^\mu W_{2\mu}]/4 + g^2\rho^2 Z^\mu Z_\mu/(2\cos\theta_W)^2 \qquad (15.8)$$

with the $W_1{}^\mu$ and $W_2{}^\mu$ and Z^μ gauge bosons acquiring masses and the electromagnetic field A^μ massless.

15.1.2 ElectroWeak Higgs Mechanism Generation of Fermion Masses

We now consider the ElectroWeak Higgs Mechanism for the eight species of fermions (four species of "normal" matter and four species of Dark Matter). We shall consider the mass terms for the four normal species which is the same as that of the four Dark species except for the values in the various species mass matrices. Therefore we define the initial 4-vector for the generations of the normal species by

$$\Psi_s = \begin{bmatrix} \psi_{11} \\ \psi_{12} \\ \psi_{13} \\ \psi_{14} \\ \dots \\ \psi_{41} \\ \psi_{42} \\ \psi_{43} \\ \psi_{44} \end{bmatrix} \tag{15.9}$$

where ψ_{ki} is the generation index for the ith generation of the kth species. ψ_{k1} is the wave function for the 1st generation, ψ_{k4} is the 4th generation member of the kth species, and we omit other indices in the interests of clarity. The normal fermion species are ordered: charged lepton ($k = 1$), up-type quark, neutral lepton, and down-type quark ($k = 4$). Other indices of these wave functions are surpressed in the interests of clarity. A 4th generation fermion of any species is yet to be found experimentally.

We assume that a "double" doublet Higgs field exists with four components:

$$\eta = \begin{bmatrix} \varphi_{1+} \\ \varphi_{10} \\ \varphi_{2+} \\ \rho_{20} \end{bmatrix} \tag{15.10}$$

with conjugate Higgs doublet

$$\eta' = \begin{bmatrix} \varphi_{10} \\ -\varphi_{1-} \\ \varphi_{20} \\ -\rho_{2-} \end{bmatrix} \qquad (15.11)$$

The Higgs sector lagrangian has the form:

$$\mathcal{L}_{EW}^{Higgs} = (\partial\eta^{\dagger}/\partial X^{\mu})(\partial\eta/\partial X^{\mu}) - \lambda(\eta^{\dagger}\eta - \rho^2)^2 + \mathcal{L}_{EW}^{Higgs}{}_{FermionMasses} \qquad (15.12)$$

where the symmetry breaking follows from the choice of unitary gauge

$$\eta = \begin{bmatrix} 0 \\ \rho_1 \\ 0 \\ \rho_2 \end{bmatrix} \qquad \eta' = \begin{bmatrix} \rho_1 \\ 0 \\ \rho_2 \\ 0 \end{bmatrix} \qquad (15.13)$$

where η is a real field quadruplet.

The lagrangian density mass term for the four normal fermion species is

$$\mathcal{L}_{EW}^{Higgs}{}_{FermionMasses} = \Sigma_{\alpha,\beta} \{\psi_{kL\alpha}\,\overline{\eta}m_{k\alpha\beta}\,\psi_{kR\beta} + \psi_{kL\alpha}\,\overline{\eta'}m'_{k\alpha\beta}\psi_{kR\beta}\} + c.c. \qquad (15.14)$$

where $m_{k\alpha\beta}$ and $m'_{k\alpha\beta}$ are complex constant matrices, where $\alpha, \beta = 1, \dots, 4$, and where the second term is the double conjugation doublet used to produce a total mass term invariant under weak hypercharge. The total fermion lagrangian mass terms are

$$\mathcal{L}^{Higgs}{}_{FermionMasses} = \mathcal{L}_{U}^{UHiggs}{}_{FermionMasses} + \mathcal{L}_{EW}^{Higgs}{}_{FermionMasses} \qquad (16.48)$$

plus the iota mass which is negligible except perhaps for neutrinos. $\mathcal{L}_{EW}^{Higgs}{}_{FermionMasses}$ is the contribution of ElectroWeak Higgs Mechanism to the fermion masses. Using the vacuum expectation value of η in eq. 15.13 we find

$$\mathcal{L}_{EW}^{Higgs}{}_{FermionMasses} = \Sigma_{\alpha,\beta} \{\overline{\psi}_{2L\alpha}\,\rho_1 m_{2\alpha\beta}\psi_{2R\beta} + \overline{\psi}_{4L\alpha}\,\rho_2 m_{4\alpha\beta}\psi_{4R\beta} + $$
$$+ \overline{\psi}_{1L\alpha}\,\rho_1 m'_{1\alpha\beta}\psi_{1R\beta} + \overline{\psi}_{3L\alpha}\,\rho_2 m'_{3\alpha\beta}\psi_{3R\beta}\} + c.c$$
$$. \qquad (15.15)$$

giving mass terms for all four species. **There is an implicit color summation over the color quarks in each generation and quark species.**

The four mass matrices m_1, … , m_4 are all complex, constant mass matrices. They can be brought to diagonal form D_k with non-negative values by U(4) matrices A_k and B_k:

$$A_k m_k B_k^{-1} = D_k \qquad (15.16)$$

or

$$m_k = A_k^{-1} D_k B_k \qquad (15.17)$$

for k= 1, …, 4.

We now note, that although, D_k has non-negative real values, down-type quarks are all tachyonic and up-type quarks are all non-tachyonic, and neutral leptons are all tachyonic and charged leptons are all Dirac non-tachyonic leptons, due to their lagrangian kinetic terms.

We further note that $m_k^\dagger m_k$ is hermitean, and A_k and B_k are members of U(4) as is D_k for k = 1, 2, 3, 4, with the result that all matrices m_k are members of the U(4) group.

We can use these U(4) transformations A_k and B_k to define the sixteen "physical" fermion fields:

$$\psi_{2L\alpha}\,\rho_1 m_{2\alpha\beta}\psi_{2R\beta} + \psi_{4L\alpha}\,\rho_2 m_{4\alpha\beta}\psi_{4R\beta} + \psi_{1L\alpha}\,\rho_1 m_{1\alpha\beta}\psi_{1R\beta} + \psi_{3L\alpha}\,\rho_2 m_{3\alpha\beta}\psi_{3R\beta} \qquad (15.18)$$

$$= (\psi_{2L}A_2^{-1})_\alpha \rho_1 D_{2\alpha\beta}(B_2\psi_{2R})_\beta + (\psi_{4L}A_4^{-1})_\alpha \rho_2 D_{4\alpha\beta}(B_4\psi_{4R})_\beta +$$
$$+ (\psi_{1L}A_1^{-1})_\alpha \rho_1 D_{1\alpha\beta}(B_1\psi_{1R})_\beta + (\psi_{3L}A_3^{-1})_\alpha \rho_2 D_{3\alpha\beta}(B_3\psi_{3R})_\beta$$

$$= \psi_{2L phys\alpha}\rho_1 D_{2\alpha\beta}\psi_{2R phys\beta} + \psi_{4L phs\alpha}\rho_2 D_{4\alpha\beta}\psi_{4R phys\beta} + \psi_{1L phys\alpha}\rho_1 D_{1\alpha\beta}\psi_{1R phys\beta} + \psi_{3L phs\alpha}\rho_2 D_{3\alpha\beta}\psi_{3R phys\beta}$$

Species:
　　Up-type quarks　　　down-type quarks　　　charged leptons　　　neutral leptons

The preceding discussion with changes in the values of constants and constant matrices holds for Dark Matter also where the Dark quarks and leptons acquire ElectroWeak mass terms. The Dark Matter species ElectroWeak mass terms, with the subscript D signifying Dark Matter, are[172]

$$\psi_{D2Lphys\alpha}\rho_{D1} D_{D2\alpha\beta}\psi_{D2Rphys\beta} + \psi_{D4Lphs\alpha}\rho_{D2} D_{D4\alpha\beta}\psi_{D4Rphys\beta} + \psi_{D1Lphys\alpha}\rho_{D1} D_{D1\alpha\beta}\psi_{D1Rphys\beta} + \psi_{D3Lphs\alpha}\rho_{D2} D_{D3\alpha\beta}\psi_{D3Rphys\beta}$$

Dark　　　　　　　　　　　　　　　　　　　　　　　　　　　　　　(15.19)
Species: Up-type quarks　　　down-type quarks　　　charged leptons　　　neutral leptons

---

[172] Dark quarks are Color SU(3) singlets by *assumption* in the Unified SuperStandard Model..

15.1.3 ElectroWeak Higgs Mechanism for Gauge Fields Including Dark Gauge Fields

In section 15.1.2 we introduced a double doublet (a quadruplet) to derive the symmetry breaking fermion mass spectrum of the four normal fermion families and saw that a similar derivation should be operative for Dark fermions by Ockham's Razor – it is the simplest possible approach to handling the broken $SU(2) \otimes U(1)$ symmetry that the Dark ElectroWeak sector has in the Unified SuperStandard Model.

Now we generalize the SU(2)$\otimes$U(1) ElectroWeak symmetry breaking via the Higgs Mechanism that gives mass to three SU(2)$\otimes$U(1) gauge bosons to include Dark SU(2)$\otimes$U(1) symmetry breaking via the Higgs Mechanism that will give mass to three Dark SU(2)$\otimes$U(1) gauge bosons and also yield a massless gauge boson analogous to the electromagnetic field. This assumption is consistent both with the Leibniz Minimax Principle and Ockham's Razor having both simplicity within the context of the Unified SuperStandard Model and achieving a maximal effect with a minimal extension of the formalism. We will therefore be constructing an $SU(2) \otimes U(1) \otimes SU(2) \otimes U(1)$ Higgs Mechanism symmetry breaking derivation using the double doublets of section 15.1.2.

We assume that a "double" doublet Higgs field exists with four components:

$$\eta = \begin{bmatrix} \varphi_{1+} \\ \varphi_{10} \\ \varphi_{2+} \\ \rho_{20} \end{bmatrix} \tag{15.20}$$

with conjugate Higgs doublet

$$\eta' = \begin{bmatrix} \varphi_{10} \\ -\varphi_{1-} \\ \varphi_{20} \\ -\rho_{2-} \end{bmatrix} \tag{15.21}$$

The Higgs sector lagrangian has the form:

$$\mathcal{L}_{EW}^{Higgs} = (\partial\eta^{\dagger}/\partial X^{\mu})(\partial\eta/\partial X^{\mu}) - \lambda(\eta^{\dagger}\eta - \rho^2)^2 + \mathcal{L}_{EW}^{Higgs}{}_{EWMasses} \tag{15.22}$$

where the symmetry breaking follows from the choice of unitary gauge (similar in form to eq. 15.13):

$$\eta = \begin{bmatrix} 0 \\ \rho_1 \\ 0 \\ \rho_2 \end{bmatrix} \qquad \eta' = \begin{bmatrix} \rho_1 \\ 0 \\ \rho_2 \\ 0 \end{bmatrix} \tag{15.23}$$

where η is a real field quadruplet.

The covariant derivative of η in the unitary gauge is

$$
D_{...\mu}\,\eta = \begin{bmatrix} \{\partial/\partial X^\mu + ... + ig\mathbf{t}\cdot\mathbf{W}_\mu + + ig't_0 W_{0\mu}\} & 0 \\ & \\ 0 & \{\partial/\partial X^\mu + ... + ig_D\mathbf{t}\cdot\mathbf{W}_{D\mu} + + ig_D't_0 W_{D0\mu}\} \end{bmatrix} \begin{bmatrix} 0 \\ \rho_1 \\ 0 \\ \rho_2 \end{bmatrix} \tag{15.24}
$$

where g and g' are coupling constants with g_D and g_D' their Dark equivalents, t_0.is a ½ the identity matrix, the $\mathbf{t}$ matrices are ½ the vector of Pauli matrices, and the zeros and derivative terms are all 2×2 submatrices.The ellipses indicate additional indices and additional terms respectively.

Then

$$
D_{...\mu}\,\eta = \begin{bmatrix} \tfrac{1}{2}ig\rho_1(W_{1\mu} - iW_{2\mu}) \\ \partial\rho_1/\partial X^\mu - \tfrac{1}{2}ig\rho_1(\cos\theta_W)^{-1}Z_\mu \\ \tfrac{1}{2}ig_D\rho_2(W_{D1\mu} - iW_{D2\mu}) \\ \partial\rho_2/\partial X^\mu - \tfrac{1}{2}ig_D\rho_2(\cos\theta_{WD})^{-1}Z_{D\mu} \end{bmatrix} \tag{15.25}
$$

where θ_{WD} is the Dark Weinberg angle and

$$
\begin{aligned}
W_3{}^\mu &= Z^\mu\cos\theta_W + A^\mu\sin\theta_W \\
W_0{}^\mu &= -Z^\mu \sin\theta_W + A^\mu\cos\theta_W \\
W_{D3}{}^\mu &= Z_D{}^\mu\cos\theta_{WD} + A_D{}^\mu\sin\theta_{WD} \\
W_{D0}{}^\mu &= -Z_D{}^\mu \sin\theta_{WD} + A_D{}^\mu\cos\theta_{WD}
\end{aligned} \tag{15.26}
$$

From eq. 15.25 we find the corresponding Higgs field kinetic terms in the lagrangian are

$$
(D_{...\mu}\,\eta)^\dagger D^\mu\eta = \partial\rho_1/\partial X^\mu\partial\rho_1/\partial X_\mu + g^2\rho_1^2[W_1{}^\mu W_{1\mu} + W_2{}^\mu W_{2\mu}]/4 + g^2\rho_1^2\, Z^\mu Z_\mu/(2\cos\theta_W)^2 +
$$
$$
+ \partial\rho_2/\partial X^\mu\partial\rho_2/\partial X_\mu + g_D{}^2\rho_2^2[W_{D1}{}^\mu W_{D1\mu} + W_{D2}{}^\mu W_{D2\mu}]/4 + g_D{}^2\rho_2^2\, Z_D{}^\mu Z_{D\mu}/(2\cos\theta_{WD})^2 \tag{15.27}
$$

with the $W_1{}^\mu$ and $W_2{}^\mu$ and Z^μ gauge bosons acquiring masses and the electromagnetic field A^μ massless, and with the Dark $W_{D1}{}^\mu$ and $W_{D2}{}^\mu$ and $Z_D{}^\mu$ gauge bosons acquiring masses and the Dark electromagnetic field $A_D{}^\mu$ massless.

Thus the SU(2)⊗U(1)⊗SU(2)⊗U(1) Higgs Mechanism symmetry breaking yields the desired results. Note it is consistent with a massless Dark electromagnetic field.

15.1.4 Higgs Mechanism for Tachyons

The Higgs mechanism is currently the favored mechanism for spontaneous symmetry breaking and to give masses to fermions and bosons. The nature of the free tachyon lagrangian terms,

$$\mathcal{L}_{free} = \psi_T{}^\dagger i\gamma^0\gamma^5(\gamma^\mu \partial/\partial x^\mu + m_0)\psi_T(x) \tag{15.28}$$

where m_0 is a possible bare mass, requires a Higgs sector that contributes to a tachyon mass through spontaneous symmetry breaking having the general form:

$$\mathcal{L}_{Higgs} = \tfrac{1}{2}\partial\phi/\partial x^\mu \partial\phi/\partial x_\mu - \psi_T{}^\dagger i\gamma^0\gamma^5\psi_T\phi - V(\phi) \tag{15.29}$$

where

$$V(\phi) = g^2(\phi^2 - (\delta m)^2)^2 \tag{15.30}$$

Note the quadratic term in ϕ – the "mass" term has the negative sign of a tachyon – again showing that tachyons are a feature of modern physics. In the present case the quartic term "stabilizes" the tachyon field which then can be shifted to the minimum of the potential.

The spontaneous symmetry breaking resulting from the potential $V(\phi)$ causes the mass of the tachyon to change to

$$m = m_0 \pm \delta m \tag{15.31}$$

A choice of vacuum state corresponding to the positive sign in eq. 15.31 causes the tachyon mass to increase. A choice of the vacuum state corresponding to the negative sign causes the tachyon mass to decrease, and could cause m to become negative. However this event would not make the tachyon into a normal particle. Rather it would essentially transform the tachyon into its antiparticle.

15.2 Generation Group Symmetry Breaking using the Higgs Mechanism

15.2.1 Generation U(4) Gauge Symmetry Breaking and Long Range Forces

Earlier we showed that there was good experimental evidence for a conserved Baryon Number B and we proceeded to develop a simple U(1) gauge theory that would imply Baryon Number conservation in a manner analogous to QED's implying electric charge conservation. We used a new local U(4) symmetry group to generalize the one generation SuperStandard

Model to a four generation SuperStandard Model based on four conserved particle numbers: B, L, B_D, and L_D.[173]

We now assumethat the four generation SuperStandard Model has a local U(4) symmetry that is broken by mass terms gewnerated by the Higgs Mechanism . Further, we will assume that the Higgs breakdown yields two massless (long range) fields which we associate with Baryon Number B and Dark Baryon Number B_D. The remaining fields acquire masses and generate short range forces.

We use the following U(4) diagonal matrices:

$$G_1 = \text{diag}(1, 1, 1, 1) \qquad (15.32)$$
$$G_2 = \text{diag}(0, 1, 0, 0)$$
$$G_3 = \text{diag}(0, 0, 1, 0)$$
$$G_4 = \text{diag}(0, 0, 0, 1)$$

The U(4) algebra has 16 hermitean matrices that satisfy

$$G_i^\dagger = G_i \qquad (15.33)$$

The particle numbers can be expressed in terms of the diagonal generators as

$$B = G_1 - G_2 - G_3 - G_4 \qquad (15.34)$$
$$B_D = G_2$$
$$L = G_3$$
$$L_D = G_4$$

The covariant derivatives have the general form *for both normal and Dark U gauge fields*:

$$D_{...\mu} = \partial/\partial X^\mu + ... - \tfrac{1}{2}ig_G\mathbf{G}\cdot U_\mu \qquad (15.35)$$

where the ellipses indicate the other details of the particular covariant derivative. We now wish to express the four gauge fields $U_i(X)$ for i = 1, 2, 3, 4 corresponding to the diagonal generators in terms of the fields of the four particle number gauge fields: B_μ, L_μ, $B_{D\mu}$, and $L_{D\mu}$.

$$U_{i\mu} = A_{ik} N_{k\mu} \qquad (15.36)$$

where A_{ik} are the elements of a matrix of constants and

---

[173] Charge, although a conserved number, is a part of the ElectroWeak sector, account of which has already been taken.

$$N_\mu = \begin{bmatrix} B_\mu(X) \\ L_\mu(X) \\ B_{D\mu}(X) \\ L_{D\mu}(X) \end{bmatrix} \tag{15.37}$$

is a column vector consisting of the gauge fields corresponding to each of the conserved particle numbers.

The matrix A must have non-zero determinant so that eq. 15.36 can be inverted to express the particle number fields in terms of the four $U_i(X)$ gauge fields:

$$N_\mu = A^{-1}U_\mu \tag{15.38}$$

Resulting in

$$B_\mu(X) = U_{1\mu} \tag{15.39}$$
$$L_\mu(X) = U_{1\mu} + U_{2\mu}$$
$$B_{D\mu}(X) = U_{1\mu} + U_{3\mu}$$
$$L_{D\mu}(X) = U_{1\mu} + U_{4\mu}$$

Then

$$D_{...\mu} = \partial/\partial X^\mu + ... - \tfrac{1}{2}ig_G\left[\sum_{i=5}^{16} G_iU_{i\mu} + BB_\mu(X) + LL_\mu(X) + B_DB_{D\mu}(X) + L_DL_{D\mu}(X)\right] \tag{15.40}$$

where the particle numbers, which are analogous to the charges Q and Q' in ElectroWeak theory, are B, L, B_D, and L_D. They are expressed in terms of U(4) generators by eqs. 15.34.

15.2.2 Higgs Mass Mechanism for U(4) Generation Gauge Fields

We now require that there are two massless fields, one coupled to Baryon number and one coupled to Dark Baryon number. The Dark sector is assumed to be analogous to the normal particle sector in this respect. There are fourteen remaining fields that acquire masses and longitudinal components. These fields become short range, ultra-weak generational forces. The masses they acquire through the Higgs Mechanism are presumably very large as these gauge particles have not been found experimentally.[174]

We assume that a scalar Higgs field exists which is a U(4) vector with four components corresponding to the fermion generations. It is an SU(2)⊗U(1)⊗SU(3) ElectroWeak scalar. Its lagrangian density is

$$\mathcal{L}_U^{UHiggs} = (\partial\eta^\dagger/\partial X^\mu)(\partial\eta/\partial X^\mu) - \lambda(\eta^\dagger\eta - \rho^2)^2 + \mathcal{L}_U^{UHiggs}{}_{FermionMasses}$$

---

[174] Section 16.4 discusses this topic in more detail.

where $\mathcal{L}_U{}^{UHiggs}{}_{FermionMasses}$ are the fermion masses produced by the U Higgs Mechanism and where we choose a unitary gauge in which

$$\eta = \begin{bmatrix} 0 \\ \rho_1 \\ 0 \\ \rho_2 \end{bmatrix} \tag{15.41}$$

where ρ_1 and ρ_2 are real fields. Then the covariant derivative of η is

$$D_{\ldots\mu}\eta = \{\partial/\partial X^\mu + \ldots - \tfrac{1}{2}ig_G[\Sigma \mathbf{G_i}U_{i\mu} + BB_\mu(X) + LL_\mu(X) + B_D B_{D\mu}(X) + L_D L_{D\mu}(X)]\} \begin{bmatrix} 0 \\ \rho_1 \\ 0 \\ \rho_2 \end{bmatrix}$$

$$\tag{1542}$$

The sum over i is from 5 through 16, and $[\mathbf{G_i}]_{jk}$ is the jk element of $\mathbf{G_i}$. Then

$$D_{\ldots\mu}\eta = \begin{bmatrix} -\tfrac{1}{2}ig_G\{\rho_1\Sigma[\mathbf{G_i}]_{12}U_{i\mu} + \rho_2\Sigma[\mathbf{G_i}]_{14}U_{i\mu}\} \\ \partial\rho_1/\partial X^\mu - \tfrac{1}{2}ig_G\rho_1 L_\mu - \tfrac{1}{2}ig_G\{\rho_1\Sigma[\mathbf{G_i}]_{22}U_{i\mu} + \rho_2\Sigma[\mathbf{G_i}]_{24}U_{i\mu}\} \\ -\tfrac{1}{2}ig_G\{\rho_1\Sigma[\mathbf{G_i}]_{32}U_{i\mu} + \rho_2\Sigma[\mathbf{G_i}]_{34}U_{i\mu}\} \\ \partial\rho_2/\partial X^\mu - \tfrac{1}{2}ig_G\rho_2 L_{D\mu} - \tfrac{1}{2}ig_G\{\rho_1\Sigma[\mathbf{G_i}]_{42}U_{i\mu} + \rho_2\Sigma[\mathbf{G_i}]_{44}U_{i\mu}\} \end{bmatrix} \tag{15.43}$$

$$= \begin{bmatrix} -\tfrac{1}{2}ig_G\Sigma\{\rho_1[\mathbf{G_i}]_{12} + \rho_2[\mathbf{G_i}]_{14}\}U_{i\mu} \\ \partial\rho_1/\partial X^\mu - \tfrac{1}{2}ig_G\rho_1 L_\mu - \tfrac{1}{2}ig_G\rho_2\Sigma[\mathbf{G_i}]_{24}U_{i\mu} \\ -\tfrac{1}{2}ig_G\Sigma\{\rho_1[\mathbf{G_i}]_{32} + \rho_2[\mathbf{G_i}]_{34}\}U_{i\mu} \\ \partial\rho_2/\partial X^\mu - \tfrac{1}{2}ig_G\rho_2 L_{D\mu} - \tfrac{1}{2}ig_G\rho_1\Sigma[\mathbf{G_i}]_{42}U_{i\mu} \end{bmatrix} \tag{15.44}$$

since the generators $\mathbf{G_i}$ have zeroes along their diagonals for $i = 5, \ldots, 16$.

From eq. 15.43 we find the corresponding Higgs field kinetic terms in the lagrangian are

$$(D_{\ldots\mu}\eta)^\dagger D_{\ldots}{}^\mu \eta = \partial\rho_1/\partial X^\mu \partial\rho_1/\partial X_\mu + \partial\rho_2/\partial X^\mu \partial\rho_2/\partial X_\mu + g_G{}^2\rho_1{}^2 L_\mu L^\mu/4 + g_G{}^2\rho_2{}^2 L_{D\mu} L_D{}^\mu/4 + \ldots$$

$$\tag{15.45}$$

Note there are differing mass squared terms for the Lepton $(g_G^2\rho_1^2/4)$ and Dark Lepton $(g_G^2\rho_2^2/4)$ gauge fields making them short range fields with the likelihood of very large masses much beyond ElectroWeak gauge field masses, and with an ultra weak coupling constant g_G as suggested by the "experimental" coupling for the Baryonic force.

The Baryonic and Dark Baryonic gauge fields are massless and thus long range although their coupling constant appears to be ultra weak – much below the gravitational coupling constant G.

We now turn to calculating the remaining terms in eq. 15.45 that determine the masses of the remaining 14 gauge fields. We begin by assigning matrix elements for the remaining hermitean U(4) generators:

$$[G_5]_{ik} = \delta_{i1}\delta_{k2} + \delta_{i2}\delta_{k1}$$
$$[G_6]_{ik} = -i\delta_{i1}\delta_{k2} + i\delta_{i2}\delta_{k1}$$
$$[G_7]_{ik} = \delta_{i1}\delta_{k3} + \delta_{i3}\delta_{k1}$$
$$[G_8]_{ik} = -i\delta_{i1}\delta_{k3} + i\delta_{i3}\delta_{k1}$$
$$[G_9]_{ik} = \delta_{i1}\delta_{k4} + \delta_{i4}\delta_{k1}$$
$$[G_{10}]_{ik} = -i\delta_{i1}\delta_{k4} + i\delta_{i4}\delta_{k1}$$
$$[G_{11}]_{ik} = \delta_{i2}\delta_{k3} + \delta_{i3}\delta_{k2}$$
$$[G_{12}]_{ik} = -i\delta_{i2}\delta_{k3} + i\delta_{i3}\delta_{k2}$$
$$[G_{13}]_{ik} = \delta_{i2}\delta_{k4} + \delta_{i4}\delta_{k2}$$
$$[G_{14}]_{ik} = -i\delta_{i2}\delta_{k4} + i\delta_{i4}\delta_{k2}$$
$$[G_{15}]_{ik} = \delta_{i3}\delta_{k4} + \delta_{i4}\delta_{k3}$$
$$[G_{16}]_{ik} = -i\delta_{i3}\delta_{k4} + i\delta_{i4}\delta_{k3}$$

(15.46)

Then completing eq. 15.45 using eq. 15.44 we find

$$(D_{\dots\mu}\eta)^{\dagger} D_{\dots}{}^{\mu}\eta = \partial\rho_1/\partial X^{\mu}\partial\rho_1/\partial X_{\mu} + \partial\rho_2/\partial X^{\mu}\,\partial\rho_2/\partial X_{\mu} + g_G^2\rho_1^2 L_{\mu}\,L^{\mu}/4 + g_G^2\rho_2^2 L_{D\mu}\,L_D{}^{\mu}/4 +$$
$$+ (g_G/2)^2\rho_1^2(U_5^2 + U_6^2) + (g_G/2)^2\rho_2^2(U_9^2 + U_{10}^2) + (g_G/2)^2\rho_1^2(U_{11}^2 +$$
$$+ U_{12}^2) + + (g_G/2)^2(\rho_1^2 + \rho_2^2)(U_{13}^2 + U_{14}^2) + + (g_G/2)^2\rho_2^2(U_{15}^2 + U_{16}^2)$$

(15.47)

up to total divergences which generate surface terms which we discard and assuming that all fields satisfy the gauge condition

$$\partial U_i{}^{\mu}/\partial X^{\mu} = 0$$

(15.48)

Note that there are no mass terms for $U_7(X)$ and $U_8(X)$ as well as $B_{\mu}(X)$ and $B_{D\mu}(X)$ due to our choice of unitary gauge eq. 15.41. Consequently there are four massless long range fields and 12 gauge fields that acquire masses of three different values: $(g_G/2)\rho_{10}$, $(g_G/2)\rho_{20}$, and $(g_G/2)(\rho_{10}^2 + \rho_{20}^2)^{\frac{1}{2}}$ where ρ_{10} and ρ_{20} ar the vacuum expectation values of ρ_1 and ρ_2 respectively. The fields $U_7(X)$ and $U_8(X)$ are not "diagonal" and thus appear in the fermion sector as terms connecting fermions in different generations within the four species of normal fermions and within the four species of Dark fermions.[175] Therefore they do not change the values of any of the four types of particle numbers.

Based on an earlier estimate (eq. 12.6) the ultra weak value of the coupling constant is

$$g_G = (4\pi\alpha_B)^{1/2} \approx 1.218 \, (Gm_H^2)^{1/2} \tag{15.49}$$

The gauge field $B_\mu(X)$ is now part of a quadruplet of long range fields. It is the massless, long range field discussed in chapter 12. The two non-diagonal long range forces, being between different generations of a species and having an ultra-weak coupling constant are not of great consequence because of the short lifetime of the higher generations of a species. Therefore, despite their long range, they have only the "shortest" time to exert an inter-generation force before a higher generation particle decays.

Since we expect the other massive fields to have very large masses (and thus very large Higgs field vacuum expectation values), and ultra weak coupling, they are not likely to be experimentally found for the foreseeable future.

15.2.3 Impact of the Generation U(4) Higgs Mechanism on Fermion Generation Masses

The fermion masses of the charged lepton, and the up-type quark, and down-type quark species' generations all show a rapid increase of mass with the generation. For example the u quark mass is a few MeV while the t quark (third generation) has a mass of about 170 GeV/c. The ratio of these masses is about 170,000. While one can account for this great difference by the judicious choices of Higgs' parameter values, when one considers the Generation group, and its associated numerical quantities: ultra weak coupling, very large U particle masses – perhaps of the order of hundreds or thousands of GeV/c, and the corresponding very large Higgs particle vacuum expectation values in this U gauge field sector[176] then the differences in fermion masses within a species become more understandable and natural from a Leibniz Principle perspective.

Thus the popular view that the ElectroWeak gauge field symmetry breaking is solely via ElectroWeak Higgs fields is not part of our SuperStandard Model unless the U(4) sector is removed. In our model there are two sets of contributions to fermion symmetry breaking: ElectroWeak Higgs particles symmetry breaking, and Generation group U(4) Higgs particles symmetry breaking. The Generation group causes each species to break into four generations.

In the conventional Standard Model the breakup of species into generations is inserted "by hand." It is not a consequence of the existence of $SU(2) \otimes U(1)$ symmetry or symmetry breaking. In our approach the U(4) Generation group causes the appearance of generations. We

---

[175] Neutral lepton, charged lepton, up-type quark and down-type quark plus the four corresponding Dark species..
[176] They are not the Higgs particles of the $SU(2) \otimes U(1)$ ElectroWeak sector.

base the existence of the Generation group[177] on the four conserved particle numbers. Leibniz' Principle and Ockham's Razor then lead to the above construction/derivation.

15.2.4 Generation Group Higgs Mechanism for Fermion Masses

We now consider the Generation group Higgs Mechanism for the eight species of fermions (four species of "normal" matter[178] and four species of Dark Matter). We shall consider the mass terms for the four normal species, which is the same in form as that of the four Dark species except for the values in the various species mass matrices. Therefore we define the initial 4-vector for the generations of the normal species by

$$\Psi_s = \begin{bmatrix} \psi_{11} \\ \psi_{12} \\ \psi_{13} \\ \psi_{14} \\ \cdots \\ \psi_{41} \\ \psi_{42} \\ \psi_{43} \\ \psi_{44} \end{bmatrix} \tag{15.50}$$

where ψ_{ki} is the generation index for the ith generation of the kth species. ψ_{k1} is the wave function for the 1st generation, ψ_{k4} is the 4th generation member of the kth species, and we omit other indices in the interests of clarity. The normal fermion species order here are: charged lepton (k = 1), up-type quark, neutral lepton, and down-type quark (k = 4). Other indices of these wave functions are suppressed in the interests of clarity. A 4th generation fermion of any species is yet to be found experimentally. The lagrangian density mass terms for the four normal fermion species are

$$\mathcal{L}_U^{UHiggs}{}_{FermionMasses} = \Sigma_{k,\alpha,\beta} \, \bar{\psi}_{kL\alpha} \, \eta_k m_{k\alpha\beta} \psi_{kR\beta} + c.c. \tag{15.51}$$

where $m_{k\alpha\beta}$ is complex constant matrix, where k labels species, and where α, $\beta = 1, \ldots , 4$. The total of fermion lagrangian mass terms is

---

[177] In earlier books we suggested the fermion generations might be the result of a wormhole to another 4-dimensional universe. The new approach is simpler and more consistent with known facts – thus more consistent with the Leibniz Minimax Principle.

[178] Not taking account of the three color quark species of normal matter yet.

$$\mathcal{L}^{Higgs}_{FermionMasses} = \mathcal{L}_U^{UHiggs}{}_{FermionMasses} + \mathcal{L}_{EW}^{Higgs}{}_{FermionMasses} \qquad (15.52)$$

where $\mathcal{L}_{EW}^{Higgs}$ is the contribution of ElectroWeak Higgs Mechanism to the fermion masses (discussed above). Using the vacuum expectation value of η in eq. 15.37 we find

$$\mathcal{L}_U^{UHiggs}{}_{FermionMasses} = \Sigma_{\alpha,\beta} \left\{ \bar{\Psi}_{2L\alpha}\, \rho_1 m_{2\alpha\beta} \Psi_{2R\beta} + \bar{\Psi}_{4L\alpha}\, \rho_2 m_{4\alpha\beta} \Psi_{4R\beta} \right\} + c.c. \qquad (15.53)$$

giving mass terms tor the up-type and down-type quark species but not to lepton species. There is an implicit color summation over the color quarks in each generation and quark species. *Qualitatively eq. 15.53 could be viewed as corresponding to the experimentally known largeness of quark masses relative to lepton masses in each generation of normal matter.*

The mass matrices $m_2 = [m_{2\alpha\beta}]$ and $m_4 = [m_{4\alpha\beta}]$ are both complex, constant mass matrices. They can be brought to diagonal form with non-negative values by U(4) matrices A_k and B_k:

$$A_2 m_2 B_2^{-1} = D_2 \qquad (15.54)$$
$$A_4 m_4 B_4^{-1} = D_4$$

or

$$m_2 = A_2^{-1} D_2 B_2 \qquad (15.55)$$
$$m_4 = A_4^{-1} D_4 B_4$$

We now note, that although, both D_2 and D_4 have non-negative real values, down-type quarks are all tachyonic and up-type quarks are all non-tachyonic due to their lagrangian kinetic terms as seen in chapter 5.

We further note that $m_2^\dagger m_2$ and $m_4^\dagger m_4$ are hermitean, and A_k and B_k are members of U(4) as is D_k for $k = 2,4$, with the result that m_2 and m_4 are also both members of the U(4) group. Thus

$$m_2^{-1} = m_2^\dagger \qquad (15.56)$$
$$m_4^{-1} = m_4^\dagger$$

We can express the mass matrices in terms of U(4) generators

$$m_2 = \Sigma G_i m_{2i} \qquad (15.57)$$
$$m_4 = \Sigma G_i m_{4i}$$

$$m_2^{-1} = m_2^\dagger = \Sigma G_i m_{2i}{}^* \qquad (15.58)$$
$$m_4^{-1} = m_4^\dagger = \Sigma G_i m_{4i}{}^*$$

since the matrices G_i are all hermitean, where $\{m_{2i}\}$ and $\{m_{4i}\}$ are each a set of sixteen complex constants.

While we do not as yet know the 4th generation fermions or their masses, the third generation quarks have masses that are far greater than the 1st and 2nd generation quarks or their sum suggesting that the trace of m_2 and m_4 is dominated by the 4th generation mass of the two quark species with a similar situation holding for the two Dark quark species.   Therefore if we take the trace of m_2 and m_4 then it seems probable based on the trend of the generations that the 4th generation mass dominates the trace:

$$D_{24} \approx \text{tr } D_2 \tag{15.59}$$
$$D_{44} \approx \text{tr } D_4$$

We can use these A_k and B_k U(4) transformations to define the eight "physical" (up to further ElectroWeak Higgs Mehanism effects) up-type and down-type quark generations fields:

$$\bar{\Psi}_{2L\alpha}\,\rho_1 m_{2\alpha\beta}\Psi_{2R\beta} + \bar{\Psi}_{4L\alpha}\,\rho_2 m_{4\alpha\beta}\Psi_{4R\beta} = (\bar{\Psi}_{2L}A_2^{-1})_\alpha \rho_1 D_{2\alpha\beta}(B_2\Psi_{2R})_\beta + (\bar{\Psi}_{4L}A_4^{-1})_\alpha \rho_2 D_{4\alpha\beta}(B_4\Psi_{4R})_\beta$$

$$= \bar{\Psi}_{2L\text{phys}\alpha}\,\rho_1 D_{2\alpha\beta}\Psi_{2R\text{phys}\beta} + \bar{\Psi}_{4L\text{phs}\alpha}\,\rho_2 D_{4\alpha\beta}\Psi_{4R\text{phys}\beta} \tag{15.60}$$

Species: up-type quarks   down-type quarks

The preceding discussion with changes in the values of constants and constant matrices holds for Dark Matter also where the Dark quarks acquire mass terms but the Dark leptons do not. The Dark Matter species mass terms, with the subscript D signifying Dark Matter, are

$$= \bar{\Psi}_{D2L\text{phys}\alpha}\,\rho_{D1} D_{D2\alpha\beta}\Psi_{D2R\text{phys}\beta} + \bar{\Psi}_{D4L\text{phs}\alpha}\,\rho_{D2} D_{D4\alpha\beta}\Psi_{D4R\text{phys}\beta} \tag{15.61}$$

Dark Species: up-type quarks   down-type quarks

15.3 Layer Groups Higgs Symmetry Breaking

The Layer groups symmetry is also broken. *In this section we use the Higgs Mechanism to implement symmetry breaking for the lowest generation of the four layers for Layer group '1'.[179] The symmetry breaking for each of the other three generations of the four layers is assumed to have a similar form.*

---

[179] We omit the designation of the Layer group and gauge fields. The presentation is only for generation '1' in this section (whose species contain e, v_e, u and d type fermions.)

15.3.1 Layer Group Higgs Mechanism Contributions to Layer Gauge Field Masses

In this section we will determine a Layer group's Higgs contributions to gauge field masses. (The fermion mass contributions from the various Higgs interactions are discussed in section 15.3.2. We will see that all[180] layers have Layer group Higgs contributions to the layer's fermion masses.) There are four Layer groups – one for each of the four fermion generations. We will consider the case of the generation '1' Layer group below.[181] The cases of other layers is assumed to be similar.

We begin by assuming that four scalar Higgs fields η_a exist, which are each U(4) Layer group 4-vectors. The η_a are SU(2)⊗U(1)⊗SU(3) ElectroWeak and Strong Interaction scalars. The lagrangian density terms for the ath η_a are[182]

$$\mathcal{L}_V{}^{Higgs} = (\partial\eta_a{}^\dagger/\partial X^\mu)(\partial\eta_a/\partial X^\mu) - \lambda(\eta_a{}^\dagger\eta_a - \rho_a{}^2)^2 + \mathcal{L}_V{}^{Higgs}{}_{FermionMasses} \qquad (15.62)$$

where $\mathcal{L}_V{}^{Higgs}{}_{FermionMasses}$ are the fermion masses produced by the Layer Higgs Mechanism and where we set the η Layer 4-vector with Higgs field components to

$$\eta_a = \begin{bmatrix} \rho_{1a} \\ \rho_{2a} \\ \rho_{3a} \\ \rho_{4a} \end{bmatrix} \qquad (15.63)$$

where ρ_{1a}, ρ_{2a}, ρ_{3a} and ρ_{4a} are real fields.[183] Then the covariant derivative of η_a is

$$D_{\dots\mu}\eta_a = \{\partial/\partial X^\mu + \dots - \tfrac{1}{2}ig_V[\Sigma_i\mathbf{G}_{Li}\mathbf{V}_{i\mu} + \mathbf{G}_{L1}\mathbf{V}_{1\mu} + \mathbf{G}_{L2}\mathbf{V}_{2\mu} + \mathbf{G}_{L3}\mathbf{V}_{3\mu} + \mathbf{G}_{L4}\mathbf{V}_{4\mu}]\} \begin{bmatrix} \rho_{1a} \\ \rho_{2a} \\ \rho_{3a} \\ \rho_{4a} \end{bmatrix}$$

$$(15.64)$$

The sum over i is from 5 through 16 (non-diagonal matrices), and $[\mathbf{G}_{Li}]_{jk}$ is the jk^{th} element of $\mathbf{G}_{Li}$. Then

---

[180] All Layers have Layer group Higgs contributions to avoid massless Layer group gauge fields. *Since transitions between fermions layers have not been found we assume all Layer groups gauge fields are very massive and/or the coupling constants are ultra-small.*

[181] We omit an index number for the Layer group and other quantities in the interest of simplicity.

[182] Again we use the standard formulation of the Higgs Mechanism because of its familiarity.

[183] Each field ρ_{ia} can be expressed as a pseudoquantum field: $\rho_{ia} = \varphi_{1ia} + \varphi_{2ia}$ where φ_{1ia} has the vacuum expectation value ρ_{i0a} for i = 1, ... , 4. Thus our pseudoquantum field theory version is implemented easily.

$$D_{...\mu}\eta_a = \begin{bmatrix} \partial\rho_{1a}/\partial X^\mu - \tfrac{1}{2}ig_V\{\rho_{1a}\mathbf{G}_{L1}\mathbf{V}_{1\mu} + \rho_{2a}\Sigma[\mathbf{G}_{Li}]_{11}\mathbf{V}_{i\mu} + \rho_{2a}\Sigma[\mathbf{G}_{Li}]_{12}\mathbf{V}_{i\mu} + \rho_{3a}\Sigma[\mathbf{G}_{Li}]_{13}\mathbf{V}_{i\mu} + \rho_{4a}\Sigma[\mathbf{G}_{Li}]_{14}\mathbf{V}_{i\mu}\} \\ \partial\rho_{2a}/\partial X^\mu - \tfrac{1}{2}ig_V\{\rho_{2a}\mathbf{G}_{L2}\mathbf{V}_{2\mu} + \rho_{1a}\Sigma[\mathbf{G}_{Li}]_{21}\mathbf{V}_{i\mu} + \rho_{2a}\Sigma[\mathbf{G}_{Li}]_{22}\mathbf{V}_{i\mu} + \rho_{3a}\Sigma[\mathbf{G}_{Li}]_{23}\mathbf{V}_{i\mu} + \rho_{4a}\Sigma[\mathbf{G}_{Li}]_{24}\mathbf{V}_{i\mu}\} \\ \partial\rho_{3a}/\partial X^\mu - \tfrac{1}{2}ig_V\{\rho_{3a}\mathbf{G}_{L3}\mathbf{V}_{3\mu} + \rho_{1a}\Sigma[\mathbf{G}_{Li}]_{31}\mathbf{V}_{i\mu} + \rho_{2a}\Sigma[\mathbf{G}_{Li}]_{32}\mathbf{V}_{i\mu} + \rho_{3a}\Sigma[\mathbf{G}_{Li}]_{33}\mathbf{V}_{i\mu} + \rho_{4a}\Sigma[\mathbf{G}_{Li}]_{34}\mathbf{V}_{i\mu}\} \\ \partial\rho_{4a}/\partial X^\mu - \tfrac{1}{2}ig_V\{\rho_{4a}\mathbf{G}_{L4}\mathbf{V}_{4\mu} + \rho_{1a}\Sigma[\mathbf{G}_{Li}]_{41}\mathbf{V}_{i\mu} + \rho_{2a}\Sigma[\mathbf{G}_{Li}]_{42}\mathbf{V}_{i\mu} + \rho_{3a}\Sigma[\mathbf{G}_{Li}]_{43}\mathbf{V}_{i\mu} + \rho_{4a}\Sigma[\mathbf{G}_{Li}]_{44}\mathbf{V}_{i\mu}\} \end{bmatrix}$$

$$(15.65)$$

$$= \begin{bmatrix} \partial\rho_{1a}/\partial X^\mu - \tfrac{1}{2}ig_V\{\rho_{1a}\mathbf{G}_{L1}\mathbf{V}_{1\mu} + \rho_{2a}\Sigma[\mathbf{G}_{Li}]_{12}\mathbf{V}_{i\mu} + \rho_{3a}\Sigma[\mathbf{G}_{Li}]_{13}\mathbf{V}_{i\mu} + \rho_{4a}\Sigma[\mathbf{G}_{Li}]_{14}\mathbf{V}_{i\mu}\} \\ \partial\rho_{2a}/\partial X^\mu - \tfrac{1}{2}ig_V\{\rho_{2a}\mathbf{G}_{L2}\mathbf{V}_{2\mu} + \rho_{1a}\Sigma[\mathbf{G}_{Li}]_{21}\mathbf{V}_{i\mu} + \rho_{3a}\Sigma[\mathbf{G}_{Li}]_{23}\mathbf{V}_{i\mu} + \rho_{4a}\Sigma[\mathbf{G}_{Li}]_{24}\mathbf{V}_{i\mu}\} \\ \partial\rho_{3a}/\partial X^\mu - \tfrac{1}{2}ig_V\{\rho_{3a}\mathbf{G}_{L3}\mathbf{V}_{3\mu} + \rho_{1a}\Sigma[\mathbf{G}_{Li}]_{31}\mathbf{V}_{i\mu} + \rho_{2a}\Sigma[\mathbf{G}_{Li}]_{32}\mathbf{V}_{i\mu} + \rho_{4a}\Sigma[\mathbf{G}_{Li}]_{34}\mathbf{V}_{i\mu}\} \\ \partial\rho_{4a}/\partial X^\mu - \tfrac{1}{2}ig_V\{\rho_{4a}\mathbf{G}_{L4}\mathbf{V}_{4\mu} + \rho_{1a}\Sigma[\mathbf{G}_{Li}]_{41}\mathbf{V}_{i\mu} + \rho_{2a}\Sigma[\mathbf{G}_{Li}]_{42}\mathbf{V}_{i\mu} + \rho_{3a}\Sigma[\mathbf{G}_{Li}]_{43}\mathbf{V}_{i\mu}\} \end{bmatrix} \quad (15.66)$$

since the generators $\mathbf{G}_i$ have zeroes along their diagonals for $i = 5, \ldots , 16$.

From eq. 15.66 we find the corresponding a[th] Higgs field kinetic terms in the lagrangian are

$$(D_{...\mu}\eta_a)^\dagger D_{...}{}^\mu\eta_a = \partial\rho_{1a}/\partial X^\mu\,\partial\rho_1/\partial X_\mu + \partial\rho_{2a}/\partial X^\mu\,\partial\rho_2/\partial X_\mu + \partial\rho_{3a}/\partial X^\mu\,\partial\rho_3/\partial X_\mu + \partial\rho_{4a}/\partial X^\mu\,\partial\rho_4/\partial X_\mu +$$
$$+ g_V^2\rho_{1a}^2\mathbf{V}_{1\mu}\mathbf{V}_1{}^\mu/4 + g_V^2\rho_{2a}^2\mathbf{V}_{2\mu}\mathbf{V}_2{}^\mu/4 + g_V^2\rho_{3a}^2\,\mathbf{V}_{3\mu}\mathbf{V}_3{}^\mu/4 + g_V^2\rho_{4a}^2\,\mathbf{V}_{4\mu}\mathbf{V}_4{}^\mu/4 + \ldots$$
$$(15.67)$$

We now turn to calculating the remaining terms in eq. 15.67 that determine the masses of the remaining 14 gauge fields. We begin by assigning matrix elements for the remaining hermitean U(4) generators:

$$[\mathbf{G}_{L5}]_{ik} = \delta_{i1}\delta_{k2} + \delta_{i2}\delta_{k1} \qquad (15.68)$$
$$[\mathbf{G}_{L6}]_{ik} = -i\delta_{i1}\delta_{k2} + i\delta_{i2}\delta_{k1}$$
$$[\mathbf{G}_{L7}]_{ik} = \delta_{i1}\delta_{k3} + \delta_{i3}\delta_{k1}$$
$$[\mathbf{G}_{L8}]_{ik} = -i\delta_{i1}\delta_{k3} + i\delta_{i3}\delta_{k1}$$
$$[\mathbf{G}_{L9}]_{ik} = \delta_{i1}\delta_{k4} + \delta_{i4}\delta_{k1}$$
$$[\mathbf{G}_{L10}]_{ik} = -i\delta_{i1}\delta_{k4} + i\delta_{i4}\delta_{k1}$$
$$[\mathbf{G}_{L11}]_{ik} = \delta_{i2}\delta_{k3} + \delta_{i3}\delta_{k2}$$
$$[\mathbf{G}_{L12}]_{ik} = -i\delta_{i2}\delta_{k3} + i\delta_{i3}\delta_{k2}$$
$$[\mathbf{G}_{L13}]_{ik} = \delta_{i2}\delta_{k4} + \delta_{i4}\delta_{k2}$$
$$[\mathbf{G}_{L14}]_{ik} = -i\delta_{i2}\delta_{k4} + i\delta_{i4}\delta_{k2}$$
$$[\mathbf{G}_{L15}]_{ik} = \delta_{i3}\delta_{k4} + \delta_{i4}\delta_{k3}$$
$$[\mathbf{G}_{L16}]_{ik} = -i\delta_{i3}\delta_{k4} + i\delta_{i4}\delta_{k3}$$

Then completing eq. 15.67 using eq. 15.66 we find

$$(D_{...\mu}\eta_a)^\dagger D_{...}{}^\mu\eta_a = \partial\rho_{1a}/\partial X^\mu\,\partial\rho_1/\partial X_\mu + \partial\rho_{2a}/\partial X^\mu\,\partial\rho_{2a}/\partial X_\mu + \partial\rho_{3a}/\partial X^\mu\,\partial\rho_{3a}/\partial X_\mu + \partial\rho_{4a}/\partial X^\mu\,\partial\rho_{4a}/\partial X_\mu +$$
$$+ g_V^2\rho_{1a}^2\mathbf{V}_{1\mu}\mathbf{V}_1{}^\mu/4 + g_V^2\rho_{2a}^2\,\mathbf{V}_2^2/4 + g_V^2\rho_{3a}^2\,\mathbf{V}_3^2/4 + g_V^2\rho_{4a}^2\,\mathbf{V}_4^2/4 +$$

$$+ (g_V/2)^2(\rho_{1a}^2 + \rho_{2a}^2)(V_5^2 + V_6^2) + (g_V/2)^2(\rho_{1a}^2 + \rho_{3a}^2)(V_7^2 + V_8^2) +$$
$$+ (g_V/2)^2(\rho_{1a}^2 + \rho_{4a}^2)(V_9^2 + V_{10}^2) + (g_V/2)^2(\rho_{2a}^2 + \rho_{3a}^2)(V_{11}^2 + V_{12}^2) +$$
$$+ (g_V/2)^2(\rho_{2a}^2 + \rho_{4a}^2)(V_{13}^2 + V_{14}^2) + (g_V/2)^2(\rho_{3a}^2 + \rho_{4a}^2)(V_{15}^2 + V_{16}^2)$$

$$(15.69)$$

up to total divergences, which generate surface terms which we discard, and also assuming that all fields satisfy the gauge condition

$$\partial V_{ia}^{\mu}/\partial X^{\mu} = 0 \qquad (15.70)$$

Eq.15.69 shows all Layer groups gauge fields have masses. The combination of an ultra-weak coupling constant and very large gauge field masses results in extremely weak interactions between the fields in each layer, which leads to almost independent layers of normal and Dark fermions. Thus the Darkness! They result in very rare decays between layers, and very weak interactions between fermions in different layers. The higher layers with presumably much more massive fermions are thus well "insulated" from our layer. Thus they are Dark to us as well.

We estimate Layer groups gauge field masses to be very large – of the order of many TeV or they would have been detected at CERN by now. Their detection must await the construction of much more powerful accelerators. *The "non-diagonal" Layer gauge fields are the means by which we may hope to eventually find fermions of the higher layers.*

15.3.2 Layer Group Higgs Mechanism Contributions to Fermion Masses

The fermion masses of the charged lepton, and the up-type quark, and down-type quark species' generations all show a rapid increase of mass with the generation. For example the u quark mass is a few MeV while the t quark (third generation) has a mass of about 170 GeV/c. The ratio of these masses is about 170,000. While one can account for this great difference by the judicious choices of Higgs' parameter values, when one considers the Layer groups and their associated numerical quantities: ultra-weak coupling, very large Layer gauge field masses – perhaps of the order of hundreds or thousands of GeV/c, then a large difference in particle masses between layers is understandable and natural.

The form of the fermion lagrangian mass terms for the charged lepton species, generation '1' Layer group L_1 due to Higgs symmetry breaking is

$$\mathcal{L}^{Higgs}_{FermionMasses,\, l\,=\,'1',e} = \Sigma_{\delta,\gamma} \bar{\Psi}_{L1'e\delta}\, \eta_{'1e}m_{'1e\delta\gamma}\Psi_{R'1'e\gamma} + \Sigma_{\delta,\gamma} \bar{\Psi}_{DL'1'e\delta}\, \eta_{D'1'e}m_{D'1'e\delta\gamma}\Psi_{DR'1'e\gamma}$$

$$(15.71)$$

The indices δ and γ label *layer* rows and columns. The fields labeled η (with subscripts) are Higgs fields that have non-zero vacuum expectation values. Replacing 'e' with another species designator gives the terms for another species in the normal or Dark sectors. Replacing '1' with another generation number '2', '3', or '4' specifies a different generation corresponding to a different one of the other three Layer groups.

15.4 The Total Fermion Mass Matrix

The total fermion mass matrix lagrangian terms are

$$\mathcal{L}^{Higgs}_{FermionMasses} = \mathcal{L}_{EWFermionMasses} + \mathcal{L}_{UFermionMasses} + \mathcal{L}_{VtotFermionMasses} + \mathcal{L}_{\Theta\text{-}SymmetryFermionMasses} +$$
$$+ \mathcal{L}_{GravSpeciesFermionMasses} + c.c. \qquad (15.72)$$

where $\mathcal{L}_{EW}^{Higgs}$ is the contribution of ElectroWeak Higgs Mechanism to the fermion masses, $\mathcal{L}_{UFermionMasses}$ is the Generation group contribution, $\mathcal{L}_{VtotFermionMasses}$ is the sum of the four Layer groups contributions, $\mathcal{L}_{\Theta\text{-}SymmetryFermionMasses}$ is the Θ-Symmetry 'Rotation of Interactions' (discussed later) contribution, and $\mathcal{L}_{GravSpeciesFermionMasses}$ is the Gravitational Species group (discussed later) contribution.

The mass matrices in eq. 15.72 are complex, constant mass matrices that can be totaled and brought to diagonal form with non-negative values by U(4) matrices. (They cannot be separately diagonalized.) The resulting diagonalized mass matrices are the mass matrices of the physical fermions.

15.5 Four Generation CKM Matrix

The four generation generalization for a single layer of the CKM matrix[184]

$$C_4 = A_2 B_4^{-1} \qquad (15.73)$$

is a constant 4×4 U(4) Generation group matrix. Redefining field phases it can be reduced to an SU(4) matrix. It appears in the charge-changing quark current

$$J^{\mu}_{charged} = \overline{\psi}_{4Lphs}\gamma^{\mu}C_4\psi_{2Lphys} \qquad (15.74)$$

Thus a specific Generation group matrix determines the mixing between the generations. This matrix may be expected to generate CP violation. C_4 can be constrcted as a product of SU(4) factors with nine arbitrary parameters.

15.6 Four Layer CKM-like Matrix

The four layer 16×16 CKM-like matrices would have been four blocks of 4×4 CKM-like matrices along the diagonal for each of the leptons, quarks, Dark leptons and Dark quarks *IF* the Layer groups did not mix the four layers of fermions. However since the Layer groups do mix the four layers of fermions, the four[185] 16×16 CKM-like matrices C_{4i} (for i = leptonic,

---

[184] M. Kobayashi and K. Maskawa, Prog. Theo. Phys. **49**, 652 (1975) and references therein.
[185] For leptons, quarks, Dark leptons and Dark quarks.

quark, Dark leptonic and Dark quark) are not block diagonal.[186] Each CKM-like matrix is a 16×16 non-block-diagonal, constant, complex matrix.

The charged ElectroWeak currents have the form

$$J^{\mu}_{i} = \overline{\psi}_{physa}\gamma^{\mu}[C_i]_{ab}\psi_{physb} \tag{15.75}$$

with sums over the indices a and b. They may be expected to generate CP violation.

---

[186] Thus attempts to guess the form of the CKM-like matrices based on simple group theory hypotheses are unlikely to be successful. In view of the contributions of five contributions to the total fermion mass matrix (eq. 15.72), and the need to diagonalize the total mass matrix, partial diagonalizations (for example, based solely on ElectroWeak symmetry breaking) are not physically meaningful.

16. Faddeev-Popov Method for Gauge Fields

16.1 Faddeev-Popov Method for 'Normal' Yang-Mills Gauge Fields

The Faddeev-Popov Method for normal Yang-Mills fields (those with real-valued parameters) has been described in numerous textbooks. For example, the reader may read Huang (1992) or Kaku (1993).

16.2 Pure Gauge Complexon Path Integral Formulation and Faddeev-Popov Method

The path integral formalism for complexon non-abelian, pure, Yang-Mills fields differs significantly from the conventional gauge field path integral formalism. The path integral for a complexon gauge field can be written symbolically as:

$$Z(J^\mu) = N\!\int\! DA_C \Delta_{FP}(A_C)\delta(F(A_C))\Delta_C(A_C)\delta(F_C(A_C))\exp\{i\!\int\! d^7y[\mathscr{L} + J^\mu(y)A_{C_\mu}(y)]\} \qquad (16.1)$$

where $\delta(F(A_C))$ specifies the gauge, $\Delta_{FP}(A_C)$ is its Faddeev-Popov determinant; and $\delta(F_C(A_C))$ specifies the complexon condition (eqs. 2.123a and 2.124) with $\Delta_C(A_C)$ the Faddeev-Popov determinant for the complexon condition. In both cases the Faddeev-Popov determinant can be calculated in the standard way.[187]

First we consider the gauge fixing delta function. Note that it can be written as a delta function in the gauge times a determinant:

$$\delta(F(A_C^\omega)) = \delta(\omega - \omega_0)|\det \delta F(A_{C_\mu}^\omega(x))/\delta\omega(x)|^{-1}\big|_{F(A_C)=0} \qquad (16.2)$$

where ω_0 is a reference gauge, where

$$\begin{aligned} A_{C\,\mu}^{a\,\omega}(x) &= A_{C\,\mu}^{a}(x) - g^{-1}D_\mu\omega^a + f^{abc}\,\omega^b(x)A_{C\,\mu}^{c}(x) \\ &= A_{C\,\mu}^{a}(x) + \delta A_{C\,\mu}^{a\,\omega}(x) \end{aligned} \qquad (16.3)$$

and

$$\text{Re } F_{C\,\mu k}^{a\,\omega} = \text{Re } F_{C\,\mu k}^{a} + f^{abc}\,\omega^b(x)F_{C\,\mu k}^{c} \qquad (16.4)$$

---

[187] See for example Huang (1992).

$$= \mathrm{Re}\, F_C{}^a{}_{\mu k} + \delta(\mathrm{Re}\, F_C{}^a{}_{\mu k}{}^\omega)$$

under an infinitesimal gauge transformation, and where

$$\Delta_{FP}(A_C) = \left| \det \delta F(A_{C\mu}{}^\omega(x))/\delta\omega(x) \right|_{F(A_C)=0,\,\omega=0} \tag{16.5}$$

We will choose the complexon Lorentz gauge to evaluate the Faddeev-Popov determinant:

$$F^a(A_C) = D_\mu A_C{}^{a\mu}(x) = 0 \tag{16.6}$$

We find

$$F^a(A_{C\mu}{}^\omega(x)) = D^\mu(A_C{}^a{}_\mu(x) - g^{-1}D_\mu\omega^a(x) + f^{abc}\omega^b(x)A_C{}^c{}_\mu(x))$$
$$= -g^{-1}D^\mu D_\mu\omega^a(x) + f^{abc}A_C{}^c{}_\mu(x)D^\mu\omega^b(x) \tag{16.7}$$

Thus

$$\delta F^a(A_{C\mu}{}^\omega(x))/\delta\omega^b(x) = -g^{-1}\,\delta^{ab}D^\mu D_\mu + f^{abc}A_C{}^{c\mu}(x)D_\mu \tag{16.8}$$

and

$$\Delta_{FP}(A_C) = \left| \det\, (g^{-1}\delta^{ab}D^\mu D_\mu - f^{abc}A_C{}^{c\mu}(x)D_\mu) \right| \tag{16.9}$$

where $| \dots |$ represent absolute value.

We can rewrite the Faddeev-Popov determinant as a path integral over anti-commuting c-number fields with a ghost Lagrangian term:

$$\Delta_{FP}(A_C) = \int D\chi^* D\chi \, \exp[\, i\!\int d^7x \, \mathscr{L}^{ghost}(x)] \tag{16.10}$$

where

$$\mathscr{L}^{ghost}(x) = \chi^{a*}(x)[\delta^{ab}D^\mu D_\mu - gf^{abc}A_C{}^{c\mu}(x)D_\mu]\chi^b(x) \tag{16.11}$$

16.3 Faddeev-Popov Application to the Complexon Condition

The complexon condition can also be implemented within the path integral formalism using the Faddeev-Popov Mechanism. Using the identity

$$1 = \int DA_C\Delta_C(A_C)\delta(F_C(A_C)) \tag{16.12}$$

we see that an infinitesimal gauge transformation yields eqs. 16.3 and 16.4. This enables us to relate $\Delta_C(A)$ to the determinant

$$\delta(F_C(A_C{}^\omega)) = \left| \det \delta F_C(A_{C\mu}{}^\omega(x))/\delta\omega(x) \right|^{-1}_{F_C(A_C)=0,\,\omega=0} \tag{16.13}$$

and

$$\Delta_C(A_C) = \left| \det \delta F_C(A_{C\mu}{}^\omega(x))/\delta\omega(x) \right|_{F_C(A_C)=0,\,\omega=0} \tag{16.14}$$

From eq. 16.31 we see

$$F_C(A_{C\mu}(x)) = A_C^{a\mu}[\partial^2 A_{C\,\mu}^a/\partial x_r^k \partial x_i^k - \partial^2 A_{C\,k}^a/\partial x_r^\mu \partial x_i^k + gf^{abc}\partial(A_{C\,\mu}^b A_{C\,k}^c)/\partial x_i^k] \quad (16.15)$$

with

$$F_C^a(A_{C\mu}) = 0 \quad (16.16)$$

Inserting eq. 16.3 and 16.4 we find

$$[\delta F_C(A_{C\mu}^{\omega}(x))/\delta\omega^a(x)]|_{F_C(A_C)=0,\ \omega=0} = \delta[\delta A_C^{b\mu\omega}\text{Re}\ \partial F_{C\,\mu k}^b/\partial x_{ik} + A_C^{b\mu}\ \partial\delta(\text{Re}\ F_{C\,\mu k}^{b\,\omega})/\partial x_{ik}]/\delta\omega^a(x)|_{\omega=0}$$

$$= -g^{-1}(\text{Re}\ \partial F_{C\,\mu k}^a/\partial x_{ik})D^\mu - f^{abc}A_C^{b\mu}(\text{Re}\ F_{C\,\mu k}^c)\partial/\partial x_{ik} \quad (16.17)$$

Thus

$$\Delta_C(A_C) = |\det (g^{-1}(\text{Re}\ \partial F_{C\,\mu k}^a/\partial x_{ik})D^\mu + f^{abc}A_C^{b\mu}(\text{Re}\ F_{C\,\mu k}^c)\partial/\partial x_{ik}| \quad (16.18)$$

where $|\ \dots\ |$ represent absolute value.

We can rewrite this Faddeev-Popov determinant as a path integral over anti-commuting c-number fields with a ghost Lagrangian:

$$\Delta_C(A_C) = \lim_{r\to\infty} \int D\chi_C^* D\chi_C\ \exp[ir^{-2}\!\int d^7x\ \mathcal{L}_C^{\text{ghost}}(x)] \quad (16.19)$$

where r is a constant that is taken to the limit ∞, and where

$$\mathcal{L}_C^{\text{ghost}}(x) = \chi_C^*(x)\{D^\mu D_\mu + r^2 t^a[(\text{Re}\ \partial F_{C\,\mu k}^a/\partial x_{ik})D^\mu + gf^{abc}A_C^{b\mu}(\text{Re}\ F_{C\,\mu k}^c)\partial/\partial x_{ik}]\}\chi_C(x) \quad (16.20)$$

where t^a is a 3×3 matrix of the $\underline{3}$ representation of color SU(3) and $\chi_C(x)$ is a three row field in the $\underline{3}$ representation. *The introduction of $D^\mu D_\mu$ is based on consistency with the complexon formalism. It is needed to establish a perturbative expansion of the path integral. Its effect vanishes in the limit $r \to \infty$ reducing ghost loops of this type to point interactions.* The reader will note that second order and third order derivative terms appear in the interaction in $\mathcal{L}_C^{\text{ghost}}(x^\mu)$ and raise the issue of non-renormalizable divergences. If one uses the Two-Tier approach to quantum field theory developed by Blaha (2005a) then all potential divergences disappear. *The Two-Tier formulation of the pure, complexon, Yang-Mills theory that we are discussing is finite. See chapter 9 for Two-Tier quantum field theory.*

The complete pure complexon, Yang-Mills path integral is

$$Z(J^\mu) = N \int DA_C D\chi^* D\chi D\chi_C^* D\chi_C \Delta_{FP}(A_C)\delta(F(A_C))\Delta_C(A_C)\delta(F_C(A_C))\exp\{i\!\int d^7y\ [\mathcal{L} + J^\mu A_{C\mu}]\} \quad (16.21)$$

where

$$\mathcal{L} = \mathcal{L}_{CG} + \mathcal{L}^{\text{ghost}} + \mathcal{L}_C^{\text{ghost}} \quad (16.22)$$

with the lagrangian terms specified by eqs. 16.11, and 16.20.

17. The 'Interaction Rotations' Θ Group

We now define an 'Interactions Rotation' group, the Θ-Symmetry group G_Θ. It rotates the 192 gauge field components of

$$\Omega = \prod_{k=1}^{4} SU(3)_k \otimes SU(2)_k \otimes U(1)_k \otimes SU(2)_k \otimes U(1)_k \otimes U(4)_k \otimes U(4)_k \qquad (17.1)$$

where k labels the layer for all factors except the last U(4) factor where it enumerates the four Layer groups defined previously[188].The factors correspond to the ElectroWeak, Strong, Generation, and Layer groups. We abbreviate Θ as

$$\Omega = [SU(3) \otimes SU(2) \otimes U(1) \otimes SU(2) \otimes U(1) \otimes U(4) \otimes U(4)]^4 \qquad (17.2)$$

Since there are 192 gauge fields in Ω, the Θ-Symmetry group G_Θ of rotations of gauge fields is U(192). There are also 192 fermions in the layered Periodic Table of Fermions seen earlier. Each layer has its own set of Standard Model-like gauge fields plus there are four Layer group gauge fields that operate between layers. There is no direct leakage between layers except for the 'Interaction Rotation' group interactions. G_Θ rotates all the components of all of the fields in Ω and the fermion fields of the Periodic Table of Fermions.

The 192 gauge fields that are rotated, using a vector notation, are

$$\mathbf{A}_I{}^\mu = \Sigma_{a,j} \, {}^a\mathbf{A}_{Ij}{}^\mu(x) \cdot {}^a\mathbf{T}_{Ij} \qquad (17.3)$$

for a = 1,2, 3, 4 which specifies the layer, where j which labels each of the interactions in ${}^a\mathbf{A}_I{}^{\mu},$[189] where the ${}^a\mathbf{T}_I$ vector of matrix generators is specified below and where the vector

$${}^a\mathbf{A}_I{}^\mu = (g^a{}_1 {}^a\mathbf{A}_{SU(3)}{}^\mu(x_C),\ g^a{}_2 {}^a\mathbf{W}^\mu(x)\ ,\ g^a{}_3 {}^a\mathbf{A}_E{}^\mu(x),\ g^a{}_4 {}^a\mathbf{W}_D{}^\mu(x),\ g^a{}_5 {}^a\mathbf{A}_{DE}{}^\mu(x),\ g^a{}_6 {}^a\mathbf{U}^\mu(x)\ ,\ g^a{}_7 {}^a\mathbf{V}^\mu(x))$$
$$(17.4)$$

Each vector ${}^a\mathbf{A}_I{}^\mu$ is a vector of the gauge fields of the respective interactions in layer a. For simplicity we label the respective layer dependent coupling constants as $g^a{}_1$, $g^a{}_2$, ... , $g^a{}_7$. The subscript 'D' labels Dark matter interactions, 'W' labels Weak fields, 'E' labels

---

[188] These groups appear in the Θ-symmetry group because they are all directly connected with defining the form of the Periodic Table of Fermions. Other groups do not specify aspects of the Periodic Table's form.

[189] In the case of the Layer groups' gauge fields ${}^a\mathbf{V}^\mu(x)$ the index 'a' merely enumerates the four Layer groups since the groups straddle all four layers.

Electromagnetic fields, $U^\mu(x)$ labels U(4) Generation group fields, and $V^\mu(x)$ labels U(4) Layer group fields.

Similarly we define a 7-vector of 48 matrix generators for each a:

$$^a T_I = (^a T_{SU(3)},\ ^a \tau_{SU(2)},\ ^a I_{U(1)},\ ^a \tau_{DSU(2)},\ ^a I_{DU(1)},\ ^a G_{U(4)},\ ^a G_{LU(4)}) \tag{17.5}$$

We put a layer index on these matrices: since, although they have the same form in all layers, the respective matrices operate in 'different spaces' for each layer since the gauge fields, and the fermion fields they operate on, differ from layer to layer.

The plenitude of interactions that we have identified and summarized in chapter 25 of Blaha (2017b) leads us to consider the possibility of unification based partly on the 'rotation of interactions' and more fully, on an interactions unification based on the Riemann-Christoffel tensor.[190]

When we use the PseudoQuantum (two fields per gauge particle) formalism[191] the part of these gauge fields interactions within a covariant derivative has the form

$$^a A_I^{1\mu}(x) + {}^a A_I^{2\mu}(x) \tag{17.6}$$

summed over a.

17.1 Θ-Symmetry: 'Interaction Rotations' for Fermions

The Θ-Symmetry gauge fields appear in Dirac-like equation covariant derivatives. We define a column vector ψ containing the 4-spinors of all 192 normal and Dark fundamental fermions in our theory based on four generations in four layers as detailed earlier and in Blaha (2016a), (2016b) and (2016c). Then the flat space-time massless Dirac equation[192] has the form:

$$\gamma_\mu D^\mu \psi = \gamma_\mu \{ \partial^\mu + i\, [g_\Theta A_\Theta^{1\mu}(x) + g_\Theta A_\Theta^{2\mu}(x) + g_S A_S^{1\mu}(x) + g_S A_S^{2\mu}(x) + \Sigma_a (^a A_I^{1\mu}(x) + $$
$$+\ {}^a A_I^{2\mu}(x))] \} \psi = 0 \tag{17.7}$$

where

$$A_I^{i\mu} = \Sigma_a\, {}^a A_I^{i\mu}(x) \cdot {}^a T_I \tag{17.8}$$

and with the Θ-Symmetry gauge field terms $A_\Theta^{1\mu}(x)$ and $A_\Theta^{2\mu}(x)$ inserted. Each of these fields consists of a gauge field multiplied by a U(192) matrix generator in the <u>192</u> representation since there are 192 fermion fields.

Then the massless Dirac equation becomes

---

[190] Most of the equations in this section first appeared in Blaha (2017a).
[191] See Blaha (2017b).
[192] Similar considerations apply to other Dirac-like equations described in Blaha (2017b).

$$\gamma_\mu D^\mu \psi = \gamma_\mu \{\partial^\mu + i \, [g_\Theta A_\Theta^{1\mu}(x) + g_\Theta A_\Theta^{2\mu}(x) + g_S A_S^{1\mu}(x) + g_S A_S^{2\mu}(x) +$$
$$+ A_I^{1\mu}(x) + A_I^{2\mu}(x)]\} \psi = 0 \qquad (17.9)$$

The $A_\Theta^{1\mu}(x)$ and $A_\Theta^{2\mu}(x)$ gauge fields each separately have a 192×192 matrix representation:

$$A_\Theta^{i\mu}(x) = \Sigma_n A_\Theta^{i\mu}{}_n(x) T_{U(192)n} \qquad (17.10)$$

for i = 1., 2 where n sums over the 192^2 generators of U(192).

The fields in $A_I^{1\mu}(x)$, and $A_I^{2\mu}(x)$ separately transform under a Θ-transformation. Under a Θ-transformation we find the Dirac equation transforms to

$$\gamma_\mu \{\partial^\mu + i \, [g_\Theta A'_\Theta{}^{1\mu}(x) + g_\Theta A'_\Theta{}^{2\mu}(x) + + g_S A'_S{}^{1\mu}(x) + g_S A'_S{}^{2\mu}(x) +$$
$$+ A'_I{}^{1\mu}(x) + A'_I{}^{2\mu}(x)]\} \psi' = 0$$

where

$$A'_\Theta{}^{1\mu}(x) = C_\Theta(x) A_\Theta^{1\mu}(x) C_\Theta^{-1}(x) - i \, C_\Theta(x)\partial^\mu C_\Theta^{-1}(x)/g_\Theta \qquad (17.11)$$
$$A'_\Theta{}^{2\mu}(x) = C_\Theta(x) A_\Theta^{2\mu}(x) C_\Theta^{-1}(x)$$
$$A'_S{}^{1\mu}(x) = C_\Theta(x) A_S^{1\mu}(x) \, C_\Theta^{-1}(x)$$
$$A'_S{}^{1\mu}(x) = C_\Theta(x) A_S^{1\mu}(x) \, C_\Theta^{-1}(x)$$
$$A'_I{}^{1\mu}(x) = C_\Theta(x) A_I^{1\mu}(x) \, C_\Theta^{-1}(x)$$
$$A'_I{}^{2\mu}(x) = C_\Theta(x) A_I^{2\mu}(x) \, C_\Theta^{-1}(x)$$
$$\psi' = C_\Theta(x)\psi$$

The effect of the Θ-transformation $C_\Theta(x)$ is to rotate the gauge fields components and the fermion fields. It is accompanied by a rotation of the generator matrices components. Together they define an equivalent formulation of the original Dirac equation and thus the fermion sector. The next section provides an explicit simple example of Θ-transformations – ElectroWeak theory.

Note that a Θ-transformation causes a change of gauge in the $A_\Theta^{1\mu}(x)$ field. The other gauge fields, $A_\Theta^{2\mu}(x)$, $A_I^{1\mu}(x)$ and $A_I^{2\mu}(x)$, are 'rotated' but do not undergo a change of gauge. (Each of the 192 gauge fields within $A_I^{1\mu}(x)$ and $A_I^{2\mu}(x)$ do undergo their own particular changes of gauge for their own transformation groups.)

17.2 Broken Θ-Symmetry

Θ-symmetry is broken in several ways. First, it is is broken by the complexon nature of color SU(3) gauge fields and quarks. More importantly it must be completely broken because effects of Θ-symmetry are not seen in Nature. *We therefore assume Θ-symmetry is completely broken via the Higgs Mechanism. Since the unbroken symmetry is U(192) there must be a 192-plet of Higgs bosons. The discussion (section 15.3) of complete symmetry breaking of the U(4)*

Layer groups shows how the complete symmetry breaking of the U(192) Θ-symmetry can be implemented.

The result is complete symmetry breaking and the creation of mass terms for all Θ-symmetry gauge fields and Θ-symmetry contributions to all fermion masses.

17.2.1 Θ-Symmetry and Higgs Fields

The Higgs boson fields η (or our PseudoQuantum alternative) also participate in Θ-symmetry rotations. Consider a composite boson field constructed from the concatenation of all Higgs fields. The term generating the gauge field masses has the form

$$(D^\mu\eta)^\dagger D^\mu\eta$$

Upon rotating gauge fields, mass terms will also correspondingly 'rotate' to maintain the covariance of the overall theory. *Thus Θ-symmetry transformations apply to gauge fields, fermions, and Higgs bosons.*

17.2.2 Θ-symmetry Breaking due to Complexon Coordinates

The color SU(3) gauge fields $A_{SU(3)}^{j\mu}(x)$ for $j = 1,2$ appearing in $A_I^{i\mu}(x)$ are complexon fields in our formulation of SuperStandard Model. Thus they are functions of complex spatial coordinates

$$x_c = (t, \mathbf{x_r} - i\mathbf{x_i})$$

Therefore color SU(3) does not admit of Θ-transformations. Consequently Θ-Symmetry is 'first' broken to $[SU(3)_{Color}]^4 \otimes U(160)$. This symmetry breaking reduces the number of Θ-Symmetry generators to 160^2. Then the overall symmetry breaking described above occurs.

17.2.3 Large A_Θ gauge field masses

The symmetry breaking mechanism may be expected to lead to large A_Θ gauge field masses. Together with the likely smallness of the g_Θ coupling constant, this leads us to expect that the A_Θ interaction, which occurs between all 192 fermions, will not be easily detected.

17.3 Example: ElectroWeak-like Theory

ElectroWeak model is an example of a global Θ-transformation which does not include the full gamut of features outlined above. Focussing on the Weak and Electromagnetic interactions we consider the covariant derivative

$$\{\partial^\mu + i\,[g\mathbf{W}^\mu\cdot\boldsymbol{\tau} + g'W_0^\mu\tau_0]\}\psi = 0 \qquad (17.12)$$

Rotating W_3^μ and W_0^μ with $C^{-1}{}_\Theta$

$$\begin{bmatrix} gW_3{}^\mu \\ g'W_0{}^\mu \end{bmatrix} \rightarrow \begin{bmatrix} g\cos\theta\ Z^\mu + g\sin\theta\ A^\mu \\ -g'\sin\theta\ Z^\mu + g'\cos\theta\ A^\mu \end{bmatrix}$$

and rotating the generator matrices by C_Θ

$$\begin{bmatrix} \tau_3 \\ \tau_0 \end{bmatrix} \rightarrow \begin{bmatrix} \cos\theta\ \tau_3 - \sin\theta\ \tau_0 \\ \sin\theta\ \tau_3 + \cos\theta\ \tau_0 \end{bmatrix}$$

and making an appropriate choice of θ_W yields the electromagnetic field A and Z we find

$$g'W_0{}^\mu \tau_0 + gW_3{}^\mu \tau_3 \rightarrow Z^\mu[(g\cos^2\theta - g'\sin^2\theta)\tau_3 - \tfrac{1}{2}\sin(2\theta)(g + g')\tau_0 + A^\mu[\tfrac{1}{2}\sin(2\theta)(g + g')\tau_3 + (g'\cos^2\theta - g\sin^2\theta)\tau_0]$$

If we choose the coefficient of A^μ be e, then

$$e(\tau_3 + \tau_0) = [\tfrac{1}{2}\sin(2\theta)(g + g')\tau_3 + (g'\cos^2\theta - g\sin^2\theta)\tau_0]$$

and thus

$$\tfrac{1}{2}\sin(2\theta)(g + g') = (g'\cos^2\theta - g\sin^2\theta) = e$$

Consequently

$$g' = -(\tfrac{1}{2}\sin(2\theta) - \sin^2\theta)/(\tfrac{1}{2}\sin(2\theta) - \cos^2\theta)$$

If we further choose $g = g'$ (since equal coupling constants is an often stated goal of theorists) for simplicity we find the angle $\theta = \pi/8$ or $22.5°$ while the usual Weinberg angle θ_W is about $30°$. Thus we find a close similarity to standard ElectroWeak theory.[193]

　　　We conclude that ElectroWeak theory can be viewed as an example of a Θ-transformation.[194]

---

[193] We can, of course, make the angles equal by adjusting the values of g ang g'.

[194] A rotation of fermions was not required in this case because the interactions are not charge changing. In the general case, a rotation of interactions also requires a rotation of the 192 fermions.

17.4 Path Integral Formulation, and the Faddeev-Popov Method

The path integral formulation of our theory with the complete set of eleven interactions is fairly straightforward with one exception. Since we use complex valued coordinates for the color SU(3) gauge theory, a somewhat different approach must be followed for it. We detailed this approach in Blaha (2015a).

In this section we will explicitly consider the Faddeev-Popov Method for the Ω gauge field $A_\Omega^{1\mu}(x)$ only – noting that the Yang-Mills gauge fields of the other interactions must also be subject to gauge fixing using the Faddeev-Popov Method.

The path integral we consider for the $A_\Omega^{1\mu}(x)$ gauge field is:[195]

$$Z(J^\mu) = N\!\int\! DA_\Omega^{1\mu}\, \Delta(A_\Omega^1)\delta(F(A_\Omega))\exp\{i\!\int\! d^4y[\mathscr{L} + J^\mu(y)\, A_\Omega{}^1{}_\mu(y)]\} \qquad (17.13)$$

where $\delta(F(A_\Omega))$ specifies the gauge, and $\Delta(A_\Omega^1)$ is its Faddeev-Popov determinant. The Faddeev-Popov determinant can be calculated in the standard way.[196]

First we consider the gauge fixing delta function. Note that it can be written as a delta function in the gauge times a determinant:

$$\delta(F(A_\Omega^\omega)) = \delta(\omega - \omega_0)|\det \delta F(A_\Omega{}^1{}_\mu{}^\omega(x))/\delta\omega(x)|^{-1}\big|_{F(A_\Omega)=0} \qquad (17.14)$$

where ω_0 is a reference gauge, where

$$A_\Omega{}^{1a}{}_\mu{}^\omega(x) = A_\Omega{}^{1a}{}_\mu(x) - g_\Omega{}^{-1}\partial_\mu\omega^a + f_\Omega{}^{abc}\,\omega^b(x)A_\Omega{}^{1c}{}_\mu(x) \qquad (17.15)$$
$$= A_\Omega{}^{1a}{}_\mu(x) + \delta A_\Omega{}^{1a}{}_\mu{}^\omega(x)$$

and where

$$F_A{}^a{}_\mu{}^\omega = F_A{}^a{}_\mu + f_\Omega{}^{abc}\,\omega^b(x)F_A{}^c{}_\mu \qquad (17.16)$$

under an infinitesimal gauge transformation where the $f_\Omega{}^{abc}$ are SU3)⊗U(160) structure constants, and a, b, and c label the generators of SU3)⊗U(160).

Also

$$\Delta(A_\Omega^1) = |\det \delta F(A_\Omega{}^1{}_\mu{}^\omega(x))/\delta\omega(x)|\big|_{F(A_\Omega) = 0,\, \omega = 0} \qquad (17.17)$$

We will choose the Lorentz gauge to evaluate the Faddeev-Popov determinant:

$$F^a(A_\Omega) = \partial_\mu A_\Omega{}^{1a\mu}(x) = 0 \qquad (17.18)$$

We find

$$F^a(A_{\Omega\mu}{}^\omega(x)) = \partial^\mu[A_\Omega{}^{1a}{}_\mu(x) - g_\Omega{}^{-1}\partial_\mu\omega^a(x) + f_\Omega{}^{abc}\,\omega^b(x)A_\Omega{}^{1c}{}_\mu(x)]$$

---

[195] The $A_\Omega^{2\mu}(x)$ transforms homogeneously and thus does not have a Faddeev-Popov determinant $\Delta(A_\Omega^2)$.
[196] See for example Huang (1992).

$$= -g_\Omega^{-1}\partial^\mu\partial_\mu\omega^a(x) + f_\Omega^{abc}A_\Omega^{1c}{}_\mu(x)\,\partial^\mu\omega^b(x) \qquad (17.19)$$

Thus

$$\delta F^a(A_{\Omega_\mu}{}^\omega(x))/\delta\omega^b(x) = -g_\Omega^{-1}\delta^{ab}\partial^\mu\partial_\mu + f_\Omega^{abc}A_\Omega^{1c\mu}(x)\partial_\mu \qquad (17.20)$$

and

$$\Delta(A_\Omega) = |\det(g_\Omega^{-1}\delta^{ab}\partial^\mu\partial_\mu - f_\Omega^{abc}A_\Omega^{1c\mu}(x)\partial_\mu)| \qquad (17.21)$$

where $|\ldots|$ represents the absolute value.

We can rewrite the Faddeev-Popov determinant as a path integral over anti-commuting c-number fields χ^a with a ghost Lagrangian:

$$\Delta(A_\Omega) = \int D\chi^* D\chi\,\exp[\,i\!\int d^4x\,\mathscr{L}_\Omega^{ghost}(x)] \qquad (17.22)$$

where

$$\mathscr{L}_\Omega^{ghost}(x) = \chi^{a*}(x)[\delta^{ab}\partial^\mu\partial_\mu - gf_\Omega^{abc}A_\Omega^{1c\mu}(x)\partial_\mu]\chi^b(x) \qquad (17.23)$$

with a, b and c ranging from 1 through 192.

The Ω gauge field lagrangian thus acquires a ghost lagrangian:

$$Z(J^\mu) = N\!\int DA_\Omega^{1\mu}\delta(F(A_\Omega))\exp\{i\!\int d^4y[\mathscr{L}_\Omega^{ghost}(y) + J^\mu(y)\,A_\Omega^1{}_\mu(y)]\} \qquad 17.24)$$

in addition to other factors for the other gauge field interactions.

17.5 Interactions of $A_\Omega^\mu(x)$ and Ghost 'Fields'

In general the $A_\Omega^\mu(x)$ field with its 192 components embodies interactions between all 192 fermions in the Fermion Periodic Table. Similarly, the 192 ghost fields also yield interactions between all 192 fermions through their affect on A_Ω^μ quanta propagated between frmions.

The $A_\Omega^\mu(x)$ field quanta has in and out states in perturbation theory. Ghost fields only exist as interactions between A_Ω^μ quanta within Feynman diagrams and do not have in or out states.

18. Unified SuperStandard Model Multi-Quark Particles, and Dark Matter 'Chemistry'

18.1 New Multi-quark Particles such as Penta-Quark Particles

The CERN LHC has found evidence for penta-quark particles – particles consisting of five quarks.[197] This discovery is a step beyond the tetraquark (four quark) particle named $Z_C(3900)$ found, and announced,[198] in 2013. The Z_C particle could be viewed as two D-mesons bound together by the strong color force in a color singlet "hadron molecule."

Similarly, the pentaquark could be viewed as a particle consisting of a two quark meson and a three quark baryon formed into a type of molecule by the strong interaction.

18.2 Multi-Quark Molecules – Quark Chemistry

These recent discoveries are the initial steps to a spectrum of multi-quark particles consisting of

$$2m + 3b$$

quarks where m is the number of "meson-like" constituents and b is the number of "baryon-like" constituents. The pentaquark has $m = b = 1$ constituents.

We are now confronted by a new type of "molecule" where the binding force is not electromagnetic but, instead, the strong color force. The tight binding of these "molecules" by the strong force makes them effectively into particles. But they are perhaps more comparable to the binding of protons and neutrons into atomic nuclei by the nuclear force.

The development of a quark molecular chemistry would, if molecules were produced in quantity, lead to quark materials with extremely important physical characteristics – quark matter – with both new and important "chemical" properties. However, quark molecules have an extraordinarily short lifetime and so the creation of a number of quark molecules, and their fabrication into materials, does not appear to be feasible. Nevertheless, their theoretical and experimental study might serve to drive conventional chemistry and materials science forward.

---

[197] Announced July 15, 2015. See arXiv.org/abs/1507.03414.
[198] June 20, 1013. Discovery by the Belle Collaboration, High Energy Accelerator Research Organization, in Tsukuba, Japan; and BESIII Collaboration at the Beijing Electron Positron Collider in China. Phys. Rev. Lett. **110**, 252001, 2013 and **110**, 252002, 2013 .

18.3 Large Multi-Quark "Molecules" and Eventually Quark Stars

One can envision the eventual creation of very "large" multi-quark particles/molecules. Taken to the limit of extremely large "molecules" of extraordinary size it appears possible to develop a dynamics of quark stars where the dominant force is not the nuclear force as we usually know it and electromagnetism but rather the strong force. Astronomy has yet to discover an unambiguous quark star.

18.4 Dark Matter Features: Particle Spectrum and Basic Chemistry of Dark Matter, Dark Matter Bodies

In chapter 39 of Blaha (2012a) we initially described Dark Matter particles, Dark atoms, the Dark Periodic Table, and Dark basic chemistry – all based on the additional $SU(2) \otimes U(1)$ symmetry in the SuperStandard Model that more or less mirrors normal ElectroWeak symmetry. The major differences were 1) that Dark quarks are assumed to be $SU(3)$ singlets as suggested by the known weakness of Dark Matter interactions with normal matter and 2) that *physical* hadron charges – both electric charge and Dark electric charge – are quantized with whole number values. (Dark quarks are the emasculated "hadrons" of this theory since they do not experience the Strong Interaction.) This chapter contains material from chapter 39 of Blaha (2012a) for completeness as well as some additional thoughts.

Another important issue is the effect of gravitation on Dark Matter—Can Dark matter aggregate under the force of gravity to form galactic clusters, galaxies, and smaller objects such as Dark suns and perhaps Dark planets? We discuss these possibilities below.

This chapter is based on the assumption of four generations.[199] Previous chapters present a strong theoretical case for four generations of fermions.

18.5 Fundamental Dark Particles

We now consider the Dark particles that are associated with our new Dark $SU(2) \otimes U(1)$ symmetry. Since Dark Matter only interacts weakly with known matter, Dark particles must be color singlets. Thus there will be 12 (in the three generation case) or 16 (in the four generation case) Dark particles (plus their antiparticles.) Recent experiments suggest Dark particles have extremely large masses of the order of 8.6 GeV/c or larger. Dark quarks are color singlets with complex 3-momenta in our Unified SuperStandard Model.

---

[199] This chapter is based on the assumption of one layer of four generations. The introduction of the Layer group and four layers of four fermion generations dramatically increases the number of Dark fermion states.

"Periodic Table" of 'Known' Fermions

NORMAL ELECTROWEAK FERMION DOUBLETS

Generation	Leptons		Color Triplet Quarks	
	Real-Valued 3-Momenta		Complex-Valued 3-Momenta	
1	e	ν_e	$u_1\ u_2\ u_3$	$d_1\ d_2\ d_3$
2	μ	ν_μ	$c_1\ c_2\ c_3$	$s_1\ s_2\ s_3$
3	τ	ν_τ	$t_1\ t_2\ t_3$	$b_1\ b_2\ b_3$
4?	ω	ν_ω	$V_1\ V_2\ V_3$	$W_1\ W_2\ W_3$

DARK SU(2)⊗U(1) FERMION DOUBLETS

Generation Quarks	Dark Leptons		Color Singlet Dark	
	Real-Valued 3-Momenta		Complex-Valued 3-Momenta	
1	e_D	ν_{eD}	u_D	d_D
2	μ_D	$\nu_{\mu D}$	c_D	s_D
3	τ_D	$\nu_{\tau D}$	t_D	b_D
4?	ω_D	$\nu_{\omega D}$	V_D	W_D

Figure 18.1 "Periodic Table" of four generations of normal and Dark fundamental fermions assuming only one layer of fermions.

In addition to Dark fermions there will be four SU(2)⊗U(1) Dark gauge bosons – also with large masses:

Dark Particle Gauge Bosons

$$\text{U(1): } W'^{\mu}_0 \qquad\qquad \text{SU(2): } W'^{\mu}_1 \qquad W'^{\mu}_2 \qquad W'^{\mu}_3$$

18.6 Dark Particle Chemistry

The chemistry of Dark Matter will be different from the chemistry of known matter due to the absence of the color interaction. Dark particles cannot combine through a strong interaction to form a hadronic spectrum, or atomic nuclei, such as we see in normal matter. Thus Dark particle atoms are like hydrogen atoms, and consist of a Dark quark particle bound to a Dark lepton particle by the Dark electric force assuming Dark charge is quantized and has equal integer absolute values for Dark quarks and leptons. More complex Dark molecules are also possible as we will see later.

Suppose all Dark charged particles have charge $\pm 1e_D$ (e_D is the Dark unit of charge which may possibly be equal to e, the electromagnetic charge), and each Dark doublet has a Dark particle with unit Dark charge and a neutral Dark particle. Then we would expect that (in the case of four generations) Dark Matter would consist of:

1. Lepton-like fundamental particles: Four Dark charged and four Dark charge neutral particles and their anti-particles.

2. Quark-like fundamental particles: Four Dark charged and four Dark charge neutral particles and their anti-particles.

3. There would be a total of eight neutral Dark quarks and leptons.

4. Atoms are composed of oppositely charged Dark particles of different types. There are 16 Dark "atoms" of the form leptoDark particle - quarkDark particle, $e_D u_D$, $e_D c_D$, $e_D t_D$, $\mu_D u_D$, $\mu_D c_D$, $\mu_D t_D$, $\tau_D u_D$, $\tau_D c_D$, $\tau_D t_D$, $e_D v_D$, $\mu_D v_D$, $\tau_D v_D$, $\omega_D v_D$, $\omega_D u_D$, $\omega_D c_D$, and $\omega_D t_D$ plus their anti-matter equivalents. There are 12 "quasi-stable" particle-anti-particle combinations: six leptoDark - antileptoDark particle combinations, and six quarkDark - antiquarkDark particle combinations. (There is no attractive nuclear force.) All of these atoms are bound by the Dark electric force.

5. Simple molecules of the type of Figs. 18.2 – 18.4 below, and so on, based on Dark dipole interactions, Dark van der Waals forces and other Dark electromagnetic interactions are possible.

After a sufficiently long time, collisions would lead perhaps to the dominance of Dark particles and the "disappearance" of antiDark particles if the number of Dark particles is overwhelmingly dominant in a fashion similar to normal matter.[200] (The other possibility is not excluded.) The Dark Periodic Table is presented in Fig. 18.2.

---

[200] The study of electron-positron production through Weak Interaction with Dark Matter by M. Aguilar et al, Phys. Rev. Letters **110**, 141102 (2013) does not seem to clarify this issue.

Periodic Table of Simple Dark Particle Atoms
(Assuming Dark anti-particles are Annihilated)

$e_D u_D$	$\mu_D u_D$	$\tau_D u_D$	$\omega_D u_D$	d_D
$e_D c_D$	$\mu_D c_D$	$\tau_D c_D$	$\omega_D c_D$	s_D
$e_D t_D$	$\mu_D t_D$	$\tau_D t_D$	$\omega_D t_D$	b_D
$e_D v_D$	$\mu_D v_D$	$\tau_D v_D$	$\omega_D v_D$	w_D

Figure 18.2 "Periodic Table" of Simple Dark atoms assuming only one layer of fermions.

There also are similar antiparticle atoms. Bound states are assumed bound, perhaps into hydrogen-like atoms, through a Dark electromagnetic force. The last column consists of Dark charge neutral quarks. Antiparticle atoms of these states might also exist or be created through the Dark ElectroWeak interactions. One could extend the table with a column of Dark neutrinos. The decays, and mixing between generations, remains to be determined in the unknown Dark Higgs sector.

The periodic table (Fig. 18.2) that we constructed is based on an analogy with the features of normal matter: quarks are much heavier than leptons and leptons "revolve" around the Dark quark nuclei. If this view is correct then one can conceive of a chemistry of Dark Matter with molecules bound by Dark electromagnetic forces. Pair bonding of Dark leptons would be possible and so one could conceive of a fairly complex Dark chemistry bounded by the fact that a Dark particle atom has only one Dark lepton. Thus there would only be 64 bound pairs of atoms[201] similar to H_2.

$$e_D$$
$$u_D \qquad\qquad u_D$$
$$e_D$$

Figure 18.3. Example of two Dark atoms binding in a manner similar to the binding of hydrogen molecules in H_2. Considering the possible combinations of quarks and leptons there are 64 bound varieties of this type. Chains of these molecules are also possible in principle.

Dark particles can be combined into chains and more complex molecules. A simple chain of Dark particles appears in Fig. 18.4.

---

[201] Plus anti-particle equivalents.

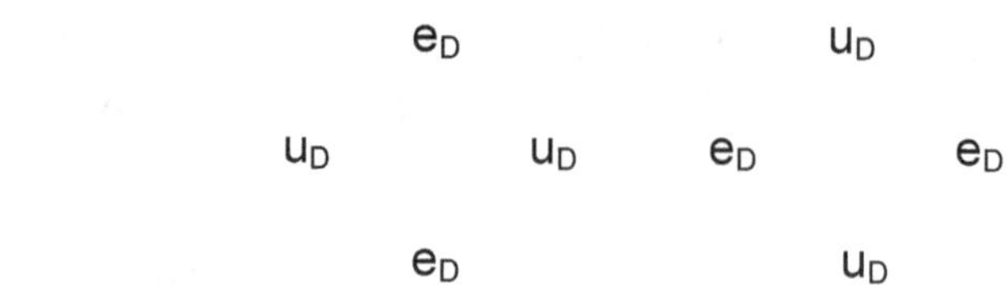

Figure 18.4. Example of a simple Dark Matter chain.

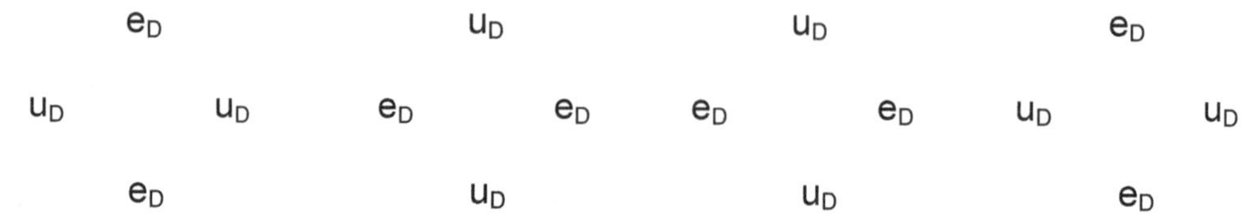

Figure 18.5. A segment of two strands interacting with each other. Other combinations are possible using other Dark charged fermions. This diagram is suggestive of a Dark form of DNA. Dark Life?

More complex chains as well as two and three dimensional bound aggregates are also possible. These considerations suggest something approaching the complexity of simple life forms may be possible.

Dark dipole effects could lead to the (weaker) binding of larger assemblages of Dark atoms. For example, if the masses of the leptonic and quark Dark particles are not too dissimilar, crystalline Dark Matter appears possible.

We thus arrive at a Dark Matter sector with much less variety than normal matter. It may, or may not, preclude the existence of Dark life forms, and possibly of Dark solids composed of Dark Matter. Solid Dark planets may exist. The detection of planet-like and star-like Dark Matter objects within the Dark Matter "cloud" surrounding a galaxy would be of great importance.

18.7 Gravitation and Dark Matter Bodies: Galactic Clusters, Galaxies

Dark Matter is thought to equal approximately 83% of the total matter in the universe. It is known to form halos around galaxies and to form filaments between galaxies in galactic clusters. It appears to influence the orbits of stars in galaxies possibly causing the rotational speed of stars at large radial distances from galactic centers to be roughly constant and independent of the radius.

When two galaxies collide it appears that the Dark Matter drags between the galaxies as they separate. Thus Dark Matter has forces as we suggested earlier.

We see these effect in the Dark Matter halos around galaxies. Galaxies are permeated with Dark Matter. Our galaxy, for example, appears to have ten times as much Dark Matter as normal matter.[202] Galaxies that are composed of predominantly Dark matter have been found.

18.8 Dark Stars, Dark Planets, Dark Globular Clusters—Or Just Darkened?

However we have not seen (as yet?) clumps of Dark Matter within our galaxy similar in size to stars and planets or globular clusters. Nor do we see the earth, or other Solar System planets, having a discrepancy in mass due to Dark Matter clumps within these bodies. Nor is there a Dark planet in our Solar System.

Yet there is no apparent reason why Dark Matter should not clump within planets (and our sun) causing a discrepancy in solar system dynamics and ostensibly the force of gravity. The ten to one dominance of Dark Matter in our galaxy creates a striking discrepancy with the apparent absence of Dark Matter gravitational effects in the Solar System.

18.8.1 Open Questions on Dark Matter in Our Solar System

1. Why does not the earth's mass have a significant contribution from the Dark Matter within the earth since our galaxy has 10 times more Dark Matter than normal matter? Does not Dark Matter clump somewhat around planet size objects?

2. Same question as 1 but applied to the sun. Where is the effect of the Dark Matter within the sun?

3. Why is there no Dark Jupiter in our Solar System with normal moons circling it?

4. Why are not the planetary dynamics of the Solar System affected by the presence of Dark Matter?

18.8.2 Open Questions on Dark(ened) Stars and Dark(ened) Globular Clusters of Stars

There are many stars in the galaxy composed of normal matter apparently. However the ten to one ratio of Dark Matter to normal matter in our galaxy should be reflected in stellar dynamics which is a supposedly well understood field. Stellar dynamics depends significantly on the mass of stars.

---

[202] Therefore one would expect more gravitational phenomena reflecting the presence of Dark Matter locally in our galaxy than its effect on galactic rotation. Since Dark Matter hardly interacts with normal matter it would intersperse with normal matter throughout the galaxy forming a hidden part of the galaxy – present but almost completely not detectable. The hidden part would occupy the same space as normal galactic matter rather like an evanescent ghost in a horror movie.

1. Why does not stellar dynamics take account of the Dark Matter contribution to the mass of stars?

2. When a star contracts generating gravitational energy why does not the contraction of the Dark Matter clumped within the star impact gravitationally on the contraction of the normal matter?

3. Why is not the evolution and structure of globular clusters of stars affected by their presumably large Dark Matter content?

4. Will we find a Dark star circled by some large normal matter planets?

5. Can we detect a Dark cluster by its gravitational effects?

18.8.3 Comments on these Questions

The only currently apparent resolution of these questions is a lack of significant Dark chemical interactions/reactions between Dark Matter atoms that may prevent the formation of Dark solids and liquids.[203] Dark nuclear decays similar to the normal nuclear reactions that power stars may also not exist. Thus no Dark stars. But Dark aggregates of the mass of the sun or more should be possible.

Dark Matter in galaxies appears to be rather like a distributed gas within a gravitational well formed by the combined masses of a galaxy. It also forms a filament between galaxies.[204] Within a galaxy Dark Matter appears dispersed in a somewhat uniform way with only very minor gravitational clumping by clusters, stars and planets.

---

[203] Thus Dark atoms may not have Dark dipole forces or other multipole forces that would induce the formation of Dark Matter liquids and solids.

[204] A galaxy composed overwhelmingly of Dark Matter has recently been found.

19. Complex General Relativity Reformulated

We have seen that Complex Special Relativity is the basis of flat space-time phenomena. Flat space-time coordinates are complex-valued in general. The real-valued coordinates that we experience in everyday life are the result of our measuring instruments: clocks and rulers. Real-valued coordinates are generated from complex-valued coordinates by Reality group transformations.

If flat space-time is governed by Complex Special Relativity then it is clear that curved 'space-time' is governed by Complex General Relativity. Here again there is a Reality group the General Relativistic Reality group – a U(4) group – that maps complex-valued General Relativity coordinates to real-valued curved coodinates.[205] There is a corresponding U(4) Internal Symmetry Reality group that we call the *Species group*. This group rotates fermions as described in chapter 20.

We can isolate the General Relativistic Reality group by factoring complex General Relativistic coordinate transformations into parts that consist of a real-valued General Coordinate transformation and complex-valued coordinate transformations. It will be apparent that the General Relativistic Reality group emerges in this discussion. The Species group is distinct from the Internal Symmetry Reality group of the SuperStandard Model: R = SU(3)⊗SU(2)⊗U(1)⊗SU(2)⊗U(1). We begin by defining the tetrad notation.

19.1 Tetrad (Vierbein) Formalism

The *vierbein* formalism begins with the Equivalence Principle that allows us to define an inertial coordinate system in the neighborhood of any point Z in space-time. We will use the notation $\varsigma^\alpha(Z)$ to denote the inertial coordinates at Z. We define a tetrad or vierbein as

$$v^\alpha_{\ \mu}(x) = (\partial\varsigma^\alpha(x)/\partial x^\mu)_{x=Z} \tag{19.1}$$

and, in a neighborhood of Z, we can invert the relation between ς and x to define an inverse

$$w^\mu_{\ \alpha}(x) = (\partial x^\mu(\varsigma)/\partial\varsigma^\alpha)_{x=X} \tag{19.2}$$

such that

$$w^\mu_{\ \alpha}(x)v^\alpha_{\ \nu}(x) = \delta^\mu_{\ \nu}$$
$$w^\mu_{\ \beta}(x)v^\alpha_{\ \mu}(x) = \delta^\alpha_{\ \beta} \tag{19.3}$$

---

[205] Much of this chapter appears in Blaha (2016h) and (2017a).

In real General Relativity all *tetrads* are real-valued. In Complex General Relativity a *tetrad* $v^\alpha_\mu(x)$ is complex-valued.

The metric at a curved space-time point X is defined in terms of *tetrads* as

$$g_{\rho\sigma}(x) = \eta_{\alpha\beta}\, v^\alpha_\rho(x)v^\beta_\sigma(x)$$
$$g^{\rho\sigma}(x) = \eta^{\alpha\beta}\, w^\rho_\alpha(x)w^\sigma_\beta(x)$$

(19.4)

The inverse of a *tetrad* transformation can also be expressed as

$$w_\beta^{\ \nu}(x) = v_\beta^{\ \nu}(x) = \eta_{\beta\alpha}g^{\nu\mu}(x)v^\alpha_\mu(x)$$

Then a *tetrad* and its inverse satisfy the relations

$$v^\alpha_\mu(x)v_\beta^{\ \mu}(x) = \delta^\alpha_\beta$$

(19.5)

and

$$v^\alpha_\mu(x)v_\alpha^{\ \nu}(x) = \delta^\nu_\mu$$

There are two general types of space-time transformations that can be performed on a tetrad.

1. A complex-valued (possibly real-valued) General Relativistic coordinate transformation:

$$v'^\alpha_\mu(x) = \partial x^\nu/\partial x'^\mu \, v^\alpha_\nu(x)$$

2. A complex-valued, local *Lorentzian transformation*

$$v'^\beta_\mu(x) = \Lambda(x)^\beta_\alpha\, v^\alpha_\mu(x)$$

where $\Lambda(x)^\beta_\alpha$ is an element of a subset of the local Complex Lorentz Group.

The local Lorentzian transformations $\Lambda(x)^\beta_\alpha$ consist of local Lorentz transformations that are real-valued, and complex-valued Lorentz transformations. Both types of transformations satisfy the orthogonality condition:

$$\eta_{\alpha\beta}\Lambda^\alpha_\rho(x)\Lambda^\beta_\sigma(x) = \eta_{\rho\sigma}$$

(19.6)

Thus the *tetrad* partakes of both local (position dependent) General Relativistic transformations and local Lorentzian transformations.

19.2 Complex General Relativistic Transformations

The General Relativistic Reality group interaction emerges from complex General Relativistic transformations. We can separate elements of the set of all complex General Coordinate transformations into a product of two factors: a real-valued General Coordinate transformation and a complex-valued General Coordinate transformation. The set of complex factors can be further factored into those that satisfy

$$\Lambda(\omega, \mathbf{u})^T G \Lambda(\omega, \mathbf{u}) = G \tag{19.7}$$

and those that do not. We then see that the set of those that do not satisfy the above equation form a curved space representation of the U(4) group under 'multiplication' of transformations.

The elements of the set of real and complex General Coordinate transformations whose flat complex space-time limit satisfy the above equation form the elements of the Complex Lorentz group.[206]

We thus find the set of all 4-dimensional complex, curved space General coordinate transformations can be visualized as in Fig. 19.1. The next section describes the interplay of the three parts displayed in Fig. 19.1.

19.3 Structure of Complex General Coordinate Transformations

Complex General Coordinate transformations can be uniquely factored into products of two terms, which will later be further factored into three factors. They have the form

$$\partial x''^{\nu}(x)/\partial x^{\mu} = U(x'')^{\nu}{}_{\beta} \, \partial x'^{\beta}(x)/\partial x^{\mu} \tag{19.8}$$

where

$$x''^{\nu}(x) = U(x'')^{\nu}{}_{\beta} x'^{\beta}$$
$$x'^{\mu}(x) = U^{-1\mu}{}_{b}(x'') \, x''^{b}$$

where $U(x')^{\nu}{}_{\beta}$ is complex and where $\partial x'^{\beta}(x)/\partial x^{\mu}$ is a purely real General Coordinate transformation.

We define

$$U(x'')^{\mu}{}_{\nu} = w^{\mu}{}_{a}(x'')\left[\exp\!\left(i \sum_{k} g_k \Phi_k(x'')\tau_k\right)\right]^{a}{}_{b} v^{b}{}_{\nu}(x'') \tag{19.9}$$

$$U^{-1}(x'')^{\mu}{}_{\nu} = w^{\mu}{}_{a}(x'')\left[\exp\!\left(-i \sum_{k} g_k \Phi_k(x'')\tau_k\right)\right]^{a}{}_{b} v^{b}{}_{\nu}(x'')$$

---

[206] It is this part of curved space-time General Relativity that becomes the flat space-time Complex Lorentz group, which leads to the SU(3)⊗SU(2)⊗U(1)⊗SU(2)⊗U(1) Standard Model Reality group.

where the constants g_k are real, and Φ_k and τ_k are hermitean. The uniqueness of the factorization follows from the Reality group (and U(4)) property that any complex 4-vector can be uniquely mapped to any specified real 4-vector.

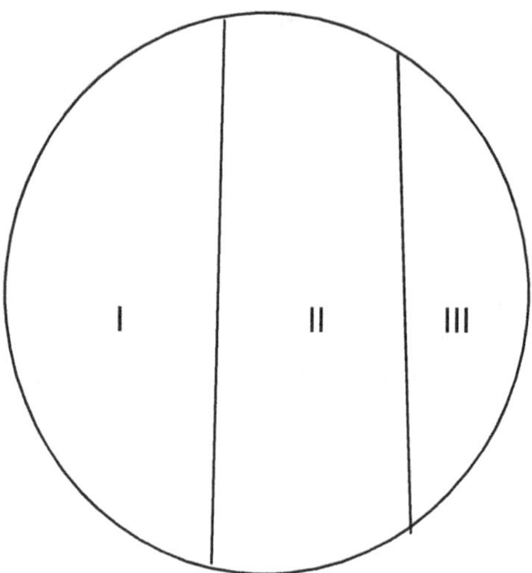

Figure 19.1. A visualization of the set of General Coordinate transformations separated into real-valued General coordinate transformations (part I), complex transformations that satisfy $\Lambda(\omega, \mathbf{u})^T G \Lambda(\omega, \mathbf{u}) = G$ (part II), and complex transformations that do not satisfy $\Lambda(\omega, \mathbf{u})^T G \Lambda(\omega, \mathbf{u}) = G$ (part III). Part I and part II combine in the limit of flat space-time to form the Complex Lorentz group. Parts II and III elements form a U(4) group that we call the General Relativistic Reality group.

Given the factorization above it becomes possible to separate the affine connection correspondingly.

19.4 Complex Affine Connection – General Relativistic Reality Group

The structure of a complex general coordinate transformation enables us to calculate its affine connection for later use in determining the covariant derivative, and the dynamic equations. First the transformation to the real-valued x' coordinates from inertial coordinates is

$$\Gamma^{\sigma}_{\lambda\mu}(x') = \partial x'^{\sigma}/\partial \varsigma^{\rho}\; \partial^2 \varsigma^{\rho}/\partial x'^{\lambda}\partial x'^{\mu} \tag{19.10}$$

Next the Reality group transformation has the affine connection

$$\Gamma^{\sigma}_{\lambda\mu}(x'') = \partial x''^{\sigma}/\partial \varsigma^{\rho}\; \partial^2 \varsigma^{\rho}/\partial x''^{\lambda}\partial x''^{\mu}$$

which can be re-expressed as

$$\Gamma^{\sigma}{}_{\lambda\mu}(x'') = \partial x''^{\sigma}/\partial x'^{\beta}\ \partial x'^{\beta}(\varsigma)/\partial\varsigma^{\rho}\ \partial/\partial x''^{\mu}[\partial\varsigma^{\rho}/\partial x'^{\alpha}\ \partial x'^{\alpha}/\partial x''^{\lambda}]$$
$$= \partial x''^{\sigma}/\partial x'^{\beta}\ \partial x'^{\alpha}/\partial x''^{\lambda}\ \partial x'^{\gamma}/\partial x''^{\mu}\ \Gamma^{\beta}{}_{\alpha\gamma}(x') + \partial x''^{\sigma}/\partial x'^{\beta}\ \partial^{2}x'^{\beta}/\partial x''^{\lambda}\partial x''^{\mu}$$

$$(19.11)$$

Next substituting the General Relativistic Reality group transformation

$$x''^{\nu}(x) = U(x'')^{\nu}{}_{\beta}x'^{\beta}$$
$$x'^{\mu}(x) = U^{-1}(x'')^{\mu}{}_{\beta}\ x''^{\beta}$$

together with

$$\partial x''^{\sigma}/\partial x'^{\beta} = \partial[U(x'')^{\sigma}{}_{\alpha}x'^{\alpha}]/\partial x'^{\beta} = U(x'')^{\sigma}{}_{\beta} + x'^{\alpha}\ \partial U(x'')^{\sigma}{}_{\alpha}/\partial x'^{\beta}$$

$$\partial x'^{\sigma}/\partial x''^{\beta} = \partial[U^{-1}(x'')^{\sigma}{}_{\alpha}x''^{\alpha}]/\partial x''^{\beta} = U^{-1}(x'')^{\sigma}{}_{\beta} + x''^{\alpha}\ \partial U^{-1}(x'')^{\sigma}{}_{\alpha}/\partial x''^{\beta}$$

we find the second term above is the Reality fields affine connection

$$\Gamma_{R}{}^{\sigma}{}_{\lambda\mu}(x'') = \partial[U(x'')^{\sigma}{}_{\alpha}x'^{\alpha}]/\partial x'^{\beta}\ \partial\{\partial[U^{-1}(x'')^{\beta}{}_{\alpha}x''^{\alpha}]/\partial x''^{\lambda}\}/\partial x''^{\mu}$$

and so we find the affine connections are approximately additive. Thus approximately

$$\Gamma^{\sigma}{}_{\lambda\mu}(x'') = \Gamma_{GR}{}^{\sigma}{}_{\lambda\mu}(x') + \Gamma_{R}{}^{\sigma}{}_{\lambda\mu}(x'')$$

if $x''^{\sigma} \simeq x'^{\sigma}$.

A complex transformation of types II and III in Fig. 19.1 has the form:

$$U(x'')^{\mu}{}_{\nu} = w^{\mu}{}_{a}(x'')[\exp(i\sum_{k}\Phi_{k}(x'')\tau_{k})]^{a}{}_{b}\ l^{b}{}_{\nu}(x'')$$
$$U^{-1}(x'')^{\mu}{}_{\nu} = w^{\mu}{}_{a}(x'')[\exp(-i\sum_{k}\Phi_{k}(x'')\tau_{k})]^{a}{}_{b}\ l^{b}{}_{\nu}(x'')$$

where τ_{k} is a U(4) generator matrix. Its infinitesimal transformation is approximately

$$U(x'')^{\nu}{}_{\beta} \approx \delta^{\nu}{}_{\beta} + i\sum_{k}\Phi_{k}(x'')[\tau_{k}]^{\nu}{}_{\beta}$$

$$(19.12)$$

$$U^{-1}(x'')^{\nu}{}_{\beta} \approx \delta^{\nu}{}_{\beta} - i\sum_{k}\Phi_{k}(x'')[\tau_{k}]^{\nu}{}_{\beta}$$

using the *vierbein* flat space-time limits

$$w^{\mu}{}_{a}(x'') \approx \delta^{\mu}{}_{a}$$
$$l^{b}{}_{\nu}(x'') \approx \delta^{b}{}_{\nu}$$

where

$$\Phi_k(x) = \int^{x} dy_\lambda \, A_{Rk}{}^\lambda(y) \tag{19.13}$$

Then

$$\Gamma_R{}^\sigma{}_{\lambda\mu} = -\tfrac{1}{2}i\{\textstyle\sum_k A_{Rk}(x'')_\mu [\tau_k]^\sigma{}_\lambda + \sum_k A_{Rk}(x'')_\lambda [\tau_k]^\sigma{}_\mu\} \tag{19.14}$$

$$= A_R{}^\sigma{}_{\mu\lambda} + A_R{}^\sigma{}_{\lambda\mu}$$

(summed over k) with the matrix $A_R{}^\sigma{}_{\mu\lambda}$ given by

$$A_R{}^\sigma{}_{\mu\lambda} = -\tfrac{1}{2}i\textstyle\sum_k A_{Rk\mu}[\tau_k]^\sigma{}_\lambda \tag{19.15}$$

with $A_R{}^\sigma{}_{\mu\lambda}$ transformable to matrix row and column numbers

$$A_{R_{flat}}{}^{\mu a}{}_b = A_{R_{flatk}}{}^\mu [\tau_k]^\sigma{}_\lambda \delta^a_\sigma \delta^\lambda_b$$

using the flat space-time vierbein values, and so $A_{R_{flat}}{}^a{}_{\mu b}$ may be written in matrix form as

$$A_{R_{flat\mu}} = -\tfrac{1}{2}i\textstyle\sum_k A_{R_{flatk\mu}}\tau_k \tag{19.16}$$

In the flat space-time limit $A_{Rk}{}^\lambda(y)$ becomes the Coordinate Species group U(4) gauge fields $A_{R_{flatk}}{}^\lambda(y)$.

19.5 PseudoQuantization of Affine Connections

Having obtained the form of the general affine connection we now PseudoQuantize them for later use in our unification program. We define

$$R^{1\beta}{}_{\sigma v\mu} = \partial_\mu H^{1\beta}{}_{\sigma v} - \partial_v H^{1\beta}{}_{\sigma\mu} + H^{1\gamma}{}_{v\sigma}H^{1\beta}{}_{\gamma\mu} - H^{1\gamma}{}_{\mu\sigma}H^{1\beta}{}_{\gamma v}$$

$$R^{2\beta}{}_{\sigma v\mu\rho} = \partial_\mu H^{2\beta}{}_{\sigma v} - \partial_v H^{2\beta}{}_{\sigma\mu} + H^{2\gamma}{}_{v\sigma}H^{2\beta}{}_{\gamma\mu} - H^{2\gamma}{}_{\mu\sigma}H^{2\beta}{}_{\gamma v} +$$

$$+ H^{1\gamma}{}_{v\sigma}H^{2\beta}{}_{\gamma\mu} - H^{1\gamma}{}_{\mu\sigma}H^{2\beta}{}_{\gamma v} + H^{2\gamma}{}_{v\sigma}H^{1\beta}{}_{\gamma\mu} - H^{2\gamma}{}_{\mu\sigma}H^{1\beta}{}_{\gamma v} \tag{19.17}$$

where

$$H^\sigma{}_{v\mu} = \Gamma_{GR}{}^\sigma{}_{v\mu} + \Gamma_{GR}{}^{2\sigma}{}_{v\mu} + \Gamma_R{}^{1\sigma}{}_{v\mu} + \Gamma_R{}^{2\sigma}{}_{v\mu} \tag{19.18}$$

and where $\Gamma_{GR}{}^\sigma{}_{v\mu}$ and $\Gamma_{GR}{}^{2\sigma}{}_{v\mu}$ are affine connections for real-valued General Relativity, and $\Gamma_R{}^{1\sigma}{}_{v\mu}$ and $\Gamma_R{}^{2\sigma}{}_{v\mu}$ are affine connections for a complex-valued set of transformations embodying a U(4) gauge group that combine with real-valued General Relativistic transformations to yield Complex General Relativistic transformations.

The affine connection is most often viewed as a derived quantity—part of the derivation of the curvature tensor in General Relativity. It is typically derived from manipulations of the metric $g_{\mu\nu}$. However, the affine connection can also be viewed as a set of independent fields that become related to the metric via dynamic equations.

Some years ago A. Einstein and H. Weyl[207] pointed out that the metric and the affine connection should be treated as independent quantities and subject to independent arbitrary infinitesimal variations:

> "In contrast to Einstein's original "metric" conception in terms of the $g_{\nu\mu}$ there was later developed, by Eddington, by Einstein himself, and recently by Schrödinger, an affine field theory operating with the components $\Gamma^{\sigma}_{\nu\mu}$ of an affine connection. But in 1925 Einstein also advocated a "mixed" formulation by means of a lagrangian in which both the $g_{\nu\mu}$ and the $\Gamma^{\sigma}_{\nu\mu}$ are taken as basic field quantities and submitted to independent arbitrary infinitesimal variations.[208] In certain respects this seems to be the most natural procedure."

Following this approach we have introduced the above affine connections for use in the construction of our unification of the eleven particle interactions.

---

[207] H. Weyl, Phys. Rev. **77**, 699 (1950).
[208] A. Einstein, Sitzungsber., Preuss. Akad. Der Wissensch. (1925), p. 414.

20. Species Group U(4) Gauge Fields

From the discussion in sections 19.4-19.5 we see the flat space-time limit of $A_{Rk}{}^{\lambda}(y)$ is a local U(4) coordinate space gauge field.There are, *by assumption,*[209] a corresponding internal symmetry gauge fields $A_{Sk}{}^{\lambda}(y)$ – the Internal Symmetry U(4) Species Group gauge fields. The mathematical features of this field is quite similar to to the U(4) Generation group fields. The interaction that appears in covariant derivatives is $g_8 A_S{}^{\mu}(x) = g_8 A_{Sk}{}^{\mu}(x)\mathbf{G}_{Sk}$ where the $\mathbf{G}_{Sk}$ are U(4) generator matrices[210] and k is summed from 1, ... , 16.

Below we will see that the effect of the Internal Symmetry Species Group is to U(4) rotate the four components of each fermion's field. Since it preserves the species of each fermion we call this group the *Species Group.* It performs a U(4) rotation of the spinor representation of each fermion

We will see that the Higgs Mechanism breakdown of the Species Group endows each fermion with a mass[211] contribution that breaks the scale invariance of the Unified SuperStandard Model.

Since the Species Group Higgs Mechanism breaking gives each fermion a 'gravity generated' mass, and since this mass sets the mass scale for each fermion, we conclude later that the principle of the equality of inertial and gravitational mass is a direct consequence. *Inertial mass equals gravitational mass.*

20.1 Species Group Covariance

A Species Group transformation on a Dirac equation must be covariant. Consider the Dirac equation under an Internal Symmetry Species Group transformation:

$$\overline{\psi}(x)[i\gamma_{\mu}(\partial/\partial x_{\mu} - ig_8 A_{Sk}{}^{\mu}(x)\mathbf{G}_{Sk}) - m]\psi(x) = 0 \tag{20.1}$$

summed over k. If we perform a Species group transformation U on lagrangian terms:

$$\overline{\psi}(x)[i\gamma_{\mu}(\partial/\partial x_{\mu} - ig_8 A_{Sk}{}^{\mu}(x)\mathbf{G}_{Sk}) - m]U^{-1}U\psi(x)$$

---

[209] In this discussion we *assume* that the Coordinate Species Reality Group with gauge fields $A_{Rk}{}^{\lambda}(y)$ has a corresponding Internal Symmetry Reality group that we call the Species Group. This assumption parallels the assumptions for the $SU(3){\otimes}SU(2){\otimes}U(1){\otimes}SU(2){\otimes}U(1)$ Internal Symmetry Reality Group presented in previous chapters.

[210] In chapter 19 the $\mathbf{G}_{Sk}$ matrices were denoted τ_k.

[211] The breakdown of Θ-symmetry also gives each fermion as mass and thus also breaks scale invariance.

or

$$\bar{\psi}(x)U^{-1}U[iU^{-1}U\gamma_\mu U^{-1}U(\partial/\partial x_\mu - ig_8 A_{Sk}{}^\mu(x)G_{Sk}) - m]U^{-1}U\psi(x)$$

we find

$$\bar{\psi}'(x)[i\gamma_\mu'U(\partial/\partial x_\mu - ig_8 A_{Sk}{}^\mu(x)G_{Sk})U^{-1} - m]\psi'(x)$$

where

$$\gamma_\mu'(x) = U\gamma_\mu U^{-1}$$

is locally equivalent to a Dirac matrix by Good's Theorem.[212] If we set

$$A'_S{}^\mu(x) = -(i/g_8)U[\partial U^{-1}/\partial x^\mu] + UA_S{}^\mu(x)U^{-1}$$

then the transformed lagrangian terms are

$$\bar{\psi}'(x)[i\gamma_\mu'(x)(\partial/\partial x_\mu - ig_8 A'_{Sk}{}^\mu(x)G_{Sk}) - m]\psi'(x) \tag{20.2}$$

They have the same form as the original terms above and thus the expression is covariant. We note the indices of the matrices G_{Sk} are spinor indices and so $G_{Sk}\gamma_\mu$ has an implicit spinor matrix summation. But the symmetry group is U(4).

The coordinate dependence of $\gamma_\mu'(x)$ introduces locality into the Dirac matrix. This locality might be viewed with concern except that an inverse Species group transformation exists that removes the locality. Thus the physical impact of this 'new' locality is eliminated.

20.2 Spontaneous Symmetry Breaking of the General Relativity U(4) Reality Group – The Species Group

We begin the discussion of the Internal Symmetry Species Group symmetry breaking by defining a Higgs field η which is a Species group 4-vector

$$\eta = \begin{bmatrix} \rho_1 \\ \rho_2 \\ \rho_3 \\ \rho_4 \end{bmatrix} \tag{20.3}$$

where ρ_1, ρ_2, ρ_3 and ρ_4 are real fields.[213] Then the covariant derivative of η (taking account only of the Species group) is

---

[212] R. H. Good, Jr., Rev. Mod. Phys., **27**, 187 (1955).

$$D_{...\mu}\,\eta \;=\; \{\partial /\partial X^{\mu} + \ldots - \tfrac{1}{2}ig_8 \Sigma\, A_{Sk}{}^{\mu}(x)G_{Sk}\} \begin{bmatrix} \rho_1 \\ \rho_2 \\ \rho_3 \\ \rho_4 \end{bmatrix}$$

$$(20.3)$$

Following steps similar to eqs. 15.4 through 15.17 for the Generation Group symmetry breaking we find with ρ_i being the vacuum expectation value of the Higgs field:

$$
(D_{...\mu}\eta)^{\dagger} D_{...}{}^{\mu}\eta \;=\; \partial\rho_1/\partial X^{\mu}\,\partial\rho_1/\partial X_{\mu} + \partial\rho_2/\partial X^{\mu}\,\partial\rho_2/\partial X_{\mu} + \partial\rho_3/\partial X^{\mu}\,\partial\rho_3/\partial X_{\mu} + \partial\rho_4/\partial X^{\mu}\,\partial\rho_4/\partial X_{\mu} +
$$
$$
+\; \tfrac{1}{4}\,g_8{}^2\{\rho_1{}^2 A_{S1}{}^2 + \rho_2{}^2 A_{S2}{}^2 + \rho_3{}^2 A_{S3}{}^2 + \rho_4{}^2 A_{S4}{}^2 +
$$
$$
+\,(\rho_1{}^2 + \rho_2{}^2)(V_5{}^2 + V_6{}^2) + \tfrac{1}{4}(\rho_1{}^2 + \rho_3{}^2)(V_7{}^2 + V_8{}^2) +
$$
$$
+\,(\rho_1{}^2 + \rho_4{}^2)(V_9{}^2 + V_{10}{}^2) + \tfrac{1}{4}(\rho_2{}^2 + \rho_3{}^2)(V_{11}{}^2 + V_{12}{}^2) +
$$
$$
+\,(\rho_2{}^2 + \rho_4{}^2)(V_{13}{}^2 + V_{14}{}^2) + \tfrac{1}{4}(\rho_3{}^2 + \rho_4{}^2)(V_{15}{}^2 + V_{16}{}^2)\}
$$

$$(20.4)$$

up to total divergences, which generate surface terms which we discard. We also assume that all fields satisfy the gauge condition

$$\partial A_{Si}{}^{\mu}/\partial X^{\mu} = 0 \qquad\qquad (20.5)$$

Eq. 20.4 shows all Species Group gauge fields have masses. Thus Species Group symmetry is completely broken. The combination of an ultra-weak coupling constant and very large gauge field masses results in extremely Species interactions.

We assume Species group gauge field masses to be very large – of the order of the Planck mass in view of its origin in Complex General Relativity.

20.3 Species Group Higgs Mechanism Contributions to Fermion Masses

The symmetry breaking of the Species Group results in a contribution to each fermion mass of all types, species, generations, and layers. The Species Group contributions to normal and Dark fermion mass terms are

$$\mathcal{L}^{Higgs}{}_{FermionMassesSpecies} = \Sigma_{s,g,l}\,\bar{\psi}_{sglL}\,\rho_s\, m_{sgl}\,\psi_{sglR} + \Sigma_{s,g,l}\,\bar{\psi}_{DsglL}\,\rho_s\, m_{Dsgl}\,\psi_{DsglR} + \text{c.c.} \qquad (20.6)$$

The η field expectation value has components labeled ρ_s.[214] The mass matrices m_{sgl} and m_{Dsgl} are the complex constant Species mass matrix contributions for normal and Dark species.

---

[213] Each field ρ_i can be expressed as a Pseudoquantum field: $\rho_i = \varphi_{1i} + \varphi_{2i}$ where φ_{1i} has the vacuum expectation value ρ_{i0} for $i = 1, \ldots, 4$. Thus our Pseudoquantum field theory version is implemented easily.
[214] The Higgs fields $\eta\ldots$ in our Pseudoquantum formulation are $\eta\ldots = \varphi_{1...}(x) + \varphi_{2...}(x)$ as described earlier.

20.4 Species Group Higgs Masses Shows Inertial Mass Equals Gravitational Mass

In Blaha (2016h) we showed that a Complex General Relativity transformation can be factored into the product of a complex-valued transformation and a real-valued General Coordinate transformation. The set of complex valued transformations form a U(4) group that we called the General Coordinate Reality group. The analogous Internal Symmetry Species Group has gauge fields that undergo spomtaneous symmetry breaking and generate contributions to all fermion masses.

Since fermion field masses are now sums of ElectroWeak Higgs contributions, Generation group Higgs contributions, Layer group Higgs contributions, and General Coordinate Species Group contributions, and since the Species Group Higgs fields appear in all fermion masses, the equality of inertial and gravitational mass is proven. The Species Group Higgs particles' equations set the mass scale of gravitational mass, and thereby of all Higgs mass contributions. The scale of inertial masses equals to the scale of gravitational masses. **Since an expression cannot mix mass scales, the gravitational mass scale must be the same as the inertial mass scale.**

Inertial mass equals gravitational mass.

We have established the equality of inertial and gravitational mass at the short distance quantum level. In our view, this explanation is far more satisfying than basing the equality on a combination of large distance phenomena and quantum phenomena. As Einstein and Weyl have pointed out, all fundamental physics phenomena should be based on a local theory.

We have mapped Complex General Relativitistic transformations consisting of U(4) transformations and Real General Relativistic transformations into transformations consisting of Internal Symmetry Species Group factors and Real General Relativistic transformations factors. The Higgs Mechanism breakdown of the Species Group has the important consequence that it prevents Species Group transformations that rotate between fermions and anti-fermions.

21. Derivation Status at this Point

At this point in our derivation of the Unified SuperStandard Model we have completed the following:

1. Derived the dimension of space-time based on the requirement that Physical phenomena must support parallel processes.
2. Assuming Complex Lorentz group symmetry we have derived the existence of exactly four fermion species and four Dark fermion species. Using the Reality group associated with the Complex Lorentz group as a guide we have postulated the existence of the Standard Model symmetry group $SU(3) \otimes SU(2) \otimes U(1) \otimes SU(2) \otimes U(1)$ with the additional $SU(2) \otimes U(1)$ factors for Dark ElectroWeak symmetry.
3. Assuming that the Unified SuperStandard Model can be described by a lagrangian we have derived conserved particle numbers such as Baryon Number. The four conserved particle numbers for baryons, leptons, Dark baryons and Dark leptons suggest a $U(4)$ symmetry that we call the Generation Group. This group's $\underline{4}$ representation gives a four generation fermion spectrum.
4. The enhanced free fermion lagrangian, now having four generations of fermions, has four new conserved fermion numbers for each of the four generations. The four conserved particle numbers for each generation are for baryons, leptons, Dark baryons and Dark leptons of each generation. These numbers suggest four $U(4)$ symmetries – one $U(4)$ symmetry for each generation. We call these symmetries Layer Groups symmetries. We then add indices to fermion fields of each generation to create a $U(4)$ $\underline{4}$ representation (Fig. 13.2). Thus we have four sets of four generations. We call each set of four generations a layer.(Fig. 13.3)
5. We assume each Layer has its own set of $SU(3) \otimes SU(2) \otimes U(1) \otimes SU(2) \otimes U(1) \otimes U(4)$ where the $U(4)$ factor is for the Generation group. We assume the gauge interactions of each layer is different since transitions between layers due to these interactions have not been found experimentally. Introducing a $U(4)$ Layer group factor for each of the generations we can specify the Unified SuperStandard Model symmetry in a compact form as

$$[SU(3) \otimes SU(2) \otimes U(1) \otimes SU(2) \otimes U(1) \otimes U(4) \otimes U(4)]^4$$

at this point in the derivation.
6. Introducing $U(192)$ gauge fields for the 'rotation of interactions' Θ-Symmetry group, G_Θ, the Unified SuperStandard symmetry group grows to

$$[SU(3) \otimes SU(2) \otimes U(1) \otimes SU(2) \otimes U(1) \otimes U(4) \otimes U(4)]^4 \otimes U(192)$$

since there are 192 gauge fields and 192 fermion fields in the Model at present.

7. Turning to Complex General Relativity we find that it factors into a U(4) group, which becomes the Internal Symmetry Species Group, combined with Real General Relativity.

8. Thus the Unified SuperStandard Model symmetry is

$$[SU(3)\otimes SU(2)\otimes U(1)\otimes SU(2)\otimes U(1)\otimes U(4)\otimes U(4)]^4\otimes U(192)\otimes U(4)$$

plus Real General Relativity.

9. We also showed how to eliminate divergences in perturbation theory calculations using the Two-Tier formulation of Quantum Field Theory. The Two-Tier theory essentially smears space-time points at ultra-small distances eliminating divergences.

10. In addition we described PseudoQuantum Field Theory which enables quantization in any coordinate system, offers a cleaner formulation of symmetry breaking and supports canonical formulations of higher derivative field theories.

11. The Higgs Mechanism for symmetry breaking and particle mass generation for all the above interactions was also described.

The appearance of the integer 4 throughout the Unified SuperStandard Model can be traced back to the axiom that physical processes can take place in parallel, which determined the dimension of space-time in our universe to be 4. The 4 generators of qubit interactions, the fundamental construct of Logic and Physics, also specified a 4-dimensional space-time based on the principle that the number of fundamental interactions determines the dimension of space-time. There is no space-time without interactions since all mass-energy can be concentrated at a point in the absence of interactions.

22. The Riemann-Christoffel Curvature Tensor and Vector Boson Part of Unified SuperStandard Model Lagrangian

22.1 The Covariant Derivative

The covariant derivative[215] which appears in fermion and gravitation equations uses

$$^a\mathbf{A}_I{}^\mu(x) = (^a g_1{}^a\mathbf{A}_{SU(3)}{}^\mu(x_C),\ ^a g_2{}^a\mathbf{W}^\mu(x)\ ,\ ^a g_3{}^a\mathbf{A}_E{}^\mu(x),\ ^a g_4{}^a\mathbf{W}_D{}^\mu(x),\ ^a g_5{}^a\mathbf{A}_{DE}{}^\mu(x),\ ^a g_6{}^a\mathbf{U}^\mu(x),\ ^a g_7{}^a\mathbf{V}^\mu(x)) \tag{22.1}$$

where a labels the layer, a = 1, 2, 3, 4. We define the sum over a and the the components of the vector $^a\mathbf{A}_I{}^\mu(x)$ labeled with i by

$$\mathbf{C}_I{}^\mu(x) = \Sigma_{a,i}\ ^a\mathbf{A}_{Ii}{}^\mu(x) + g_8\mathbf{A}_S{}^\mu(x) + g_\Theta\mathbf{A}_\Theta{}^\mu(x) \tag{22.2}$$

For simplicity we label the respective coupling constants in layer a as $^a g_1$, $^a g_2$, …, $^a g_7$. In the equation above: the subscript 'D' labels Dark matter interactions, 'W' labels Weak fields, 'E' labels Electromagnetic fields, $\mathbf{U}^\mu(x)$ labels U(4) Generation group fields, and 'V' labels U(4) Layer group fields. $\mathbf{A}_S$ labels the U(4) Species Group fields.

The symmetry is $[SU(3)\otimes SU(2)\otimes U(1)\otimes SU(2)\otimes U(1)\otimes U(4)\otimes U(4)]^4\otimes U(4)\ \otimes U(192)$. The number of gauge fields for each of the elements of the vectors in each layer is 8. 3, 1, 3, 1, 16, and 16, totalling 48 components. For four layers there are 4×48 = 192 elements plus 16 Species gauge fields and 192^2 Θ-Symmetry fields.

We will consider the case of one layer of vector bosons below omitting the ^{a} superscript. The generalization to all layers is straightforward.

Using the above definitions the PseudoQuantum covariant derivative of a 4-vector is

$$D_\nu V_\mu = (\partial_\nu + iF_\nu)V_\mu - H^\sigma{}_{\nu\mu}V_\sigma \tag{22.3}$$
$$= [g^\sigma{}_\mu\partial_\nu + ig^\sigma{}_\mu F_\nu - H^\sigma{}_{\nu\mu}]V_\sigma$$
$$= [g^\sigma{}_\mu\partial_\nu + iD^\sigma{}_{\mu\nu}]V_\sigma$$

where[216]

---

[215] This section has equations obtained from Blaha (2017d).

[216] We will omit the insertion of the spinor coupling constants of $B^{1\mu}$ and $B^{2\mu}$ in eq. 22.2 in the interests of simplifying expressions.

$$F^\mu = C_I^{1\mu}(x) + \mathbf{C}_I^{2\mu}(x) + B^{1\mu} + B^{2\mu} \tag{22.4}$$

and

$$H^\sigma{}_{\nu\mu} = \Gamma_{GR}{}^\sigma{}_{\nu\mu} + \Gamma_{GR}{}^{2\sigma}{}_{\nu\mu} \tag{22.5}$$

$$D^\sigma{}_{\mu\nu} = g^\sigma{}_\mu F_\nu + iH^\sigma{}_{\nu\mu}$$

where we have abstracted the complex part of the complex affine connection into the U(4) gauge field $A_S{}^\mu$. $H^\sigma{}_{\nu\mu}$ is the real-valued part of the complex affine connection.

Commutators of the vector fields in F_μ are implicit when the covariant derivative is applied to vectors and tensors such as V_σ.

22.2 The Curvature Tensor

The curvature tensor applied to a 4-vector is

$$R'^\beta{}_{\sigma\nu\mu}V_\beta = g^\alpha{}_\mu(\partial_\nu + iF_\nu)g^\beta{}_\sigma(\partial_\alpha + iF_\alpha)V_\beta - H^\alpha{}_{\mu\nu}g^\beta{}_\sigma(\partial_\alpha + iF_\alpha)V_\beta + \tag{22.6}$$
$$+ H^\alpha{}_{\mu\nu}H^\beta{}_{\sigma\alpha}V_\beta - g^\alpha{}_\mu(\partial_\nu + iF_\nu)H^\beta{}_{\sigma\alpha}V_\beta - H^\gamma{}_{\nu\sigma}\{g^\alpha{}_\gamma(\partial_\mu + iF_\mu)V_\alpha - H^\alpha{}_{\gamma\mu}V_\alpha\} -$$
$$- \{\mu \leftrightarrow \nu\}$$

$$= ig^\beta{}_\sigma(\partial_\nu F_\mu - \partial_\mu F_\nu - i[F_\nu, F_\mu])V_\beta + (\partial_\mu H^\beta{}_{\sigma\nu} - \partial_\nu H^\beta{}_{\sigma\mu} + H^\gamma{}_{\nu\sigma}H^\beta{}_{\gamma\mu} - H^\gamma{}_{\mu\sigma}H^\beta{}_{\gamma\nu})V_\beta$$

$$= ig^\beta{}_\sigma(F_E{}^1{}_{\nu\mu} + F_E{}^2{}_{\nu\mu} + F_W{}^1{}_{\nu\mu} + F_W{}^2{}_{\nu\mu} + F_{DE}{}^1{}_{\nu\mu} + F_{DE}{}^2{}_{\nu\mu} + F_{DW}{}^1{}_{\nu\mu} + F_{DW}{}^2{}_{\nu\mu} + F_{SU(3)}{}^1{}_{\nu\mu} +$$
$$+ F_{SU(3)}{}^2{}_{\nu\mu} + F_U{}^1{}_{\nu\mu} + F_U{}^2{}_{\nu\mu} + F_V{}^1{}_{\nu\mu} + F_V{}^2{}_{\nu\mu} + F_S{}^1{}_{\nu\mu} + F_S{}^2{}_{\nu\mu} + F_\Theta{}^1{}_{\nu\mu} + F_\Theta{}^2{}_{\nu\mu} + F_B{}^1{}_{\nu\mu} + F_B{}^2{}_{\nu\mu})V_\beta +$$
$$+ (\partial_\mu H^\beta{}_{\sigma\nu} - \partial_\nu H^\beta{}_{\sigma\mu} + H^\gamma{}_{\nu\sigma}H^\beta{}_{\gamma\mu} - H^\gamma{}_{\mu\sigma}H^\beta{}_{\gamma\nu})V_\beta$$

$$= R'_E{}^\beta{}_{\sigma\nu\mu}V_\beta + R'_{SU(2)}{}^\beta{}_{\sigma\nu\mu}V_\beta + R'_{DE}{}^\beta{}_{\sigma\nu\mu}V_\beta + R'_{DSU(2)}{}^\beta{}_{\sigma\nu\mu}V_\beta + R'_{SU(3)}{}^\beta{}_{\sigma\nu\mu}V_\beta + R'_U{}^\beta{}_{\sigma\nu\mu}V_\beta +$$
$$+ R'_V{}^\beta{}_{\sigma\nu\mu}V_\beta + R'_S{}^\beta{}_{\sigma\nu}V_\beta + R'_\Theta{}^\beta{}_{\sigma\nu}V_\beta + R'_B{}^\beta{}_{\sigma\nu}V_\beta + R'_G{}^\beta{}_{\sigma\nu\mu}V_\beta$$

where all $F_{...}{}^1{}_{\nu\mu}$ and $F_{...}{}^2{}_{\nu\mu}$ terms have summations over the four layers (see below) except the terms $F_S{}^1{}_{\nu\mu} + F_S{}^2{}_{\nu\mu} + F_\Theta{}^1{}_{\nu\mu} + F_\Theta{}^2{}_{\nu\mu} + F_B{}^1{}_{\nu\mu} + F_B{}^2{}_{\nu\mu}$, and where

$$R'_{SU(3)}{}^\beta{}_{\sigma\nu\mu} = ig^\beta{}_\sigma(F_{SU(3)}{}^1{}_{\nu\mu} + F_{SU(3)}{}^2{}_{\nu\mu}) \tag{22.7}$$
$$R'_{SU(2)}{}^\beta{}_{\sigma\nu\mu} = ig^\beta{}_\sigma(F_W{}^1{}_{\nu\mu} + F_W{}^2{}_{\nu\mu})$$
$$R'_E{}^\beta{}_{\sigma\nu\mu} = ig^\beta{}_\sigma(F_E{}^1{}_{\nu\mu} + F_E{}^2{}_{\nu\mu})$$
$$R'_U{}^\beta{}_{\sigma\nu\mu} = ig^\beta{}_\sigma(F_U{}^1{}_{\nu\mu} + F_U{}^2{}_{\nu\mu})$$
$$R'_V{}^\beta{}_{\sigma\nu\mu} = ig^\beta{}_\sigma(F_V{}^1{}_{\nu\mu} + F_V{}^2{}_{\nu\mu})$$
$$R'_{DSU(2)}{}^\beta{}_{\sigma\nu\mu} = ig^\beta{}_\sigma(F_{DW}{}^1{}_{\nu\mu} + F_{DW}{}^2{}_{\nu\mu})$$
$$R'_{DE}{}^\beta{}_{\sigma\nu\mu} = ig^\beta{}_\sigma(F_{DE}{}^1{}_{\nu\mu} + F_{DE}{}^2{}_{\nu\mu})$$
$$R'_S{}^\beta{}_{\sigma\nu\mu} = ig^\beta{}_\sigma(F_S{}^1{}_{\nu\mu} + F_S{}^2{}_{\nu\mu})$$

$$R'_{\Theta}{}^{\beta}{}_{\sigma\nu\mu} = ig^{\beta}{}_{\sigma}(F_{\Theta}{}^{1}{}_{\nu\mu} + F_{\Theta}{}^{2}{}_{\nu\mu})$$

$$R'_{B}{}^{\beta}{}_{\sigma\nu\mu} = ig^{\beta}{}_{\sigma}(F_{B}{}^{1}{}_{\nu\mu} + F_{B}{}^{2}{}_{\nu\mu})$$

and

$$R'_{G}{}^{\beta}{}_{\sigma\nu\mu} = \partial_{\mu}H^{1\beta}{}_{\sigma\nu} - \partial_{\nu}H^{1\beta}{}_{\sigma\mu} + H^{1\gamma}{}_{\nu\sigma}H^{1\beta}{}_{\gamma\mu} - H^{1\gamma}{}_{\mu\sigma}H^{1\beta}{}_{\gamma\nu} + \partial_{\mu}H^{2\beta}{}_{\sigma\nu} - \partial_{\nu}H^{2\beta}{}_{\sigma\mu} + \tag{22.8}$$

$$+ H^{2\gamma}{}_{\nu\sigma}H^{2\beta}{}_{\gamma\mu} - H^{2\gamma}{}_{\mu\sigma}H^{2\beta}{}_{\gamma\nu} + H^{1\gamma}{}_{\nu\sigma}H^{2\beta}{}_{\gamma\mu} - H^{1\gamma}{}_{\mu\sigma}H^{2\beta}{}_{\gamma\nu} + H^{2\gamma}{}_{\nu\sigma}H^{1\beta}{}_{\gamma\mu} - \Gamma^{2\gamma}{}_{\mu\sigma}\Gamma^{\beta}{}_{\gamma\nu}$$

$$= R^{1\beta}{}_{\sigma\nu\mu} + R^{2\beta}{}_{\sigma\nu\mu}$$

with

$$H^{\beta}{}_{\sigma\nu\mu} = \partial_{\mu}H^{\beta}{}_{\sigma\nu} - \partial_{\nu}H^{\beta}{}_{\sigma\mu} + H^{\gamma}{}_{\nu\sigma}H^{\beta}{}_{\gamma\mu} - H^{\gamma}{}_{\mu\sigma}H^{\beta}{}_{\gamma\nu} \tag{22.9}$$

$$R^{1\beta}{}_{\sigma\nu\mu} = \partial_{\mu}H^{1\beta}{}_{\sigma\nu} - \partial_{\nu}H^{1\beta}{}_{\sigma\mu} + H^{1\gamma}{}_{\nu\sigma}H^{1\beta}{}_{\gamma\mu} - H^{1\gamma}{}_{\mu\sigma}H^{1\beta}{}_{\gamma\nu}$$

$$R^{2\beta}{}_{\sigma\nu\mu p} = \partial_{\mu}H^{2\beta}{}_{\sigma\nu} - \partial_{\nu}H^{2\beta}{}_{\sigma\mu} + H^{2\gamma}{}_{\nu\sigma}H^{2\beta}{}_{\gamma\mu} - H^{2\gamma}{}_{\mu\sigma}H^{2\beta}{}_{\gamma\nu} +$$

$$+ H^{1\gamma}{}_{\nu\sigma}H^{2\beta}{}_{\gamma\mu} - H^{1\gamma}{}_{\mu\sigma}H^{2\beta}{}_{\gamma\nu} + H^{2\gamma}{}_{\nu\sigma}H^{1\beta}{}_{\gamma\mu} - H^{2\gamma}{}_{\mu\sigma}H^{1\beta}{}_{\gamma\nu}$$

and

$$H^{1\sigma}{}_{\nu\mu} = \Gamma_{GR}{}^{\sigma}{}_{\nu\mu} \tag{22.10}$$

$$H^{2\sigma}{}_{\nu\mu} = \Gamma_{GR}{}^{2\sigma}{}_{\nu\mu}$$

and where with summations over four layers indicated by Σ (and layer numbers on fields not shown to avoid clutter) we have

$$F_{SU(3)}{}^{1}{}_{\varkappa\mu} = \Sigma \{\partial A_{SU(3)}{}^{1}{}_{\mu}/\partial x^{\varkappa} - \partial A_{SU(3)}{}^{1}{}_{\varkappa}/\partial x^{\mu} + ig_{1}[A_{SU(3)}{}^{1}{}_{\varkappa}, A_{U(3)}{}^{1}{}_{\mu}]\} \tag{22.11}$$

$$F_{W}{}^{1}{}_{\varkappa\mu} = \Sigma \{\partial W^{1}{}_{\mu}/\partial x^{\varkappa} - \partial W^{1}{}_{\varkappa}/\partial x^{\mu} + ig_{2}[W^{1}{}_{\varkappa}, W^{1}{}_{\mu}]\}$$

$$F_{E}{}^{1}{}_{\varkappa\mu} = \Sigma \{\partial A_{E}{}^{1}{}_{\mu}/\partial x^{\varkappa} - \partial A_{E}{}^{1}{}_{\varkappa}/\partial x^{\mu}\}$$

$$F_{DW}{}^{1}{}_{\varkappa\mu} = \Sigma \{\partial W_{D}{}^{1}{}_{\mu}/\partial x^{\varkappa} - \partial W_{D}{}^{1}{}_{\varkappa}/\partial x^{\mu} + ig_{4}[W_{D}{}^{1}{}_{\varkappa}, W_{D}{}^{1}{}_{\mu}]\}$$

$$F_{DE}{}^{1}{}_{\varkappa\mu} = \Sigma \{\partial A_{DE}{}^{1}{}_{\mu}/\partial x^{\varkappa} - \partial A_{DE}{}^{1}{}_{\varkappa}/\partial x^{\mu}\}$$

$$F_{U}{}^{1}{}_{\varkappa\mu} = \Sigma \{\partial U^{1}{}_{\mu}/\partial x^{\varkappa} - \partial U^{1}{}_{\varkappa}/\partial x^{\mu} + ig_{6}[U^{1}{}_{\varkappa}, U^{1}{}_{\mu}]\}$$

$$F_{V}{}^{1}{}_{\varkappa\mu} = \Sigma \{\partial V^{1}{}_{\mu}/\partial x^{\varkappa} - \partial V^{1}{}_{\varkappa}/\partial x^{\mu} + ig_{7}[V^{1}{}_{\varkappa}, V^{1}{}_{\mu}]\}$$

$$F_{S}{}^{1}{}_{\varkappa\mu} = \partial A_{S}{}^{1}{}_{\mu}/\partial x^{\varkappa} - \partial A_{S}{}^{1}{}_{\varkappa}/\partial x^{\mu} + ig_{8}[A_{S}{}^{1}{}_{\varkappa}, A_{S}{}^{1}{}_{\mu}]$$

$$F_{\Theta}{}^{1}{}_{\varkappa\mu} = \partial A_{\Theta}{}^{1}{}_{\mu}/\partial x^{\varkappa} - \partial A_{\Theta}{}^{1}{}_{\varkappa}/\partial x^{\mu} + ig_{\Theta}[A_{\Theta}{}^{1}{}_{\varkappa}, A_{\Theta}{}^{1}{}_{\mu}]$$

$$F_{B}{}^{1}{}_{\varkappa\mu} = \partial B^{1}{}_{\mu}/\partial x^{\varkappa} - \partial B^{1}{}_{\varkappa}/\partial x^{\mu} + i[B^{1}{}_{\varkappa}, B^{1}{}_{\mu}]$$

$$F_{SU(3)}{}^{2}{}_{\varkappa\mu} = \Sigma \{\partial A_{SU(3)}{}^{2}{}_{\mu}/\partial x^{\varkappa} - \partial A_{SU(3)}{}^{2}{}_{\varkappa}/\partial x^{\mu} + ig_{1}[A_{SU(3)}{}^{2}{}_{\varkappa}, A_{SU(3)}{}^{2}{}_{\mu}] + \tag{22.12}$$

$$+ ig_{1}[A_{SU(3)}{}^{1}{}_{\varkappa}, A_{SU(3)}{}^{2}{}_{\mu}] + ig_{1}[A_{SU(3)}{}^{2}{}_{\varkappa}, A_{SU(3)}{}^{1}{}_{\mu}]\}$$

$$F_{W}{}^{2}{}_{\varkappa\mu} = \Sigma \{\partial W^{2}{}_{\mu}/\partial x^{\varkappa} - \partial W^{2}{}_{\varkappa}/\partial x^{\mu} + ig_{2}[W^{2}{}_{\varkappa}, W^{2}{}_{\mu}] + ig_{2}[W^{1}{}_{\varkappa}, W^{2}{}_{\mu}] + ig_{2}[W^{2}{}_{\varkappa}, W^{1}{}_{\mu}]\}$$

$$F_{E}{}^{2}{}_{\varkappa\mu} = \Sigma \{\partial A_{E}{}^{2}{}_{\mu}/\partial x^{\varkappa} - \partial A_{E}{}^{2}{}_{\varkappa}/\partial x^{\mu}\}$$

$$F_{DW}{}^2{}_{\varkappa\mu} = \Sigma \ \{\partial W_D{}^2{}_\mu/\partial x^\varkappa - \partial W_D{}^2{}_\varkappa/\partial x^\mu + ig_4[W_D{}^2{}_\varkappa, W_D{}^2{}_\mu] + ig_4[W_D{}^1{}_\varkappa, W_D{}^2{}_\mu] +$$
$$+ \ ig_4[W_D{}^2{}_\varkappa, W_D{}^1{}_\mu]\}$$

$$F_{DE}{}^2{}_{\varkappa\mu} = \Sigma \ \{\partial A_{DE}{}^2{}_\mu/\partial x^\varkappa - \partial A_{DE}{}^2{}_\varkappa/\partial x^\mu\}$$

$$F_U{}^2{}_{\varkappa\mu} = \Sigma \ \{\partial U^2{}_\mu/\partial x^\varkappa - \partial U^2{}_\varkappa/\partial x^\mu + ig_6[U^2{}_\varkappa, U^2{}_\mu] + ig_6[U^1{}_\varkappa, U^2{}_\mu] + ig_6[U^2{}_\varkappa, U^1{}_\mu]\}$$

$$F_V{}^2{}_{\varkappa\mu} = \Sigma \ \{\partial V^2{}_\mu/\partial x^\varkappa - \partial V^2{}_\varkappa/\partial x^\mu + ig_7[V^2{}_\varkappa, V^2{}_\mu] + ig_7[V^1{}_\varkappa, V^2{}_\mu] + ig_7[V^2{}_\varkappa, V^1{}_\mu]\}$$

$$F_S{}^2{}_{\kappa\mu} = \partial A_S{}^2{}_\mu/\partial x^\kappa - \partial A_S{}^2{}_\kappa/\partial x^\mu + ig_8[A_S{}^2{}_\kappa, A_S{}^2{}_\mu] + ig_8[A_S{}^1{}_\kappa, A_S{}^2{}_\mu] +$$
$$+ \ ig_8[A_S{}^2{}_\kappa, A_S{}^1{}_\mu]$$

$$F_\Theta{}^2{}_{\kappa\mu} = \partial A_\Theta{}^2{}_\mu/\partial x^\kappa - \partial A_\Theta{}^2{}_\kappa/\partial x^\mu + ig_\Theta[A_\Theta{}^2{}_\kappa, A_\Theta{}^2{}_\mu] + ig_\Theta[A_\Theta{}^1{}_\kappa, A_\Theta{}^2{}_\mu] + ig_\Theta[A_\Theta{}^2{}_\kappa, A_\Theta{}^1{}_\mu]$$

$$F_B{}^2{}_{\varkappa\mu} = \partial B^2{}_\mu/\partial x^\varkappa - \partial B^2{}_\varkappa/\partial x^\mu + i[B^2{}_\mu, B^2{}_\varkappa] + i[B^1{}_\mu, B^2{}_\varkappa] + i[B^2{}_\mu, B^1{}_\varkappa]$$

Note that $R'^\beta{}_{\sigma\nu\mu}$ factorizes into $[U(1){\otimes}SU(2){\otimes}U(1){\otimes}SU(2){\otimes}SU(3){\otimes}U(4){\otimes}U(4)]^4{\otimes}U(4){\otimes}U(192)$ parts and a Riemann-Christoffel Gravitational curvature tensor part. For later use in defining a lagrangian we define

$$R'^\beta{}_{\sigma\nu\mu} = R'_E{}^{1\beta}{}_{\sigma\nu\mu} + R'_E{}^{2\beta}{}_{\sigma\nu\mu} + R'_{SU(2)}{}^{1\beta}{}_{\sigma\nu\mu} + R'_{SU(2)}{}^{2\beta}{}_{\sigma\nu\mu} + R'_{DE}{}^{1\beta}{}_{\sigma\nu\mu} + R'_{DE}{}^{2\beta}{}_{\sigma\nu\mu} + R'_{DSU(2)}{}^{1\beta}{}_{\sigma\nu\mu} +$$
$$+ \ R'_{DSU(2)}{}^{2\beta}{}_{\sigma\nu\mu} + R'_{SU(3)}{}^{1\beta}{}_{\sigma\nu\mu} + R'_{SU(3)}{}^{2\beta}{}_{\sigma\nu\mu} + R'_U{}^{1\beta}{}_{\sigma\nu\mu} + R'_U{}^{2\beta}{}_{\sigma\nu\mu} + R'_V{}^{1\beta}{}_{\sigma\nu\mu} + R'_V{}^{2\beta}{}_{\sigma\nu\mu} +$$
$$+ \ R'_S{}^{1\beta}{}_{\sigma\nu\mu} + R'_S{}^{2\beta}{}_{\sigma\nu\mu} + R'_\Theta{}^{1\beta}{}_{\sigma\nu\mu} + R'_\Theta{}^{2\beta}{}_{\sigma\nu\mu} + R'_B{}^{1\beta}{}_{\sigma\nu\mu} + R'_B{}^{2\beta}{}_{\sigma\nu\mu} + R^{1\beta}{}_{\sigma\nu\mu} + R^{2\beta}{}_{\sigma\nu\mu}$$

$$(22.13)$$

where

$$R'_E{}^{1\beta}{}_{\sigma\nu\mu} = ig^\beta{}_\sigma F_E{}^1{}_{\nu\mu} \qquad\qquad (22.14)$$
$$R'_E{}^{2\beta}{}_{\sigma\nu\mu} = ig^\beta{}_\sigma F_{DE}{}^2{}_{\nu\mu}$$

$$R'_{DE}{}^{1\beta}{}_{\sigma\nu\mu} = ig^\beta{}_\sigma F_E{}^1{}_{\nu\mu}$$
$$R'_{DE}{}^{2\beta}{}_{\sigma\nu\mu} = ig^\beta{}_\sigma F_{DE}{}^2{}_{\nu\mu}$$

$$R'_{SU(2)}{}^{1\beta}{}_{\sigma\nu\mu} = ig^\beta{}_\sigma F_W{}^1{}_{\nu\mu}$$
$$R'_{SU(2)}{}^{2\beta}{}_{\sigma\nu\mu} = ig^\beta{}_\sigma F_{DW}{}^2{}_{\nu\mu}$$

$$R'_{DSU(2)}{}^{1\beta}{}_{\sigma\nu\mu} = ig^\beta{}_\sigma F_W{}^1{}_{\nu\mu}$$
$$R'_{DSU(2)}{}^{2\beta}{}_{\sigma\nu\mu} = ig^\beta{}_\sigma F_{DW}{}^2{}_{\nu\mu}$$

$$R'_{SU(3)}{}^{1\beta}{}_{\sigma\nu\mu} = ig^\beta{}_\sigma F_{SU(3)}{}^1{}_{\nu\mu}$$
$$R'_{SU(3)}{}^{2\beta}{}_{\sigma\nu\mu} = ig^\beta{}_\sigma F_{SU(3)}{}^2{}_{\nu\mu}$$

$$R'_U{}^{1\beta}{}_{\sigma\nu\mu} = ig^\beta{}_\sigma F_U{}^1{}_{\nu\mu}$$

$$R'_{U}{}^{2\beta}{}_{\sigma\nu\mu} = ig^{\beta}{}_{\sigma}F_{U}{}^{2}{}_{\nu\mu}$$

$$R'_{V}{}^{1\beta}{}_{\sigma\nu\mu} = ig^{\beta}{}_{\sigma}F_{V}{}^{1}{}_{\nu\mu}$$

$$R'_{V}{}^{2\beta}{}_{\sigma\nu\mu} = ig^{\beta}{}_{\sigma}F_{V}{}^{2}{}_{\nu\mu}$$

$$R'_{S}{}^{1\beta}{}_{\sigma\nu\mu} = ig^{\beta}{}_{\sigma}F_{S}{}^{1}{}_{\nu\mu}$$

$$R'_{S}{}^{2\beta}{}_{\sigma\nu\mu} = ig^{\beta}{}_{\sigma}F_{S}{}^{2}{}_{\nu\mu}$$

$$R'_{\Theta}{}^{1\beta}{}_{\sigma\nu\mu} = ig^{\beta}{}_{\sigma}F_{\Theta}{}^{1}{}_{\nu\mu}$$

$$R'_{\Theta}{}^{2\beta}{}_{\sigma\nu\mu} = ig^{\beta}{}_{\sigma}F_{\Theta}{}^{2}{}_{\nu\mu}$$

$$R'_{B}{}^{1\beta}{}_{\sigma\nu\mu} = ig^{\beta}{}_{\sigma}B^{1}{}_{\nu\mu}$$

$$R'_{B}{}^{2\beta}{}_{\sigma\nu\mu} = ig^{\beta}{}_{\sigma}B^{2}{}_{\nu\mu}$$

The total Ricci tensor is

$$R'_{\sigma\mu} = R'^{\beta}{}_{\sigma\beta\mu} \tag{22.15}$$

$$= iF_{E}{}^{1}{}_{\sigma\mu} + iF_{E}{}^{2}{}_{\sigma\mu} + iF_{W}{}^{1}{}_{\sigma\mu} + iF_{W}{}^{2}{}_{\sigma\mu} + iF_{DE}{}^{1}{}_{\sigma\mu} + iF_{DE}{}^{2}{}_{\sigma\mu} + iF_{DW}{}^{1}{}_{\sigma\mu} + iF_{DW}{}^{2}{}_{\sigma\mu} + iF_{SU(3)}{}^{1}{}_{\sigma\mu} + iF_{SU(3)}{}^{2}{}_{\sigma\mu} +$$
$$+ iF_{U}{}^{1}{}_{\sigma\mu} + iF_{U}{}^{2}{}_{\sigma\mu} + iF_{V}{}^{1}{}_{\sigma\mu} + iF_{V}{}^{2}{}_{\sigma\mu} + iF_{S}{}^{1}{}_{\sigma\mu} + iF_{S}{}^{2}{}_{\sigma\mu} + iF_{\Theta}{}^{1}{}_{\sigma\mu} + iF_{\Theta}{}^{2}{}_{\sigma\mu} + iF_{B}{}^{1}{}_{\sigma\mu} + iF_{B}{}^{2}{}_{\sigma\mu} +$$
$$+ \partial_{\mu}H^{1\beta}{}_{\sigma\beta} - \partial_{\beta}H^{1\beta}{}_{\sigma\mu} + H^{1\gamma}{}_{\beta\sigma}H^{1\beta}{}_{\gamma\mu} - H^{1\gamma}{}_{\mu\sigma}H^{1\beta}{}_{\gamma\beta} +$$
$$+ \partial_{\mu}H^{2\beta}{}_{\sigma\beta} - \partial_{\beta}H^{2\beta}{}_{\sigma\mu} + H^{2\gamma}{}_{\beta\sigma}H^{2\beta}{}_{\gamma\mu} - H^{2\gamma}{}_{\mu\sigma}H^{2\beta}{}_{\gamma\beta} + H^{1\gamma}{}_{\beta\sigma}H^{2\beta}{}_{\gamma\mu} - H^{1\gamma}{}_{\mu\sigma}H^{2\beta}{}_{\gamma\beta} + H^{2\gamma}{}_{\beta\sigma}H^{1\beta}{}_{\gamma\mu} - H^{2\gamma}{}_{\mu\sigma}H^{1\beta}{}_{\gamma\beta}$$

$$= R'_{E}{}^{1}{}_{\sigma\mu} + R'_{E}{}^{2}{}_{\sigma\mu} + R'_{SU(2)}{}^{1}{}_{\sigma\mu} + R'_{SU(2)}{}^{2}{}_{\sigma\mu} + R'_{DE}{}^{1}{}_{\sigma\mu} + R'_{DE}{}^{2}{}_{\sigma\mu} + R'_{DSU(2)}{}^{1}{}_{\sigma\mu} + R'_{DSU(2)}{}^{2}{}_{\sigma\mu} + R'_{SU(3)}{}^{1}{}_{\sigma\mu} +$$
$$+ R'_{SU(3)}{}^{2}{}_{\sigma\mu} + R'_{U}{}^{1}{}_{\sigma\mu} + R'_{U}{}^{2}{}_{\sigma\mu} + R'_{V}{}^{1}{}_{\sigma\mu} + R'_{V}{}^{2}{}_{\sigma\mu} + R'_{S}{}^{1}{}_{\sigma\mu} + R'_{S}{}^{2}{}_{\sigma\mu} + R'_{\Theta}{}^{1}{}_{\sigma\mu} + R'_{\Theta}{}^{2}{}_{\sigma\mu} + R'_{B}{}^{1\beta}{}_{\sigma\beta\mu} +$$
$$+ R'_{B}{}^{2\beta}{}_{\sigma\beta\mu} + R^{1}{}_{\sigma\mu} + R^{2}{}_{\sigma\mu}$$
$$= R'^{1}{}_{\sigma\mu} + R'^{2}{}_{\sigma\mu}$$

where

$$R'^{1}{}_{\sigma\mu} = R'_{E}{}^{1}{}_{\sigma\mu} + R'_{SU(2)}{}^{1}{}_{\sigma\mu} + R'_{DE}{}^{1}{}_{\sigma\mu} + R'_{DSU(2)}{}^{1}{}_{\sigma\mu} + R'_{SU(3)}{}^{1}{}_{\sigma\mu} + R'_{U}{}^{1}{}_{\sigma\mu} + R'_{V}{}^{1}{}_{\sigma\mu} + R'_{S}{}^{1}{}_{\sigma\mu} +$$
$$+ R'_{\Theta}{}^{1}{}_{\sigma\mu} + R'_{B}{}^{1\beta}{}_{\sigma\beta\mu} + R^{1}{}_{\sigma\mu} \tag{22.16}$$

$$R'^{2}{}_{\sigma\mu} = R'_{E}{}^{2}{}_{\sigma\mu} + R'_{SU(2)}{}^{2}{}_{\sigma\mu} + R'_{DE}{}^{2}{}_{\sigma\mu} + R'_{DSU(2)}{}^{2}{}_{\sigma\mu} + R'_{SU(3)}{}^{2}{}_{\sigma\mu} + R'_{U}{}^{2}{}_{\sigma\mu} + R'_{V}{}^{2}{}_{\sigma\mu} + R'_{S}{}^{2}{}_{\sigma\mu} +$$
$$+ R'_{\Theta}{}^{2}{}_{\sigma\mu} + R'_{B}{}^{2\beta}{}_{\sigma\beta\mu} + R^{2}{}_{\sigma\mu} \tag{22.17}$$

with

$$R'_{E}{}^{1}{}_{\sigma\mu} = iF_{E}{}^{1}{}_{\sigma\mu} \tag{22.18}$$

$$R'^2_{E\ \sigma\mu} = iF^2_{E\ \sigma\mu}$$

$$R'^1_{SU(2)\ \sigma\mu} = iF^1_{W\ \sigma\mu}$$

$$R'^2_{SU(2)\ \sigma\mu} = iF^2_{W\ \sigma\mu}$$

$$R'^1_{DE\ \sigma\mu} = iF^1_{DE\ \sigma\mu}$$

$$R'^2_{DE\ \sigma\mu} = iF^2_{DE\ \sigma\mu}$$

$$R'^1_{DSU(2)\ \sigma\mu} = iF^1_{DW\ \sigma\mu}$$

$$R'^2_{DSU(2)\ \sigma\mu} = iF^2_{DW\ \sigma\mu}$$

$$R'^1_{SU(3)\ \sigma\mu} = iF^1_{SU(3)\ \sigma\mu}$$

$$R'^2_{SU(3)\ \sigma\mu} = iF^2_{SU(3)\ \sigma\mu}$$

$$R'^1_{U\ \sigma\mu} = iF^1_{U\ \sigma\mu}$$

$$R'^2_{U\ \sigma\mu} = iF^2_{U\ \sigma\mu}$$

$$R'^1_{V\ \sigma\mu} = iF^1_{V\ \sigma\mu}$$

$$R'^2_{V\ \sigma\mu} = iF^2_{V\ \sigma\mu}$$

$$R'^1_{S\ \sigma\mu} = iF^1_{S\ \sigma\mu}$$

$$R'^2_{S\ \sigma\mu} = iF^2_{S\ \sigma\mu}$$

$$R'^1_{\Theta\ \sigma\mu} = iF^1_{\Theta\ \sigma\mu}$$

$$R'^2_{\Theta\ \sigma\mu} = iF^2_{\Theta\ \sigma\mu}$$

$$R'^1_{B\ \sigma\mu} = iF^1_{B\ \sigma\mu}$$

$$R'^2_{B\ \sigma\mu} = iF^2_{B\ \sigma\mu}$$

with the further definition of $R''^1_{\sigma\mu}$ and $R''^2_{\sigma\mu}$:

$$R''^1_{\sigma\mu} = R'^1_{SU(3)\ \sigma\mu} + R^1_{\sigma\mu}$$
$$R''^2_{\sigma\mu} = R'^2_{SU(3)\ \sigma\mu} + R^2_{\sigma\mu} \tag{22.19}$$

$R''^1_{\sigma\mu}$ is the Ricci tensor. An additional Ricci-like tensor is

$$H_{\sigma\mu} = H^\beta_{\ \sigma\beta\mu} \tag{22.20}$$

The curvature scalar is

$$R' = g^{\sigma\mu}R'_{\sigma\mu} = + \partial^\sigma H^{1\beta}_{\ \sigma\beta} - \partial_\beta H^{1\beta\ \sigma}_{\ \sigma} + H^{1\gamma}_{\ \beta\sigma}H^{1\beta\ \sigma}_{\ \gamma} - H^{1\gamma}_{\ \mu\sigma}H^{1\beta}_{\ \gamma\beta} + \partial^\sigma H^{2\beta}_{\ \sigma\beta} - \partial_\beta H^{2\beta\ \sigma}_{\ \sigma} +$$
$$+ H^{2\gamma}_{\ \beta\sigma}H^{2\beta\ \sigma}_{\ \gamma} - H^{2\gamma\sigma}_{\ \sigma}H^{2\beta}_{\ \gamma\beta} + H^{1\gamma}_{\ \beta\sigma}H^{2\beta\ \sigma}_{\ \gamma} - H^{1\gamma\sigma}_{\ \sigma}H^{2\beta}_{\ \gamma\beta} + H^{2\gamma}_{\ \beta\sigma}H^{1\beta\ \sigma}_{\ \gamma} - H^{2\gamma\sigma}_{\ \sigma}H^{1\beta}_{\ \gamma\beta}$$
$$\tag{22.21}$$

$$= g^{\sigma\mu}(R^{1\beta}_{\ \sigma\beta\mu} + R^{2\beta}_{\ \sigma\beta\mu})$$

22.3 Vector Boson and Graviton Lagrangian Terms

The vector boson and gravitational part of the lagrangian of the Unified SuperStandard (with the Higgs sector and the Faddeev-Popov terms gauge sector not displayed here) is:

$$\mathcal{L} = \mathrm{Tr}\ \sqrt{g}[MD_\nu R''^1_{\sigma\mu}D^\nu R''^{2\sigma\mu} + aR'^1_{\sigma\mu}R'^{2\sigma\mu} + bR' + cg^{\sigma\mu}g^2_{\sigma\mu} + c'g^{2\sigma\mu}g^2_{\sigma\mu} - dA_{SU(3)}{}^2_\mu A_{SU(3)}{}^{2\mu}] \tag{22.22}$$

where M, a, b, c, c', and d are constants, and $R''^i_{\sigma\mu}$ for i = 1, 2 determined above.[217]

This higher derivative lagrangian maintains the locality of the theory but does entail a modest modification in the derivation of the Euler-Lagrange equations of motion. It also requires the use of principal value propagators rather than ordinary Feynman propagators for gluon and graviton interactions. Thus the Strong Interaction sector, and the Gravitation sector are Action-at-a-Distance theories that are similar in spirit to Wheeler-Feynman Electrodynamics. The two U(1) Electromagnetic sectors, the Generation group U(4) gauge field sector, the Layer group U(4) gauge field sector, the two SU2) Weak sectors, the U(4) A_S gauge field sector, the spinor connection sector, and the Θ-interaction sector may, or may not, be Action-at-a-Distance fields. They are not constrained to be Action-at-a-Distance by the present considerations.

Since we wish to apply our theory cosmologically, and within hadrons, where the gravitational spinor connections are negligible due to the smallness of the gravitational constant G and the 'smallness' of B spin on the cosmological scale, we set $F^1_{\nu\mu} = F^2_{\nu\mu} = 0$ and find[218]

---

[217] One may ask why $R''^1_{\sigma\mu}$ and $R''^2_{\sigma\mu}$ appear in the first term of the lagrangian, and not other interaction terms. We believe the primary reason is: "The extended vierbein $l^{\mu ai}(x)$ can be viewed as located at a point in a higher dimensional complex-valued space.

$$l^{\mu ai}(x) = (\partial\xi_X{}^{ai}(x)/\partial x_\mu)_{X=h(x)}$$

where $\xi_X{}^{ai}$ is a set of locally inertial coordinates located at point X, and x = h(x) is a 4-dimensional point in a tangent subspace of the higher dimensional space:

$$X = h(x)$$

The relation between complex 4-dimensional coordinates x and the higher dimensional coordinates X is an embedding of a 4-dimensional surface within the higher dimensional complex space when account is taken of the range of possible x values. We have considered such embeddings in Blaha (2015a), and in earlier books, and developed a theory of a higher dimensional complex-valued space (the *Megaverse*) that contains our universe and probably many other universes." Thus SU(3) and Gravitation have a special role in our particle dynamics based on geometry. The second reason is the common feature of color SU(3) and real-valued General Relativity is that they are the only interactions that do not participate in 'rotations of interactions' as described earlier and in chapter 31 of Blaha (2017b). The third, practical reason is the experimental reality that the Strong Interaction and Gravitation are known to have 'anomalous' features that will be seen to be remedied by these insertions while the other interactions are 'conventional.'

[218] The constants have the dimensions: M has the dimension of inverse mass squared, b has dimension mass squared, a is dimensionless, c and c' have dimension mass, and d has dimension mass squared.

$$\mathcal{L} = \mathrm{Tr}\ \sqrt{g}[MD_v(R'^1_{SU(3)\sigma\mu} + R^1_{\sigma\mu})D^v(R'_{SU(3)}{}^{2\sigma\mu} + R^{2\sigma\mu}) + \\ + aR'^1_{\sigma\mu}R'^{2\sigma\mu} + bR' + cg^{\sigma\mu}g^2_{\sigma\mu} + c'g^{2\sigma\mu}g^2_{\sigma\mu} - dA_{SU(3)}{}^2_{\mu}A_{SU(3)}{}^{2\mu}] \qquad (22.23)$$

Since there are no strong interaction fields in 'empty' space and gravity is negligible within hadrons,[219] we can drop the interaction terms between the Strong interaction and the Gravity interaction. However, we cannot drop the interaction terms amongst Electromagnetism, the Weak interaction, the Strong Interaction, the Generation group U(4) interaction, the U(4) Layer groups interactions, the U(4) Species group interaction, and the U(192) Θ-interaction – within, and between, hadrons. The interaction terms between Electromagnetism and Gravitation are important cosmologically.

The above lagrangian terms can therefore be expressed as:[220]

$$\mathcal{L} = \mathcal{L}_E + \mathcal{L}_{SU(2)} + \mathcal{L}_{DE} + \mathcal{L}_{DSU(2)} + \mathcal{L}_{SU(3)} + \mathcal{L}_U + \mathcal{L}_V + \mathcal{L}_S + \mathcal{L}_\Theta + \mathcal{L}_G + \mathcal{L}_{int} \qquad (22.24)$$

where taking traces of $\mathcal{L}$s terms is understood, and with coupling constants not displayed below to avoid clutter,

$$\mathcal{L}_E = \mathrm{Tr}\ \sqrt{g}\{M\{[\partial_v + i(A_E{}^1_v + A_E{}^2_v)]F^1_{E\sigma\mu}[\partial^v + i(A_E{}^{1v} + A_E{}^{2v})]F^2_E{}^{\sigma\mu}\} + aF_E{}^1_{\sigma\mu}F_E{}^{2\sigma\mu}\} \qquad (22.25)$$

$$\mathcal{L}_{SU(2)} = \mathrm{Tr}\ \sqrt{g}[aF_W{}^1_{\sigma\mu}F_W{}^{2\sigma\mu}]$$

$$\mathcal{L}_{DE} = \mathrm{Tr}\ \sqrt{g}\{M\{[\partial_v + i(A_{DE}{}^1_v + A_{DE}{}^2_v)]F^1_{DE\sigma\mu}[\partial^v + i(A_{DE}{}^{1v} + A_{DE}{}^{2v})]F_{DE}{}^{2\sigma\mu}\} + aF_{DE}{}^1_{\sigma\mu}F_{DE}{}^{2\sigma\mu}\}$$

$$\mathcal{L}_{DSU(2)} = \mathrm{Tr}\ \sqrt{g}[aF_W{}^1_{\sigma\mu}F_W{}^{2\sigma\mu}]$$

$$\mathcal{L}_{SU(3)} = \mathrm{Tr}\ \sqrt{g}\{M[\partial_v + i(A_{SU(3)}{}^1_v + A_{SU(3)}{}^2_v)]F_{SU(3)}{}^1_{\sigma\mu}[\partial^v + i(A_{SU(3)}{}^{1v} + A_{SU(3)}{}^{2v})]F_{SU(3)}{}^{2\sigma\mu} + \\ + aF_{SU(3)}{}^1_{\sigma\mu}F_{SU(3)}{}^{2\sigma\mu} - dA_{SU(3)}{}^2_{\mu}A_{SU(3)}{}^{2\mu}\}$$

$$\mathcal{L}_U = \mathrm{Tr}\ \sqrt{g}[aF_U{}^1_{\sigma\mu}F_U{}^{2\sigma\mu}]$$

$$\mathcal{L}_V = \mathrm{Tr}\ \sqrt{g}[aF_V{}^1_{\sigma\mu}F_V{}^{2\sigma\mu}]$$

$$\mathcal{L}_S = \mathrm{Tr}\ \sqrt{g}[aF_S{}^1_{\sigma\mu}F_S{}^{2\sigma\mu}]$$

$$\mathcal{L}_\Theta = \mathrm{Tr}\ \sqrt{g}[aF_\Theta{}^1_{\sigma\mu}F_\Theta{}^{2\sigma\mu}]$$

$$\mathcal{L}_G = \mathrm{Tr}\ \sqrt{g}[MD_vR^1_{\sigma\mu}D^vR^{2\sigma\mu} + aR^1_{\sigma\mu}R^{2\sigma\mu} + bg^{\sigma\mu}(R^{1\beta}_{\sigma\beta\mu} + R^{2\beta}_{\sigma\beta\mu}) + cg^{\sigma\mu}g^2_{\sigma\mu} + c'g^{2\sigma\mu}g^2_{\sigma\mu}] \\ = \mathrm{Tr}\ \sqrt{g}[MD_vR^1_{\sigma\mu}D^vR^{2\sigma\mu} + aR^1_{\sigma\mu}R^{2\sigma\mu} + bH + cg^{\sigma\mu}g^2_{\sigma\mu} + c'g^{2\sigma\mu}g^2_{\sigma\mu}]$$

$$\mathcal{L}_{int} = \mathcal{L} - (\mathcal{L}_E + \mathcal{L}_{SU(2)} + \mathcal{L}_{DE} + \mathcal{L}_{DSU(2)} + \mathcal{L}_{SU(3)} + \mathcal{L}_U + \mathcal{L}_V + \mathcal{L}_S + \mathcal{L}_\Theta + \mathcal{L}_G) \qquad (22.26)$$

---

[219] We show gravity weakens at very short distances using our Two-Tier Quantum Field Theory formalism. See Appendix A, and Blaha (2003) and (2005a) among other books by the author.
[220] We only consider the gauge field lagrangian terms.

with appropriate sums over layers. Thus $\mathcal{L}_{SU(3)}$, $\mathcal{L}_{SU(2)}$, $\mathcal{L}_E$, $\mathcal{L}_{DE}$, $\mathcal{L}_{DSU(2)}$, $\mathcal{L}_U$, $\mathcal{L}_V$, $\mathcal{L}_S$, $\mathcal{L}_\Theta$, and parts of $\mathcal{L}_{int}$ are the dominant interactions within hadrons, and $\mathcal{L}_G$, $\mathcal{L}_E$ and parts of $\mathcal{L}_{int}$ are the dominant interactions in space within the framework of this discussion.

The $D_\nu R^1{}_{\sigma\mu}$ and $D^\nu R^{2\sigma\mu}$ terms have the form:

$$D_\nu R^i{}_{\sigma\mu} = + \partial_\nu R^i{}_{\sigma\mu} - H^{1\beta}{}_{\sigma\nu} R^i{}_{\beta\mu} - H^{1\beta}{}_{\nu\mu} R^i{}_{\sigma\beta} \tag{22.27}$$

for i = 1, 2.

22.4 New Vector Boson Interactions

The above lagrangian can be broken up into pieces in the following manner:

$$\mathcal{L}_E = \mathrm{Tr}\,\sqrt{g}\{M\{[\partial_\nu + i(A_E^1{}_\nu + A_E^2{}_\nu)]F^1{}_{E\sigma\mu}[\partial^\nu + i(A_E^{1\nu} + A_E^{2\nu})]F^2{}_E{}^{\sigma\mu}\} + aF_E^1{}_{\sigma\mu}F_E^{2\sigma\mu}\} \tag{22.28}$$

$$\mathcal{L}_{SU(2)} = \mathrm{Tr}\,\sqrt{g}[aF_W^1{}_{\sigma\mu}F_W^{2\sigma\mu}]$$

$$\mathcal{L}_{DE} = \mathrm{Tr}\,\sqrt{g}\{M\{[\partial_\nu + i(A_{DE}^1{}_\nu + A_{DE}^2{}_\nu)]F^1{}_{DE\sigma\mu}[\partial^\nu + i(A_{DE}^{1\nu} + A_{DE}^{2\nu})]F_{DE}^{2\sigma\mu}\} + aF_{DE}^1{}_{\sigma\mu}F_{DE}^{2\sigma\mu}\}$$

$$\mathcal{L}_{DSU(2)} = \mathrm{Tr}\,\sqrt{g}[aF_W^1{}_{\sigma\mu}F_W^{2\sigma\mu}]$$

$$\mathcal{L}_{SU(3)} = \mathrm{Tr}\,\sqrt{g}\{M[\partial_\nu + i(A_{SU(3)}^1{}_\nu + A_{SU(3)}^2{}_\nu)]F_{SU(3)}^1{}_{\sigma\mu}[\partial^\nu + i(A_{SU(3)}^{1\nu} + A_{SU(3)}^{2\nu})]F_{SU(3)}^{2\sigma\mu} +$$

$$+ aF_{SU(3)}^1{}_{\sigma\mu}F_{SU(3)}^{2\sigma\mu} - dA_{SU(3)}^2{}_\mu A_{SU(3)}^{2\mu}\} \tag{22.29}$$

$$\mathcal{L}_U = \mathrm{Tr}\,\sqrt{g}[a\Gamma_U^1{}_{\sigma\mu}\Gamma_U^{2\sigma\mu}]$$

$$\mathcal{L}_V = \mathrm{Tr}\,\sqrt{g}[aF_V^1{}_{\sigma\mu}F_V^{2\sigma\mu}]$$

$$\mathcal{L}_S = \mathrm{Tr}\,\sqrt{g}[aF_S^1{}_{\sigma\mu}F_S^{2\sigma\mu}]$$

$$\mathcal{L}_\Theta = \mathrm{Tr}\,\sqrt{g}[aF_\Theta^1{}_{\sigma\mu}F_\Theta^{2\sigma\mu}]$$

$$\mathcal{L}_G = \mathrm{Tr}\,\sqrt{g}[MD_\nu R^1{}_{\sigma\mu}D^\nu R^{2\sigma\mu} + aR^1{}_{\sigma\mu}R^{2\sigma\mu} + bg^{\sigma\mu}(R^{1\beta}{}_{\sigma\beta\mu} + R^{2\beta}{}_{\sigma\beta\mu}) + cg^{\sigma\mu}g^2{}_{\sigma\mu} + c'g^{2\sigma\mu}g^2{}_{\sigma\mu}]$$

$$= \mathrm{Tr}\,\sqrt{g}[MD_\nu R^1{}_{\sigma\mu}D^\nu R^{2\sigma\mu} + aR^1{}_{\sigma\mu}R^{2\sigma\mu} + bH + cg^{\sigma\mu}g^2{}_{\sigma\mu} + c'g^{2\sigma\mu}g^2{}_{\sigma\mu}]$$

$$\mathcal{L}_{int} = \mathcal{L} - (\mathcal{L}_E + \mathcal{L}_{SU(2)} + \mathcal{L}_{DE} + \mathcal{L}_{DSU(2)} + \mathcal{L}_{SU(3)} + \mathcal{L}_U + \mathcal{L}_V + \mathcal{L}_S + \mathcal{L}_\Theta + \mathcal{L}_G) \tag{22.30}$$

again with appropriate sums over layers and with coupling constants not displayed to avoid clutter. Thus $\mathcal{L}_{SU(3)}$, $\mathcal{L}_{SU(2)}$, $\mathcal{L}_E$, $\mathcal{L}_{DE}$, $\mathcal{L}_{DSU(2)}$, $\mathcal{L}_U$, $\mathcal{L}_V$, $\mathcal{L}_S$, $\mathcal{L}_\Theta$, and parts of $\mathcal{L}_{int}$ are the dominant interactions within hadrons, and $\mathcal{L}_G$, $\mathcal{L}_E$ and parts of $\mathcal{L}_{int}$ are the dominant interactions in space within the framework of this discussion. The terms of $\mathcal{L}_{int}$ have 'new' interactions between gauge fields that are described in some detail in Blaha (2017b). These interactions are not in the conventional Standard Model. They lead to modifications of gravity, the Strong Interactions, spin dynamics and so on.

23. Gravitational Potential on the Three Distance Scales

This chapter[221] and the next chapter describe some of the possible results of the unified lagrangian terms in eq. 22.22. Here we pull together results on the gravitational potential that were previously found in Blaha (2016g), (2016) and (2017a). We note that a new experimental study of 33,000 galaxies indicate that the gravitational potential at inter-galactic distances deviates significantly from the Newtonian potential G/r. In Blaha (2017a) we showed that such a deviation occurs in our theory. This experimental result was not known at the time of its writing.[222]

23.1 Total Gravitational Potential

The gravitational potential generated from the real-valued affine connection term bR' in the lagrangian eq. 22.22 was shown in chapter 6 of Blaha (2016e) (See Appendix L for the contents of chapter 6.) to be

$$V^{tot}_{G}(\mathbf{r}) = -G/r - a_1 Ge^{-m_{G}r}/r + a_2 G\cos(m_G r)/r \tag{23.1}$$

if a massless graviton term were required, where

$$m_G^2 = 2b/a \cong 10^{55}\ ev^2 = 10^{27}\ GeV^2 \tag{23.2}$$
$$a_1 = \tfrac{1}{2}$$
$$a_2 = \tfrac{1}{2}$$

The gravitational potential contribution due to the gravitational gauge field $A_S^{1\mu}$ was shown to be:

$$V_S(\mathbf{r}) \equiv V_{GA1}(\mathbf{r}) = -[1/(96\pi a)][1/r - e^{-m_{A}r}/r] \tag{23.3}$$

where

$$m_A = (a/M)^{1/2} = m_{SI} \cong 10^{-71}\ GeV \tag{23.4}$$

---

[221] Most of this chapter appears in Blaha (2017a) and earlier books by the author.

[222] The gravitational potential was found to be greater than G/r at inter-galactic distances in a survey of 33,000 galaxies by M. Brouwer and colleagues at the Leiden Observatory (The Netherlands) in an announcement on December 18, 2016.

Since a $\cong$ 1 the coupling constant

$$1/(96\pi a) \cong 0.0033 \tag{23.5}$$

In comparison the electromagnetic fine structure constant is

$$\alpha \cong 0.0073$$

Thus the Species Group A_S field coupling constant is approximately ½ of the fine structure constant.

The total gravitational potential due to both sources of gravity is

$$V(\mathbf{r}) = -G/r - a_1 Ge^{-m_G r}/r + a_1 Gcos(m_G r)/r - [1/(96\pi a)][1/r - e^{-m_A r}/r] \tag{23.6}$$

We now examine $V(\mathbf{r})$ at short distances (within the solar system), distances of tens of thousands of light years (intra-galactic distances), and distances between galaxies (hundreds of thousands to millions of light years and beyond).

23.2 Intra-Solar System Distance Scale

Since m_G is very large and m_A is extremely small, the gravitational potential at distances of up to at least several light years is approximately

$$V(\mathbf{r}) \cong -G/r \tag{23.7}$$

to well within feasible experimental limits.

23.3 Galactic Distance Scale

At distances of several tens of thousands of light years up to perhaps 100,000's of light years we find eq. 23.6 becomes approximately

$$V(\mathbf{r}) \cong -[G + 1/(96\pi a)]/r + \tfrac{1}{2}Gcos(m_G r)/r \tag{23.8}$$

Note that the approximate expansion of the terms in eq. 23.8 to third order yields

$$V(\mathbf{r}) \sim -[G + 1/(96\pi a)]/r + a_1 Gm_G^3 r^2 - a_1 Gm_G^2 r + constants \tag{23.9}$$

with the resultant force

$$F = \nabla V^{tot}_G(\mathbf{r}) \sim [G + 1/(96\pi a)]\mathbf{r}/r^3 + 2a_1 Gm_G^3 \mathbf{r} - a_2 Gm_G^2 \mathbf{r}/r + ... \tag{23.10}$$

This result is to be compared to the MoND force of A. Balakin et al, Phys. Rev. **D70**, 064027 (2004):[223]

$$F = -\lambda Gm[M/r^2 - |\,\Pi_c\,|\,r/c^2]$$
(23.11)

and the vector form suggested by H-S Zhao et al, Phys. Rev. **D82**, 103001 (2010):

$$\partial\Phi/\partial\mathbf{r} = Gm\mathbf{r}/r^3 + (Gm)^{1/2}\mathbf{r}/r^2$$
(23.12)

Recent studies of 153 galaxies confirm the MoND discrepancy from Newtonian gravitation.[224]

23.4 Intergalactic Distance Scale

We can estimate the gravitational potential of $V(\mathbf{r})$ in eq. 23.1 for large distances of the order of many hundreds of thousands of light years, and beyond, we find

$$V(\mathbf{r}) \cong -[G + 1/(96\pi a)]/r$$
(23.13)

to good approximation since $m_G^2 = 2b/a \cong 10^{55}$ ev$^2 = 10^{27}$ GeV^2 sets a distance scale of the order of tens of thousands of light years causing the oscillating term to 'wash out,' and the $a_1 Ge^{-m_G r}/r$ term to be negligible. The $e^{-m_A r}/r$ term whose distance scale is short range is also negligible.

Consequently we find a deeper potential and thus a larger attractive gravitational force at inter-galactic distances in agreement with the 33,000 galaxy survey of M. Brouwer and colleagues.

23.5 Theoretical Agreement With Gravitational Data at all Known Distances

We conclude our unified theory agrees with known gravitational data at the three distance scales.

---

[223] The constant c in eq. 28.11 is the speed of light, and M is the mass (not the M used in our lagrangian equations.)
[224] S. S. McGaugh, F. Lelli, and J. M. Schombert, arXiv: 1609.0591 (2016).

24. The Strong Interaction Sector

In this chapter[225] we describe some of the implications of the Strong Interaction sector of our theory. We will find a linear quark potential and quark confinement follows.

24.1 Strong Interaction Lagragian Terms

The flat space-time Strong Interaction gauge field lagrangian terms (eq. 22.9) is

$$\mathcal{L}_{SU(3)} = \mathrm{Tr}\ \sqrt{g}\{M[\partial_\nu + ig_1(A_{SU(3)}{}^1{}_\nu + A_{SU(3)}{}^2{}_\nu)]F_{SU(3)}{}^1{}_{\sigma\mu}[\partial^\nu + i\ g_1(A_{SU(3)}{}^{1\nu} + A_{SU(3)}{}^{2\nu})]F_{SU(3)}{}^{2\sigma\mu} +$$
$$+ aF_{SU(3)}{}^1{}_{\sigma\mu}F_{SU(3)}{}^{2\sigma\mu} - dA_{SU(3)}{}^2{}_\mu A_{SU(3)}{}^{2\mu}\} \tag{22.29}$$

Dropping the subscript $_{SU(3)}$ for added clarity and adding color fermion terms we obtain[226]

$$\mathcal{L}_{SU(3)} = \mathrm{Tr}\ \{MD_\nu F^1{}_{\sigma\mu}D^\nu F^{2\sigma\mu} + aF^1{}_{\sigma\mu}F^{2\sigma\mu} - dA^2{}_\mu A^{2\mu}\} + \bar{\psi}[i\overleftrightarrow{\nabla} + g_1(A^1 + A^2) - m]\psi \tag{24.1}$$

where (for $j = 1, 2$)

$$D_\nu F^j{}_{\sigma\mu} = \partial_\nu F^j{}_{\sigma\mu} + ig_1[A^1{}_\nu, F^j{}_{\sigma\mu}] + ig_1[A^2{}_\nu, F^j{}_{\sigma\mu}] \tag{24.2}$$

The lagrangian is equal to eq. 17 of S. Blaha, Phys. Rev. **D11**, 2921 (1974) (Appendix J) except for additional terms $MD_\nu F^1{}_{\sigma\mu}D^\nu F^{2\sigma\mu}$ and $[A^2{}_\varkappa, A^2{}_\mu]$; and the following changes in parameters:

$$a = -\tfrac{1}{2} \qquad d = \tfrac{1}{2}\lambda^2$$

Since that paper essentially contains a complete description of our Strong Interaction theory (modulo the additional terms) we refer the reader to it and to its predecessor paper referenced therein. There are a few additional changes required to bring the 1974-5 papers into agreement with our current theory:

1. Eq. 30 of the above referenced paper must be modified to

$$F^2{}_{\varkappa\mu} = \partial A^2{}_\mu/\partial x^\varkappa - \partial A^2{}_\varkappa/\partial x^\mu + ig_1[A^2{}_\varkappa, A^2{}_\mu] + ig_1[A^1{}_\varkappa, A^2{}_\mu] + ig_1[A^2{}_\varkappa, A^1{}_\mu] \tag{24.3}$$

---

[225] Most of this chapter appears in Blaha (2016h) and earlier books by the author.
[226] We note the constant a, that appears in this chapter is NOT the Charmonium constant a.

with the addition of the 'bolded' term if$[A^2_{\varkappa}, A^2_{\mu}]$. There is also a trivial change of notation of coupling constant from 'g_1' to 'f'.

2. Eqs. 6 and 18 should have the interaction term expanded to

$$g_1 A^1 \;\rightarrow\; g_1(A^1 + A^2) \quad \text{```} \tag{24.4}$$

and similarly in eq. 20. Eqs. 38 – 41 directly show that the additional interaction term leads to a gluon propagator[227] $<A^1 + A^2, A^1 + A^2> = 2<A^1, A^2> + <A^1, A^1>$, and introduces a 1/r term in the potential part of the gluon propagator.

As a result the effective gluon propagator in the theory, **if the $Mf^2 D_v F^1_{\sigma\mu} D^v F^{2\sigma\mu}$ term is neglected**, combines eqs. 38 and 39 to give the *short-distance*[228] gluon propagator between quarks:

$$g_{\mu v}\delta_{ab}P[\lambda^2/k^4 - 1/k^2] \tag{24.5}$$

up to a constant factor.[229]

These changes *explicitly* leads to a Strong Interaction potential of the form

$$V(r) = -2f^2/r + f^2\lambda^2 r \tag{24.6}$$

Naturally one can expect perturbative corrections to eq. 24.6 in higher order in f. However, as will be seen later, the apparent relative smallness of f suggests eq. 24.6 is a good approximation to the **short-distance, inter-quark interaction.**

24.1.1 Canonical Equal Time Commutation Relations

The Euler-Lagrange equations of motion, eqs. 27 – 31 in the 1975 paper (Appendix J), are modified most significantly by the $Mf^2 D_v F^1_{\sigma\mu} D^v F^{2\sigma\mu}$ lagrangian term in eq. 24.1. In order to follow the canonical method to obtain the contributions to the equations of motion of this higher derivative term we will use integration by parts and discard surface terms, as is usually done in quantum field theory. We should start with eq. 24.1. However, with a view towards perturbation theory which appears reasonable in view of the smallness of the strong interaction coupling constant $f^2/4\pi = 0.024$ seen below, we will abstract a quadratic expresssion in the fields from eq.

---

[227] Eqs. 40-41 in the above referenced paper.

[228] We anticipate that the $Mf^2 D_v F^1_{\sigma\mu} D^v F^{2\sigma\mu}$ term will affect the short-distance behavior of the inter-quark interaction. The equivalent term in the gravitation sector influences the long-distance form of the gravity potential and leads to a MoND-like behavior. See chapter 23.

[229] This propagator is taken in Principal value to avoid potential unitarity problems. This topic is described in detail in earlier papers and books by the author.

24.1 and then proceed to develop gluon propagators and the strong interaction potential. The 'free' Strong Interaction lagrangian that we use is

$$\mathcal{L}_{SU(3)F} = Tr\{MD_{Fv}F_F{}^{1a}{}_{\sigma\mu}D_F{}^vF_F{}^{2a\sigma\mu} + aF_F{}^{1a}{}_{\sigma\mu}F_F{}^{2a\sigma\mu} - dA^{2a}{}_\mu A^{2a\mu}\} + \bar{\psi}[i\nabla + f(A^1 + A^2) - m]\psi$$

(24.7)

where

$$F^{1a}{}_{\mu\varkappa} = \partial A^{1a}{}_\mu/\partial x^\varkappa - \partial A^{1a}{}_\varkappa/\partial x^\mu$$
$$F^{2a}{}_{\mu\varkappa} = \partial A^{2a}{}_\mu/\partial x^\varkappa - \partial A^{2a}{}_\varkappa/\partial x^\mu$$

(24.8)

and

$$D_{Fv} = \partial_v$$

(24.9)

The conjugate momenta to $A^{1a}{}_\mu$ and $A^{2a}{}_\mu$ are respectively

$$\pi^{1a}{}_\mu = \partial\mathcal{L}_{SU(3)F}/(\partial A^{1a}{}_\mu/\partial t) = aF_F{}^{2a\mu t}$$
$$\pi^{2a}{}_\mu = \partial\mathcal{L}_{SU(3)F}/(\partial A^{2a}{}_\mu/\partial t) = aF_F{}^{1a\mu t}$$

(24.10)

The non-zero, equal time commutation relations are

$$[\pi^{ia}{}_\mu(\mathbf{x}, t), A^{jb}{}_v(\mathbf{y}, t)] = i(1 - \delta^{ij})\delta^{ab}\delta^{G(\mu v)}(\mathbf{x} - \mathbf{y})$$

(24.11)

where i and j label the fields, and $G(\mu v)$ indicates the gauge[230] G and the associated index expressions, with

$$\delta^{G(\mu v)}(\mathbf{x} - \mathbf{y}) = \int d^4k \, exp(-ik{\cdot}x)b^G{}_{\mu v}(k)/(2\pi)^4$$

(24.12)

where $b^G{}_{\mu v}(k)$ is a polynomial in k with a δ-function factor restricting the integration over k.

24.1.2 Dynamical Equations

After performing partial integrations on the $MD_{Fv}F_F{}^1{}_{\sigma\mu}D_F{}^vF_F{}^{2\sigma\mu}$ term (and discarding surface terms at 'infinity') the Euler-Lagrange dynamical equations (in the Landau gauge) due to independent variations with respect to $A^{1a}{}_\mu$ is

$$2M\square^2A^{2a}{}_\mu - 2a\square A^{2a}{}_\mu = -g_1\bar{\psi}T^a\gamma_\mu\psi$$

(24.13)

and, with respect to $A^{2a}{}_\mu$, is

---

[230] Not the gravitational coupling constant.

$$2M\Box^2 A^{1a}{}_{\mu} - 2a\Box A^{1a}{}_{\mu} - 2dA^{2a}{}_{\mu} = -g_1\,\bar{\psi}\,T^a\gamma_{\mu}\psi \qquad (24.14)$$

where T^a is an SU(3) generator. Subtracting the equations we find

$$2M\Box^2 A^{1a}{}_{\mu} - 2a\Box A^{1a}{}_{\mu} - 2M\Box^2 A^{2a}{}_{\mu} + 2a\Box A^{2a}{}_{\mu} - 2dA^{2a}{}_{\mu} = 0$$

or

$$A^{2a}{}_{\mu} = [2M\Box^2 - 2a\Box + 2d]^{-1}[2M\Box^2 A^{1a}{}_{\mu} - 2a\Box A^{1a}{}_{\mu}] \qquad (24.15)$$

with the result

$$\{2M\Box^2 - 2a\Box - 2d[2M\Box^2 - 2a\Box + 2d]^{-1}[2M\Box^2 - 2a\Box]\}A^{1a}{}_{\mu} = -g_1\bar{\psi}\,T^a\gamma_{\mu}\psi$$

or

$$\{2M\Box^2 - 2a\Box - 2d[2M\Box^2 - 2a\Box + 2d]^{-1}[2M\Box^2 - 2a\Box]\}A^{1a}{}_{\mu} = -g_1\bar{\psi}\,T^a\gamma_{\mu}\psi$$

$$\{2M\Box^2 - 2a\Box - 2d + 4d^2[2M\Box^2 - 2a\Box + 2d]^{-1}\}A^{1a}{}_{\mu} = -g_1\bar{\psi}\,T^a\gamma_{\mu}\psi \qquad (24.16)$$

Eq. 24.16 leads to the Principal Value (Feynman) propagator:

$$D^{11}{}_{\mu\nu}(x-y) = P -i<0|T(A^1{}_{\mu}(x),\,A^1{}_{\nu}(y)|0>$$
$$= P \int d^4k \, \exp(-ik\cdot x)b_{\mu\nu}(k)D_1(k)/(2\pi)^4 \qquad (24.17)$$

where $b_{\mu\nu}(k)$ is a Landau gauge polynomial in k, and

$$D_1(k) = \{2Mk^4 - 2ak^2 - 2d + 4d^2[2Mk^4 - 2ak^2 + 2d]^{-1}\}^{-1}$$
$$= [2Mk^4 - 2ak^2 + 2d](2Mk^4 - 2ak^2)^{-2}$$

Thus

$$D^{11}{}_{\mu\nu}(x-y) = P \int d^4k \, \exp(-ik\cdot x)b_{\mu\nu}(k)[2Mk^4 - 2ak^2 + 2d]/[(2\pi)^4(2Mk^4 - 2ak^2)^2]$$
$$= P \int d^4k \, \exp(-ik\cdot x)b_{\mu\nu}(k)[2Mk^4 - 2ak^2 + 2d]/[(2\pi)^4 k^4(2Mk^2 - 2a)^2] \qquad (24.18)$$

indicating a linear potential r term as well as terms of lower powers in r and Yukawa-like terms with a mass of $(a/M)^{\frac{1}{2}}$. We will describe the resulting effective Strong Interaction potential in more detail later.

Eq. 24.11 leads to the other propagator:

$$D^{12}{}_{\mu\nu}(x-y) = P -i<0|T(A^1{}_{\mu}(x),\,A^2{}_{\nu}(y)|0>$$
$$= P \int d^4k \, \exp(-ik\cdot x)b_{\mu\nu}(k)D_2(k)/(2\pi)^4 \qquad (24.19)$$

where

$$D_2(k) = [2Mk^4 - 2ak^2]^{-1} \tag{24.20}$$

Thus

$$D^{12}{}_{\mu\nu}(x-y) = P \int d^4k \, \exp(-ik\cdot x) b_{\mu\nu}(k)/[(2\pi)^4 k^2(2Mk^2 - 2a)] \tag{24.21}$$

indicating a 1/r potential term plus a Yukawa term with a mass of $(a/M)^{\frac{1}{2}}$.

Due to the form of the interaction with quarks (eq. 24.7) the total effective gluon interaction between quarks is

$$\begin{aligned}
D^{tot}{}_{\mu\nu}(x-y) &= D^{11}{}_{\mu\nu}(x-y) + 2D^{12}{}_{\mu\nu}(x-y) \\
&= P \int d^4k \, \exp(-ik\cdot x) b_{\mu\nu}(k)\{[3Mk^4 - 3ak^2 + 2d]/[(2\pi)^4 k^4(2Mk^2 - 2a)^2]\}
\end{aligned} \tag{24.22}$$

24.2 Strong Interaction Potential

Eq. 24.22 leads to the form of the total Strong Interaction potential. We note that the $\mu = \nu = 0$ part of the Feynman propagator for transverse gluons has the form:

$$D^{tot}{}_{00}(x-y) = \ldots - \int d^4k \, V_{SI}(\mathbf{k}) \exp(-ik\cdot(x-y))/(2\pi)^4 = \ldots + V_{SI}(\mathbf{x}-\mathbf{y})\delta(x_0 - y_0) \tag{24.23}$$

where

$$V_{SI}(\mathbf{x}) = - \int d^3k \, \exp(i\mathbf{k}\cdot\mathbf{x}) V_{SI}(\mathbf{k})/(2\pi)^3$$

with

$$\begin{aligned}
V_{SI}(\mathbf{k}) &= [3Mk^4 + 3ak^2 + 2d]/[\mathbf{k}^4(2Mk^2 + 2a)^2] \\
&= (2M)^{-2}\{2d(M/a)^2/\mathbf{k}^4 + [3a(M/a)^2 - 4d(M/a)^3]/\mathbf{k}^2 \\
&\quad + 2d(M/a)^2/(\mathbf{k}^2 + a/M)^2 + [-3a(M/a)^2 + 4d(M/a)^3]/(\mathbf{k}^2 + a/M)\}
\end{aligned} \tag{24.24}$$

Letting

$$m_{SI} = (a/M)^{\frac{1}{2}} \tag{24.25}$$

we find

$$V_{SI}(\mathbf{k}) = (2a)^{-2}\{2d/\mathbf{k}^4 + [3a - 4dm_{SI}^{-2}]/\mathbf{k}^2 + 2d/(\mathbf{k}^2 + m_{SI}^2)^2 + [4dm_{SI}^{-2} - 3a]/(\mathbf{k}^2 + m_{SI}^2)\} \tag{24.26}$$

The constant, a, is dimensionless and of order 1. The constant M has the dimension of inverse mass squared. We anticipate M will be extremely large resulting in a very small gluon mass m_{SI}.

There are massless gluon terms that generate color confinement reducing the impact of the massive gluon terms to a negligible effect outside hadronic regions.

We also see that the value of the inverse of graviton masses is of the order of the average galactic radius (the average galactic radius is large) and thus generate a Modified Newtonian potential (MoND) as seen in chapter 23.

Substituting eq. 24.26 we obtain a sum of massless and Yukawa-like potentials. The Yukawa potential is

$$V_Y(\mathbf{r}) = \int d^3k \, \exp(i\mathbf{k}\cdot\mathbf{r})/[(2\pi)^3(\mathbf{k}^2 + m^2)] = \exp(-mr)/[4\pi r] \qquad (24.27)$$

Thus we obtain

$$V_{SI}(\mathbf{r}) = -(2a)^{-2}\{2d(dV_Y(\mathbf{r})/dm^2)|_{m=0} + [3a - 4dm_{SI}^{-2}]/(4\pi r) - 2d(dV_Y(\mathbf{r})/dm^2|_{m=m_{SI}}) + [4dm_{SI}^{-2} -3a]V_Y(\mathbf{r})|_{m=m_{SI}}\}$$

$$(24.28)$$

with the form

$$V_{SI}(\mathbf{r}) = \alpha_1 r + \alpha_2/r + \alpha_3 e^{-m_{SI}r}/(4\pi m_{SI}) + \alpha_4 e^{-m_{SI}r}/(4\pi r) \qquad (24.29)$$

where the constants α_i are:

$$\alpha_1 = -(2a)^{-2}d/(8\pi) \quad \text{(up to an infrared divergent constant)} \qquad (24.30)$$
$$\alpha_2 = -(2a)^{-2}[3a - 4dm_{SI}^{-2}]/(4\pi)$$
$$\alpha_3 = -(2a)^{-2}d$$
$$\alpha_4 = -(2a)^{-2}[4dm_{SI}^{-2} -3a]$$

Thus we find the form of the potential of eq. 24.29 is linear plus Yukawa-like terms with small mass m_{SI} – perhaps near zero. As a result the form of the effective potential is

$$V(r) = -2g^2/r + g^2\lambda^2 r \qquad (24.31)$$

24.3 Charmonium and the Strong Interaction

In 1974 a bound state of a charmed and an anti-charmed quark was discovered by two experimental groups. Since charmed quarks are quite massive theoretical attempts were made to understand the charmed quark bound states within the framework of non-relativistic quantum mechanics. The "Cornell group" developed a fairly satisfactory[231] charmed quark bound state spectrum in 1974-5 using a combination of a linear and inverse 1/r potential as the strong interaction. In a recent fit[232] they gave the potential energy:

$$V(r) = -\kappa/r + r/a^2 \qquad (24.32)$$

where $\kappa = 0.61$, $a = 2.38$ GeV^{-1} and the charmed quark mass was 1.84 GeV.

---

[231] As did a Harvard group.

[232] E. J. Eichten, K. Lane, and C. Quigg, arXiv:hep-ph/ 0206018 (2002). See this paper for references to earlier work by the "Cornell group" and the "Harvard group" as well as papers by other researchers.

24.4 The Origin of the Linear Potential

The linear potential appears to have originated in a suggestion of Feynman in the Spring of 1974. This author proposed[233] a non-Abelian gauge quantum field theory, which yielded a linear potential. These papers, which had 4th order dynamic equations for the gauge fields, showed how to avoid the problems previously associated with higher derivative theories by using principal-value gauge field propagators that were similar in concept to the action-at-a-distance propagators used by Feynman and Wheeler in the late 1940's to formulate action-at-a-distance Quantum Electrodynamics.

Thus a non-Abelian quantum field theory of the strong interaction yielding a linear potential was created. In parallel with this development, Professor Kenneth Wilson (later a Nobelist) was developing lattice gauge theory. Because lattice lines focus the field of gauge boson, lattice gauge theory exhibited a linear potential as well between quarks. Thus it offered an alternative to our gauge theory. However, the linearity of the lattice potential was "built-in" by the lattice theory formulation and thus was an artifact of the lattice formulation. This approach, and other proposed approaches, all share the problem that the linear potential that they produce cannot be proven to truly be a consequence. Rather the linear potential is the "likely result."

On the other hand our higher dimensional theory produces the linear potential if the standard rules of quantum field theory are followed with the proviso that gauge field propagators are principal-value propagators.

This author had several discussions with Professor Wilson in late 1974 in the author's office and while walking to lunch at the Cornell Faculty Club. Professor Wilson proposed possible flaws in the author's theory on almost a daily basis. The author was able to show these suggested flaws were not flaws but physically acceptable. The final discussion with Wilson ended with Wilson stating words to the effect, "Your theory may be a correct phenomenological approximation to my theory of the strong interaction and quark confinement. But my theory is the correct one. Your theory is only a phenomenology." In the forty plus years since this concluding discussion no one has proved that the conventional strong interaction theory truly has a linear potential and quark confinement although some approximations suggest it does.

In the absence of a demonstration of a linear potential in the standard strong interaction model we suggest our theory is a viable alternative. Since the linear potential appears to fairly successfully describe much of the charmonium spectrum we feel our theory with its explicit derivation of a linear potential is worthy of interest – especially because it is in agreement with experiment as far as we know. *An experimentally completely correct phenomenology is a theory.*

---

[233] S. Blaha, Phys. Rev. **D10**, 4268 (July, 1974) and Phys. Rev. **D11**, 2921 (December, 1974) – Appendix J. These papers appeared before the charmonium calculations of the Cornell and Harvard groups in 1975.

24.5 Comparison Between Our Theory and the Charmonium Analysis

Comparing our above results with the Charmonium calculation we find[234]

$$g^2\lambda^2 = -(2a)^{-2}d/(8\pi) \tag{24.33}$$
$$-2g^2 = -(2a)^{-2}[3a - 4dm_{SI}^{-2}]/(4\pi) \tag{24.34}$$

where $g = \sqrt{(\kappa/2)} = 0.552$ and $\lambda = 0.761$ GeV (the result of Charmonium analysis). Thus

$$\lambda^2 = d/[4dm_{SI}^{-2} - 3a] \tag{24.35}$$

Note the Strong 'fine structure' constant found by the Cornell group

$$g^2/(4\pi) = 0.024 \tag{24.36}$$

is small. The electromagnetic fine structure constant α is 0.0073, a factor of about 3 lower than the Strong 'fine structure' constant. Thus it seems perturbation theory in the Strong interactions is not necessarily unreasonable.

---

[234] We note the constant, a, that appears in this chapter is NOT the Charmonium constant, a.

25. Other Effects of New Interactions Between Bosons

25.1 Missing Nucleon Spin Puzzle

The estimates of nucleon spin that are obtained from parton analyses of deep inelastic electron – nucleon interactions are woefully short of the spin expected in quark models of nucleons. The missing spin has been attributed to a number of causes. However the Missing Spin Puzzle remains.

From eq. 22.28 - 22.30 it is clear that there are important new interaction terms between the Electromagnetic and Strong interaction fields. After taking traces we find

$$\mathcal{L}_{intEM\text{-}S} = -\mathrm{Tr}\ iM\{(A_E{}^1{}_\nu + A_E{}^2{}_\nu)\ F_{SU(3)}{}^1{}_{\sigma\mu}D^\nu F_{SU(3)}{}^{2\sigma\mu} + iD_\nu F_{SU(3)}{}^1{}_{\sigma\mu}(A_E{}^{1\nu} + A_E{}^{2\nu})F_{SU(3)}{}^{2\sigma\mu}\} \quad (25.1)$$

$\mathcal{L}_{intEM\text{-}S}$ generates a combined photon-gluon vertex insertion in gluon interactions between quarks within a hadron. Figs. 25.1 and 25.2 show two simple possible vertex insertions in a gluon line.

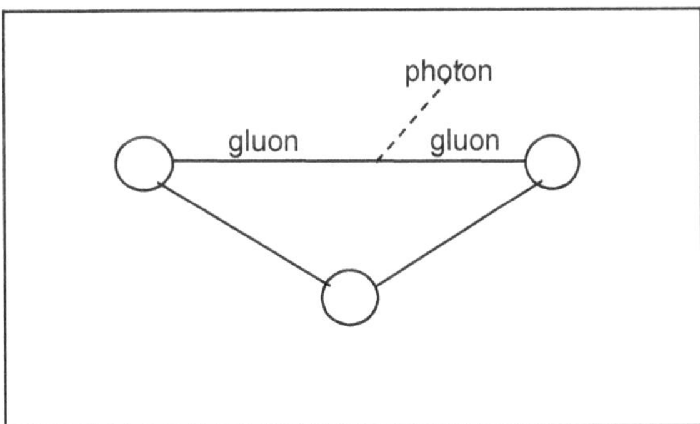

Figure 25.1. A single 'outgoing' photon vertex insertion in a gluon line. Only single gluon lines between the three quarks are displayed.

The gluon line, by itself, has $1/k^4$ and $1/k^2$ momentum space propagator terms. *The insertion of the vertex in Fig.25.1 generated by $\mathcal{L}_{intEM\text{-}S}$ yields a combined momentum factor of $k^3(k^4k^4)^{-1} = k^{-5}$ which would make it (summed over all gluon lines) a significant contribution to* *the proton spin determination in deep inelastic electron-nucleon scattering.*[235] *The insertion of the vertex in Fig. 25.2 generated by $\mathcal{L}_{intEM-S}$ yields a combined momentum factor of $k^2(k^4k^4)^{-1} = k^{-6}$ which may have a less significant effect.*

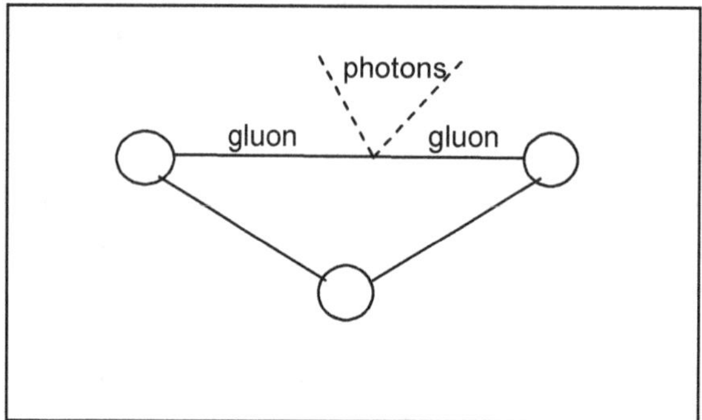

Figure 25.2. An 'outgoing' two photon vertex insertion in a gluon line. Only single gluon lines between the three quarks are displayed.

Thus our unified theory may help solve the nucleon spin puzzle. The interactions in Figs. 25.1 and 25.2 introduce a new direct connection between photons and spin one gluons. Thus their contributions to the summations of proton spin interactions in parton models may account for the 'missing' two-thirds of proton spin. Our unified theory has a new gluon-photon interaction that is not found in the conventional Standard Model.

25.2 Discrepancies between Proton Radius Measurements

Recently experiments have confirmed that the radius of a proton in a muonic hydrogen atom is smaller than the proton radius measured in a conventional hydrogen atom composed of a proton encircled by an electron. The lagrangian in eq. 22.23 indicates that there are direct interactions between gluons and Generation group gauge fields such as

$$U^{1i}_{\ \nu}U^{2iv}\partial_\alpha A_{SU(3)\mu}^{\ \ \ 1}\partial^\alpha A_{SU(3)}^{\ \ \ 2\mu}$$

These interactions result in Feynman diagrams that modify the electromagnetic field between a proton and a circling muon or electron. Fig. 25.3 shows the simplest forms of this interaction for a muon and a quark within a proton.

---

[235] See C. A. Aidala, S. D. Bass, D. Hasch, and G. K. Mallot, arXiv: 1209.2803v2 (2013) and references therein for a review of the 'missing' nucleon spin puzzle.

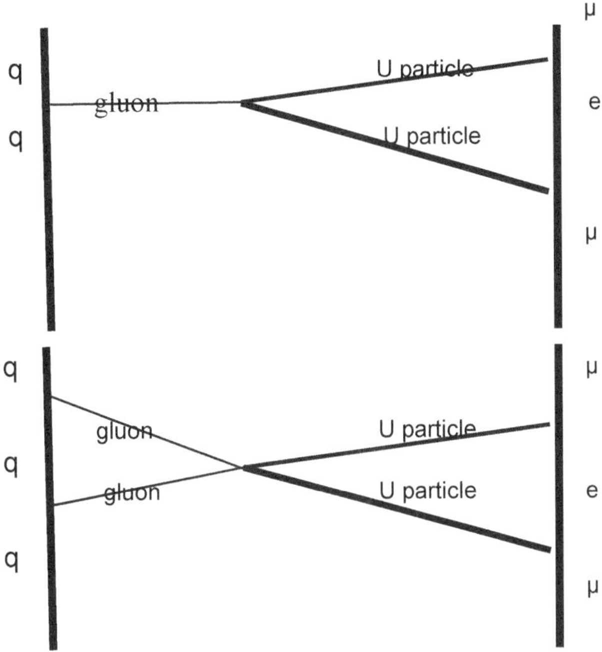

Figure 25.3. Gluon-Generation U gauge particle interactions between a muon and quark (within a proton).

The net effect of this additional interaction is to increase the force between the quark and muon beyond what one would expect using the electromagnetic potential, from the usual

$$V = - e^2/r$$

to

$$V' = - (e + \delta)^2/r$$
$$= - e^2/r'$$

(25.2)

where δ is small compared to e, and where

$$r' = r[e/(e + \delta)]^2 < r$$

giving an apparently smaller radial distance than the actual muon radial distance. This change would cause an energy shift to a lower value (more negative) in the muonic energy levels and, consequently, the proton radius would appear smaller for muonic atoms – as experiment confirms.[236]

Thus the new 'interactions between interactions' of our unified theory have significant effects that are already qualitatively verified by experiment.

---

[236] Fig. 25.3 is not of importance for 'normal' hydrogen since the electron is so much lighter than the muon.

26. Quantum Gravity and the Wheeler-DeWitt Equation Extended to Complex Coordinates

26.1 Introduction

In this chapter we will consider Quantum Gravity as embodied in the Wheeler-DeWitt[237,238] equation. This equation has many noteworthy points that we will consider below – particularly its extension to complex space-time. It also raises important basic quantum questions: such as "Who is the Observer?" that we addressed in Blaha (2014a) where most of this chapter first appeared.

26.2 Analytically Continued Wheeler-DeWitt Equation to Complex Metrics under a Faddeev-Popov Method Restriction

In this section we extend the Wheeler-DeWitt equation for Quantum Gravity to complex coordinates and metrics <u>by analytic continuation</u> (piece-wise if necessary) and impose the condition that metrics must be real-valued using the Faddeev-Popov Method. Our procedure will be to take the Wheeler-DeWitt equation for real-valued metrics (and coordinates), **analytically continue it to the case of complex metrics and coordinates,**[239] and then impose the condition that physically acceptable metrics must be real-valued through use of a Reality group transformation implemented via the Faddeev-Popov Method. Our motivation is two-fold: we have shown that the form of The Standard Model of Particles can be derived from complex space-time considerations demonstrating that we exist in a "masked" complex space-time; and we have shown the imposition of a restriction on a gauge theory such as gravitation[240] can be implemented using the Faddeev-Popov Method or equivalent.

We start by noting the canonical decomposition of a <u>real-valued</u> metric $g_{\mu\nu}$ is defined by:

---

[237] DeWitt, B. S., Phys. Rev. **160**, 1113 (1987).
[238] Hartle and Hawking also derive the Wheeler-DeWitt equation from a path integral formalism for quantum gravity.
[239] The piecewise analytic continuation of general relativity to complex coordinates and metrics is described in some detail in Blaha (2004). The gist of the continuation is that all equations have the same form after analytic continuation due to a basic theorem of complex mathematics that the analytic extension of equations to complex values from real values is unique.
[240] Such as real-valuedness.

$$g_{\mu\nu}(x) = \eta_{\alpha\beta}\, \partial\omega^{\alpha}/\partial x^{\mu}\; \partial\omega^{\beta}/\partial x^{\nu} \qquad (26.1)$$

with

$$g_{\mu\nu} = g_{\nu\mu}$$

and with inverse

$$g^{\mu\nu} = \eta^{\alpha\beta}\, \partial x^{\mu}/\partial\omega^{\alpha}\; \partial x^{\nu}/\partial\omega^{\beta} \qquad (26.2)$$

The decomposition of the real-valued metric is

$$g_{\mu\nu} = \begin{bmatrix} -\alpha^{2}\beta_{k}\beta^{k} & \beta_{j} \\[2ex] \beta_{i} & \gamma_{ij} \end{bmatrix} \qquad (26.3)$$

$$g^{\mu\nu} = \begin{bmatrix} -\alpha^{-2} & \alpha^{-2}\beta^{j} \\[2ex] \alpha^{-2}\beta^{i} & \gamma^{ij} - \alpha^{2}\beta^{i}\beta^{j} \end{bmatrix} \qquad (26.4)$$

where

$$\gamma_{ik}\,\gamma^{kj} = \delta_{i}^{j} \qquad\qquad \beta^{i} = \gamma^{ij}\,\beta_{j} \qquad (26.5)$$

The Wheeler-DeWitt equation <u>for real-valued metrics</u> is

$$(G_{ijkl}\, \delta/\delta\gamma_{ij}\, \delta/\delta\gamma_{kl} + \gamma^{\frac{1}{2}\,(3)}R + 2\lambda\, \gamma^{\frac{1}{2}\,(3)})\Psi(^{(3)}\mathcal{G}) = 0 \qquad (26.6)$$

where λ is the cosmological constant, and where the Wheeler-DeWitt metric is

$$G_{ijkl} = \tfrac{1}{2}\, \gamma^{-\frac{1}{2}}(\,\gamma_{ik}\gamma_{jl} + \gamma_{il}\gamma_{jk} - \gamma_{ij}\gamma_{kl}) \qquad (26.7)$$

The functional derivatives $\delta/\delta\gamma_{ij}$ have several interpretations that are presumably equivalent. DeWitt characterizes them as coordinate independent specifications of the 3-metric. The wave function $\Psi(^{(3)}\mathcal{G}) = \Psi(\gamma_{ij})$, where $^{(3)}\mathcal{G}$ is a geometry, is *not* coordinate dependent. It is invariant under coordinate changes. $^{(3)}\mathcal{G}$ is a discrete infinity of independent invariants constructed from products of the Riemann tensor and its covariant derivatives.

Hartle and Hawking[241] derive the Wheeler-DeWitt equation from a path integral formalism for quantum gravity. Their path integral can be represented as

$$Z = N \int \delta g(x)\, \exp(iS_{E}[g]) \qquad (26.8)$$

---

[241] Hartle, J. B. and Hawking, S. W., Phys. Rev. D **28**, 2960 (1983).

where S_E is the classical action for gravity and the functional integral is an integral over all 4-geometries. Changing to DeWitt's notation based on the spatial metric γ_{ij} and expressing eq. 26.8 in a more explicit form for use in conjunction with the Faddeev-Popov Method we have

$$Z = N \int \sum_{i,j} \prod_x d\gamma_{ij}(x) \, \exp(iS_E[\gamma]) = N \int D\gamma \, \exp(iS_E[\gamma]) \qquad (26.9)$$

The integrand, being a functional integral over all space, is independent of the coordinates.

The Wheeler-DeWitt equation applies to real-valued metrics $\gamma^{ij}(x)$. We now extend this equation to apply to complex-valued metrics using the local Reality group. In doing this we realize that there are an infinite number of complex-valued metrics in the orbit corresponding to each real-valued metric.

This redundancy can be resolved by realizing that the physical measurement of an invariant interval, and the coordinates from which it is derived, are always real-valued. Yardsticks and clocks can only measure real-valued numbers. Based on this physical principle we can generalize quantum gravity to complex coordinates and metrics by using the Faddeev-Popov method to constrain the set of paths in the quantum gravity path integral eq. 26.9. Using the Faddeev-Popov Method the constraint can be expressed in terms of an infinitesimal transformation of the metric to a complex value. Using an infinitesimal Reality group transformation V:

$$[\exp(ia_j(x'',x')U_j)]^{\alpha}{}_{\mu} = S(x'',x')^{\alpha}{}_{\mu} = \partial x''^{\alpha}/\partial x'^{\mu} \qquad (26.10)$$

$$V = \exp(ia_k(x'',x')U_k) \cong I + ia_k(x'',x')U_k \qquad (26.11)$$

$$V^{\dagger} = [\exp(ia_k(x'',x')U_k)]^{\dagger} \cong I - ia_k(x'',x')U_k \qquad (26.12)$$

where $a_j(x'',x')$ is the jth real-valued local infinitesimal parameter, and U_k is one of the 16 hermitean generators of the Reality group. (We treat the index on a_k and U_k as lower case and sum on k.) We find the condition fixing the metric to a physical *real* value is:

$$F^{ij}(\gamma(x)) = \text{Im } \gamma^{ij}(x) \equiv -\tfrac{1}{2} i(\gamma^{ij}(x) + \gamma^{ij}(x)^*) = 0 \qquad (26.13)$$

where

$$F^{aij}(\gamma^a(x)) = \text{Im}\{(\delta^i{}_m + ia_k(x',x)U_k{}^i{}_m)(\delta^j{}_p + ia_k(x',x)U_k{}^j{}_p)\gamma^{mp}(x)\} \qquad (26.14)$$

using the infinitesimal form.

The Reality condition eq. 26.13 is implemented within the path integral formalism with the Faddeev-Popov Method identity

$$1 = \int D\gamma \, \Delta(F(\gamma)) \, \delta(\delta F^{aij}(\gamma^a(x))/\delta a_n(x',x)) = \int D\gamma \, \Delta(F(\gamma)) \, \delta(F(\gamma^{ij})) \qquad (26.15)$$

Eq. 26.14 yields

$$\delta F^{aij}(\gamma^a(x))/\delta a_n(x',x)|_{a=0} = \text{Im}\{\delta_{kn}iU_{k\,m}^{\,i}\gamma^{mp}(x)\delta_{\,p}^{j} + \delta_{kn}\delta_{\,m}^{i}i\gamma^{mp}(x)U_{k\,p}^{\,j}]\}$$
$$= \text{Re}\{\gamma^{mp}[[U_{n\,m}^{\,i}\delta_{\,p}^{j} + \delta_{\,p}^{i}U_{n\,m}^{\,j}]\}$$
$$= \gamma^{mp}(x)\text{Re}\{U_{n\,m}^{\,i}\delta_{\,p}^{j} + \delta_{\,p}^{i}U_{n\,m}^{\,j}\} = \gamma^{mp}(x)\xi^{ij}_{\,nmp} \qquad (26.16)$$

since $\gamma^{mp}(x)$ is made real by the $\delta(\text{Im }\gamma^{mp}(x))$ where

$$\xi^{ij}_{\,nmp} = \text{Re}\{U_{n\,m}^{\,i}\delta_{\,p}^{j} + \delta_{\,p}^{i}U_{n\,m}^{\,j}\} \qquad (26.17)$$

Note $\xi^{ij}_{\,nmp}$ is symmetric in i and j, and becomes effectively symmetric in m and p when combined with $\gamma^{mp}(x)$ in eq. 26.16. Calculating $\Delta(F(\gamma))$ we obtain

$$\Delta(F(\gamma)) = [\det \delta F^{aij}(\gamma^a(x))/\delta a_n(x',x)|_{a=0}]^{-1} \qquad (26.18)$$

We can rewrite this Faddeev-Popov determinant as a path integral over an anti-commuting c-number scalar field χ with a ghost Lagrangian:

$$\Delta(F(\gamma)) = \int D\chi^* D\chi \, \exp[i\int d^4x \, \gamma^{\frac{1}{2}}\mathcal{L}_\gamma^{\,ghost}(x)] \qquad (26.19)$$

where

$$\mathcal{L}_\gamma^{\,ghost}(x) = \chi^*(x)[U_{nij}\xi^{ij}_{\,nmp}\gamma^{mp}(x)]\chi(x)$$
$$= \chi^*(x)(\text{Re }U_{nij})\xi^{ij}_{\,nmp}\gamma^{mp}(x)]\chi(x) \qquad (26.20)$$
$$= \chi^*(x)\xi_{mp}\gamma^{mp}(x)\chi(x) \qquad (26.21)$$

since $\xi_{mp}\gamma^{mp}(x)$ is a c-number:

$$\xi_{mp} = (\text{Re }U_{nij})\xi^{ij}_{\,nmp} = 2\,\text{Re }U_{npi}\,\text{Re }U_{n\,m}^{\,i} \qquad (26.22)$$

using $\text{Re }U_{n\,m}^{\,i} = \text{Re }U_{n\,i}^{\,m}$ for the U(4) generators in its fundamental representation $\underline{4}$.

We then find ξ_{mp} is a diagonal matrix and has the value

$$\xi_{mp} = \xi_m\delta_{mp} \qquad (26.22a)$$

where the diagonal elements are

$$\xi_0 = 8$$
$$\xi_1 = 8$$
$$\xi_2 = 8$$
$$\xi_3 = 8 \qquad (26.22b)$$

and the non-diagonal elements are zero making ξ_{mp} considered as a matrix a multiple of the Identity matrix..

The Faddeev-Popov generated terms when added to the Einstein Action appear to have important ramifications – particularly with respect to the Cosmological Constant. We explore that issue next.

26.3 Possible Source of the Cosmological Constant in the Complex Space-time – Faddeev-Popov Constraint Term

The Wheeler-DeWitt equation

$$(G_{ijkl}\, \delta/\delta\gamma_{ij}\, \delta/\delta\gamma_{kl} + \gamma^{\frac{1}{2}\,(3)}\, R + 2\lambda\, \gamma^{\frac{1}{2}\,(3)})\Psi(^{(3)}g) = 0 \tag{26.23}$$

has the cosmological constant, Λ, as one of its terms. This equation is derived from the Hamiltonian constraint that ultimately follows from the Einstein action. Inserting the Faddeev-Popov term of the previous section in the Einstein lagrangian yields

$$S_E(\gamma) = -(16\pi G)^{-1} \int d^4x\, \gamma^{\frac{1}{2}}\{\, R(x) - 2\lambda + \chi^*(x)\chi(x)\xi_{mp}\gamma^{mp}(x)\} \tag{26.24}$$

The constant matrix ξ_{mp} is a product of parts of the generators of U(4) given by eq. 26.22a:

$$\xi_{mp} = 2\,(Re\, U_{npi})\,(Re\, U^i_{n\,m}) \tag{26.25}$$

We will now show that the term

$$\Pi = \gamma^{\frac{1}{2}}\chi^*(x)\chi(x)\xi_{mp}\gamma^{mp}(x) \tag{26.26}$$

upon variation of the metric $\delta\gamma_{\mu\nu}$, gives a term which is approximately a constant cosmological term assuming $\chi(x)$ is approximately constant (with its implied divergence eliminated by renormalization of the path integral), and assuming an almost flat space-time $\gamma_{\mu\nu} \cong \eta_{\mu\nu}$. Varying the metric for Π yields

$$\delta\Pi = \delta\gamma_{\mu\nu}\,\{\tfrac{1}{2}\,\gamma^{\frac{1}{2}}\chi^*(x)\chi(x)\xi_{mp}\gamma^{mp}(x)\gamma^{\mu\nu} - \gamma^{\frac{1}{2}}\chi^*(x)\chi(x)\xi_{mp}\gamma^{m\mu}\gamma^{p\nu}\} \tag{26.27}$$

The resulting modified Einstein field equation is

$$R^{\mu\nu} - \tfrac{1}{2}\,g^{\mu\nu}R + \lambda\,\gamma^{\mu\nu} + \tfrac{1}{2}[\gamma^{mp}(x)\gamma^{\mu\nu} - \gamma^{m\mu}\gamma^{p\nu}]\xi_{mp}\chi^*(x)\chi(x) = -8\pi T^{\mu\nu} \tag{26.28}$$

The terms $\tfrac{1}{2}[\gamma^{mp}(x)\gamma^{\mu\nu} - \gamma^{m\mu}\gamma^{p\nu}]\xi_{mp}\chi^*(x)\chi(x)$ appearing above is, or is a contribution to, the cosmological constant assuming a nearly flat universe as our universe seems to be. Thus $\gamma_{\mu\nu} \cong$

$\eta_{\mu\nu}$ to good approximation and $\chi^*(x)\chi(x)$ can be taken to be constant since the time derivative of $\chi(x)$ does not appear in the lagrangian. Consequently the total cosmological constant term is

$$\lambda_{tot}{}^{\mu\nu} \cong \lambda g^{\mu\nu} + 4\,\chi^*(x)\chi(x)g^{\mu\nu} = (\lambda + \lambda_{F\text{-}P})\,g^{\mu\nu} \tag{26.29}$$

by eq. 26.22b.

Given the somewhat problematic state of our understanding of the cosmological constant it is not impossible that the complexity of space-time leading to $\lambda_{F\text{-}P}$ may be the sole origin of the cosmological constant. In evaluating eq. 26.29 we may normalize $\chi^*(x)\chi(x) = 1/4$ by adjusting the overall normalization of the path integral since $\chi(x)$ is time independent. Then if the "bare" cosmological constant is zero we obtain

$$\lambda_{tot} = \lambda_{F\text{-}P} = 1 \tag{26.30}$$

by eq. 26.29.

If this is true then we have achieved a space-time origin for the cosmological constant rather than an ad hoc origin. The modified Einstein field equation is then

$$R^{\mu\nu} - \tfrac{1}{2}\,g^{\mu\nu}R + \lambda_{tot}\gamma^{\mu\nu} = -8\pi T^{\mu\nu} \tag{26.31}$$

by eqs. 26.28 and 26.30.

26.4 Impact of the Faddeev-Popov Complexity Term on the Wheeler-DeWitt Equation

The Faddeev-Popov term that arises because of the restriction of the metrics and coordinates to real values also impacts on the Wheeler-DeWitt equation since it is derived from the lagrangian via the Hamiltonian it generates. The Wheeler-DeWitt equation changes to

$$(G_{ijkl}\,\delta/\delta\gamma_{ij}\,\delta/\delta\gamma_{kl} + \gamma^{1/2\,(3)}R + 2\lambda\,\gamma^{1/2\,(3)})\Psi(^{(3)}\mathcal{G}) = 0 \tag{26.32}$$

The functional Wheeler-DeWitt equation of eq. 26.32 resembles a Klein-Gordon equation.[242] Solutions of this equation can be expressed as path integrals:

$$\Psi(^{(3)}\mathcal{G},\,\mathcal{L}_F) = N \int_{\mathcal{C}} \delta g(x)\,\exp(-I(g,\,\mathcal{L}_F)) \tag{26.33}$$

where $I(g,\,\mathcal{L}_F)$ is the effective total Euclidean action for the open universe case. See Hartle and Hawking for a detailed study in the case of a symmetric cosmological constant.

It does not appear that the issues of the Wheeler-DeWitt equation are resolved by the extended Extended Wheeler-DeWitt equation presented here:

---

[242] Hartle, J. B. and Hawking, S. W., Phys. Rev. D **28**, 2960 (1983).

- Divergences in integrals in inner products, thus requiring renormalization.
- Negative probabilities in inner products,
- Issues with the requirement of space-like surfaces,
- The frontier divergence singularity.

Later we consider another form of the Wheeler-DeWitt equation expressed in terms of Megaverse geometry. We will reconsider the issues of the above formulation again in the Megaverse.

27. SuperSymmetric Aspect of the Unified SuperStandard Model

The Unified SuperStandard Model has a SuperSymmetric aspect embedded within it. It has equal numbers of fermions and vector bosons -192 of each.[243] These vector bosons[244] are directly related to the form of the fermion particles' structure – the Fermion Periodic Table. The 192 fermions form a <u>192</u> representation of the Θ-Symmetry group, and the 192 vector bosons form another <u>192</u> representation of the U(192) Θ-Symmetry group.

27.1 The Θ-Symmetry Group

The U(192) Θ-Symmetry group has 192^2 gauge fields and generators. Elements of the group algebra form a closed set under commutation. The fundamental representation has 192 dimensions.

27.2 Fermionic Operators

One can define two sets[245] of 192^2 fermionic operators $\{A_{\Theta F}{}^{\mu}{}_{nm}(x)\}$ and $\{A_{\Theta B}{}^{\mu}{}_{nm}(x)\}$ that transform fermions into bosons between the <u>192</u> Θ-Symmetry fermion and boson representations; and that also transform bosons into fermions between the <u>192</u> Θ-Symmetry fermion and boson representations respectively:

$$A_{\Theta F}{}^{\mu}{}_{nm}(x)\psi_m(x) \to A^{\mu}{}_n$$
$$A_{\Theta B}{}^{\mu}{}_{nm}(x)A_{\mu m}(x) \to \psi_n$$

$$27.1)$$

where n = 1, ... , 192 and m = 1, ... , 192 label multiplet fields irrespective of internal symmetries.[246] Each (n, m) pair label one of 192^2 fermionic operators.

The fermionic operators are closed under anticommutation. One simple 'free' field representation of these operators in terms of the fermionic and bosonic fields is

---

[243] The 16 Species group vector bosons 'do not count' because they follow from Complex General Relativity.

[244] There are also Higgs bosons and Θ-Symmetry vector bosons. These vector bosons do not affect the number or form of the Fermion Periodic Table and thus can be viewed as a separate issue. In fact we will show the Θ-Symmetry group plays a role in establishing a SuperSymmetry-like formalism in the Unified SuperStandard Model.

[245] While these operators form a Jordan algebra they do not have group representations.

[246] The ability to 'rotate between interactions' using the Θ-Symmetry group enables the fermionic operators to be transformed in a lagrangian covariant manner.

$$\mathbf{A}_{\Theta F}{}^{\mu}{}_{nm}(x) = \mathbf{A}^{\mu}{}_{n}(x)\psi_{m}(x) \tag{27.2}$$

$$\mathbf{A}_{\Theta B}{}^{\mu}{}_{nm}(x) = \psi_{n}(x)\mathbf{A}^{\mu}{}_{m}(x)^{\dagger} \tag{27.3}$$

where n and m are labels ranging from 1, ... 192 for fermions and bosons, $\mathbf{A}^{\mu}{}_{n}(x)$ is the nth of the 192 vector fields, $\psi_{m}(x)$ is the mth of the 192 fermion fields in eq. 27.2-27.3; and $\mathbf{A}^{\mu}{}_{m}(x)^{\dagger}$ is the hermitean conjugate of the mth vector field of the 192 vector fields and $\psi_{n}(x)$ is the nth fermion field of the 192 fermion fields. Spatial integrals of the above operators have simpler anticommutation relations.

Elements of the Θ-Symmetry group can be used to map the fermionic operators to different forms. For example if U_{Θ} is an element of the Θ-Symmetry group we can redefine fermion operators as a different map between fermions and bosons with expresions like

$$\mathbf{A}'_{\Theta F}{}^{\mu}{}_{nm} = U_{\Theta}\mathbf{A}_{\Theta F}{}^{\mu}{}_{nm}U_{\Theta}{}^{-1} \tag{27.4}$$

$$\mathbf{A}'_{\Theta B}{}^{\mu}{}_{nm} = U_{\Theta}\mathbf{A}_{\Theta B}{}^{\mu}{}_{nm}U_{\Theta}{}^{-1} \tag{27.5}$$

Thus one can SuperSymmetric map between the fermion and vector boson multiplets. We have not displayed internal symmetry indices on the fields. Clearly the difference between the internal symmetry fundamental representation of the fermions and the adjoint representation of the vector bosons enables the fermionic transformations to 'sluff' over internal symmetries.

27.3 Relation of the Θ-Symmetry Fields and the Fermionic Fields

The fermionic field operators that we have defined can be related to the U(192) Θ-Symmetry field operators which we will denote $\mathbf{A}_{\Theta}{}^{\mu}{}_{nm}(x)$ where n = 1, ... , 192 and m = 1, ... , 192 are labels on the <u>192</u> fermion or boson fields. The following relations hold:

1.　The product of $\mathbf{A}_{\Theta F}{}^{\mu}{}_{nm}(x)$ and $\mathbf{A}_{\Theta F}{}^{\mu}{}_{mp}(x)^{\dagger}$ can be expressed as an equivalent sum of Θ-Symmetry field operators.

2.　The product of $\mathbf{A}_{\Theta B}{}^{\mu}{}_{nm}(x)^{\dagger}$ and $\mathbf{A}_{\Theta B}{}^{\mu}{}_{mp}(x)$ can be expressed as an equivalent sum of Θ-Symmetry field operators.

3.　The product of $\mathbf{A}_{\Theta F}{}^{\mu}{}_{nm}(x)$ and $\mathbf{A}_{\Theta B}{}^{\mu}{}_{mp}(x)$ can be expressed as an equivalent sum of Θ-Symmetry field operators.

Thus we find that the U(192) Θ-Symmetry field operators can be obtained from products of the fermionic operators.

We also note that the product of a fermionic operator and a Θ-Symmetry field operator is a fermionic operator.

These relations establish a close connection between fermionic field operators and the Θ-Symmetry field operators.

We conclude that there is a SuperSymmetry relation between the fermion and vector boson multiplets in the Unified SuperStandard Model. The benefits of SuperSymmetric transformations in this context remain to be determined. However there is a hint of a significant role in the case of the Megaverse as we see later.

28. Unified SuperStandard Model and its Hidden SuperString Aspect

The Unified SuperStandard Model that we have developed is a type of Quantum Field Theory that contains the known Standard Model and Gravitation as well as significant new features. It uses extensions of Quantum Field Theory: a Two-Tier PseudoQuantum formulation that makes the theory finite to all orders in perturbation theory and can be quantized in any coordinate system.

In this chapter we will show that the theory has a String aspect that, together with its SuperSymmetric aspect, enables us to characterize it as a hidden variant of SuperString theory.

28.1 String Theory vs. Two-Tier Quantum Field Theory

We begin by comparing the Gervais-Sakita[247] SuperString lagrangian (GS) with a simple Two-Tier Quantum Field Theory lagrangian:[248]

$$\mathcal{L} = \overline{\psi}(x)(i\gamma^{\mu}\partial/\partial x^{\mu} - m)\psi(x) - \partial_a X_{\mu}\partial^a X^{\mu} \qquad \text{(GS)}$$

$$\mathcal{L} = \overline{\psi}(X_T)(i\gamma^{\mu}\partial/\partial X_T^{\mu} - m)\psi(X_T) + \tfrac{1}{4}\, M_c^{\,4} F^{\mu\nu}(X_T)F_{\mu\nu}(X_T) \qquad \text{(TT)}$$

where

$$X_T^{\mu}(y) = y^{\mu} + i\, Y^{\mu}(y)/M_c^{\,2}$$

and

$$F^{\mu\nu} = \partial X^{\mu}/\partial y_{\nu} - \partial X^{\nu}/\partial y_{\mu}$$
$$\equiv i\,(\partial Y^{\mu}/\partial y_{\nu} - \partial Y^{\nu}/\partial y_{\mu})/M_c^{\,2}$$

The Gervais-Sakita lagrangian differs from the above Two-Tier lagrangian by the introduction of the 'string' X^{μ} in the fermion part of the Two-Tier lagrangian.

We note that the Two-Tier quantum field Y_{μ} has two degrees of freedom like the string field. Thus the Two-Tier fermion field is a function of a string[249] whereas the SuperString lagrangian treats the fermion and the string as independent. As a result the Two-Tier quantum

---

[247] Gervais, J. L. and Sakita, B., Nucl. Phys. **B34**, 632 (1971).
[248] Blaha (2002) and (2005a).
[249] As Blaha (2002) and (2005a) show there is a form of the Calculus of Variations that supports a canonical derivation of the equations of motion.

field theory embedded in the Unified SuperStandard Model is a variant on SuperString theory if one also takes account of the SuperSymmetry attributes of the Unified SuperStandard Model.

28.2 Strings in Particles? or Strings Made into Particles?

A comparison of the two two types of theories is made more pointed by the form of Feynman propagators in Two-Tier theories. Fig. 9.1, which is discussed in detail in section 9.6 and displayed below because of its relevance, shows that a 'dressed' free fermion particle has an associated cloud of 'strings' that are the source of the finiteness of perturbation theory calculations. Thus Two-Tier Unified SuperStandard Model calculations are finite to all orders in perturbation theory by putting 'strings' in particles unlike SuperString theories which make particles from strings.

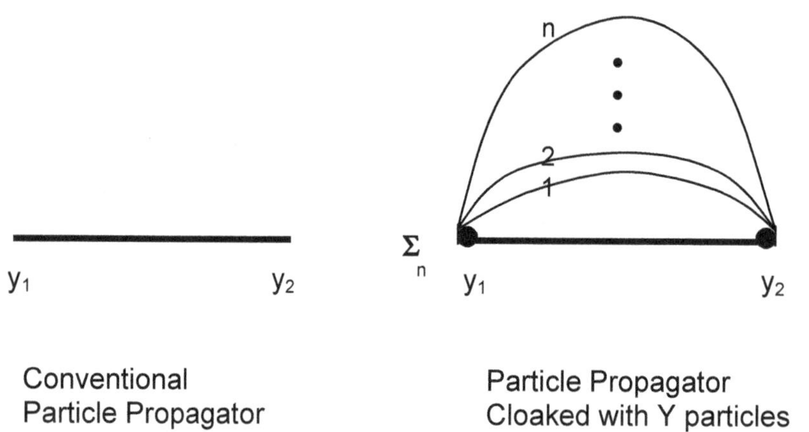

Figure 28.1. Feynman diagram for conventional and cloaked Two-Tier propagators.

28.3 Further Evidence of the String-like Substructure of the Unified SuperStandard Model

Quantum Dimensions $X_T{}^\mu(y)$ endow a particle with an extended structure that resembles to some extent the extended structure seen in bosonic string and Superstring theories. For example, Bailin (1994) use the operator[250]

$$V_\Lambda(k) = \int d^2\sigma \sqrt{-h}\, W_\Lambda(\tau, \sigma)\, e^{-ik\cdot X}$$

where $X_T{}^\mu$ is a quantized fourier expansion of the string fields (see eq. 7.22 of Bailin (1994)).

---

[250] D. Bailin and A. Love, *Supersymmetric Gauge Field Theory and String Theory* (Institute of Physics Publishing, Philadelphia, PA, 1994) page 272.

We note our $X_T{}^\mu$ coordinate-field has two transverse degrees of freedom due to gauge invariance, which also invites comparison to the bosonic string. A point of difference is that we will create a well-defined quantum field theoretic formulation in conventional space-time that has the Standard Model as its "large distance" behavior thus introducing a note of reality that is not (yet?) very apparent in Superstring theories. We see that the interacting quantum field theories based on this approach also have good, finite, short distance behavior just as string theories.

Scalar, and other particles, Feynman propagators can be viewed as describing the propagation of a particle cloaked (accompanied) by a cloud of Y particles.

If one considers high energy hadron-hadron collisions then the beclouded quarks within the hadrons can interact in a 'string-like' manner through the degrees of freedom afforded by the Y particle clouds. Then the behavior of the colliding hadrons could well simulate the types of effects that gave rise to String theory in the 1970's.

29. Status of the Derivation at this Point

Since the derivation progress report of chapter 21 we have extended the derivation to include:

1. The calculation of the Riemann-Christoffel Curvature tensor for bosons.
2. The specification of the vector boson and gravitation terms of the Unified SuperStandard Model lagrangia.
3. A demonstration of modified gravitation and confining strong interaction consequences of these lagrangian terms as well as other possible effects on the proton spin and radius puzzles.
4. A discuusion of the consequences of Quantum Gravity using the Wheeler-DeWitt equation including a Faddeev-Popov Mechanism generation of the Cosmological Constant.
5. A discussion of SuperSymmetric aspects of The Unified SuperStandard Model.
6. A discussion of a SuperString view of the The Unified SuperStandard Model.

Having presented a Unified SupeStandard Theory that uccessfully accounts for confirmed features of elementary particles and gravitation (possibly excepting new CP-related phenomena) we now turn to evidnce for phenomena beyond our universe – The Megaverse.

30. Evidence for Entities Beyond Our Universe

30.1 Theoretical and Experimental Support

Why are we not content with one universe given its enormous size and variety? It appears that there are important theoretical reasons, and some important experimental observations, that suggest that there is more than our universe 'out there.'

In this chapter[251] we will discuss theoretical reasons and experimental suggestions of a larger space—that we call the *Megaverse*—that contains our universe and, most likely, other universes. The existence of a Megaverse resolves theoretical issues and may address some important astronomical puzzles that have appeared in recent years.

The theoretical issus, which have been subjects of discussion for many years, are:

1. The need for a 'clock' to measure 'time' knowing that it is to some extent relative and local.
2. The need for a 'quantum observer' to complete the understanding of quantum gravity as described by the Wheeler-DeWitt equation and in other efforts to develop a quantum gravity.
3. The need for other universes to provide theoretical measuring platforms for quantities beyond the charge and mass of the universe. We think here of the other quantum numbers of particles and particle number operators such as Baryon number.
4. The need for an ultimate source of mass and inertia in our universe.

In Blaha (2015a) and earlier books we have suggested that there are weighty reasons to believe that other universes exist.[252] The existence of other universes is a solution to these problems.

These problems have a source in Quantum Gravity and the interpretation of the Wheeler-DeWitt equation in particular. See chapter 26 for a discussion of the Wheeler-DeWitt equation and its implications. We now consider the issues raised above.

---

[251] Most of this chapter appears in Blaha (2015a) and in earlier books by the author.

[252] In Blaha (2013a), before the Higgs particle was discovered at CERN we suggested an alternate mechanism was possible if a sister universe existed (making the existence of other universes a reasonable possibility. The Higgs discovery makes the sister universe mechanism unlikely.

30.1.1 Universe Clocks

Asynchronous Logic provides the equivalent of a clock for the synchronization of processes within large electrical systems such as VLSI chips. Similarly there is a need for a universal clock for our universe. As DeWitt[253] points out in his studies of quantum gravity,

'"The variables ... [of the quantized Friedmann model] because of their lack of hermiticity, are not rigorously observable and hence cannot yield a measure of proper time which is valid under all circumstances. ... . It is for this reason that we may say that "time" is only a phenomenological concept ... If the principle of general covariance is truly valid then the quantum mechanics of everyday usage with its dependence on the Schrödinger equations ... is only a phenomenological theory. For the only "time" which a covariant theory can admit is an intrinsic time defined by the contents of the universe itself. Any intrinsically defined time is necessarily non-Hermitean, which is equivalent to saying that there exists no clock, whether geometrical or material, which can yield a measure of time which is operationally valid under *all* circumstances, and hence there exists no operational method for determining the Schrödinger state function with arbitrarily high precision."

The lack of a clock within our universe invalidates quantum mechanics in principle and Quantum Gravity in particular. DeWitt concludes, "Thus [quantum gravity] will say nothing about time unless a clock to measure time is provided."

Unruh[254] also has an issue with the source of time:

"One of the key problems is that of time. We see and experience the world in terms of time. We see things grow, develop, and change. However, time does not enter into the Euclidean formulation of quantum gravity directly. In the usual Hamiltonian formulation, the Hamiltonian for quantum gravity is made up of densities which are the generators, not only of spatial coordinate transformations, but also of temporal coordinate transformations. The content of four of Einstein's equations, namely, the 6 „components, is that these generators are zero. Thus all wave functions are invariant under all spatial and all temporal coordinate transformations. There is nothing in the wave function or the amplitudes which refers to the coordinate t, or the corresponding points of the manifold in any way. How then do we recover the indubitable and ubiquitous experience we have of time? The standard answer is that our experience of time is actually an experience of different correlations between physical quantities in the world. Time is replaced by the readings of clocks. I know that time has changed, not through any direct experience with time, but because the hands of my watch have changed.

Although the implementation of this idea is actually extremely difficult in practice, and although I personally believe that one should formulate one's quantum theory of gravity so as to contain time explicitly, let us nevertheless pursue the consequences of this idea of time as

---

[253] DeWitt, B. S., Phys. Rev. **160**, 1113 (1987).
[254] Unruh, W. G., Phys. Rev. D **40**, 1053 (1989).

defined internally, as the "reading" of a dynamic variable. For an observer inside the theory, his "time" is not the coordinate t. Rather his time is some one of the given dynamic variables of the theory: y or P. Thus although the coupling to the baby universes via the effective action S,. is independent of the coordinates t or x, that does not mean that the observer inside the theory will experience the interactions as being independent of time. For him and/or her, time is one of the dynamic variables and so it can depend on the various dynamic variables of the theory, even if it does not depend on the time coordinate t. In general one would expect the observer to see what looks to him like a time-dependent interaction with the baby universes. At one time, some one of the baby universes may couple strongly to the large universe, while at some other time, another of the baby universes will couple more strongly."

In Blaha (2015a) and earlier books, we suggested the existence of other universes provides a 'clock' in principle for our universe. And being universes, these other universes are an excellent clock. DeWitt points out,

"Because every clock has a "one-sided" energy spectrum, its ultimate accuracy must necessarily be inversely proportional to its rest mass. When the whole universe is cast in the role of a clock, the concept of time can of course be made fantastically accurate (at least in principle) ... "

Setting a mass scale using other universes, also sets[255] a time scale and resolves the issue of a clock for our universe. *In principle the existence of other universes validates the role of time in the Copenhagen interpretation of Quantum Mechanics.*

30.1.2 Quantum Observer

Attempts to create a quantum gravity theory have to confront the need for an *Observer* in any quantum theory within the context of the Copenhagen interpretation. DeWitt points out,

"The Copenhagen view depends on the assumed a priori existence of a classical level to which all questions of observation may ultimately be referred. Here, however, the whole universe is the object of inspection; there is no classical vantage point, and hence the interpretation question must be re-argued from the beginning. While we do not wish to stress this point unduly, since, after all, the Friedmann model ignores the vast complexities of the real universe, it is nevertheless clear that the quantum theory of space-time must ultimately force a deviation from the traditional Copenhagen doctrine."

And Unruh states

---

[255] For example the Planck time value is set by the Planck mass.

"One of the key features in the interpretation of such transition amplitudes, or wave functions, is the idea that we, as observers are also a part of the Universe as a whole. We, as physical observers, must be describable from within the theory and not as observers external to the theory as in usual quantum mechanics. In usual quantum mechanics, the interpretation is usually given in terms of observers that are outside of the theory. There one makes a split, with the quantum world at one side of the split, and the observer on the other. von Neumann argued that the predictions of quantum mechanics, at least under certain assumptions, are independent of the exact location of that split, but Bohr argued adamantly for the necessity of such a split (classical observers and quantum world). *There is a great difficulty in setting up such a split for physical observers contained within and influenced by a quantum universe,* [itallics added] and for the Universe as a whole, especially including gravity, one cannot argue that the predictions will be independent of where one puts the split. Since all energies interact gravitationally, and our observations are surely energetic phenomenon, the treatment of the energetics of observation as classical would lead to different predictions than if they were treated quantum mechanically. One is therefore forced to devise an interpretation of quantum mechanics in which the observer is part of the quantum system, rather than outside the quantum system.

This means that the interpretation of these transition amplitudes becomes somewhat non-intuitive. One must ask what the system looks like from within, from the viewpoint of an observer who is part of that world, rather than being able to interpret them directly in terms of probabilities for observations made by an external observer."

While the *Observer* question is addressed by a number of authors, the proposed answers are not entirely convincing. *The existence of other universes provides macroscopic Quantum Observers for our universe.* And our universe provides a macroscopic quantum observer for other universes. Thus the quantum observer issue is resolved.

These considerations lead us to view the existence of other universes as a critical solution to the above problems.

30.1.3 The Higgs Mechanism is Explainable by Extra Dimensions

The Higgs Mechanism 'explains' (generates) fermion nd boson masses. However the Higgs potential contains a quadratic term with a constant with the dimensions of [mass]. In a sense the Higgs Mechanism trades one mass for another. From where do the Higgs potentials' masses come?

A further explanation is needed is to determine the origin of the "dimensionful" mass terms in the Higgs' particle equations themselves. At present little if any thought has been given to the origin of these terms. We suggested that, excluding a "deus ex machina" source, the only known way to generate these mass terms in the Higgs' equations is through the separation of equations technique of differential equations. This technique requires additional parameters which can only be the coordinates of *extra unknown dimensions*. The best example of the generation of mass terms appears in the Schwarzschild solution of General Relativity where a separation constant, often denoted M, appears that has the dimension of [mass].

Thus extra space-time dimensions would resolve the origin of Higgs potentials' masses. Given extra dimensions it is reasonable to expect that these extra dimensions contain universes. Thus the Megaverse!

30.1.4 Possible Accretion of Megaverse Matter to Fuel Expansion of Our Universe

If matter is distributed outside of universes in the Megaverse, and if this matter can be accreted to universes by gravitational attraction, then the apparent increasing expansion of our universe may be due to this accretion. In chapter 14 of Blaha (2017c) we presented a model in which this possibility is realized. If true, then we would have tangible evidence of the residence of our universe in the Megaverse.

30.1.5 Asynchronous Logic is a Requirement of Universes

By establishing Asynchronous Logic principles[256] as the basis for the existence of universes and for setting the number of dimensions in each universe – four; and basis of fermion particles - iotas – we have found deeper principles of organization for the foundations of physics. The principles built on this foundation serve to enable the coordination of complex physical processes.

Usually we look at particle processes primarily from a space-time perspective: particles collide and produce new particles. We primarily think of the incoming and outgoing particles in a collision. However, considering the set of fundamental particles – and the particle transforming interactions in themselves – neglecting space-time and momentum considerations – leads us to view particles as constituting an alphabet and their interactions as a type of computer grammar.[257] Then the Asynchronicity Principles enable us to bring in space-time in a way that gives us the maximum complexity with the most minimal assumptions. As Leibniz[258] points out our universe has maximal complexity with minimal assumptions. (See chapter 1.)

30.1.6 The Meaning of Total Quantities of a Universe

The 'external' properties of a universe are normally questioned—for the simple reason that it is assumed that there is no 'outside' of our universe. For example, Misner (1973) asserts:[259]

---

[256] The basis of this section is described in detail in Blaha (2015a). That book places Physics within a logical framework that is a possible deeper ground for fundamental Physics theory.

[257] This conceptual approach was first described in Blaha (1998) who went on to characterize our universe as one enormous word evolving in time.

[258] See Rescher (1967).

[259] Pp. 457 - 458.

'There is no such thing as "the energy (or angular momentum, or charge) of a closed universe," according to general relativity, and this for a simple reason. To weigh something one needs a platform on which to stand to do the weighing.'

Misner et al presumes no such platform exists. If there is but one closed universe as most currently believe then one canot measure any totals of a closed universe (which ours may be to be). Yet if we take a more general view that our universe is only one of many then it becomes possible to measure total mass, charge, angular momentum, baryon number, and many other quantities of interest. Indeed, the existence of other universes (within the encompassing Megaverse) opens the door to an understanding of time, mass, energy, and all the other quantities necessary to develop a dynamical theory of universes.

Our new 'rotations of interactions' formalism (described previously) enables us to rotate measurable quantities. These quantities (quantum numbers) furnish a set of totals for our universe such as baryon numbers (normal and Dark), lepton numbers, angular momentum and so on that characterize our universe.

Later we will also see that one can then treat universes as 'particles', and develop 'universe dynamics', which might explain knotty problems such as the Big Bang and its precursor (if any). We will do this in subsequent chapters after first considering the possible structure of universes in general in the Megaverse.

30.2 Possible Experimental Evidence for the Megaverse

At first glance it would seem impossible to produce evidence for the existence of other universes. However there are subtle means by which we can 'sense' experimentally 'nearby' universes should they exist. The mechanism would appear to be gravitational effects exerted on objects within our universe by unseen objects of enormous mass. Currently there appears to be three experimental suggestions of the existence of 'nearby' universes and one theoretical argument based on an influx of mass-energy from the Megaverse that may support an understanding of the expansion of our universe.

30.2.1 Great Attractors

One potential support is the discovery of the Great Attractor (at the center of the Laniakea Galaxy Supercluster), and the more massive Shapley Attractor (centered in the Shapley Supercluster)[260]. These attractors contain massive numbers of galaxies and are drawing galaxies over a distance of millions of light years towards them.

If another universe(s) is 'near' our universe it could act as a 'gravitational magnet' and draw galaxies within our universe towards it to form one or more superclusters which could then act as attractors. Thus attractors might indirectly reveal the presence of other nearby

---

[260] Tully, R. Brent; Courtois, Helene; Hoffman, Yehuda; Pomarède, Daniel, "The Laniakea Supercluster of galaxies". Nature (4 September 2014). 513 (7516): 71–73; arXiv:1409.0880.

universes—contrary to the expected large scale uniformity of the universe. The only other apparent source of superclusters is chance. Chance seems an unsatisfactory possibility in the present case.

30.2.2 Bright Bumps in Universe Sugesting Collision with Another Universe

A recent study[261] of the residual brightness of parts of the accessible universe found that bright patches appeared if a model of the CMB (Cosmic Microwave Background) with gases, stars and dust was 'subtracted' from the PLANCK map of the entire sky. After the subtraction one would expect only noise spread throughout the sky. However, bright patches were seen in a certain range of frequencies. These anomalies are thought to be a result of our universe colliding with another object – presumably another universe in the Megaverse.

30.2.3 Cold Spot in Universe Suggesting Collision with Another Universe

Another recent study[262] of a huge cold region of the universe spanning billions of light years revealed that this region is not a relatively empty region but rather is similar to in its distribution of galaxies to the rest of the universe. Previous the Cold Spot (an area where cosmic microwave background radiation – the leftover Big Bang radiation is weak – making it significantly colder (0.00015C colder) than the average temperature in the universe 31.1.73C above absolute zero.)

An analysis of 7,000 galaxy redshifts using new high-resolution data has now shown that the Cold Spot is similar to the rest of the universe. The Durham University group suggested that the Cold Spot might have been caused by a collision between our universe and another Universe. They further suggested that there is only a 1 in 50 chance that it could explained by standard cosmology. could produce this feature

Thus we have another important piece of circumstantial evidence in favor of other universes and thus the Megaverse.

30.2.4 Megaverse Energy-Matter Infusion into Our Universe

In chapter 14 of Blaha (2017c) we presented a model for an influx of mass-energy from the Megaverse to support the Bondi-Gold-Hoyle-Narlikar Steady State Cosmology, which was originally based on the 'continuous creation of mass-energy' by Hoyle and Narliker. This model explains why the value of Ω makes the universe close to flat. If this model is correct then we would have concrete support for a Megaverse with a low mass-energydensity leaking mass-energy into our universe. *More generally, it suggests that universes are surfaces of high mass-energy density in a Megaverse of low mass-energy density – with a ratio of mass-energy densities of the other of 10^{30}.*

---

[261]   Ranga-Ram Chary, arXiv.org:/1510.00126 (2015).
[262] T. Shanks et al, Durham University (Australia), Monthly Notices of the Royal Astronomical Society, 2016 .

30.2.5 Conclusion

We conclude that data is beginning to emerge favoring multiple universes and a physical Megaverse in support of the theoretical justifications presented earlier.

30.3 Historical Trend Towards Larger Space-Time Structures

Looking back through the history of Mankind's view of the universe we see a clear progression to a larger and larger view. Before the 16th century the earth was the universe. In the 16th century Giordano Bruno (and possibly others) suggested that the stars were suns with many worlds circling them. So our view of the universe expanded to include stars.

Then over time it was noticed that nebulae existed in space. The astronomer, Edwin Hubble, studied the Andromeda nebula with the 'new' Mt. Wilson telescope and in 1929 announced that it was a galaxy composed of stars. Now the universe was conceptually similar to our current view.

Now we seem to have significant theoretical considerations and some suggestive experimental data that lead us to consider the possibility that our universe is not alone—that our universes is but one of many universes in a space we call the Megaverse. This book (and Blaha (2017c)) pulls together much of our earlier work and adds new insights into the nature of the Megaverse. We shall take the Megaverse as fact, extrapolate the form of our universe into the form of other possible universes, and develop a fairly detailed theory of the Megaverse and its resident universes. We will consider escaping our universe into the Megaverse—mindful that such travel will not happen until the very distant future. There are many technical bridges to cross before we can travel to other universes.

Given the vastness of our universe one might ask Why travel? We considered reasons in some detail in Blaha (2017c). For now, it suffices to say, Because they are there. Mankind has always grown and prospered through exploration and exploitation of new territories. Indeed the eminent Historian, Arnold Toynbee, stated that new 'turf' is the source of growth in all civilizations. Eventually, in the very distant future, we may need new turf in the Megaverse.

30.4 Other Megaverses? Will It Ever End?

Does the trend to larger and larger expanses of space suggest that our Megaverse may be but one of many duplicate Megaverses of the same number of dimensions (but diferent orientations)? One cannot decide this question in our present, or likely near future, state of knowledge.

However, based on our estimate of the dimension of the Megaverse, and its basis in the geometry and group structure of its interactions, it is reasonable to conjecture duplicate Megaverses would have the same number of dimensions as our Megaverse, and thus have universes with the same number of dimensions as our universe, and thus have a Physics in each other Megaverses' universes similar to the Physics of our universe.

This scenario would appear to be unlikely in the author's view as it appears uneconomical and gives rise to the question How could a plethora of Megaverses arise?

Another scenario, in which Megaverses appear within Megaverses like the toy Chinese nested boxes, also appears unlikely. For it would require a chain of Megaverses of ever increasing dimension, and raise the question of its origin—a question that would never be answerable. Nor would the nested set of Megaverses be experimentally accessible.

So we are content with one Megaverse.

31. Megaverse Dimension

The determination of the dimensions of the Megaverse can only be described as guesswork unless a principle is consistently used to specify the dimensions. In this chapter we will use the known interactions of our universe to determine the Megaverse's dimension.

We will assume that interactions have a dual role in fundamental physics: they determine the dynamics of particles, and they act to determine the dimensions of the Megaverse. The first role is evident within the universe and is therefore obvious from experiment. The second role is external to our universe. It acts to determine the dimensions of the Megaverse.

31.1 Role of Interactions in Determining the Megaverse Dimension D

The motivation[263] for the second role can be discerned from considering a 2-dimensional space, and introducing a simple 1/r potential such as:

$$V = g^2/(x^2 + y^2)^{-\frac{1}{2}}$$

where g is a coupling constant.

One views V as the potential of a force. However the values of V suitably extended to therange $[-\infty, +\infty]$ can be viewed as a third dimension.

With this alternate perspective in mind we take the Unified SuperStandard Model vector interactions: $E = [SU(3)\otimes SU(2)\otimes U(1)\otimes SU(2)\otimes U(1)\otimes U(4)\otimes U(4)]^4$ which has 192 generators and set the Megaverse dimension to[264]

$$D = 192 \qquad\qquad (31.1)$$

It is important to note that the fields of the Unified Standard Model extend beyond universe boundaries since each point of a universe is urrounded by Megaverse points. This feature is described in detail in the following chapter. Under these circumstances we require all fields be equally describable by the coordinates of the universe *and* Megaverse coordinates. Then continuity requires smooth transitions between values at universe points and Megaverse points in the neighborhood of universe points. Thus fields in our universe are intimately connected to their Megaverse counterparts, partially justifying our determination of the Megaverse dimension above. We provide firther support for eq. 31.3 when we consider the 'rotation of symmetries' U(192) Θ-Group later.

---

[263] We discuss the origin of our 4-dimensioal space-time along similar lines in chapter 1.

[264] We will use the symbol D to denote the dimension of the Megaverse in the following chapters.

Lastly, we note for consistency, the Megaverse must have complex-valued coordinates since it encompasses the complex-valued coordinates of our universe (although more complicated embeddings can be visualized) We opt for simplicity. In addition the metric of the Megaverse must be Euclidean with a flat-space metric tensor limit consisting entirely of -1's along the diagonal (using our conventions). For good reason we view the Megaverse as having at best a very low mass-energy density compared to that of our universe. Thus the Megaverse is essentially flat with 'universe bubbles.'

32. The Embedding of a Universe in a Higher Dimension Megaverse – Surface Tension

32.1 Universes as Mass-Energy Islands

In developing the theory of the Megaverse we view universes as islands of mass-energy that maintain their 'integrity' as surfaces due to gravitational force. Gravity holds universes together rather like molecular forces within a water droplet hold water molecules together. Molecular attractive forces gives droplets cohesion as they (perhaps) descend through the earth's atmosphere. They are the origin of *surface tension* in water.

Similarly gravity holds higher density[265] mass-energy universes together and gives rise to gravity surface tension. In a model presented in Blaha (2017c) for the Big Bang and the expansion of the universe within the Megaverse we found the mass-energy density of the universe was a factor of 10^{30} more than the density in the surrounding Megaverse space.

Thus we have good reason to study the surface tension of universes as a result of gravitational attraction within universes.

32.2 Boundary of a Universe within the Megaverse

In this chapter we describe the embedding of universes within the Megaverse. Much of this chapter appears in several earlier books by the author such as Blaha (2015a).

As stated earlier, we define a universe to be a closed or open surface in the Megaverse with a much higher mass-energy density than the Megaverse.

There are two types of boundaries for a universe embedded in a space of larger dimensions. First there is a boundary of the universe determined by treating the universe as a surface in the space. Secondly, there is another type of universe boundary defined by the observation that any neighborhood – not strictly within the universe – of every point of the universe has an infinite number of points of the enclosing Megaverse space.[266] Or, every point has neighborhoods with Megaverse points within it. Thus *each point of a universe is on a boundary of the universe due to the larger dimensions of the Megaverse space* within which it

---

[265] Higher density in comparison to the much lower density of the inter-universe space of the Megaverse.

[266] Any neighborhood of any point in the universe – with all its points strictly within the universe – has an infinite number of points within the universe. We assume the neighborhood is so small that the curvature of the universe's space can be neglected.

resides. Fig. 32.1 schematically illustrates these neighborhoods for any universe point for a universe contained within a higher dimensional Megaverse.

32.3 Confinement of Universes due to 'Surface Tension'

We will assume that other universes have the same physics as our universe with the possible differences that they may have differing interaction coupling constants and particle masses. As we will discuss later, every point of a universe in a higher dimensional Megaverse has Megaverse points in any neighborhood of the point (with the exception of neighborhoods strictly within the universe). Thus we confront the question: what keeps mass-energy at points in a universe or is there leakage from the universe into the Megaverse?

If there is little or no leakage into the Megaverse then, since each point in a universe is part of a Megaverse surface, one can only assume that there is a barrier to movement into the Megaverse. Taking a note from fluid dynamics, and viewing Megaverse space as one 'material' and the universe as a different 'material,'[267] we view the barrier as 'surface tension.'[268] The Megaverse appears to "exert a force" confining the contents of the universe to within itself.[269]

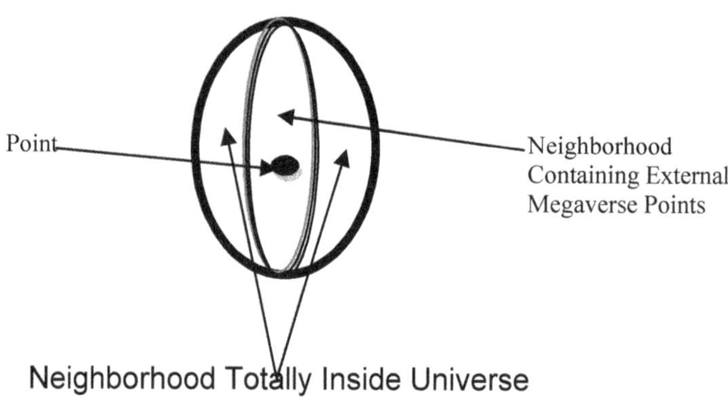

Figure 32.1. Schematic diagram of a 3-dimensional projection of 'orthogonal' neighborhoods of a point within a universe with one neighborhood strictly within the universe and the other neighborhood containing both universe and external Megaverse points in general. The 'orthogonal' circles around the point differentiate between the two types of neighborhoods.

---

[267] Meaning material with much higher mass-energy density and consequently larger internal gravitational attraction.
[268] See Landau (1987).
[269] Although in actuality it is the universe that holds itself together by gravitation.

The surface tension[270] of a universe γ satisfies the relation

$$\gamma = W/\Delta A \qquad (32.1)$$

where γ is expressed in erg/cm^2, W is the Work, and ΔA is the Area upon which the work is exerted. The pressure Δp exerted by the surface tension for a 'spherical' surface area is

$$\Delta p = 2\gamma/R \qquad (32.2)$$

where R equals the radius of curvature of the surface. The above equations embody the concept that the surface tension force equals the pressure difference at the surface.

If the universe is flat then the surface pressure approaches ∞ giving confinement of fields and particles to the universe:

$$R \to 0 \quad \text{implies } \Delta p \to \infty \qquad (32.3)$$

Thus we have the theorem:

Theorem: A universe has no leakage of fields or particles into a higher dimensional space if the universe is exactly flat.

This theorem is particularly interesting in the case of our universe. It appears to be flat (or very close to flat). The flatness of our universe may be the reason no leakage of fields or particles from our universe has been detected to high accuracy.

If a universe is found with a non-zero radius of curvature then one can expect that some fields and particles may emerge from it into the Megaverse.

While a zero radius of curvature prevents the exit of fields and particles from a universe, it does not prevent the entry of mass-energy into the universe from the Megaverse. Thus our Continuous Creation Model of chapter 14 in Blaha (2017c) may be relevant. Entry is possible; exit is forbidden in this case.

Eqs. 32.1 – 32.2 are reminiscent of the 'four laws for Black Holes' which are stated later.

32.4 Quantum Fields 'Emanating' from a Universe

If the curvature of the universe is zero (open universe) then no fields emanate from it. If the curvature of the universe is non-zero (closed universe) then fields may 'leak' into the

---

[270] A useful analogy: the Megaverse is a pool of water; a universe is a denser oil bubble within it. Surface tension caused by the cohesiveness of the oil molecules in the bubble makes it spherical (confines it to a spherical shape). Similarly a universe (denser than the Megaverse) is 'confined' within the Megaverse.

Megaverse. Then continuity conditions between a universe field and its Megaverse counterpart becomes of interest. We discuss this in detail later.

32.5 Universe Confinement by Conservation Laws

Every point in our universe is "infinitely" close to points of the Megaverse.

A universe occupies a region within the Megaverse. However because it is a lower dimension surface within the Megaverse the neighborhood of every point within a universe has an infinite number of Megaverse points that are not within the universe.

One might think that particles within a universe could then 'slip' into the Megaverse outside the universe with ease. However that is not the case. The law of momentum conservation compels particles and interactions within a universe to be confined to the universe. More importantly, the Megaverse surface tension of a flat universe confines particles and fields within a universe.[271]

The only possible ways that a particle could exit from a universe are 1) if the particle collides with a particle with a momentum, some of whose components are in Megaverse dimensions extraneous to the universe's dimensions, or 2) a particle within the universe experiences forces with components in Megaverse dimensions extraneous to the universe. We shall consider the second possibility later when we consider a mechanism for a starship to exit our universe (chapter 34). The first possibility exists if the Megaverse has a very low matter density outside of universes. The fact that this phenomena has not been observed implies the Megaverse matter density is extremely low.

Thus conservation of momentum for particles and interactions, and surface tesion, effectively confines particles within a universe even though the neighborhood of every point of a particle's trajectory contains an infinity of Megaverse points exterior to the universe. Similarly every interaction within a universe is confined when expanded in a fourier series (assuming free fields) in universe coordinates.

It is also possible for a Megaverse particle to enter[272] a universe through perhaps a collision that results in the particle being within the universe with momentum also solely within the universe. Thus the 'point boundary' of universes is porous. Particles can enter/exit a universe under appropriate conditions.

We conclude the 'point boundary' of a universe is not a barrier although surface tension force controls the entry/exit of particles and fields. We consider an exit mechanism from a universe in chapter 34.

---

[271] A useful analogy: the Megaverse is a pool of water; a universe is a denser, oil bubble within it. Surface tension caused by the cohesiveness of the oil molecules in the bubble makes it spherical (confines it to a spherical shape).

[272] In chapter 14 of Blaha (2017c) we consider a model that supplies a mechanism for the Hoyle-Narlikar continuous creation theory for the expansion of our universe. This model 'creates' mass-energy as an inflow from the Megaverse.

32.6 Objects Straddling a Universe-Megaverse Boundary

When a starship, or some other extended object, is entering/exiting a universe at some velocity the question of the state of the object arises It is partially in and partially out of the universe. We know that the object being 4-dimensional will continue to be 4-dimensional, barring effects of forces that might "twist" parts of the object into additional dimensions.

There is also the more subtle quantum effects on the object due to the possibility of different quantizations of the particles in a universe and the Megaverse. Quantizations in different coordinate systems might result in different physical interpretations of matter. We resove this issue by our PseudoQuantization Method described earlier. In Appendix A we show that one can quantize using a form of Pseudoquantization that preserves (unitarily equivalent) particle interpretations in a universe and the Megaverse.

Thus extended objects can be partly in a universe, and partly in the external Megaverse without issues..

33. General Properties of the Megaverse

If one wishes to have a depiction of the Megaverse it seems likely that the universes within it would be scattered in a fashion similar to galaxies within our universe although on a much larger distance scale and in multiple dimensions. In this chapter we overview properties of the Megaverse and its universes. Much of this material previously appeared in Blaha (2017c) and in earlier Physics and starship travel books.

33.1 Megaverse Size, Lifetime, and Universe Separation

The first questions that naturally arise are the age, size, and general of universes within the Megaverse. In the absence of any experimental detail we will assume the relative size of the entities (galaxies) in the universe equals the relative size of entities (universes) in the Megaverse:[273]

(Average Galaxy Size)/(Universe Size) = (Average Universe Size)/(Megaverse Size)　　(33.1)

Taking the average diameter of galaxies to be 400,000 light years, the age of the universe to be 13,800,000,000 light years, and the diameter of the universe to be 91.4 billion light years[274] (the estimated diameter of last scattering surface) we find the diameter of the Megaverse

$$\text{Diameter}_{\text{Megaverse}} = 2 \times 10^{16} \text{ light years} \qquad (33.2)$$
$$= 228,500 \times \text{Diameter}_{\text{Universe}}$$

And, using the 228,500 scale factor, we find the Megaverse age since the Megaverse 'Big Bang' to be

$$\text{Age}_{\text{Megaverse}} = 3 \times 10^{15} \text{ years} \qquad (33.3)$$
$$= 3 \text{ million billion years}$$

If the average separation between galaxies in our universe is 3,000,000 light years, then assuming distance scaling by 228,500, the average separation between universes in the Megaverse would be

---

[273] We assume the Megaverse is homogeneous at large distance scales. We also assume a proportionality between the average entity size (galaxies in the case of unbiverses; umiverses in the case of the Megaverse) in part based on the relatively long lifetimes of universes and the Megaverse.

[274] There are much larger estimates of the universe's diameter based on the Inflation theory of A. Guth and others.

$$\text{Separation}_{\text{Universes}} = 228{,}500 \times 3{,}000{,}000 \text{ ly} = 7 \times 10^{11} \text{ light years} \qquad (33.4)$$
$$= 700 \text{ billion light years}$$

If we now assume the mass of a universe equals the total mass-energy of our universe (including Dark mass and Dark energy) which is estimated to be $m_{\text{universe}} = 3 \times 10^{54}$ kg then the gravitational potential energy between two such universes separated by 700 billion light years is

$$V = G\ m_{\text{universe}}^{2}/\text{Separation}_{\text{Universes}} \qquad (33.5)$$
$$= 9 \times 10^{70} \text{ kgm}^{2}\text{s}^{-2}$$

The gravitational force is

$$F = G\ m_{\text{universe}}^{2}/\text{Separation}_{\text{Universes}}^{2} \qquad (33.6)$$
$$= 1.35 \times 10^{43} \text{ kgms}^{-2}$$

and the resulting gravitational acceleration of universes is

$$a = 4.5 \times 10^{-12} \text{ m/s}^{2} \qquad (33.7)$$

The Baryonic and Leptonic forces associated with the Generation group will slightly modify the force between the universes. We view the small acceleration between universes due to gravity as Physically acceptable. In a billion years the universe velocity would be $v = 1.4 \times 0^{5}$ m/s with $v/c = 0.0005$ and the distance traveled equal to 2.2×10^{21} m = 236,540 light years – negligible compared to the $\text{Separation}_{\text{Universes}}$. Universes would only make contact after extraordinary long times. Thus we have a coherent view of Megaverse size and distance parameters.

33.2 Likely Features of the Megaverse

There are a number of features of the Megaverse that appear to be true:

33.2.1 Megaverse Curvature

The Megaverse has gravitation. Gravitation appears to be weak in the Megaverse so it is close to a flat space. The sources of Megaverse gravitation are the mass-energy of the universes within it and the density of mass-energy of particles outside of universes. Universes have a larger relative force of gravity due to a higher mass-energy density. Universes therefore have more curvature.

33.2.2 Megaverse Time Dimension

We will assume that the Megaverse has one complex time dimension denoted y^{D} for the simple reason that the absence of a time dimension would make make the Megaverse static.

33.2.3 Megaverse Forces

In addition to Megaverse gravitation, the Megaverse has the forces in the Unified SuperStandard Model. These forces satisfy continuity conditions at universe boundaries.

33.2.4 Megaverse Parameters

Megaverse physical constants and particle masses have the same values as in our universe due to continuity.

33.2.5 Megaverse Vacuum Fluctuations

Megaverse Vacuum fluctuations may be a source of the generation of universes and particles. Vacuum fluctuations might account for the Big Bang. The time scale for the persistence of universes generated by a vacuum fluctuation is likely to be an extrapolation of vacuum fluctuation persistence within our universe.

33.2.6 Megaverse Matter and Chemistry

The existence of many more dimensions in the Megaverse suggest that multi-dimensional forms of matter and energy could exist between universes. As a result Megaverse atoms, compounds and Chemistry will be very different and much more varied than in our universe. If such matter exists in the Megaverse then 'mining' such matter for use in our universe—would give us exotic new compounds and Chemistry that would be partially inside, and partially outside, of our universe.

This possibility makes venturing into the Megaverse economically and scientifically desirable since such materials cannot be created within our universe.

33.3 Features of Universes Within the Megaverse

We know of our universe from the 'inside.' However the features of our universe from a Megaverse perspective are not at all certain. Im this section we will describe the Megaverse view of a universe's properties.

We shall assume a universe is a closed or open surface within the Megaverse of much higher mass-energy density than the Megaverse's mass-energy density by perhaps as much as a factor of 10^{30}. Chapter 14 of Blaha (2017c) describes a model of Megaverse mass-energy inflow into our universe exemplifying this feature.

33.3.1 Universe Area and Mass

The mass of a universe is an important property since mass is one of the sources of Megaverse gravitation, and interaction between universes.

While a universe is not believed to be a black hole (although Hawking has recently jokingly? suggested that our universe may be a black hole, and even more recently suggested black holes are not quite black holes – grey?), there are general qualitative similarities that lead us to consider the possibility that the four laws of black holes[275] may apply in part (or their entirety) to universes. In particular the 2nd law states

$$dM = \kappa dA/8\pi + \Omega dJ \tag{33.8}$$

where dM is the change in "mass/energy," A is the area of the Black Hole (universe), Ω is its angular velocity and J is the angular momentum.[276] From eq. 33.8 it appears we can reasonably define a "mass" for a universe in terms of a universe's area:

$$M = \kappa A/8\pi \tag{33.9}$$

This definition seems to capture the physics of universes that could be used in developing a dynamics of universes as we do later. It allows us to escape the dilemma of having zero total energy for universes that would preclude treating universes as particles in the Megaverse and developing a Megaverse dynamics of universe-particles. Later we will also show how to define a mass for a universe that is time dependent.

33.3.2 Relation Between Universe and Megaverse Vector Fields

Earlier in this book we defined the interactions of The Unified SuperStandard Model. These interactions and their groups also exist in the Megaverse. The fourier expansions of the fields in the Megaverse are different. They must be expressed in terms of the D coordinates of the Megaverse.

Within a universe, since each point of the universe is surrounded by Megaverse points, a field has both an expression in universe coordinates, x, and an expression in Megaverse coordinates, denoted y. For a vector field in the universe $A_U^\mu(x)$ there is an equivalent representation of the field in Megaverse coordinates $A_M^i(y)$. These representations are related by a coordinate transformation. If we define a map from universe coordinates to Megaverse coordinates with

$$y^i = f^i(x) \tag{33.10}$$

then the field representations are related by[277]

---

[275] Wald, R. M., "The Thermodynamics of Black Holes", *Living Reviews in Relativity* **4** (6): 12119 (2001).

[276] Although the angular momentum of a universe is not measurable if there is only one universe (as DeWitt argued in a quote earlier), the existence of multiple universes within the Megaverse enables the relative angular momentum of a universe to be determined.

[277] Implicit in eq. 33.10 is an inverse relation $x^\mu = f^{-1}(y)$ which is necessarily based on a restriction of the y-coordinates to obtain a 1:1 relation between the y and x coordinates. The restriction is best implemented by requiring the domain of y coordinates be restricted to those y coordinates within the universe surface. The result is a 1:1 relation between the y-domain coordinates and the x universe coordinates.

$$A_M^{\ i}(y) = \partial y^i/\partial x^\mu \, A_U^{\ \mu}(x)$$
$$= \partial y^i/\partial x^\mu \, A_U^{\ \mu}(f^{-1}(y))$$

(33.11)

in the domain of the universe. The values of the field at Megaverse points in a neighborhood of a universe point are determined by continuity.

Outside the domain of the universe the Megaverse field value is determined by its sources.

At the boundary of the universe there must be continuity in the expectation values of the Megaverse fields.

33.3.3 Megaverse Gravitation and Free Matter

We assume that a generalization of Einstein's theory of Gravity exists in the Megaverse.

Just as our universe has matter and radiation between galaxies, it seems reasonable to assume that 'free' matter and radiation exists in the Megaverse outside of universes. Such mass-energy would have two roles: to gravitationally affect the dynamics of the Megaverse and the motion of universes within it, and to possibly fuel the expansion of universes. The expansion of our universe may be due to an influx of matter and energy from the external Megaverse. Many years ago Hoyle and Narlikar considered the possibility of 'continuous creation of matter.' We suggest that an influx of Megaverse matter may be the actual source. We consider this possibility in chapter 14 of Blaha (2017c), which contains a paper by this author written approximately seven years ago. (unpublished)

Thus we arrive at a view of the Megaverse of matter and universes that is analogous to our universe of galaxies.

33.3.4 Expansion of Universes

Our universe expanded from the Big Bang to its current size and is still expanding. It is likely that other universes have undergone similar expansions. According to chapter 32 there is likely an infinite surface tension at the boundary of our universe. How can the universe have expanded, and continue expanding, under such conditions. We see a two phase expansion of the universe.

For a period of time after the Big Bang the universe did not have an infinite surface tension preventing expansion into the Megaverse due to an effect discovered by Eőtvos – a temperature dependence of the surface tension force. Eőtvos pointed out a critical temperature T_c existed that caused the surface tension force to decline as the temperature increased:

$$\gamma V^{2/3} = k(T_c - T)$$

(33.12)

where k is the Eőtvos constant, and V is the volume of the universe (the liquid 'drop') Assuming a spherical universe, the volume is

$$V = 4\pi R^3/3. \tag{33.13}$$

Thus

$$\gamma = (4\pi/3)^{-2/3} k(T_c - T)R^{-2} \tag{33.14}$$

For very high temperatures such as existed after the Big Bang $T > T_c$ and thus γ would be negative indicating that there was an outward pressure from the universe into the Megaverse promoting expansion. Thus in the high temperature period after the Big Bang the surface tension force favors expansion of the universe.

After this phase, the surface tension γ is positive. The Megaverse is then superficially 'impeding' expansion. However, the surface tension pressure of the Megaverse causes leakage *into* the universe fom the Megaverse causing its mass to increase, and its radius and volume to increase – Expansion! – due to the accretion of Megaverse mass-energy.

The above scenario is supported by the two phase model suggested by section 14.15 consisting of a Big Bang expansion model (chapters 11 – 13), and a mass-energy accretion model (chapter 14) – all of Blaha (2017c). Note as the radius of curvature goes to zero, eq. 33.14 suggests an increasing surface tension pressure 'pushing' particles into the universe.

Thus a complete universe expansion scenario is evident based on surface tension physics.

33.3.5 Universe Generation from Vacuum Fluctuations

Vacuum fluctuations could generate universe-antiuniverse pairs. Antiuniverses would have certain 'negative quantum numbers.' We view universes/antiuniverses, when created (Big Bangs), as 'ultra-small' particles of mass-energy that can proceed to expand to great size. If they are generated by a vacuum fluctuation they will, after possibly a certain time, recombine into the vacuum.

33.3.6 Universes as Black Holes

It is conceivable that a universe could be so dense and confined that it would be effectively a Black Hole. In this case the internal quantum numbers of the Black Hole universe would be inaccessible and only its mass, velocity and angular momentum would be observables.

33.3.7 Life in Other Universes

It is likely that life exists in some if not all of the universes of the Megaverse. If the physical constants, laws, and masses of the interior of a universe are similar if not identical to ours then the possibility of intelligent species, even human-like species, is very likely. This possibility is an important motivation for humanity to reach for the Megaverse as we discuss later.

34. On Exiting Our Universe

The first question on realizing that one is in a confined spaxe is How do I get out? This question takes on a dramatic new aspect when we ask How do we exit from our universe? Like tadpoles in a pool of water it is a question of imporatance since universes, like pools of water, may 'evaporate' – a possibility raised from time to time by cosmolgists.

In this chapter we present a mechanism for exiting our universe realizing that it could not be developed and used for perhaps tens of thousands of years.[278]

34.1 Looking for an Exit Point

Due to our discussion of surface tension at the boundary of the universe the first issue that we will address is whether there are 'points' where egress from our universe is facillated. We will see that the vicinity of bodies with large gravitational fields such as neutron stars and Black Holes seriously weaken the nearby universe surface tension and are thus the best locations for exiting our universe into the Megaverse.

34.2 Neutron Star (Black Hole) Exit Point Surface Tension

Near a neutron star, or similar small, very massive body, the curvature of the universe (which is close to zero – flat) becomes significant. Then, as we saw in chapter 32, the surface tension force becomes possibly much smaller (not infinite!) and the force keeping mass-energy within the universe decreases enabling a starship to conveniently exit into the Megaverse.

In chapter 32 we saw the pressure exerted by the surface tension for a 'spherical' surface area is

$$\Delta p = 2\gamma/R \tag{34.1}$$

where R equals the radius of curvature of the surface. R = 0 for a flat space. A non-zero value for R causes the infinite force of flat space-time to become a finite force which the starship can overcome to exit.

Later in this chapter we will describe a slingshot maneuver for a starship in which a starship 'wips' around a neutron star using its gravity to twist into the Megaverse. The trajectory

---

[278] This time frame could be shortened if it were possible to use faster-than-light starships to execute the maneuvers described in this chapter around a Black Hole without the destruction of the starship and its passengers. If faster than light starships become a reality then exits from the universe may become feasible only thousands of years afterwards.

is depicted in Fig. 34.1a. We call starships with this capability *uniships* since their ultimate goal is to travel to other universes.[279]

A neutron star (Black Hole) slingshot has an additional advantage for transit to the Megaverse. This advantage is enhanced by an ultra-fast uniship speed. An extremely large speed causes the uniship to appear 'hot' using the conversion between velocity and temperature. A high speed will create a 'hotspot' in the universe boundary surface that could substantially lower the surface tension force.[280] Eőtvos showed that the surface tension force has a critical temperature T_c which satisfies:

$$\gamma V^{2/3} = k(T_c - T) \tag{34.2}$$

where k is the Eőtvos constant and V is the volume of the universe (the liquid 'drop'). As the uniship speed increases to an equivalent temperature beyond T_c, the Megaverse surface tension force will drop to zero, and the uniship can freely enter the Megaverse.

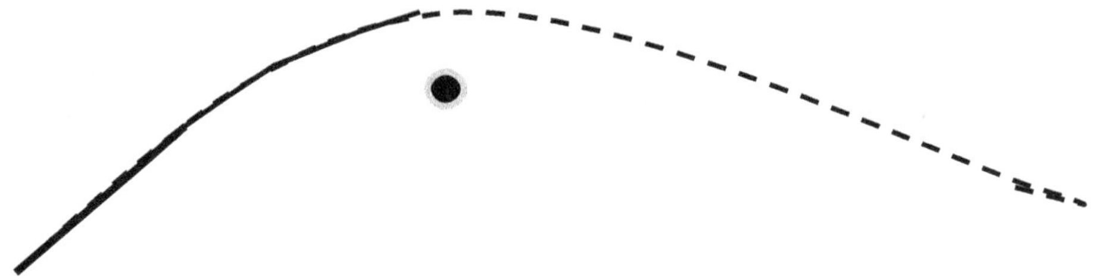

Figure 34.1a. The trajectory of a uniship in a slingshot maneuver with a neutron star or Black Hole (the dark circle). The repulsive baryonic force causes the "turn" away from the star/Black Hole into the Megaverse. The solid line corresponds to the time in which the uniship is wholly within the universe and dominated by gravity. The dotted line reflects the transition of the uniship into the Megaverse.

Therefore the Megaverse – universe interface can be viewed as a local surface bubbling that thins out at high speed to nothing – no barrier force.

---

[279] The design and human factors of uniships: living conditions, travel times, communications from the Megverse, long-lived ship materials, suspended animation, and so on are described in Blaha (2017c) and earlier books.

[280] Much more importantly is the temperature in the region around a neutron star. An idea of the temperatures involved is the magnitude of the temperature in a region $1 - 7$ km in size in a colliding neutron stars event. The temperature was estimated to be the extraordinary temperature 350 GK (gigakelvins) by R. Oechslin et al, arXiv:astro-ph/0507099 (2006). Temperatures of the order of 10^{12} K would seem to severely lessen or eliminate surface tension.

We can estimate the critical velocity v_c as a function of the critical temperature, at which the surface tension force goes to zero, using the velocity maximum of the superluminal Maxwell-Boltzmann distribution:[281]

$$v_c \approx c + \tfrac{1}{2}\, m^2 c^5/(2kT_c)^2 \qquad (34.3)$$

where m is the mass of the uniship (hundreds of thousands of metric tons presumably). The critical temperature can also be expected to be extremely high. As a result the ratio, the critical velocity, is likely to be large. At sufficiently high superluminal velocity there is no barrier to entry to the Megaverse.

The precise determination of the critical temperature remains to be done. Chapter 14 of Blaha (2017c) estimates the critical temperature for our models in chapters 11 - 14.

34.3 Uniship Slingshot into the Megaverse

In this section we will consider a baryonic force mechanism for trips into the Megaverse using a slingshot around a neutron star. Black Holes may be a better choice than neutron stars – especially a spinning Black Hole.

34.3.1 How may We Evade Euclid's Proof of Three Dimensions and Enter Other Dimensions?

We will now consider a mechanism to escape from a lower dimensional space to a higher dimensional space containing the lower dimensional space as a subspace. We will see that the key factor in this escape mechanism is *a force in the higher dimensional space* that boosts an object in the lower dimensional space into the extra dimensions in the higher dimensional space.

A simple example illustrating this mechanism is a two dimensional flat space (often called *Flatland*) within three dimensional space. We will take the Flatland to be the x-y plane and the z-axis to be the other dimension of the three dimensional space.

Suppose a charged particle is moving with speed v in the positive x direction. Suppose further that a constant magnetic force B points in the positive y direction. Then due to the magnetic vector force law for an electric charge q

$$\mathbf{F} = (q/c)\, \mathbf{v} \times \mathbf{B}$$

the force on the charge will be in the positive z direction – popping the charge out of Flatland into the full three dimensional space. Note that the magnetic force is in the Flatland but the 3-

---

[281] Blaha (2017c) section 1.4.3.

dimensional form of the force law causes the force on the charge to propel it from the Flatland into three dimensional space.[282]

34.3.2 Escaping Our Universe using the Baryonic Gauge Field

The baryonic force[283] field around a neutron staris extremely large. Since the universe points and the Megaverse points in any region of the universe are interspersed, we expect that continuity would require the baryon force fields to be partly at our universe's points *and* also at adjoining Megaverse points. Also, the surface tension of the universe should be almost zero near a neutron star allowing universe fields, such as the baryon gauge fields, to 'leak' into the Megaverse. *Consequently we must express baryon field values in Megaverse coordinates. By doing so, we are led to D = 192 baryon field components which are functions of D = 192 coordinates. Those force field components beyond our four dimensions open the uniship door to the Megaverse.*

The preceding example illustrates the general form of the mechanism for particles and uniships to escape from our four dimensional universe into the D-dimensional Megaverse. The baryonic gauge force field will 'fuel' the escape. It can be used to "pop" baryons and uniships (which are primarily composed of baryons) into the Megaverse.

We have focused on the baryonic force in this chapter. The reason is the long range possibility of travel into the Megaverse to other universes using the baryonic force to exit our universe into D-dimensional space using a slingshot orbit around a neutron star. We discuss this possibility below and refer the interested reader to Blaha (2014c) for another detailed discussion.

34.3.3 Neutron Stars Overview

An important factor in the nature of neutron stars is their unusually large rotation rates ranging up to over 700 rotations per second. The rotation of a neutron star (assuming a baryonic gauge force exists as we do) generates a *baryonic* magnetic field as well as a baryonic electric field. However we shall see that the neutron star spin-generated baryonic magnetic and electric fields only impart a force to a uniship within our universe. Thus the *Coulomb baryonic force* is the key to slingshoting a uniship into the Megaverse as we pointed out in earlier books.

The important neutron star parameters for the determination of the slingshot trajectory into the Megaverse are its mass, its radius, and its spin. From these parameters we can determine its baryon number, its baryonic current density, and its baryonic electric and magnetic fields. Then the trajectory of a uniship into the Megaverse can be determined.

---

[282] We assume the escape from the two dimensional subspace is not impeded by a "horizon."
[283] The field associated with Baryon Number Conservation.

34.3.4 The Baryonic Field of a Rapidly Spinning Neutron Star

Spinning neutron stars, and they all spin at varying rates except in extreme old age, generate baryonic fields, which *might* have provided a mechanism to escape from our universe into the Megaverse.[284] In fact, we shall see that *the spin generated force does not* have a component in the direction of Megaverse spatial coordinates. Thus they cannot be used to exit into the Megaverse.

The baryonic Coulomb force does have components in Megaverse directions outside our universe and can slingshot uniships into the Megaverse.

This section calculates the spin-dependent baryonic fields for a neutron star of mass M, internal mass density $\rho(r)$, radius R, volume V, and rotation rate Ω measured in rotations per second. We assume the neutron star is rigid and rotates uniformly. We believe this is a reasonable assumption in view of the close packing of the neutron star throughout most of its body.

34.3.5 Baryonic Current of a Neutron Star

We assume that a neutron star is spherical to good approximation although a rapidly spinning neutron star will deviate slightly from a sphere. We also assume that the neutron star can be treated as point-like since its radius, of the order of 10 km, is very small compared to the closest point of approach of a uniship which will be, at minimum, thousands of kilometers distant.

The baryonic current of a spinning neutron star, if the coordinate system is oriented so that it spins around the z-axis, yields a current J_φ in the φ direction of spherical coordinates (r, θ, φ). The current is to good approximation

$$J_\varphi = \int dr d\theta d\varphi\, r^2 \beta_B \Omega \rho(r)/m_n = (\beta_B \Omega/m_n)\int dr d\theta d\varphi\, r^2 \rho(r) = \beta_B \Omega M/m_n \qquad (34.4)$$

where β_B is the baryonic charge (analogous to e in electromagnetism), where m_n is the mass of a neutron, $\rho(r)/m_n$ is the neutron (nucleon) number density at radius r, and where the current at a radial distance r is $\beta_B \Omega \rho(r)/m_n$.

As discussed in Blaha (2014) the baryonic "Coulomb" potential in Megaverse coordinates is

$$\phi(y_1, y_2, \ldots, y_{(D-1)}) = (\beta_B^2/4\pi)N_1 N_2/(y_1^2 + y_2^2 + \ldots + y_{(D-1)}^2)^{\frac{1}{2}} \qquad (34.5)$$

---

[284] It is possible that other types of stars such as white dwarfs and perhaps even ordinary stars might be used by uniships to escape from a universe into the Megaverse. The enormous mass, and extremely small size, of neutron stars make them more favorable for slingshot escape from the universe.

where $\alpha_B = (\beta_B{}^2/4\pi)$ is the equivalent of the electromagnetic fine structure constant α. Earlier we estimated the order of magnitude of α_B using the differences[285] between various experiments to determine the gravitational constant G. We found the order of magnitude to be

$$\alpha_B = \beta_B{}^2/4\pi \simeq 0.118 \; Gm_H{}^2 \qquad (34.6)$$

where m_H is the mass of a hydrogen atom.

We conclude this subsection by determining the baryonic current from the above equations:

$$J_\varphi = (4\pi \cdot 0.118G)^{\frac{1}{2}}\Omega M \qquad (34.7)$$
$$= 1.22 \; G^{\frac{1}{2}}\Omega M$$

using the approximation $m_H = m_n$.

34.3.6 The D Component Baryonic Vector Potential

Earlier we described some of the features of the baryonic vector potential, which we recapitulate here for the reader's convenience. As in electromagnetism there is an antisymmetric tensor of the second rank that appears in the free part of the baryonic field $F_{Bu\mu\nu}(y)$ lagrangian:

$$\mathscr{L}_{Bu} = -\tfrac{1}{4} \; F_{Bu}{}^{ij}(y)F_{Buij}(y) \qquad (34.8)$$

where

$$F_{Buij}(y) = \partial B_{ui}(y)/\partial y^j - \partial B_{uj}(y)/\partial y^i \qquad (34.9)$$

and i, j = 1, 2, ... , D. The Dth coordinate corresponds to the time coordinate. While the coordinates are complex in general we will treat the (D − 1) spatial coordinates as real and the Dth coordinate as pure imaginary with the resulting invariant interval

$$ds^2 = dy_1{}^2 + dy_2{}^2 + ... + dy_{(D-1)}{}^2 - c^2dy_D{}^2 \qquad (34.10)$$

which is invariant under D-dimensional Lorentz transformations. The coordinates can be transformed into complex-valued coordinates using the Reality group defined earlier.

The tensor F_{Buij} is conveniently separated into a baryon electric part and a baryon magnetic part in a manner similar to the separation of the electromagnetic fields into electric and magnetic fields. However the (D − 1) spatial dimensions change the forms of the baryon fields. Analogously to electromagnetism the baryonic force is given by

---

[285] The recent experiment by T. Quinn et al, Phys. Rev. Lett. **111**, 101102 (2013) differs significantly from the 2010 CODATA world average of previous experiments. See P. J. Mohr, B.N. Taylor, and D. B. Newell, Rev. Mod. Phys. **84**, 1527 (2012).. We attribute the difference to the baryonic force between masses.

$$f_i = F_{Buij}(y)J_B^{\,j}/c \tag{34.11}$$

where $J_B^{\,j}$ is the jth baryonic current.

The baryon "electric" field is

$$E_{Bui} = -F_{Bui0}(y)/c \tag{34.12}$$

while the baryon "magnetic" field is

$$B_{Bui} = \varepsilon_{ijk}F_{Bu}^{\,jk}(y) \tag{34.13}$$

where i, j, k = 1, 2, ... , (D – 1) and where ε_{ijk} is a totally anti-symmetric tensor with component values ±1. If i < j < k then ε_{ijk} is +1. Even permutations of these three indices yield a value of +1 for the tensor components. Odd permutations of these three indices yield a value of –1. For example, $\varepsilon_{246} = +1$, $\varepsilon_{426} = -1$, $\varepsilon_{642} = -1$, $\varepsilon_{264} = -1$, $\varepsilon_{462} = +1$, $\varepsilon_{624} = +1$.

With these definitions of the $\mathbf{E_{Bu}}$ and $\mathbf{B_{Bu}}$ fields we derive the D-dimensional generalization of the *Lorentz force law* for a baryon of charge q and (D – 1)-velocity v_j:

$$F_i = qE_{Bui} + q\varepsilon_{ijk}v_jB_{Buk}/c \tag{34.14}$$

for i = 1, 2, ... , (D – 1). One important difference from the 4-dimensional case is the forms of the $\mathbf{E_{Bu}}$ and $\mathbf{B_{Bu}}$ fields

$$E_{Bui} = -F_{Bui0}(y)/c = [-\partial B_{uD}(y)/\partial y^i - \partial B_{ui}(y)/\partial y^D]/c \tag{34.15}$$

or, expressed as a (D – 1)-vector,

$$\mathbf{E_{Bu}} = [-\nabla_{(D-1)}\phi(y) - \partial\mathbf{B_u}(y)/\partial y^D]/c \tag{34.16}$$

where ϕ is the baryonic Coulomb potential $B_{uD}(y)$, $\nabla_{(D-1)}$ is the (D – 1)-dimensional grad operator, and $\mathbf{B_u}(y)$ is the baryonic (D – 1)-vector potential.

The (D – 1)-dimensional baryon magnetic field has the form of eqn. 34.13. The baryon magnetic field exhibits more complexity than the 3-dimensional magnetic field of electromagnetism:

$$B_{Bu1} = \varepsilon_{1jk}F_{Bu}^{\,jk}(y)/c = [F_{Bu}^{\,23}(y) + F_{Bu}^{\,24}(y) + ... + F_{Bu}^{\,2,(D-1)}(y) + F_{Bu}^{\,34}(y) + F_{Bu}^{\,35}(y) +$$
$$... + F_{Bu}^{\,3,(D-1)}(y) + F_{Bu}^{\,45}(y) + ... + F_{Bu}^{\,(D-2),(D-1)}(y)]/c \tag{34.17}$$

34.3.7 The Baryonic Electric and Magnetic Field Strengths

In this section we will calculate the baryonic electric and magnetic field strengths for a neutron star due to its spin. The spatial field strengths are determined by the dynamical equation:

$$\partial F_{Bu}{}^{ij}(y)/\partial y^i = J^j \qquad (34.18)$$

The current for a neutron star is well approximated by the constant current in the φ direction (in the spherical coordinates)

$$J_\varphi = \beta_B \Omega M/m_n \qquad (34.19)$$

due to the neutron star's small size. Expressing J_φ in rectangular coordinates assuming the rotation is along the z-axis we find

$$J_x = -\sin \varphi \, J_\varphi \qquad (34.20)$$

$$J_y = \cos \varphi \, J_\varphi \qquad (34.21)$$

We will use the relative flatness of space, and the small size of the neutron star neighborhood, to identify the x, y, and z of our universe with the Megaverse coordinates y^1, y^1, and y^3. Inserting eqns. 34.20 and 34.21 in eq. 34.18 yields D – 1 equations:

$$\begin{aligned}
\partial F_{Bu}{}^{ix}(y)/\partial y^i &= -\sin \varphi \, J_\varphi \\
\partial F_{Bu}{}^{iy}(y)/\partial y^i &= \cos \varphi \, J_\varphi \\
\partial F_{Bu}{}^{ij}(y)/\partial y^i &= 0
\end{aligned} \qquad (34.22)$$

for j = 4, ... , (D – 1). These (D – 1) equations have a solution that gives a magnetic force that is solely within the three spatial dimensions of our universe. *Thus they cannot participate in a slingshot into the Megaverse.*

34.3.8 The Baryonic Coulomb Force Slingshot into the Megaverse

The D-dimensional version of the Lorentz force is

$$F_i = qE_{Bui} + q\varepsilon_{ijk}v_j B_{Buk}/c \qquad (34.23)$$

where

$$\mathbf{E_{Bu}} = [-\nabla_{(D-1)}\phi(y) - \partial \mathbf{B_u}(y)/\partial y^D]/c \qquad (34.24)$$

We have seen that the baryonic force slingshot is wholly derived from the baryonic Coulomb force:

$$\phi(y_1, y_2, \ldots , y_{(D-1)}) = (\beta_B{}^2/4\pi)N_1 N_2/(y_1{}^2 + y_2{}^2 + \ldots + y_{(D-1)}{}^2)^{\frac{1}{2}} \qquad (34.25)$$

between two baryon masses with baryon numbers N_1 and N_2. The baryonic Coulomb force is:

$$F_i = \nabla_{(D-1)i}\phi(y) \tag{34.26}$$

where $\nabla_{(D-1)i}$ is the ith component of the $(D-1)$-dimensional grad operator $\nabla_{(D-1)}$.
The part of the Lorentz force that slingshots a uniship into the Megaverse is

$$\begin{aligned}F_{isling} &= \partial\phi(y)/\partial y^i \\ &= (\beta_B^2/4\pi)N_1 N_2\, y_i\,/(y_1^2 + y_2^2 + \dots + y_{(D-1)}^2)^{3/2}\end{aligned} \tag{34.27}$$

for $i = 4, 5, \dots, (D-1)$.

A uniship will have a baryonic "Coulomb" force directing it out of our universe into the Megaverse. The baryonic force between the uniship and the neutron star is repulsive since they are both composed of a majority of baryons.

This force will undoubtedly be small compared to the gravitational forces during a slingshot maneuver. Thus we can see that the uniship will slowly ease out of our universe. For a short time it will be partly in and partly out of the universe creating a physical situation not hitherto encountered in physics. It has some advantages since, for example, it allows an "umbrella" of thrust tubes that originally are in 3-dimensional space to widen into a $(D-1)$-dimensional umbrella of thrust tubes that would enable the uniship to travel in any direction in the Megaverse – Megaverse maneuverability. We shall consider this possibility later.

In directions into the Megaverse, gravitation forces being absent, the baryonic force will be the sole force.

Thus we find a clear division: gravitation dominates in directions within the universe; baryonic force dominates in directions out of our universe. Fig. 34.1a depicts the trajectory of a uniship in a slingshot maneuver. Note the attractive hyperbolic motion due to the dominance of gravity in our universe.

34.3.9 Point-like Uniship Slingshots

The major problems of a close approach of perhaps 100,000 km to a neutron star are the large gravity of the neutron star, its strong tidal gravitation effects (stresses) on the uniship's structure, its strong magnetic field, and its emission of large amount of primarily x-ray/gamma rays. These properties of a neutron star neighborhood would appear to significantly affect the structural integrity of the uniship, and, more importantly, seriously impact on the safety and life of its human crew.

Fortunately there is a saving grace in this physical environment. A uniship approaching the neutron star could resolve these issues by traveling at extremely high speed[286] so that the time spent in the "danger" zone of the neutron star would be very small thus sharply reducing its deleterious effects.

---

[286] The starship would approach at perhaps c/2 and acquire an additional speed due to gravity of c/3. Combining these speeds using the rules of special relativity for the addition of velocities yields an approach speed of 0.7c near the neutron star.

34.3.9 The Uniship Slingshot Trajectory

The slingshot trajectory of a uniship is approximately a hyperbola in our universe due to the dominance of gravity. As it approaches a neutron star, with a distance of closest approach of perhaps 100,000 km, a uniship could be programed to take perhaps 3 seconds or so to circle around the neutron star. Consequently the uniship need only spend a minimal time near the neutron star with little gravitational tidal stress, magnetic field exposure, and radiation exposure issues.

Since the difficulties of a neutron star slingshot are surmountable, we will now turn to the issue of a uniship escape from our universe. We are fortunate that the neutron star's force on the uniship can be conveniently divided into two parts: gravitation which only influences the spatial motion of the uniship in our universe's coordinates, and the baryonic force which only significantly influences the uniship's other $D - 4$ spatial Megaverse coordinates. The weakness of the baryonic force compared to gravity makes its impact on motion in our universe's spatial coordinates negligible.

The baryonic force generated by the interaction of a uniship's baryons with the baryons within the neutron universe causes the uniship's course to be deflected into the Megaverse.

34.3.10 Uniship Neutron Star Slingshot Dynamics

In this subsection we will describe the dynamics of the neutron star slingshot. We also show that the high gravity of the neutron star causes the universe to have a curvature 'bubble' from flat. This non-zero curvature lowers the 'surface tension' of the universe locally, and thus reduces the confining pressure to a finite value[287] (from infinity), enabling a starship to more easily escape the universe into the Megaverse. The starship goes through the thinner 'fabric' of the universe.

We will assume a flat space-time in our universe with the understanding that space-time may be significantly curved in the immediate vicinity of the neutron star. We assume the uniship will not enter that region. We will define the neutron star to be at the origin of the Megaverse coordinates. Thus $y_i = 0$ for $i = 1, \ldots , (D - 1)$. We will assume the universe spatial coordinates x_i to be equal to the first three Megaverse coordinates:

---

[287] See chapter 32 for a detailed discussion of the surface tension and confining pressures of universes.

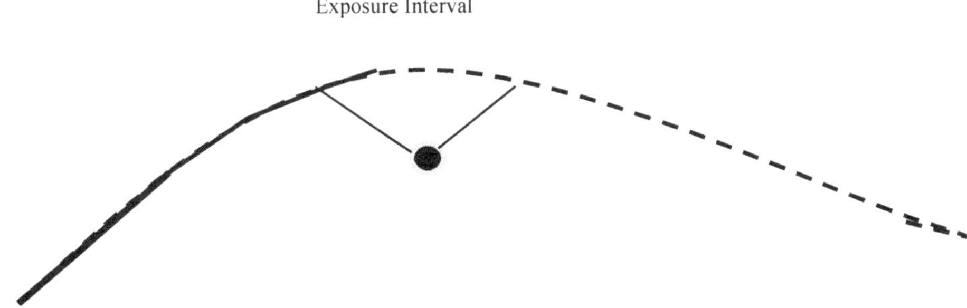

Figure 34.1b. Depiction of the uniship exposure interval of closest approach to a neutron star. The time interval spent in this region might be about three second or so. The uniship speed as it approaches might be about 0.7c or so to avoid capture by the neutron star. When the uniship totally exits our universe it disappears from view since electromagnetic radiation (light) from the uniship (or any object) "cannot penetrate"through the boundary of our universe from the Megaverse. The dashed line indicates the partial exit of the uniship from our universe into the Megaverse with Megaverse velocity and momentum components.

$$y_i = x_i \qquad (34.28)$$

for $i = 1, 2, 3$. We are allowed to do this by the flat space-time assumption for the universe. The total potential energy of the uniship in the neutron star's reference frame is[288]

$$V_{tot} = -GM_1M_2/r_u + (\beta_B^2/4\pi)\, N_1N_2/r_F \qquad (34.29)$$

where M_1 and M_2 are the masses of the neutron star and uniship, N_1 and N_2 are their baryon numbers, and where

$$r_u^2 = x_1^2 + x_2^2 + x_3^2 = y_1^2 + y_2^2 + y_3^2 \qquad (34.30)$$
$$r_F^2 = y_1^2 + \ldots + y_{(D-1)}^2 = r_u^2 + y_4^2 + \ldots + y_{(D-1)}^2 \qquad (34.31)$$

The force is the gradient of the potential

$$F^i_{slingshot} = \partial\, V_{tot}/\partial y_i$$
$$= GM_1M_2\, (\delta^{i1} + \delta^{i2} + \delta^{i3})y^i/r_u^{3/2} - (\beta_B^2/4\pi)N_1N_2\, y^i/r_F^{3/2} \qquad (34.32)$$

---

[288] Again we note that we are assuming a uniship trajectory in flat space-time so that we may use special relativistic potentials and dynamic equations.

where the Kronecker delta functions restrict the gravity force to the spatial coordinates of the universe.

The dynamic equation of the uniship motion[289] is

$$dp^i/d\tau = F^i_{slingshot} \tag{34.33}$$

where τ is the invariant interval. We now consider the initial phase of the escape to the Megaverse in which

$$r_u \gg r_F - r_u \tag{34.34}$$

The universe spatial distance is much greater than the purely Megaverse spatial distance $r_F - r_u$. In this case eq. 34.1b has different forms to good approximation for i = 1, 2, 3 and the remaining Megaverse spatial coordinates:

$$dp^i/d\tau \simeq GM_1M_2y^i/r_u^{3/2} \tag{34.35}$$

for i = 1, 2, 3 and

$$dp^i/d\tau = -(\beta_B^2/4\pi)N_1N_2\,y^i/r_F^{3/2} \tag{34.36}$$

for i = 4, ... , (D − 1). Eq. 34.35 yields a solution of the central force problem which in the present case is an approximately hyperbolic trajectory of the form depicted in Fig. 34.1a.

Due to our well-justified assumption that the distance into the Megaverse will be small compared to the distance of the unish from the neutron star we can approximate

$$r_F^{-3/2} \simeq r_u^{-3/2}[1 - (3/2)(r_F^2 - r_u^2)/r_u^2] \tag{34.37}$$

Then the momentum becomes approximately

$$dp^i/d\tau \simeq -(\beta_B^2/4\pi)N_1N_2\,y^i[1 - (3/2)(r_F^2 - r_u^2)/r_u^2]/r_u^{3/2} \tag{34.38}$$

or

$$dp^i/d\tau \simeq -(\beta_B^2/4\pi)N_1N_2\,y^i[1 - (3/2)\Sigma y_k^2/r_u^2]/r_u^{3/2} \tag{34.39}$$

where the sum is from k = 4, ..., (D − 1). The above equations yield an initially exponential-like trajectory into the Megaverse to leading order.

Thus we have shown that the neutron star slingshot clearly drives the uniship out of the universe into the Megaverse. The uniship passes into the Megaverse. For a small time it is partly in and partly out of the universe.

---

[289] The neutron star is assumed to be stationary due to the largeness of its mass relative to the uniship.

34.4 Umbrella-Shaped Uniship Slingshots

We take it for granted that we can move in one spatial direction or another with ease. However when one enters a higher dimensional space from a space of lower dimension, movement in the additional dimensions, which requires the expenditure of force in those dimensions, becomes an issue. A lower dimension object does not automatically have forces within it in the additional directions and so it cannot move itself, or part of itself in those directions.

In the previous subsections we saw how to use the baryonic force to escape from our universe to the higher dimension Megaverse. In this subsection we will show how to give "wings" to a uniship so that it will have the ability to maneuver in any direction in the Megaverse. To achieve this capability we will have to design the uniship so that it will expand in all Megaverse directions as it enters the Megaverse. This will require an umbrella-like configuration that will use the baryonic force to open the umbrella in all Megaverse 'spatial' directions. The spokes of the umbrella will be long thrust tubes through which uniship thrust can be directed to move the uniship in a desired direction. Fig. 34.2 is a depiction of the simplest form of umbrella uniship.

34.4.1 Scenario for the Opening of a Uniship Umbrella

As an umbrella uniship slingshots around a neutron star the body of the ship including fuel tanks is rigid and moves as a unit. The spokes of the umbrella, the thrust tubes, are moveable and will each move differently because of their differing average distance from the neutron star. As a result they will point in different directions in the Megaverse and can be further moved within the Megaverse relative to each other to deliver thrust in any Megaverse direction. After positioning they can be locked in place and used to maneuver the uniship towards another universe.

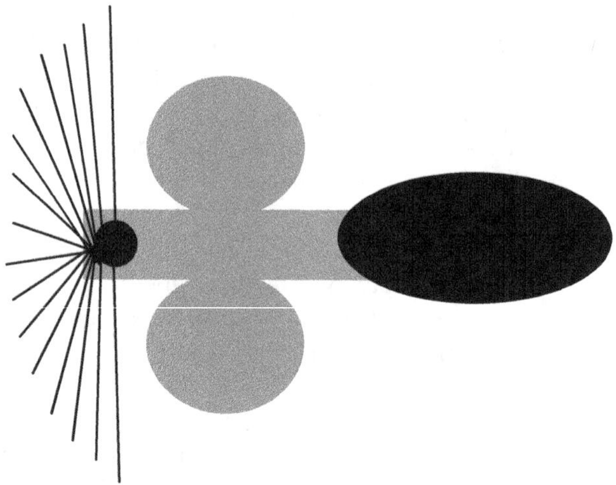

Figure 34.2. Tentative umbrella-like uniship design with the spokes of the umbrella forming a fan. The thrust tubes (D − 1 umbrella spokes − not fully shown) extend kilometers from the thrust power generator(s) core to enable the baryonic force to maneuver them in the (D − 1) different Megaverse directions. The thrust tubes are able to swivel into all (D − 1) spatial directions in response to the baryonic force as the uniship enters the Megaverse. Two fuel spheres are depicted under the assumption that one holds hydrogen and the other holds anti-hydrogen since they are presently the most powerful known possible energy source. The black forward part is for crew, supplies, and cargo. The rear gray part holds the engine apparatus and other related engine components.

34.4.2 Equations for the Motion of the Thrust Tubes Entering the Megaverse

Earlier we developed the equations for the slingshot mechanism for a compact rigid uniship. In this section we will extend the equations to the case of a uniship with an umbrella with a fan shape (an array of spokes as pictured in Fig. 34.2. The array is flat with each spoke in the plane of the ship's trajectory.) A uniship with a true umbrella of spokes (thrust tubes) is another possibility that may be of importance. This is a technical question that we will not address.

We define radial distances from the neutron star center for a "fan-shaped" umbrella with the nth spoke end[290] at radial distance r_{un}. Then we find

$$r_{Fn}^{-3/2} \simeq r_{un}^{-3/2}[1 - (3/2)(r_{Fn}^2 - r_{un}^2)/r_{un}^2] \qquad (34.40)$$

---

[290] The spoke center of mass actually.

for the end of each spoke and we then find

$$dp_n^{\,i}/d\tau \simeq -(\beta_B^{\,2}/4\pi)N_1 N_2\, y^i[1 - (3/2)(r_{Fn}^{\,2} - r_{un}^{\,2})/r_{un}^{\,2}]/r_{un}^{\,3/2} \qquad (34.41)$$

or

$$dp_n^{\,i}/d\tau \simeq -(\beta_B^{\,2}/4\pi)N_1 N_2\, y^i[1 - (3/2)\Sigma y_k^{\,2}/r_{un}^{\,2}]/r_{un}^{\,3/2} \qquad (34.42)$$

for the nth spoke where the sum is from k = 4, …, (D – 1). For each spoke radial distance r_{un} these equations yield different initial trajectories into the Megaverse to leading order.

Thus the spokes will enter the Megaverse in different Megaverse directions since they are not rigidly attached to the uniship body. Upon entry they can change each other's direction giving a final configuration with full Megaverse maneuverability. Uniships can now move in any Megaverse spatial directions.

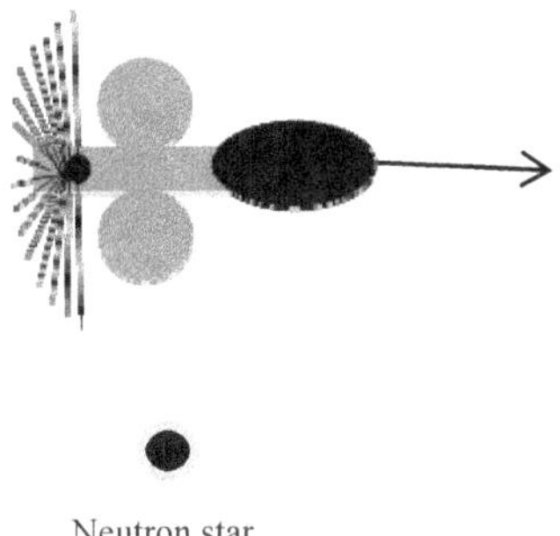

Figure 34.3. Uniship slingshot past neutron star. Note the umbrella spokes which are attached to the uniship but moveable will respond differently to the baryonic force since they are at different radial distances from the neutron star and thus feel differing amounts of force. The baryonic force will twist them in different directions. They then can be re-oriented by the uniship computer to provide mobility in all Megaverse directions.

34.4.3 Equations for the Motion of the Thrust Tubes Entering the Megaverse

Earlier we developed the equations for the slingshot mechanism for a compact rigid uniship. In this subsection we will extend the equations to the case of a rigid uniship with an umbrella with a fan shape (a flat array of spokes as picture in Fig. 34.2. The array is flat with each spoke in the plane of the ship's trajectory.) A uniship with a true umbrella of spokes (thrust tubes) is another possibility that may be of importance. This is a technical question that we will not address.

We define radial distances from the neutron star center for a "fan-shaped" umbrella with the nth spoke end[291] at radial distance r_{un}. Then eq. 34.37 becomes

$$r_{Fn}^{-3/2} \simeq r_{un}^{-3/2}[1 - (3/2)(r_{Fn}^2 - r_{un}^2)/r_{un}^2] \qquad (34.43)$$

for the end of each spoke and then we find

$$dp_n^i/d\tau \simeq -(\beta_B^2/4\pi)N_1 N_2\, y^i[1 - (3/2)(r_{Fn}^2 - r_{un}^2)/r_{un}^2]/r_{un}^{3/2} \qquad (34.44)$$

or

$$dp_n^i/d\tau \simeq -(\beta_B^2/4\pi)N_1 N_2\, y^i[1 - (3/2)\Sigma y_k^2/r_{un}^2]/r_{un}^{3/2} \qquad (34.45)$$

for the nth spoke where the sum in eq. 34.42 is from $k = 4, ..., (D - 1)$. For each spoke radial distance r_{un} these equations yield different initial trajectories into the Megaverse to leading order.

Thus the spokes will enter the Megaverse in different Megaverse directions since they are not rigidly attached to the uniship body. Upon entry they can change each other's direction giving a final configuration with full Megaverse maneuverability. *Uniships can now move in any Megaverse spatial directions.*

34.5 Alternate Approaches to Escaping our Universe

There are two other possible approaches to escaping from our universe into the Megaverse.[292] One approach sends a uniship through a massive ring or cylinder and uses the baryonic force to enter the Megaverse. The other approach uses a rotating uniship, which of course is primarily made of baryons, to rapidly spin relative to a baryon concentration in its neighborhood thus generating a baryonic force that enables the uniship to enter Megaverse.

Both approaches have some merit. But they also have the drawback of requiring large nearby baryon masses of the order of the sun's mass as well as other drawbacks that make them less desirable than the neutron star slingshot approach described in the previous chapter. Again we defer detailed discussion.

34.5.1 Rotating Baryon Ring/Cylinder Mechanism for Escape from Universe

The weakness of the baryonic force requires a large mass, or a gateway ring or a cylinder mechanism to escape from the universe. Alternate mechanisms probably should be located near a large gravitational field – perhaps revolving around a black hole.

We will now consider a rotating ring or cylinder mechanism consisting of a rotating ring or, more likely, a rotating cylinder of great size and length. Not knowing the strength[293]

---

[291] The spoke center of mass actually.
[292] This chapter appeared in Blaha (2014c).
[293] We estimate the coupling constant earlier. It is extremely small relative to the coupling constants of other forces.

(coupling constant) of the baryonic force we can only be certain that a baryonic ring or cylinder must be of great size, and large baryonic mass. (Most solid substances are overwhelmingly composed of baryons by mass.) Making a ring, or cylinder, rotate at high rotational velocity will require an extremely large amount of energy. But once set in motion the ring or cylinder will continue to rotate indefinitely. The rotating ring/cylinder must be electrically neutral. The rotating baryon ring/cylinder does have a baryon charge which generates a baryonic field just as a rotating charged ring generates an electromagnetic field.

The baryon numbers required for a baryon ring or cylinder would have to be extremely large (thus large masses) to generate the required sizeable baryonic "magnetic" field. Thus these mechanisms for uniship entry into the Megaverse are at least many thousands of years into the future.

34.5.2 Rotating Ring/Cylinder/Mass Configuration/Baryon Cloud Baryonic Force for Escape to Megaverse

A rotating ring, cylinder or mass configuration has a baryon current that generates baryon fields and forces (in a manner analogous to a rotating configuration of electric charges generates electromagnetic fields and forces). This baryon current generates a baryon field that will cause a uniship to spiral out of our universe.

The current in the rotating source creates forces between baryon current loops similar to those of the Biot-Savart law in electromagnetism.

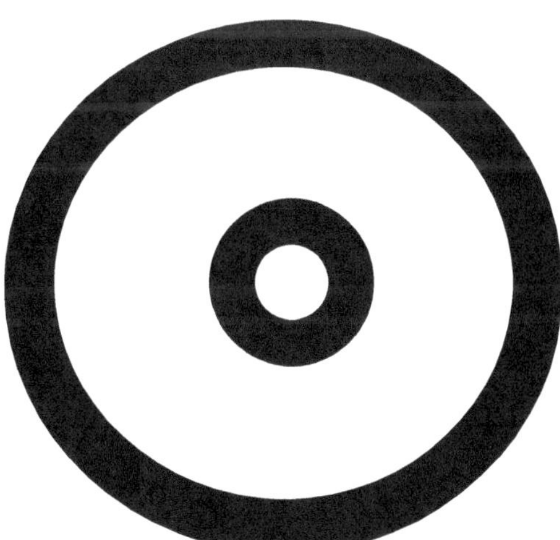

Figure 34.4. A cross section of an outer baryon current loop (in a ring, cylinder, mass configuration, or baryonic cloud) and an inner baryon current loop generated by a large spinning uniship. A "ring within a ring" configuration.

The baryons in the outer "loop" of baryons (ring, cylinder, mass configuration, or baryonic cloud) are effectively fixed in solid masses or slightly moving in a baryon cloud. Thus baryon currents are unlike electromagnetic currents which are generated by moving electrons or ions.

The baryon current generated by the uniship is from the combined rotation of the uniship baryons. It is approximately equivalent to a ring if the uniship is approximately cylindrical.

We have assumed the inner and outer rings are within a region with a large gravitational field such as a region near a black hole or massive star. The presence of a large gravitational field causes the coordinates of the universe x^μ to differ significantly from Megaverse coordinates y^i. Temporarily ignoring quantum aspects, universe coordinates are related to Megaverse coordinates by

$$y_i = f_i(x) \tag{34.46}$$

Because of the disparity in coordinates the baryonic Biot-Savart force implies a baryonic Ampère force between the currents of the form

$$\mathbf{F} = (I_{b1}I_{b2}/c^2) \oint \oint \mathbf{dy_1 \cdot dy_2} \; \mathbf{y}/|\mathbf{y}|^3 \tag{34.47}$$

where I_{b1} and I_{b2} are the Megaverse baryonic currents in the rings, $\mathbf{dy_1 \cdot dy_2}$ is the spatial $(D-1)$-dimensional inner product between the line elements of each ring *in Megaverse coordinates*, and $\mathbf{y}$ is the $(D-1)$-dimensional spatial vector distance between the line elements.

$\mathbf{F}$ clearly has components in Megaverse coordinates that will propel the uniship out of our universe into the Megaverse for transit to other universes. Note the current loops in our universe's coordinates must also be loops in Megaverse coordinates. These loops which might appear circular in a strong gravitational field in our universe will appear to be distorted closed loops in Megaverse coordinates – a simple topological consequence.

Both the uniship and the ring, or other mass configuration, will experience $\mathbf{F}$ in opposite directions. The uniship, being of much less mass than the enclosing ring, will experience a large acceleration into the Megaverse while the enclosing ring will experience negligible acceleration.

Thus these other mechanisms for escaping our universe are feasible in principle.[294] However the construction of large rings and cylinders in the neighborhood of a black hole or other large mass is a major technological issue that clearly will not happen (if it does) for many millennia. It requires a new massive scale of technology far beyond our present technology.

The neutron star slingshot mechanism appears to be the most promising approach – especially when one realizes that other universes will not have mass configurations similar to

---

[294] The rate of spin of the uniship has to be acceptable from a human factors point of view.

those mentioned above except possibly for a large baryon cloud with a central gap located near a black hole or some very massive sun or other massive object.

The major challenge for the neutron star slingshot is the avoidance of difficulties associated with a close approach. This seems possible due to the very short time spent close to the neutron star.

34.5.3 Spinning Baryon Ring Gateway to the Megaverse

A spinning baryon ring generates a baryonic field that causes a moving baryon uniship in the "center" to spiral into the Megaverse. As soon as the uniship is in the Megaverse, outside our universe, it is no longer visible. Once in the Megaverse the uniship can use its own built-in propulsion system to travel to other universes.

One problem with this approach is the return to our universe. There is no spinning ring in other universes and the construction of one would be costly time-wise, and technology-wise, not to mention the large amount of mass that would have to be assembled to make a ring or cylinder. An alternative approach would be to use the slingshot mechanism for the return to our universe or the spinning uniship mechanism depicted in Fig. 34.7 below.

//

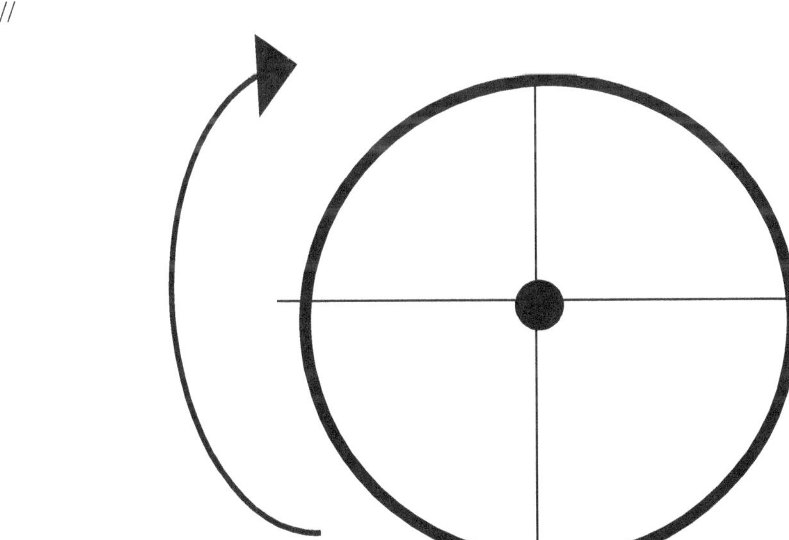

Figure 34.5. A clockwise rotating thick baryonic ring with a uniship at its center moving "straight out of the page."

34.5.4 Spinning Baryon Cylinder Gateway to the Megaverse

A uniship enters a baryonic cylinder and under the baryonic force begins to spiral from our universe into the Megaverse disappearing from view. It then can use its engines to traverse the Megaverse to other universes. It can also view other universes in the Megaverse shining with "baryonic light." The return to our universe would require a baryonic cylinder (or ring or spinning uniship (see below) or slingshot maneuver) in the Megaverse to reenter our universe. Sending a uniship, with a large positive baryonic charge along the axis (center) of the cylinder will cause the uniship to spiral into the Megaverse. The uniship can then travel to other universes.

34.5.5 Rotating Baryon Mass Configuration Gateway to the Megaverse

One can also envision other baryon mass configurations that can be used to enable a uniship to escape from our universe into the Megaverse. One example of a mass configuration is four baryon spheres of large baryon number rotating uniformly. A uniship enters at high speed along their central axis and then spirals into the Megaverse by means of the baryonic force.

34.5.6 Disadvantages of Ring/Cylinder/Mass Configurations to Escape our Universe

The basic problem with all the possible above gateways is that they offer one way transit out of our universe and do not provide for a return. The approaches below provide a capability for a return to our universe.

All of these mechanisms require extremely large masses and are beyond our capabilities in the foreseeable future.

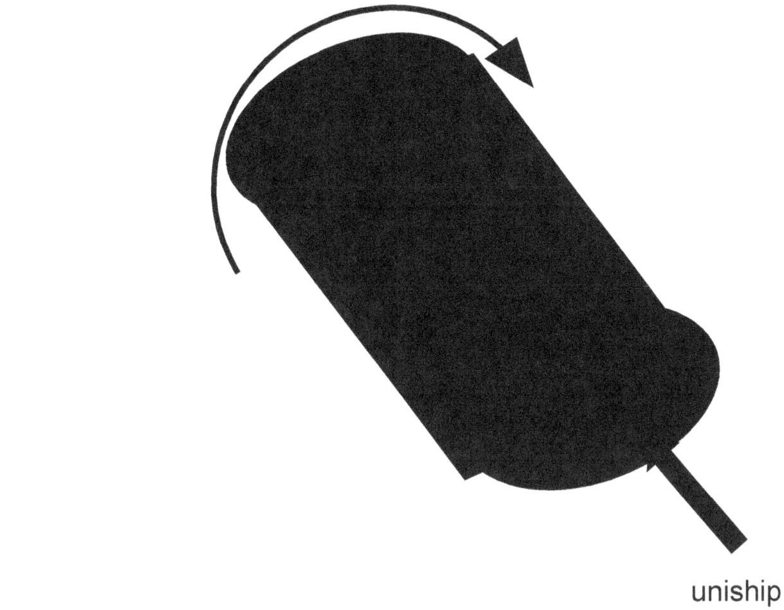

Figure 34.6. Depiction of a uniship entering a spinning baryonic cylinder along the cylinder's central axis.

This type of large masses configuration would be difficult to assemble and rotate in synchronization. However it might require less total mass than the ring or cylinder gateways.

34.6 Escape using a Rapidly Rotating Uniship

Above we explored using external large masses to cause a uniship to enter the Megaverse. In this section we consider a rotating uniship moving at high speed through a void in a massive cloud of baryons. (Fig. 34.8) It is necessary to move through a void to avoid the severe erosion that a high speed uniship would experience moving through a dense cloud of matter.

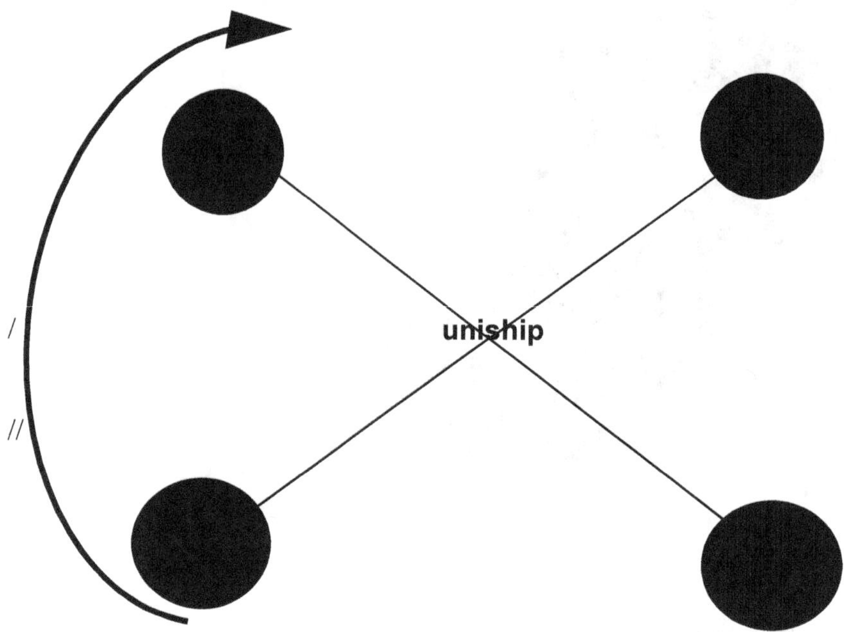

Figure 34.7. Depiction of a configuration of four large masses rotating around a common center. A uniship goes through the center ("out of the page") and enters the Megaverse.

As the fast rotating uniship moves through a void within the massive cloud of baryons it will experience a baryonic force that will cause it to spiral out of the universe into the surrounding Megaverse. A rotating uniship in a relatively static baryon cloud is equivalent to a non-rotating uniship within a rotating ring or cylinder or mass configuration.

This uniship escape mechanism could presumably be used in any large universe and could also be used to reenter our universe. It avoids the need to create gigantic mass configurations relying on Nature to provide massive clouds within a universe.

The major problem of the rotating uniship is the centripetal force that the crew would experience.

/

Figure 34.8. A rotating uniship traveling outward from page within a void in the midst of a massive baryon cloud.

34.7 Uniship Design Topics

This section discusses some uniship design topics. It appeared in Blaha (2014c).

34.7.1 How Many Dimensions Exist Inside an Escaped Uniship?

When a uniship escapes from a universe into the Megaverse the questions arise, What is the dimensionality of the space within the uniship? How do the uniship occupants and equipment interface with the $(D-1)$ spatial dimensions outside the uniship?

The first issue that must be considered in answering those questions is the physical reality of the Megaverse. Are its coordinates a mathematical fiction? Or are they physically real? The answer seems to lie in the existence of a quantum field that depends directly on the D dimensions of the Megaverse. The existence of such a field gives Megaverse coordinates a physical reality since the quantum field, being a physical construct, could not be defined without the physical reality of the Megaverse. An additional requirement for a physical Megaverse is an interaction between the quantum field and the physical masses of our experience. The baryon gauge field $B_{ui}(y)$ that we have postulated previously fulfills both of these requirements. (See Blaha (2014) and earlier parts of this volume.) Thus the Megaverse which is the "ocean" of the set of "island universes" is a physically real entity.

The second issue is the relation of Megaverse coordinates and the coordinates of a universe. We note that every point in a universe's coordinates has a corresponding point in the Megaverse due to the relation:

$$y_i = f_i(x) \tag{34.48}$$

Thus every point in a universe has two sets of coordinates. This also holds also for the interior of a uniship.

In the uniship the occupants will live in a 4-dimensional flat space "bubble"[295] within the $(D - 1)$-dimensional Megaverse. The second question we raised concerned how the uniship occupants could interface with $(D - 1)$-dimensional space within and without its confines. This is considered in the next section.

34.7.2 How Does a Uniship Control Direction and Maneuver in the D-dimensional Megaverse?

Transiting from a subspace with one set of dimensions to a larger set of dimensions in a larger space is a physically challenging task. One cannot simply use Euclid's algorithm for proving three spatial dimensions. For it assumes that we can move in the direction of any physical dimension. But if the available physical force to move in a given direction is not present then the dimension is physically inaccessible.

In the present case we enable a uniship to escape from our universe using, for example, the slingshot around a neutron star mechanism. After entering the Megaverse it will have some momentum in Megaverse directions beyond the three spatial dimensions of our universe. But that is not enough. A uniship must have the capability to move in any or all $D - 1$ directions in the Megaverse.

Thus we are led to propose a uniship designed having $D - 1$ thrust exhausts with each direction of thrust pointing in a direction in $(D - 1)$-space. Thrust would be generated in a central chamber and then directed, using (perhaps) bending magnets, through one or more exhaust ports accelerating the uniship in a specified direction. In our universe all $D - 1$ exhaust ports and the direction mechanism would point in a variety of directions in our 3-space. As a uniship slingshots around the neutron star it will experience strong gravitational and baryonic tidal forces. With a correct alignment of the $D - 1$ exhaust ports and central chamber in three dimensions, the exhaust ports and chamber will open like an umbrella in all the directions of the $(D - 1)$ Megaverse dimensions. Then the uniship will be able to accelerate in any direction in the Megaverse. Fig. 34.9 depicts the uniship before and after the slingshot maneuver.

---

[295] An analogous situation is a liquid sphere in zero gravity that splits into two spheres. Both spheres retain the same topology and other characteristics.

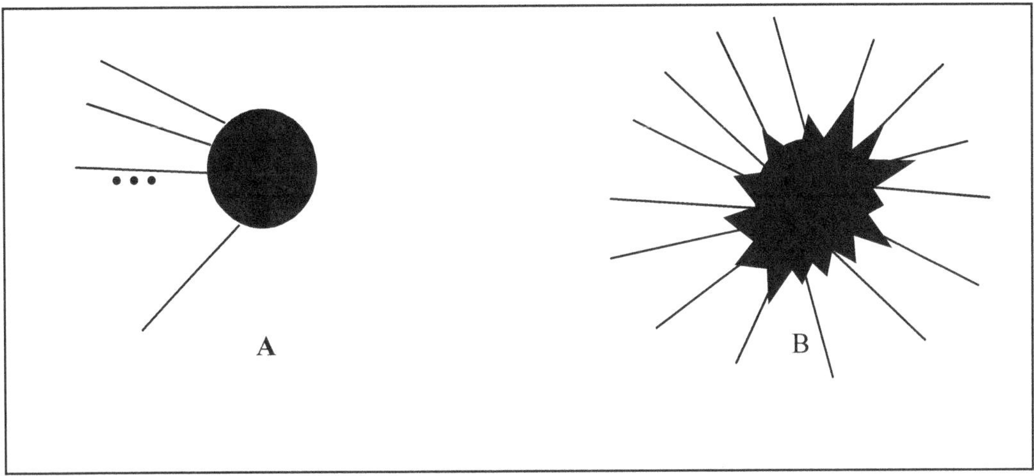

Figure 34.9. Symbolic depiction of the uniship thrust exhaust ports before (A) in 3 dimensions, and in (D − 1) dimensions after (B) the slingshot maneuver to exit our universe.

34.7.3 Power Source for Uniships

The distance scale of starship travel in our universe (and probably in other universes) between galaxies is of the order of millions of light years. Consequently starships will have to be powered by particle-antiparticle annihilation and/or fusion energy for longer trips, and nuclear energy, at minimum, for intra-galaxy trips.[296] Part of the power requirements are necessitated by the need for short travel times of a few years at most so that interstellar commerce, and migration, becomes feasible. For these reasons we wish to have starships capable of speeds up to tens of thousands times the speed of light.

The distance scale of travel in the Megaverse is likely to be trillions of light years. We anticipate universes are separated by trillions of light years. Due to these enormous distances, and the fast transportation time that we wish, uniships will require proton-antiproton annihilations on a large scale to power uniships. Proton-antiproton annihilation transforms mass entirely into energy unlike nuclear and fusion reactions where only a fraction of the masses of the particles are converted into energy yielding of the order of 1,000 the energy production of fusion.

34.7.4 Possible Uniship Design

We therefore require a large sphere containing protons that are densely packed. (An alternative that would avoid electromagnetic repulsion problems associated with densely packed

---

[296] Blaha (2013a) and earlier books.

protons would be a densely packed sphere of hydrogen atoms.) We also need a large sphere containing antiprotons that are densely packed. (Again an alternative that would avoid electromagnetic repulsion problems associated with densely packed antiprotons would be a densely packed sphere of anti-hydrogen atoms.) In addition to being the most powerful energy source available these spheres would form a baryon dipole that could play a role in the escape of a uniship from a universe. We therefore start with a uniship design of the form of Fig. 34.10.

Figure 34.10. An initial depiction of the overall design of a uniship. The upper and lower ovals contain a dense mass of protons and of antiprotons respectively. The antiproton sphere requires a confinement mechanism to avoid premature particle annihilation. (The ovals could alternatively contain liquid hydrogen and anti-hydrogen to avoid electromagnetic repulsion issues.) The module in front contains the crew quarters, equipment and cargo. The tail section contains a central region that controls the particle-anti-particle combustion process directing the thrust through one or more of the (D − 1) thrust chambers to power the uniship in the desired direction in the Megaverse.

34.7.5 Intra-Universe Starship Designs Summary

Starship design has a number of requirements to satisfy for trips within the galaxy and a more stringent set of requirements for travel to other galaxies in our universe. Uniships have a yet stricter set of requirements for travel to other universes within the Megaverse. The discussion in this section sets the background, based on our starship designs, for the consideration of important design details for uniships presented in succeeding chapters. Most of this section is abstracted from Blaha (2013a) and earlier books on nuclear ships, space guns, and starships. Appendix A of Blaha (2013a) briefly overviews the designs in these earlier books for mass transit from the earth to space, travel in the solar system, and, most importantly, starship travel within our galaxy and to other galaxies.

34.7.5.1 Types of Starships

There are two general types of starships due to the nature of starship drives: starships based on circular accelerators and starships based on linear accelerators. In both cases we assume the process that creates the quark-gluon plasma thrust is a stream of collisions of spherules of some material.

34.7.5.2 Starship Engine Energy Sources

The starship engine that we have designed requires massive amounts of energy. At this point in time the only feasible energy source for starships is nuclear energy. It is reasonable to expect that fusion energy, a more concentrated energy source, will become a reality within the next thirty years. In either case the starship will need an energy source to drive the spherule accelerator rings and associated devices of the engine for periods up to perhaps a few months or years, then turn off for perhaps many years, and then resume operations for further maneuvers.

In the extreme case of travel to another galaxy, the energy source will need to turn off for up to millions of years of starship time. While the energy source is turned off, a residual "battery" will need to operate to support monitoring the progress of time, activating the startup of the main energy source, and possibly to detect and monitor objects ahead of the starship in the line of flight. This battery source may well be a plutonium (or longer lived) source similar to those used in current space probes.[297]

The main energy source, if it is a nuclear reactor of some kind, will probably have to be a reactor that is different from current nuclear reactors. Chapter 5 of Blaha (2013a) describes long shelf life nuclear reactors that could meet this need. Since the startup process from a battery driven state needs to be gradual due to a "small" battery, it appears the set of nuclear reactors would be composed of perhaps five reactors of increasing size. The battery starts the smallest reactor by concentrating its nuclear fuel. The smallest reactor then generates the energy to concentrate that fuel, and start the second smallest reactor, and so on until the main reactor(s) is started. At this point the accelerators power up and starship thrust begins.

If the source of the energy is fusion energy then the startup process might begin in small stages in the boot up of the fusion reaction through fusing larger and larger amounts of (perhaps) ^{3}He with increasingly powerful laser beams a la the tokamak approach.

During the coasting period of a starship, nuclear reactors should be powered down to conserve nuclear fuel. When powering down a nuclear reactor power source the nuclear material (U^{235} or plutonium) (the reactor fuel) residing in the medium of the reactors would be diluted to sharply reduce fission reactions to "near zero" using the energy of the next largest reactor. The smallest reactor would be powered down by a battery. This battery would retain enough energy to bring the smallest reactor back up after the coasting period ends. Then the reactors would boot up in turn to provide energy to the vehicle. (The battery would be at extremely low temperature during a coasting phase and thus not lose a significant amount of electrical power.)

In the case of a fusion power source a battery could be used to initiate the fusion power. A gradual turnoff process could execute at the start of a coasting phase to bring the fusion process to zero in such a way that the battery could initiate the boot up process for the fusion power source at the end of a coasting period.

---

[297] We note that a natural nuclear reactor existed in the Congo Region of Central Africa for millions of years. (Parenthetical note: Could this be the stimulus for the rapid evolution of species in Africa including very early Mankind?)

34.7.5.3 An Alternate Starship Accelerator Ring Engine Design

In an earlier work Blaha (2009b) we proposed an alternate accelerator ring design for quark-gluon fluid acceleration. It appears that this design is not feasible with current or near term (one hundred years) technology. The reasons are:

1. The quark-gluon fluid ring is not feasible because fireballs cannot be injected and accelerated to form a ring in a few fm/c - the time available.

2. A quark-gluon fluid ring would be similar to fusion tokamaks but much more challenging in its requirements for confinement and stability due to the much higher density and temperature of a quark-gluon fluid ring.

34.7.5.4 Some Starship Configurations

There are a variety of possible configurations for faster than light starships. In this subsection we categorize the starship configurations by their thrusters.

34.7.5.4.1 Rear Thrust Exhausts on Circular Accelerator Starships

In Blaha (2010a), and later books, quarks and gluons were shown to have complex 3-momenta. The real part of the 3-momenta of each quark was orthogonal to the imaginary part of its 3-momenta.

If we "add" pairs of exiting quarks to create a complex 3-momentum to the rear, then a complex total pair 3-momentum is created that generates a rearward thrust. The complex thrust can be used in a one thrust exhausts, or two thrust (or multi-thrust) exhausts as shown in Figs. 34.11 and 34.12.

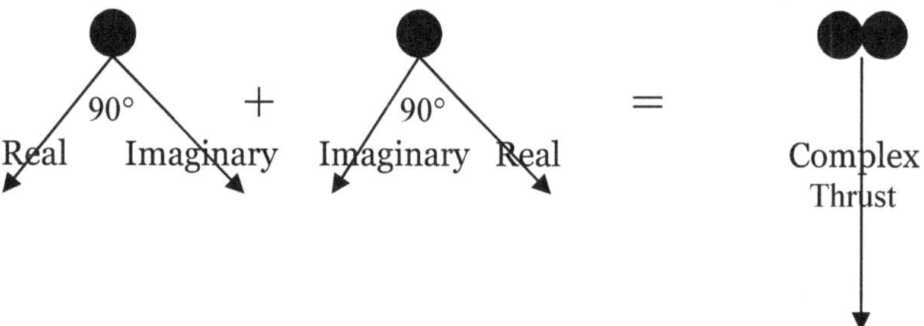

Figure 34.11. A circular accelerator starship with one thrust exhaust.

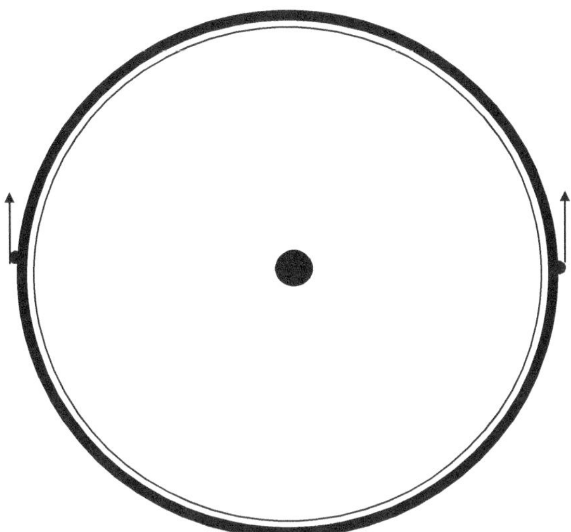

Figure 34.12. Top view of a circular starship with two complex thrust exhausts.

These types of starships generate thrust that only moves the starship forward in space. They do not cause the starship to rotate. The next subsection introduces the use of the imaginary perpendicular part of a starship's thrust to cause the starship to rotate.

34.7.5.4.2 ROTATING CIRCULAR AND CYLINDRICAL STARSHIPS DUE TO IMAGINARY PART OF THRUST

In the case of disc-shaped and cylindrical starships the imaginary part of the quark-gluon thrust can be directed by magnets to be tangent to the circular edge of the starship. The resulting tangential force causes the starship to rotate through an imaginary angle. The rotation creates a *negative* centripetal force in the starship – "artificial gravity" that varies with the distance from the central axis of the starship. Fig. 34.13 shows a "top view" of a starship.

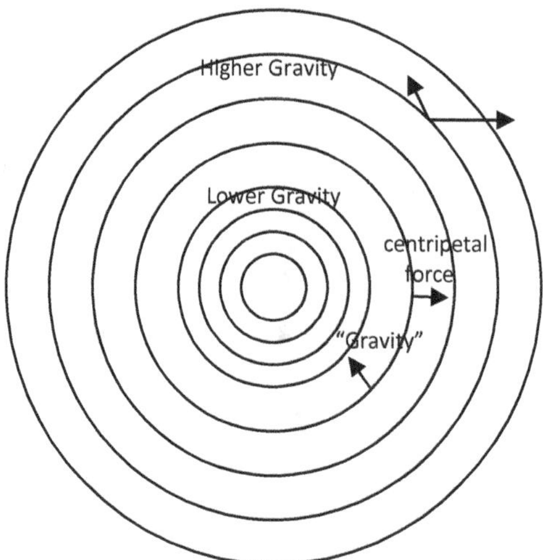

Figure 34.13. Top View of a Disc-like or Cylindrical Starship. View of cargo/people levels of inner hub. The center has lower "gravity." Circles indicate levels of equal artificial gravity in a disc-shaped or cylindrical starship. The outer parts have higher "gravity." The "gravity" force is inward towards the center on all levels. The centripetal force is outward from the center as shown by the arrow in the diagram and the discussion below (opposite to the direction of conventional centripetal force.) The artificial "gravity" force is thus inward to the center.

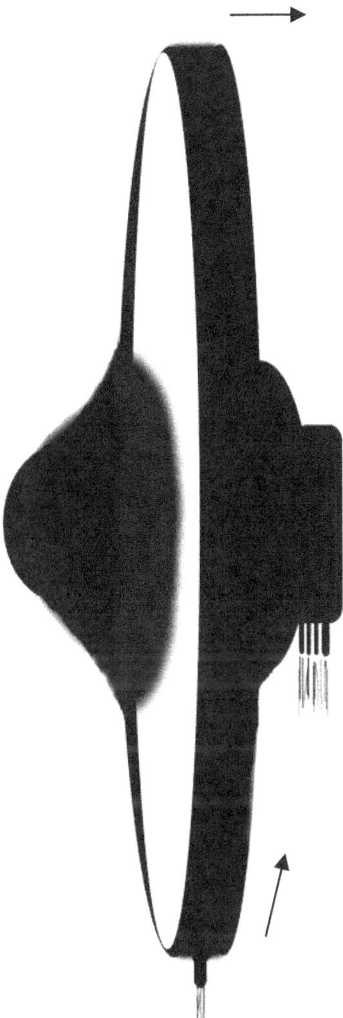

Figure 34.14. A disc-like starship with thrust consisting of a real and imaginary part. The real part drives the ship to the left. The imaginary part is tangent to the edge of the starship causing it to rotate counterclockwise.

Fig. 34.14 is an example of a disc-shaped starship with the thrust direction displayed. The horizontal thrust arrows represent the real part of the quark-gluon thrust. It accelerates the starship to the left in the figure. The vertical arrows represent the imaginary part of the thrust. It rotates the starship in a counterclockwise direction.

34.7.5.4.3 ARTIFICIAL GRAVITY LEVELS - ROTATING CIRCULAR AND CYLINDRICAL STARSHIPS

In rotating starships the real part of the thrust propels a starship. The combination of the real and imaginary parts of the thrust enables a starship to exceed the speed of light. The imaginary part of the thrust causes the starship to spin around its central axis.

We will examine the case of a cylindrical starship and see how the "artificial gravity" emerges as a result of the imaginary part of the thrust. Fig. 34.15 shows a cylindrical starship and Fig. 34.16 shows the cylindrical coordinate system used to calculate its motion.

The force generated by the quark-gluon thrust has the form

$$\mathbf{F} = g_r \check{\mathbf{z}} + i g_i \mathbf{\emptyset} \tag{34.49}$$

in the starship's cylindrical coordinates rest frame where $\check{z}$ is a unit vector in the positive z direction and $\emptyset$ is a unit vector in the positive $\emptyset$ angle direction. g_r and g_i are constants specifying the thrust. g_r is the starship's mass times its real-valued acceleration in the $\check{z}$ direction. g_i is the starship's mass times its imaginary-valued acceleration in the $\emptyset$ direction. The time derivative of the momentum $\mathbf{p}$ of a small part of mass m_0 at the edge of the starship is

$$d\mathbf{p}/dt = m_0(\gamma z')'\check{\mathbf{z}} + m_0[(\gamma\rho\emptyset')' + \gamma\rho'\emptyset')]\mathbf{\emptyset} + m_0[-\gamma\rho(\emptyset')^2 + (\gamma\rho')']\mathbf{\rho} \tag{34.50}$$

/where $\mathbf{\rho}$ is the unit vector in the positive ρ direction and the $'$ symbol indicates a time derivative.

Figure 34.15. Cigar shaped starship with "horizontal accelerator ring(s). The real part of the thrust points downward. The imaginary part of the thrust is horizontal and tangent to the cigar surface. The fins are for supplementary nuclear maneuvering rockets.

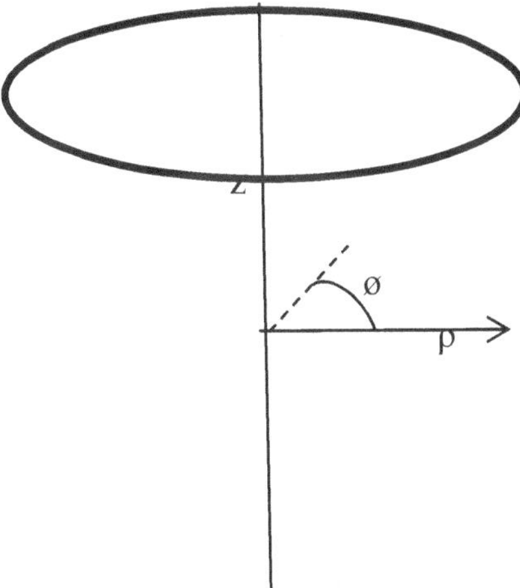

Figure 34.16. Starship cylindrical coordinate system. The diamond-shaped Starship in Fig. 34.17 is centered on the z axis.

34.7.5.4.4 STARSHIPS BASED ON LINEAR PARTICLE ACCELERATORS

Starships based on linear accelerators can have a variety of forms. Fig. 34.17 contains a diamond-shaped starship. Generally starships based on linear accelerators need great length – of the order of miles for the primary linear accelerator tubes unless a very rapid linear accelerator mechanism is developed. The angles between the linear accelerator tubes are determined by maximizing the efficiency, and amount, of thrust generation. Thus the shape of the *lower half* of the diamond-shaped starship is more or less determined. The lower part of the starship can be sharp edged diamond shaped or rounded diamond shaped. The upper part of the starship should minimize the effects of space dust.

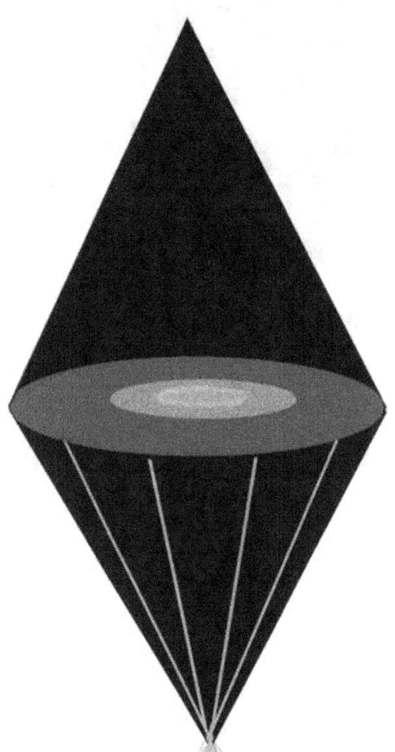

Figure 34.17. A diamond shaped ship powered by four linear accelerators. The lower part has the accelerators, magnets, nuclear reactors, and propellant. The upper part contains the crew, cargo, shielding and nuclear shuttles for exploring a solar system.

34.8 Megaverse Communications

Quantum communication is a rapidly developing field. At this point in time it has only been demonstrated for short ranges. Eventually it should mature as a long distance communication device such as that described in Blaha (2014c). Its great point is its instantaneous nature. Instantaneous communication at very great distances becomes feasible.

Thus one can hope to communicate across the universe. One can also hope for instantaneous communication with ships in the Megaverse. Sending a uniship into the Megaverse should not interfere with quantum communication using a detector at home and another that traversed into the Megaverse on a uniship. The difference in the number of dimensions should not affect the communication between the quantum devices. We expect that quantum states straddling the universe-Megaverse border will not go "out of sync."

Therefore quantum communication offers the possibility of instantaneous communications in our universe, and with uniships/colonies in the Megaverse and other Megaverse universes.

34.8.1 Rapid Interstellar Communication

Recent work on quantum entanglement suggests that instantaneous communication may be possible using this mechanism if an advanced long range form of this laboratory phenomenon can be developed. Quantum entanglement can transcend the borders of universes since it is based on coordinated parts of a quantum state. Its instantaneous nature, which has been verified to great accuracy in recent experiments, makes it the ideal mechanism for communication over trillions of light years. We will describe the application of this concept in more detail in the following sections.

34.8.2 "Instantaneous" Interstellar Communication

Once a uniship capability is achieved it will clearly necessitate a very rapid, if not instantaneous, means of communication. All electromagnetic means of communication are limited by the speed of light and are thus insufficient for multi-million light year communication. If neutrinos are tachyons (faster than light) then they could provide a communications channel except that neutrino detection is very difficult, not reliable, and would require massive detectors that would be an unacceptable addition to the mass of a uniship. More importantly, because neutrinos are extremely light particles their speed is not much more than the speed of light at best.

The only possible method appears to be a quantum entanglement mechanism – currently a subject of intense scientific interest. Based on current thinking about this form of quantum communication it will have the following very desirable features:

1. It is a 1:1 form of communication with no possibility of being intercepted by others.

2. It requires a small amount of power no matter what the distance.

3. It is instantaneous and thus gives direct real time communications over any distance – even trillions of light years.

If history is any guide, the development of inter-universe communications will be similar to the development of telecommunications over the past 150 years, but on a much longer development time scale. Thus we anticipate that it will begin with a primitive Morse code equivalent, and progress eventually to fast digital transmission of images and data. We anticipate bilateral switchboards initially that eventually will lead to communications with uniships beyond our universe in the Megaverse or other universes. Obviously this capability would be needed for exploration – particularly by the initial robot-driven uniships, and for communications between colonies and earth scientific or commercial reasons in other universes.

The basic mechanism will consist of a bilateral quantum entanglement setup that begins as two electrons[298] of opposite spins in a quantum state with total spin zero. Each electron is nudged into a magnetic bottle that does not affect their joint spin state.[299] One bottle is retained on earth; the other bottle is placed in a uniship. As the uniship travels the state of the electron spin within its bottle can be periodically sampled but without changing its state.[300] This can also be done on the earth based electron in its bottle. If either electron's spin is flipped the spin of the other electron flips instantaneously no matter what the distance. Thus instantaneous communication of one computer bit takes place.

Eight such bottle pairs allow us by flipping bits to exchange bytes of data. Because of the time contraction associated with much faster than light uniships the byte change must be almost instantaneous for effective communication between a uniship and earth. This fast exchange can be done by ultrafast computers.

Eventually arrays of "bottles" can transmit bytes in bulk in support of large data and image transfer. One can envision electronic switchboards eventually linking arrays of bottles to form a network with a set of uniships and/or colonies. The thought processes and designs are similar to those used in telecommunications.

It is important to note that quantum communications does not require powerful transmitters Thus quantum communications is energy efficient.

---

[298] Protons would be another reasonable alternative.

[299] Several experimental groups have recently been able to detect parts within quantum states without affecting the overall quantum state.

[300] 2012 Nobel Prize winner Serge Haroche of France developed ways of detecting the state of particles without disturbing their quantum state.

34.8.3 Experimental Support for Instantaneous Quantum Entanglement Data Transfer

One might ask if instantaneous quantum data transfer is possible. Both quantum theory and numerous experiments have shown that instantaneous data transfer via entangled pairs works at large distances.[301]

34.8.4 Interstellar Communications and SETI

If our (and others) suggestion that quantum communication is the only reasonable way for communications at large distances, then this might be the reason for SETI's failure to find communications by alien civilizations. Aliens may very well not be communicating by radio or laser waves.

It is important to note that quantum communication, as we have proposed it, is inherently private 1:1 communication with no visible manifestations for others to detect of which we are aware.

34.8.5 Quantum Entanglement and the Megaverse

In 1935 Einstein, Podolsky, and Rosen,[302] and shortly afterwards Erwin Schrödinger, began the discussion of what has become known as *quantum entanglement*. It has become a subject of growing, widespread interest since its validity has been demonstrated in numerous experiments.

Quantum entanglement can be described as a phenomenon in which a quantum state of two or more particles evolves to a point where the particles are separated by a distance such that there is no (hitherto) known mechanism by which the particles can communicate. The particles are separated by a space-like distance and thus cannot classically "communicate" at speeds at or below the speed of light. Yet a measurement of a property of one of the particles causes an instantaneous change in the other particles initially entangled with it.

Einstein's unhappiness with quantum entanglement is succinctly expressed with his often quoted description of quantum entanglement, "spooky action at a distance." Yet, since the 1930's, experiments have repeatedly shown that quantum entanglement is correct.

The bothersome issue which many physicists, starting with Einstein, are concerned is the mechanism by which instantaneous quantum effects happen. Dynamically we know that the evolution of a quantum system in the usual coordinates of our universe (whether a flat or curved space-time) yields quantum entanglement. However, a quantum measurement is outside the dynamical equations being a sharp transition between the quantum state before and after the measurement – only regulated by continuity conditions between the initial quantum state and the post-measurement quantum state.

---

[301] Matson, John, Quantum Teleportation Achieved Over Record Distances, Nature, 13 August 2012.
[302] Einstein A, Podolsky B, and Rosen N, "Can Quantum-Mechanical Description of Physical Reality Be Considered Complete?", Phys. Rev. **47**, 777 (1935).

In the case of a two particle system in an initial quantum state the invariant distance (squared) between the parts of the system when they are widely separated is

$$\Delta s^2 = \Delta t^2 - \Delta \mathbf{x}^2 \qquad (34.51)$$

with $\Delta t = 0$ at equal time. The coordinates are real-valued and governed by the transformations of the Lorentz group in the usual discussions of quantum entanglement. Thus the quandary of instantaneous changes in the parts of a system separated by a space-like distance.

We now suggest that quantum measurements should be viewed within the framework of complex-valued 4-dimensional space-time and the complex Lorentz group which we have shown in prior books leads to the form of The Standard Model of Elementary Particles (and thus quantum theory).[303]

If complex coordinates are the proper coordinates for observers then quantum entanglement is understandable. If a state is created and then parts of the state separate, measurement of part of the state at a certain time can set the other part of the state instantaneously. The invariant distance between the parts are space-like separated in real-valued space-time at equal time:

$$\Delta s^2 = -\Delta \mathbf{x}^2 = -(\Delta x^2 + \Delta y^2 + \Delta z^2) < 0 \qquad (34.52)$$

But they can be transformed by a complex Lorentz group transformation $R^i_{\ j}$

$$[\Delta \mathbf{x'}]^i = R^i_{\ j} [\Delta \mathbf{x}]^j \qquad (34.53)$$

to zero spatial distance giving

$$\Delta s'^2 = -\Delta \mathbf{x'}^2 = 0 \qquad (34.54)$$

In this new coordinate system the time difference is not zero – it is a pure imaginary number $\Delta t' = i|\Delta t'|$ thus maintaining the invariance of the full invariant interval:

$$\Delta s^2 = \Delta t^2 - \Delta \mathbf{x}^2 = \Delta s'^2 = \Delta t'^2 - \Delta \mathbf{x'}^2 \qquad (34.55)$$

---

[303] We also showed that the Wheeler-DeWitt equation of quantum gravity when generalized to complex-valued General Relativity explains many of the features seen in the universe such as the web of clusters of galaxies and the lopsided nature of the universe. Thus we conclude that the Megaverse and its universes have complex-valued coordinates. Some relevant references are Blaha (2011c), (2012a), (2012b), (2013a), (2013b), and (2014a). Much of the content of these books were introduced in earlier books by the author.

Thus, using *complex* coordinates,[304] which are truly the fundamental coordinates of space-time, all spatial distances can be mapped to zero and the apparent separation suggesting quantum entanglement at a distance is a result of the map of complex-valued coordinates to the real-valued coordinates. The complex coordinates of our universe are related to the coordinates of the complex D-dimensional Megaverse by

$$y_i = f_i(x) \tag{34.56}$$

for $i = 1, \ldots , D$ where x is a complex 4-vector in our universe and the y_i are the corresponding coordinates in the Megaverse. The Quantum observer can make a measurement in either set of coordinates in a reference frame where the spatial location of the apparently separated parts of the quantum state coincide.

Spatial separations at equal time can thus be mapped to zero for observations (measurements) in complex coordinates while the dynamical evolution of the quantum state is expressed in real-valued coordinates. These real-valued coordinates can be related to the complex coordinates of our universe or the Megaverse.

We conclude that there is no mysterious "spooky action at a distance" in quantum entanglement. The mystery is explained if we use the right coordinate system for the observer: complex-valued coordinates.

Thus all parts of a quantum state can be viewed as being located at the same spatial point in the right reference frame. Properly viewed, there is no mystery to quantum entanglement.

Returning to our subject: Despite the disappearance of mystery in quantum entanglement, it still remains an exciting topic in view of its potential applications such as communicating instantaneously at interstellar, intergalactic, and Megaverse distances which we have discussed elsewhere.

34.8.6 An Improbable Megaverse Connection to Spirits, the "Spirit World", and UFOs

Ghosts, spirits, and spiritual phenomena have been associated with "the fourth dimension" and other extra dimensions since the 1850s. UFOs are also often described as coming from or through other dimensions. Some scientists have also associated physical phenomena with other dimensions.

We do not believe these characterizations of phenomena as artifacts of other dimensions to be correct in the manner in which they are described.

However, we do think that one can simulate spiritual-like phenomena and UFO-like phenomena using the Megaverse. This section will briefly outline these possibilities. The hope is to forestall attempts to use our Megaverse theory to bolster support for these phenomena.

---

[304] As shown in our earlier books complex-valued coordinates in any coordinate system can be mapped into the real-valued physical coordinates that we experience using the Reality group. See Blaha (2011c) and other books by the author for more details.

34.8.6.1 "Spiritual" Phenomena and the Megaverse

Spiritual phenomena have several types: the appearance of visions of people or things of one sort or another; material objects passing through solid objects; unseen voices; and so on. All of these phenomena could be simulated using the Megaverse to make things appear from the Megaverse or to go from within the universe through the Megaverse and reappear in our universe again.

Some examples on Megaverse simulations are:

1. Using a "Sidestep" into the Megaverse to Circumvent Solid Obstacles in our Universe.

2. Going through a Solid One Dimensional Wall. A 1-dimensional example is

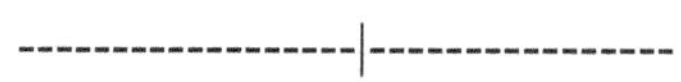

3. Persons or things going through Solid Three Dimensional Walls.

While these "partly in and partly out of the universe" phenomena can be simulated by a Megaverse agent, the agent's nature and purpose are not determinable from physical considerations of the Megaverse.

34.8.6.2 UFO Phenomena and the Megaverse

UFOs have been "seen" in the skies in many regions of the earth. They are often characterized as having high speeds, extremely high accelerations, and the ability to make abrupt changes in direction at high speeds. All of these phenomena are understandable if we are seeing objects in the Megaverse where modest changes in speed, acceleration and direction in Megaverse coordinates can map to the UFO movements that we see in our universe's coordinates.

Beyond stating these phenomena can be understood from a Megaverse perspective we can say no more. They may be real. Their purpose, if they are real, cannot be discerned. Megaverse physics cannot enlighten us on these subjects.

34.9 Traveling and Navigating in the Megaverse

I saw Eternity the other night,
Like a great ring of pure and endless light,
All calm as it was bright;
"The World" – Henry Vaughan

Upon entering the Megaverse the first uniship will want to scan for universes and simply create charts of "nearby" interesting universes rather like astronomers on earth began by charting the galaxies in the heavens and looking for novel features and phenomena. Subsequently, the first uniship, and the following uniships, will travel to universes of interest and begin their exploration. What they will encounter is anyone's guess. But we have good reasons to believe that other universes will have similar physical laws although they may have different values for physical constants such as elementary particle masses and coupling constants. The reasons are:

1. We believe the dimensionality of universes is fixed by the principles of Asynchronous Logic that make four space-time dimensions the minimal acceptable number of dimensions. (We note that higher dimensional space-times are not excluded by these principles. But Nature tends to favor extremums – a minimal number of dimensions in the present case. And Nature also tends to repeat successful designs.)

2. We have shown that complex four dimensional space-time leads to The Standard Model directly (particularly the form of the fermion mass spectrum) with an additional $SU(2){\otimes}U(1)$ that we associate with the Dark Matter sector.

3. Since all space and time measurements yield real-valued numbers we require a Reality group to map complex space-time coordinates to real-valued coordinates. The Reality group gives part of the group structure of an extended Standard Model $SU(3){\otimes}SU(2){\otimes}U(1){\otimes}SU(2){\otimes}U(1)$. Thus the form of the Standard Model (somewhat extended) is directly based on Logic and geometry.

4. Gravitation is also based on geometry.

Thus we have the form of physical theory based on geometry although the numerical constants in the theory remain to be specified.

We conclude almost every universe has a similar fundamental physics theory possibly with different physical constants. We say "almost every" because the possibility of irregular universes cannot be ruled out. Nature can choose the bizarre at times. In Blaha (2014) we describe the types of universes that can occur. In earlier books, such as those listed at the beginning of this book, we developed the geometric theory of Physics described briefly above.

We have described the necessary structure of universes realizing that they may contain unusual phenomena and may contain alien civilizations with which we can have scientific and cultural exchanges. Consequently it behooves us to explore the Megaverse to extend the range of human knowledge and possibly find reasons for establishing commerce.

34.9.1 Uniship Range and Travel Times

Ideally we would like uniships capable of traveling large distances of the order of trillions of light years in relatively short times of the order of years or, at least, decades. These wishes cannot be accomplished with current or foreseeable technology. Crews can only withstand modest accelerations up to 8g at present and probably not much more with foreseeable space medicine advances. Thus accelerating to millions of times the speed of light will take a considerable amount of time.

Given the acceleration time requirements trips between universes become lengthy – perhaps decades of earth years. On a uniship, at these very high speeds the occupants and equipment will see time progress rapidly – at a rate equal to the speed measured in units of c (the speed of light) times earth time because of relativitistic dilation. A uniship traveling at 1,000,000c will experience time increasing at 1,000,000 times earth time. An earth year thus becomes a million years of uniship time. This time dilation effect places stringent requirements on uniship equipment and requires suspended animation for the crew. We will discuss these issues in more detail later.

On the positive side, if the uniship makes a roundtrip at that speed only two earth years will have elapsed, and if they were in perfect suspended animation, the crew's body clocks will have aged only two years in sync with earth time.

The preceding discussion has ignored one major point – the energy required to accelerate to millions of c is enormous. That is why we have to use the most powerful "concentrated" energy available. We hope that energy derived in quantity from particle annihilation will be available 50,000 years hence if not much sooner. Otherwise travel into the Megaverse becomes analogous to an ant swimming across the Pacific Ocean.

The above comments cannot be viewed as encouraging. But with scientific and technological advances in the next 50,000 years there is hope. And there is good reason to begin thinking of that ultimate goal and how we may eventually reach it.

34.9.2 Uniship Baryonic D-Dimensional Observation/Seeing Techniques

If a uniship enters the Megaverse of the Megaverse there should be countless universes in view if there is a method for seeing them. However the ability to detect universes is very limited. We cannot detect gravitational waves from universes because their gravity is confined to their universes.

More importantly we cannot detect electromagnetic waves emerging from universes because they are confined to the universe within which they were created.

Consequently, the radiation from univereses that we can expect to encounter in the Megaverse is baryonic or electromagnetic radiation. There are no other known long range types of radiation emitted by universes. This creates a quandary that we can only begin to address at our present state of knowledge. Observing baryonic radiation will be a difficult task – not just because of the weakness of the baryonic field coupling constant but also for several important reasons:

1. We can't see baryonic radiation either visually or through technology at present. No detectors.

2. We cannot focus on the source(s) of baryonic radiation so as to distinguish their direction and distribution.

3. We have no mechanism to magnify the pattern of incoming baryonic radiation. No lenses, telescopes or other viewing mechanisms with "zoom" capabilities. And no technology to amplify baryonic radiation.

4. Baryonic radiation is $(D - 1)$-dimensional. We would have to be able to create 3-dimensional hologram projections that provide an intelligible view. The holographic images would have to be manipulated to enable navigation.

5. Baryonic radiation has $(D - 2)$ polarizations which would provide information on the nature of universes. The detection and analysis of these polarizations is currently beyond our capabilities.

Our inability to detect gravity waves (except in recent studies of the Big Bang – BICEP2) shows the difficulty of detecting and analyzing baryonic radiation.

34.9..2.1 Relativistic Effects on Radiation

The baryonic view of the Megaverse that a uniship crew "sees", when the uniship is traveling faster than the speed of light, is very different from its view when traveling at low speeds of a few tens of miles per second.

An observer on a uniship traveling at a relativistic speed near, but below, the speed of light will detect universes with baryonic or electromagnetic radiation compressed to within a cone in the frontal direction of the uniship (Fig.34.18). The baryonic radiation cone becomes narrower as the speed of light is approached due to aberration and in the limit as the speed approaches the speed of light becomes a point directly ahead of the uniship.

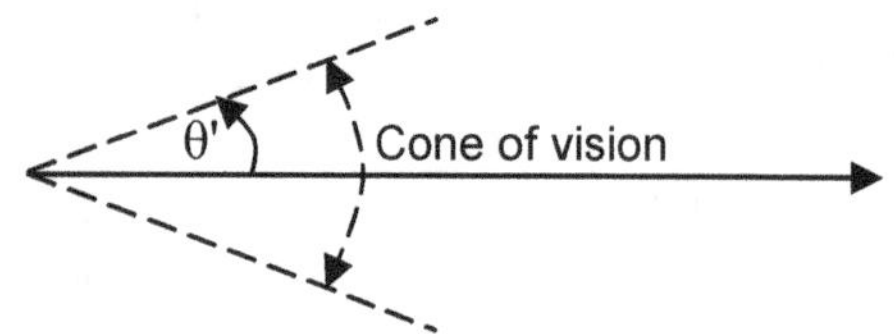

Figure 34.18. Baryonic (and electromagnetic) radiation cone of visibility around direction of uniship motion in the uniship coordinate system with the angle θ' determined by eq. 34.57 for sublight uniship speeds.

The relativistic equation for baryonic radiation aberration is

$$\cos \theta' = (\cos \theta + \beta)/(1 + \beta \cos \theta) \tag{34.57}$$

where θ is the angle of a universe relative to the uniship's direction of motion as measured in the Megaverse coordinate system and θ' is the angle of a universe relative to the uniship's direction of motion as measured in the uniship's coordinate system.

The inverse relation is

$$\cos \theta = (\cos \theta' - \beta)/(1 - \beta \cos \theta') \tag{34.58}$$

34.9.2.1.1 SUBLIGHT CASE: B < 1

As $\beta \rightarrow 1$ (the speed of light) eq. 20.1 indicates $\theta' \rightarrow 0°$ showing the entire view of the multiuniverse baryonic radiation is compressed to the forward direction. Fig. 34.18 shows the cone of visibility for a uniship traveling near the speed of light at perhaps .6c - .9c. The cone angle θ' satisfies

$$\cos \theta' > \beta \tag{34.59}$$

The rest of the field of view of the uniship is total baryonic radiation blackness except the point in the directly rearward direction ($\theta' = 180°$) for any object at $\theta = 180°$.

34.9.2.1.2 SUPERLUMINAL CASE: B > 1

For $\beta > 1$ eqns. 7.1 and 7.2 still hold and there is a cone of baryonic radiation visibility similar to that depicted in Fig. 34.18. However the cone angle θ' for superluminal speeds, $\beta > 1$, satisfies the relation

$$\cos \theta' > 1/\beta \tag{34.60}$$

The rest of the field of view of the uniship is total baryonic radiation blackness, as in the sublight speed case, except the point in the directly rearward direction ($\theta' = 180°$). We note that as β gets very large the cone of visibility becomes larger. At β = ∞ the cone of visibility becomes the angular region between θ' = 0° and θ' = 90° (the forward hemisphere).

34.9.2.1.3 SUPERLUMINAL UNISHIP VISIBILITY

As a result, "visual" baryonic radiation navigation at high superluminal speeds becomes difficult unless we develop an electronic imaging system that "undoes" the effects of aberration and enables "visual" baryonic radiation navigation.

A further problem is the location of a destination. If we send a uniship to a far universe we have to project the location of the universe at the time the uniship arrives based on the universe's current motion. If the motion of the universe is modified by forces exerted by nearby universes as baryonic radiation from the destination universe travels to the uniship, or if the universe's motion is not accurately determined, a uniship could arrive at a point that is still some distance from the destination universe. Thus navigation to a far universe is a significant issue.

34.9.2.1.4 EFFECT OF DOPPLER SHIFT AT SUPERLUMINAL SPEEDS

A uniship traveling at relativistic sublight speeds will see universes having their baryonic radiation spectrum (color) changed significantly due to the Doppler Shift effect. At superluminal speeds the Doppler Shift will also change the "colors" of objects "seen" by the uniship.

This issue is again surmountable if we use electronic imaging techniques to "undo" the Doppler shift and thus display universes with their true baryonic radiation spectrum in the uniship's reference frame.

The relativistic Doppler shift for sublight speeds of a baryonic radiation wave of frequency ν is given by

$$\nu = \nu_0 (1 - \beta^2)^{1/2} / (1 - \beta \cos \theta') \tag{34.61}$$

where ν_0 is the frequency of the baryonic radiation emitted by the source universe and θ' is the angle of the source relative to the uniship's velocity (Fig. 20.2).

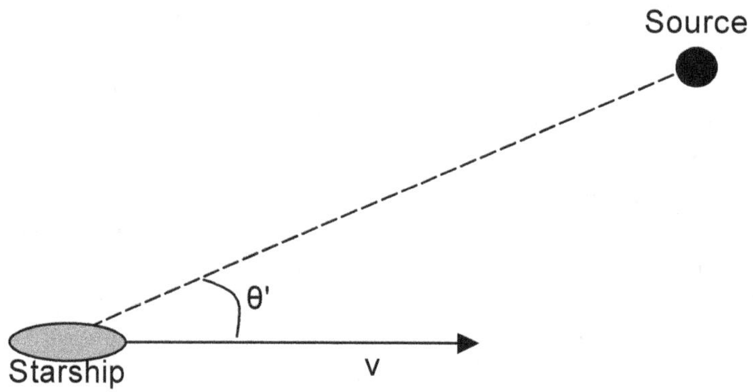

Figure 34.19. The angle of a universe θ' with respect to the uniship's velocity v.

The Doppler shift for superluminal speeds is

$$v = v_0(\beta^2 - 1)^{\frac{1}{2}}/(\beta \cos \theta' - 1) \qquad (34.62)$$

This can be seen by considering a baryonic plane wave, which is a combination of

$$\cos[(k\cdot x - vt)/2\pi] \qquad \text{and} \qquad \sin[(k\cdot x - vt)/2\pi] \qquad (34.63)$$

Upon transforming from the uniship coordinate system, for example, to a coordinate system moving in the "x-direction" at a speed faster than light, both the energy v' (up to a constant) and the time t' obtain a factor of i (that cancel each other) so eq. 34.62 is the correct frequency in the superluminal (faster than light) frame. The sign of the frequency is always positive by convention due to the form of baryonic waves and eq. 34.60 dictates the form of the denominator in eq. 34.62.

For large $\beta \gg 1$ eq. 34.62 becomes approximately

$$v \approx v_0/\cos \theta' \qquad (34.64)$$

In the forward direction $\theta' = 0$ the Doppler shift goes to zero. Due to eq. 34.60 the maximum value of the Doppler shift for large β in the field of vision is

$$v \approx \beta v_0 \qquad (34.65)$$

So the "wide" angle baryonic waves are shifted to large frequency.

Eq.34.62, and the discussion that follows, suggest that frequency shifts will be substantial for extremely fast uniships. The result will be a distorted view of the Megaverse.

However electronic imaging techniques can again be implemented to restore the "correct" view of the baryonic radiation. The combined effects of aberration and the Doppler shift on the view of the Megaverse from the uniship can be electronically corrected to give a "normal" view of the Megaverse. In addition a projection system, probably based on holograms, is required to transform $(D - 1)$-dimensional views of the Megaverse into sets of 3-dimensional depictions of parts of the $(D - 1)$-dimensional view.

34.9.3 Uniship Navigation

Navigating on earth and in space is often a difficult task. First one must know where one is and then one must know where the destination is, and how to get there. In the Megaverse all three items are challenging to discern. For we are in a D-dimensional space where our intuition, based as it is on three dimensional space, fails. We are thus at the mercy of technology to detect these three things with only electronic baryonic eyes to see and guide our motion.

Baryon detectors on board detect other universes through their baryonic radiation just as we detect stars and galaxies by their electromagnetic radiation currently. Within the next 50,000 years we anticipate that baryonic radiation optics will develop and mature to the point where it can provide visual capabilities similar to electromagnetic light that we use at present. Then we can develop universe maps for the Megaverse just as we have star maps currently.

An important issue is the ability to distinguish anti-matter universes from universes dominated by matter such as our universe. We do not want uniships to enter anti-universes unless we can properly shield them from disintegrating under particle-antiparticle interactions.

The navigation system, using 3-dimensional views (holograms) obtained from $(D - 1)$-dimensional pictures of the Megaverse, can then plot courses to universes of interest for exploration.

The courses selected then direct the $(D - 1)$-directional thrust system to execute the correct combination of accelerations to travel to the selected universe.

Most of the technology that we have discussed remains to be created. However the rapid progress of technology, if it continues, would seem to be able to provide the needed components in the future.

34.9.4 Uniship Exit from a Universe

We have seen that only baryonic radiation can penetrate a universe horizon. We must now amend that conclusion to consider the exit or entry of uniships from a universe. In the slingshot mechanism for exiting, and the other possible exit methods that we considered, a key role was played by the baryonic interaction in all cases. The role is easily visualized by considering the example of escaping from 2-dimensional flatland using magnetic force.

However in the current situation (4-dimensional object exiting into a D-dimensional space) we have to inquire into the nature and results of the exit. It is clear that the escape of a uniship must be view as a bubble (the 4-dimensional universe) subdividing into two 4-dimensional bubbles in (D − 1)-dimensional space in a continuous process rather like biological cell division. One bubble is of course the universe. The other bubble is the uniship which is also a surface from the viewpoint of the Megaverse. Fig. 34.20 symbolically depicts an intermediate stage in the subdivision process.

Figure 34.20. Symbolic depiction of an intermediate stage in the subdivision representing the exit of a uniship from a universe. The uniship (small circle) emerges from the universe (large circle) into the Megaverse. (Not drawn to scale.)

We thus see that a uniship is a 4-dimensional entity but with D - 1 thrust ports extruding into (D − 1)-dimensional space. (The thrust ports were generated by tidal effects in the slingshot process.) Thus it is a hybrid with 4-dimensional and D-dimensional parts. The crew, the equipment and the fuel modules will be in 3-dimensional space. The combustion chamber and thrust ports and viewing mechanism for navigation will be in (D − 1)-dimensional space. The combined parts will constitute the uniship. (We note that the 3-dimensional parts are in a subspace of (D – 1)-dimensional space just as a piece of paper is a 2-dimensional object in 3-dimensional space.) Therefore there are no connection problems between the 3-dimensional and (D – 1)-dimensional parts of the uniship.

34.9.5 Uniship Entry into a Universe

We now consider the entry of a uniship into a universe, which we will assume is 4-dimensional like our universe for reasons given earlier. The uniship has both 3-dimensional and (D – 1)-dimensional spatial parts. If we position the uniship 3-dimensional part within the universe and retract or rotate the (D – 1)-dimensional part into the 3-dimensions of the universe, then the uniship will be entirely within the universe giving a successful entry.

Retracting (D – 1)-dimensional parts is easy to visualize but requires a mechanism for retraction in (D – 4) of the (D – 1) dimensions. This mechanism can be a simple retraction mechanism but must exist in each of the dimensions outside the universe being entered. Rotating thrust exhausts also is easy to visualize but, again, the rotation device must be in the (D – 4) external dimensions and rotate into the three spatial dimensions of the universe being entered. In both cases we are faced with Cheshire cat situations: after the retraction or rotation the (D – 4)-dimensional devices that perform these chores will still exist and thus not be fully within the universe.

Whether the uniship can explore a new universe, being partly outside it, is an open question. This author feels that the uniship will be able to successfully navigate in the new universe rather like a shark swims the sea with its fin sticking out of the water. The (D – 4)-dimensional "fin" will add to the mass of the uniship but will not be affected by gravitation because it is out of the universe.

34.10 Issues for Life on a Megaverse Uniship

Life on a uniship that travels to universes is similar to life on a starship that travels to stars and galaxies except it has significantly more stringent requirements. Travel within the universe is measured in light years ranging to millions of light years. Travel within the Megaverse is most likely in hundreds of billions to trillions of light years in starship time. This results in major large requirements for fuel, materials strength and longevity, and in human suspended animation time, among other requirements.

Starship requirements are described in Blaha (2013a), 2014b) and (2014c). In this chapter we will describe some uniship requirements. The following chapters describe other uniship requirements.

34.10.1 Long Distance Starship Requirements for Travel to Far Stars and Galaxies

If we wish to travel to long distances – up to trillions of light years eventually, then critical advances are necessary that could take up to 50,000 years.

We see the uniship effort as a long term exploration and colonization program in an ever widening ring around our universe. In this chapter we discuss many of the advances that would be needed.

A major problem of uniships is the rapid progress of time on a much faster than light uniship. If a uniship has a speed that is much faster than the speed of light, then the progress of time in the uniship is much faster than the progress of time on earth.[305] For example if the uniship is traveling 5,000,000 times the speed of light, then the increase in time on the uniship is 5,000,000 times the increase in time on earth. In an interval of one year of earth time, 5,000,000 years will have passed on the uniship.

The extraordinarily fast passage of time on a very fast uniship requires materials, equipment and engines to continue to work effectively for long periods of uniship time which is just as real on a uniship as earth time is real on earth. One cannot avoid the fact that the distance traveled by a uniship measured in light years is equal to the time of flight to cover that distance measured in years.[306]

Thus we come to the first important long distance uniship requirement – very long lifetime equipment and uniship superstructure. Other requirements follow in this chapter.

34.10.2 Long-Lived Materials

The materials that we use today to build large vehicles such as oil tankers, submarines and aircraft carriers are meant to last up to, at most, a century and often much less. Many of these materials age, deteriorate, rust, migrate within computer chips over time, and actually slowly flow like a liquid in many cases.

Not many materials keep their original characteristics over long periods of time. In the past fifty years there has been much progress in developing new harder, stronger and age resistant materials and metals. But uniships requirements are extraordinarily larger.

Uniships, in which time moves quickly so that thousands and perhaps millions of years of uniship time elapse, must be composed of materials with a very long stable lifetime. An important part of the R&D for a uniship is the development and use of age tolerant materials. The examination of materials used hundreds of thousands of years ago such as tools and dwellings shows the ravages of age. A uniship should have an initial goal of tens of millions of years of stability without aging. Ultimately one would hope that uniships that don't age in trillions of years could be built to travel to other universes.

These design requirements are far ahead of current technology.

---

[305] See p. 15 of Blaha (2011c): *All The Universe.*

[306] Neglecting the time required to accelerate the starship at the beginning and the time required to decelerate back to a "normal" speed of a few miles per second.

34.10.3 Long-Lived Machinery and Electronics

If one has materials that preserve their composition, shape and performance characteristics over millions of years or more, then one can construct machinery and electronic gear such as computers that can last similar periods of time. Long-lived machinery and electronic gear then can be used when a uniship travels the Megaverse.

Long distance uniships need materials and machinery that last "nearly forever" – exactly the opposite of the intent of Earth industries.

34.10.4 Long Shelf Life Nuclear Reactors and Nuclear Shuttles

Several types of nuclear reactors are required for a uniship:

1. A continuous running reactor that can run for up to hundreds of millions of years to provide power to a uniship in flight to a distant location. This reactor may be a low power reactor. It should have a very long lifetime. That this is possible is suggested by the natural nuclear reactor that ran for millions of years about a billion years ago in the Congo.[307]
2. Long shelf life reactors that are not activated until a uniship destination is reached. These would power the uniship inside universes and their solar systems, and nuclear shuttles for travel and landings within a solar system.

34.10.5 Suspended Animation for Long Trips

It is necessary to have suspended animation available for crews on uniship journeys. With suspended animation a crew could go on a journey lasting hundreds of millions of years, or more, of uniship time, and, upon return to earth, have aged physiologically only a short time while out of suspended animation exploring distant universes and their star systems. The round trip travel time will not have aged them. When they return to earth they may be some months older, but their families and friends (having aged only by the earth travel time) will still be roughly contemporary with them.

A mechanism for long term suspended animation is thus a major requirement. Any suspended animation mechanism must take account of three important facts: 1) suspended animation must reduce human body temperatures to a low value to "halt" life processes and bodily decay; 2) lowering body temperatures will cause cells to rupture due to the expansion of water upon freezing; 3) the entry into suspended animation and the reentry to a normal bodily state must be rapid and uniform throughout the body.[308]

---

[307] The author suspects that the vast diversity of life in Africa may in part be due to genetic changes caused by this natural reactor. The development of mammalian life may in part also be due to this reactor and the radiation from its waste products and radioactive deposits in the Congo region over the millennia.
[308] One cannot "unfreeze" part of a human body and have the rest still frozen.

A mechanism to achieve these goals is not presently known. The current approaches to suspended animation (which all include lowering body temperature) are:

1. Replacing part or all of the blood in an organism with an "antifreeze" solution that will prevent cells and body tissue from bursting when the temperature is lowered. Revival takes place by raising the temperature of the organism while returning blood to the organism's circulatory system. This approach has been successfully applied to dogs that have been put into suspended animation for three hours. Unfortunately some of the dogs had nerve and coordination problems after revival.[309]

2. An organism can have a chemical injected or absorb a chemical while breathing that will counteract the tendency of water to expand when body temperature is lowered and/or lower the metabolic rate of the organism.

3. NASA and other groups have studied the possibility of placing humans into hibernation. Since hibernating organisms do age – perhaps more slowly – this approach is not true suspended animation.

4. A combination of electromagnetic "vibration" of a body having an innocuous chemical dispersed in the body (while awake or in suspended animation) might allow bodily temperatures to be lowered to a stable "frozen" state without cell rupturing. Turning off the electromagnetic vibration combined with a revival jolt might be an effective way to exit suspended animation procedure.

34.10.6 Robotic Driven Uniships

The initial uniship flights could be manned by robots rather than humans. This approach would be useful to test uniships, and their components, without endangering a crew. The robot guidance systems would, of course, have to be constructed of long-lived components. If it is successful then a robotic trip would also help demonstrate the long term reliability of long-lived computer equipment.

Robotic flights would be especially useful if a method of rapid faster than light, or instantaneous, communication between the uniship and earth existed.

34.10.7 Long-Life Computer Chips

Computers hardened for battle and bad weather conditions currently exist. A long distance uniship would require computers with working lifetimes of between thousands and hundreds of millions of years. In time periods of these lengths computer chips would be subject to aging processes such as the intermixing of the metals composing the various chips of the

---

[309] At the University of Pittsburgh's Safar Center for Resuscitation Research.

computer and the aging of the wiring of the computer. Since new materials of greater strength and other superior properties are being discovered fairly frequently one can hope that the required types of metals and materials will eventually be found.

34.10.8 Space Dust

The effect of dust and gas molecules in space on uniships are of great importance. These effect should be detectable in "short" distance uniship voyages. If it is important, as it seems to be, then the design of shielding for long distance uniships should incorporate appropriate "armor" to protect the uniship and crew.

34.10.9 Length Dilation Effect

Lengths on a uniship, traveling at high speed much greater than the speed of light, are significantly dilated. A length measured on a uniship will appear to be larger to an observer on earth by a factor of the speed measured in terms of the speed of light than the length on earth. For example if a uniship is moving at 5,000 times the speed of light then a 2 meter long stick on the uniship would appear to be 10,000 meters long to an earth observer.[310]

Does this length dilation phenomenon affect the contents of the speeding uniship? No. It is an illusion that the earth observer "sees." An occupant of the uniship would not notice a difference and would see the stick as still two meters in length. Due to length dilation, a starship traveling at speeds much beyond the speed of light, would appear to be enormously long.

34.10.10 QFT Acceleration Thermal Vacuum Heating – Particle Baths

Recently questions have been raised about spaceships accelerating at a high rate in 'empty' space. Based on an analysis of Unruh and others[311] it has been suggested that the ship would see an incoming wave of particles and a 'heated' vacuum due to its motion. The basis of this claim and the work of Unruh (and others) is standard quantum field theory, which in turn is based on Special Relativity.

Since accelerating reference frame transformations and other exotic reference frame transformations are outside the framework of Special Relativity we suggest that this spaceship phenomena merely reflects a flaw in the choice of second quantization. Earlier we described a more general formulation of quantum field theory in which particle states are unchanged under transformations between accelerating reference systems as well as under transformations to exotic coordinate systems. Thus the particle 'baths' suggested by conventional quantum field theory do not take place for accelerating spaceships nor does the vacuum 'heat up' and spew particles.

---

[310] This discussion assumes that the stick and the starship motion are parallel. For a detailed discussion of this length contraction phenomena see Blaha (2011c) p. 17.

[311] See the relevant references in Appendix A.

34.11 Uniship Particle Annihilation Drive

This section (from Blaha (2011a)) epitomizes the possibility that the CERN LHC can be used as a *test instrument for the development* of a starship ion drive capable of travel to the stars (after the LHC is retired from its investigation of elementary particle phenomena).

We have shown in I that quarks and gluons have complex-valued velocities that enable them to travel faster-than-light (tachyonic), which we showed in I enables quarks to generate more massive particles after collisions than the initial particles. Recently, new experiments (announced in April, 2017[312]) at the LHC have shown that ultra-high energy proton-proton collisions have producd an abundance of particles similar to that produced in collisions of atomic nuclei. Thus initial quarks in protons effectively 'multiply' to have atomic nuclei characteristics – a feature we attribute to their tachyonic nature.

The tachyonic attributes that seem to explain these new results suggest that a faster-than-light ion starship engine could be created. CERN LHC seems able to do design studies to aid in the development of this new type of starship engine at now accessible energies.

34.12 Voyages into the Megaverse

Traveling in in the Megaverse between universes places extraordinary demands on uniships. Speed, acceleration, and fuel requirements are the primary issues because the distance between universes is vast and human lifetimes are short. We should like to be able to travel fairly quickly between universes. If one wishes colonization, commerce, and timely exploration then the trip between two universes, on average, should be perhaps six months to a year of earth time. (Time on a uniship proceeds much, much more rapidly than earth time. But we can circumvent this potential problem using suspended animation for passengers and very long-live machinery for the uniship and its contents.)

In this chapter we will consider Uniship distance and speed related requirements for "short" distance travel in the Megaverse.

34.12.1 Starship Distance and Speed Requirements in Our Universe

In our book *All the Universe!* (and in earlier books) we developed the theory of faster-than-light starships for travel between stars and galaxies in our universe. For the reader's convenience we reproduce part of *All the Universe!* It appears that speeds of 60,000c give acceptable travel times (up to a year) within our galaxy, and *speeds of a few million c give acceptable travel times to nearby universes where c denotes the speed of light.*

At 3,000,000,000c a universe that is 2,000,000,000 light years away could be reached in eight months (neglecting acceleration and deceleration times) – an acceptable time for trade, exploration and possibly colonization. The fundamental problem is to develop an energy source that can fuel such enormous speeds – a reason for anticipating a 50,000 years of necessary development time.

---

[312] ALICE Collaboration, reported in the journal Nature Physics (April, 2017).

Destination	Distance (ly)	Approximate Travel Time (years)
To the other end of the Milky Way Galaxy	100,000	3
To the Center of the Milky Way	30,000	1
Large Magellenic Galaxy	150,000	5
Small Magellenic Galaxy	200,000	7
Andromeda Galaxy	2,000,000	70

Figure 34.21. "Coasting" part of travel time to various destinations at a real velocity of 30,000c.

34.12.2 Uniship Distance and Speed Requirements for "Short" Distances in the Megaverse

In this section we will consider some issues associated with uniship "short distance" trips in the Megaverse such as a three trillion light year trip to a nearby universe.[313] The primary issue is the energy (and fuel) required to make a trip at high speed so that the travel time is of the order of months. Another important issue is the acceleration times that are required to attain high speeds.

34.12.2.1 Energy Required to Attain High Velocities

The energy E required to reach a high velocity much greater than the speed of light is not a simple mathematical expression in the speed v because the energy of an object moving much faster than the speed of light approaches zero. (The momentum approaches mc where m is the mass of the object. Thus $E^2 - c^2p^2 = -m^2c^4$ for $v > c$. The object is tachyonic.)

We will calculate the energy required to reach a speed $v > c$ in the earth's rest frame using the development presented earlier in this book. The energy E expended in the spatial x interval from x_0 to x is defined as

$$E = \int_{x_0}^{x} F dx = \int_{t_0}^{t} F v dt \qquad (34.66)$$

where t_0 is the initial time and t is the final time. From eq. 34.66 we obtain

$$E = \int_{t_0}^{t} F v dt = \int_{t_0}^{t} g \gamma v dt \qquad (34.67)$$

---

[313] We will use the results found earlier for acceleration and velocity in the x direction. The general case follows directly from this special case.

Eq. 34.67 can be transformed to the form:

$$E = (m/2) \int_{v_0}^{v} dv \; \gamma^{-1} \, d[(\gamma v)^2]/dv \qquad (34.68)$$

The exact form of the integral is:

$$E = [m\gamma(v^2 + c^2)/2 + mc^2/(2\gamma)] - [m\gamma_0(v_0^2 + c^2)/2 + mc^2/(2\gamma_0)] \qquad (34.69)$$

where $\gamma_0 = (1 - v_0^2/c^2)^{-1/2}$. Since the velocity is in general complex-valued, v and γ are also complex-valued as is E.

Using the Reality group that we introduced in previous books, the physical velocity is the absolute value of v, |v|, and the physical value of E is |E|. |E| is calculated by first substituting the complex values of v, γ, v_0 and γ_0 in eq. 25.2 and then taking the absolute value of the resulting complex quantity E.

We begin by defining

$$E_{part}(v) = m\gamma(v^2 + c^2)/2 + mc^2/(2\gamma) \qquad (34.70)$$

The integral in eq. 34.67, although superficially real-valued has complex quantities in the integrand. The result of the integration from a speed below the speed of light to a speed greater than the speed of light can be written as

$$E_{tot} = E_{part}(c - i\varepsilon) - E_{part}(v_0) + E_{part}(v) - E_{part}(c + i\varepsilon) \qquad (34.71)$$

as $\varepsilon \to 0$ due to the singularity at v = c. Since we are interested in the energy required to reach ultra-high speeds of the order of tens of thousands to trillions of times the speed of light, we see that

$$E_{tot} \to \lim_{v \to \infty} E_{part}(v) = -imcv/2 \qquad (34.72)$$

to leading order in v.   Thus

$$|E_{tot}| \to mc|v|/2 \qquad (34.73)$$

or, expressed in a more convenient way:

$$|E_{tot}| \to \tfrac{1}{2} E_{rest} |v|/c \qquad (34.74)$$

as $v \to \infty$ in the earth's reference frame where $E_{rest} = mc^2$ is the rest mass-energy of the entire ship including the fuel. Knowing the desired cruising speed v of a starship or uniship the ship must have an energy supply equal to $2|E_{tot}| = mc|v|$ when it leaves the earth's reference frame to provide for acceleration *and deceleration* plus the additional energy needed for other ship functions. For a round trip an energy supply of $4|E_{tot}| = 2mc|v|$ would be needed for the ship's engines.

34.12.2.2 Acceleration Time Required to Reach Extremely High Speed

We can determine the acceleration time required to reach velocity v from eq. 34.69 of the previous section:

$$v = c\{1 - 2/(1 + ((c + v_0)/(c - v_0))\exp[2g(t - t_0)/(mc)])\} \qquad (34.75)$$

Inverting eq. 34.75 above yields the acceleration time interval required to achieve a speed v in the earth's reference frame:

$$t(v) - t_0 = (mc/2g)\ln\{[(c - v_0)/(c + v_0)]\,[(c + v)/(c - v)]\} \qquad (34.76)$$

where v_0 is the speed at t_0. For large v in the earth reference frame the time interval approaches

$$|t(v) - t_0| \rightarrow (mc/2g)\ln[(c - v_0)/(c + v_0)] + mc^2/(gv) \qquad (34.77)$$

to leading order in v. As $v \rightarrow \infty$ the interval becomes a constant:

$$|t(v) - t_0| \rightarrow (mc/2g)\ln[(c - v_0)/(c + v_0)] \qquad (34.78)$$

The reason for this limit on the earth reference frame time interval can be understood when one realizes that the corresponding ship time interval approaches infinity. Ship time increases much faster at high speeds greater than $\sqrt{2}c$. We shall see this in the next subsection.

34.12.2.3 Speed, Distance and Acceleration Time in the Starship/Uniship Reference Frame

In this subsection we will calculate dynamical quantities from the point of view of the starship/uniship reference frame.

In case examined in the previous section we found the acceleration in the earth reference frame to be $\gamma g/m$ by eq. 34.68. In the ship reference frame the acceleration is g/m.

The transformation law from the earth reference frame (unprimed coordinates) to the starship/uniship reference frame (primed coordinates) in which the ship is moving at instantaneous speed v in the x direction is

$$\begin{aligned}
t' &= \gamma(t - \beta x/c) \\
x' &= \gamma(x - \beta ct) \\
y' &= y \\
z' &= z
\end{aligned} \qquad (34.79)$$

Substituting

$$x = x_0 + (mc^2/g)\ln[(1 - v_0/c)/(1 - v/c)] - c(t - t_0) \qquad (34.80)$$

from the previous section and

$$t(v) - t_0 = (mc/2g)\ln\{[(c - v_0)/(c + v_0)]\,[(c + v)/(c - v)]\} \qquad (34.81)$$

from this section with $t_0 = 0$ and $x_0 = 0$, we obtain the ship time to be

$$t' = \gamma\{(m(c - v)/2g)\ln\{[(c - v_0)/(c + v_0)]\,[(c + v)/(c - v)]\} - (mv/g)\ln[(c - v_0)/(c - v)]\}$$
$$(34.82)$$

Eq. 34.82 gives starship time as a function of speed. Thus the faster a ship speeds the more quickly time passes on the ship. As a result ships should have long lifetime equipment and place passengers in suspended animation for long periods of time.

For large v, the absolute value of the ship time approaches

$$|t'| \rightarrow |mc/g)\ln v| \qquad (34.83)$$

Thus the ship time approaches infinity as the ship speed approaches infinity. This situation explains the limit on earth time in eq. 34.78. It corresponds to ship time approaching infinity. (The ship speed approaches infinity in this limit as well.) Of course the limit is never reached because it would require an infinite amount of thrust and fuel. However it allows very, very large velocities to be reached making starship and uniship travel in reasonable time frames possible.

If the conditions in the above section hold, then it is possible to reach "infinite" speed in a finite time in the earth's reference frame. To reach this limiting speed will require an infinite amount of starship time, and fuel.

35. Unified Super Standard Model in the Megaverse

Our view of the Megaverse consists of a vast number of universes in a much vaster space. The space consists of a very low density of radiation, particles and dust (perhaps accreted into bodies of matter.) Our models[314] suggest that the Megaverse mass-energy density between universes is perhaps 10^{-30} of the average density in our universe.

35.1 Particle Interactions in the Space Between Universes

We assume the interactions between particles in the Megaverse space between universes are described by the Unified SuperStandard Model, suitably generalized to D dimensions, that was presented earlier in this book. We further assume, based on continuity considerations, that coupling constants, masses and other parameters of the Unified Standard Model have the same values in the Megaverse as they do in our universe.

Fermion and boson free fields would have fourier expansions in D-dimensional Megaverse coordinates. Using Two-Tier Megaverse coordinates all Megaverse perturbation theory calculations are convergent. There are no infinities in Two-Tier calculations irrespective of the space-time dimension.

35.2 Example of a Megaverse Baryonic Gauge Field – The Planckton Field

The conservation of baryon number has been repeatedly investigated by experimenters and found to be true to extremely high accuracy. For decades theorists have suggested that the conservation law follows from the existence of a gauge field in a manner much like electric charge conservation follows from the properties of the electromagnetic abelian gauge field.[315]

We will therefore assume a baryonic gauge field exists that is similar to the electromagnetic field except for features due to its definition and existence in the D-dimensional Megaverse. This field will couple extremely weakly to individual baryons as well as universe particles with non-zero baryon number. We will call the baryonic gauge field particle a *planckton*. Its electromagnetic analogue is the photon. We described Megaverse baryonic gauge field quantization in Blaha (2014a), (2015a) and earlier books.

---

[314] See chapters 11 – 14 of Blaha (2017c).
[315] See Gell-Mann, M. and Levy, M. *Nuovo Cimento* 16, 705 (1960) for a proof.

Plancktons propagate in the Megaverse, both within universes, and in the Megaverse external to universes. So we will define the planckton field in D-dimensional Megaverse coordinates. They will interact with baryons within a universe with Megaverse coordinates mapped to the curvilinear coordinates in the universe.

Since a planckton field in D-dimensional quantization in conventional coordinates would lead to divergences we will use quantum coordinates:

$$Y^i(y) = y^i + i\, Y_u^{\ i}(y)/M_u^{D/2} \tag{35.1}$$

with quantum coordinate derivatives defined by

$$\partial_i = \partial/\partial Y^i(y) = \partial/\partial(y^i - Y_u^{\ i}(y)/M_u^{D/2}) \tag{35.2}$$

to obtain a completely finite theory of planckton interactions with elementary particles and later with universe particles.

Plancktons, fermion particle fields, gravitation, other gauge fields, Higgs fields, and the $Y_u^{\ i}(y)$ field of quantum coordinates are the only fields in the space between universes in the Megaverse. Universe.

35.3 Planckton Second Quantization

We begin by noting that Megaverse quantum coordinates are defined by eqs. 35.1 and 35.2 above. The lagrangian density terms for the free planckton $B_u^{\ i}(Y(y))$ fields is

$$\mathcal{L}_{Bu} = -\tfrac{1}{4}\, F_{Bu}^{\ \mu\nu}(Y(y))F_{Bu\mu\nu}(Y(y)) \tag{35.3}$$

with $Y(y)$ given by eq. 35.1 and the free planckton lagrangian is

$$L_{Bu} = \int d^{D-1}y\, \mathcal{L}_{Bu}(Y(y)) \tag{35.4}$$

with

$$F_{Bu\mu\nu} = \partial B_{u\mu}(Y(y))/\partial Y^\nu(y) - \partial B_{u\nu}(Y(y))/\partial Y^\mu(y) \tag{35.5}$$

where the values of μ and ν range from 1 to D.

The equal time commutation relations, derived in the usual way, are:

$$[B_u^{\ \mu}(Y(y, y^0)), B_u^{\ \nu}(Y(y', y^0))] = [\pi_u^{\ \mu}(Y(y, y^0)), \pi_u^{\ \nu}(Y(y', y^0))] = 0 \tag{35.4}$$

$$[\pi_{uj}(Y(y, y^0)), B_{uk}(Y(y', y^0))] = -i\, \delta^{D-1\ tr}_{jk}(Y(y,0) - Y(y',0)) \tag{35.5}$$

where

$$\pi_u^{\ k} = \partial \mathcal{L}_{Bu}(B_u(Y(y)))/\partial B_{uk}'(Y(y)) \tag{35.6}$$

$$\pi_u{}^0 = 0 \tag{35.7}$$

and

$$\delta^{D-1\,tr}{}_{jk}(\mathbf{y} - \mathbf{y}') = \int d^{D-1}k\; e^{i\,\mathbf{k}\bullet(Y(y,0) - Y(y',0))}\,(\delta_{jk} - k_j k_k/\mathbf{k}^2)/(2\pi)^{D-1} \tag{35.8}$$

$$B_{uk}{}'(Y(y)) = \partial B_{uk}(Y(y))/\partial y^D \tag{35.9}$$

for $j, k = 1, 2, \ldots, D - 1$.

　　If we choose the Coulomb gauge for $B_{uk}(Y(y))$:

$$B_u{}^{16}(Y(y)) = 0$$
$$\partial B_u{}^{j}(Y(y))/\partial Y^j(y) = 0$$

for $j = 1, 2, \ldots, D-1$ then D - 2 degrees of freedom are present in the vector potential.[316] The Fourier expansion of the vector potential $B_u{}^i(Y(y))$ is:[317]

$$B_u{}^i(Y(y)) = \int d^{D-1}k\, N_{0B}(k) \sum_{\lambda=1}^{D-2} \varepsilon^i(k, \lambda)[a_B(k,\lambda) :e^{-ik\cdot Y(y)}: + a_B{}^\dagger(k,\lambda) :e^{ik\cdot Y(y)}:] \tag{35.10}$$

for $i = 1, \ldots, D - 1$ where

$$N_{0B}(k) = [(2\pi)^{D-1}\, 2\omega_k]^{-\frac{1}{2}} \tag{35.11}$$

and (since the field is massless)

$$k^D = \omega_k = (\mathbf{k}^2)^{\frac{1}{2}} \tag{35.12}$$

where k^D is the energy, and where the $\varepsilon^i(k, \lambda)$ are the polarization unit vectors for $\lambda = 1, \ldots, D - 2$ and $k^\mu k_\mu = (k^D)^2 - \mathbf{k}^2 = 0$.

　　The commutation relations of the Fourier coefficient operators are:

$$[a_B(k,\lambda), a_B{}^\dagger(k',\lambda')] = \delta_{\lambda\lambda'}\,\delta^{D-1}(\mathbf{k} - \mathbf{k}') \tag{35.13}$$
$$[a_B{}^\dagger(k,\lambda), a_B{}^\dagger(k',\lambda')] = [a_B(k,\lambda), a_B(k',\lambda')] = 0 \tag{35.14}$$

and the polarization vectors satisfy

$$\sum_{\lambda=1}^{D-2} \varepsilon_i(k, \lambda)\varepsilon_j(k, \lambda) = (\delta_{ij} - k_i k_j/\mathbf{k}^2) \tag{35.15}$$

The $B_u{}^\mu$ Feynman propagator is

____

[316] Note we use the Coulomb gauge for Y(y) also.
[317] Normal ordering of the exponentials is required.to avoid spurious infinities. It is analogous to the normal ordering seen in perturbation theory. See Blaha (2005a) for additional details.

$$iD_F^{\text{trTT}}(y_1 - y_2)_{jk} = <0|T(B_{uj}(Y(y_1))B_{uk}(Y(y_2)))|0> \qquad (35.16)$$

$$= -\,ig_{jk} \int \frac{d^D k \; e^{-ik\cdot(y_1 - y_2)} \, R(\mathbf{k}, y_1 - y_2)}{(2\pi)^D \, (k^2 + i\varepsilon)} \qquad (35.17)$$

where g_{jk} is the D-dimensional Lorentz metric and where $R(\mathbf{k}, y_1 - y_2)$ is given by

$$R(\mathbf{k}, y_1 - y_2) = \exp[-\,k^i k^j \Delta_{Tij}(y_1 - y_2)/M_u^D] \qquad (35.18)$$
$$= \exp\{-k^2[A(v) + B(v)\cos^2\theta] \,/\, [(2\pi)^{D-2}M_u^4 z^2]\}$$

with

$$z^\mu = y_1{}^\mu - y_2{}^\mu$$
$$z = |\mathbf{z}| = |\mathbf{y_1} - \mathbf{y_2}|$$
$$k = |\mathbf{k}|$$
$$v = |z^0|/z$$
$$A(v) = (1 - v^2)^{-1} + .5v \, \ln[(v - 1)/(v + 1)]$$
$$B(v) = v^2(1 - v^2)^{-1} - 1.5v \, \ln[(v - 1)/(v + 1)]$$
$$\mathbf{k\cdot z} = kz \cos\theta$$

where $|\mathbf{k}|$ denoting the length of a spatial $(D - 1)$-vector $\mathbf{k}$ while $|z^0|$ is the absolute value of $z^0 \equiv z^D$. Thus the k^2 factor in the exponential damps all spatial components.

As eq. 35.18 indicates, the Gaussian damping factor $R(k, z)$ for all large spatial momentum k^j is the same for both the positive and negative frequency parts of the (two-tier) B_u Feynman propagator. We are assuming the spatial momentum is real-valued in this discussion. It is also important to note that $R(k, z)$ does not depend on $k^0 = k^D$ (in the B_u and Y_u Coulomb gauges) and thus the integration over k^0 proceeds in the usual way to produce time-ordered positive and negative frequency parts.

The Gaussian exponential factor in eqs. 35.17 – 35.18 causes the Feynman propagator to be finite and, together with the Gaussian factor in universe particle propagators, causes all perturbation theory calculations when interactions are introduced to be finite as we have seen earlier in The Unified SuperStandard Model.

For small momentum much less than M_u then $R(\mathbf{k}, y_1 - y_2) \rightarrow 1$ and the Feynman propagator is the "normal" propagator of conventional D-dimensional quantum field theory. For large momentum the corresponding potential approaches $r^{(D-3)}$ due to the k^2/z^2 factor in the exponential in eq. 35.18.[318] Thus the B_u potential is highly non-singular at large energies.

---

[318] See section 35.5 below or eqs. 4.11 – 4.24 in Blaha (2005a). Blaha (2005a) has a complete discussion of Two-Tier perturbation theory that directly generalizes to the Megaverse.

35.4 Planckton Interactions with Universe Particles and Individual Baryons

Section 15.1 of Blaha (2015a) describes the second quantization of plancktons in 16 dimensions. In this section we will develop an interacting theory in D dimensions of universe particles and plancktons from a model lagrangian of universe particles and plancktons using quantum coordinates. The universe particle – planckton lagrangian is:

$$\mathcal{L} = \overline{\psi}(Y(y))[i\gamma^\mu \partial/\partial y^\mu - e_B\gamma^\mu B_{u\mu}(Y(y)) - m(t)]\psi(Y(y)) - \tfrac{1}{4} F_{Bu}{}^{\mu\nu}(Y(y))F_{Bu\mu\nu}(Y(y)) - $$
$$- \tfrac{1}{4} F_u{}^{\mu\nu}(y)F_{u\mu\nu}(y) \tag{35.19}$$

where $\mu, \nu = 1, 2, \ldots, D$ and where

$$\overline{\psi} = \psi^\dagger \gamma^D$$

$$F_{Bu\mu\nu} = \partial B_{u\mu}(Y(y))/\partial Y^\nu(y) - \partial B_{u\nu}(Y(y))/\partial Y^\mu(y) \tag{35.20}$$
$$F_{u\mu\nu} = \partial Y_\mu/\partial y^\nu - \partial Y_\nu/\partial y^\mu$$
$$Y^i(y) = y^i + i\, Y_u^i(y)/M_u^{D/2}$$

$$e_B = e_{B0}/M_u^{D/2 - 2}$$

with e_{B0} a dimensionless coupling constant, and with μ and ν ranging from 1 through D.

The corresponding lagrangian is

$$L = \int d^{(D-1)}y\, \mathcal{L} \tag{35.21}$$

Note the dimensions of the fields in the D dimensional space are:

$$Y^\mu \;\sim\; [\text{mass}]^{D/2 - 1} \tag{35.22}$$
$$B_{u\mu} \;\sim\; [\text{mass}]^{D/2 - 1}$$
$$\psi \;\sim\; [\text{mass}]^{(D-1)/2}$$

as can be seen from the above lagrangian as well as earlier equations. Note also that the mass and thus the size of universe particles is time dependent in general. They can expand or contract with time depending on their internal characteristics (gravitation and effects of elementary particle interactions) which are not embodied in this lagrangian. As a result, this model theory does not conserve energy unless m(t) is constant.

The lagrangian generates baryonic interactions of universe particles using Two–Tier quantum coordinates which prevent infinities in perturbation theory calculations.

The interaction of baryon elementary particles with the baryonic field requires terms in Unified SuperStandard Model covariant derivatives specifying the baryon field interaction baryons with the form

$$e_B \gamma^\mu B_{u\mu}(Y(y)) \qquad (35.23)$$

35.5 Megaverse Free Scalar Fields

In this section we describe a Megaverse scalar field $\phi(Y)$ (spin 0) which could be viewed as a prototype for a Higgs scalar field. We will use Two-Tier coordinates defined by eqs. 35.1 and 35.2. The scalar $\phi(Y)$ lagrangian density terms are

$$\mathscr{L}_F = \tfrac{1}{2}[\,(\partial\phi/\partial Y)^2 - m^2\phi^2\,] \qquad (35.24)$$

The free field fourier expansion is

$$\phi(Y) = \int d^{\,D-1}p\, N_m(p)\, [a(p){:}e^{-ip\cdot Y}{:} + a^\dagger(p){:}e^{ip\cdot Y}{:}] \qquad (35.25)$$

$$= \int d^{D-1}p\, N_m(p)\, [a(p){:}e^{-ip\cdot(y + iY/M_u^{\,2})}{:} + a^\dagger(p){:}e^{ip\cdot(y + iY/M_u^{\,2})}{:}]$$

We note the equal time commutation relations of ϕ and the conjugate momentum π_ϕ are the same as the conventional equal time commutation relations of a scalar field despite the fact that Y^μ is itself a quantum field since $[Y^\mu(\mathbf{y}, y^0), Y^\nu(\mathbf{y}', y^0)] = 0$ for $\mathbf{y} \neq \mathbf{y}'$. In addition, we note the ϕ and π_ϕ fields are not hermitean.

The Fourier expansion of ϕ in eq. 3.25 does require one refinement – the exponential terms in X^μ must be *normal ordered* to avoid infinities in the unequal time commutation relations:

Since the hamiltonian as well as other quantities are normal ordered in quantum field theory the additional requirement of normal ordering in the field operator is merely an extension of a standard procedure to a more complex situation and is not disturbing. The unequal time commutation relation of the normal ordered ϕ field is:

$$[\phi(Y^\mu(y_1)), \phi(Y^\mu(y_2))] = i\Delta(y_1 - y_2) + O(1/M_u^{\,2}) \qquad (35.26)$$

where

$$\Delta(y_1 - y_2) = -i \int d^{D-1}k\, (e^{-ik\cdot(y_1 - y_2)} - e^{ik\cdot(y_1 - y_2)})/[(2\pi)^3 2\omega_k] \qquad (35.27)$$

is a familiar c-number invariant singular function. The additional terms in eq. 35.26 are q-number terms that become significant at very short distances of the order M_u^{-1}. Thus precise measurements of field strengths at larger distances are limited by standard quantum effects as indicated by the commutation relation.

The principle of *microscopic causality* is violated at extremely short distances of the order M_u^{-1} since the commutator is non-zero, in general, for space-like distances of the order of M_u^{-1} due to the q-number terms. This violation is not experimentally measurable now – and for the foreseeable future – and reflects a type of non-locality at extremely short distances.

The short distance behavior of two-tier quantum field theory leads to the elimination of divergences resulting in finite interacting quantum field theories.

35.5.1 Vacuum Fluctuations

While the expectation value of a *conventional* free scalar field $\phi_{conv}(Y)$ is zero in a conventional quantum field theory:

$$<0|\phi_{conv}(Y)|0> = 0 \qquad (35.28)$$

the vacuum fluctuations of *conventional* scalar quantum field theory are quadratically divergent:

$$<0|\phi_{conv}(Y)\phi_{conv}(Y)|0> = \int d^{D-1}p/[(2\pi)^3 2\omega_p] \qquad (35.29)$$

In Two-Tier quantum field theory we find the vacuum expectation value of a free field is zero *and the expectation value of the square of the field is also zero:*

$$<0|\phi(Y)\phi(Y)|0> = \int d^{D-1}p\, e^{-p^i p^j \Delta_{Tij}(0)/Mu^4}/[(2\pi)^3 2\omega_p] = 0 \qquad (35.30)$$

since the exponential factor in the integral is $-\infty$. The exponent contains

$$\Delta_{Tij}(z) = \int d^{D-1}k\, e^{-ik\cdot z}\,(\delta_{ij} - k_i k_j/\mathbf{k}^2)/[(2\pi)^3 2\omega_k] \qquad (35.31)$$

where "T" is for "Two-Tier". Thus *vacuum fluctuations are zero in Two-Tier quantum field theory.* Correspondingly, we will see that renormalization constants are finite in the Two-Tier Unified SuperStandard Model .

35.5.2 The Feynman Propagator

The Feynman propagator for a Two-Tier free scalar quantum field is:

$$i\Delta_F^{TT}(y_1 - y_2) = <0|T(\phi(Y(y_1)),\phi(Y(y_2)))|0> \qquad (35.32)$$
$$\equiv <0|\phi(Y(y_1))\phi(Y(y_2))|0>\,\theta(y_1^0 - y_2^0) + \phi(Y(y_2))\phi(Y(y_1))|0>\,\theta(y_2^0 - y_1^0)$$

Since $Y^0 = y^0$ in the Coulomb gauge of the Y^μ field there is no ambiguity in the choice of the relevant time variable. A straightforward calculation shows:

$$i\Delta_F^{TT}(y_1 - y_2) = i \int d^4p \, e^{-ip\cdot(y_1 - y_2)} \, R(\mathbf{p}, y_1 - y_2)/[(2\pi)^4(p^2 - m^2 + i\varepsilon)] \qquad (35.33)$$

where $R(\mathbf{p}, y_1 - y_2)$ is given by eq. 35.18.

35.5.3 Large Distance Behavior of Two-Tier Theories

The large distance behavior of the two-tier Feynman propagator approaches the behavior of the conventional Feynman propagator since

$$R(\mathbf{p}, y_1 - y_2) \rightarrow 1 \qquad (35.34)$$

when $(y_1 - y_2)^2$ becomes much larger than M_u^{-2}. Thus the behavior of a conventional quantum field theory naturally emerges at large distance. We will see that a conventional SuperStandard Model is the large distance limit of the Two-Tier SuperStandard Model thus *realizing a form of Correspondence Principle for Quantum Field Theory*. Some features of a conventional SuperStandard Model that depend specifically on the existence of divergences, such as the axial anomaly, will not be divergent in the Two-Tier SuperStandard Model since it is a divergence-free theory.

35.5.4 Short Distance Behavior of Two-Tier Theories

At short distances the Gaussian factor dominates and radically changes the behavior of the Feynman propagator eliminating its short distance singular behavior, and thus paving the way to finite quantum field theories. Near the light cone, $M_u^{-2} \gg -(y_1 - y_2)^2 \rightarrow 0$, we can approximate eq. 35.33 with

$$i\Delta_F^{TT}(y_1 - y_2) \approx \int d^{D-1}p \, [N(p)]^2 \, R(\mathbf{p}, y_1 - y_2) \qquad (35.35)$$

since $e^{-ip\cdot(y_1 - y_2)}$ is approximately unity for small $(y_1 - y_2)$. We assume the mass of the ϕ particle is zero or is negligible at high energies so we set $m = 0$ to study the high energy behavior of eq. 35.35. Upon performing the integrations in eq. 35.35 for space-like $(y_1 - y_2)^2$ (and analytically continuing to the time-like regions[319,320]) we find that as $(y_1 - y_2)^2 \rightarrow 0$ from the space-like or time-like side of the light cone we find eq. 35.35 becomes:

---

[319] See S. Blaha, "Relativistic Bound State Models with Quasi-Free Constituent Motion", Phys. Rev. **D12**, 3921 (1975) and references therein.

[320] It should be noted that A and B in eq. 35.18 have the same sign for $0 \leq v < 1.1243$ thus making for easy analytic continuation across the light cone (which corresponds to $v = 1$ in eqs. 35.31 and 35.32).

$$i\Delta_F{}^{TT}(y_1 - y_2) \sim |(y_1 - y_2)^\mu (y_1 - y_2)_\mu|^{(D-2)/2} \qquad (35.36)$$

Eq. 35.36 has several noteworthy points:

1. The propagator is well behaved on the light cone and approaches zero smoothly from both space-like and time-like directions. This good behavior near the light cone will be seen later for other particle propagators with the net result that the usual infinities found in conventional quantum field theory are absent in two-tier quantum field theories.

2. The quadratic form of the propagator in eq. 35.36 is suggestive of attempts to formulate a relativistic harmonic oscillator model of elementary particles[321] and more recent attempts to achieve quark confinement. The fact that the absolute value of the quadratic term appears in eq. 35.36 neatly avoids the common pitfall seen in fully relativistic harmonic oscillator attempts.

3. The behavior *in coordinate space* of the propagator at short distances (q. 35.36) is equivalent to a high-energy behavior of

$$p^{(2-2D)} \qquad (35.37)$$

in momentum space. Thus we get the equivalent *of a higher derivative theory* in two-tier quantum field theory at high energies while retaining a positive definite energy spectrum. The problems of negative metric states that have plagued conventional higher derivative quantum field theories are avoided.[322]

　　　This concludes the basic description of scalar Unified SuperStandard Model particles such as Higgs bosons in the Megaverse.

35.6 Megaverse Free Fermion Fields

　　　The fundamental fermions of the Unified SuperStandard Model that exist in the Megaverse have a basic description very similar to that of fermion universe particles.[323] Fundamental SuperStandard Model fermions in the space between universes have the same masses, internal symmetries and interactions, that they do in our universe. They have four species just as universe particles. They have corresponding fourier expansions.

---

[321] H. Yukawa, H., Phys. Rev. **91**, 416 (1953); Y. S. Kim and M. E. Noz, Phys. Rev. **D8**, 3521 (1973) and references therein.

[322] S. Blaha, Phys.Rev. **D10**, 4268 (1974); S. Blaha, Phys.Rev. **D11**, 2921 (1975); S. Blaha, Nuovo Cim. **A49**, :113 (1979); S. Blaha, "Generalization of Weyl's Unified Theory to Encompass a Non-Abelian Internal Symmetry Group" SLAC-PUB-1799, Aug 1976; S. Blaha, "Quantum Gravity and Quark Confinement" Lett. Nuovo Cim. **18**, 60 (1977); Nakanishi, N., Suppl. Prog. Theo. Phys. **51**, 1 (1972); and references therein.

[323] Universe particle are particle formulatlions for universes that we describe in chapter 36.

Due to the similarity of fundamental SuperStandard Model fermions to universe particles we will not describe their Megaverse form of SuperStandard Model fermions here but refer the reader to chapter 36.

36. Quantum Field Theory of the Megaverse and Universes

In this chapter we describe the Megaverse as a quantum entity within the framework of the Wheeler-DeWitt equation suitably generalized.[324] In addition we describe the relation between a particle Quantum Field Theory in the Megaverse and in a universe within the Megaverse. *We also describe the needed generalizations of Quantum Field Theory to adequately and consistently describe a universe quantum field's relation to its Megaverse counterpart. These generalizations are: the local definition of asymptotic particle states, Bogoliubov transformations between a universe quantum field and its Megaverse counterpart, Two-Tier QFT to eliminate perturbation theory infinities, and Pseudoquantum Field Theory to accommodate Higgs vacuum expectation values and higher derivative interactions.*

36.1 Quantum Gravity in the Megaverse and its Universes

Since our universe is described by Quantum Gravity, and other universes are also, it is reasonable to assume the Megaverse is described by D-dimensional classical and Quantum Gravity. The source of Quantum Gravity in a universe is the mass-energy within the universe including particle masses, the Baryonic and Dark Baryonic gauge fields, the Two Tier Y^μ field, and interaction gauge fields.

The sources of classical Gravitation and Quantum Gravity in the Megaverse are analogously the "masses" of universe particles, SuperStandard fermions, the Baryonic and Dark Baryonic gauge fields, other gauge fields, Higs particles, the Ω-group gauge interactions, the mass-energy in the Megaverse outside of universes,[325] and the Megaverse Two-Tier Y^μ field.

The surface of a universe in Quantum Gravity is fuzzy and determined by the quantum universe's wave function. This wave function is a solution of the universe Wheeler-Dewitt equation. The Wheeler-DeWitt equation assumes a single 4-dimensional universe with real-valued coordinates and metrics.

Since our theory requires complex coordinates and thus complex-valued metrics, we extended the Wheeler-DeWitt equation to complex metrics. Now we must define a Wheeler-DeWitt for the Megaverse's complex-valued coordinates and metric.

---

[324] Most of this chapter appeared in Blaha (2017b), Blaha (2015a) and earlier books by the author.
[325] We believe the Megaverse has a very low density mass-energy between universes.

36.2 Megaverse Wheeler-Dewitt Equation

The Megaverse Wheeler-DeWitt equation inside a universe is

$$\left(G_{ijkl}\left\{\prod_x\sum_m\{[\delta^m_{\ j}\,\partial f_n/\partial x^i + \delta^m_{\ i}\,\partial f_m/\partial x^j](\partial^2 f_n/\partial x^{m2})\}^{-1}\partial/\partial x^m\right\}\left\{\prod_x\sum_m\{[\delta^m_{\ k}\,\partial f_n/\partial x^l + \right.$$

$$\left.+ \delta^m_{\ l}\,\partial f_n/\partial x^k](\partial^2 f_n/\partial x^{m2})\}^{-1}\partial/\partial x^m\right\} + \gamma^{\frac{1}{2}\,(3)}R + 2\lambda\gamma^{\frac{1}{2}\,(3)}\right)\Psi(^{(3)}G, L_F) = 0$$

$$(36.2)$$

with universe coordinates x. In the Megaverse the Wheeler-DeWitt equation, in terms of Megaverse coordinates y_n is

$$0 = \left(G_{ijkl}\left\{\prod_x\sum_m\{[\delta^m_{\ j}\,\partial y_n/\partial x^i + \delta^m_{\ i}\,\partial y_m/\partial x^j](\partial^2 y_n/\partial x^{m2})\}^{-1}\partial/\partial x^m\right\}\left\{\prod_x\sum_m\{[\delta^m_{\ k}\,\partial y_n/\partial x^l + \right.\right.$$

$$\left.\left.+ \delta^m_{\ l}\,\partial y_n/\partial x^k](\partial^2 y_n/\partial x^{m2})\}^{-1}\partial/\partial x^m\right\} + \gamma^{\frac{1}{2}\,(D-1)}R + 2\lambda\,\gamma^{\frac{1}{2}\,(D-1)}\right)\Psi(^{(D-1)}G, L_F)$$

$$(36.3)$$

where the sums over n and m in each pair of {} are done independently. All references to the metric are expressed in terms of Megaverse quantiyties.

Due to the appearance of products over all coordinates, the Megaverse expression for $\delta/\delta\gamma_{ij}$ is independent of x, and of Megaverse coordinates, in accordance with the space-time independence of the original Wheeler-DeWitt equation.

The Megaverse form of the Wheeler-DeWitt equation also directly relates the metric of a universe to the Megaverse. Every universe has two sets of coordinates: universe coordinates, usually labeled x, embodying the curvature of the universe induced by gravitation, and Megaverse coordinates usually labeled y.

Outside of universes the Megaverse Wheeler-DeWitt equation becomes

$$(G_{ijkl}\,\delta/\delta\gamma_{ij}\,\delta/\delta\gamma_{kl} + \gamma^{\frac{1}{2}\,(D-1)}R + 2\lambda_M\,\gamma^{\frac{1}{2}\,(D-1)})\Psi(^{(D-1)}\mathcal{G}) = 0 \qquad (36.4)$$

with a cosmological constant, λ_M which is assumed to be present based on its presence in the universe case above and the uniformity of Nature. We assume a representation of the D-dimensional metric with a "spatial" sub-metric part γ_{ij00}. $\gamma^{\frac{1}{2}\,(D-1)}$ is the determinant of γ_{ik}.

Due to the products over all coordinates, the Megaverse expression for $\delta/\delta\gamma_{ik}$ is independent of Megaverse coordinates in analogy with the space-time independence of the original Wheeler-DeWitt equation. But the solutions of the Wheeler-DeWitt equations for a universe must be related to the solutions of the Wheeler-DeWitt equations of the Megaverse within the universe. This relationship must constrain the universe solutions at the boundary of a universe.

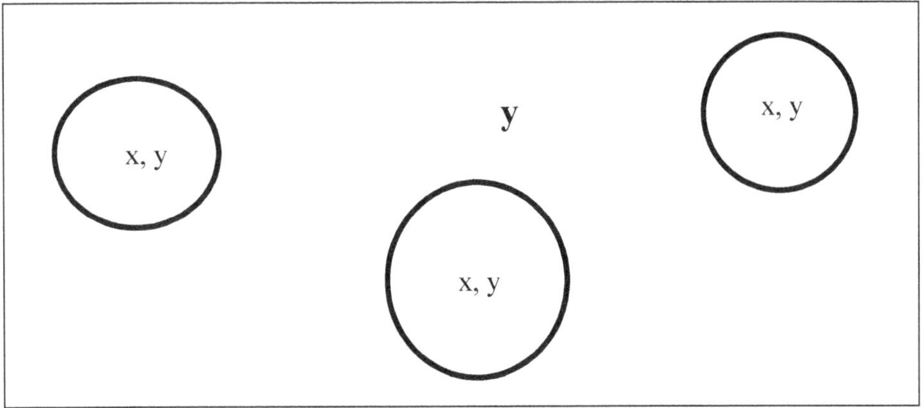

Figure 36.1. A symbolic view of part of the Megaverse with universes depicted as circles. Each universe has its own curvilinear coordinate 4-vector denoted x. The Megaverse coordinates are a D-vector y. A Wheeler-DeWitt equation applies within each universe and a comprehensive equation describes the entire Megaverse transitioning to each universe's wave function within its domain.

36.3 Megaverse Metric Functional Integral in a Universe

Corresponding to the functional derivative in eq. 36.3 is an explicit Megaverse form of the functional integral for a universe

$$\int D\gamma \equiv \int \prod_x \sum_{i,j} d\gamma_{ij}(x) = \int \prod_x \sum_{i,j,k} d(\partial f_k/\partial x^i \, \partial f_k/\partial x^j) \tag{36.5}$$

$$= \int \prod_x \sum_{i,j,k} [d(\partial f_k/\partial x^i) \, \partial f_k/\partial x^j + d(\partial f_k/\partial x^j) \, \partial f_k/\partial x^i]$$

$$= 2\int \prod_x \sum_{i,j,k} d(\partial f_k/\partial x^i) \, \partial f_k/\partial x^j$$

$$= 2\int \prod_x \sum_{i,j,k} dx(\partial^2 f_k/\partial x^{i2}) \, \partial f_k/\partial x^j \equiv 2 \int \delta f(x) \tag{36.6}$$

$$= 2\int \prod_x \sum_{i,j,k} dx(\partial^2 y_k/\partial x^{i2}) \, \partial y_k/\partial x^j \equiv 2 \int \delta f(x) \tag{36.7}$$

The third line is due to the summation over i and j. The factor of 2 can be absorbed in the normalization factor. Eqs. 36.5 – 36.7, like the Wheeler-DeWitt equation, are independent of the coordinates due to the product over coordinates x. Thus the solution of the Wheeler-DeWitt equation takes the form

$$\Psi(^{(D-1)}\mathcal{G}, \mathcal{L}_F) = N \int \delta f(x) \exp(-I(g, \mathcal{L}_F)) \qquad (36.8)$$

upon absorption of the factor of 2 into the normalization constant N

The Wheeler-DeWitt equation depends on the metric γ_{ij} which is a $(D-1)^2$ components construct with $D \cdot (D-1)/2$ independent components due to its symmetry.

36.4 Megaverse Wheeler-DeWitt Solutions

36.4.1 Tachyonic Solutions of Wheeler-DeWitt Equation

The original Wheeler-DeWitt equation, and its Megaverse equivalent, have a form that is similar in many respects to the Klein-Gordon equation. In particular, as DeWitt[326] noted, it resembles "a Klein-Gordon equation with $-\gamma^{\frac{1}{2}\,(D-1)}R$ (our notation) playing the role of a mass-squared term. An important difference, however, is that $^{(D-1)}R$ can be either positive or negative, and hence the wave propagation of the state functional is not confined to time-like directions." DeWitt proceeds later in his paper to exclude consideration of negative "mass" squared terms.

In our view the negative "mass" squared cases represent tachyonic solutions of the Wheeler-DeWitt equation and should be considered as having the same validity as the positive mass squared solutions.

In a universe G_{ijkl} can be regarded as the contravariant metric of a 6-dimensional Riemannian manifold M with hyperbolic signature (-1, 1, 1, 1, 1, 1) with a time-like coordinate. Tachyonic solutions are part of the set of solutions of the Wheeler-DeWitt equation. In the Megaverse formulation the tachyonic solutions are indicators of tachyonic universes in the Megaverse.

If we use the Megaverse form of the Wheeler-DeWitt equation then the changes in time (dilations of γ_{ij} or equivalently dilations of y_n) can be viewed as specifying the overall motion of entire universes. We see now that we can have "normal" or tachyonic motion of universes.

Clearly we are building towards a particle view of universes.[327]

36.4.2 Problems in the Solutions of the Wheeler-DeWitt Equation

There are a number of problem areas associated with the original and our Megaverse Wheeler-DeWitt equations. DeWitt identified most of them in his paper, referenced below. While one could view these apparent problems as negatives, we will take the view that they are indicators of a deeper structure of universes below the Wheeler-DeWitt solutions just as the Dirac equation resolves analogous difficulties with the Klein-Gordon equation.

---

[326] DeWitt, B. S., Phys. Rev. **160**, 1113 (1987).p. 1124.

[327] The Reality groups of 4-dimensional universes, and of the Megaverse, play a role in the physical interpretation of Megaverse phenomena.

36.4.2.1 Negative Frequencies and Probabilities – Anti-Universes

DeWitt noted the existence of negative frequencies and negative probabilities associated with the solutions of the Wheeler-DeWitt equation. The rather close analogy to the Klein-Gordon equation in whose solutions similar issues appear is suggestive.

It appears that two general types of universes are embodied in the Wheeler-DeWitt equation. One type, which we call "normal", consists of universes like ours which have an excess of baryons (and consequently electrons to make an overall charge neutral universe). The other type of universe we will call *anti-universes*. These universes have an excess of anti-baryons (and positrons). We suggest such pairs can result from vacuum fluctuations in the Megaverse.

In addition we have "normal" universes and tachyonic universes whose motions in the Megaverse are analogous to the motions of fermions within a universe. In direct analogy with the Klein-Gordon equation the negative frequency and negative probability issues are thereby resolved.

One can speculate that the Wheeler-Dewitt equation, which is effectively second order in the "time" derivative, can be factored into first derivative equations – perhaps through the introduction of more degrees of freedom in a fashion similar to Dirac's introduction of spinors – to achieve first order equations in "time." Then we would have to face the issue of interpreting "Dirac-like" Wheeler-DeWitt equations.[328] We could again fall back on the rationale for introducing spinors for Standard Model fermions as in Blaha (2015a) – Asynchronous Logic – and suggest that universes embody a multi-valued Logic – although the concept of universes as logic values, at first glance, appears strange. We suggest that a universe's logic value is its spin.[329] Tied up with spin is "handedness." Our universe may be 'left-handed' as many phenomena within the universe seem to favor left-handedness. Therefore we can see a meaning for spin in universe internal properties. The "left handedness" of our universe suggests other universes with "right handedness" may exist.

We will be content, for the present, to assume both universes and anti-universes exist. An anti-universe will be presumed to be a universe in the Megaverse where anti-particles predominate. We will postulate that universes, and anti-universes like protons and anti-protons, are always bodies with extension and not point-like. Blaha (2013) and later chapters in this book provide an example of a Big Bang where a universe begins with extension and is not point-like.

Later we describe a proposed dynamics of universes and anti-universes based on Megaverse gauge fields.

36.5 Quantum Aspects of the Megaverse

The Megaverse contains quantized universes. Being quantized, the horizon (surface) of a universe is not precisely defined but undergoes presumably mild quantum smearing. In

---

[328] We will not consider factoring the Wheeler-DeWitt equation in this book.
[329] Fermionic universe spin is defined later.

particular the surface of a quantum universe in the Megaverse must be defined, as particle positions are defined in quantum mechanics, by a wave function whose "square" at each Megaverse point is the probability of a part of the universe being there. Physically we would expect the probability to fall sharply shortly beyond the classical horizon (surface) of the universe.

So in a Megaverse with a low density of universes most of the Megaverse will have zero probability of a universe being present.

There are however four sources of quantum phenomena in the Megaverse: 1) gauge fields that provides interactions between universes as well as quantum fluctuations that create universes; 2) universe particles; 3) Megaverse mass-energy density; and 4) the Y^μ field that appears within universes and in the exterior Megaverse that quantizes the coordinates of each universe. A Y^μ fields existence in the Megaverse makes perturbation theory computations finite to all orders.

36.6 Quantum Field Theory in the Megaverse

The conventional form of Quantum Field Theory does not meet the requirements of a Physical theory in a Megaverse of more than four dimensions. Perturbation theory calculations will yield infinities – not predictions susceptible to physical interpretation. To remedy this flaw in Quantum Field Theory, and for other reasons cited below, we have introduced three enhancements to Quantum Field theory over the past fourteen years. We briefly describe these enhancements within the framework of the Megaverse in this section and refer the reader to previous books that provide detailed descriptions.

36.6.1 Two Tier Quantum Field Theory in the Megaverse

Two Tier Quantum Field Theory,[330] which was based on a new method in the Calculus of Variations, uses two 'layers' of fields to introduce quantum coordinates. We shall consider this technique, which applies to all fields, for the specific case of a massless vector field $V^i(y)$ analogous to the electromagnetic field.

Since a field, quantized in D-dimensional conventional coordinates (D > 4), would lead to divergences in perturbation theory calculations we use D-dimensional quantum coordinates:

$$Y^i(y) = y^i + i\, Y_u^{\ i}(y)/M_u^{\ D/2} \tag{36.9}$$

where $Y_u^{\ i}(y)$ for i = 1, ..., D is a D-dimensional free gauge field and M_u is a mass of the order of the Planck mass or greater. The $Y_u^{\ i}(y)$ term adds a quantum field to the D coordinates making them a set of quantum coordinates. Quantum coordinate derivatives are defined by

---

[330] See Blaha (2005a), and Blaha (2002), for discussions of this new method to eliminate infinities in quantum field theory calculations.

$$\partial_i = \partial/\partial Y^i(y) = \partial/\partial(y^i - Y_u^{\ i}(y)/M_u^{D/2}) \qquad (36.10)$$

The use of these coordinates to quantize particle fields leads to a completely finite perturbation theory. We applied them earlier to create a finite fundamental theory of mater. We will apply them to fields in the Megaverse to achieve a finite theory of Megaverse dynamics for elementary particles and universe particles.

The second quantization of the vector gauge field, $V^i(y)$ is analogous to the second quantization of the electromagnetic field. The lagrangian density terms for the free $V^i(Y(y))$ fields is

$$\mathscr{L}_{Vu} = -\tfrac{1}{4}\, F_{Vu}^{\ ij}(Y(y))F_{Vuij}(Y(y)) \qquad (36.11)$$

The lagrangian is

$$L_{Vu} = \int d^D y\, \mathscr{L}_{Vu}(Y(y)) \qquad (36.12)$$

with

$$F_{Vuij} = \partial V_i(Y(y))/\partial Y^j(y) - \partial V_j(Y(y))/\partial Y^i(y) \qquad (36.13)$$

where the values of i and j range from 1 to D in this section.

The equal time commutation relations, using the Dth coordinate as the time coordinate, are specified in the usual way:

$$[V^i(Y(\mathbf{y}, y^0)), V^j(Y(\mathbf{y}', y^0))] = [\pi^i(Y(\mathbf{y}, y^0)), \pi^j(Y(\mathbf{y}', y^0))] = 0 \qquad (36.14)$$
$$[\pi_j(Y(\mathbf{y}, y^0)), V_k(Y(\mathbf{y}', y^0))] = -i\,\delta^{(D-1)tr}_{\ jk}(Y(\mathbf{y},0) - Y(\mathbf{y}',0)) \qquad (36.15)$$

where

$$\pi_u^{\ k} = \partial \mathscr{L}_{Vu}(V(Y(y)))/\partial V_k'(Y(y)) \qquad (36.16)$$
$$\pi_u^{\ D} = 0 \qquad (36.17)$$

for $k = 1, \dots, (D-1)$, and

$$\delta^{(D-1)tr}_{\ jk}(\mathbf{y} - \mathbf{y}') = \int d^{(D-1)}k\, e^{i\,\mathbf{k}\cdot(Y(\mathbf{y},0) - Y(\mathbf{y}',0))} (\delta_{jk} - k_j k_k/\mathbf{k}^2)/(2\pi)^{D-1} \qquad (36.18)$$
$$V_k'(Y(y)) = \partial V_k(Y(y))/\partial y^{1D} \qquad (36.19)$$

for j, k = 1, 2, ... , (D – 1).

If we choose the Coulomb gauge for $V_k(Y(y))$:

$$V^D(Y(y)) = 0$$
$$\partial V^j(Y(y))/\partial Y^j(y) = 0$$

for $j = 1, 2, \ldots, (D-1)$ then $(D-2)$ degrees of freedom (polarizations) are present in the vector potential.[331] The Fourier expansion of the vector potential $V^i(Y(y))$ is:

$$V^i(Y(y)) = \int d^{(D-1)}k \, N_{0V}(k) \sum_{\lambda=1}^{D-2} \varepsilon^i(k, \lambda)[a_V(k,\lambda) :e^{-ik\cdot Y(y)}: + a_V^\dagger(k,\lambda) :e^{ik\cdot Y(y)}:] \quad (36.20)$$

for $i = 1, \ldots, (D-2)$ where

$$N_{0V}(k) = [(2\pi)^{(D-1)} 2\omega_k]^{-\frac{1}{2}} \quad (36.21)$$

and (since the field is massless)

$$k^D = \omega_k = (\mathbf{k}^2)^{\frac{1}{2}} \quad (36.22)$$

where k^D is the energy, and where the $\varepsilon^i(k, \lambda)$ are the polarization unit vectors for $\lambda = 1, \ldots, (D-2)$ and $k^\mu k_\mu = k^{D\,2} - \mathbf{k}^2 = 0$.

The commutation relations of the Fourier coefficient operators are:

$$[a_V(k,\lambda), a_V^\dagger(k',\lambda')] = \delta_{\lambda\lambda'}\delta^{D-1}(\mathbf{k} - \mathbf{k}') \quad (36.23)$$
$$[a_V^\dagger(k,\lambda), a_V^\dagger(k',\lambda')] = [a_V(k,\lambda), a_V(k',\lambda')] = 0 \quad (36.24)$$

and the polarization vectors satisfy

$$\sum_{\lambda=1}^{D-2} \varepsilon_i(k, \lambda)\varepsilon_j(k, \lambda) = (\delta_{ij} - k_i k_j/\mathbf{k}^2) \quad (36.25)$$

The V^μ Feynman propagator is

$$iD_F^{trTT}(y_1 - y_2)_{jk} = \langle 0|T(V_j(Y(y_1))V_k(Y(y_2)))|0\rangle \quad (36.26)$$

$$= -ig_{jk} \int \frac{d^D k \, e^{-ik\cdot(y_1 - y_2)} R(\mathbf{k}, y_1 - y_2)}{(2\pi)^{16} (k^2 + i\varepsilon)} \quad (36.27)$$

where g_{jk} is the D-dimensional Lorentz metric and where $R(\mathbf{k}, y_1 - y_2)$ is given by qeq. 35.17.

---

[331] Note we use the Coulomb gauge for $Y(y)$ also.

The Gaussian damping factor R(k, z) for *all* large spatial momentum k^j is the same for both the positive and negative frequency parts of the (Two Tier) V Feynman propagator. We are assuming the spatial momentum is real-valued in this discussion. It is also important to note that R(k, z) does not depend on $k^0 = k^D$ (in the V and Y_u Coulomb gauges) and thus the integration over k^0 proceeds in the usual way to produce time-ordered positive and negative frequency parts.

The Gaussian exponential factor in *all* spatial coordinates causes the Feynman propagator to be finite and, together with the Gaussian factor in universe particle propagators, causes all perturbation theory calculations when interactions are introduced to be finite as we have seen in I.

For small momentum much less than M_u then $R(\mathbf{k}, y_1 - y_2) \to 1$ and the Feynman propagator is the "normal" propagator of conventional D-dimensional quantum field theory. For large momentum the corresponding potential approaches r^{D-3}. The V potential is highly non-singular at large energies.

Thus using Two-Tier Quantum Field Theory we can perform perturbation theory caluculations in the Megaverse that always yield a finite result.[332] This is not true if conventional Quantum Field is used.

36.6.2 Pseudoquantum Field Theory in the Megaverse

Pseudoquantum Field Theory which we developed in a series of books[333] also can be formulated in the Megaverse. Thus we can use it in the Megaverse to implement the Higgs Mechanism to generate particle masses and symmetry breaking.

In this section we generalize Pseudoquantum field theory to the Megaverse for a scalar field. It can be implemented for other particle fields in an analogous manner.

We will now Pseudoquantize a scalar particle field in the Megaverse that will become a Higgs particle with a non-zero vacuum expectation value.[334] We begin by defining two fields that correspond to the scalar particle:[335] $\varphi_1(x)$ and $\varphi_2(x)$ where x is a D-dimensional vector. These fields will be assumed to have the equal time commutators

$$[\varphi_a(x), \pi_b(y)] = i(1 - \delta_{ab})\delta^{(D-1)}(\mathbf{x} - \mathbf{y}) \qquad (36.28)$$
$$[\varphi_a(x), \varphi_b(y)] = 0$$

---

[332] In particular, the fermion triangle divergence (anomaly) does not occur in our Two-Tier Quantum Field Theory of the fermion sector. Thus there is no requirement for axion-like particles in the Megaverse (or in universes) although the possible existence of this type of particle is not ruled out.

[333] See I for the discussion of the Pseudoquantum field theory formalism for Higgs particles in our Extended Standard Model.

[334] Much of this chapter appears in Blaha (2016c), and earlier books, as well as in S. Blaha, Phys. Rev. **D17**, 994 (1978) (Appendix I). The case of fermion Pseudoquantization is also discussed in S. Blaha, Il Nuovo Cimento **49A**, 35 (1979) (Appendix A).

[335] The subscripts on the fields are not gauge symmetry indices but simply identifiers distinguishing the fields from each other.

$$[\pi_a(x), \pi_b(y)] = 0$$

where δ_{ab} is the Kronecker δ and where $\pi_a(x)$ is the canonically conjugate momentum to $\varphi_a(x)$. The fields $\varphi_1(x)$ and $\pi_1(y)$ will be observable classical fields. The fields $\varphi_2(x)$ and $\pi_2(y)$ will not be observables so that $\varphi_1(x)$ and $\pi_1(y)$ can both be sharp on the set of physical states.

We now specify the lagrangian density for a scalar Megaverse Klein-Gordon particle:

$$\mathcal{L} = \partial\varphi_1/\partial x_\mu \partial\varphi_2/\partial x^\mu \qquad (36.29a)$$

with hamiltonian density

$$\mathcal{H} = \pi_1\,\pi_2 + \partial\varphi_1/\partial x_i \partial\varphi_2/\partial x^i \qquad (36.29b)$$

where μ labels 16-dimensional coordinates, i labels Megaverse spatial coordinates ($(D-1)$-dimensional), and $\pi_1 = \partial\varphi_2/\partial t$ and $\pi_2 = \partial\varphi_1/\partial t$ with $t = x^D$. Eqs. 36.30 are without a potential or mass term.

The lagrangian and hamiltonian for a massive scalar particle are

$$\mathcal{L} = \partial\varphi_1/\partial x_\mu \partial\varphi_2/\partial x^\mu - m^2\,\varphi_1\varphi_2 \qquad (36.30a)$$

with hamiltonian density

$$\mathcal{H} = \pi_1\,\pi_2 + \partial\varphi_1/\partial x_i \partial\varphi_2/\partial x^i + m^2\,\varphi_1\varphi_2 \qquad (36.30b)$$

The massless fields can be fourier expanded in terms of creation and annihilation operators:

$$\varphi_i(\mathbf{x}, t) = \int d^{(D-1)}k\,[a_i(k)f_k(x) + a_i^\dagger(k)f_k^*(x)] \qquad (36.31)$$

for $i = 1, 2$ where

$$f_k(x) = N(k)e^{-ik\cdot x}$$

with $N(k)$ being a normalization factor.

The creation and annihilation operators satisfy the commutation relations:

$$[a_a(k), a_b^\dagger(k')] = (1 - \delta_{ab})\delta^{(D-1)}(\mathbf{k} - \mathbf{k'}) \qquad (36.32)$$
$$[a_a(k), a_b(k')] = 0$$
$$[a_a^\dagger(k), a_b^\dagger(k')] = 0$$

for $a, b = 1, 2$.

In this formulation the defining properties of a physical state are:

$$\varphi_1(x)|\Phi, \Pi\rangle = \Phi(x)|\Phi, \Pi\rangle \qquad (36.33)$$
$$\pi_1(x)|\Phi, \Pi\rangle = \Pi(x)|\Phi, \Pi\rangle$$

where $\Phi(x)$ and $\Pi(x)$ are sharp on the states and thus classical fields with

$$\Phi(\mathbf{x}, t) = \int d^{(D-1)}k \, [\alpha(k)f_k(x) + \alpha^*(k)f_k^*(x)] \qquad (36.34)$$

and correspondingly for $\Pi(x)$.

To implement the mass generation mechanism we set Φ equal to a constant. We can define a set of states satisfying

$$a_1(k)|\alpha\rangle = \alpha(k)|\alpha\rangle$$
$$a_1^\dagger(k)|\alpha\rangle = \alpha^*(k)|\alpha\rangle$$

and correspondingly a set of coherent states

$$|\alpha\rangle = C\exp\left\{\int d^3k \, [\alpha(k)a_2^\dagger(k) + \alpha^*(k)a_2(k)]\right\}|0\rangle \qquad (36.35)$$

where C is a normalization constant and where the vacuum state $|0\rangle$ satisfies

$$a_1(k)|0\rangle = a_1^\dagger(k)|0\rangle = 0 \qquad (36.36a)$$

$$a_2(k)|0\rangle \neq 0 \qquad\qquad\qquad a_2^\dagger(k)|0\rangle \neq 0 \qquad (36.36b)$$

The dual vacuum state satisfies

$$\langle 0|a_2(k) = \langle 0|a_2^\dagger(k) = 0 \qquad (36.37a)$$
$$\langle 0|a_1(k) \neq 0 \qquad\qquad\qquad \langle 0|a_1^\dagger(k) \neq 0 \qquad (36.37b)$$

With this coherent state formalism, which gives purely classical fields and yet also has quantum fields through the use of φ_2 and its creation and annihilation operators, we now have the machinery to define a mass mechanism without the introduction of a potential whose origin can only be described as dubious.

For we can define a coherent state for some k as

$$|\Phi, \Pi\rangle = C\exp\{[2N(k)]^{-1}\Phi[a_2^\dagger(k) + a_2(k)]\}|0\rangle \qquad (36.38)$$

where C is a normalization constant, that yields a non-zero vacuum expectation value:

$$\varphi_1(x)|\Phi, \Pi\rangle = \Phi|\,\Phi, \Pi\rangle \qquad (36.39)$$

where Φ is a constant. Evaluating a fermion interaction term we find a mass term emerges[336]

---

[336] When matrix elements with a "vacuum state" are taken.

$$\overline{\psi}(\varphi_1 + \varphi_2)\psi \;\;\rightarrow\;\; \overline{\psi}(\Phi + \varphi_2)\psi \qquad\qquad (36.40)$$

It generates a mass for an interaction with a gauge field of the form

$$A^{\mu}(\varphi_1 + \varphi_2)^2 A_{\mu} \;\;\rightarrow\;\; A^{\mu}(\Phi + \varphi_2)^2 A_{\mu} \qquad\qquad (36.41)$$

It also yields a quantum field theoretic interaction that would result in the production of ElectroWeak particles from these scalar fields. The production of Higgs particles that decay into ElectroWeak gauge particles has recently been found at CERN.

Thus our Pseudoquantum formalism is well-adapted to generate particle masses and symmetry breaking. Section 36.6.3 below shows that we can define Quantum Field Theories that support quantization for arbitrary timelike directions with local definitions of asymptotic states.

36.6.3 Local Definition of Asymptotic Particle States

The local definition of particle states is a significant point of interest for the Megaverse. Given the need for a quantum formulation of particle theory, we must address the issue of particle field quantization in a universe vs. quantization in the Megaverse. A particle state of a field quantized in one coordinate system is, in general, a superposition of particle states if the particle field is quantized in a different coordinate system. This is true within a universe. It is also true if one quantizes a field within a universe's coordinate system, and also quantizes the field in a Megaverse coordinate system. Since a universe can be described in a universe coordinate system, or in Megaverse coordinates, the problem of field quantization in coordinate systems, and the interpretation of particle states, becomes more critical when Megaverse coordinates are brought into consideration.

Some years ago this author developed a formulation[337] of quantum field theory in which the particle interpretation of particle states was unambiguous: an n particle state in one field quantization's coordinate system was an n particle state for the field quantized in any other coordinate system. Thus the particle interpretation of states was independent of the coordinate system chosen for second quantization.

In this section we overview this method of quantization since it relates to the very real issue of the transition of particles between the Megaverse and a universe. If one envisions a particle (or a starship!) traveling between a universe and the Megaverse, then the fate of the particle (starship) after the transition is directly related to the possible(?) quantum field theoretic change of particle state(s).

---

[337] S. Blaha, "The Local Definition of Asymptotic Particle States", IL Nuovo Cimento **49A**, 35 (1979). This paper is reprinted in Appendix A for the reader's convenience. Also the paper S. Blaha, "New Framework for Gauge Field Theories", IL Nuovo Cimento **49A**, 113 (1979) applies this quantization method to non-Abelian gauge theories. It is reprinted in Appendix B.

The second quantization method described in this section is a generalization of the Pseudoquantization procedure described earlier in this section. The method was developed in the late 1970's by the author to provide a quantization procedure which supports a unique particle interpretation of states in arbitrary non-static space-times where no global timelike coordinate (Killing vector) exists. An N particle state in one quantization is an N particle state in other quantizations. Physical particle states of different quantizations are related by a unitary Bogoliubov transformation that preserves the particle number of the states. (The particle number operator commutes with the operator generating the unitary transformation.) See Appendix A for a detailed discussion.[338] We shall assume the reader has read Appendix A and extend the discussion to Megaverse quantization vs. universe quantization below.

We will consider the case of a scalar particle. Charged scalars, fermions and gauge fields are considered in Appendices A and B. These other cases are completely analogous.

36.6.3.1 Second Quantization and the Definition of Particle States in a Universe

Let us consider the case of a scalar particle that we second quantize in some fashion based on a timelike Killing vector

$$\varphi(x) = \sum_{\alpha} \chi_\alpha(x)A_\alpha + \chi_\alpha^*(x)A_\alpha^\dagger \tag{36.42}$$

where the $\chi_\alpha(x)$ are positive frequency with respect to a definition of positive frequency within a universe – following the notation of Appendix A.

36.6.3.2 Second Quantization and the Definition of Particle States in the Megaverse

Consider now the case of the same scalar particle that we second quantize in the Megaverse based on a timelike Megaverse Killing vector

$$\varphi(y) = \sum_{\beta} \psi_\beta(y)b_\beta + \psi_\beta^*(y)b_\beta^\dagger \tag{36.43}$$

where the $\psi_\beta(x)$ are positive frequency with respect to a Megaverse definition of positive frequency.

---

[338] The discussions in the papers of Apendices A and B were predicated on an assumption of one universe. The generalization to Megaverse-universe quantizations is straightforward. We note, as Appendix A points out, that differences in the quantizations of two relatively accelerating observers *do cause* different numbers of particles to be evident in corresponding states – both physically, and in the quantized theories particle states. The quantizations described here are for *one* observer using different coordinate systems. The case of *two* relatively accelerating observers is different as noted in Appendix A.

36.6.3.3 Relation Between the Definitions of Quantized Fields and Particle States

Comparing eqs. 36.42 and 36.43 we note the difference in the definition of the coordinates used in the field expansions as well as the implicit difference in the definitions of positive frequency. Therefore, to relate the quantizations to each other *solely within a universe*, we must use the relation between Megaverse coordinates y and universe coordinates x:

$$y_i = f_i(x) \tag{36.44}$$

or, in vector form,

$$y = f(x) \tag{36.45}$$

for $i = 1, 2, \ldots, D$. Thus

$$\varphi(f(x)) = \sum_{\beta} \psi_{\beta}(f(x))b_{\beta} + \psi_{\beta}^{*}(f(x))b_{\beta}^{\dagger} \tag{36.46}$$

Inverting the above equations to obtain the relation of the fourier coefficient operators we see:

$$A_{\alpha} = \sum_{\beta} [C_{\alpha\beta} \, b_{\beta} + C'_{\alpha\beta} \, b_{\beta}^{\dagger}] \tag{36.47}$$

where $C_{\alpha\beta}$ and $C'_{\alpha\beta}$ are c-number functions of α and β:

$$\begin{aligned} C_{\alpha\beta} &= (\chi_{\alpha}(x), \varphi(f(x)) \\ C'_{\alpha\beta} &= (\chi_{\alpha}^{*}(x), \varphi(f(x))) \end{aligned} \tag{36.48}$$

with eq 36.46 substituted for $\varphi(f(x))$ in the inner products, which are integrals over the universe coordinates x.

Eqs. 36.48 imply an N particle state in a universe will appear as a superposition of states of various numbers of particles in Megaverse coordinates IF THE STANDARD QUANTUM FIELD THEORY FORMULATION IS USED. A practical implication of this formalism is that a mouse in a universe is a superposition of protoplasm in the Megaverse – an unpleasant prospect for manned travel out of a universe into the Megaverse.

To REMEDY this situation – which we take to be unphysical[339] – we must reformulate quantum field theory in a manner similar to the Pseudoquantum formulation presented earlier.

The new formulation associates two fields with a particle in a manner very similar to that of Pseudoquantization field theory discussed earlier. The scalar particle case is discussed in Appendix A between eqs 6 – 31. The reader is directed to read that section.

The conclusions of that section, and the sections following it, in Appendix A are:

---

[339] A mouse is a mouse whether in the universe or Megaverse since the transition between them does not involve a physical change in the mouse. Earlier we suggested the transition is smooth since all neighborhoods of every point of a universe contains an infinite number of exterior Megaverse points.

1. One can define corresponding particle states in a Megaverse quantization or in a universe quantization with the same number of particles.
2. The fourier coefficient operators of the two quantizations are related by Bogoliubov transformations and are unitarily equivalent.
3. The group of the local Bogoiubov transformations is an infinite tensor product of $SU_{1,1}$ groups.
4. The vacua of the particle are invariant under Bogoliubov transformations that relate the the Megaverse and the universe quantizations.
5. Unitarily equivalent perturbation theories of both the Megaverse and the universe quantizations can be defined.

The equations of Appendix A (and B) can be taken to apply to a universe quantization vs. a Megaverse quantization with the proviso that a map of Megaverse coordinates to universe coordinates must be used to calculate fourier coefficient operators such as in eqs. 36.46 – 36.48 above.

We thus have shown that our generalized Pseudoquantization formalism can be used to relate Megaverse quantum field theory in Megaverse coordinates to universe quantum field theory in universe coordinates.

36.7 SuperStandard Particles in Universe and Megaverse Coordinates

There are two aspects to the relation of universe fields and Megaverse fields: first the transformation of fields between universe coordinates and Megaverse coordinates; secondly the relation between the fourier expansions of the quantum fields in the respective coordinate systems.

36.7.1 Coordinate Transformation of Fields

The first issue of transforming between universe and Megaverse coordinates must be mindful that universe points are also Megaverse points. The values of a field at Megaverse points in a neighborhood of a universe point are determined by continuity. If we define a map from universe coordinates to Megaverse coordinates with

$$y^i = f^i(x) \qquad (36.49)$$

then the vector field representations of a universe vector field $A_U{}^\mu(x)$ and a Megaverse vector field $A_M{}^i(y)$ are related by[340]

---

[340] Implicit in eq. 33.10 is an inverse relation $x^\mu = f^{-1}(y)$ which is necessarily based on a restriction of the y-coordinates to obtain a 1:1 relation between the y and x coordinates. The restriction is best implemented by requiring the domain of y coordinates be restricted to those y coordinates within the universe surface. The result is a 1:1 relation between the y-domain coordinates and the x universe coordinates.

$$A_M^i(y) = \partial y^i/\partial x^\mu \, \dot{A}_U^\mu(x)$$
$$= \partial y^i/\partial x^\mu \, A_U^\mu(f^{-1}(y)) \qquad (36.50)$$

in the domain of the universe. Note the Megaverse vector field has D components.

For scalar fields such as Higgs fields the relationship is

$$\varphi_M^i(y) = \dot{A}_U^\mu(x)$$
$$= \varphi_U^\mu(f^{-1}(y)) \qquad (36.51)$$

Fermion field relations are somewhat more intricate. There must be a map from a 4-spinor universe fermion field $\psi_{Ua}(x)$ where a = 1, 2, 3, 4 to the $2^{D/2}$-spinors of a Megaverse $\psi_{Ma}(y)$:[341]

$$\psi_{Ma}(y) = \sum_b D_{ab} \, \psi_{Ub}(x)$$

$$= \sum_b D_{ab} \, \psi_{Ub}(f^{-1}(y)) \qquad (36.52)$$

with Megaverse spinor components labeled b. The transformation matrix D_{ab} produces the many spinor component equivalent to a universe fermion field.

36.7.2 Quantum Field Transformation Between Coordinate Systems

The above transformation between universe and Megaverse fields is suitable for classical fields. However, in the case of quantum fields the fourier representation of the fields requires Bogoliubov transformation between the fourier components to guarantee the the unitary equivalence of asymptotic particle states in each coordinate system. Chapter 10 describes the PseudoQuantum formalism to implement this change of coordinate systems for the case of scalar particles. Appendix A describes the procedure for the case of vector particles and fermions.

36.8 Unruh Bath for Accelerating Starships?

The preceding discussion in section 36.6.3 eliminates the issue of a starship moving with large acceleration encountering a thermal bath of incoming particles and a 'hot' vacuum.

36.9 Integrity of Starships Passing into the Megaverse

The preceding discussion in section 36.6.3 also eliminates problems with the transition of starships from a universe into the Megaverse. Particles are particles – within and without the Megaverse.

---

[341] See eqs. 37.16 and 37.17 in chapter 37 for Megaverse spinor details.

37. Introduction to Universe Particles

37.1 Megaverse Dynamics due to Universe Fields

The particles and fields that we have found in our universe will also exist in the Megaverse but the form of their fourier expansions will be 192-dimensional. There will also be universe particles – quantum fields for entire universes.[342]

The universe particles within a Megaverse have dynamical motions in the Megaverse due to forces between them as well as Megaverse gravity.

In Blaha (2017c) as well as earlier books we described features of the Wheeler-DeWitt equation that suggested that universes could be viewed as particles or anti-particles, or tachyons. The solutions of their equations are wave functions on a manifold that are analogous to the solutions of the Klein-Gordon and Dirac equations. The issues of negative probabilities, possible tachyonic solutions, and negative frequency solutions suggest a need for an appropriate particle interpretation of universes that can possibly resolve these problems.

Some physicists have taken the Wheeler-DeWitt equation as the starting point for a theory of a universe as a particle. The Wheeler-DeWitt equation describes the interior of a universe in a quantum framework.

We will take a different approach using the Megaverse as the environment of universe particles that internally have Quantum Gravity, and externally have Megaverse Quantum Gravity.

We view a universe as an extended particle and begin by ignoring the detailed inner structure of universes. This approach is similar to the historical treatment of hadrons such as the proton as particles and developing a theory of them as fundamental particles using form factors, structure functions and so on to approximate their inner structure. Afterwards, as detailed data became available, the detailed investigation of the internal structure of hadrons using quark-parton models followed. We will pursue a similar theoretical development beginning with a theory of universes as extended particles in the 192-dimension Megaverse.[343] The internal structure of the particle universes will eventually be specified by the Wheeler-DeWitt equation expressed in Megaverse coordinates.

---

[342] Discussed later.

[343] Of course we follow a different course here – looking from within universe particles 'outwards' whereas in our universe we look 'into' particles from 'outside.'

The two simplest choices for the nature of universes are "spin 0" *bosonic universes* and *fermionic universes* with odd half integer spin, s_M.[344] We will first consider the possibility of fermionic universes, and then consider "spin 0" *bosonic* universes.

The first issue of fermionic universes (reminiscent of the discussions of spin in the 1920's) is the interpretation of spin states. We suggest that the upper $2^{D/2-1}$ components (with $2^{D/2-2}$ "spin up" and $2^{D/2-2}$ "spin down" states) of a fermionic universe wave function represent a left-handed universe with an excess number of baryons. The lower $(2^{D/2-1})$ components lead to right-handed anti-universes where there is an excess of anti-baryons. These associations are analogous to the interpretations of the Dirac electron wave function.[345]

The universe particle "spin up" and "spin down" states are distinguished by their interactions with gauge fields in a manner analogous to quantum electrodynamics.

37.2 "Free Field" Dynamics of Fermionic Universe Particles

We now consider fermionic universes as extended particles with an odd half integer spin – *fermionic universe particles* - in the D-dimensional Megaverse. In the Megaverse there are D 'Dirac' matrices with $2^{D/2}$ rows and $2^{D/2}$ columns that are the equivalent of the four Dirac matrices in four dimensions. We will denote these D matrices as γ_M^i for i = 1, 2, ... , D. They satisfy the anti-commutation relations:

$$\{\gamma_M^i, \gamma_M^j\} = 2\,\delta^{ij} \qquad (37.1)$$

and thus form a Clifford algebra. We will choose the coordinate y^D to be the time coordinate and thus make it pure imaginary with a Reality group transformation. (The D-dimensional Megaverse space is a complex Euclidean space.) Therefore γ^D will be hermitean $((\gamma^D)^2 = 1)$, and the γ^i matrices for i = 1, ... , (D – 1) will be anti-hermitean with $(\gamma^i)^2 = -1$. The number of linearly independent matrices in D dimensions is 2^D.

The Megaverse metric is (by use of the Reality group) chosen to be

$$g^{ij} = -\delta^{ij}, \qquad g^{D,D} = 1 \qquad (37.2)$$

for i, j = 1, 2, ... , (D – 1); and zero otherwise.

Except for the additional dimensions, fermion dynamics is quite similar to the 4-dimensional case. The free universe particle Dirac equation is

---

[344] The Megaverse is assumed to be D-dimensional where we have assumed provisionally to be D = 192. The spin of fermionic universe particles was shown to be s_M in Blaha (2017c) in chapter 3.

[345] It is known that phenomena in our universe tend to be left-handed. If this feature of our universe's phenomena is also a property of the universe itself, then, since handedness is an attribute of spin, the treatment of a universe as having spin is not unreasonable.

$$(i\gamma^i \partial_i + m)\psi(y) = 0 \tag{37.3}$$

summed over i = 1, 2, ... , D where the mass is assumed to be constant, and set below. The derivative operator, is based on the use of quantum coordinates[346]

$$X^i(y) = y^i + i\, Y_u^{\ i}(y)/M_u^{\ D/2} \tag{37.4}$$

for i = 1, ... , D and is defined to be

$$\partial_i = \partial/\partial X^i(y) = \partial/\partial(y_i - Y_{ui}(y)/M_u^{\ D/2}) \tag{37.5}$$

where *we assume $M_u = M_c$ with M_c being a very large mass scale of perhaps the order of the Planck mass.*

$Y_u^{\ i}$ is a D-dimensional Megaverse gauge field equivalent of the universe $Y^\mu(x)$ used in Two Tier renormalization:

$$X^\mu(z) = z^\mu + i\, Y^\mu(z)/M_c^{\ 2} \tag{37.6}$$

where $Y^\mu(z)$ is a free QED-like field. The $Y^i(y)$ quantum coordinates will be used in the Megaverse to eliminate potential divergences, in a manner similar to the case of our universe when universe particle interactions are introduced later. In a high dimension space the potential divergences in calculations are much greater. *Two-Tier coordinates eliminate these potential divergences.*

37.3 Four Types of Fermionic Universe Particles

Assuming universe energies are real-valued,[347] there are four possible types of fermionic universe particles in the Megaverse. They are analogous to the four species of fermions. Two of these types are tachyonic. It is important to note that DeWitt points out that the Wheeler-DeWitt equation has tachyonic solutions since the mass-like term dependent on $^{(3)}R$ can be positive or negative.[348] A negative mass is an indication of tachyonic behavior wherein the wave propagation of the state functional is not necessarily in time-like directions and is thus tachyonic.

Eq. 37.3 is a a D-dimensional Dirac equation. There are three other general types of universe particle equations. (By assumption fermionic universes come in four species like

---

[346] Giving Two-Tier renormalization without infinities.

[347] The energy of universe particles need not be real-valued since universes can 'decay' – unlike elementary particles which are not subject to decay, by definition, since they are assumed to be *fundamental*. We choose to consider the case of universes with real-valued energies. The case of universes with complex-valued energies is a simple extension of the real-value cases considered here.

[348] DeWitt, B. S., Phys. Rev. **160**, 1113 (1967) p. 1124.

fermions.) The derivation of the four types of universe particles (chapter 38) is similar to the derivation of fermion types discussed previously. We will now consider the D-dimensional equivalent for universe particles in the Megaverse.

The general form of a pure D-dimensional complex Lorentz group[349] boost can be expressed in terms of a complex relative (D – 1)-velocity $\mathbf{v_c}$ between inertial reference frames. A D-dimensional coordinate boost has the form

$$\Lambda_C(\mathbf{v_c}) \equiv \Lambda_C(\omega, \mathbf{v_c}) = \exp[i\omega\hat{\mathbf{w}}\cdot\mathbf{K}] \tag{37.7}$$

where

$$\omega = (\omega_r^2 - \omega_i^2 + 2i\omega_r\omega_i\,\hat{\mathbf{u}}_r\cdot\hat{\mathbf{u}}_i)^{\frac{1}{2}} \tag{37.8}$$

and

$$\hat{\mathbf{w}} = (\omega_r\hat{\mathbf{u}}_r + i\omega_i\hat{\mathbf{u}}_i)/\omega \tag{37.9}$$

with all vectors being (D – 1)-dimensional spatial vectors. We define the real and imaginary unit vectors $\hat{\mathbf{u}}_r\cdot\hat{\mathbf{u}}_r = 1 = \hat{\mathbf{u}}_i\cdot\hat{\mathbf{u}}_i$ with the result

$$\hat{\mathbf{w}}\cdot\hat{\mathbf{w}} = 1 \tag{37.10}$$

The complex relative velocity is

$$\mathbf{v_c} = \hat{\mathbf{w}}\,\tanh(\omega) \tag{37.11}$$

The free dynamical equations of the four universe particle species will be generated by D-dimensional Lorentz boosts of the free Dirac equation of a universe particle at rest with the *requirement that the time variable* $(t = y^D)$ *and energy are real in the resulting field equations.*[350] The procedure can most easily be performed in D-dimensional momentum space with the Megaverse coordinate space version of the generated equation determined from the momentum space version.

The result is a number of different types of fermion universes that mirror the particle species that we have seen earlier.

- Dirac-like Universe Particle
- Tachyon Universe particle
- Left-Handed Tachyonic Universe Particles
- Right-Handed Tachyonic Universe Particles
- "Up-Quark-like" Universe Particles
- Left-Handed "Down-Quark-like" Tachyonic Universe Particles

---

[349] The D-dimensional Complex Lorentz group has similar features to the 4-dimensional Complex Lorentz group. We shall only discuss it to the extent needed for our universe particle type's derivation. See Weinberg (1995) for the 4-dimensional Lorentz group – the D-dimensional Lorentz group generalizes directly from the features of the 4-dimensional Lorentz group.

[350] The D-dimensional "energy" must be real since it relates to the area of the universe – a real number.

- Right-Handed Down-Quark-like Tachyonic Universe Particles

Free lagrangians, which are similar to the corresponding fermion lagragians seen earlier, follow for these universe particles.

37.4 Embedding Universes Within the Megaverse

Earlier we developed a view of the general nature of universes in the Megaverse. We now examine the general nature of the Megaverse itself. Consistency *suggests that complex-valued coordinates universes, such as our universe, require an embedding in a complex-valued Megaverse* space. In chapter 31 we determined the provisional dimension of the Megaverse, D = 192, from the dimensions implied by interactions within our universe, and from a need to extend Ω-group symmetry to Megaverse coordinates.

We now assume the Megaverse has an analytic, complex, symmetric metric g_{ik} which satisfies

$$g_{ij} = g_{ji} \tag{37.12}$$

for i, j = 1, 2, ... , D.

The embedding equations for our curved complex-valued universe {x} within the complex, D-dimensional Megaverse {y} are:

$$z_i = f_i(x) \tag{37.13}$$

where x^μ are the complex-valued 4-dimensional coordinates of a universe.

The infinitesimal Megaverse invariant distance is

$$ds^2 = g_{ik}dz^i dz^k \tag{37.14}$$

where the z_i are the complex-valued D-dimensional coordinates of the Megaverse. g_{ik} is the complex-valued Megaverse metric tensor, ds is the D-dimensional invariant Megaverse distance,[351] and the complex-valued, 4-dimensional metric of a universe has the form

$$g_{\mu\nu} = \partial f_j/\partial x^\mu \, \partial f_j/\partial x^\nu \tag{37.15}$$

with an implied sum over the subscript j.

The picture that we paint of the Megaverse is that of a complex, D-dimensional space containing (perhaps) countless universes, some of which may be (almost) flat like ours, and some of which may be curved, open or closed 4-surfaces. *We assume universes are closed, or curved and open, or flat.*

---

[351] We note that ds^2 is complex and this might be found troubling. A Reality group transformation would yield the absolute value of ds^2 and thus $ds_{physical} = ||ds^2|^{1/2}$ would be the physical real-valued invariant distance. This remark also applies to 4-dimensional complex General Relativity.

The Megaverse does have a gravitational field due to universe masses, particle masses, the Two-Tier[352] field Y^μ, the Unified SuperStandard Model fields. *We will assume all universes are complex, 4-dimensional although universes with a larger number of dimensions are possible and will be discussed below.*

37.5 Possible Dimensions of Universes

In Blaha (2011c) (and in this book) we determined the dimensionality of our universe based on principles of Asynchronous Logic that suggested a 4-valued logic that could be embodied in a 4-dimensional spinor matrix formulation. This 4-dimensional spinor formulation led to a 4-dimensional space-time.

The requirement that the speed of light is the same in all inertial reference frames, and that transformations between reference frames in faster than light relative motion are physically allowed, led to the requirement of complex coordinates and the Complex Lorentz transformation group (as was found necessary in Axiomatic Quantum Field Theory studies.) The reality of all physical time and distance measurements led to the introduction of the Reality group that mapped complex quantities to real physical values. This chain of logic is in accord with Leibniz's minimax principle: nature uses the simplest means to create complex physical phenomena.

While the preceding paragraph applies very nicely to our universe, and presumably to other universes, the question of the existence of other universes within the Megaverse with different dimensions naturally arises. Stars and galaxies have many varieties. Why should all universes be of dimension four?

Having developed the fundamental nature of our universe from Logic (the only sure requirement of any physical theory) it seems reasonable to classify possible universes based on their fundamental logic. In Blaha (2011c) we developed matrix formulations for many-valued logics. Assuming no separate clock mechanism to synchronize parts of complex processes, we developed an n × n matrix formalism for n-valued logic.

Therefore we were able to develop a principal sequence of types of universes based on n-valued logic. We summarize the small n-valued cases in Table 37.1.

Fant (2005) points out that VLSI circuits with spatially separated parts, which require time synchronization of activity without clocks, need a 4-valued logic at minimum. Thus for a complex universe such as ours, the minimum space-time dimensionality is 4. For a smaller number of dimensions the complexity of physical processes is much diminished as the many solvable models of low space-time dimensionality show: easily solved – not very complex phenomena!

Smaller dimensioned universes may well exist – but not with the richness of complexity that leads to our type of universe's phenomena such as life.

Larger dimension universes may well also exist. They would have an excess of phenomena that might preclude life as we know it, or engender new forms of life.

n-Valued Logic	Matrix Representation Size	Spinor Components	Space-Time Dimensionality[353]
1	1×1	1	1
2	2×2	2	2
3	3×3	2	3
4	4×4	4	4
5	5×5	4	5
6	6×6	8	6

Table 37.1. Space-time dimensionality and number of spinor components corresponding to various n-valued logics. It would seem that the minimal acceptable number of dimension is 4 if one is to have a physically acceptable universe as we understand it.

The general tendency of physical phenomena to be largely based on extrema suggests that 4-dimensional space-time based on Leibniz's minimax principle is the "logical" choice for all universes.The case of 6-dimensional universes also appears attractive for a number of reasons.

The above classification scheme for universes is based on logic. Another important consideration is size. It appears that universes can have differing sizes and in fact can also grow or diminish in size (expansion or contraction). We have considered the size issue earlier and will again when we consider the expansion/contraction of a universe due to a time-dependent mass. Other possible differentiating factors between universes will also be considered.

37.6 Fermion Spins in the Megaverse for D = 192

The spinors in free field expansions of $\psi(y)$ have $2^{D/2}$ column entries.[354] The D 'Dirac' γ-matrices have $2^{D/2} \times 2^{D/2}$ rows and columns. The lowest fermion spin is

$$s_M = (2^{D/2 - 1} - 1)/2 \tag{37.16}$$

For D = 192 these quantities are (rounded off – they are half-integer values in fact):

---

[353] Weinberg (1995) p. 216 exhibits an equation that relates the number of components of a spinor to the dimension of its space-time.
[354] See Weinberg (1995) p. 216.

Number of spinor components:
$$N_{MRC} = 2^{D/2} = 8 \times 10^{28} \qquad (37.17)$$

Spin:
$$s_M = 4 \times 10^{28}$$

Spin Range
$$s_{Mz} = -4 \times 10^{28}, -4 \times 10^{28} + 1, \ldots, +4 \times 10^{28}$$

At first glance the range of spins gives one pause for thought. However, universes are not small generally. Consequently a great range of spins is understandable.

38. Fermionic and Bosonic Universe Particles

The Wheeler-DeWitt equation, because of its similarity to the Klein-Gordon equation, has led to numerous proposals to view universes as particles.[355]

In this chapter[356] we will consider a possible particle interpretation of universes that, while consistent with the spirit of the Wheeler-DeWitt equation and the Megaverse, goes far beyond our current experimental knowledge, although some recent astronomical data tends to support it. It can only be justified in this century by its generality and simplicity. It just looks right.

38.1 The Hierarchy of the Cosmos

In our universe we have seen that natural phenomena form a hierarchy ranging from the simplest to the largest/most complex phenomena. One current view of the hierarchy of levels of physical phenomena is:

Elementary particles: leptons, quarks, gluons, gauge bosons, and Higgs particles

Hadrons: protons, neutrons, …

Molecules

Agglomerations of molecules

Macroscopic objects

Planets

Stars

Galaxies

Clusters of galaxies

Supergalaxies

The Universe

Each level generally has a set of "simplified" physical laws that describe its phenomena.[357] For example molecules have quantum mechanical laws and regularities that help to understand the phenomena at the molecular level.

---

[355] Some suggestions of this interpretation are: DeWitt, B. S., Phys. Rev. **160**, 1113 (1967); Robles-Perez, S. J., arXiv:1212.4598 (2012); and references therein.

[356] This chapter is obtained from Blaha (2015a) with a change of dimension from 16 to D.

[357] This point was often made by Nobelist Kenneth Wilson of Cornell and Ohio State Universities.

Interestingly, while all phenomena at each level should be explainable by the laws at lower levels, and ultimately, all phenomena should be explainable at the level of elementary particles, connecting phenomena at different levels is often quite difficult and, in many cases, impossible.

Consequently, while we believe physical phenomena are ultimately reducible to the lowest level, the problem of relating phenomena at different levels is largely unresolved.

In this book we introduce new levels in the hierarchy of nature: the level of multiple universes, and the level of the all-encompassing Megaverse. In doing this, we seek to maintain what we know of our universe, as embodied in our Extended Standard Model and Quantum Gravity. We will now turn to a discussion of the universes level and a portrayal of universes as extended particles.

38.2 The Particle Interpretation of Extended Wheeler-DeWitt Equation Solutions

In earlier chapters we described features of the Wheeler-DeWitt equation that suggested that universes could be viewed as particles or anti-particles, or tachyons. The solutions of this equation are scalar wave functions on a manifold that are analogous to the solutions of the Klein-Gordon equation. The issues of negative probabilities, possible tachyonic solutions, and negative frequency solutions suggest a need for an appropriate particle interpretation of universes that can possibly resolve these problems.

Some physicists have taken the Wheeler-DeWitt equation as the starting point for a theory of a universe as a particle. The Wheeler-DeWitt equation describes the interior of a universe in a quantum framework.

We will take a different approach using the Megaverse as the environment of universe particles that internally have Quantum Gravity, and externally have Megaverse Quantum Gravity.

We view a universe as an extended particle and begin by ignoring the detailed inner structure of universes. This approach is similar to the historical treatment of hadrons such as the proton as particles and developing a theory of them as fundamental particles using form factors, structure functions and so on to approximate their inner structure. Afterwards, as detailed data became available, the detailed investigation of the internal structure of hadrons using quark-parton models followed. We will pursue a similar theoretical development beginning with a theory of universes as extended particles in the D-dimension Megaverse. The internal structure of the particle universes will eventually be specified by the Wheeler-DeWitt equation expressed in Megaverse coordinates.

The two simplest choices for the nature of universes are "spin 0" *bosonic universes* and *fermionic universes* with odd half integer spin, s_M.[358] We will first consider the possibility of fermionic universes, and then briefly consider "spin 0" *bosonic* universes.

---

[358] Since the Megaverse is D-dimensional, the spin of fermionic universe particles was shown to be s_M in chapter 37.

The first issue of fermionic universes (reminiscent of the discussions of spin in the 1920's) is the interpretation of spin states. We suggest that the upper $2^{D/2-1}$ components (with $2^{D/2-2}$ "spin up" and $2^{D/2-2}$ "spin down" states) of a fermionic universe wave function represent a left-handed universe with an excess number of baryons. The lower $(2^{D/2-1})$ components lead to right-handed anti-universes where there is an excess of anti-baryons. These associations are analogous to the interpretations of the Dirac electron wave function.[359]

The universe particle "spin up" and "spin down" states are distinguished by their interactions with gauge fields in a manner analogous to quantum electrodynamics.

38.3 "Free Field" Dynamics of Fermionic Universe Particles

We now consider universes as extended particles with an odd half integer spin – *fermionic universe particles* - in the D-dimensional Megaverse. In the Megaverse there are D 'Dirac' matrices with $2^{D/2}$ rows and $2^{D/2}$ columns that are the equivalent of the four Dirac matrices in four dimensions. We will denote these D matrices as $\gamma_M{}^i$ for i = 1, 2, ... , D. They satisfy the anti-commutation relations:

$$\{\gamma_M{}^i, \gamma_M{}^j\} = 2\,\delta^{ij} \tag{38.1}$$

and thus form a Clifford algebra. We will choose y^D to be the time coordinate and thus make it pure imaginary with a Reality group transformation. (The D-dimensional Megaverse space is a complex Euclidean space.) Therefore γ^D will be hermitean $((\gamma^D)^2 = 1)$, and the γ^i matrices for i = 1, ... , (D − 1) will be anti-hermitean with $(\gamma^i)^2 = -1$. The number of linearly independent matrices in D dimensions is 2^D.

The Megaverse metric is (by use of the Reality group) chosen to be

$$g^{ij} = -\delta^{ij}, \qquad g^{D,D} = 1 \tag{38.2}$$

for i, j = 1, 2, ... , (D − 1); and zero otherwise.

Except for the additional dimensions, fermion dynamics is quite similar to the 4-dimensional case. The free universe particle Dirac equation is

$$(i\gamma^i\partial_i + m)\psi(y) = 0 \tag{38.3}$$

summed over i = 1, 2, ... , D where the mass is assumed to be constant, and set by eq. 38.119 below. The derivative operator, is based on the use of quantum coordinates[360]

---

[359] It is known that phenomena in our universe tend to be left-handed. If this feature of our universe's phenomena is also a property of the universe itself, then, since handedness is an attribute of spin, the treatment of a universe as having spin is not unreasonable.
[360] Giving Two Tier renormalization.

$$Y^i(y) = y^i + i\, Y_u{}^i(y)/M_u{}^{D/2} \qquad (38.4)$$

For i = 1, …, D and is defined to be

$$\partial_i = \partial/\partial Y^i(y) = \partial/\partial(y_i - Y_{ui}(y)/M_u{}^{D/2}) \qquad (38.5)$$

where *we assume* $M_u = M_c$ *with* M_c *being a very large mass scale of perhaps the order of the Planck mass.*

$Y_u{}^i$ is a D-dimensional Megaverse gauge field equivalent of the universe $Y^\mu(x)$ used in Two Tier renormalization (discussed in I):

$$Y^\mu(y) = y^\mu + i\, Y^\mu(y)/M_c{}^2$$

where $Y^\mu(y)$ is a free QED-like field. The $Y^i(y)$ quantum coordinates will be used in the Megaverse to eliminate potential divergences, in a manner similar to the case of our universe when universe particle interactions are introduced later.

38.3.1 Four Types of Fermionic Universe Particles

Assuming universe energies are real-valued,[361] there are four possible types of fermionic universe particles in the Megaverse that are analogous to the four species of fermion described in Blaha (2010b) for The Extended Standard Model. Two of these types are tachyonic. It is important to note that DeWitt points out that the Wheeler-DeWitt equation has tachyonic solutions since the mass-like term dependent on $^{(3)}R$ can be positive or negative.[362] A negative mass is an indication of tachyonic behavior wherein the wave propagation of the state functional is not necessarily in time-like directions and is thus tachyonic.

Eq. 38.3 is a Dirac-type D-dimensional Dirac equation. There are three other general types of universe particle equations. (By assumption fermionic universes come in four species like fermions.) The derivation of the four types of universe particles is similar to the derivation of fermion types in the Extended Standard Model in 4-dimensional complex space-time given in Blaha (2010b). We will now consider the D-dimensional equivalent for universe particles in the Megaverse.

The general form of a pure D-dimensional complex Lorentz group[363] boost can be expressed in terms of a complex relative $(D-1)$-velocity $\mathbf{v_c}$ between inertial reference frames. A D-dimensional coordinate boost has the form

---

[361] The energy of universe particles need not be real-valued since universes can 'decay' – unlike elementary particles which are not subject to decay, by definition, since they are assumed to be *fundamental*. We choose to consider the case of universes with real-valued energies. The case of universes with complex-valued energies is a simple extension of the real-value cases considered here.

[362] DeWitt, B. S., Phys. Rev. **160**, 1113 (1967) p. 1124.

[363] The D-dimensional complex Lorentz group has similar features to the 4-dimensional complex Lorentz group. We shall only discuss it to the extent needed for our universe particle type's derivation. See Weinberg (1995) for the 4-

$$\Lambda_C(\mathbf{v_c}) \equiv \Lambda_C(\omega, \mathbf{v_c}) = \exp[i\omega\hat{\mathbf{w}}\cdot\mathbf{K}] \tag{38.6}$$

where

$$\omega = (\omega_r{}^2 - \omega_i{}^2 + 2i\omega_r\omega_i\,\hat{\mathbf{u}}_r\cdot\hat{\mathbf{u}}_i)^{\frac{1}{2}} \tag{38.7}$$

and

$$\hat{\mathbf{w}} = (\omega_r\hat{\mathbf{u}}_r + i\omega_i\hat{\mathbf{u}}_i)/\omega \tag{38.8}$$

with all vectors being $(D-1)$-dimensional spatial vectors. We define the real and imaginary unit vectors $\hat{\mathbf{u}}_r\cdot\hat{\mathbf{u}}_r = 1 = \hat{\mathbf{u}}_i\cdot\hat{\mathbf{u}}_i$ with the result

$$\hat{\mathbf{w}}\cdot\hat{\mathbf{w}} = 1 \tag{38.9}$$

The complex relative velocity is

$$\mathbf{v_c} = \hat{\mathbf{w}}\,\tanh(\omega) \tag{38.10}$$

The free dynamical equations of the four universe particle species will be generated by D-dimensional Lorentz boosts of the free Dirac equation of a universe particle at rest with the *requirement that the time variable* $(t = y^D)$ *and energy are real in the resulting field equations.*[364] The procedure can most easily be performed in D-dimensional momentum space with the Megaverse coordinate space version of the generated equation determined from the momentum space version.

38.3.1.1 Dirac-like Equation – Type I universe Particle

A positive energy plane wave solution of the Dirac equation eq. 38.3 for a universe particle at rest is

$$\psi(y) = \exp[-imt]w(0) \tag{38.11}$$

where we set $\partial_t = \partial/\partial y_D$ while temporarily ignoring the $Y_u{}^i(y)/M_u{}^{D/2}$ term. $w(0)$ is a $2^{D/2}$ component spinor column vector. The solution $\psi(y)$ satisfies the momentum space Dirac equation for a particle at rest:

$$(m\gamma^D - m)\psi(y) = 0 \tag{38.12}$$

The $2^{D/2} \times 2^{D/2}$ spinor matrix form of a D-dimensional Lorentz boost with a relative real velocity $\mathbf{v}$ of the Dirac matrices is[365]

$$S^{-1}(\Lambda(\mathbf{v}))\gamma^\nu S(\Lambda(\mathbf{v})) = \Lambda^\nu{}_\mu(\mathbf{v})\gamma^\mu \tag{38.13}$$

---

dimensional Lorentz group – the D-dimensional Lorentz group generalizes directly from the features of the 4-dimensional Lorentz group.

[364] The D-dimensional "energy" must be real since it relates to the area of the universe – a real number.

[365] **The indices ν and μ from this point in this chapter have values: 1, 2, … , D.**

where $\Lambda^{\nu}{}_{\mu}(\mathbf{v})$ is a D-dimensional Lorentz boost. $S(\Lambda(\mathbf{v}))$ has the form

$$S(\Lambda(\mathbf{v})) = \exp(-\omega\gamma^D\gamma\cdot\mathbf{v}/(2|\mathbf{v}|))$$

$$= \cosh(\omega/2)I + \sinh(\omega/2)\gamma^D\gamma\cdot\mathbf{p}/|\mathbf{p}| \qquad (38.14)$$

with *real* $\omega = \operatorname{arctanh}(|\mathbf{v}|)$ and *real* $\mathbf{v}$. $|\mathbf{p}|$ is the magnitude of the spatial $(D-1)$-vector. Also

$$S^{-1}(\Lambda(\mathbf{v})) = \gamma^D S^{\dagger}(\Lambda(\mathbf{v}))\gamma^D = \exp(\omega\gamma^D\gamma\cdot\mathbf{v}/(2|\mathbf{v}|))$$

$$= \cosh(\omega/2)I - \sinh(\omega/2)\gamma^D\gamma\cdot\mathbf{p}/|\mathbf{p}| \qquad (38.15)$$

If we now apply $S(\Lambda(\mathbf{v}))$ to the momentum space Dirac equation of a particle at rest (eq. 38.12) we find

$$0 = S(\Lambda(\mathbf{v}))(m\gamma^D - m)\,\psi(y)$$
$$= [mS(\Lambda(\mathbf{v}))\gamma^D S^{-1}(\Lambda(\mathbf{v})) - m]S(\Lambda(\mathbf{v}))w(0)$$

A straightforward evaluation shows

$$mS(\Lambda(\mathbf{v}))\gamma^D S^{-1}(\Lambda(\mathbf{v})) = g_{D\mu\nu}p^{\mu}\gamma^{\nu} = \displaystyle{\not}p \qquad (38.16)$$

where p is a momentum D-vector. In addition we define the D-dimension spinor ($2^{D/2}$ components)

$$S(\Lambda(\mathbf{v}))w(0) = w(p) \qquad (38.17)$$

which can be viewed as a "positive energy D Dirac spinor". The Dirac equation in momentum space has the familiar form:

$$(\displaystyle{\not}p - m)\exp[-ip\cdot y]w(p) = 0 \qquad (38.18)$$

Eq. 38.18 implies the free, coordinate space Dirac equation:

$$(i\gamma^{\mu}\partial/\partial y^{\mu} - m)\psi(y) = 0 \qquad (38.19)$$

We identify this equation as the dynamical equation of a type 1 universe particle. It corresponds to the free charged lepton elementary particle species Dirac equation in particle physics.

38.3.1.2 Complex Boosts

The form of the D-dimensional spinor boost transformation corresponding to the coordinate transformation eq. 38.6 is:

$$S_C(\omega, \mathbf{v_c}) \equiv S_C = \exp(-\omega\gamma^D\gamma\cdot\hat{\mathbf{w}}/2)$$
$$= \cosh(\omega/2)I + \sinh(\omega/2)\gamma^D\gamma\cdot\hat{\mathbf{w}} \qquad (38.20)$$

with *complex* $\mathbf{v_c}$ and $\hat{\mathbf{w}}$ defined by eqs. 38.10 and 38.8 respectively. The inverse transformation is

$$S_C^{-1}(\omega, \mathbf{v_c}) = \exp(\omega\gamma^D\gamma\cdot\hat{\mathbf{w}}/2)$$

$$= \cosh(\omega/2)I - \sinh(\omega/2)\gamma^D\gamma\cdot\hat{\mathbf{w}} \qquad (38.21)$$

Note that S_C is not unitary just as in the 4-dimensional case.

We now apply a spinor boost to the Dirac equation for a particle at rest in this more general case of complex ω and $\hat{\mathbf{w}}$.

$$0 = S_C(\omega, \mathbf{v_c}))(m\gamma^D - m)\exp[-imt]w(0)$$
$$= [mS_C\gamma^D S_C^{-1} - m]\exp[-imt]S_C w(0) \qquad (38.22)$$

where $S_C = S_C(\omega, \mathbf{v_c})$. After some algebra we find

$$mS_C\gamma^D S_C^{-1} = m[\cosh(\omega)\gamma^D - \sinh(\omega)\gamma\cdot\hat{\mathbf{w}}] \qquad (38.23)$$

We will use these *complex* boosts to generate the other species' Dirac-like equations.

38.3.1.3 Tachyon Universe particle Dirac Equation

The development of the complex spinor boost transformation (subsection 38.3.1.2 above) leads to two possible forms of the tachyon Dirac-like equation. One form will lead to a lagrangian dynamics for left-handed universe particles. The other form leads to a lagrangian dynamics for right-handed universe particles.

38.3.1.4 Type IIa Case: Left-Handed Tachyonic Universe Particles

If the real and imaginary relative vectors parts of $\hat{\mathbf{w}}$, namely $\hat{\mathbf{u}}_r$ and $\hat{\mathbf{u}}_i$, are parallel, then $\hat{\mathbf{u}}_r\cdot\hat{\mathbf{u}}_i = 1$ and

$$\omega = \omega_r + i\omega_i \qquad (38.24)$$

Eqs. 38.23 and 38.24 then imply

$$mS_C\gamma^D S_C^{-1} = m[\cosh(\omega_r)\cos(\omega_i) + i\sinh(\omega_r)\sin(\omega_i)]\gamma^D -$$
$$- m[\sinh(\omega_r)\cos(\omega_i) + i\cosh(\omega_r)\sin(\omega_i)]\gamma\cdot\hat{\mathbf{u}}_r \qquad (38.25)$$

or

$$mS_C\gamma^D S_C^{-1} = \cos(\omega_i)\gamma\cdot p_r + i\sin(\omega_i)\gamma\cdot p_i \qquad (38.26)$$

where

$$p_r^{\,0} = m\cosh(\omega_r) \qquad p_i^{\,0} = m\sinh(\omega_r) \qquad (38.27)$$

and

$$\mathbf{p}_r = m\hat{\mathbf{u}}_r\sinh(\omega_r) \qquad\qquad \mathbf{p}_i = m\hat{\mathbf{u}}_r\cosh(\omega_r) \qquad (38.28)$$

If $\omega_i = 0$, then we recover the momentum space Dirac-like equation. If $\omega_i = \pi/2$, then we obtain the left-handed momentum space tachyon equation:

$$mS_C\gamma^D S_C^{-1} = i\gamma\cdot p_i \qquad (38.29)$$

and the tachyon energy and momentum expressions

$$\mathbf{p} = m\mathbf{v}\gamma_s \qquad\qquad E = m\gamma_s \qquad (38.30)$$

where $\sinh(\omega) = \gamma_s = (\beta^2 - 1)^{-\frac{1}{2}}$ with $\beta = v/c > 1$. v is the absolute value of the $(D-1)$ component spatial velocity. Also

$$S_C w(0) = w_C(p) \qquad (38.31)$$

is a tachyon spinor.

The momentum space tachyonic Dirac-like equation is

$$(i\not{p} - m)\exp[-ip\cdot y]w_T(p) = 0 \qquad (38.32)$$

where $p\cdot y = p^D y^D - \mathbf{p}\cdot\mathbf{y}$ after performing a corresponding boost in the exponential factor. If we apply $i\not{p}$ to eq. 38.32 we find the tachyon mass condition is satisfied

$$-E^2 + \mathbf{p}^2 = m^2 \qquad (38.33)$$

Transforming back to coordinate space we obtain the "left-handed" *tachyonic Dirac-like equation*:

$$(\gamma^\mu \partial/\partial y^\mu - m)\psi_T(y) = 0 \qquad (38.34)$$

38.3.1.5 Type IIb Case: Right-Handed Tachyonic Universe Particles

If the real and imaginary relative vectors parts of $\hat{\mathbf{w}}$, $\hat{\mathbf{u}}_r$ and $\hat{\mathbf{u}}_i$, are anti-parallel $\hat{\mathbf{u}}_r = -\hat{\mathbf{u}}_i$, then $\hat{\mathbf{u}}_r\cdot\hat{\mathbf{u}}_i = -1$ and

$$\omega = \omega_r - i\omega_i \tag{38.35}$$

then

$$mS_C\gamma^D S_C^{-1} = m[\cosh(\omega_r)\cos(\omega_i) - i\sinh(\omega_r)\sin(\omega_i)]\gamma^D -$$
$$- m[\sinh(\omega_r)\cos(\omega_i) - i\cosh(\omega_r)\sin(\omega_i)]\gamma\cdot\hat{\mathbf{u}}_r \tag{38.36}$$

or

$$mS_C\gamma^D S_C^{-1} = \cos(\omega_i)\gamma\cdot p_r - i\sin(\omega_i)\gamma\cdot p_i \tag{38.37}$$

where

$$p_r^{\ D} = m\cosh(\omega_r) \qquad p_i^{\ D} = m\sinh(\omega_r) \tag{38.38}$$

and

$$\mathbf{p}_r = m\hat{\mathbf{u}}_r\sinh(\omega_r) \quad \mathbf{p}_i = m\hat{\mathbf{u}}_r\cosh(\omega_r) \tag{38.39}$$

If $\omega_i = \pi/2$, then we obtain the right-handed momentum space tachyon equation.[366]

$$(-\gamma^\mu\partial/\partial y^\mu - m)\psi_T(y) = 0 \tag{38.40}$$

38.3.1.6 Type III Case: "Up-Quark-like" Universe Particles

There are two other cases where we can obtain fermion dynamical equations with a *real* time variable and real energy.[367] In one case we set $\hat{\mathbf{u}}_r\cdot\hat{\mathbf{u}}_i = 0$ and have a real ω.

If the real and imaginary relative vectors parts of $\hat{\mathbf{w}}$, namely $\hat{\mathbf{u}}_r$ and $\hat{\mathbf{u}}_i$, are perpendicular, $\hat{\mathbf{u}}_r\cdot\hat{\mathbf{u}}_i = 0$, then

$$\omega = (\omega_r^{\ 2} - \omega_i^{\ 2})^{1/2} \tag{38.41}$$

Thus ω is either pure real ($\omega_r \geq \omega_i$) or pure imaginary ($\omega_r < \omega_i$).

The momentum space equation generated by the corresponding spinor boost is

$$\{m\cosh(\omega)\gamma^D - m\sinh(\omega)\gamma\cdot(\omega_r\hat{\mathbf{u}}_r + i\omega_i\hat{\mathbf{u}}_i)/\omega - m\}\exp[-imt]w_c(p) = 0 \tag{38.42}$$

Defining the momentum 4-vector

$$p = (p^D, \mathbf{p}) \tag{38.43}$$

where

$$p^D = m\cosh(\omega) \qquad \mathbf{p} = \mathbf{p}_r + i\mathbf{p}_i \tag{38.44}$$

---

[366] We note that $\gamma_s = (\beta^2 - 1)^{-1/2}$, *if expressed in terms of ω, has a branch cut extending from $<-\infty, +\infty>$ in the complex ω plane. Thus values of ω with positive imaginary parts are physically different from values of ω with negative imaginary parts.*

[367] The requirement of a real energy for a universe is not strict. For a fundamental free particle the energy must be real or the particle would be subject to decay – contrary to its assumed fundamental nature. Universes can 'decay' to 'smaller' universes. Therefore the requirement for real energy can be violated. Nevertheless the requirement for real energy is appealing since it leads to four species of universes strengthening the analogy of universes to elementary particles.

with

$$\mathbf{p_r} = m\omega_r\hat{\mathbf{u}}_r \sinh(\omega)/\omega \qquad \mathbf{p_i} = m\omega_i\hat{\mathbf{u}}_i \sinh(\omega)/\omega \qquad (38.45)$$
$$\mathbf{p_r \cdot p_i} = 0 \qquad (38.46)$$

then we obtain a positive energy Dirac-like equation

$$[\mathbf{p \cdot \gamma} - m]\exp[-imt]w_c(p) = 0$$

or

$$[p^D\gamma^D - (\mathbf{p_r} + i\mathbf{p_i})\cdot\mathbf{\gamma} - m]\exp[-ip\cdot y]w_c(p) = 0 \qquad (38.47)$$

with a complex 3-momentum $\mathbf{p}$ and the 4-momentum mass shell condition:

$$p^2 = (p^D)^2 - \mathbf{p_r \cdot p_r} + \mathbf{p_i \cdot p_i} = m^2 \qquad (38.48)$$

Note

$$|\mathbf{v_c}| = |\mathbf{p}|/p^D = [(\mathbf{p_r} + i\mathbf{p_i})\cdot(\mathbf{p_r} + i\mathbf{p_i})]^{1/2}/p^D = \tanh(\omega) \qquad (38.49)$$

and so the Lorentz factor is

$$\gamma = \cosh(\omega) \qquad (38.50)$$

Eq. 38.47 is the momentum space equivalent of the wave equation[368]

$$[i\gamma^D\partial/\partial t + i\mathbf{\gamma}\cdot(\nabla_r + i\nabla_i) - m]\psi_u(t, \mathbf{y_r}, \mathbf{y_i}) = 0 \qquad (38.51)$$

where $\mathbf{y} = \mathbf{y_r} - i\mathbf{y_i}$, and where the grad operators ∇_r and ∇_i are with respect to $\mathbf{y_r}$ and $\mathbf{y_i}$ respectively. Since $\hat{\mathbf{u}}_r \cdot \hat{\mathbf{u}}_i = 0$ we see that there is a subsidiary condition on the wave function

$$\nabla_r \cdot \nabla_i \, \psi_u(t, \mathbf{y_r}, \mathbf{y_i}) = 0 \qquad (38.52)$$

We note eq. 38.52 can be put into covariant form as the difference of two vectors squared (which is a real D-dimensional Lorentz group invariant):

$$[\gamma^D\partial/\partial t + i\mathbf{\gamma}\cdot(\nabla_r + i\nabla_i)]^2 - [\gamma^D\partial/\partial t + i\mathbf{\gamma}\cdot(\nabla_r - i\nabla_i)]^2 = 4\nabla_r \cdot \nabla_i.$$

We identify eq. 38.51 as the dynamical equation of an "up-quark-like" universe particle.

---

[368] The gradient operators ∇_r and ∇_i are 191-dimensional spatial gradient operators.

38.3.1.7 Type IVa Case: Left-Handed "Down-Quark-like" Tachyonic Universe Particles

In this case we set $\hat{\mathbf{u}}_r \cdot \hat{\mathbf{u}}_i = 0$. Then by eq. 38.7

$$\omega = (\omega_r^2 - \omega_i^2)^{\frac{1}{2}}$$

Thus ω again starts out either pure real (if $\omega_r \geq \omega_i$) or pure imaginary (if $\omega_r < \omega_i$). In this case we also choose ω real, and then change ω to

$$\omega = (\omega_r^2 - \omega_i^2)^{\frac{1}{2}} \rightarrow \omega' = (\omega_r^2 - \omega_i^2)^{\frac{1}{2}} + i\pi/2 = \omega + i\pi/2$$

by adding $i\pi/2$ to ω since ω is a free parameter. We then proceed as we did in the prior tachyon case.[369]. The resulting Lorentz boost

$$\Lambda_C = \exp[i((\omega_r^2 - \omega_i^2)^{\frac{1}{2}} + i\pi/2)(\omega_r\hat{\mathbf{u}}_r + i\omega_i\hat{\mathbf{u}}_i)\cdot\mathbf{K}/\omega] \qquad (38.53)$$

becomes a left-handed "quark-like" boost. The tachyon dynamical equation is[370]

$$[\gamma^{\mathbf{D}}\partial/\partial t + \gamma\cdot(\nabla_r + i\nabla_i) - m]\psi_d(y) = 0 \qquad (38.54)$$

with the constraint equation

$$\nabla_r \cdot \nabla_i \, \psi_d(t, \mathbf{y}_r, \mathbf{y}_i) = 0 \qquad (38.55)$$

We will call the universe particles satisfying eqs. 38.54 and 38.55 left-handed *tachyonic quark-like universe particles.*

38.3.1.8 Type IVb Case: Right-Handed Down-Quark-like Tachyonic Universe Particles

In this case we set $\hat{\mathbf{u}}_r \cdot \hat{\mathbf{u}}_i = 0$. Then by eq. 38.7

$$\omega = (\omega_r^2 - \omega_i^2)^{\frac{1}{2}}$$

Thus ω again starts out either pure real (if $\omega_r \geq \omega_i$) or pure imaginary (if $\omega_r < \omega_i$). In this case we also choose ω real, and then change ω to

$$\omega = (\omega_r^2 - \omega_i^2)^{\frac{1}{2}} \rightarrow \omega' = (\omega_r^2 - \omega_i^2)^{\frac{1}{2}} - i\pi/2 = \omega - i\pi/2$$

---

[369] Here again the choice of ω in eq. 38.53 leads to a "left-handed" universe particle while the choice $\omega' = \omega - i\pi/2$ leads to a right-handed one.
[370] The gradient operators ∇_r and ∇_i are $(D-1)$-dimensional spatial gradient operators.

since ω is a free parameter and proceed as we did in the prior case. The resulting Lorentz boost

$$\Lambda_C = \exp[i((\omega_r^2 - \omega_i^2)^{\frac{1}{2}} - i\pi/2)(\omega_r\hat{\mathbf{u}}_r + i\omega_i\hat{\mathbf{u}}_i)\cdot\mathbf{K}/\omega] \tag{38.56}$$

becomes a right-handed quark-like boost. The resulting tachyon dynamical equation is

$$[-\gamma^D\partial/\partial t - \gamma\cdot(\nabla_r + i\nabla_i) - m]\psi_d(y) = 0 \tag{38.57}$$

with the constraint equation

$$\nabla_r\cdot\nabla_i\,\psi_d(t, \mathbf{y}_r, \mathbf{y}_i) = 0 \tag{38.58}$$

We will call the universe particles satisfying eqs. 38.57 and 38.58 right-handed *tachyonic quark-like universe particles.*

38.3.2 Lagrangians

In this section we will develop a lagrangian formalism for each of the four types of fermionic universe particles noting that a tachyonic universe particles have two forms: left-handed and right-handed (discussed later in section 38.3.5).

The various types of universe particles described in section 38.3.1 correspond to universes with differing internal characteristics and motion in the Megaverse. The equations are all free field equations. Internal potentials and interactions must be introduced in these equations to complete the universe dynamical equations. A connection to the Wheeler-DeWitt description of their internal quantum structure also remains to be established (section 38.3.6).

In defining the lagrangians for the four fermionic universe types that yield their dynamical equations in a canonical manner, we require the conventional quantum field theory feature that the hamiltonian derived from the lagrangian is hermitean. We will develop a separate lagrangian for each type.

38.3.2.1 Type I Universe Particle Lagrangian

The Universe particle Dirac equation lagrangian is

$$\mathcal{L}_u = \bar{\psi}(i\gamma^\mu\partial/\partial y^\mu - m)\psi(y) \tag{38.59}$$

where

$$\bar{\psi} = \psi^\dagger\gamma^D$$

and $\psi^\dagger$ is the hermitean conjugate of ψ.

38.3.2.2 Type II Tachyon Universe Particle Lagrangian

This lagrangian includes both left-handed and right-handed cases. It can be separated into lagrangian terms for each case using parity projection operators.

$$\mathcal{L}_{uT} = \psi_T^{\,S}(\gamma^\mu \partial/\partial y^\mu - m)\psi_T(y) \tag{38.60}$$

where

$$\psi_T^{\,S} = \psi_T^{\,\dagger}\, i\gamma^D \gamma^5 \tag{38.61}$$

with γ^5 being the D-dimensional equivalent for γ^5 in 4 dimensions. The peculiar form of the tachyon universe lagrangian is necessitated by the hermiticity of the hamiltonian calculated from it.

38.3.2.3 Type III "Up-Quark-like" Universe Particle Lagrangian

The lagrangian density of a free "up-quark-like" universe particle is

$$\mathcal{L}_u = \bar{\psi}_u(i\gamma^\mu D_\mu - m)\psi_u(y) \tag{38.62}$$

where $\bar{\psi}_u = \psi_u^{\,\dagger}\gamma^D$ and

$$\psi_u^{\,\dagger} = [\psi_u(\mathbf{y}_r, \mathbf{y}_i)]^\dagger\big|_{\mathbf{y}_i = -\mathbf{y}_i} \tag{38.63}$$

$$\begin{aligned} D_D &= \partial/\partial y^D \\ D_k &= \partial/\partial y_r^{\,k} + i\,\partial/\partial y_i^{\,k} \end{aligned} \tag{38.64}$$

for $k = 1, 2, \dots , (D-1)$. The action

$$I = \int d^{(D-1)}y\,\mathcal{L}_u \tag{38.65}$$

It is easy to show that this action is also real.

38.3.2.4 Type IV "Down-Quark-like" Tachyon Universe Particle Lagrangian

The lagrangian density of a free "down-quark-like" universe particle is

$$\mathcal{L}_d = \psi_d^{\,C}(y)(\gamma^D \partial/\partial t + \boldsymbol{\gamma}\cdot(\nabla_r + i\nabla_i) - m)\psi_d(y) \tag{38.66}$$

where

$$\psi_d^{\,C}(y) = [\psi_d(y)]^\dagger\big|_{\mathbf{y}_i = -\mathbf{y}_i}\, i\gamma^D\gamma^5 \tag{38.67}$$

In words, eq. 38.67 states: take the hermitean conjugate of $\psi_d(y)$; change y_i to $-y_i$; and then post-multiply by the indicated factors.

The action is

$$I = \int d^{(D-1)}y \, \mathcal{L}_d \tag{38.68}$$

The action is real. The lagrangian can also be separated into left-handed and right-handed parts using projection operators.

38.3.3 Form of The Megaverse Quantum Coordinates Gauge Field

The discussions of sections 38.3.1 and 38.3.2 assumed the coordinates were Megaverse coordinates and their derivatives. Prior to those discussions we indicated we would use quantum coordinates in the Megaverse of the form[371]

$$Y^i(y) = y^i + i \, Y_u^{\ i}(y)/M_u^{\ D/2} \tag{38.4}$$

and their derivatives

$$\partial_i = \partial/\partial Y^i(y) = \partial/\partial(y^i - Y_u^{\ i}(y)/M_u^{\ D/2}) \tag{38.5}$$

for $i = 1, 2, \ldots , D$ to eliminate divergences in quantum field theory. The subscript "u" signifies universes. The mass constant for the Megaverse, M_u, may be the same as the mass constant M_c appearing in the Two Tier mechanism for our universe.

In this section we define the gauge fields $Y_u^{\ i}(y)$ of the Megaverse.[372] They are similar to the $Y^\mu(y)$ fields of our New Standard Model.[373] The $Y_u(y)$ D-dimensional vector gauge field, in the absence of external sources, will be defined in a D-dimensional Coulomb gauge:

$$Y_u^{\ D}(y) = 0 \tag{38.69}$$
$$\partial Y_u^{\ j}(y)/\partial y^j = 0$$

where the sum over j is over the $D - 1$ spatial y coordinates. We follow a procedure similar to Blaha (2003) but for D-dimensional space. The lagrangian density for the free $Y_u^{\ j}(y)$ fields is

$$\mathcal{L}_u = -\tfrac{1}{4} F_u^{\ \mu\nu} F_{u\mu\nu} \tag{38.70}$$

and the lagrangian is

$$L_u = \int d^{(D-1)}y \, \mathcal{L}_u \tag{38.71a}$$

with

---

[371] The denominator $M_u^{\ D/2}$ is necessitated by the dimension of $Y_u^{\ i}(y)$ which is $[m]^{D/2-1}$. Eqs. 38.78 and 38.81 below imply this conclusion.

[372] This choice implies that the Megaverse Y mass $M_u = M_C$, its universe mass.

[373] See Blaha (2005a) for details.

$$F_{u\mu\nu} = \partial Y_{u\mu}/\partial y^{\nu} - \partial Y_{u\nu}/\partial y^{\mu} \qquad (38.71b)$$

The equal time commutation relations, derived in the usual way, are:

$$[Y_u^{\mu}(\mathbf{y}, y^0), Y_u^{\nu}(\mathbf{y}', y^0)] = [\pi_u^{\mu}(\mathbf{y}, y^0), \pi_u^{\nu}(\mathbf{y}', y^0)] = 0 \qquad (38.72)$$

$$[\pi_u^{j}(\mathbf{y}, y^0), Y_{uk}(\mathbf{y}', y^0)] = -i\, \delta^{(D-1)\text{tr}}_{jk}(\mathbf{y} - \mathbf{y}') \qquad (38.73)$$

for $\mu, \nu, j, k = 1, 2, \ldots, (D-1)$ where

$$\pi_u^{k} = \partial \mathcal{L}_u/\partial Y_{uk}' \qquad (38.74)$$

$$\pi_u^{0} = 0 \qquad (38.75)$$

and

$$\delta^{\text{tr}}_{jk}(\mathbf{y} - \mathbf{y}') = \int d^{(D-1)}k\, e^{i\, \mathbf{k}\bullet(\mathbf{y} - \mathbf{y}')}\, (\delta_{jk} - k_j k_k/\mathbf{k}^2)/(2\pi)^{D-1} \qquad (38.76)$$

$$Y_{uk}' = \partial Y_{uk}/\partial y^{D} \qquad (38.77)$$

The Coulomb gauge indicates $D - 2$ degrees of freedom are present in the vector potential. The Fourier expansion of the vector potential is:

$$Y_u^{i}(y) = \int d^{(D-1)}k\, N_0(k) \sum_{\lambda=1}^{D-2} \varepsilon^{i}(k, \lambda)[a(k,\lambda)\, e^{-ik\cdot y} + a^{\dagger}(k,\lambda)\, e^{ik\cdot y}] \qquad (38.78)$$

where

$$N_0(k) = [(2\pi)^{(D-1)}2\omega_k]^{-\frac{1}{2}} \qquad (38.79)$$

and (since the field is massless)

$$k^{D} = \omega_k = (\mathbf{k}^2)^{\frac{1}{2}} \qquad (38.80)$$

where k^{D} is the energy, and where the $\varepsilon^{i}(k, \lambda)$ are the polarization unit vectors for $\lambda = 1, \ldots, (D - 2)$ and $k^{\mu}k_{\mu} = k^{D\,2} - \mathbf{k}^2 = 0$.

The commutation relations of the Fourier coefficient operators are:

$$[a(k,\lambda), a^{\dagger}(k',\lambda')] = \delta_{\lambda\lambda'}\delta^{(D-1)}(\mathbf{k} - \mathbf{k}') \qquad (38.81)$$

$$[a^{\dagger}(k,\lambda), a^{\dagger}(k',\lambda')] = [a(k,\lambda), a(k',\lambda')] = 0 \qquad (38.82)$$

and the polarization vectors satisfy

$$\sum_{\lambda=1}^{D-2} \varepsilon_i(k, \lambda)\varepsilon_j(k, \lambda) = (\delta_{ij} - k_i k_j/\mathbf{k}^2) \qquad (38.83)$$

It will be convenient to divide the Y field into positive and negative frequency parts:

$$Y_{u}{}^{+}{}_{i}(y) = \int d^{(D-1)}k \, N_0(k) \sum_{\lambda=1}^{D-2} \varepsilon_i(k,\lambda) \, a(k,\lambda) \, e^{-ik\cdot y} \qquad (38.84)$$

and

$$Y_{u}{}^{-}{}_{i}(y) = \int d^{(D-1)}k \, N_0(k) \sum_{\lambda=1}^{D-2} \varepsilon_i(k,\lambda) \, a^{\dagger}(k,\lambda) \, e^{ik\cdot y} \qquad (38.85)$$

For later use we note the commutator between the positive and negative frequency parts is:

$$[\, Y_{u}{}^{-}{}_{j}(y_1), Y_{u}{}^{+}{}_{k}(y_2)] = - \int d^{(D-1)}k \, e^{ik\cdot(y_1 - y_2)} \, (\delta_{jk} - k_j k_k/\mathbf{k}^2)/[(2\pi)^{D-1} \, 2\omega_k] \quad (38.86)$$

38.3.3.1 Y^{μ} Fock Space Imaginary Coordinate States

States can also be defines for the quantized Y^{μ} field. These states will be similar in form to electromagnetic photon states but play a different role in our approach since they are in fact coordinate excitation states for the imaginary part of $Y^i(y)$ (eq. 38.4). Thus universe particles (and other fields) will exist in a real D-dimensional space with quantum excitations into imaginary Quantum Dimensions. These excitations become significant at high energies. At low energies space appears as c-number complex; at very high energies space becomes slightly q-number complex.

There are two types of imaginary coordinate excitations: 1.) Quantum excitations into Fock states consisting of a superposition of states with a definite finite number of Y_u "particles" and 2.) Imaginary coordinate excitations into coherent Y_u states with an "infinite" number of particles. Coherent states can be viewed as representing "classical" fields.

In this section we will consider Y_u field states with a definite number of excitations ("particles"). The raising and lowering operators of the Y_u field can be used to define free particle states. For example a one particle state can be defined by

$$|k, \lambda\rangle = a^{\dagger}(k, \lambda)|0\rangle \qquad (38.87)$$

with corresponding bra state

$$\langle k, \lambda| = \langle 0|a(k, \lambda) \qquad (38.88)$$

where the "coordinate vacuum" is defined as usual:

$$a(k, \lambda)|0\rangle = 0 \qquad (38.89)$$

$$\langle 0|a^{\dagger}(k, \lambda) = 0 \qquad (38.90)$$

Multi-particle states can also be defined in the conventional way with products of the raising and lowering operators applied to the vacuum. The set of all states containing a finite number of "particles" constitutes a Fock space.

A state with a finite number of Y_u "particles" represents a quantum fluctuation into imaginary Quantum Dimensions.

38.3.3.2 Y_u Coherent Imaginary Coordinate States

Coherent Y_u states bring us closer what we might consider to be "classical" imaginary dimensions – dimensions that we can, in principle, experience as we do normal dimensions. Let us define the coherent state[374]

$$| y, p> = e^{-\mathbf{p} \cdot Y_u^-(y)/M_u^{D/2}}|0> \tag{38.91}$$

This state is an eigenstate of the coordinate operator $Y_u^+(y')$:

$$Y_{u\;j}^{+}(y_1)\,|y_2, p> = -[Y_{u\;j}^{+}(y_1), \mathbf{p} \cdot Y^-(y_2)]/M_u^{D/2}|y, p> \tag{38.92}$$

$$= - \int d^{D-1}k\,[N_0(k)]^2\, e^{ik \cdot (y_2 - y_1)}\,(p_j - k_j \mathbf{p} \cdot \mathbf{k}/\mathbf{k}^2)/M_u^{D/2}|y, p>$$

$$= p^i \Delta_{Tij}(y_1 - y_2)/M_u^{D/2}|y, p> \tag{38.93}$$

where $p^i \Delta_{Tij}(y_1 - y_2)/M_u^{D/2}$ is the eigenvalue of $Y_{u\;j}^{+}(y_1)$. As we will see later, the eigenvalue of Y_u^+ becomes large as $(y_1 - y_2)^2 \to 0$. Thus the imaginary Quantum Dimensions become significant at very short distances, and then significantly modifies the high-energy behavior of quantum field theories. In particular, Quantum Dimensions have a significant effect when

$$(y_1 - y_2)^2 \lesssim (2^{D-2}\pi^{D-2}M_u^2)^{-1} \tag{38.94}$$

We assume the mass scale $M_u = M_C$ is very large – perhaps of the order of the Planck mass (1.221×10^{19} GeV/c^2).

38.3.3.3 Quantization of the Type I Free Universe Particle Dirac Field

The quantization procedure is formally identical to that of a conventional Dirac particle. The standard equal time anti-commutation relations for a D-dimensional fermion field are:

$$\{\psi_\alpha(Y), \psi_\beta(Y')\} = \{\pi_{\psi\alpha}(Y), \pi_{\psi\beta}(Y')\} = 0 \tag{38.95}$$

$$\{\pi_{\psi\alpha}(Y), \psi_\beta(Y')\} = i\,\delta_{\alpha\beta}\,\delta^{D-1}(Y - Y') \tag{38.96}$$

---

[374] Coherent states are well known in the physics literature. See for example T. W. B. Kibble, J. Math. Phys. **9**, 315 (1968) and references therein; V. Chung, Phys. Rev. **140**, B1110 (1965); J. R. Klauder, J. McKenna, and E. J. Woods, J. Math. Phys. **7**, 822 (1966) and references therein.

where α and β are the spinor indices ranging from 1 to $N_{MRC} = 2^{D/2}$ and where

$$\pi_{\psi a}(Y) = i\,\psi_a^\dagger(Y) \tag{38.97}$$

The field can be expanded in a fourier series:

$$\psi(Y(y)) = \sum_s \int d^{D-1}p\; N^d{}_m(p)\, [b(p,s)u(p,s)\,:e^{-ip\cdot(y\,+\,iYu/M_u{}^{D/2})}: + d^\dagger(p,s)v(p,s)\,:e^{ip\cdot(y\,+\,iYu/M_u{}^{D/2})}:] \tag{38.98}$$

$$\psi^\dagger(Y(y)) = \sum_s \int d^{D-1}p\; N^d{}_m(p)\, [b^\dagger(p,s)\bar{u}(p,s)\gamma^0\,:e^{+ip\cdot(y\,+\,iYu/M_u{}^{D/2})}: + d(p,s)\bar{v}(p,s)\gamma^0\,:e^{-ip\cdot(y\,+\,iYu/M_u{}^{D/2})}:]$$

$$\tag{38.99}$$

where

$$N^d{}_m(p) = [m/((2\pi)^{D-1}E_p)]^{\frac{1}{2}} \tag{38.100}$$

and

$$E_p = p^D = (\mathbf{p}^2 + m^2)^{\frac{1}{2}} \tag{38.101}$$

with : ... : signifying normal ordering. The commutation relations of the Fourier coefficient operators are:

$$\{b(p,s),\, b^\dagger(p',s')\} = \delta_{ss'}\delta^{D-1}(\mathbf{p}-\mathbf{p}') \tag{38.102}$$
$$\{d(p,s),\, d^\dagger(p',s')\} = \delta_{ss'}\delta^{D-1}(\mathbf{p}-\mathbf{p}') \tag{38.103}$$
$$\{b(p,s),\, b(p',s')\} = \{d(p,s),\, d(p',s')\} = 0 \tag{38.104}$$
$$\{b^\dagger(p,s),\, b^\dagger(p',s')\} = \{d^\dagger(p,s),\, d^\dagger(p',s')\} = 0 \tag{38.105}$$
$$\{b(p,s),\, d^\dagger(p',s')\} = \{d(p,s),\, b^\dagger(p',s')\} = 0 \tag{38.106}$$
$$\{b^\dagger(p,s),\, d^\dagger(p',s')\} = \{d(p,s),\, b(p',s')\} = 0 \tag{38.107}$$

The spinors $u(p,s)$ and $v(p,s)$ are defined in a conventional way (as in Bjorken and Drell). However their form is different from the 4-dimensional case. If one takes the $N_{MRC} \times N_{MRC} \equiv 2^{D/2} \times 2^{D/2}$ $\gamma \cdot p$ matrix, then the first $2^{D/2\,-\,1}$ columns give $u(p,s)$ up to a normalization for the free particle case, the remaining $2^{D/2\,-\,1}$ columns give $v(p,s)$ up to a normalization.

Since there are $2^{D/2-1}$ possible spin values, using the equation $2s + 1 =$ total number of spin values, we see that the spin of a fermionic universe particle is

$$s_M = 2^{D/2\,-\,2} - \tfrac{1}{2}$$

The possible universe particle spin values are:

Up spin values: $+1/2^{D/2\,-\,1}, +2/2^{D/2\,-\,1}, \ldots, +2^{D/2\,-\,2}/2^{D/2\,-\,1}$

$$\text{Down spin values: } -2^{D/2-2}/2^{D/2-1} = -\tfrac{1}{2}, \ -2^{D/2-2}/2^{D/2-1} + 1, \ \dots, \ -1/2^{D/2-1}$$

This enormous number of possible spins is reasonable considering the number of dimensions D and the enormous variety in spins one should expect in universes – given their large size and complexity.

38.3.3.4 Feynman Propagators for the Type I Free Universe Particle Dirac Field

The form of the fermionic universe particle Feynman propagator differs from a conventional fermion propagator by having a Gaussian factor $R(\mathbf{p}, z)$ in its fourier expansions. This follows from using quantum Megaverse coordinates (eq. 38.4).

$$iS_F^{TT}(y_1 - y_2) = <0|T(\bar{\psi}(Y(y_1))\psi(Y(y_2)))|0> \tag{38.108}$$

where the time ordering is with respect to y_1^D and y_2^D. Expanding the free fields leads to the fourier representation:

$$iS_F^{TT}(y_1 - y_2) = i \int \frac{d^D p \, e^{-ip\cdot(y_1 - y_2)} \, (\mathbf{p} + m) \, R(\mathbf{p}, y_1 - y_2)}{(2\pi)^D (p^2 - m^2 + i\varepsilon)} \tag{38.109}$$

where

$$R(\mathbf{p}, y_1 - y_2) = \exp[-p^i p^j \Delta_{Tij}(y_1 - y_2)/M_u^D] \tag{38.110}$$
$$= \exp\{-p^2[A(v) + B(v)\cos^2\theta] / [(2\pi)^{D-2} M_c^4 z^2]\} \tag{38.111}$$

(Note p^2 is the square of the spatial $(D-1)$-vector.) with

$$z^\mu = y_1^\mu - y_2^\mu \tag{38.112}$$
$$z = |\mathbf{z}| = |\mathbf{y_1} - \mathbf{y_2}| \tag{38.113}$$
$$p = |\mathbf{p}| \tag{38.114}$$
$$v = |z^0|/z \tag{38.115}$$
$$A(v) = (1 - v^2)^{-1} + .5v \ln[(v-1)/(v+1)] \tag{38.116}$$
$$B(v) = v^2(1 - v^2)^{-1} - 1.5v \ln[(v-1)/(v+1)] \tag{38.117}$$
$$\mathbf{p}\cdot\mathbf{z} = pz \cos\theta \tag{38.118}$$

and $|\mathbf{p}|$ denoting the length of a spatial $(D-1)$-vector $\mathbf{p}$ while $|z^0|$ is the absolute value of $z^0 \equiv z^D$.

As eq. 38.109 indicates, the Gaussian damping factor[375] $R(p, z)$ for large spatial momentum p is the same for both the positive and negative frequency parts of the Two-Tier

---

[375] Note the Gaussian damping is for all $D-1$ spatial momentum integrations.

Feynman propagator. We are assuming the spatial momentum is real-valued in this discussion. It is also important to note that R(p, z) does not depend on $p^0 = p^D$ (in the Y Coulomb gauge) and thus the integration over p^0 proceeds in the usual way to produce time-ordered positive and negative frequency parts.

38.3.3.5 Feynman Propagators for the Types II, III, and IV Free Universe Particle Dirac Fields

These propagators differ in details from the Type I propagator. The differences modulo the change in dimension appear in Blaha (2011c). See also Blaha (2005a) for a detailed discussion of 4-dimensional spin ½ particle propagators.

38.3.4 Expanding and Contracting Universes: Impact of Time Dependent Universe Particle Masses

Our discussions of the dynamics of universe particles assumed their masses were constant. However the definition of mass in terms of the area of a universe based on the physics of black holes is

$$M = \kappa A/8\pi \tag{38.119}$$

where A is the area of the black hole shows that *the mass of a universe particle is time dependent* because the area of a universe is generally time dependent. For example, our universe is expanding and its surface area is thus growing with time.

Eqs. 38.11 (and subsequent fermionic dynamic equations) must then be modified from

$$\psi(y) = \exp[-imt]w(0) \tag{38.11}$$

to a covariant form:

$$\psi(y) = \exp[-i\int_0^{w\cdot y} m(t')dt']w(0) \tag{38.120}$$

where w is a unit D-vector in the time (y^D) direction $(w^2 = 1)$. The lower bound on the integral, 0, is the time of the beginning of the universe particle – its Big Bang. Thus the cumulative change in the mass of the universe particle may be significant. It is interesting to note that the Wheeler-Dewitt equation also has a variable value mass term R that also depends on the evolution of the universe.

Eq. 38.120 satisfies the free covariant Dirac-like universe particle field dynamic equation

$$[i\gamma^i\partial/\partial y^i - m(w\cdot y)]\psi(y) = 0 \tag{38.121}$$

In contrast to the constant mass equation eq. 38.19. Substituting eq. 38.120 in eq. 38.121 we find

$$(\gamma^i w_i\, m(w{\cdot}y) - m(w{\cdot}y))\psi(y) = 0 \qquad (38.122)$$

or

$$(\gamma^i w_i - 1)\psi(y) = 0 \qquad (38.123)$$

Upon performing a D-dimensional Lorentz boost (of the type of eqs. 38.13 – 38.16) on eq. 38.123 we obtain

$$(\gamma_i p^i/m_0 - 1)\psi(y) = 0$$

or

$$(\gamma_i p^i - m_0)\psi(y) = 0 \qquad (38.124)$$

where p^i is a momentum D-vector with $p^2 = m_0^2$. Eq. 38.123 is the constant mass momentum space dynamic equation. It determines the spinor in $\psi(y)$. After taking account of the quantum coordinates the quantum Dirac-like universe particle wave function has the form

$$\psi(Y(y)) = \sum_s \int d^{(D-1)}p\, N^d_m(p)\, [b(p,s)u(p,s) : \exp[-iG(p, Y(y))]: + d^\dagger(p,s)v(p,s) \cdot$$
$$\cdot :\exp[+iG(p, Y(y))]:\} \qquad (38.125)$$

$$\psi^\dagger(Y(y)) = \sum_s \int d^{(D-1)}p\, N^d_m(p)\, \{b^\dagger(p,s)\bar{u}(p,s)\gamma^0 :\exp[+iG(p, Y(y))]: + d(p,s)\bar{v}(p,s)\gamma^0 \cdot$$
$$\cdot \exp[-iG(p, Y(y))]:\} \qquad (38.126)$$

where : ... : denotes normal ordering and

$$G(p, Y(y)) = \int_0^{p\cdot Y(y)/\lambda} m(t')dt' \qquad (38.127)$$

with $\lambda = m_0$, and $N^d_m(p)$ a normalization constant. Contrast eqs. 38.125-38.126 to the constant mass case eqs. 38.98-38.101. The *constant mass case* simply sets $m(t') = m_0$.

If we examine the integral eq. 38.127 for a short time interval δt in the particle's rest frame then $G(p, Y(y)) \approx m(0)\delta t$ and so we define $m(0) = m_0$. Based on the formula for universe particle mass (eq. 38.119) we anticipate that m_0 might be as large as the Planck mass or larger – thus an extremely short radius. Blaha (2013) describes a quantum Big Bang model in which the initial radius of the universe is $O(EM_{Planck}^{-2})$ where E is of the order of 1 and has the dimensions of [mass].

Thus we have a closed form definition of a quantum universe particle wave function for universe particles of type I. A similar procedure can be followed for universe particles of types II, III, and IV.

The Feynman propagator for type I quantum fields is *not* eq. 38.109 but now has a form reflecting the Y(y) dependence of the quantum fields in eqs. 38.125 and 38.126:

$$iS_F^{TT}(y_1, y_2) = i \frac{\int d^D p \; \{ <0|\theta(y_{1D} - y_{2D})G(y_1, y_2) + \theta(y_{2D} - y_{1D})G(y_2, y_1)\}0>}{(2\pi)^D (p - m_0)}$$

(38.128)

where p^D is the energy and

$$G(y_1, y_2) = \; : \exp[-iG(p, Y(y_1))]: \; :\exp[+iG(p, Y(y_2))]:$$

(38.129)

Let

$$G_{tot}(y_1, y_2) = <0|\theta(y_{1D} - y_{2D})G(y_1, y_2) + \theta(y_{2D} - y_{1D})G(y_2, y_1)\}0>$$

$$= <0|\theta(y_{1D} - y_{2D}):\exp[-iG(p, Y(y_1))]::\exp[+iG(p, Y(y_2))]: \; +$$

$$+ \theta(y_{2D} - y_{1D}) :\exp[-iG(p, Y(y_2))]::\exp[+iG(p, Y(y_1)):]|0>$$

(38.130)

$$= <0|\theta(y^D_1 - y^D_2): \exp[-i\int_0^{p \cdot Y(y1)/\lambda} m(t')dt']::\exp[+i\int_0^{p \cdot Y(y2)/\lambda} m(t')dt']: \; +$$

$$+ \theta(y^D_2 - y^D_1):\exp[-i \exp[+i\int_0^{p \cdot Y(y2)/\lambda} m(t')dt']::\exp[+i\int_0^{p \cdot Y(y1)/\lambda} m(t')dt']:|0>$$

with $\lambda = m_0$ then

$$iS_F^{TT}(y_1, y_2) = i \frac{\int d^D p \; G_{tot}(y_1, y_2)}{(2\pi)^D (p - m_0)}$$

(38.131)

Except for the case of a constant mass, where $m(t) = m_0$, the Feynman propagator is not a function of $y_1 - y_2$. The evaluation of eq. 38.130 in the general case of a variable mass is straightforward but cumbersome. For the special case of a linear time dependence of the mass, $m(t) = at$, we find eq. 38.130 gives

$$G_{tot}(y_1, y_2) = <0|\theta(y^D_1 - y^D_2):\exp[-ia(p \cdot Y(y_1)/m_0)^2/2]::\exp[+ia(p \cdot Y(y_2)/m_0)^2/2]: \; +$$

$$+ \theta(y^D_1 - y^D_2):\exp[-ia(p \cdot Y(y_2)/m_0)^2/2]::\exp[+ia(p \cdot Y(y_1)/m_0)^2/2]:|0>$$

(38.132)

yielding a complex function of p, y_1, and y_2. *Note that the lower bound of the integrals in the Feynman propagator cancel and thus the need for an understanding of the beginning of a universe is removed in this case.*

We have shown that universe particle theory can handle the case of a variable universe mass m(t). Expanding or contracting (or oscillating) universe particles correspond to expanding and contracting (or oscillating) universes.

38.3.5 Left-Handed and Right-handed Universe Particles

In sections 38.3.1 and 38.3.2 we found that left-handed and right-handed tachyonic universe particles existed. The tachyonic nature of the universe particles indicates that their speed in the universe exceeds the "speed of light" of the Megaverse. The physical meaning of the handedness of these types of universes is an interesting issue. When we consider our universe we see left-handedness in the weak interactions of elementary particles. In addition it appears that organic molecules overwhelmingly favor left-handedness on earth although right-handed molecules exist in outer space and can be created in the laboratory. Right-handed molecules transform into left-handed molecules in watery media through electromagnetic effects.

Why nature favors left-handedness is an open question. It has given rise to speculations that gravitation, especially quantum gravitons, may be left-handed. The European Space Agency's Planck telescope will study polarization effects in the cosmos and may well be able to show that the gravitons starting from the beginning of the universe, and magnified by inflation in the universe's expansion, may be left-handed.

If handedness of gravitation is verified experimentally, then our theory of left-handed/right-handed universe particles would be supported. *Our universe would then be tachyonic and probably left-handed. We, in the universe, would, of course, not know of the velocity of the universe in the Megaverse.*

38.3.6 Internal Structure of Universe Particles

We have treated universes as particles in the preceding discussion taking an extremely large view of Megaverse particles just as elementary particle theory viewed nucleons at low energies (large distances). Now we develop a more detailed view of universe particles in a manner analogous to the high energy view of the internal dynamics of nucleons that led to the quark-parton model of nucleons. In the present case we shall see that high energy Baryonic and other field probes of universe particles can yield a model of the internal structure of universes.

We know that universes are composed of matter and radiation. We believe that there is at least one possible accessible interaction between universes dependent on baryon number – a baryonic, D-dimensional gauge field. There are also other particle number gauge fields – but these are less likely to be significant since Dark matter has yet to be found except through its gravitational effects. In this section we will discuss the use of a baryonic gauge field to probe the baryon structure of universe particles.

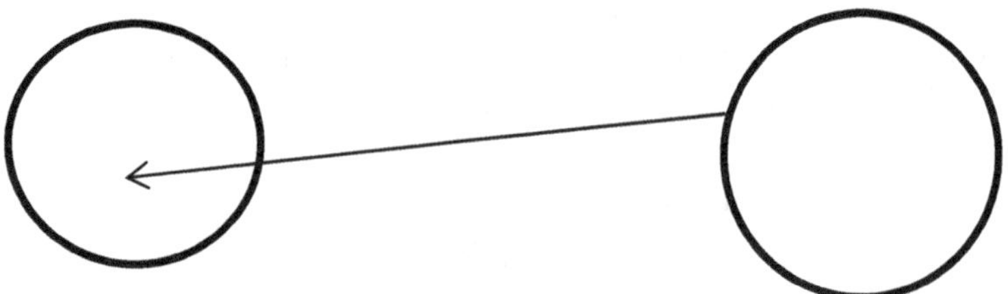

Figure 38.1. A symbolic view of a high resolution (high energy) probe from a universe to a specific baryonic part of another universe.

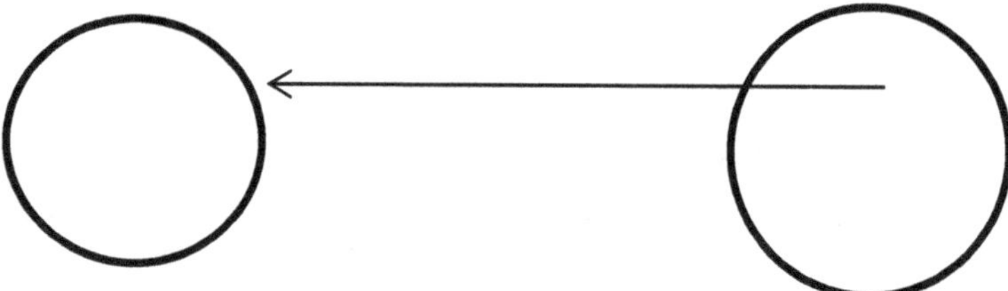

Figure 38.2. A symbolic view of a low resolution (lower energy) probe from a universe to an entire universe.

There appears to be two types of probes[376] of a universe: 1) a series of high energy probes to specific small regions inside another universe for the purpose of mapping its internal structure; 2) a low energy probe of another universe to get a global view of its structure.[377] The first type of probe corresponds to deep inelastic (high energy) electron-nucleon scattering which led to the quark-parton model of nucleons. The second type of probe corresponds to low energy electron-nucleon scattering to get a "global" view of a nucleon. In both case an electromagnetic (gauge) field particle (photon) was the probe particle.

Besides the inherent scientific interest in such experiments it is possible that they may be of use in the very distant future if Mankind is able to develop Megaverse starships that can travel in the Megaverse to other universes. Then the baryonic gauge field may become the "eyes" of the starship in addition to the electromagnetic field (light) are the eyes of current spaceships. We considered the possibility of universe starships in the book entitled, *All the Megaverse! II* in detail.

---

[376] It would appear that the probes are in Megaverse coordinates, not universe coordinates, in order to 'bridge' the distance between universes. Chapter 35 shows the manner of the mapping between Megaverse coordinates and universe coordinates for quantum fields.

[377] **We note these probes, being limited by the speed of light, are at best theoretical speculations since the travel time between universes is so very large.**

38.5 Bosonic Universe Particles

The previous sections has described fermionic universe particles. In this section we will briefly describe aspects of bosonic (spin 0) universe particles. First it is important to note that the Wheeler-DeWitt equation being second order like the Klein-Gordon equation seems to suggest that universe particles can be bosonic – like Klein-Gordon equation particles.[378] The Wheeler-DeWitt equation has a mass-like term R that can be positive or negative. If the mass term is negative then the wave-like propagation of the state functional (wave equation solution) can be in space-like directions implying a tachyonic solution. Thus the Wheeler-DeWitt equation supports "normal" state functionals that propagate in time-like directions as well as tachyonic propagation.

For this reason we suggest that bosonic universe particles can be either normal or tachyonic. Tachyonic bosonic universe particles can fission in a manner similar to tachyonic fermionic universe particles. The fission equations of section 38.5.2 also apply to tachyonic bosonic universe particles.

The quantum field theory of normal and tachyonic bosonic universe particles is similar to that of ordinary bosons. See Blaha (2005a) for the boson case discussion that is paralleled by our universe particle formalism.

38.6 Physical Meaning of Universe Particle Spin

The physical meaning of spin is a continuing discussion topic. We have suggested[379] that spin states are in essence logic states with changes in spin an analogous to changes in logical values in a discourse or computer program. Since the matrix formalism for spin ½ and higher spin states is formally similar to the formalism for angular momentum, one can combine spin and angular momentum as we do in quantum theory.

In the case of universe particles, one can also associate universe particles with "true" and "false" values. Fermionic universe states have $2^{D/2-1}$ truth values and correspond to a multi-valued logic. The numerousness of truth values is due to the D-dimensional space within which universe particles reside.

Naturally one would like to know the physical differences between these $2^{D/2-1}$ types of universe particles. Does the difference reside in different shapes of the universe particles? Or is the difference somehow a consequence of the global mass-energy distribution of the universe that we have not been able to discern since we only know of one universe?

The physical meaning of spin for elementary particles is also somewhat elusive. It does not reflect the flow of charge within a particle. For if it did reflect physical spinning of a particle, the outer edges of a particle such as an electron would be traveling at a speed faster than light. So spin is not a mechanical property of the internal structure of an elementary

---

[378] One should remember that the Wheeler-DeWitt equation is not in space but in a 6-dimensional manifold, denoted M, of metrics with one "time" dimension – having hyperbolic signature – + + + + + when the metric is positive definite. See DeWitt's paper.

[379] Blaha (2011c) and subsequent books.

particle. We have suggested that it is a truth value in the matrix formulation of a 4-valued logic called Asynchronous Logic. Thus it has no certain tangible physical basis.

In the case of universe particles the situation is unclear at present. It could be taken to be an indirect reflection of the structure of mass-energy within a universe. This view would be contrary to our proposed view of elementary particle spin as truth values. So we can only assert that a logic interpretation is the only sensible one (based on our present knowledge or our lack thereof). The physical role of universe particle spin is only evident in interactions between universe particles via gauge fields. Thus one must simply view it as a construct for the present.

Other than our mapping of spin values to logic values in Asynchronous Logic there is little anyone can say about the physical origin of elementary particle spin. Specifying a symmetry group as the origin is not sufficient.

39. Megaverse Interactions

39.1 Unified SuperStandard Model Interactions in the Megaverse

The particles and interactions of the Unified SuperStandard Model exist in the Megaverse space between universes. They are, of course, required to be specified in terms of Megaverse coordinates.

Boson representations are described in Megaverse coordinates in sections 33.3.2, 34.3.6, chapter 35, and section 36.6.

Chapter 38, which describes fermion universe particles, also provides a description of fermion particle fields in the Megaverse.

39.2 Megaverse Lorentz Group

The Megaverse Lorentz group is SU(191,1) – a direct generalization of the SU(3, 1) Lorentz group for Special Relativity in our universe. The apparently extremely low density of the Megaverse suggests the Megaverse is most likely a flat space-time. In our work we have chosen y^D transformed to a real value as Megaverse time. This assumption is required in order for the Megaverse to be a dynamic entity. We assume the time variable is real since clocks measure real-valued time. A Megaverse Reality group transformation can map complex time coordinates to real values with no loss of generality.

39.3 Megaverse Gravitation

Megaverse Complex General Relativity and Gravitation is also a direct generalization of the 4-dimensional case with one significant exception: the Einstein field equations change to:

$$R_{\mu\nu} - [1/(D - 2)]g_{\mu\nu}R = -8\pi GT_{\mu\nu}$$

where the metric is such that $g_\mu{}^\mu = 190$, $R_{\mu\nu}$ is the Megaverse Ricci tensor, and R is the Megaverse curature scalar. (The overall form of these quantities is the same as that of 4-dimensional space-time.)

39.4 Megaverse SuperSymmetry

The discussions of the Θ-symmetry and SuperSymmetry in our 4-dimensional space-time were presented earlier. A similar discussion of Θ-symmetry and SuperSymmetry with additional ramifications will now be considered. It was discussed to some extent in Blaha (2017c).

An important new feature relating to Θ-symmetry and SuperSymmetry is the dimension of the Megaverse D = 192. Due to the value of this dimension we can define a U(192) unitary Reality group that transforms the complex-valued coordinates of the Megaverse to real-valued coordinates. We also have the Θ-symmetry for interactions. Lastly we can define 192 antisymmetry coordinates that supports Supersymmetric field theory in the Megaverse.

This section discusses these possibilities. Much of the discussion parallels the discussions in Blaha (2017c).

39.4.1 Θ-SymmetryTransformations in the Megaverse

The choice of dimension, 192, enables us to extend the Θ-Symmetry to Megaverse coordinates. Let $C_\Theta(y)$ be a local U(4) Θ-Symmetry transformation that rotates interactions in Megarverse space y. Then applying a Θ-Symmetry transformation to the Dirac equation in Megaverse space:[380]

$$\gamma_\mu D^\mu \psi \;=\; \gamma_\mu \{ \partial^\mu + i\,[g_\Theta A_\Theta^{1\mu}(y) + g_\Theta A_\Theta^{2\mu}(y) + \mathbf{A}_I^{1\mu}(y) + \mathbf{A}_I^{2\mu}(y)] \}\psi(y) = 0$$

we find it transforms to

$$\gamma_\mu \{ \partial^\mu + i\,[g_\Theta A'_\Theta{}^{1\mu}(y) + g_\Theta A'_\Theta{}^{2\mu}(y) + \mathbf{A}'_I{}^{1\mu}(y) + \mathbf{A}'_I{}^{2\mu}(y)] \}\psi'(y') = 0$$

where

$$A'_\Theta{}^{1\mu}(y') = C_\Theta(y)A_\Theta^{1\mu}(y)C_\Theta^{-1}(y) - i\,C_\Theta(y)\partial^\mu C_\Theta^{-1}(y)/g_\Theta$$
$$A'_\Theta{}^{2\mu}(y') = C_\Theta(y)A_\Theta^{2\mu}(y)C_\Theta^{-1}(y)$$
$$\mathbf{A}'_I{}^{1\mu}(y') = C_\Theta(y)\mathbf{A}_I^{1\mu}(y)\,C_\Theta^{-1}(y)$$
$$\mathbf{A}'_I{}^{2\mu}(y') = C_\Theta(y)\mathbf{A}_I^{2\mu}(y)\,C_\Theta^{-1}(y)$$
$$\psi'(y') \;= C_\Theta(y)\psi(y)$$

The Θ-Symmetry interaction is experienced by all particles (fields) in the Megaverse. It is also experienced by all universes in the Megaverse and thus affects universe dynamics.

39.9 Megaverse SuperSpace

The Megaverse offers an exciting new platform for SuperSymmetry studies of the Unified SuperStandard Model and possible SuperSymmetric generalizations using Megaverse Superspace with antisymmetric coordinates in addition to normal coordinates.

We can define a generalized set of Megaverse coordinates that associate 192 antisymmetric coordinates with the 192 normal Megaverse coordinates:

$$\{y^\mu,\, \Theta^\nu\}$$

---

[380] We omit the Species group interaction.

where $\mu = 1, \ldots, 192$ *and* also $\nu = 1, \ldots, 192$. We now have an equal numbers of normal and antisymmetric coordinates in Megaverse superspace. This was not the case in 4-dimensional space-time.

Supersymmetric transformations on superspace coordinates have the infinitesimal form:

$$y^\mu \rightarrow y^\mu + i\,\varepsilon\,\Gamma^\mu\Theta$$
$$\Theta^\nu \rightarrow \Theta^\nu + \varepsilon^\nu$$

where Γ^μ is a 192×192 matrix generalization of the Dirac γ matrices.[381] There are 192 Γ matrices.

Using Megaverse superspace coordinates one can proceed to construct a superfields formalism with many features analogous to the 4-dimensional formulation but of significantly greater complexity. We defer doing this in view of the finiteness of Megaverse Two-Tier Quantum Field Theory which, like the 4-dimensional case, yields totally finite results in all orders of perturbation theory. We note the major benefit of 4-dimensional SuperField Theory was the finiteness of calculations in perturbation theory. Given the complexity of Megaverse SuperField perturbation theory it would appear the simplicity of Two-Tier Quantum Field Theory is preferable.

---

[381] See Blaha (2017c) for their definition for arbitrary dimension D.

40. Universe Particle Dynamics

This chapter describes aspects of the interactions of universe particles due to gravitation, baryon number forces, and collisions between universes. It also the describes the genesis of universes due to vacuum fluctuations, the fission of universes, and the internal distortion of universes due to acceleration and the presence of 'nearby' universes..

40.1 The Internal Distortion of Universes

In the absence of external forces univereses are considered to be uniform in the large. However the acceleration of a universe in the Megaverse can distort the universe. Also the existence of a nearby universe(s) could cause the uniformity of a universe to be lost due to gravitation and baryonic forces.[382]

40.1.1 Impact of Universe Particle Acceleration – Lopsided Internal Structure of Universe

Universes can accelerate within the Megaverse due to external Megaverse forces. Universe acceleration should be detectable within a universe as a "lopsidedness" – there would be a shift of parts of the universe away from the direction of acceleration resulting in a difference in the features of the universe "in front" compared to those "in back" – an acceleration effect just as one sees when a jet accelerates.

Interestingly new data from the Planck observatory of the European Space Agency confirms and extends earlier data from NASA's WMAP observatory that one side of the universe appears different from the other side. There are temperature differences and mass distribution differences – just as one might expect if the universe were accelerating as a unit.

Thus we see the beginning of data suggesting our universe may be accelerating through the Megaverse. Some Planck observatory scientists have suggested their data is a preliminary indication of the Megaverse.

40.1.2 Impact of External Forces on Universe Structure

The presence of a nearby universe could cause a universe to lose its large scale uniformity and the mass-energy of the universe to drift over time to the 'nearby' side of the universe due to gravitation and baryonic forces.

---

[382] The baryonic forces, the Baryonic force and the Dark Baryonic force, on a universe are large due to their additivity in the universe and nearby universes.

40.2 Universes in Collision

We can assume that the dynamics of universes in collision will be analogous to that of galaxies in collision since gravity is a dominant force in both cases. Colliding galaxies have often been observed. Their dynamics should provide guidance for the case of universes in collision.[383]

It is clear in the case of colliding galaxies, and of colliding large nuclei (gold and lead typically) that there are several types of collisions with differing results. These types of universe collisions can be qualitatively classified as

1. Clean collisions in which universes nudge each other but retain their identity. These are extreme peripheral collisions. If the universes overlap slightly then the typically spherical symmetry of the universes may become distorted and they may become lopsided.[384]

2. Peripheral collisions in which the universes retain their identity but are connected by a trailing string of mass-energy. Eventually the string breaks and the universes separate. Subsequently the pieces of trailing string in each universe contract due to their universe's gravitational effects.

3. Two universes can collide and produce multiple universes.

4. Two universes can collide in a "central" collision and amalgamate into one universe. They can intermix with both the baryonic, gauge, and gravitational forces causing a redistribution of their masses. They may separate afterwards or may coalesce into a single universe. One result of this may be lopsided universes. Our universe appears to be lopsided. Some cosmologists believe this is due to a near collision of our universe with another shortly after the Big Bang.

40.3 Creation of Universes through Gauge Field Fluctuations

One of the most exciting questions in Cosmology is the origin of our universe. The conventional view is that it originated in a Big Bang from an infinitesimal point in space. The source of the Big Bang and the prior state of the Cosmos, if there was one, is the subject of much speculation. Based on the particle interpretation of the Wheeler-DeWitt equation, the possibility of a baryonic force strongly supported by conservation of baryon number, and the

---

[383] The high energy collision of atomic nuclei at Brookhaven, CERN and other laboratories also is analogous in overall detail with universes in collision.

[384] The Wilkinson Microwave Anisotropy Probe (WMAP) and the Planck European Space Agency satellite have been accumulating data since 2001 that suggests the universe may be lopsided with hot and cold spots on opposite sides of the universe differing from those on the other side being hotter and colder respectively. *Perhaps the result of a collision when the universe was young.*

Megaverse concept, it is reasonable to consider the possibility that the universe originated in a vacuum fluctuation.

In this case there would be two Big Bangs one for our universe and one for an anti-universe. One would expect that they would have opposite corresponding features: one with baryon dominance – one with anti-baryon dominance, and one left-handed – one right-handed.

Our formulation of universe particle theory provides for the generation of a universe particle and anti-particle as a vacuum fluctuation. We view a universe particle as having a substantial excess of baryons, N, as we see in our universe. Its anti-universe at the time of creation (the Big Bang point) is its "mirror image" having the "same" number of anti-baryons (baryon number –N) so that baryon number is conserved by the fluctuation event. Thus the excesses of one universe are compensated by the excesses of the other.

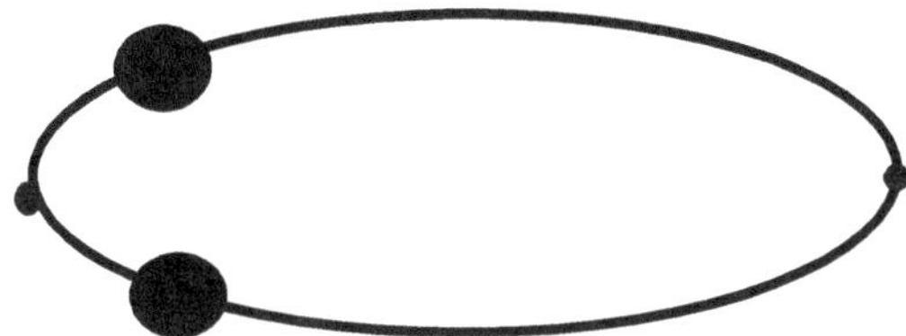

Figure 40.1. Generation of a universe – anti-universe pair as a vacuum fluctuation.

The small value of the coupling constant should lead to an extremely long lifetime for the universes generated by the fluctuation. Thus the 13.7 billion year life of our universe is not unreasonable. The probability of the creation of universes by vacuum fluctuations should be correspondingly small.

40.4 Fission of Universes

Under certain circumstances the distribution of matter in the universe may lead to the fission of the universe into two separate universes. Our theory supports this possibility for universe particles. The detailed mechanism of the fission process is not specified by the model.

40.4.1 Fission of Normal Universes

The fission of universe particles in our universe particle model is depicted in the Feynman diagram in Fig. 40.2.

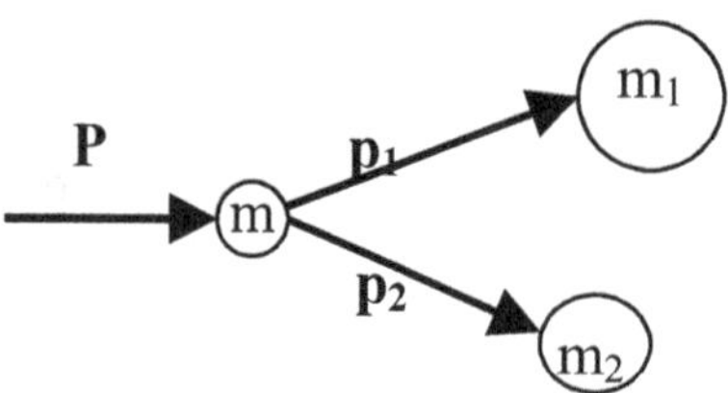

Figure 40.2. Fission of a universe particle into two universe particles.

The sum of the masses of the output universe particles is usually less than the original universe particle mass. However if the fission takes a long time and the masses are time dependent then the produced universe particles combined masses may exceed the original universe's mass.

40.4.2 Tachyon Universe Particle Fission to More Massive Universe Particles

In Blaha (2007a) we showed that a tachyonic (faster than light) particle could fission into particles of larger mass through the conversion of momentum into mass. In this section we show that a tachyonic universe particle may fission into two more massive universe particles.[385] This phenomenon is of particular interest because it enables tachyonic universes to spawn in a new novel way not previously considered in discussions of the origin of universes.

A simple model lagrangian for a tachyonic universe particle is

$$\mathcal{L}_{\|} = \overline{\psi}_{T}{}^{S}(Y(y))[\gamma^{\mu}\partial/\partial y^{\mu} - e_{B}\gamma^{\mu}B_{u\mu}(Y(y)) - m(t)]\psi_{T}(Y(y)) - \tfrac{1}{4}F_{Bu}{}^{\mu\nu}(Y(y))F_{Bu\mu\nu}(Y(y)) -$$
$$- \tfrac{1}{4}F_{u}{}^{\mu\nu}(y)F_{u\mu\nu}(y)$$

We assume m(t) is constant.

When a particle or a universe particle fissions (decays) one normally expects that the masses of the particles or universe particles produced by the decay to be smaller than the mass of the original particle or nucleus. In the case of tachyonic (faster-than-light) elementary particles, or universe particles, a much different possibility is present: a tachyon universe can decay into heavier tachyons (perhaps through a distortion of the universe internally into two 'lumps'.) We will consider the specific case of a tachyon universe particle decaying into two universe particles whose total mass is greater than the original. (See Fig. 40.3.)

---

[385] We will use the term mass here to denote mass-energy. Since we identified mass as a multiple of area earlier the comments here would appear to apply to universe area as well.

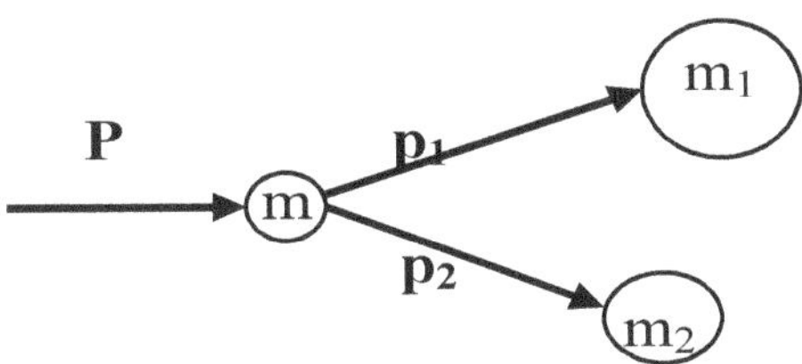

Figure 40.3. Two universe particle decay of a tachyon universe particle.

We will assume the initial tachyon universe particle has zero energy ($p^D = 0$) and thus the tachyons universe particles emerging from the decay also have total universe particle energy zero. The analysis is based on conservation of total universe energy and momentum in Megaverse space outside of universes. The below discussion applies to D-dimensional space with (D – 1)-dimensional spatial coordinates.

Momentum conservation implies

$$\mathbf{P} = \mathbf{p_1} + \mathbf{p_2}$$

Since all energies are zero

$$(cP)^2 = (c\mathbf{P})^2 = m^2$$
$$(cp_1)^2 = (c\mathbf{p_1})^2 = m_1^2$$
$$(cp_2)^2 = (c\mathbf{p_2})^2 = m_2^2$$

where $P = |\mathbf{P}|$, $p_1 = |\mathbf{p_1}|$, and $p_2 = |\mathbf{p_2}|$. If we now square the above equation for **P** and then use the above three equations we obtain

$$m^2 = m_1^2 + m_2^2 + 2m_1 m_2 \cos\theta$$

where θ is the opening angle between the emerging universe particles momenta $\mathbf{p_1}$ and $\mathbf{p_2}$. There are a number of interesting cases:

Case $\theta = 0$:

$$m = m_1 + m_2$$

The masses of the outgoing universe particles sum to the mass of the original tachyon universe particle.

Case $\theta = \pi/2$:

$$m^2 = m_1^2 + m_2^2$$

The masses of each outgoing universe particle tachyon are less than the mass of the original tachyon universe particle.

Case $\theta = \pi$:

$$m^2 = (m_1 - m_2)^2$$

In this case either $m_1 > m$ or $m_2 > m$. Thus one of the outgoing tachyon universe particles has a greater mass than the original tachyon universe particle. Mass is effectively created from the spatial momentum of the initial universe particle. This process is the inverse of normal particle and universe particle fission where the sum of the outgoing masses is always less than the original particle's mass and the difference is mass converted into energy in the form of additional photons.

This last case, where one of the outgoing universe particles is more massive than the original universe particle, is not just for $\theta = \pi$. Since

$$\cos\theta = (m^2 - m_1^2 - m_2^2)/(2m_1 m_2)$$

we see that the sum of the outgoing universe particle masses is always greater than the original tachyon universe particle *mass (except when $\theta = 0$)* since

$$\cos\theta = 1 + [m^2 - (m_1 + m_2)^2]/(2m_1 m_2) \leq 1$$

and thus

$$[m^2 - (m_1 + m_2)^2]/(2m_1 m_2) \leq 0$$

Note $m = m_1 + m_2$ only if $\theta = 0$.

Since we can transform the above discussion to the case of universe particle tachyons having non-zero Megaverse energy using an ordinary D-dimensional Lorentz transformation the discussion in this subsection is general.

We therefore conclude that when a tachyon universe particle decays into two tachyon universe particles the sum of the masses of the produced tachyon universe particles is greater than the mass of the original tachyon universe particle except if the angle between the momenta of the produced tachyon universe particles is zero. In that case the sum of the masses of the produced tachyon equals the mass of the original tachyon universe particle and the produced universe particles overlap.

41. Universe Particle – Planckton Interactions

While the gravitational force between universe particles is a simple generalization of 4-dimensional gravitation, the baryonic forces, normal and Dark, between universe particles have some points of difference.[386]

They are quite similar to Two-Tier electromagnetic interactions except that universe particles have time-dependent masses, and that the space is D-dimensional.

The time dependence of the universe particle masses is illustrated by Fig. 41.1: the mass of a universe particle after a baryonic interaction vertex is the same as it was before the interaction assuming the point-like interaction specified in the lagrangian.

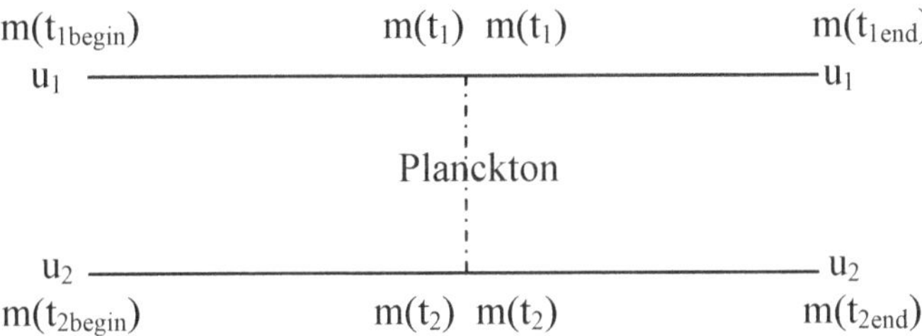

Figure 41.1. A Feynman diagram illustrating the continuity of a universe particle mass through a Planckton interaction.

The reader may verify this by writing the perturbation theory equivalent. A universe particle vertex corresponds to

$$iS_F^{TT}(y_1, y_2)\gamma^\mu iS_F^{TT}(y_2, y_3)$$

The universe particle mass is the same on either side of the interaction vertex.

41.1 Internal Structure of Universe Particles

Planckton field theory gives interactions between baryons. This theory is applicable to universe-universe interactions.[387] It also yields baryon particle – baryon particle interactions as well as baryon particle – universe particle interactions.

---

[386] Since Plancktons travel at the speed of light this section is only of theoretic interest at best.

[387] Again of theoretic interest only due to the vast differences between universes and the limitation of their speed to c.

It is possible for a planckton to be emitted in one universe and interact with a baryon elementary particle in another universe. This type of "probe" must be a high energy probe just as a photon probe of the internal structure of a nucleon[388] must be a high energy photon to bring out the nucleon's internal structure (parton model).

In this section we will discuss planckton probes of other universes, and the internal structure of a universe as a mass distribution governed by gravitation as it relates to universe particles.

41.1.1 Planckton Probes

Plancktons can be generated in one universe and be used to probe the baryon distribution of another universe. Since the planckton propagator is expressed in Megaverse coordinates the baryon distribution in the target universe will be a distribution in Megaverse coordinates. Megaverse coordinates can be expressed in terms of the curved space-time coordinates of a universe x^μ. However the inversion of the map between universe coordinates and Megaverse coordinates

$$x^\mu = f^{-1\mu}(y)$$

is not 1:1 since x^μ is 4-dimensional and y is a D-dimensional vector. The universe coordinates x^μ are each individually determined up to a subspace. One might be concerned about this situation but the determination of the distribution in Megaverse coordinates gives a more direct picture not convoluted by the curvature of the target universe.

The detailed probing of a target universe requires high energy plancktons. The similarity of this procedure to deep inelastic electron-nucleon scattering is obvious to the high energy physicist. But in doing this planckton probe experiment one obtains a picture of a different universe – something that is not possible to do with electromagnetic or graviton probes.

The great problem of this approach is the limitation of planckton velocity to the speed of light.

41.1.2 Internal Structure of a Universe Particle

The development of the theory of universe particles which resulted in the lagrangian appearing earlier does not fully describe universe particles since it neglects the internal structure of a universe particle. The internal structure of a universe particle is primarily determined by gravitation, electromagnetic effects and nuclear physics.

Consequently the full lagrangian of a universe particle has the form

$$\mathcal{L}_{tot} = \mathcal{L}_{internal} + \mathcal{L}$$

---

[388] Deep inelastic electron-nucleon scattering.

where $\mathscr{L}$ is determined above. As a result the complete quantum wave function of a universe particle has the form

$$\psi_{tot} = \psi_{internal}(Y)\psi_{ext}(Y)$$

where $\psi_{internal}(Y)$ is the internal wave function and $\psi_{ext}(Y)$ is external wave function. It seems reasonable to have a separable equation except when universes collide. In that situation a perturbative mixing of the universes and their wave functions applies and it may be possible to calculate the collision output universes by introducing a further interaction between the internal and external aspects of the universe particles.

41.2 Ultra-high LHC Proton-Proton Collisions Resemble Heavy Nuclei Collisions

Our earlier discussion of tachyonic universes interacting, or decaying, to produce heavier universes is buttressed by new particle interaction data from CERN. New 7 TeV proton-proton collision experiments by the CERN ALICE Collaboration[389] have revealed that the end products of these collisions resemble the end products of heavy nuclei collisions.

Tachyonic particles can transform momentum into mass as pointed out earlier. Since d and s[390] quarks are tachyonic and are present in proton-proton collisions, it is likely that the tachyonic mass production mechanism, at least partly, causes pp collisions to simulate heavy nuclei collisions. The quasi-free nature of the quark-gluon plasma generated in ultra-high energy pp collisions lends weight to this interpretation.

The ALICE results may indicate tachyonic mass creation from the momentum of d and s quarks. Normal u and c quarks do not have a tachyonic nature and thus cannot generate mass from momentum. Their collisions can only generate mass from energy.

---

[389] J. Adam et al, Nature Physics (2017). DOI: 10.1038/nphys4111. Data from LHC run 1.
[390] Enhanced production of s quarks is another feature of ultra-high energy pp collisions according to the ALICE collaboration.

42. Megaverse Hadrons, Atoms, Chemistry, and Materials

The higher dimensionality of the Megaverse causes profound changes in the structure of matter. We assume that the particles of the Unified SuperStandard Model are the fundamental particles in the very low density Megaverse space between universes. These particles exist in D dimensions and have corresponding fourier expansions. They have the Unified SuperStandard Model interactions plus additional Megaverse interactions described in chapter 39. Based on these assumptions we can reasonably project the forms of matter that will result over time due to interactions and colisions.

42.1 Hadronic Species

Megaverse quarks must combine in ways satisfying the dictates of SU(3) color confinement. As a result the usual structure of hadrons should be produced as indicated in chapter 18: protons, neutrons, …, charmonium, …, pentaquarks, …

Due to the large Megaverse dimension we can expect that additional varieties of hadrons will exist. The additional spatial dimensions will permit new 'semi-stable' configurations of quarks and antiquarks as these particles disperse in the extra dimensions.

42.2 New Type of Atomic Nuclei

The stable hadrons such as protons, neutrons, and so on will combine using the Megaverse form of the nuclear force to produce atomic nuclei (and using electrons to form atoms.) Atomic nuclei are fairly well modeled by liquid drop models and shell models. The Megaverse equivalents of these models will yield tighter binding and perhaps enhanced stability of isotopes (and a wider variety of isotopes as well.) For example we expect the magic numbers of atomic nucei shells to increase. The result will be new forms of matter. These new types of matter could yield materials with extraordinary properties of density, hardness, pliability, resilience, conductivity, and so on.

42.3 Megaverse Chemistry

They will be the basis of a vast new Megaverse chemistry with complex molecular configurations of atoms due to the large number of new dimensions.

42.4 The Key Factor in the Megaverse

The key factor that profoundly influences the hadrons, atomic nuclei, chemistry, and life forms in the Megaverse is the enhanced propinquity of the components, which leads to stronger forces amongst them. This feature is simply illustrated by considering a one-dimensional array of four constituents composing an entity:

compared to a two-dimensional array:

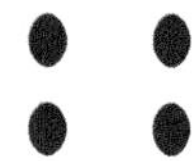

In the two-dimensional case the average spacing between constituents, and thus the force between them, is stronger resulting in tighter binding and denser entities.

In the case of the 192-dimensional Megaverse the effect can only be much more pronounced.

42.5 New Forms of Megaverse Life Beyond Organic

They may also open the possibility of new forms of living creatures beyond organic and hypothesized silicon-based life. See chapter 43.

42.6 Can these New Entities Exist Half in and Half Out of the Universe?

Based on our knowledge of the earth, cosmic rays, and the universe these possibilities do not appear to have been seen. It is conceivable that in regions in the universe, where gravitation and thus spatial curvature is very large, the surface tension of the universe may be near zero. Under these conditions hadrons, atomic nuclei, and atoms might exist with parts in our universe and parts in the Megaverse. Then the question arises: Will the parts that are in our universe be recognizable as in this state? It may be reasonable to search for 'fragments' of these types in spectra and in cosmic rays.

Some spectacular cosmic ray particle events such as the Niu particle event (seen some years ago) may be the result of the entry of a higher dimensional Megaverse particle into our universe.

43. Life Forms in the Megaverse

We have seen that there is evidence for the existence of the Megaverse . We have also seen that it is quaite likely that matter and energy exist in Megaverse space between universes. It is possible that Megaverse matter might have accreted due to gravity and other forces to form stars and planets. If so, then possibly life exists in the Megaverse outside of universes. This possibility raises many provocative questions. We will call Megaverse life *Megaversian.*

We know that life on earth exists in many niches that we would consider hostile: deserts, high altitudes, deep oceans, intense cold and darkness, and even kilometers below the surface of the earth. Life is prevalent on earth.

We have found other solar system bodies have environments that could support life. And we have found planets around distant stars that appear to be capable of supporting life.

Given these facts it sems reasonable to inquire as to the form and nature of life within the D-dimensional Megaverse space.

43.1 Why Life?

If life is prevalent within, and without, our universe then there must be a fundamental reason for Nature's tendency to produce life in just about any 'hospitable' environment. We believe the reason is that *"Life is maximally entropic."* Entropy almost always increases with time. Sometimes quickly; sometimes more slowly. In any given situation where life exists we believe that entropy increases more quickly than it would if life were not present. This hypothesis, of course, needs comprehensive study. But we will assume it as a working hypothesis in our investigation.

43.2 A Necessary Condition for Life

As part of our fundamental hypothesis for the existence of life, we require that energetic processes must exist in an environment where life appears.[391] Living things need energy to develop and grow as well as for motion should they be mobile.

If one considers the simple fact that giant living objects do not pop into existence, but rather start from the small, we see another requirement for life is that its origin is local – life originates in the small. On earth it went through an evolutionary process starting from viruses to cells to multi-cell creatures – eventuslly becoming the small and large creatures of our experience.

---

[391] See Feinberg (1980) for a study of energy processes that could support life.

But one must remember the beginning of all life is in the small and based on local sources of energy that could fuel life in the small and enable it to grow and evolve over time. *We conclude the beginnings of life require a local source of energy.In the earliest stages of the development of life forms, the seed ('virus'?) must utilize an energy source in its immediate vicimity. Only at 'later' stages can life forage for energy over a distance.*

43.3 Types of Known Life

A great deal is known about life on earth. However there is still much to learn. We will pursue the question of Megaverse life based on our knowledge of aspects of earth life.

43.4 Fundamental Aspects of Megaverse Life

It would be nice if the physical laws, coupling constants, and particle masses of the Megaverse were the same as those of our 4-dimensional universe. And it may well be so, since our Continuous Creation Model based on an inflow from the Megaverse (Chapter 14 of Blaha (2017c)) would suggest it.

We shall assume that this is the case but, where possible, not   rely on precise details based on phenomena in our universe. Our focus will be on the impact of D dimensions on features of Megaverse life.

We will assume, therefore, that the stable nuclei of atoms are the same in our universe and the Megaverse. However the arrangement of electron shells, and their energies, around the nuclei of atoms will be different due to $(D - 1)$-dimensional space. As a result the chemistry of molecules and, most particularly, of the equivalent of DNA (if there is one) will differ substantially.

Thus we can expect life in Megaverse space to be quite different from life within our universe. Nevertheless, we can make some 'conclusive' statements based on general considerations.

43.5 Megaversian DNA

DNA is the common denominator of all known forms of life on earth. We assume an equivalent to DNA exists as the basis of life in $(D - 1)$-dimensional Megaverse space (Megaverse 'time' is the additional dimension to make the D-dimensional Megaverse.) Based on earth life, we assume Megaversian DNA, *Mega-DNA*, has the following properties:

1. It consists of long strands of linked nucleotides (the equivalents of cytosine, guanine, adenine, and thymine) as well as equivalents of deoxyribose and a phosphate group.

2. There are two strands, one of which may be viewed as a 'back up' copy, to provide stability and for use in DNA replication.

The purpose of the *two* strands suggests that the number of strands is independent of the dimensionality of space.

The large number of dimensions of Megaverse space has a number of important implications:

1. The long Mega-DNA strands can be compacted to better fit into cells as DNA strands are compacted in balls in cells on earth. Earth DNA strands range up to 2 meters in length. Mega-DNA strands, perhaps of length of the order of

$$2(\text{Megaverse-cell radius})^{(D-4)} \text{ meters}$$

can be fit within Megaverse-life cells.

2. The additional Megaverse dimensions suggest more 'flexibility' in the Mega-DNA to provide more epigenetic adaptability and more capacity for mutations.

3. Mega-DNA should support a greater variety of cell and multi-cell life than found on earth.

Thus we can envision 'fantastic' creatures able to rapidly adapt to changing environmental conditions and to possibly have the feature called 'shape shifting' in popular SciFi movies.

43.6 Megaversian Brain Size

Perhaps the most important effect of a large number of dimensions is intelligence level. In our universe, assuming electromagnetic connections in the brain, the size of brains is constrained by the time it takes for signals to go between parts of the brain. (Currently many investigators believe that consciousness, and possibly abstract thought, are a result of total brain 'collaboration.') The maximum brain size is usually estimated to be slightly larger than that of the human brain.

In the Megaverse we do not have three space dimensions but rather $D - 1$ spatial dimensions. Thus the volume of a Megaverse brain is not proportional to r^3 where r is the radius of the brain. Rather it is proportional to $r^{(D-1)}$. Thus a Megaverse brain of the same radius as an earthly brain has incomparably more volume and contents than an earthly brain of the same radius. Thus if a Megaverse intelligent species should exist one can expect it would have massively more memory and analytical power.

Another aspect of Megaversian brain structure is the possibility of distributed brain power. We are familiar with networks of computers uniting to do massive computations. Some earthly species such as the octopus have a distributed brain structure: an octopus has a central brain and a 'sub-brain' in each of its eight limbs. Distributed brain power in Megaverse creatures could result in formidable computational and analytical abilities.

43.7 Megaversian Locomotion

In our universe, particularly on earth, larger animals tend to have two or four legs. Smaller creatures such as insects may have many more legs. Two-legged (and four-legged) animals appear to have two (or four) legs to be able to turn effectively in three dimensions. They do not have three or five legs because the added mobility is outweighed by the added stability requirements placed on the brain and nervous system.

In the Megaverse we would expect that creatures would have $D - 1$ legs (appendages) to maneuver in $D - 1$ spatial dimensions. It is possible that appendages would have 'sub-brains' like the octopus to off load processing and control of limbs. Whether they have hands and arms is open to question since legs could also play the role of arms.

43.8 Megaversian Vision

In $D - 1$ dimensions it would appear reasonable to have compound eyes like insect species on earth. Processing the data coming from the eyes would be a significant burden on the Megaverse brain.

43.9 Composition of Megaversian Life

If the elements present in Megaverse matter are similar to those in our universe the composition of creatures may be similar to that of life in our universe. However the many Megaverse dimensions, and the expected differences in the electronic structure of atoms and compounds might lead to a different chemical composition of life. Perhaps arsenic-based life might exist. Organic chemicals would be different in composition and structure.

These differences might lead to changes in the environments suitable for life: temperature extremes, atmospheres, and food requirements among other things.

43.10 Megaversian Societies and Civilizations

The average density of matter in the Megaverse may be expected to be low (section 14.14) and, consequently, the number of planets and stars per unit volume will be correspondingly small. On some of these objects life may develop and, on a much smaller number, societies and possibly civilizations may appear. After all, on earth we find complex societies of ants, bees, and so on. Their societies are similar to human socities in many respects[392] despite the vast difference between Mankind and insects. *The key factor in the growth of civilizations is the availability of large amounts of surplus energy.*

---

[392] See Blaha (2010c) for a comprehensive study of human societies and almost all the known civilizations of Mankind. It shows that the 'ups and downs' of civilizations and societies is based on energetics (Thermodynamics).

On this basis we suggest that Megaverse societies and civilizations are a likely possibility given the existence of Megaverse life forms.

43.11 Extension of Human Sight to the Megaverse

If a human entered the Megaverse, his/her sight would be limited to three dimensions since bodily motions would be so limited.[393] To see in all 191 spatial directions a human would need a device (similar to a periscope in concept) that could be oriented to other of the 191 directions. This device would be Megaversian in the sense that part of it would be in the three dimensional subspace defined by the human's orientation, and the rest would be in the remaining 188 dimensions into which the human would want to look. Of course rotating the device to other directions would require force components in those directions. Thus the viewing device would be no simple piece of equipment.

Perhaps the most analogous viewing 'device' in our universe is a multi-faceted insect eye. Each facet provides a view of the surroundings. The corresponding Megaverse 'eye' would have one facet for each spatial direction.

43.12 Megaversian Life

If Megaverse bodies are at all habitable, then one can expect life to exist, and yet to be very different due to the D Megaverse dimensions. Yet despite the differences one can expect certain similarities to life in our universe.

---

[393] One can visualize this situation by considering a 2-dimensional flatworld where the inhabitants could only see things within the flatworld but could not see the things in a third (vertical) dimension. To see things above the flatworld they would need a device that would partly be in the third dimension. These thoughts generalize to the Megaverse case.

44. Unified SuperStandard Model Map to Reality

We have derived a finite Unified SuperStandard Model from basic assumptions and examined some of its consequences. We have gone beyond our universe and considered the possibility of a Megaverse of universes.

Our derivation gave us an unchanging theory of Reality conceptually similar to the concept of Parmenides. The question now becomes: How is that Reality imposed on the material universe that we see – everywhere with 'infinite precision?' There appear to be three possible answers to that question:

1. Our primitives, axioms and derivation are the only possible choices (modulo minor variations). They are naturally adapted to describe physical reality by construction. This view is similar to that of Professor Hawking who said he did not see a necessity for God in the workings of the universe.

2. There are other possible explanations of the features of the material universe such as a SuperString theory. These explanation(s) may lead to our theory.

3. There is an 'Unmoved Mover' that causes Reality (our theory) to be imposed on the material universe (and Megaverse.) Some see this entity as God.

In any case it is clear that our theory maps directly to the material universe that we see although there are questions that remain to be addressed.

Appendix A. The Local Definition of Asymptotic Particle States

This appendix reprints S. Blaha, "The Local Definition of Asymptotic Particle States", IL Nuovo Cimento **49A**, 35 (1979).[394] It describes the PseudoQuantization of boson and fermion field theories for use in the quantization of fields in universes and the Megaverse.

---

[394] © Copyright Stephen Blaha 1978.

IL NUOVO CIMENTO   VOL. 49 A, N. 1   1 Gennaio 1979

The Local Definition of Asymptotic Particle States (*).

S. BLAHA

Physics Department, Williams College - Williamstown, Ma. 01267

(ricevuto il 28 Luglio 1978)

Summary. — A generalization of quantum field theory is described which has a unique particle interpretation — even in space-times where no global timelike co-ordinate exists. The formulation is described in detail for the case of scalar bosons and spin–one-half fermions in flat space-time. We show that it is possible to construct a model in our approach which is physically equivalent to any given model in the usual formulation. In addition, a new class of models can be constructed which are not possible in the usual formulation. This class includes quantum action-at-a-distance models which can be used to develop models with higher-derivative field equations which are unitary. Our formulation allows some latitude in the choice of boundary conditions, so that one can opt for a continuum of possible Green's functions ranging from Feynman propagators to principal-value propagators (half advanced-half retarded).

1. – Introduction.

Our experience in flat space-time has fostered the opinion that a given action leads to a unique quantum field theory upon implementation of the canonical quantization procedure. This is apparently not true in general. A given action corresponds to an infinity of physically inequivalent quantum field theories in nonstatic space-times where no timelike Killing vector exists ([1,2]). The origin of this plurality of quantum theories can be seen in free field the-

---

(*) Supported in part by grants from the National Science Foundation, and Research Corporation.
([1]) S. A. FULLING: *Phys. Rev. D*, **7**, 2850 (1973); C. SOMMERFIELD: *Ann. Phys.*, **84**, 285 (1974).
([2]) B. DEWITT: *Phys. Rep.*, **19**, 295 (1975).

ories (cf. FULLING [1]). The usual quantization procedure is based on a definition of positive frequency which selects an acceptable complete orthonormal set of field equation solutions to use in field quantization. In nonstatic space-times no unique criterion exists for defining positive frequency. As a result, there is no restriction on the choice of complete orthonormal set of field equation solutions used to Fourier-expand fields. Having different choices leads to unitarily (and physically) inequivalent representations of the field algebra. The set of physical particle states in one quantization is generally not unitarily related to the set of physical states in another quantization [1].

The absence of a criterion to select the « correct » quantum-field theory in the usual formulation has led us to consider a generalization of quantum field theory. In this generalization we introduce extra degrees of freedom in such a way that quantizations based on differing definitions of positive frequency are unitarily equivalent. Thus for a given action there is one resulting quantum field theory up to unitary equivalence.

In particular the physical particle states of different quantizations are related by a unitary transformation. Since the particle number operator is invariant under this transformation, a N-particle state in one quantization is a superposition of N-particle states in any other quantization. This is made possible by a local definition of particle states in the Fourier-transformed space (momentum space in the case of flat space-time).

It is important to note that the plethora of inequivalent quantizations in the usual formulation is faced by *one* observer. It is not a question of quantizations in different co-ordinate systems corresponding to different observers. The differences in the quantizations of two relatively accelerating observers, for example, are physically real and, in fact, also exist within the framework of our formulation. Relatively accelerating observers will, in general, « see » different numbers of particles.

Sections **2** and **3** contain our formulation of a free-scalar-boson field theory and a free spin–one-half fermion field theory in flat space-time. Significant differences exist between our formulation and the usual formulation. However, models exist in our formulation which make predictions which are identical to those of conventional field theory models, *e.g.*, quantum electrodynamics. Models also exist in our formulation which are completely outside the framework of the usual formulation. For example, a choice of boundary conditions is possible in our formulation which allows for virtual particles to propagate via non-Feynman propagators. In general, our particle propagator has the form

$$(1) \qquad G = \sin^2 \theta\, G_F + \cos^2 \theta\, C G_F^* C^{-1} ,$$

where θ is an arbitrary angle, G_F the usual Feynman propagator with G_F^* its complex conjugate, and where C is the relevant charge conjugation matrix. G is a Feynman propagator if $\theta = \pi/2$.

If $\theta = \pi/4$ then G is a principal-value propagator (half advanced-half retarded). This type of Green's function has appeared in classical action-at-a-distance theories. Our formulation thus encompases quantum action-at-a-distance theories. The use of principal-value propagators allows a substantial enlargement of the class of unitary, renormalizable field theory models. For example, models with higher-derivative field equations cannot simultaneously satisfy the requirements of positive probabilities and unitarity, if Feynman propagators are used. But if principal-value propagators are used, both requirements can be consistently satisfied [3]. This has allowed us to previously construct a unitary, higher-derivative non-Abelian model of the strong interactions with a manifest linear potential and quark confinement [4]. Of course, the use of principal-value propagators leads to a different type of analytic structure for amplitudes. We take the view that analyticity is an experimental question rather than a fundamental requirement on field theory [5]. It is amusing to note that confinement of color in this model serves to sharply dampen if not eliminate the potential nonanalyticity.

We will discuss our formulation of non-Abelian field theories in detail in a subsequent paper [6].

2. – Boson quantization.

In flat space-time a timelike co-ordinate exists and as a result the Hamiltonian occupies a privileged position in defining positive frequency. In a non-static space-time, with no global timelike co-ordinate, no corresponding operator exists and the definition of « positive frequency » appears to be arbitrary. Consider a free scalar-field theory in such a situation. The field equation has an infinite number of possible complete orthonormal sets of solutions which span the space of solutions. Consider two possible sets: $\{\chi_\alpha, \chi_\alpha^*\}$ and $\{\psi_\beta, \psi_\beta^*\}$, where the χ_α are positive frequency with respect to one definition of positive frequency, and ψ_β are positive frequency with respect to a different definition. Then mode expansions of the scalar field

$$(2) \qquad \varphi(x) = \sum_\alpha [\chi_\alpha(x) A_\alpha + \chi_\alpha^*(x) A_\alpha^\dagger],$$

$$(3) \qquad \varphi(x) = \sum_\beta [\psi_\beta(x) b_\beta + \psi_\beta^*(x) b_\beta^\dagger],$$

---

[3] S. BLAHA: Phys. Rev. D, **10**, 4268 (1974).

[4] S. BLAHA: Phys. Rev. D, **11**, 2921 (1975).

[5] R. E. CUTKOWSKY, P. V. LANDSHOFF, D. I. OLIVE and J. C. POLKINGHORNE: Nucl. Phys., **12** B, 281 (1969); T. D. LEE: in Quanta-Essays in Theoretical Physics Dedicated to Gregor Wentzel, edited by P. G. O. FREUND, C. J. GOBEL and Y. NAMBU (Chicago, Ill., 1970); H. RECHENBERG and E. C. G. SUDARSHAM: Nuovo Cimento, **14** A, 299 (1973).

[6] S. BLAHA: Nuovo Cimento, **49** A, 58 (1978).

can be inverted to relate the Fourier coefficient operators

$$(4) \qquad A_\alpha = \sum_\beta [C_{\alpha\beta} b_\beta + \tilde{C}_{\alpha\beta} b_\beta^\dagger] ,$$

where $C_{\alpha\beta}$ and $\tilde{C}_{\alpha\beta}$ are c-number functions of α and β. Equation (4) shows that A_α is related to b_β and $b_\beta^\dagger$ through a local Bogoliubov transformation. As a result, the quantizations are, in general, not unitarily equivalent, have different vacua, and different particle interpretations ([1,2]). The basis of this difficulty is the noncommutativity of Fourier coefficient operators and their Hermitian conjugates

$$(5) \qquad [b_\beta, b_{\beta'}^\dagger] = \delta_{\beta\beta'} .$$

We shall propose a generalization of quantum field theory in which (in the free-field case) the Fourier coefficient operators and their Hermitian conjugates commute. In order to maintain the quantum character of the theory a sup plementary field and the corresponding Fourier coefficient operators will be introduced. We shall confine our discussion to flat space-time in this section, and in sect. **3** which deals with free spin–one-half fermion quantization. In sect. **4** we discuss the particle interpretation of the formulation in nonstatic space-time.

Let us provisionally introduce the Lagrangian

$$(6) \qquad \mathscr{L} = \partial_\mu \varphi_1 \partial^\mu \varphi_2 - \tfrac{1}{2} \partial_\mu \varphi_1 \partial^\mu \varphi_1 - m^2 \varphi_1 \varphi_2 + \tfrac{1}{2} m^2 \varphi_1^2 .$$

Following canonical procedures, we obtain the field equations

$$(7) \qquad (\Box + m^2)\varphi_i = 0 ,$$

for $i = 1, 2$ and the canonical momenta

$$(8) \qquad \pi_1 = \dot{\varphi}_2 - \dot{\varphi}_1$$

and

$$(9) \qquad \pi_2 = \dot{\varphi}_1 ,$$

which are taken to satisfy the canonical equal-time commutation relations

$$(10) \qquad [\varphi_i(x), \pi_j(y)] = i\delta_{ij}\delta^3(\boldsymbol{x} - \boldsymbol{y}) ,$$

$$(11) \qquad [\varphi_i(x), \varphi_j(y)] = [\pi_i(x), \pi_j(y)] = 0 ,$$

for $i, j = 1, 2$. Equations (9) and (10) imply

$$\text{(12)} \qquad [\varphi_1(x), \dot{\varphi}_1(y)] = 0 \, ,$$

$$\text{(13)} \qquad [\varphi_1(x), \dot{\varphi}_2(y)] = i\delta^3(\boldsymbol{x} - \boldsymbol{y}) \, ,$$

$$\text{(14)} \qquad [\varphi_2(x), \dot{\varphi}_2(y)] = i\delta^3(\boldsymbol{x} - \boldsymbol{y}) \, ,$$

at equal times. The most general form for the mode expansion of the fields is

$$\text{(15)} \qquad \varphi_1(x) = \int \mathrm{d}^3k \, [(C_{11}A_{1k} + C_{12}A_{2k})f_k(x) + (\tilde{C}_{11}A_{1k}^\dagger + \tilde{C}_{12}A_{2k}^\dagger)f_k^*(x)] \, ,$$

$$\text{(16)} \qquad \varphi_2(x) = \int \mathrm{d}^3k \, [(C_{21}A_{1k} + C_{22}A_{2k})f_k(x) + (\tilde{C}_{21}A_{1k}^\dagger + \tilde{C}_{22}A_{2k}^\dagger)f_k^*(x)] \, ,$$

where $(2\pi)^{\frac{3}{2}}(2\omega_k)^{\frac{1}{2}}f_k(x) = \exp[-ik\cdot x]$ and where C_{ij} and $\tilde{C}_{ij}$ are a set of constants. In view of the afore-mentioned difficulties stemming from the non-commutativity of a Fourier-coefficient operator and its Hermitian conjugate, we are led to impose the commutation relations

$$\text{(17)} \qquad [A_{ik}, A_{jk'}] = [A_{ik}^\dagger, A_{jk'}^\dagger] = 0$$

and

$$\text{(18)} \qquad [A_{ik}, A_{jk'}^\dagger] = (1 - \delta_{ij})\delta^3(\boldsymbol{k} - \boldsymbol{k'}) \, ,$$

for $i, j = 1, 2$. We define two vacua (which are in fact related) $|0\rangle_1$ and $|0\rangle_2$ by

$$\text{(19)} \qquad A_{1k}|0\rangle_2 = A_{1k}^\dagger|0\rangle_2 = 0 \, ,$$

$$\text{(20)} \qquad A_{2k}|0\rangle_1 = A_{2k}^\dagger|0\rangle_1 = 0$$

with

$$\text{(21)} \qquad A_{2k}|0\rangle_2 \neq 0 \, , \qquad A_{2k}^\dagger|0\rangle_2 \neq 0 \, ,$$

and

$$\text{(22)} \qquad A_{1k}|0\rangle_1 \neq 0 \, , \qquad A_{1k}^\dagger|0\rangle_1 \neq 0 \, ,$$

for all k. These definitions are motivated by the need for vacua which would be invariant under Bogoliubov transformations—a necessary requirement if the difficulties of particle interpretation caused by relations of the form of eq. (4) are to be avoided. Let us define the local Bogoliubov transformation

$$\text{(23)} \qquad A_{ik}(\lambda_1, \lambda_2) = B_{\lambda_1\lambda_2}A_{ik}B_{\lambda_1\lambda_2}^{-1} =$$

$$= \exp[i\lambda_1]\cosh\lambda_2 A_{ik} + \exp[-i\lambda_1]\sinh\lambda_2 A_{ik}^\dagger \, ,$$

where λ_1 and λ_2 are functions of the momentum k. The operator B has the form

$$(24) \qquad B_{\lambda_1 \lambda_2} = \exp\left[2i\int d^3k\,\lambda_1(k)\,\Gamma_{3k}\right] \exp\left[2i\int d^3k\,\lambda_2(k)\,\Gamma_{2k}\right],$$

where

$$(25) \qquad \Gamma_{3k} = (A_{2k}^\dagger A_{1k} + A_{2k}A_{1k}^\dagger)/2,$$

$$(26) \qquad \Gamma_{2k} = i(A_{2k}^\dagger A_{1k}^\dagger - A_{2k}A_{1k})/2.$$

If we also define

$$(27) \qquad \Gamma_{1k} = -(A_{2k}^\dagger A_{1k}^\dagger + A_{2k}A_{1k})/2,$$

then these operators satisfy the commutation relations of a $SU_{1,1}$ algebra:

$$(28) \quad [\Gamma_{1k},\Gamma_{2k'}] = -i\delta_{kk'}\Gamma_{3k}, \qquad [\Gamma_{2k},\Gamma_{3k'}] = i\delta_{kk'}\Gamma_{1k}, \qquad [\Gamma_{3k},\Gamma_{1k'}] = i\delta_{kk'}\Gamma_{2k}.$$

Thus the group of local Bogoliubov transformations is an infinite tensor product of $SU_{1,1}$ groups. It should be noted that $|0\rangle_2$ and $|0\rangle_1$ are invariant under this group. The equations of motion and equal-time commutation relations are also invariant under this group. These properties will enable us to show the uniqueness of the particle interpretation of our formulation in nonstatic space-time in sect. 4. The Casimir operator for the k-th $SU_{1,1}$ algebra,

$$(29) \qquad \Gamma_k^2 = \Gamma_{3k}^2 - \Gamma_{1k}^2 - \Gamma_{2k}^2,$$

$$(30) \qquad \Gamma_k^2 = N_k(N_k + 2),$$

allows us to identify the particle number operator (cf. sect. 4 below for its derivation)

$$(31) \qquad N = \int d^3k\,N_k = \int d^3k\,[A_{2k}^\dagger A_{1k} + A_{2k}A_{1k}^\dagger],$$

which is left invariant by the Bogoliubov transformations.

If we compare our formulation to the usual one at this stage, we see that the enlargement of the field algebra has allowed us to define a group of local Bogoliubov transformations which is unitary and leaves the vacuum invariant —two properties not possible in the usual approach.

We now define inner products in our formalism. The structure of the commutation relations eq. (17) and (18) together with the nature of the defined vacua suggest that kets can be taken to have the form

$$(32) \qquad |\alpha\rangle = A_{2k_1}^\dagger A_{2k_2}^\dagger \dots A_{2q_1} A_{2q_2} \dots |0\rangle_2$$

and that bras should have the form

$$(33) \qquad \langle\alpha| = {}_1\langle 0|A_{2k_1}A_{1k_2}\dots A_{1q_1}^\dagger A_{1q_2}^\dagger \dots.$$

(We could have chosen to construct kets using $|0\rangle_1$, and bras using $_2\langle 0|$ with no change in consequences.) The form of the commutation relations (which are used to reduce the inner products to a multiple of $_1\langle 0|0\rangle_2$) imply that the dual of the ket space is not its Hermitian conjugate. In our case the algebra reduces inner products to $_1\langle 0|0\rangle_2$ which we define to be unity.

We can relate the dual of a ket to its Hermitian conjugate through the introduction of a Dirac metric operator [8]. We define the operator, γ, by

$$\gamma^{-1}A_{1k}\gamma = A_{2k}, \qquad \gamma^{-1}A_{2k}\gamma = A_{1k}, \tag{34}$$

$$\gamma|0\rangle_1 = |0\rangle_2 . \tag{35}$$

We find

$$\gamma = \exp\left[-\frac{i\pi}{2}\int d^3k\,[A_{2k}^\dagger A_{2k} + A_{1k}^\dagger A_{1k} - A_{2k}^\dagger A_{1k} - A_{2k}A_{1k}^\dagger]\right], \tag{36}$$

which implies that γ satisfies the necessary conditions for a metric operator, $\gamma = \gamma^\dagger = \gamma^{-1}$. The norm of a state $|\alpha\rangle$ can thus be defined by

$$(|\alpha\rangle)^\dagger \gamma |\alpha\rangle \tag{37a}$$

and inner products will generally have the form

$$(|\beta\rangle)^\dagger \gamma |\alpha\rangle . \tag{37b}$$

The adjoint operator is defined by

$$A^* = \gamma^{-1}A^\dagger \gamma . \tag{38}$$

Physical observables must be self-adjoint, $A^* = A$. Self-adjoint operators play the same role as Hermitian operators do in the usual formulation. In particular the Hamiltonian must be self-adjoint, if we are to have conservation of norm. Because q_2 satisfies a Jordan-Pauli commutation relation, we shall introduce interactions in our model using only $q_2(x)$. As a result q_2 must also be self-adjoint if the Hamiltonian is to be self-adjoint.

---

(7) Earlier two field formalisms have been considered by G. MIE: *Ann. Phys. Lpz.*, **37**, 511 (1912); P. A. M. DIRAC: *Comm. Dublin Inst. Advanced Studies*, **180 A**, 1 (1942); W. PAULI: *Rev. Mod. Phys.*, **15**, 175 (1943); M. FROISSART: *Suppl. Nuovo Cimento*, **14**, 197 (1959); T. D. LEE and G. C. WICK: *Phys. Rev. D*, **2**, 1033 (1970). Our motivation and formulation differ substantially from them. Ref. (3,4) above do describe models which can be directly incorporated within the framework of our formulation.
(8) W. PAULI: *Rev. Mod. Phys.*, **15**, 175 (1943).

The energy-momentum tensor is defined by

$$(39) \qquad T^{\mu\nu} = -g^{\mu\nu}\mathscr{L} + \frac{\delta\mathscr{L}}{\delta\partial_\mu\varphi_1}\partial^\nu\varphi_1 + \frac{\delta\mathscr{L}}{\delta\partial_\mu\varphi_2}\partial^\mu\varphi_2$$

with the Hamiltonian given by the 0-0 component. It is easy to verify that the requirements of Poincaré invariance, and the Schwinger commutation relations for $T^{\mu\nu}$ are met.

The equal-time commutation relations, and the self-adjointness of H and φ_2 place six constraints on the constants C_{ij} and $\tilde{C}_{ij}$ in eqs. (15) and (16). After some algebra we find that we are able to express the field operators in the form

$$(40) \qquad \varphi_1(x) = \int d^3k \left[\left(\frac{\cos(\theta_1-\theta_2)}{\sin\theta_1} A_{1k} + \frac{\sin(\theta_1-\theta_2)}{\sin\theta_1} A_{2k} \right) f_k(x) + \right.$$
$$\left. + \left(\frac{\cos(\theta_1-\theta_2)}{\cos\theta_1} A_{1k}^\dagger - \frac{\sin(\theta_1-\theta_2)}{\cos\theta_1} A_{2k}^\dagger \right) f_k^*(x) \right],$$

$$(41) \qquad \varphi_2(x) = \int d^3k \left[(\cos\theta_2 A_{2k} + \sin\theta_2 A_{1k}) f_k(x) + (\sin\theta_2 A_{2k}^\dagger + \cos\theta_2 A_{1k}^\dagger) f_k^*(x) \right],$$

where θ_1 and θ_2 are arbitrary constants which fix the boundary conditions of the Green's functions. (They are *not* related to the Bogoliubov transformations defined above.) We also find

$$(42) \qquad H = \int d^3k\, \omega_k (A_{2k}^\dagger A_{1k} + A_{2k} A_{1k}^\dagger) = 2\int d^3k\, \omega_k \Gamma_{3k}$$

in the free-field case independent of θ_1 and θ_2.

The theory is not invariant under Bogoliubov transformations due to their noncommutativity with H. This is consonant with the absence of any evidence in nature for such an invariance (and related constants of motion). The point of our formulation is to ensure that representations of the field algebra and dynamics, which are related to each other by Bogoliubov transformations, are unitarily equivalent. In the case of flat space-time the unitary equivalence is a moot point, since a unique generator of the dynamics, the Hamiltonian, is apparent. In nonstatic space-times, where no unique generator of the dynamical motion is determined, the unitary equivalence is necessary in order to have an unambiguous quantum field theory (given the action).

Different choices for the generator of the dynamics lead to representations which can be related by Bogoliubov transformations. These representations are unitarily equivalent in our formulation, but not equivalent in the usual formulation. We return to this issue in sect. 4.

The role of θ_1 and θ_2 is evident in the Green's functions. As usual we define the Green's functions as the vacuum expectation values of the time-ordered product of the field operators:

$$(43) \qquad iG_{ij}(x - y) = {}_1\langle 0|T(\varphi_i(x)\varphi_j(y))|0\rangle_2 \,.$$

Equation (41) implies

$$(44) \qquad G_{22}(x - y) = \sin^2 \theta_2 G_{\rm F}(x - y) + \cos^2 \theta_2 G_{\rm F}^*(x - y) \,,$$

where $G_{\rm F}(x - y)$ is the usual Feynman propagator. G_{12} and G_{11} also depend on θ_1 and θ_2, but their precise expressions will not be of use in our presentation.

We shall now show that a model exists within our formulation which is physically equivalent to any conventional scalar quantum field theory with interaction $\mathscr{L}_{\rm I}(\varphi)$. Our model Lagrangian is given by eq. (6) plus the interaction Lagrangian $\mathscr{L}_{\rm I}(\varphi_2)$, where $\mathscr{L}_{\rm I}(\varphi_2)$ is the same function of φ_2 as $\mathscr{L}_{\rm I}(\varphi)$ is of φ. In order to have Feynman propagators, it is necessary to choose the boundary condition $\theta_2 = \pi/2$. We shall demonstrate that an asymptotic state exists in our formulation which corresponds to any asymptotic state of the usual formulation, and then show that S-matrix elements between corresponding states in the two models are equal in any order of perturbation theory.

The construction of asymptotic fields and states in our model is based on the renormalized quadratic part of the Lagrangian (eq. (6)). Therefore the previous development of this section can be used if appropriate subscripts « in » or « out » are appended to the operators. In particular, since $\theta_2 = \pi/2$, we have

$$(45a) \qquad \varphi_{2\rm in}(x) = \int {\rm d}^3k \, [f_k(x) A_{1k\rm in} + f_k^*(x) A_{2k\rm in}^\dagger]$$

by eq. (41). We shall express the in-field operator of the usual formulation by

$$(45b) \qquad \varphi_{\rm in}(x) = \int {\rm d}^3k \, [f_k(x) A_{k\rm in} + f_k^*(x) A_{k\rm in}^\dagger] \,.$$

Note that the form of $\varphi_{2\rm in}$ and $\varphi_{\rm in}$ is identical except for the subscripts « 1 » and « 2 » on the operators. In addition, the commutation relations of the field operators and the Fourier-coefficient operators are also identical except for numerical subscripts. Furthermore, the application of the field operators to the vacua is also identical in effect (except for subscripts)

$$(45c) \qquad \begin{cases} \varphi_{\rm in}(x)|0\rangle = \varphi_{\rm in}^{(-)}(x)|0\rangle \,, & \varphi_{2\rm in}(x)|0\rangle_2 = \varphi_{2\rm in}^{(-)}(x)|0\rangle_2 \,, \\ \langle 0|\varphi_{\rm in}(x) = \langle 0|\varphi_{\rm in}^{(+)}(x) \,, & {}_1\langle 0|\varphi_{2\rm in}(x) = {}_1\langle 0|\varphi_{2\rm in}^{(+)}(x) \,, \end{cases}$$

where the superscript « + » labels positive-frequency parts of the field operator and « − » labels negative-frequency parts. This close parallel in properties between our model and the model of the usual formulation implies the identity

$$(45d) \qquad \langle 0|\mathscr{P}(\varphi_{\rm in})|0\rangle = {}_1\langle 0|\mathscr{P}(\varphi_{2\rm in})|0\rangle_2 \,,$$

where $\mathscr{P}(\varphi_{in})$ is any polynomial in the field φ_{in}. Later we shall use this identity to demonstrate the equality of the S-matrices in our model and the given model of the usual formulation. (Note that a straightforward application of eq. (45d) implies that the propagator G_{22} in our formulation equals the time-ordered propagator of the usual formulation.)

We now state the rule associating asymptotic particle states in our formulation with those of the usual formulation: given an in or out ket of the usual formulation, the corresponding ket in our formulation is obtained by appending the subscript « 2 » to every Fourier-coefficient operator (and to the vacuum) (e.g. $A^{\dagger}_{kin}|0\rangle \Leftrightarrow A^{\dagger}_{2kin}|0\rangle_{2}$). Given an in or out bra of the usual formulation, the corresponding bra in our formulation is obtained by appending « 1 » to each Fourier-coefficient operator (and to the vacuum) (e.g. $\langle 0|A_{kin} \Leftrightarrow {}_{1}\langle 0|A_{1kin}$). It is easily seen that energy-momentum eigenstates in the usual formulation correspond to energy-momentum eigenstates in our formulation. Thus we have identified the set of physical states in our model and find a detailed correspondence to those of the usual formulation.

The development of the perturbation theory of our model is completely analogous to the usual development. The S-matrix relates in and out fields: $\varphi_{2in}(x) = S\varphi_{2out}(x)S^{-1}$. LSZ reduction formulae are derived in the same manner as in the usual formulation. We find the reduction formula for a particle from an in-state and from an out-state to be, respectively,

$$(46) \quad \left|\begin{aligned} &\langle \beta\,\text{out}|\alpha\,p\,\text{in}\rangle = \\ &\quad = \langle \beta - p\,\text{out}|\alpha\,\text{in}\rangle + \frac{i}{\sqrt{Z}}\int d^{4}x\, f_{p}(x)(\overrightarrow{\Box + m^{2}})\langle \beta\,\text{out}|\varphi_{2}(x)|\alpha\,\text{in}\rangle , \\ &\langle \beta k\,\text{out}|\alpha\,\text{in}\rangle = \\ &\quad = \langle \beta\,\text{out}|\alpha - k\,\text{in}\rangle + \frac{i}{\sqrt{Z}}\int d^{4}x\, f_{k}^{*}(x)(\overrightarrow{\Box + m^{2}})\langle \beta\,\text{out}|\varphi_{2}(x)|\alpha\,\text{in}\rangle , \end{aligned}\right.$$

where $\varphi_{2}(x)$ is the interacting field and where we use the notation of ref. ([9]). The reduction of several particles leads to expressions which are identical to corresponding expressions of the usual model if the subscript « 2 » is appended to $\varphi(x)$.

Just as in the conventional model, we can formally develop a perturbation theory based on the U-matrix. The U-matrix relates the interacting and asymptotic field operator

$$(47a) \qquad \varphi_{2}(\boldsymbol{x}, t) = U^{-1}(t)\varphi_{2in}(\boldsymbol{x}, t)U(t)$$

([9]) We follow the conventions and notation of J. D. BJORKEN and S. D. DRELL: *Relativistic Quantum Fields* (New York, N. Y., 1965).

and is easily shown to satisfy the differential equation

$$(47b) \qquad i\frac{\partial U}{\partial t} = -\left[\int \mathrm{d}^3x\, \mathscr{L}_1(\varphi_{2\mathrm{in}})\right] U .$$

Defining $U(t, t') = U(t) U^{-1}(t')$ and solving eq. (47b) gives

$$(48) \qquad U(t, t') = T \exp\left[i\int_{t'}^{t} \mathrm{d}^4x\, \mathscr{L}_1(\varphi_{2\mathrm{in}})\right].$$

The LSZ procedure defined above reduces the calculation of S-matrix elements to the evaluation of time-ordered products of the interacting fields, $_1\langle 0|T(\varphi_2(x_1)\varphi_2(x_2)\ldots \varphi_2(x_N))|0\rangle_2$. The U-matrix can then be used to reduce this quantity to the ratio of matrix elements involving only in-fields:

$$(49) \qquad \frac{_1\langle 0|T(\varphi_{2\mathrm{in}}(x_1)\ldots \varphi_{2\mathrm{in}}(x_N)\exp[i\int\mathrm{d}^4x\,\mathscr{L}_1(\varphi_{2\mathrm{in}})])|0\rangle_2}{_1\langle 0|T(\exp[i\int\mathrm{d}^4x\,\mathscr{L}_1(\varphi_{2\mathrm{in}})])|0\rangle_2} .$$

Expanding to any order in the interaction in eq. (49) gives matrix elements of polynomials in $\varphi_{2\mathrm{in}}$ which are equal—term by term—to corresponding matrix elements of the perturbation theory of the model of the conventional formulation by eq. (45d). Thus S-matrix elements between corresponding states are identically equal in the conventional model and our corresponding model.

It should be noted that only a subset of the possible asymptotic states in our model are identified as physical particle states which correspond to states in the usual model. The operator $A_{2k\mathrm{in}}$ can also be used to create in-kets (and $A^\dagger_{1k\mathrm{out}}$ to create out bras), but the S-matrix elements between physical kets and any ket (or bra) in which these operators appear is zero. (This follows from the fact that $[\mathscr{L}_1(\varphi_{2\mathrm{in}}), A_{2k\mathrm{in}}] = [\mathscr{L}_1(\varphi_{2\mathrm{in}}), A^\dagger_{1k\mathrm{in}}] = 0$ and $_1\langle 0|A_{2k\mathrm{in}} = 0 = A^\dagger_{1k\mathrm{in}}|0\rangle_2$.) Thus the S-matrix is block diagonal in our model. The part of it corresponding to the physical state sector is identical to the S-matrix of the given model of the conventional formulation.

The expression for the vacuum expectation value from which S-matrix elements may be calculated, eq. (49), can be used to show the unitary equivalence of representations which are related by a Bogoliubov transformation. Suppose that we had not used the representation of eq. (45a), but instead the Bogoliubov-transformed representation

$$(50) \quad \varphi^B_{2\mathrm{in}}(x) = \int \mathrm{d}^3k\,[f_k(x)(A_{1k\mathrm{in}}\cosh\lambda + A^\dagger_{1k\mathrm{in}}\sinh\lambda) +$$
$$+ f^*_k(x)(A^\dagger_{2k\mathrm{in}}\cosh\lambda + A_{2k\mathrm{in}}\sinh\lambda)] \equiv B_{0\lambda}\varphi_{2\mathrm{in}}(x)B^{-1}_{0\lambda} .$$

The canonical nature of the transformation guarantees that the canonical commutation relations will be maintained. If we follow the development of the perturbation theory given by eqs. (45)-(49) with φ_{2in}^B replacing φ_{2in} and $\varphi_2^B = U^{-1}(t)\varphi_{2in}^B U(t)$ replacing φ_2, then we find that S-matrix elements are calculated from vacuum expectation values involving only φ_{2in}^B fields:

$$(51) \qquad \frac{{}_1\langle 0 \, | \, T(\varphi_{2in}^B(x_1) \ldots \varphi_{2in}^B(x_N) \exp\left[i\int d^4x \, \mathscr{L}_I(\varphi_{2in}^B)\right]) \, | \, 0\rangle_2}{{}_1\langle 0 \, | \, T(\exp\left[i\int d^4x \, \mathscr{L}_I(\varphi_{2in}^B)\right]) \, | \, 0\rangle_2} \, .$$

Since $B_{0\lambda}^{-1}|0\rangle_2 = |0\rangle_2$ and ${}_1\langle 0|B_{0\lambda} = {}_1\langle 0|$ we find that eq. (51) is equal to eq. (49). Thus the unitary equivalence of representations of the quantum field theory differing by a Bogoliubov transformation is demonstrated. (One can formally define Bogoliubov transformations for interacting fields $B^{int} = UBU^{-1}$, but B^{int} is not unitary due to the well-known difficulties of the U-matrix in the conventional formulation which are also present in our formulation. We circumvent this problem by working with the definition of the S-matrix in terms of vacuum expectation values of asymptotic in-fields, where the unitary equivalence under Bogoliubov transformation can be unambiguously shown to hold.)

A comparison of our formulation with the usual formulation shows a certain similarity of form at the Lagrangian level if our Lagrangian is put in the form

$$(52) \qquad \mathscr{L} = -\frac{1}{2}\partial_\mu(\varphi_1-\varphi_2)\partial^\mu(\varphi_1-\varphi_2) + \partial_\mu\varphi_2\partial^\mu\varphi_2 +$$
$$+ \mathscr{L}_I(\varphi_2) + \frac{m^2}{2}(\varphi_1-\varphi_2)^2 - \frac{m^2}{2}\varphi_2^2 \, .$$

In the usual approach $\varphi_3 = \varphi_1 - \varphi_2$ is an ignorable field and it would not have been surprising that we found equal S-matrix elements above. However, our formulation differs from the usual formulation in two respects—first, the field operators are both expanded in type « 1 » and « 2 » Fourier coefficient operators and, more importantly, the vacuum is defined in a way which correlates the φ_3 and φ_2 sectors. In the $\theta_2 = \pi/2$ case the first difference can be eliminated by a relabeling of Fourier-coefficient operators. However, for other values of θ_2 both differences are present and lead to a very different theory from the usual formulation. While it is clear that the correlation between the φ_3 and φ_2 sectors can be implemented in free field theory, one might ask if this remains true in the interacting case. Certainly the correlation can be implemented in the asymptotic fields and states, since that is free field theory. One can also *formally* implement the correlation for interacting fields through eq. (47a). But the implementation of the correlation in the interacting case is actually based on the reduction of the S-matrix element to the vacuum expectation value of products of asymptotic fields. Since the correlation can be maintained for the asymptotic fields and states, the physical quantities of the models, S-matrix elements, embody the correlation. (The value of the correlation we introduce is twofold: first, it is necessary in order to obtain the unitary equivalence of Bogoliubov rotated representations, and secondly, it widens the range of allowed flat-space-time quantum field theories to include those with principal-value propagators ([6]).)

We conclude this section with a brief discussion of our formulation of the charged-scalar-particle case. In the usual approach, the free-charged-scalar-particle Lagrangian may be expressed in terms of complex fields, $q(x)$ and $q^*(x)$ or in terms of two real fields $q_a(x)$ and $q_b(x)$ with

$$q(x) = [\varphi_a(x) + i\varphi_b(x)]/\sqrt{2} \,. \tag{53}$$

If we follow the same procedure as above for the real fields, double their number and use the Lagrangian form of eq. (6), we are led to the complex field expression of the Lagrangian:

$$\mathcal{L} = \partial_\mu \tilde{\varphi}_2 \partial^\mu \varphi_1 + \partial_\mu \varphi_2 \partial^\mu \tilde{\varphi}_1 - \partial_\mu \tilde{\varphi}_1 \partial^\mu \varphi_1 - m_2 \tilde{\varphi}_2 \varphi_1 - m^2 \varphi_2 \tilde{\varphi}_1 + m^2 \tilde{\varphi}_1 \varphi_1 \,. \tag{54}$$

where

$$q_i(x) = [q_{ia}(x) + i q_{ib}(x)]/\sqrt{2} \tag{55}$$

and

$$\tilde{q}_i(x) = [q_{ia}(x) - i q_{ib}(x)]/\sqrt{2} \,. \tag{56}$$

We require q_{ia} and q_{ib} to embody the same boundary conditions, so that the expansion of q_{ia} and q_{ib} utilizes the same constants, c_{i} and $\tilde{c}_{i}$. Consequently

$$q_i(x) = \int \mathrm{d}^3k [(c_{i1} A_{+1k} + c_{i2} A_{-2k}) f_k + (\tilde{c}_{i1} A^\dagger_{-1k} + \tilde{c}_{i2} A^\dagger_{-2k}) f_k^*] \,, \tag{57}$$

$$\tilde{q}_i(x) = \int \mathrm{d}^3k [(c_{i1} A_{-1k} + c_{i2} A_{-2k}) f_k + (\tilde{c}_{i1} A^\dagger_{+1k} + \tilde{c}_{i2} A^\dagger_{+2k}) f_k^*] \,, \tag{58}$$

where q_i and $\tilde{q}_i$ are related via the charge conjugation operator ([9]). Following the quantization pattern discussed above, with only minor changes due to the presence of two types of Fourier coefficient operators: positive charge, A_{+ik}, and negative charge, A_{-ik}, leads eventually to the following Green's function:

$$G_{22}(x-y) = G_F(x-y) \sin^2 \theta_2 + G_F^*(x-y) \cos^2 \theta_2 \,. \tag{59}$$

Note that it has the same form as eq. (1). (Lagrangian interaction terms are expressed solely in terms of q_2 and $\tilde{q}_2$.) As a result we require $\gamma \tilde{q}_2 \gamma^{-1} = q_2^\dagger$, the Hermitian conjugate of q_2, where γ is the metric operator so that

$$C q_2 C^{-1} = \gamma q_2^\dagger \gamma^{-1} \,, \tag{60}$$

where C is the charge conjugation operator.

3. – Fermion quantization.

In this section we describe our formulation of spin–one-half fermion quantum field theory. Again we are motivated by the need for a unique particle interpretation in nonstatic space-time. The formulation for fermions has close similarities to boson quantization.

Two fields are needed to describe a spin–one-half particle. The Lagrangian is

$$(61) \qquad \mathscr{L} = \tilde{\psi}_2 \gamma^0 (i\nabla - m)\psi_1 + \tilde{\psi}_1 \gamma^0 (i\nabla - m)\psi_2 \quad \tilde{\psi}_1 \gamma^0 (i\nabla - m)\psi_1 .$$

We follow the conventions and notation of ref. (^{9}). The fields $\tilde{\psi}_i$ will be related to the transpose of the charge conjugate field via

$$(62) \qquad \tilde{\psi}_i = \psi_i^{cT} \gamma^0 C^T$$

for $i = 1, 2$. (In the usual formulation $\psi^\dagger = \tilde{\psi}$ would hold.) The equations of motion are

$$(63) \qquad (i\nabla - m)\psi_i = 0 , \quad \tilde{\psi}_i(i\overleftarrow{\nabla} - m) = 0 ,$$

for $i = 1, 2$. The momentum conjugate to ψ_1 is

$$(64) \qquad \pi_1 = i(\tilde{\psi}_2 - \tilde{\psi}_1)$$

and the conjugate to ψ_2 is

$$(65) \qquad \pi_2 = i\tilde{\psi}_1 .$$

The canonical equal-time anticommutation relations imply

$$(66) \qquad \{\psi_{1\alpha}(x), \tilde{\psi}_{1\beta}(y)\} = 0 ,$$

$$(67) \qquad \{\psi_{1\alpha}(x), \tilde{\psi}_{2\beta}(y)\} = \delta_{\alpha\beta}\delta^3(\boldsymbol{x} - \boldsymbol{y})$$

and

$$(68) \qquad \{\psi_{2\alpha}(x), \tilde{\psi}_{2\beta}(y)\} = \delta_{\alpha\beta}\delta^3(\boldsymbol{x} - \boldsymbol{y}) .$$

The most general form for the mode of expansion of the fields is (^{9})

$$(69) \qquad \psi_i = \sqrt{2m} \sum_s \int d^3k \left[(c_{i1}b_{1ks} + c_{i2}b_{2ks}) f_k(x) u_{ks} + (\tilde{c}_{i1}d_{1ks}^\dagger + \tilde{c}_{i2}d_{2ks}^\dagger) f_k^*(x) v_{ks} \right] .$$

Just as in the charged scalar case, we develop our formulation in such a way that the even and odd charge conjugation combinations, $\psi_i \pm \psi_i^c$, implement the same boundary conditions. Therefore

$$(70) \qquad \tilde{\psi}_i = \sqrt{2m} \sum_s \int d^3k \left[(c_{i1} d_{1ks} + c_{i2} d_{2ks}) f_k(x) v_{ks}^\dagger + (\tilde{c}_{i1} b_{1ks}^\dagger + \tilde{c}_{i2} b_{2ks}^\dagger) f_k^*(x) u_{ks}^\dagger \right] .$$

The nonzero Fourier-coefficient anti-commutation relations are

$$(71) \qquad \{ d_{iks}, d_{jk's'}^\dagger \} = \{ b_{iks}, b_{jk's'}^\dagger \} = (1 - \delta_{ij}) \delta_{ss'} \delta^3(k - k')$$

for $i, j = 1, 2$. The definition of states and inner products mirror the boson case. The vacua are defined by

$$(72) \qquad b_{1ks}|0\rangle_2 = b_{1ks}^\dagger|0\rangle_2 = d_{1ks}|0\rangle_2 = d_{1ks}^\dagger|0\rangle_2 = 0$$

and

$$(73) \qquad b_{2ks}|0\rangle_1 = b_{2ks}^\dagger|0\rangle_1 = d_{2ks}|0\rangle_1 = d_{2ks}^\dagger|0\rangle_1 = 0$$

and are related by a metric operator η:

$$(74) \qquad \eta|0\rangle_1 = |0\rangle_2$$

which satisfies $\eta = \eta^\dagger = \eta^{-1}$. We conventionally choose to construct kets from $|0\rangle_2$ and define their dual as their Hermitian conjugate multiplied by the metric operator. Thus inner products have the form

$$(75) \qquad \langle \alpha | \beta \rangle = (|\alpha\rangle)^\dagger \eta |\beta\rangle .$$

Physical observables must be self-adjoint, $A = A^* = \eta^{-1} A^\dagger \eta$, in order to have real eigenvalues. The Hamiltonian must be self-adjoint in order to conserve the norm. In view of eq. (68) we only use ψ_2 and $\tilde{\psi}_2$ in interaction terms and therefore require $\tilde{\psi}_2 = \eta^{-1} \psi_2^\dagger \eta$, so that

$$(76) \qquad C \psi_2 C^{-1} = \eta^{-1} C \bar{\psi}_2^T \eta ,$$

which bears comparison with eq. (15.112) of ref. ([9]) and also eq. (60) above. The equal-time anticommutation relations, eq. (76), and the adjointness of H restrict the constants c_{ij} and $\tilde{c}_{ij}$ so that

$$(77) \qquad \psi_1 = \sqrt{2m} \sum_s \int d^3k \left[(\cos(\theta_1 - \theta_2) b_{1ks} + \sin(\theta_1 - \theta_2) b_{2ks}) f_k(x) u_{ks} \sin\theta_1 + \right.$$
$$\left. + (\cos(\theta_1 - \theta_2) d_{1ks}^\dagger - \sin(\theta_1 - \theta_2) d_{2ks}^\dagger) f_k^*(x) v_{ks} \cos\theta_1 \right]$$

and

$$(78) \qquad \psi_2 = \sqrt{2m} \sum_s \int d^3k \left[(\sin\theta_2 b_{1ks} + \cos\theta_2 b_{2ks}) f_k(x) u_{ks} + \right.$$
$$\left. + (\cos\theta_2 d_{1ks}^\dagger + \sin\theta_2 d_{2ks}^\dagger) f_k^*(x) v_{ks} \right] .$$

The Hamiltonian is

$$(79) \qquad H = \sum_{s} \int d^3k\, \omega_k (b^\dagger_{2ks} b_{1ks} - b_{2ks} b^\dagger_{1ks} + d^\dagger_{2ks} d_{1ks} - d_{2ks} d^\dagger_{1ks}) \, .$$

In contrast to the usual formulation, we see that our Hamiltonian does not have an infinite vacuum energy with respect to $|0\rangle_2$. It is not positive definite, but we will be able to develop a unitary S-matrix theory in the space of positive-energy asymptotic states, if we choose $\theta_2 = \pi/2$. This is evident from an examination of the Green's function

$$(80) \qquad S_{22}(x-y) = - i_1 \langle 0| T\big(\psi_2(x)\, \tilde\psi_2(y)\gamma_0\big)|0\rangle_2 =$$

$$(81) \qquad = \sin^2\theta_2 S_F(x-y) + \cos^2\theta_2\, C\gamma^0 S_F^*(C\gamma^0)^{-1} \, ,$$

which gives $S_{22} = S_F$, the usual Feynman propagator, if $\theta_2 = \pi/2$. As in the boson case, we introduce interactions only through type-2 operators, and use type-2 operators to LSZ reduce in and out particles. The result is a perturbation theory which, in the Fermion sector, only involves time-ordered products of ψ_2 and $\tilde\psi_2$. Thus we can establish a model in our formulation which is equivalent, so far as S-matrix elements are concerned, to any given model of the usual formulation. In particular, our model electrodynamics has the Lagrangian

$$(82) \qquad \mathcal{L} = \quad \tfrac{1}{2} F^1_{\mu\nu} F^{2\mu\nu} + \tfrac{1}{4} F^1_{\mu\nu} F^{1\mu\nu} + \tilde\psi_2\gamma^0(i\nabla - m)\psi_1 +$$

$$+ \tilde\psi_1\gamma^0(i\nabla - m)\psi_2 - \tilde\psi_1\gamma^0(i\nabla - m)\psi_1 - e_0\tilde\psi_2\gamma^0 \hat{A}_2\psi_2$$

with

$$(83) \qquad F^i_{\mu\nu} = \partial_\nu A_{i\mu} - \partial_\mu A_{i\nu}$$

for $i = 1, 2$. While we will discuss this model more fully elsewhere ([6]) two things are worth noting. First the interaction is expressed solely in terms of fields of type 2—both for fermions and the photon. Following our quantization procedure leads to S-matrix expressions which are term-by-term equal to corresponding expressions in QED. Secondly, the model is gauge invariant. The gauge transformation is

$$(84) \qquad \psi_2 \rightarrow \exp[i\Lambda]\psi_2 \, ,$$

$$(85) \qquad \tilde\psi_2 \rightarrow \exp[-i\Lambda]\tilde\psi_2 \, ,$$

$$(86) \qquad A_{2\mu} \rightarrow A_{2\mu} - \frac{1}{e}\partial_\mu\Lambda \, ,$$

$$(87) \qquad \psi_1 \rightarrow \psi_1 + \big(\exp[i\Lambda] - 1\big)\psi_2 \, ,$$

$$(88) \qquad \tilde\psi_1 \rightarrow \tilde\psi_1 + \big(\exp[-i\Lambda] - 1\big)\tilde\psi_2 \, ,$$

and its associated conserved current is

$$(89) \qquad J_\mu = -i \frac{\delta \mathscr{L}}{\delta \partial^\mu \psi_1} \psi_2 - i \frac{\delta \mathscr{L}}{\delta \partial^\mu \psi_2} \psi_2 \, ,$$

$$(90) \qquad J_\mu = \tilde{\psi}_2 \gamma^0 \gamma_\mu \psi_2 \, .$$

We close our discussion of fermions by considering the case $\theta_2 = \pi/4$ which, by eq. (81), gives the principal-value propagator

$$(91) \qquad S_{22}(x-y) = \int \frac{\mathrm{d}^4 k}{(2\pi)^4} \exp\left[-ik\cdot(x-y)\right](k+m)\frac{P}{k^2-m^2} \, ,$$

where

$$(92) \qquad \frac{P}{k^2-m^2} = \frac{1}{2}\left(\frac{1}{k^2-m^2+i\varepsilon} + \frac{1}{k^2-m^2-i\varepsilon}\right).$$

Thus a quantum action-at-a-distance model of fermions can be constructed within the framework of our formulation.

4. – Particle interpretation.

In this section we shall show that the particle interpretation of our formulation of quantum field theory is well defined for the case of a free scalar particle in a nonstatic space-time where no global timelike co-ordinate exists. We assume an action of the form

$$(93) \qquad S = \int \mathrm{d}^4 x \left[\varphi_2 D\varphi_1 - \tfrac{1}{2}\varphi_1 D\varphi_1\right] + \text{(surface terms)} \, ,$$

which under the variation of S gives the field equations

$$(94) \qquad D\varphi_1 = D\varphi_2 = 0 \, .$$

The self-adjointness of D implies

$$(95) \qquad \int_V \left[f^* Dg - (Df)^* g\right] \mathrm{d}^4 x = \int_{\Sigma_v} f^* \overleftrightarrow{D}{}^\mu g \, \mathrm{d}\Sigma_\mu \, ,$$

where Σ_v is the surface bounding V, $\mathrm{d}\Sigma_\mu$ is an outward directed surface element of Σ_v and D^μ is a two-edged vector differential operator. If Σ is a spacelike complete Cauchy hypersurface for the field equations (we assume they exist), then an inner product for complex solutions of the field equations can be de-

52

fined by

$$(96) \qquad (v_1, v_2) = i\int_{\Sigma} v_1^* \overleftrightarrow{D}^\mu v_2 \, d\Sigma_\mu \,.$$

We now choose an arbitrary complete orthonormal set of pairs of complex conjugate solutions of eq. (94), $\{V_\alpha, V_\alpha^*\}$, satisfying

$$(97) \qquad (V_\alpha, V_{\alpha'}) = - (V_\alpha^*, V_{\alpha'}^*) = \delta_{\alpha\alpha'} \,,$$

$$(98) \qquad (V_\alpha, V_{\alpha'}^*) = 0 \,,$$

and use them in the mode expansion of the field operators

$$(99) \qquad \varphi_i = \sum_\alpha [(c_{i1} A_{1\alpha} + c_{i2} A_{2\alpha}) V_\alpha + (\tilde{c}_{i1} A_{1\alpha}^\dagger + \tilde{c}_{i2} A_{2\alpha}^\dagger) V_\alpha^*] \,,$$

where c_{ij} and $\tilde{c}_{ij}$ are real c-numbers. The Fourier coefficient operators satisfy

$$(100) \qquad [A_{i\alpha}, A_{j\alpha'}^\dagger] = (1 - \delta_{ij})\delta_{\alpha\alpha'}$$

with all other commutators equal to zero. The commutativity of Fourier-coefficient operators and their Hermitian conjugate allows us to define the vacua $|0\rangle_1$ and $|0\rangle_2$ by

$$(101) \qquad A_{1\alpha}|0\rangle_2 = A_{1\alpha}^\dagger|0\rangle_2 = A_{2\alpha}|0\rangle_1 = A_{2\alpha}^\dagger|0\rangle_1 = 0 \,.$$

We choose to construct states from $|0\rangle_2$. The one-particle ket corresponding to the Fourier transform variable α is

$$(102) \qquad |\alpha\rangle = - (v_\alpha^*, \varphi_2)|0\rangle_2/\tilde{c}_{22}$$

and the one-particle bra dual to it is

$$(103) \qquad \langle\alpha| = (|\alpha\rangle)^\dagger \gamma \,,$$

where γ is a metric operator satisfying $\gamma = \gamma^{-1} = \gamma^\dagger$ and

$$(104) \qquad \gamma^{-1} A_{i\alpha} \gamma = \varepsilon_{ij} A_{j\alpha}$$

with $\varepsilon_{11} = \varepsilon_{22} = 0$ and $\varepsilon_{12} = \varepsilon_{21} = 1$. The further development of this quantization proceeds along the lines of sect. 2. In particular φ_2 is self-adjoint.

We now introduce a quantization of the particle described by S which parallels the above development in every detail except that a different complete orthonormal set of field equation solutions $\{W_\beta, W_\beta^*\}$ is used in the mode expansion of the fields

$$(105) \qquad q_i = \sum_\beta \left[(c_{i1} A_{1\beta} + c_{i2} A_{2\beta}) W_\beta + (\tilde{c}_{i1} A_{1\beta}^\dagger + \tilde{c}_{i2} A_{2\beta}^\dagger) W_\beta^* \right].$$

The question arises: how are we to relate the two quantizations? In the usual formulation only one answer is apparent—the field operators are to be identified [1,2], since they are uniquely determined by the field equations and the canonical commutation relations. But in the present case, the field operators are not uniquely determined, so that the identification of the fields in eq. (105) and (99) is not required. The relation between the quantizations must obviously be well defined (in the sense that every operator and state in one quantization can be uniquely expressed in terms of operators and states of the other representations). More importantly, it must only relate operators whose properties are fixed by the field equation and the canonical commutation relations; and whose expectation values are uniquely specified by purely geometrical restrictions on their support and do not embody a definition of positive frequency. In our formalism, the operators which satisfy these requirements are linear combinations of

$$(106) \qquad q_{iv}^{\mathrm{II}} = \sum_\alpha \left(A_{i\alpha} V_\alpha + A_{i\alpha}^\dagger V_\alpha^* \right),$$

$$(107) \qquad q_{iw}^{\mathrm{II}} = \sum_\beta \left(A_{i\beta} W_\beta + A_{i\beta}^\dagger W_\beta^* \right),$$

for $i = 1, 2$ in the respective quantizations we can restrict the discussion to these quantities. In particular, the vacuum expectation value

$$(108) \qquad {}_1\langle 0| q_1^{\mathrm{II}}(x) q_2^{\mathrm{II}}(y)|0\rangle_2 = \tfrac{1}{2}\,{}_1\langle 0|[q_1^{\mathrm{II}}(x), q_2^{\mathrm{II}}(y)]|0\rangle_2 ,$$

$$(109) \qquad {}_1\langle 0| q_1^{\mathrm{II}}(x) q_2^{\mathrm{II}}(y)|0\rangle_2 = \frac{i}{2}\,\Delta(x - y) ,$$

where $\Delta(x - y)$, the commutator function, has the well-defined geometrical property that it vanishes at spacelike distances.

Identifying q_{iw}^{II} with q_{iv}^{II}, for $i = 1, 2$, leads to the relations

$$(110) \qquad A_{i\beta} = (W_\beta, q_i^{\mathrm{II}}) = \sum_\alpha \left[(W_\beta, V_\alpha) A_{i\alpha} + (W_\beta, V_\alpha^*) A_{i\alpha}^\dagger \right],$$

for $i = 1, 2$ plus Hermitian-conjugate expressions. The form of the inner products on the right-hand side of eq. (110) is determined by requiring that the definition of positive frequency implicit in the separation of the orthonormal set $\{W_\beta, W_\beta^*\}$ into complex conjugate pairs of solutions can also be implemented by linear combinations of V_α and V_α^*. Specifically, we assume that a complete orthonormal set of pairs of complex conjugate functions, $\{V_\alpha, V_\alpha^*\}$, exists which satisfies

$$(111) \qquad V_\alpha = c_1 \tilde{V}_\alpha + c_2 \tilde{V}_\alpha^* ,$$

$$(112) \qquad V_\alpha^* = c_1^* \tilde{V}_\alpha^* + c_2^* \tilde{V}_\alpha ,$$

for all α, where c_1 and c_2 are c-number functions of α only with $|c_1| > |c_2|$, and where

$$(113) \qquad (W_\beta, \tilde{V}_\alpha^*) = 0 ,$$

$$(114) \qquad (W_\beta^*, \tilde{V}_\alpha) = 0 ,$$

for all β. The orthogonality conditions imply

$$(115) \qquad c_1 = \exp[i\lambda_1] \cosh \lambda_2 ,$$

$$(116) \qquad c_2 = \exp[i\lambda_1] \sinh \lambda_2 ,$$

where λ_1 and λ_2 are solely functions of α. The substitution of eqs. (111) and (112) in eq. (110) and use of eqs. (113) and (114) gives

$$(117) \qquad A_{i\beta} = \sum_\alpha (W_\beta, \tilde{V}_\alpha)[\exp[i\lambda_1] \cosh \lambda_2 A_{i\alpha} + \exp[-i\lambda_1] \sinh \lambda_2 A_{i\alpha}^\dagger]$$

for $i = 1, 2$. Note that the bracketed term on the right-hand side of the equation has the same form as the Bogoliubov rotated Fourier-coefficient operator given in eq. (23). In the present case we can rewrite eq. (117) in the form

$$(118) \qquad A_{i\beta} = \sum_\alpha (W_\beta, \tilde{V}_\alpha) B_{\lambda_1\lambda_2} A_{i\alpha} B_{\lambda_1\lambda_2}^{-1}$$

with

$$(119) \qquad B_{\lambda_1\lambda_2} = \exp\left[2i \sum_\alpha \lambda_1(\alpha) \Gamma_{3\alpha}\right] \exp\left[2i \sum_\alpha \lambda_2(\alpha) \Gamma_{2\alpha}\right] ,$$

where $\Gamma_{3\alpha}$ and $\Gamma_{2\alpha}$ are obtained from eqs. (25) and (26) by replacing the subscripts k with α.

The particle interpretations of the two quantizations will now be shown to be identical. First we note that the vacuum $|0\rangle_2$ of the « α » quantization is invariant under $B_{\lambda_1\lambda_2}$, so that it may be taken to be identical with the $|0\rangle_2$ vacuum of the « β » quantization. Next we note that the canonical commutation relations and the vacuum expectation value of any product of field operators are invariant under B:

$$(120) \qquad {}_1\langle 0| \varphi_{i_1}(x_1) \varphi_{i_2}(x_2) \dots |0\rangle_2 = {}_1\langle 0| B^{-1} \varphi_{i_1}(x_1) \varphi_{i_2}(x_2) \dots B |0\rangle_2 .$$

This implies that we could replace A_{ix} with $B_{\lambda_1\lambda_2}A_{ix}B_{\lambda_1\lambda_2}^{-1}$ in the mode expansions, eq. (107), with no change in physical consequences. In particular, this applies to the definition of particle kets. Equation (102) becomes

$$(121) \qquad |\alpha\rangle = (\exp[-i\lambda_1]\cosh\lambda_2 A_{2\alpha}^\dagger + \exp[i\lambda_1]\sinh\lambda_2 A_{2\alpha})|0\rangle_2 .$$

Consequently, $A_{2\beta}^\dagger|0\rangle_2$ is a superposition of one-particle states in the « α » quantization. In general, the N-particle state in the « β » quantization is a superposition of N-particle states in the « α » quantization.

The invariance of particle number under Bogoliubov transformations is reflected in the relation between the particle number operator,

$$(122) \qquad N = \sum_\alpha (A_{2\alpha}^\dagger A_{1\alpha} - A_{2\alpha}A_{1\alpha}^\dagger),$$

which is invariant under Bogoliubov transformations, and related to the Casimir operator of the Bogoliubov group (cf. eqs. (29)-(31)). Our identification of N as the particle number operator is based, as most charge and number operators are, on an invariance of the action under a global change of phase of fields. In our case we note that the action of eq. (93) is invariant under the infinitesimal phase change

$$(123) \qquad \varphi_1 \to \varphi_1 + i\varepsilon\varphi_1 , \qquad \varphi_2 \to \varphi_2 + i\varepsilon(\varphi_1 - \varphi_2) .$$

The corresponding conserved-number operator is given by

$$(124) \qquad N = i\int_{\Sigma_v} \varphi_1 \overleftrightarrow{D}{}^\mu \varphi_2 \, d\Sigma_\mu .$$

Because φ_1 and φ_2 implicitly embody a definition of positive frequency, we are led to replace them with Hermitian operators:

$$(125) \qquad N = i\int_{\Sigma_v} \varphi_1^H D^\mu \varphi_2^H \, d\Sigma_\mu$$

with φ_i^H given in eq. (106). Equation (125) can be evaluated by using eq. (96) and (97) to give eq. (122). Thus our definition of number operator is physically motivated. It is also consistent with our expectations of a number operator.

We shall now summarize our picture of second quantization in curved space-time where no global timelike Killing vector is present. Consider a complete spacelike Cauchy hypersurface. At each point on the surface there is a local timelike direction. There is, in general, a class of operators, which will locally generate a displacement in the timelike direction, but which globally generate very different motions. Due to the absence of a global timelike Killing vector, no member of the class of potential generators of the dynamics is physically

selected as the generator of the dynamics. One is free to choose any member
as the generator of the dynamics locally. Each choice implies a different definition of positive and negative frequency when field operators are represented
by Fourier expansions.

In the usual formulation of quantum field theory any choice of generator
of the dynamics (and thus Fourier representation of field operators) is unitarily
inequivalent to any other choice in general. As a result each choice gives a *different physical theory* and the second quantization of a theory is not unique. Prac-
tically, this means that 1) a one-particle state in one quantization is a many-
particle state in any other quantization (particle number is ambiguous), 2) in
general one can construct a one-particle state which is an eigenstate of a generator in one representation, but one cannot construct a one-particle state in
another representation which is an eigenstate of the same generator (the space
of states is different), and 3) (if interactions are introduced) the S-matrix differs from quantization to quantization. Obviously, there are only two acceptable alternatives in this situation; either some new principle selects one representation as the correct physical representation, or a modification of quantum
field theory is necessary. In the absence of a new physical principle, we have
formulated a modification of quantum field theory.

Our formulation allows one to quantize a field theory with any of the potential generators of the dynamics and yet to have a physically unique theory.
Different quantizations can be related by Bogoliubov transformations and, in
our formulation, are unitarily equivalent. Consequently, particle number is
invariant—N-particle states in one representation are superpositions of N-par-
ticle states in any other representation; the set of states in one representation is
unitarily equivalent to the set of states in any other representation; and the
S-matrix is uniquely determined in the case of an interacting theory (the proof
is analogous to that of the flat space-time case discussed in sect. **2**). Our formulation associates a unique physical theory with any given action. In a sense,
it implements an equivalence principle in the space of solutions to the field
equations—any complete orthonormal set of solutions to the field equations can
be used in the Fourier expansion of field operators and a unique physical theory
results (cf. eqs. (111)-(114)).

In conclusion, we note that the problem we have addressed relates to one
observer and the ambiguities of conventional quantum field theory he must face.
Different observers in relatively accelerating frames will not see the same number
of particles in our formulation. Neither is particle creation near black holes
precluded in our formulation.

* * *

I am grateful to the Aspen Center for Physics for its hospitality while part
of this work was being done, and to M. A. B. BEG, S. MANDELSTAM and D. PARK
for stimulating conversations.

● RIASSUNTO (*)

Si descrive una generalizzazione della teoria quantistica dei campi che ha un'unica interpretazione particellare — anche negli spazio-tempo in cui non esiste alcuna coordinata globale di tipo tempo. La formulazione è descritta in dettaglio per i casi di bosoni scalari e di fermioni con spin $\frac{1}{2}$ nello spazio-tempo piatto. Si mostra che è possibile costruire un modello nel nostro approccio che è fisicamente equivalente a un qualsiasi modello nella solita formulazione. Inoltre, si può costruire una nuova classe di modelli che non sono possibili nella solita formulazione. Questa classe comprende modelli quantici di azione a distanza che possono essere usati per sviluppare modelli con equazioni di campo a derivata più alta che sono unitarie. La nostra formulazione permette ampia scelta delle condizioni limite, cosicché si può optare per un continuo di possibili funzioni di Green che vanno dai propagatori di Feynman a propagatori del valore principale (mezzo avanzato-mezzo ritardato).

(*) *Traduzione a cura della Redazione.*

Локальное определение асимптотических состояний частиц.

Резюме (*). — Описывается обобщение квантовой теории поля, которая имеет единую частичную интерпретацию — даже в пространстве и времени, где не существует глобальной времениподобной координаты. Подробно описывается формулировка для случаев скалярных бозонов и фермионов со спином половина в плоском пространстве-времени. Мы показываем, что имеется возможность сконструировать модель в нашем подходе, которая физически эквивалентна любой заданной модели в обычной формулировке. Кроме того, может быть сконструирован новый класс моделей, которые являются невозможными в обычной формулировке. Этот класс включает модели квантового действия на расстоянии, которые могут быть использованы для развития моделей с полевыми уравнениями с высшими производными, которые являются унитарными. Наша формулировка допускает некоторую свободу в выборе граничных условий, так что имеется возможность выбрать континуум возможных гриновских функций, от фейнмановских пропагаторов до пропагаторов главных значений (наполовину опережающая - наполовину запаздывающая).

(*) *Переведено редакцией.*

Appendix B. New Framework for Gauge Field Theories

This appendix reprints S. Blaha, "New Framework for Gauge Field Theories", IL Nuovo Cimento **49A**, 113 (1979).[395] It describes the PseudoQuantization of gauge field theories for the purposes of defining higher derivative field theories and for use in the quantization of fields in universes and the Megaverse.

---

[395] © Copyright Stephen Blaha 1978.

IL NUOVO CIMENTO   Vol. 49 A, N. 1   1 Gennaio 1979

New Framework for Gauge Field Theories (*).

S. BLAHA

Physics Department, Williams College - Williamstown, Ma. 01267

(ricevuto l'8 Agosto 1978)

Summary. — We formulate gauge theories within the framework of a generalization of quantum field theory. In particular, we discuss models of electrodynamics and of Yang-Mills theories, a model of the strong interaction with higher-order derivatives and quark confinement and a renormalizable model of pure quantum gravity with Einstein Lagrangian. In the case of electrodynamics we show that two models are possible: one with predictions which are identical to QED and one which is a quantum action-at-a-distance model of electrodynamics. In the case of Yang-Mills theories we can construct a model which is identical in predictions to any conventional model, or a quantum action-at-a-distance model. In the second case it is possible to eliminate all loops of Yang-Mills particles (in all gauges) in a manner consistent with unitarity. A variation of Yang-Mills models exists in our formulation which has higher-order derivative field equations. It is unitary and has positive probabilities. It can be used to construct a model of the strong interactions which has a linear potential and manifest quark confinement. Finally we show how to construct an action-at-a-distance model of pure quantum gravity (whose classical limit is the dynamics of the Einstein Lagrangian) coupled to an external classical source. The model is trivially renormalizable.

1. – Introduction.

Because of the absence of an acceptable physical interpretation of conventional quantum field theory in the case of curved nonstatic space-time, we

---

(*) Supported in part by Research Corporation and the National Science Foundation.

recently developed a modified formulation of quantum field theory [1]. We showed it had a unique physical particle interpretation in nonstatic space-time. We described the flat–space-time formulation for the cases of scalar particles and spin–one-half fermions. In both cases we found that it was possible to construct a model which was equivalent in predictions to any model of the usual formulation. However, it was also possible to construct models which had no analogue in the usual formulation. In these models the quantum exchange of a « particle » did not take place via a Feynman propagator. Rather, the propagator G could be chosen to be any of a continuum of possibilities ranging from the Feynman propagator to the principal-value (half advanced-half retarded) propagator. The choice is parametrized by an angle θ, whose specification is equivalent to a choice of boundary condition

$$(1) \qquad G(x-y) = \sin^2 \theta \, G_F(x-y) + \cos^2 \theta \, C G_F^* C^{-1},$$

where G_F is the usual Feynman propagator with G_F^* its complex conjugate and C is an appropriate charge conjugation matrix.

The new degree of freedom, represented by θ, leads to new possibilities for the formulation of flat–space-time quantum field theory models which will be especially evident in the case of gauge theories.

In sect. **2** we shall explore models of quantum electrodynamics which occur within the framework of our formulation. We shall show that one model is completely equivalent to QED in its predictions. In addition, we shall show that a quantum action-at-a-distance electrodynamics is also possible.

In sect. **3** we describe model Yang-Mills theories and show that a model equivalent to any conventional model can be formulated as well as a quantum action-at-a-distance model. In the second case all non-Abelian boson loops can be eliminated (in all gauges) in a manner which is consistent with unitarity.

In sect. **4** we describe a non-Abelian model of quark confinement based on higher-derivative field equations for which a unitary physical S-matrix is obtained by the choice of principal-value propagators. This demonstrates the utility of principal-value propagators for widening the class of unitary quantum field theories to include higher-derivative theories.

In sect. **5** we describe a model of pure quantum gravity based on the Einstein Lagrangian which is trivially renormalizable, if gravitons propagate via principal-value propagators. (No graviton loops.) This illustrates the potential of the principal-value propagator to ameliorate renormalization problems by allowing one to limit the self-interactions of quantum fields.

In view of the unfamiliar features of principal-value propagators, we shall

---

[1] S. BLAHA: *Nuovo Cimento*, **49** A, 35 (1978).

devote the remainder of this section to a discussion of their properties in relation to causality and analyticity.

Among the first appearances of principal-value, or half advanced-half retarded, propagators was in Feynman's space-time approach to quantum electrodynamics [2], where he observed that one is naturally led to such a propagator for the photon, if one considers a quantum one-photon exchange between two electrons wherein the photons propagate, as in classical electrodynamics, via retarded propagators. Since one cannot distinguish between the process where a photon propagates from electron A to B and the process where the photon propagates from B to A for short time intervals one must sum the two amplitudes and a principal-value propagator results. This observation illustrates the absence of a direct relation between the propagator of the classical field and the propagator of the corresponding quantum field. In particular, it is perfectly possible for a quantum field to have a principal-value propagator, and yet have the classical field propagate via a retarded propagator. This can be understood within the framework of the absorber model of Feynman and Wheeler [3]. For a quantum process, where a finite number of quanta is exchanged, the effective propagator of the quanta cannot have this aspect of its character changed. But in a classical situation, where infinite numbers of quanta are involved, the effective propagator of the quanta can be changed through interaction with the absorber in the manner outlined in ref. [3]. We therefore conclude that the propagation of quanta via principal-value propagators does not imply that macroscopic causality (as represented by retarded propagators) is lost. This remark is relevant to our models of the strong interaction and quantum gravity discussed later.

So far as microscopic causality is concerned, it will be seen that principal-value propagators are completely consistent with the vanishing of commutators of field operators for spacelike distances. Thus neither macroscopic nor microscopic causality is necessarily inconsistent with the use of quantum principal-value propagators.

The use of principal-value propagators does lead to a different analytic structure of S-matrix amplitudes. Amplitudes are now piecewise analytic. However, several authors [4] have shown that such amplitudes are not necessarily inconsistent with experiment—even for the case of the pion-nucleon dispersion relations. In particular, considering the absence of any direct relation between the analytic properties of quark-quark scattering amplitudes,

[2] R. FEYNMAN: *Phys. Rev.* **76**, 769 (1949), footnote [5].
[3] J. WHEELER and R. FEYNMAN: *Rev. Mod. Phys.*, **17**, 157 (1945); **21**, 425 (1949); F. HOYLE and J. NARLIKAR: *Ann. of Phys.*, **54**, 207 (1969); **62**, 44 (1971).
[4] R. E. CUTKOWSKY, P. V. LANDSHOFF, D. I. OLIVE and J. C. POLKINGHORNE: *Nucl. Phys.*, **12** B, 281 (1969); T. D. LEE and G. C. WICK: *Phys. Rev. D*, **2**, 1033 (1970); M. GUNDZIK and E. C. G. SUDARSHAN: *Phys. Rev. D*, **6**, 798 (1972); H. RECHENBERG and E. C. G. SUDARSHAN: *Nuovo Cimento*, **14** A, 299 (1973).

and the analytic properties of the scattering amplitudes of their bound states one cannot rule out the possibility of principal-value propagators for color gluons. In the case of quantum gravity, another potential application of principal-value propagators, the analytic structure of scattering amplitudes due to graviton exchange is unknown and likely to remain so. In the absence of such information, the renormalizability of pure quantum gravity, and the compatibility with Mach's principle, resulting from the use of principal-value propagators for gravitons is encouraging.

Perhaps the most important question facing field theory models with principal-value propagators is unitarity. The appendix contains a detailed discussion of this issue. The physical S-matrix is shown to be unitary. In addition the value of using infinite-momentum frame variables to compute S-matrix elements is pointed out.

2. – Model electrodynamics.

In this section we shall describe a model electrodynamics which is identical to quantum electrodynamics in its predictions. We shall also discuss a model for quantum action-at-a-distance electrodynamics.

The Lagrangian density of our models [1] is

$$(2) \qquad \mathcal{L} = -\tfrac{1}{2} F^1_{\mu\nu} F^{2\mu\nu} - \tfrac{1}{4} F^1_{\mu\nu} F^{1\mu\nu} + \bar{\psi}_2 \gamma^0 \big(i(\gamma \cdot \nabla) - e_0(\gamma \cdot A_2) - m \big) \psi_2$$
$$- (\tilde{\psi}_1 - \tilde{\psi}_2) \gamma^0 \big(i(\gamma \cdot \nabla) - m \big)(\psi_1 - \psi_2) ,$$

where we have introduced two electromagnetic fields, $A_{1\mu}$ and $A_{2\mu}$, so that

$$(3) \qquad F^i_{\mu\nu} = \partial_\nu A_{i\mu} - \partial_\mu A_{i\nu} .$$

for $i = 1, 2$, and two fields, ψ_1 and ψ_2 for electrons. The field $\tilde{\psi}_i$ is related to the transpose of the charge conjugate of ψ_i by [1]

$$(4) \qquad \tilde{\psi}_i = \psi_i^{cT} \gamma^0 C^T ,$$

where C is a charge conjugation matrix. We follow the notation and conventions of ref. [5]. The free field theory for fermions has been developed in detail in ref. [1].

---

(5)   J. BJORKEN and S. DRELL: *Relativistic Quantum Fields* (New York, N. Y., 1965).

Before discussing the free field theory of the photons we note the invariance of $\mathscr{L}$ under a restricted gauge transformation of the second kind,

$$\psi_2 \to \psi_2 \exp[i\Lambda], \tag{5}$$

$$\tilde{\psi}_2 \to \tilde{\psi}_2 \exp[-i\Lambda], \tag{6}$$

$$e_0 A_{2\mu} \to e_0 A_{2\mu} - \partial_\mu \Lambda, \tag{7}$$

$$\psi_1 \to \psi_1 + [\exp[i\Lambda] - 1]\psi_2, \tag{8}$$

$$\tilde{\psi}_1 \to \tilde{\psi}_1 + [\exp[-i\Lambda] - 1]\tilde{\psi}_2. \tag{9}$$

The associated conserved current is

$$J_\mu = \tilde{\psi}_2 \gamma^0 \gamma_\mu \psi_2. \tag{10}$$

It should be noted that the Lagrangian is also invariant under an independent gauge transformation, $A_{1\mu} \to A_{1\mu} - \partial_\mu \Lambda_1$.

Upon varying the action associated with $\mathscr{L}$, we find the field equations

$$\partial^\mu F^1_{\mu\nu} = \partial^\mu F^2_{\mu\nu}, \tag{11}$$

$$\partial^\nu F^2_{\mu\nu} = e_0 \tilde{\psi}_2 \gamma^0 \gamma_\nu \psi_2, \tag{12}$$

$$(i(\gamma \cdot \nabla) - m)\psi_1 = (i(\gamma \cdot \nabla) - m)\psi_2, \tag{13}$$

$$(i(\gamma \cdot \nabla) - e_0(\gamma \cdot A_2) - m)\psi_2 = 0. \tag{14}$$

Equations (12) and (14) demonstrate the equivalence of our model to the usual model of electrodynamics at the level of c-number fields.

We now turn to the case of free photons. From the Lagrangian we can identify the canonical momentum conjugate to $A_{1\mu}$ as

$$\pi_{1\mu} = F^2_{0\mu} - F^1_{0\mu}, \tag{15}$$

while the momentum conjugate to $A_{2\mu}$ is

$$\pi_{2\mu} = F^1_{0\mu}. \tag{16}$$

Since the free electromagnetic Lagrangian is invariant under independent gauge transformations of $A_{1\mu}$ and $A_{2\mu}$ the choice of gauges for $A_{1\mu}$ and $A_{2\mu}$ are also independent. We shall second quantize the fields in the joint Coulomb gauge

$$\vec{\nabla} \cdot \vec{A}_1 = \vec{\nabla} \cdot \vec{A}_2 = 0. \tag{17}$$

The resulting equal-time commutation relations are

$$(18) \qquad [F^1_{0i}(\vec{x}, t), A_{1j}(\vec{y}, t)] = 0 ,$$

$$(19) \qquad [F^1_{0i}(\vec{x}, t), A_{2j}(\vec{y}, t)] = i\delta^{tr}_{ij}(\vec{x} - \vec{y}) ,$$

$$(20) \qquad [F^2_{0i}(\vec{x}, t), A_{2j}(\vec{y}, t)] = i\delta^{tr}_{ij}(\vec{x} - \vec{y}) ,$$

$$(21) \qquad [F^2_{0i}(\vec{x}, t), A_{1j}(\vec{y}, t)] = -i\delta^{tr}_{ij}(\vec{x} - \vec{y}) ,$$

with

$$(22) \qquad \delta^{tr}_{ij}(\vec{x} - \vec{y}) = \int \frac{d^3k}{(2\pi)^3} \exp[-i\vec{k}\cdot(\vec{x} - \vec{y})](\delta_{ij} - k_i k_j/|\vec{k}|^2) .$$

The field operator expansions are completely analogous to the expansion of a free scalar field discussed in ref. [1]. The major difference is the necessary presence of polarization vectors in the electromagnetic-field expansions

$$(23) \qquad \vec{A}_1(x) = \int d^3k \sum_{\lambda=1}^{2} \vec{\varepsilon}(k, \lambda)[f_k(x)(\cos(\theta_1 - \theta_2)a_{1k\lambda} + \sin(\theta_1 - \theta_2)a_{2k\lambda})/\sin\theta_1 +$$
$$+ f_k^*(x)(\cos(\theta_1 - \theta_2)a^\dagger_{1k\lambda} - \sin(\theta_1 - \theta_2)a^\dagger_{2k\lambda})/\cos\theta_1]$$

and

$$(24) \qquad \vec{A}_2(x) = \int d^3k \sum_{\lambda=1}^{2} \vec{\varepsilon}(k, \lambda)[f_k(x)(\cos\theta_2 a_{2k\lambda} + \sin\theta_2 a_{1k\lambda}) +$$
$$+ f_k^*(x)(\sin\theta_2 a^\dagger_{2k\lambda} - \cos\theta_2 a^\dagger_{1k\lambda})] ,$$

where θ_1 and θ_2 are arbitrary angles. As discussed in ref. [1] the choice of these angles represents a choice of boundary conditions for the free-field Green's functions. If the Fourier coefficient operators satisfy

$$(25) \qquad [a_{ik\lambda}, a_{jk'\lambda'}] = [a^\dagger_{ik\lambda}, a^\dagger_{jk'\lambda'}] = 0 ,$$

$$(26) \qquad [a_{ik\lambda}, a^\dagger_{jk'\lambda'}] = (1 - \delta_{ij})\delta_{\lambda\lambda'}\delta^3(\vec{k} - \vec{k}') ,$$

for $i, j = 1, 2$, then the equal-time commutation relations will be satisfied for arbitrary θ_1 and θ_2. In addition the Hamiltonian which is the 0-0 component of the energy-momentum tensor derived from the Lagrangian, has the form

$$(27) \qquad H = \int d^3k \sum_{\lambda=1}^{2} \omega_k(a^\dagger_{2k\lambda}a_{1k\lambda} + a_{2k\lambda}a^\dagger_{1k\lambda}) ,$$

independent of θ_1 and θ_2. There is an analogy [6] between the quantization

---

[6] S. BLAHA: *Phys. Rev. D*, **17**, 994 (1978).

of the mode amplitudes, eqs. (25) and (26), and the co-ordinates of an assembly of one-dimensional harmonic oscillators just as in the second quantization of conventional field theory. We shall define photon vacua in the present case in a manner consistent with the procedure of ref. [1] and in analogy with the simple harmonic-oscillator model of ref. [6]:

$$a_{1k\lambda}|0\rangle_2 = a_{1k\lambda}^\dagger|0\rangle_2 = 0 , \tag{28}$$

$$a_{2k\lambda}|0\rangle_2 \neq 0 , \qquad a_{2k\lambda}^\dagger|0\rangle_2 \neq 0 , \tag{29}$$

for one vacuum, $|0\rangle_2$, and

$$a_{2k\lambda}|0\rangle_1 = a_{2k\lambda}^\dagger|0\rangle_1 = 0 , \tag{30}$$

$$a_{1k\lambda}|0\rangle_1 \neq 0 , \qquad a_{1k\lambda}^\dagger|0\rangle_1 \neq 0 , \tag{31}$$

for the other vacuum, $|0\rangle_1$, for all k and λ. The vacua are related to each other by a Dirac-metric operator in a manner familiar from the scalar case. The metric operator γ is necessary in view of the commutation relations of the Fourier-coefficient operators, eqs. (25) and (26). A one-photon ket is defined by

$$|k\lambda\rangle = a_{2k\lambda}^\dagger|0\rangle_2 , \tag{32}$$

while the dual one-photon bra is defined by

$$\langle k\lambda| = {}_1\langle 0|a_{1k\lambda} , \tag{33}$$

so that

$$\langle k'\lambda'|k\lambda\rangle = \delta_{\lambda\lambda'}\delta^3(\vec{k} - \vec{k}') , \tag{34}$$

by means of eq. (26) and ${}_1\langle 0|0\rangle_2 = 1$. The metric operator is introduced in order to relate the dual of a ket to its Hermitian conjugate:

$$\langle k\lambda| = \left(|k\lambda\rangle\right)^\dagger\gamma . \tag{35}$$

Consequently the metric operator is defined to have the properties

$$\gamma|0\rangle_1 = |0\rangle_2 , \tag{36}$$

$$\gamma^{-1}a_{1k\lambda}\gamma = a_{2k\lambda} , \tag{37}$$

$$\gamma^{-1}a_{2k\lambda}\gamma = a_{1k\lambda} \tag{38}$$

with $\gamma = \gamma^\dagger = \gamma^{-1}$. It has a form similar to the metric operator of the scalar case which is given in eq. (36) of ref. [1]. The definition of many-photon bras and kets is a direct generalization of eqs. (32), (33) and (35).

The time-ordered propagator which will later be of use in the development of perturbation theory is

$$(39) \qquad G^{22}_{\mu\nu}(x - y) = -i\langle 0|T(A_{2\mu}(x)A_{2\nu}(y))|0\rangle_2$$
$$= \sin^2\theta_2 G_{F\mu\nu}(x-y) - \cos^2\theta_2 G^*_{F\mu\nu}(x - y),$$

where $G_{F\mu\nu}(x - y)$ is the usual Feynman propagator for photons, while G^*_ν is its complex conjugate. Another nonzero free-field propagator, $G^{12}_{\mu\nu}$ is defined as the vacuum expectation value of the time-ordered product of $A_{1\mu}(x)$ and $A_{2\nu}(y)$. It depends on both θ_1 and θ_2. It does not appear in perturbation theory.

Equation (39) shows that $G^{22}_{\mu\nu}$ is a Feynman propagator if $\theta_2 = \pi/2$, and a principal-value propagator if $\theta_2 = \pi/4$. Thus

$$(40) \qquad G^{22}_{\mu\nu}(x - y)\big|_{\theta_2=\pi/4} = g_{\mu\nu}\int \frac{d^4k}{(2\pi)^4}\exp[-ik\cdot(x - y)]\frac{P}{k^2},$$

where

$$(41) \qquad \frac{P}{k^2} = \frac{1}{2}\left(\frac{1}{k^2 + i\varepsilon} + \frac{1}{k^2 - i\varepsilon}\right),$$

in the Feynman gauge. This Green's function has previously appeared in action-at-a-distance formulations [3] of classical electrodynamics. The generality of our formulation of quantum field theory allows it to appear as a special case. This does not seem to be possible within the framework of the conventional formulation [7]. If we choose $\theta_2 = \pi/4$ and proceed to develop perturbation theory for the full interacting theory then we obtain a quantum-action-at-a-distance theory. We obtain quantum exchange of energy and momentum to which we associate lines in Feynman diagrams. The only difference is in the analytic structure associated with the «epsilontics» of the pole locations. Scattering amplitudes are piecewise analytic in this case [4]. More importantly, quanta associated with the electromagnetic field do not appear in asymptotic states, if unitarity is to be maintained. Thus the electromagnetic field does not have its own degrees of freedom, but is constrained to have its source in the charged matter fields which are present. The physical expectations of a quantum action-at-a-distance electrodynamics are therefore satisfied. Except for the absence of «photons» from asymptotic states, and the principal-value nature of the electromagnetic-field propagator, the perturbation theory rules, diagrams and combinatorics are identical to those of QED, although they must be unitarized using the method described in the appendix.

[7] R. FEYNMAN: Science, 153, 699 (1966).

We now turn to the Feynman case ($\theta_2 = \pi/2$). The field $\vec{A}_2$ takes the form

$$(42) \qquad \vec{A}_2 = \int d^3k \sum_{\lambda=1}^{2} \vec{\varepsilon}(k,\lambda) \{ f_k(x) a_{1k\lambda} + f_k^*(x) a_{2k\lambda}^\dagger \} \,,$$

in the Coulomb gauge. Note that the form of eq. (42) and the commutators, eq. (25) and (26) imply the free-field identity

$$(43) \qquad {}_2\langle 0 | A_{2i_1}(x_1) A_{2i_2}(x_2) \dots A_{2i_n}(x_n) | 0 \rangle_2 = \langle 0 | A_{i_1}(x_1) A_{i_2}(x_2) \dots A_{i_n}(x_n) | 0 \rangle$$

for all spatial indices $i_1, i_2, \dots, i_n$, and all n for which the right-hand side is the vacuum expectation value of n free fields in the usual formulation. This identity is the basis of our claim that the Feynman case of our formulation is *completely* equivalent to QED in its predictions. An important part of establishing the correspondence is the LSZ reduction formulae for photons: for an in photon $(k\lambda)$ we have

$$(44) \qquad \langle \beta \text{ out} | \alpha(k\lambda) \text{ in} \rangle = \langle \beta \text{ out} | a_{2k\lambda \text{ in}}^\dagger | \alpha \text{ in} \rangle =$$
$$= \langle \beta - (k\lambda) \text{ out} | \alpha \text{ in} \rangle - \frac{i}{\sqrt{Z_3}} \int d^4x \, A_{k\lambda}^\mu(x) \langle \beta \text{ out} | A_{2\mu}(x) | \alpha \text{ in} \rangle$$

and for an out photon $(k\lambda)$ we have

$$(45) \qquad \langle \beta(k\lambda) \text{ out} | \alpha \text{ in} \rangle = \langle \beta \text{ out} | a_{1k\lambda \text{ out}} | \alpha \text{ in} \rangle =$$
$$= \langle \beta \text{ out} | \alpha - (k\lambda) \text{ in} \rangle - \frac{i}{\sqrt{Z_3}} \int d^4x \, \langle \beta \text{ out} | A_{2\mu}(x) | \alpha \text{ in} \rangle \, A_{k\lambda}^{\mu*}(x)$$

in the notation of ref. ([5]).

The second quantization of the free-electron field is described in detail in ref. ([1]). We assume an in and out electron field formalism is developed along those lines. Our formulation of spin–one-half fermion theory also allows one to choose Feynman propagators or principal-value propagators for fermions. In order to establish a model equivalent to QED we choose Feynman propagators for the electrons.

The perturbation theory for our model is based on the interaction Lagrangian

$$(46) \qquad \mathcal{L}_{\text{int}} = -e_0 \, \bar{\psi}_2 \gamma^0 (\gamma \cdot A_2) \psi_2 \,.$$

We shall need the LSZ reduction formulae for electrons

$$(47) \qquad \langle \beta \text{ out} | (ps)\alpha \text{ in} \rangle = \langle \beta - (ps) \text{ out} | \alpha \text{ in} \rangle -$$
$$- \frac{i}{\sqrt{Z_2}} \int d^4x \, \langle \beta \text{ out} | \bar{\psi}_2(x) \gamma^0 | \alpha \text{ in} \rangle \, (-i\overleftarrow{\nabla} - m) \, U_{ps}(x)$$

and

$$(18) \qquad \langle \beta(ps) \text{ out} | \alpha \text{ in} \rangle = \langle \beta \text{ out} | \alpha \ (ps) \text{ in} \rangle -$$

$$- \frac{i}{\sqrt{Z_2}} \int d^4x \ \overline{U}_{ps}(x)(i\overrightarrow{\nabla} - m) \langle \beta \text{ out} | \psi_2(x) | \alpha \text{ in} \rangle$$

in the notation of ref. (^{5}). To remove a positron from an in state, we use

$$(49) \qquad \langle \beta \text{ out} | (ps)\alpha \text{ in} \rangle = \langle \beta - (ps) \text{ out} | \alpha \text{ in} \rangle +$$

$$+ \frac{i}{\sqrt{Z_2}} \int d^4x \ \overline{V}_{\overline{ps}}(x)(i\overrightarrow{\nabla} - m) \langle \beta \text{ out} | \psi_2(x) | \alpha \text{ in} \rangle$$

and, to remove a positron from an out state, we use

$$(50) \qquad \langle \beta(ps) \text{ out} | \alpha \text{ in} \rangle = \langle \beta \text{ out} | \alpha \ -(ps) \text{ in} \rangle +$$

$$+ \frac{i}{\sqrt{Z_2}} \int d^4x \langle \beta \text{ out} | \tilde{\psi}_2 \gamma^0 | \alpha \text{ in} \rangle (-i\overleftarrow{\nabla} - m) \ V_{\overline{ps}}(x) \ .$$

In view of the interaction Lagrangian and eqs. (47)-(50) it is clear that only time-ordered products of $\psi_{2\text{in}}$ and $\tilde{\psi}_{2\text{in}}$ appear in the perturbation theory expansion. The Wick theorem, which applies in the present case, reduces all vacuum expectation values of time-ordered products of $\psi_{2\text{in}}$ and $\tilde{\psi}_{2\text{in}}$ to products of the time-ordered two-point function

$$(51) \qquad iS(x-y) = {}_2\langle 0 | T\big(\psi_2(x) \, \tilde{\psi}_2(y)\gamma^0\big) | 0 \rangle_2$$

$$(52) \qquad = i \int \frac{d^4p}{(2\pi)^4} \exp\left[-ip\cdot(x-y)\right] \frac{p+m}{p^2 - m^2 + i\varepsilon} \ ,$$

where eqs. (51) is in the notation of ref. (^{1}).

The above considerations and the familiar development* of the S-matrix as an expansion in vacuum expectation values of time-ordered products of in-field operators lead us to assert that our model electrodynamics is identical to QED in its predictions. In particular

$$(53) \qquad {}_2\langle 0 | T\Big(A_{2\mu_1\text{in}}(x_1) A_{2\mu_2\text{in}}(x_2) \dots \psi_{2\text{in}}(y_1) \psi_{2\text{in}}(y_2) \dots \tilde{\psi}_{2\text{in}}(z_1) \tilde{\psi}_{2\text{in}}(z_2) \dots$$

$$\dots \exp\left[i \int_{-\infty}^{\infty} dt \, L_{\text{int}} \right] \Big) | 0 \rangle_2 =$$

$$= {}_2\langle 0 | T\Big(A_{\mu_1\text{in}}(x_1) A_{\mu_2\text{in}}(x_2) \dots \psi_{\text{in}}(y_1) \psi_{\text{in}}(y_2) \dots \bar{\psi}_{\text{in}}(z_1) \bar{\psi}_{\text{in}}(z_2) \dots \exp\left[i \int_{-\infty}^{\infty} dt \, L_{\text{int}}^{\text{QED}} \right] \Big) | 0 \rangle \ ,$$

where the right-hand side is the corresponding time-ordered product of the conventional formulation of QED with

$$(54) \qquad L_{\text{int}}^{\text{QED}} = -i : \int d^3 x \, c_0 \, \bar{\psi}_{\text{in}} (\gamma \cdot A_{\text{in}}) \psi_{\text{in}} : .$$

Thus, the perturbation theory of our model electrodynamics is term by term equivalent to QED—the equivalent QED term being obtained by ignoring the subscripts 1 and 2 on fields and vacua, and letting $\tilde{\psi}_2 \gamma^0 \to \bar{\psi}$. Feynman diagrams can be appropriated without modification for use in our formulation. We conclude with a quote from Feynman [7]: « It always seems odd to me that the fundamental laws of physics can appear in so many different forms that are not apparently identical at first.... Theories of the known which are described by different physical ideas may be equivalent in all their predictions and hence scientifically indistinguishable. However they are not psychologically identical when one is trying to move from that base into the unknown. For different views suggest different modifications ... » In that spirit we turn to the investigation of non-Abelian theories.

3. – Yang-Mills models.

The models we have hitherto investigated are members of a class whose Lagrangians have the form

$$(55) \qquad \mathscr{L} = \mathscr{L}_0(q_2) + \mathscr{L}_1(q_1 \, q_2) .$$

where $\mathscr{L}_0$ has a form which is essentially identical to a Lagrangian of the conventional formulation. For example, in the case of electrodynamics the Lagrangian of eq. (2) can be put in the form of eq. (55), if we let

$$(56) \qquad \mathscr{L}_0(A_{2\mu}, \psi_2) = -\tfrac{1}{4} F_{\mu\nu}^2 F^{2\mu\nu} + \tilde{\psi}_2 \gamma^0 \big(i(\gamma \cdot \nabla) - c_0(\gamma \cdot A_2) - m \big) \psi_2$$

and

$$(57) \qquad \mathscr{L}_1(A_{1\mu} \, A_{2\mu}, \psi_1 \, \psi_2) =$$
$$= \tfrac{1}{4}(F_{\mu\nu}^1 - F_{\mu\nu}^2)^2 + (\tilde{\psi}_2 \, \tilde{\psi}_1)\gamma^0 \big(i(\gamma \cdot \nabla) - m \big)(\psi_1 \, \psi_2) .$$

At the level of c-number fields it is clear from eqs. (56) and (57) that $A_{2\mu}$ and ψ_2 reproduce the usual electrodynamics. At the level of quantum fields one can opt to have a model reproducing the conventional quantum theory—we call it the Feynman case. It maintains the implied decoupling of $A_{1\mu} - A_{2\mu}$ and $\psi_1 - \psi_2$ from $A_{2\mu}$ and ψ_2. On the other hand, we can take advantage of the larger space of states in the larger-manifestly-indefinite-metric theory and couple the fields together by taking advantage of degrees of freedom in the

Fourier expansion of asymptotic fields and the definition of the vacuum. Since
we « perturb around » the asymptotic fields, perturbation theory will reflect
this coupling. Our principal-value or quantum action-at-a-distance models
are based on this possibility.

We shall use the form of eq. (55) as an ansatz to generate Yang-Mills models
in our formulation. This leads to the gauge field Lagrangian

$$(58) \qquad \mathscr{L} = -\tfrac{1}{4} \boldsymbol{F}_{2\mu\nu} \cdot \boldsymbol{F}_2^{\mu\nu} + \tfrac{1}{4}(\boldsymbol{G}_{1\mu\nu} - \boldsymbol{G}_{2\mu\nu}) \cdot (\boldsymbol{G}_1^{\mu\nu} - \boldsymbol{G}_2^{\mu\nu}),$$

where

$$(59) \qquad \boldsymbol{F}_{2\mu\nu} = \partial_\mu \boldsymbol{A}_{2\nu} - \partial_\nu \boldsymbol{A}_{2\mu} - g\boldsymbol{A}_{2\mu} \times \boldsymbol{A}_{2\nu}$$

and

$$(60) \qquad \boldsymbol{G}_{i\mu\nu} = \partial_\mu \boldsymbol{A}_{i\nu} - \partial_\nu \boldsymbol{A}_{i\mu}$$

for $i = 1, 2$. Under a local gauge transformation, $S = S(x)$,

$$(61) \qquad A_{2\mu} \to A'_{2\mu} = S^{-1} A_{2\mu} S + \frac{i}{g} S^{-1} \partial_\mu S,$$

$$(62) \qquad A_{1\mu} \to A'_{1\mu} = A_{1\mu} + A'_{2\mu} - A_{2\mu},$$

the Lagrangian $\mathscr{L}$ is invariant, where $A_{i\mu} = \boldsymbol{A}_{i\mu} \cdot \boldsymbol{T}$ for $i = 1, 2$ with $\boldsymbol{T}$ a
vector composed of matrices representing generators of the gauge group. There
is an additional invariance under the local transformation, $A_{1\mu} \to A_{1\mu} + \partial_\mu \Lambda$,
which is a straightforward generalization of the Abelian gauge transformation.

The field equations derived from eq. (58), upon varying the action with
respect to $\boldsymbol{A}_{1\mu}$ and $\boldsymbol{A}_{2\mu}$, are

$$(63) \qquad \partial^\mu(\boldsymbol{G}_{1\mu\nu} - \boldsymbol{G}_{2\mu\nu}) = 0$$

and

$$(64) \qquad (\partial^\mu + g\boldsymbol{A}_2^\mu \times)\boldsymbol{F}_{2\mu\nu} = 0 \,.$$

The canonical momentum conjugate to $\boldsymbol{A}_{2\mu}$ is

$$(65) \qquad \boldsymbol{\Pi}_{2\mu} = \boldsymbol{F}_{2\mu 0} + \boldsymbol{G}_{1\mu 0} - \boldsymbol{G}_{2\mu 0}$$

and the momentum conjugate to $\boldsymbol{A}_{1\mu}$ is

$$(66) \qquad \boldsymbol{\Pi}_{1\mu} = \boldsymbol{G}_{2\mu 0} - \boldsymbol{G}_{1\mu 0} \,.$$

We choose the Coulomb gauge to implement the field quantization. Due to
the form of the Lagrangian, we can treat $\boldsymbol{A}_{2\mu}$ and $\boldsymbol{A}_{3\mu} = \boldsymbol{A}_{1\mu} - \boldsymbol{A}_{2\mu}$ as inde-

pendent fields. Choosing the Coulomb gauge for $A_{2\mu}$, $\vec{\nabla} \cdot \vec{A}_2 = 0$, we can isolate the independent field quantities and establish their equal-time commutation relations in the well-known manner [*]. We also can choose to work in the Coulomb gauge of $A_{3\mu}$, $\vec{\nabla} \cdot \vec{A}_3 = 0$ which is equivalent to $\vec{\nabla} \cdot \vec{A}_1 = 0$. This possibility follows from the invariance of $\mathscr{L}$ under the local « Abelian » transformation of $A_{3\mu}$ or $A_{1\mu}$ mentioned above. The resulting equal-time commutation relations are equivalent to

$$(67) \qquad [G^{\mathrm{T}}_{\chi 0ia}(x), A_{\beta jb}(y)] = i(1 - \delta_{\chi 1}\delta_{\beta 1})\delta_{ab} J^{\mathrm{tr}}_{ij}(x - y),$$

where $\chi, \beta = 1, 2$; i and j are spatial- and a and b are internal-symmetry indices. G^{T} is the transverse part of G. One can now proceed to linearize the field equations. The Fourier expansions of the solutions of the linearized equations have the form of eqs. (23) and (24), if an internal-symmetry index is appended to the Fourier component operators. The discussion then reduces to the Abelian case considered in the last section. In particular the Green's function of the linearized model which is relevant to perturbation theory is

$$(68) \qquad G^{22}_{ab\mu\nu}(x - y) = - i \langle 0 | T(A_{2\mu a}(x) A_{2\nu b}(y)) | 0 \rangle_2.$$

It can be expressed as a linear combination of a Feynman propagator and its complex conjugate as in eq. (39). The arbitrary angle can be chosen to give Feynman propagators or principal-value propagators.

The perturbation theory is based on the linearized model. In the Feynman case the LSZ reduction of asymptotic non-Abelian quanta is essentially given in eqs. (44) and (45), if appropriate internal-symmetry indices are introduced. Only fields of « type 2 » are introduced by the LSZ reduction. Furthermore, the cubic and quartic terms of the Lagrangian « interaction » term also involve only fields of type 2. If we restrict all couplings to other fields—such as fermions—to involve only $A_{2\mu}$, then only time-ordered products of $A_{2\mu}$ appear in the perturbation theory expansion of the S-matrix. The consequence of this development is that one has constructed a model which makes predictions which are identical to a conventional formulation based on a Lagrangian of the usual type

$$(69) \qquad \mathscr{L} = - \tfrac{1}{4} F_{\mu\nu} \cdot F^{\mu\nu}.$$

This can be verified in a path integral approach based on the Lagrangian of eq. (58).

The point of our formulation is that one can make other choices—such as the principal-value propagator choice. (Note that the path integral formulation

---

(*) E. ABERS and B. W. LEE: *Phys. Rep.*, **9** C, 1 (1973).

only has a formal validity in these cases, since the Wick rotation to Euclidean co-ordinates is highly nontrivial.) The choice of principal-value propagators has several important consequences.

In the appendix we show how to calculate the unitary physical S-matrix in this case. We show that loops of principal-value propagators do not contribute to the absorptive parts of S-matrix elements. As a result Faddeev-Popov ghost loops are not needed to maintain unitarity (in any gauge). We also point out a radical possibility: a unitary S-matrix can be defined in which all non-Abelian boson loops are excluded. Thus the non-Abelian boson sector consists solely of tree diagrams. This is a quantum analogue of the absence of self-interactions in classical action-at-a-distance theories.

4. – Non-Abelain model of the strong interactions.

Several years ago a non-Abelian model of the strong interaction was constructed (^{9}) which had manifest quark confinement and a linear potential. The model was based on higher-order field equations. Such equations lead to well-known unitarity problems, if the usual quantization program is implemented. The author chose to avoid the unitarity problems through an *ad hoc* quantization procedure which gave principal-value propagators. The result was a model having unconventional analyticity properties. Despite the successful confinement scheme in this model, there was some reason for uneasiness, due to the unusual form of the Lagrangian (where two sets of fields were associated with the gluons) and the *ad hoc* method of quantization.

The new framework we have developed in ref. (^{1}) and this paper eliminates these sources of concern. Our formulation is based on fundamental ground—the need for an acceptable physical particle interpretation of quantum field theory in flat and curved space-time. We shall see that the two-field Lagrangian of the strong-interaction model lies within the general framework we have established. The principal-value propagators, required to have a unitary model, will also be obtained in a straightforward manner.

Since the strong-interaction model is described in some detail in ref. (^{9}), we shall only outline the aspects necessary to connect that model with the present work.

The strong-interaction Lagrangian departs slightly from the ansatz of eq. (55) (mainly in the replacement $G^1_{\mu\nu} \cdot G^{1\mu\nu} \to \lambda^2 A_{1\mu} \cdot A^\mu_1$):

$$(70) \qquad \mathcal{L} = -\tfrac{1}{2} F^1_{\mu\nu} \cdot F^{2\mu\nu} - \tfrac{1}{2}\lambda^2 A_{1\mu} \cdot A^\mu_1 + \bar\psi_2 \gamma^0\big(i(\gamma \cdot \nabla) + g(\gamma \cdot A_2) - m\big)\psi_2 -$$
$$(\bar\psi_1 + \bar\psi_2)\gamma^0\big(i(\gamma \cdot \nabla) - m\big)(\psi_1 + \psi_2).$$

(^{9}) S. BLAHA: *Phys. Rev. D*, **10**, 4268 (1974); **11**, 2921 (1975). It should be noted that the claim that loops of principal-value propagators are identically zero is not true. In the case of one-loop diagrams only the real part is zero.

where ψ_1 and ψ_2 are the quark fields and

$$F^2_{\mu\nu} = \partial_\mu \boldsymbol{A}_{2\nu} - \partial_\nu \boldsymbol{A}_{2\mu} + g\boldsymbol{A}_{2\mu} \times \boldsymbol{A}_{2\nu} \tag{71}$$

and

$$F^1_{\mu\nu} = D_\mu \boldsymbol{A}_{1\nu} - D_\nu \boldsymbol{A}_{1\mu} \tag{72}$$

with the covariant derivative defined by

$$D_\mu = \partial_\mu + g\boldsymbol{A}_{2\mu} \times . \tag{73}$$

The Lagrangian is invariant under the local color SU_3 gauge transformation

$$\psi_2 \to S^{-1}\psi_2 , \tag{74}$$

$$\psi_1 \to \psi_1 + (S^{-1} - I)\psi_2 , \tag{75}$$

$$A_{1\mu} \to S^{-1}A_{1\mu}S , \tag{76}$$

$$A_{2\mu} \to S^{-1}A_{2\mu}S + \frac{i}{g}S^{-1}\partial_\mu S , \tag{77}$$

with $A_{i\mu} = \boldsymbol{A}_{i\mu}\cdot\boldsymbol{T}$, where $\boldsymbol{T}$ is a vector composed of matrices representing SU_3 generators. The equations of motion and the Coulomb gauge quantization are discussed in detail in ref. [9]. (The subscripts, 1 and 2, on the gluon fields are reversed in ref. [9] relative to the present development.)

The essential feature of the model can be seen in the linearized field equations

$$\Box \boldsymbol{A}_{2\mu} = -\Lambda^2 \boldsymbol{A}_{1\mu} , \tag{78}$$

$$\Box \boldsymbol{A}_{1\mu} = -g\boldsymbol{J}_\mu , \tag{79}$$

where $\boldsymbol{J}_\nu$ is the color quark current. Equations (78) and (79) show an apparent mass term, $\Lambda^2(\boldsymbol{A}_{1\mu})^2$, actually leads to a higher-order derivative field equation. The implications of this feature for quark confinement are discussed in ref. [9]. The discussion is based on the assumption that all gluon propagators are principal-value propagators. We shall now show that the propagators for color gluons in the linearized model can be principal-value propagators.

We work in the Coulomb gauge, $\vec{\nabla}\cdot\vec{\boldsymbol{A}}_2 = 0$. A mode expansion of the gluon fields must be consistent with the free linearized field equations and the equal-time commutation relations,

$$[F^{a\alpha}_{0i}(x), A_{\beta bj}(y)] = i\delta_{ab}(1-\delta_{\alpha\beta})\delta^{tr}_{ij}(\vec{x}-\vec{y}) \tag{80}$$

(cf. eqs. (65) and (66) of ref. [9]). These conditions and the requirement of a unitary theory imply the Fourier expansions

$$(81) \qquad \vec{A}_{1\lambda}(x) = \int \frac{d^3k}{\sqrt{2}} \sum_{\lambda=1}^{2} \vec{\varepsilon}(k,\lambda)[(a_{1k\lambda} + a_{2k\lambda})f_k(x) + (a_{1k\lambda}^\dagger + a_{2k\lambda}^\dagger)f_k^*(x)],$$

$$(82) \qquad \vec{A}_{2a}(x) = \int \frac{d^3k}{\sqrt{2}} \sum_{\lambda=1}^{2} \vec{\varepsilon}(k,\lambda)[(a_{2k\lambda} - a_{1k\lambda})f_k(x) + (a_{2k\lambda}^\dagger + a_{1k\lambda}^\dagger)f_k^*(x)] +$$

$$+ A^2 \int \frac{d^4k}{\sqrt{2}} 2\omega_k \theta(k_0)\,\delta'(k^2) \sum_{\lambda=1}^{2} \vec{\varepsilon}(k,\lambda)[(a_{1k\lambda} + a_{2k\lambda})f_k(x) + (a_{1k\lambda}^\dagger - a_{2k\lambda}^\dagger)f_k^*(x)],$$

where $\delta'(k^2) = \partial\delta(k^2)/\partial k^2$ and $\omega_k = |\vec{k}|$. From eqs. (81) and (82) we can determine the Green's functions of the linearized model:

$$(83) \qquad iG^{12}_{\mu\nu ab}(x-y) = {}_2\langle 0|T(A_{1\mu a}(x)A_{2\nu b}(y))|0\rangle_2 =$$
$$= -i\delta_{ab}\int \frac{d^4k}{(2\pi)^4}\exp[-ik\cdot(x-y)]r^{12}_{\mu\nu}P\frac{1}{k^2}$$

and

$$(84) \qquad iG^{22}_{\mu\nu ab}(x-y) = {}_2\langle 0|T(A_{2\mu a}(x)A_{2\nu b}(y))|0\rangle_2 =$$
$$= i A^2\delta_{ab}\int \frac{d^4k}{(2\pi)^4}\exp[-ik\cdot(x-y)]r^{22}_{\mu\nu}P\frac{1}{k^4},$$

where

$$(85) \qquad P\frac{1}{k^{2N}} = \frac{1}{2}\left[\frac{1}{(k^2+i\varepsilon)^N} + \frac{1}{(k^2-i\varepsilon)^N}\right]$$

and where $r^{12}_{\mu\nu}$ and $r^{22}_{\mu\nu}$ are gauge-dependent tensors. While the path integral formulation is only of formal value for the present model, it can be used to determine the form of the Green's functions for any choice of gauge. It can also be used to generate the perturbation theory rules for the model. The character of the perturbation theory is described in ref. (9). Confinement arises through the Schwinger mechanism in a manner reminiscent of the two-dimensional Schwinger model. The mass scale is set by A. The Lagrangian of eq. (70) leads to an interaction between quarks which has the linear potential r as its Coulomb potential in lowest order. The r potential has had notable success in the explication of the charmonium system. A r^{-1} term may also be needed to successfully describe charmonium. Such a term can be introduced in our model by adding another interaction term

$$(86) \qquad \mathcal{L}_r = g'\bar{\psi}_2\gamma^0(\gamma\cdot A_1)\psi_2$$

to eq. (70). Since $A_{1\mu}$ transforms homogeneously under a gauge transformation, the gauge invariance of the Lagrangian is not altered by the addition of this term.

In conclusion, we have established the basis for a manifestly quark-confining model of the strong interaction within the framework of our formulation of quantum field theory. This example illustrates the value of our formulation in the construction of unitary quantum field theories with higher-order derivative field equations. Another interesting application of this approach would be to quantum gravity which may have its renormalization problems ameliorated by introducing higher-order derivatives. Certain formal similarities, and this possibility, have led the author to propose a unified model of vierbein gravity and the strong interaction [10]. The higher-order derivative gravity terms which lead to power counting renormalizability for the quantum gravity sector are linked to the higher-order derivative strong-interaction sector terms which lead to quark confinement. The recent discovery of important spin effects in high-energy p-p scattering [11] is suggestive in view of the presence of strong spin-spin interactions in unified models of the strong-interaction and vierbein gravity.

5. – Model of quantum gravity.

In this section we develop a trivially renormalizable model of quantum gravity (based on the Einstein Lagrangian) which is coupled to an external classical source. In addition to obtaining a physically interesting model, we shall see the utility of principal-value propagators in ameliorating divergence problems. The self-interactions of the « gravitons » will be reduced if principal-value propagators are used. This corresponds to the absence of self-interactions in classical action-at-a-distance models.

The model which we shall develop lies within the framework established above. It can be described as quantum action-at-a-distance gravity. We begin with the Einstein action

$$(87) \qquad I = - 2\varkappa^{-2} \int \mathrm{d}^4 x \, g^{\frac{1}{2}} R$$

with $\varkappa = 32\pi G$, where G is Newton's constant, $g = \det g_{\mu\nu}$, $R = g^{\mu\nu} R^{\lambda}_{\mu\nu\lambda}$ and where

$$(88) \qquad R^{\lambda}_{\mu\nu\lambda} = \partial_\nu \Gamma^{\alpha}_{\mu\lambda} + \Gamma^{\alpha}_{\nu\beta} \Gamma^{\beta}_{\mu\lambda} - (\nu \Leftrightarrow \lambda)$$

and

$$(89) \qquad \Gamma^{\alpha}_{\mu\nu} = \frac{1}{2} g^{\alpha\beta} [\partial_\mu g_{\nu\beta} + \partial_\nu g_{\mu\beta} - \partial_\beta g_{\mu\nu}] \, .$$

[10] S. Blaha: Lett. Nuovo Cimento, 18, 60 (1977).
[11] J. R. O'Fallon, L. G. Ratner, P. F. Schultz, K. Abe, R. C. Fernow, A. D. Krisch, T. A. Mulera, A. J. Satthouse, B. Sandler, K. M. Terwilliger, D. G. Crabb and P. H. Hansen: Phys. Rev. Lett., 39, 733 (1977).

Our approach is analogous to Feynman's covariant quantization around a flat background field [12]. We therefore let

$$(90) \qquad g_{\mu\nu} = \eta_{\mu\nu} + \varkappa h_{\mu\nu}^2,$$

where $\eta_{\mu\nu} = (-1, 1, 1, 1)$. We then introduce a second tensor field, $h_{\mu\nu}^1$; and choose the action for our model to be (cf. eq. (55))

$$(91) \qquad I = \int \mathrm{d}^4 x \big[\tfrac{1}{2}(h_{\mu\nu,\sigma}^3)^2 - (h_{\mu}^3)^2 - \tfrac{1}{2}(h_{,\mu}^3)^2 - 2\varkappa^{-2}g^{\frac{1}{2}}R \big],$$

where $h_{\mu\nu}^3 = h_{\mu\nu}^1 - h_{\mu\nu}^2$, $h_{\mu}^3 = h_{\mu\nu,\nu}^3$, $h_{,\mu}^3 = h_{\nu\nu,\mu} = \partial_\mu h$ and where the last term on the right-hand side of eq. (91) is expanded in $h_{\mu\nu}^2$ by using eq. (90). The gauge invariance of the Lagrangian is maintained by requiring $h_{\mu\nu}^1$ and $h_{\mu\nu}^2$ to satisfy the same transformation law

$$(92) \qquad h_{\mu\nu}^i \rightarrow h_{\mu\nu}^{i\prime} = h_{\mu\nu}^i - \partial_\nu \varepsilon_\mu - \partial_\mu \varepsilon_\nu,$$

where ε_μ are four small, but otherwise arbitrary, functions of the co-ordinates. We introduce a de Donder harmonic gauge fixing term

$$(93) \qquad \mathscr{L}_\mathrm{D} = -\frac{1}{2\gamma}\left(h_\mu^2 - \frac{1}{2}h_{,\mu}^2 \right)^2 + \frac{1}{2\gamma}\left(h_\mu^3 - \frac{1}{2}h_{,\mu}^3 \right)^2,$$

in order to obtain a regular kinetic matrix from the quadratic terms of the augmented Lagrangian. The quadratic terms are (if we let $\gamma = \tfrac{1}{2}$)

$$(94) \qquad \mathscr{L}_\mathrm{quad} = -h_{\alpha\beta,\mu}^1 h_{\alpha\beta,\mu}^2 + \tfrac{1}{2} h_{,\mu}^1 h_{,\mu}^2 + \tfrac{1}{2}(h_{\alpha\beta,\mu}^1)^2 - \tfrac{1}{4}(h_{,\mu}^1)^2.$$

These terms plus the higher-order « interaction » terms can be introduced into a suitable path integral [13] expression and the perturbation theory rules can be developed. The path integral approach is only of formal significance, in general, in our models. It gives the correct algebraic expressions and combinational rules. The nontriviality of Wick rotation to Euclidean co-ordinates in any case, but the Feynman case, precludes attributing anything more than formal value to the path integral approach. Of course, the applicability of the path integral formalism to the Feynman case, and the absence of any difference at the algebraic level (before integrations) between the Feynman and principal-value cases, means that the path integral formalism also describes the combinatorics of the principal-value case.

---

[12] R. FEYNMAN: *Acta Phys. Polonica*, **24**, 697 (1963).
[13] V. N. POPOV and L. FADEEV: Kiev Report. No. ITP 67-36 (1967).

In order to have a nontrivial model in the principal-value case, we shall assume the gravitational field is coupled to an external classical source. Note that covariance requires that only $h^2_{\mu\nu}$ couples to that source. This fact, plus the absence of $h^1_{\mu\nu}$ in higher-order terms, implies that the only free-field propagator necessary to develop perturbation theory is the vacuum expectation value of the time-ordered product of free $h^2_{\mu\nu}$ fields:

$$(95) \qquad iG^{22}_{\mu\nu,\varrho\sigma} = {}_2\langle 0 | T\big(h^2_{\mu\nu}(x)h^2_{\varrho\sigma}(y)\big) | 0 \rangle_2 =$$

$$= -\frac{i}{2}\,(\eta_{\mu\varrho}\eta_{\nu\sigma} + \eta_{\mu\sigma}\eta_{\nu\varrho} - \eta_{\mu\nu}\eta_{\varrho\sigma})\int \frac{\mathrm{d}^4k}{(2\pi)^4}\,\frac{\exp\left[-ik\cdot(x-y)\right]}{k^2}\,.$$

If one chooses the propagator in eq. (95) to be a Feynman propagator, then the usual model of quantum gravity results with its attendant renormalizability problems.

We shall consider the case, in which G^{22} is a principal-value propagator. One can establish this case by following a canonical quantization procedure which is completely analogous to those of models considered earlier. The propagation of « gravitons » by principal-value propagators has important consequences for perturbation theory. All graviton loops can be excluded from the physical S-matrix without leading to unitarity problems (cf. appendix). As a result, only tree diagrams occur and unitarity requires that all gravitons are absorbed—either within trees or on the external classical source. Thus we have a model quantum gravity with no renormalization problem. In the classical limit the model becomes Einstein's classical theory of gravity ([14]). There is, of course, the question of whether the classical limit is a retarded model of gravity or not. The answer lies in cosmology—is the absorber mechanism operative?

An action-at-a distance model of gravity with absorber seems to be very much in the spirit of Mach's principle. The usual approach—where the gravitational field is treated as having its own degrees of freedom—almost invariably leads to the surrepitious introduction of absolute space. The reason is simple. The solution of the field equations for a localized mass distribution is asymptotically flat. The only apparent way to avoid this problem is to say Mach restricts us to the class of closed-universe solutions.

In an action-at-a-distance gravity ([15]) the metric field exists only in the presence of matter, so that any asymptotic flatness of solutions is a mathematical

---

([14]) There is an opinion which is sometimes expressed that a tree diagram model is equivalent to a classical theory. This is not true. The correct view is that the classical limit of a tree diagram model is the corresponding classical theory. Cf. S. DESER's paper in *Quantum Gravity*, edited by C. J. ISHAM, R. PENROSE and D. W. SCIAMA (Oxford, 1975).

([15]) A. WHITEHEAD: *The Principle of Relativity* (Cambridge, 1922).

artifact. The metric properties of space are a consequence of the presence of mass-energy. (Consistent with this view, we note that mass-energy must be present to measure the metric properties of space.)

The ultimate realization of Mach's principle from the present viewpoint requires that inertia be the result of the gravitational effects of distant matter. Preliminary work [16] in this direction can be easily incorporated within the framework of an action-at-a-distance model of gravity. In fact, Sciama's model of the inertial effects of the distant stars displays a close analogy to the Feynman-Wheeler absorber model.

In closing we remark that the introduction of quantum matter fields in our action-at-a-distance quantum gravity appears to result in a nonrenormalizable model. It seems that a higher-order derivative Lagrangian of the type mentioned in the last section may be necessary for a fully renormalizable quantum gravity. Of course, that would also be a quantum action-at-a-distance model.

APPENDIX

In this appendix we define a unitary physical S-matrix for action-at-a-distance quantum field theories. The usual formal definition of the S-matrix is unacceptable because it introduces negative-metric states which lead to negative probabilities.

We begin by defining the set of physical asymptotic states to be those states of positive metric which do not contain quanta of action-at-a-distance fields. In an action-at-a-distance electrodynamics the set of physical states includes all electron and positron states not containing «action at a distance» photons. The problem is to define a unitary S-matrix taking «in» physical states to «out» physical states.

We choose to use a variation of Bogoliubov's procedure [17] for defining a unitary physical S-matrix as implemented by SUDARSHAN and co-workers [18]. The starting point is the expansion of the S-matrix in «old fashioned» perturbation theory:

$$(A.1) \qquad S_{\beta\alpha} = \delta_{\beta\alpha} - i(2\pi)^4 \delta^4(P_\beta - P_\alpha) T_{\beta\alpha}$$

with

$$(A.2) \qquad T_{\beta\alpha} = \langle \beta|H_I|\alpha\rangle + \sum_N \frac{\langle\beta|H_I|N\rangle\langle N|H_I|\alpha\rangle}{E_\alpha - E_N + i\varepsilon} + \dots ,$$

[16] D. W. SCIAMA: Roy. Astron. Soc. Month. Not., 113, 34 (1953).

[17] N. N. BOGOLIUBOV: Annales of Invitational Conference on High-Energy Physics, CERN (1958).

[18] E. C. G. SUDARSHAN: Fields and Quanta, 2, 175 (1972); C. A. NELSON: Louisiana State University preprint (1972); J. L. RICHARD: Phys. Rev. D, 7, 3617 (1973); C. C. CHIANG: University Göteborg preprint (1972), and references therein.

where H_1 is the interaction Hamiltonian. We wish to modify eq. (A.2) so that states with «action-at-a-distance quanta» do not contribute to the absorptive part of amplitudes. This would allow unitarity to be maintained under the restriction of unitarity sums to the set of physical states. SUDARSHAN and co-workers [18] have shown that this can be done by taking the energy denominator factor of unphysical states in principal value:

$$(A.3) \qquad T_{\beta\alpha} = \langle \beta | H_1 | \alpha \rangle + \sum_p \frac{\langle \beta | H_1 | p \rangle \langle p | H_1 | \alpha \rangle}{E_\alpha - E_p + i\varepsilon}$$
$$+ \sum_u \langle \beta | H_1 | u \rangle \langle u | H_1 | \alpha \rangle P \frac{1}{E_\alpha - E_u} + \dots ,$$

where p labels physical states and u labels unphysical states (*i.e.* those containing action-at-a-distance quanta in our case). Equation (A.3) defines the physical T-matrix in our models. It could serve as the basis for calculating S-matrix elements.

However, we would like to re-express this S-matrix in terms of covariant perturbation theory. In order to do this we shall take advantage of Weinberg's study [19] of the infinite-momentum frame limit of old-fashioned perturbation theory, and Chang and Ma's realization of infinite-momentum frame results by a change of variable [20].

Since the difference between the physical T-matrix and the unmodified T-matrix lies only in the character of the singularity in the energy denominator, the algebraic development of Weinberg, and his power counting arguments, in particular, can be taken over without change to our case. (We also will ignore possible complications due to interchanging the order of integration and the $P \to \infty$ limit. Actually we define our T-matrix to be the result of this procedure.) The result is that Weinberg's rules apply in the present case also —except that Weinberg's rule (d) must be modified, so that an energy denominator is taken in principal value, if the corresponding intermediate state is unphysical—*i.e.* contains action-at-a-distance quanta.

CHANG and MA showed that it was not necessary to go to the infinite-momentum frame limit in order to realize Weinberg's rules. They showed that the expression of the usual Feynman rules in terms of infinite momentum frame variables, $\eta = p^0 + p^3$ and $s = p^0 - p^3$ for each momentum p^μ, led to the same results as Weinberg's rules. This fact can be used to advantage in the case of action-at-a-distance models. S-matrix elements can be calculated in our models from the usual Feynman rules in the following way: i) for a given diagram follow the usual Feynman rules to obtain a S-matrix contribution using infinite-momentum frame variables *à la* Chang and Ma—in particular, associate Feynman propagators with action-at-a-distance particle lines; ii) evaluate the «energy» integrals by complex integration; iii) the result will be a series of terms corresponding to different intermediate states in old-fashioned perturbation theory language. Those denominators corresponding to intermediate states with principal-value quanta should be taken in principal value. Those denominators corresponding to physical intermediate states should remain unmodified.

[19] S. WEINBERG: *Phys. Rev.*, **150**, 1313 (1966).
[20] S. J. CHANG and S. K. MA: *Phys. Rev.*, **180**, 1506 (1969).

Let us examine the results of this procedure for some simple cases. First, consider the pole diagram for two-particle–to–two-particle scattering. If the pole corresponds to a principal-value particle the amplitude will be proportional to

$$(A.4) \qquad P \frac{1}{q^2 - m^2},$$

using the above rules [21]. This result also follows from explicit evaluation of the diagram in canonical perturbation theory, using eq. (1) of the text for $\theta = \pi/4$.

Next consider a second-order self-energy correction due to the emission and absorption of a principal-value quantum by a particle (propagating via Feynman propagators). Our modified Feynman rules require us to evaluate an integral

$$(A.5) \qquad I = i \int \mathrm{d}^4 k \, \frac{1}{k^2 - m^2 + i\varepsilon} \, \frac{1}{(p-k)^2 - \mu^2 + i\varepsilon},$$

using-infinite-momentum frame variables:

$$(A.6) \qquad \eta' = k^0 + k^3 \qquad s' = k^0 - k^3 \qquad \vec{q} = (k^1, k^2)$$

and $p = (s, \eta, \vec{0})$. The result is

$$(A.7) \qquad I = \frac{\pi}{2} \int \mathrm{d}^2 q \int_0^{\eta} \frac{\mathrm{d}\eta'}{\eta'(\eta - \eta')} \, \frac{1}{s - (\vec{q}^2 + \mu^2)/(\eta - \eta') - (\vec{q}^2 + m^2)/\eta' + i\varepsilon}$$

which must be modified to

$$(A.8) \qquad I' = \frac{\pi}{2} \int \mathrm{d}^2 q \int_0^{\eta} \frac{\mathrm{d}\eta'}{\eta'(\eta - \eta')} \, P \, \frac{1}{s - (q^2 + \mu^2)/(\eta - \eta') - (q^2 + m^2)/\eta'}.$$

In this case we see that $2I' = I + I^*$. The propagation of the particle which would ordinarily propagate via Feynman propagators is manifestly different in states where virtual action-at-a-distance quanta are present. This is necessary in order to define a unitary physical S-matrix. Thus the physical S-matrix only agrees with the canonically defined S-matrix for one–principal-value particle intermediate states and states without principal value particles.

If we consider a non-Abelian action-at-a-distance gauge theory coupled to an external classical source, it is possible to define the physical S-matrix in

---

[21] Compare to eqs. (5) and (6) of ref. [9].

a different manner from the above. Consider the usual definition of the S-matrix and in particular the set of tree diagrams with no external principal-value particle lines. It is easy to verify that the absorptive part of any tree diagram is zero *in the physical region* since

$$(A.9) \qquad \text{Abs} \left[P \, \frac{1}{k^2 - m^2} \right] = 0 \; .$$

As a result the set of tree diagrams defines a unitary (gauge invariant) S-matrix. (This should be contrasted with the usual theory where taking the absorptive part of a tree diagram introduces states containing non-Abelian bosons which in turn introduce loops and Fadeev-Popov ghosts.) In the present case no Fadeev-Popov ghost loops are needed to maintain unitarity [22]. Thus we wind up with a loopless, tree diagram model. The possibility of limiting the self-interaction of fields in this way allows us to develop a renormalizable model of quantum gravity.

Finally we note that the lack of contributions to unitarity sums from intermediate states containing principal-value particles allows us to avoid the introduction of Fadeev-Popov ghosts in *all* non-Abelian action-at-a-distance models.

---

[22] Cf. ref. [12] and B. W. LEE and J. ZINN-JUSTIN: *Phys. Rev. D*, **5**, 3121 (1972).

● RIASSUNTO (*)

Si formulano teorie di gauge nel sistema di una generalizzazione della teoria quantistica dei campi. In particolare si discutono modelli di teorie di elettrodinamica e di Yang-Mills, un modello dell'interazione forte con derivate di ordine più alto e confinamento dei quark, e un modello rinormalizzabile di gravità quantistica pura con una Lagrangiana di Einstein. Nel caso dell'elettrodinamica si mostra che due modelli sono possibili: uno con predizioni che sono identiche a QED e uno che è un modello quantistico di azione a distanza dell'elettrodinamica. Nel caso delle teorie di Yang-Mills si può costruire un modello che è identico per quanto riguarda le predizioni a qualsiasi modello convenzionale o modello di azione a distanza. Nel secondo caso è possibile eliminare tutti i cappi di particelle di Yang-Mills (in tutti i gauge) in una maniera consistente con l'unitarietà. Esiste una variazione dei modelli di Yang-Mills nella nostra formulazione che ha equazioni di campo con derivate ad ordine più alto. È unitario ed ha probabilità positive. Può essere usato per costruire un modello d'interazioni forti che ha un potenziale lineare e manifesta confinamento dei quark. Infine si mostra come costruire un modello di azione a distanza della gravità quantistica pura (il cui limite classico è la dinamica della Lagrangiana di Einstein) accoppiato ad una sorgente esterna classica. Il modello è grossolanamente rinormalizzabile.

(*) *Traduzione a cura della Redazione.*

Новый подход к калибровочным теориям поля.

Резюме (*). – Мы формулируем калибовочные теории в рамках обобщения квантовой теории поля. В частности, мы обсуждаем модели электродинамики и теорий Янга-Миллса, модель сильных взаимодействий с производными высших порядков и удержанием кварков и перенормируемую модель для чистой квантовой гравитации с Лагранжианом Эйнштейна. Мы показываем, что в случае электродинамики возможны две модели: одна модель имеет предсказания, которые идентичны предсказаниям квантовой электродинамики, и другая модель, которая представляет модель электродинамики с квантовым действием на расстоянии. В случае теорий Янга-Миллса мы можем сконструировать модель, которая идентична по предсказаниям любой общепринятой модели или модели с квантовым действием на расстоянии. Во втором случае имеется возможность исключить все петли для частиц Янга-Миллса (во всех калибровках). Наша формулировка содержит изменение моделей Янга-Миллса, которое включает уравнения поля с производными высших порядков. Наша модель является унитарной и имеет положительные вероятности. Наша формулировка может быть использована для конструирования модели сильных взаимодействий, которая имеет линейный потенциал и обеспечивает удержание кварков. Мы показываем, как сконструировать модель действия на расстоянии для квантовой гравитации, связанной с внешним классическим источником. Предложенная модель является перенормируемой.

(*) *Переведено редакцией.*

Direttore responsabile: CARLO CASTAGNOLI

Stampato in Bologna dalla Tipografia Compositori coi tipi della Tipografia Monograf
Questo fascicolo è stato licenziato dai torchi il 17-I-1979

Appendix C. The Three Players in Creating a Fundamental Theory of Physical Reality

When one considers constructing a theory, or model, of fundamental 'real world' physical reality, the usual course of action is to make some assumptrions such as a symmetry group, or a construct like strings, and then proceed to develop their implications.[396] Next, the authors of the theory, and experimenters, study the experimental implications of their theory. A tacit assumption, which is sometimes stated, is that their mathematical-mental construct is somehow automatically implemented by Reality. The possibility that their construction is merely a mental exercise, and not Reality, is seldom considered except by the philosophically inclined. And for good reason. No one has put forward a concept that joins thought to Reality in a convincing way except as 'word play.'

In this chapter we will seek to carefully characterize the process of constructing a fundamental physical theory.[397] In doing so, we identify 'three players' that play decisive roles in bringing a fundamental theory from conception to Reality. First is the construction process which is a mathematical-mental effort that leads to a complete theory fundamental theory. Secondly, there is the conceptual mapping of the theory to physical Reality. Thirdly, there is the 'agency' by which Reality is imprinted with the theory. The third 'player', the 'imprinter,' is often alluded to in the statement that it is surprising how well Reality is described by mathematics. This statement is not good enough! We will examine it in some detail in this chapter together with the other two players in the 'game' of constructing the fundamental theory of everything.

C.1 Construction of a Fundamental Physics Theory of Everything

The construction process of a fundamental theory is a mental exercise consisting of defining primitive constructs, specifying assumptions (axioms), and then proceeding to develop a theory from the assumptions.

C.1.1 Euclidean Geometry Model

Euclid's theory of geometry is perhaps the best model of the process, and the components of the process. First Euclid identifies primitive unspecified constructs: straight

---

[396] In the author's view this approach is satisfactory for non-fundamental theories and models where the underlying physics is presumably well understood. But at the fundamental physics level a more careful analysis is required. This chapter delves into this issue in some detail.

[397] We discuss the metaphysical and logical basis of our approach in more detail in Blaha (2011c).

lines, curves, and angles (which are viewed in no need of definition.) He then states five axioms and proceeds to state and prove theorems based on the unspecified, but understood, properties of the constructs. As the set of theorems grows, subsidiary, but unspecified, properties of geometric figures are used in proofs. These properties include the definition of angles, the construction of simple geometric figures such as rectangles, triangles, trapezoids, and so on. The more advanced developments in Euclidean geometry such as the geometry of Ptolomaic astronomy[398] with cycles and epicycles also have implicit principles for construction.

We see from Euclid that the theory construction process requires:

1. The definition of primitive terms.
2. The specification of a set of axioms.
3. The derivation of theorems from the axioms and further ad hoc but necessary assumptions.
4. The identification of the 'complete' theory with Reality.

Euclid was fortunate in the creation of Euclidean geometry because he knew the method with which to construct geometry, and its primitive terms with which to define his axioms, and then his theorems. The development of a fundamental physical theory is not so fortunate.

C.1.2 Logic Applied to Physics

Perhaps the most important aspect of constructing a physical theory is the use of Logic to derive results from the fundamental axioms. Questions have been raised about the validity of Logic. Attention is often drawn to logical 'paradoxes' and particularly to Gődel's Theorem. In Blaha (2015a) we showed the problems of these paradoxes and Gődel's Theorem were due to the choice of subjects in a statement that were not in the subject domain of the statement. Statements are a verbal form of functions. Like functions they have a domain (set of valid subjects) and a range.

Most scientists, Logicians and Mathematicians are familiar with mathematical functions that have numerical arguments and calculate a numerical value or set of numerical values. However, those familiar with Computer Science know that a function, in general, has two types of output: its numeric value(s), and a status value indicating whether the computation that the function performed was successfully done or failed. Typically the status value is a zero or one, but it is understood as true or false also. (True – computation successful; false – computation failed for some reason) Thus we categorize functions as being in one of three broad categories:

---

[398] Ptolomaic Astronomy is a very-well developed, and mathematically sound, branch of Euclidean geometry. Its flaw is its misidentification with the physical reality of the Solar System. In this regard it very well illustrates our point that the matching of theory with Reality is an important part of the creation of a fundamental theory.

Types of Functions

1	2	3
A Mathematical Function Producing A Value(s) Only;	A Function Producing a Value(s) and a Status Value;	A Function Producing a Status Value Only (T or F);
Examples:		
sin function	Programming Function	Logic Statement

Thus Logic is unsullied by paradoxes and, besides being the only way to 'do' Physics theory, is correct and consistent.

C.1.1 Primitive Terms of Fundamental Physics Theories

The primitive terms of the fundamental theory of Physics is an open question. There are various theories that have been proposed: Twistor theory, SuperString theories, SuperStandard Model theory, and so on. They differ significantly in their definition of primitive terms and their axioms (or their equivalent to axioms).

Under the circumstances probably the most significant aspects of terms are:

1. Consistency with what we know of Reality – they are somehow evident in Reality.
2. Simplicity.
3. Utility in defining meaningful axioms. This feature implies a forward-looking approach that looks to the development of the theory to guide the choice of primitives.

Euclid's specification of primitive terms agrees with these considerations.

C.1.2 Physical Axioms of a Fundamental Theory

The choice of axioms is very dependent on the eventual theory. Generally it seems that most theories choose a lagrangian and/or symmetry group, and proceed to 'back-engineer' a set of axioms if they consider defining axioms at all.

If one wishes to create a fundamental theory using a top-down approach then the set of axioms, like primitive terms, should satisfy the same three conditions plus Decision Axioms (described below):

1. Consistency with what we know of Reality – they are somehow evident in Reality.
2. Simplicity.

3. Utility in generating the fundamental theory. This feature also implies a forward-looking approach that looks to the development of the theory to guide the choice of axioms.

4. The choice of axioms should be in agreement with the below Decision axioms.

C.1.3 Decision Axioms for a Fundamental Theory

The concept of Decision Axioms does not seem to have been formally considered previously. Decision axioms guide a choice between alternatives in the construction of fundamental theory. Given the complex nature of fundamental theories choices between alternatives frequently occur. SuperString theory, for example, has millions of choices in the determination of the 'correct' SuperString theory.

What rules (Decision Axioms) can guide a choice?

In this subsection we identify six rules that provide guidance for making the proper choice.

C.1.3.1 Ockham's Razor

William of Ockham proposed a Law of Parsimony that is called *Ockham's Razor* which states that the simplest choice is to be preferred in a multiple choice situation. This principle is often stated as 'the simplest solution is usually the correct solution.'[399]

The best rationale for this principle is that it generally reduces the complexity that follows such a choice. Since many physics calculations are extremely difficult, picking the simplest choice.would generally tend to make subsequent theorems/and calculations less difficult. This point of view might be thought to be ad hoc or anthropomorphic. But it reflects the reality of scientific calculation and of theory construction.

Thus we will assume Ockham's Law of Parsimony in the construction of our theory with the proviso that Leibniz's Minimax Principle (below) takes precedence if there is a conflict in the implications of the choices.

C.1.3.2 Leibniz's Minimax Principle for Physics Theories

Leibniz[400] developed a Minimax Principle that can be phrased for our purposes as, "The universe is based on the smallest set of properties or features that lead to the greatest variety of phenomena." This principle reflects the spirit of the minimum/maximum criteria of the Calculus of Variations[401] that plays a central role in many physics theories. This principle somewhat overlaps Ockham's Law of Parsimony. Given a choice of possible theoretical lines of

---

[399] William of Ockham – Law of Parsimony – "Pluralitas non est ponenda sine necessitate" or "Plurality should not be posited without necessity." Ockham's Law was first stated by Durand De Saint-Pourçain (1270-1334 A.D.). In simple terms the principle states the simplest solution to a problem is most likely to be the correct solution.
[400] See Rescher (1967).
[401] Leibniz was one of the founders of the Calculus of Variations.

construction there is a possibility that the Law of Parsimony and the Minimax Principle would suggest different choices. Fortunately, we will see that the construction of the SuperStandard Model does not seem to present this potential dilemma.

An important, unremarked, aspect of Leibniz's Principle is the decision between a set of choices depends on the future part of the construction or theory. Thus future constructs determine past constructs in minimax decisions.[402]

C.1.3.3 Conputationally Minimal

Given a choice between two alternatives that appear equally likely from general considerations and the view of the preceding two Decision Axioms, the preferred choice is the one leading to the most minimal (simplest) computations in the full theory.

C.1.3.4 Experimentally Verifiable Directly or Indirectly

Given a choice between two alternatives, one of which is subject to experimental verification, directly or indirectly although perhaps in the future, and the other choice is not, then the first alternative is preferred.

C.1.3.5 Level of Rigor

Physics is mathematical in nature. Mathematics strives for rigor and will not be satisfied without rigorous results except for conjectures and speculations. Physics prefers rigorously derived results as wel. However, there is 'physical rigor,' which is a level of rigor that is rigorous to the extent that it is mathematically rigorous. This circuitous statement reflects the historical fact that Physics has often gone beyond the mathematics of the time. The most significant example is Newtonian physics, which until the mid-19th century used *derivatives* asserting that they were, or would become, rigorous mathematically. After a careful analysis of derivatives, and advances in the understanding of continuity and the various types of infinities, mathematicians were able to put derivatives on a rigorous footing justifying the use of derivatives by physicists for almost four centuries.

Today, the most interesting part of physics in search of mathematical rigor is the path integral formulation in quantum theory and the Faddeev-Popov Method, in particular. The problem here is again an understanding of infinities in functional derivatives and path integrals. Again, mathematical rigor is lackng. Yet physicists use these techniques with the strong belief that their results will be eventually justified rigorously.

In view of the nature of rigor in Physics it seems reasonable to state the Decision Axiom: *In the case of a situation where several possible approaches are possible, one should*

---

[402] The knowledgable reader will remember Feynman's speculation that the physical universe may be evolving from the future into the past. Quantum field theories support such an interpretation of their mathematics. The similarity of this fundamental minimax principle's feature with the corresponding feature of physics theories encourages support for the minimax principle as a 'design' law of fundamental physical theory.

choose the approach that is most rigorous – should one exist. If all approaches are at an 'equal' level of rigor then the approach that appears 'most' physical should be chosen, taking account of the other Decision Axioms.

C.1.3.6 Replication Principle - Nature Tends to Repeat Successful Strategies

In many branches of Science one sees that successful strategies and techniques are repeated in several areas – perhaps with some variations and changes. Consequently, when a physical phenomena is similar to another physical phenomena, and all other axioms that were stated above are not relevant for the phenomena, then one should model the new phenomena in a manner similar to the successfully modeled phenomena.

C.1.3.7 Principle of Maximal Serendipity

A correct physical theory will evolve from fundamental axioms with maximal simplicity to attain better-than-expected ends.

C.1.4 The Body of the Derived Theory –Jeopardy Points

After the primitives and the axioms have been specified, the body of the theory can be developed mathematically (logically). At points in the development process decisions may have to be made between alternatives. We call these points – *jeopardy points*. The above Decision Axioms can provide guidance if a strictly physical decision is not evident.

Clearly the fewer the jeopardy points in a theoretical development, the better. The existence of many such points raises the question of whether the axiomatic basis needs to be strengthened.

The general lack of jeopardy points suggests the axiomatic basis of the theory is well-founded.

C.1.5 The Fundamental Theory as a Mental Construct

If one creates a fundamental theory it is a mental construct, and although physicists are quick to identify it with Reality, it is not Reality until a map to Reality is specified. Parmenides (front cover) pointed out this distinction in a way in his only surviving poem *On Nature*. He said the fundamental nature of the universe was an unchanging unity of nature (The Way of Truth); but our view of the universe was that of change and seeming (The Way of Appearance). Translated into the framework of our discussion we see the fundamental theory as an unchanging mental construct, but the Reality to which it maps as ever changing.

In the next section we discuss the mapping realizing that most physical theories make the leap from concept to Reality without effort.

C.2 Conceptual Mapping of a Fundamental Theory to Physical Reality

Generally the fundamental theories of Physics that have been proposed are so close to Reality that the mapping process is straight-forward – pick a symmetry group, specify a lagrangian, find the physical particle spectrum (which was put there by the choice of symmetry group anyway), and calculate experimentally verifiable numbers.

However in some cases the mapping is not so clear. For example, one might find an extremely large particle spectrum with experimentally measurable masses, and the particles are not there! Then one has the dilemma of redesigning the theory.

C.2.1 Mapping of the Fundamental Physical Axioms

One could possibly redesign the set of axioms of the theory if they were at fault and if the axioms had been specified. The new theory that emerges might then be mappable to experimental reality.

C.2.2 Mapping of the Body of the Derived Fundamental Theory

If the body of the fundamental theory does not agree with a mapping to Reality, the creator(s) of the theory often change the parameters or otherwise tinker with its development. One sees this in some current theories.

C.3 The 'Agency' by which Reality is Imprinted with a Fundamental Theory

Mental processes such as the thought embodying a physical theory are clearly different from physical reality. A question that has been a subject of discussion for over 2500 years is How does thought become embodied in Reality? A further question that has been of great interest for the past five hundred years is How does the mathematics of physical theory so well explain/correspond to Reality?

Addressing the second question we note that there are two possibilities: Reality is chaotic or Reality follows physical laws (which are necessarily mathematical). If Reality were chaotic but through some principle of self-organization transformed itself into an ordered system, then we would see order but there would be pockets of chaos in the universe and on earth that we have not seen. Further the transition to order would have to repeat with every change of state of a physical system – an effect which is also not seen. Thus 'self-organization' in any form does not appear to be the source of Physical Law.

Thus we must conclude that nature is ordered and conforms to physical laws. Physical laws are intertwined through mathematics. The fact that we can define theories that embody all physical laws – rather than have a host of unrelated physical laws is another marvel.

So we are left with the issue of who imprinted the concepts of a fundamental physical theory on Reality. The philosophers (Presocratic and more recent) addressed this problem and came to the conclusion that there was an 'Unmoved Mover' – an entity that implemented the imprinting of physical law on Reality.

Most philosophers (and theologians) identified this entity as God. And they called this the proof of the existence of God.

Appendix D. New Paradigm: Composition of Extrema in the Calculus of Variations

This Appendix appeared originally in Blaha (2005a) and earlier books.

D.1 A New Paradigm in the Calculus of Variations

The Calculus of Variations has a long and venerable history in Physics and Mathematics. Many problems in Physics and Mathematics have been treated with approaches based on techniques in the Calculus of Variations.[403] In this book we have developed a unified quantum field theory of the known forces of nature based on a new type of problem, or paradigm, in the Calculus of Variations. One way of viewing the spectrum of problems in the calculus of variations is the following progression.

D.1.1 A Classification of Variational Problems

1. Variational problems in a Euclidean, or Minkowski, flat space such as the minimal distance between two points or the extrema of a field theory Lagrangian.

2. Variational problems seeking extrema on a curved surface such as the shortest distance between points on the surface of a sphere.

The development in this book suggests a third and fourth, possibility, that to the author's knowledge, has not been addressed in the literature:

3. Variational problems where the extrema are determined on a surface that is itself defined as an extremum. The discussions in this book exemplify this pardigm.

4. Variational problems where the extrema are determined on a surface that is itself defined as an extremum that depends on the extrema on the surface. More simply put the extrema, and the surface upon which they are defined, are jointly determined and are interrelated. Fortunately, our unified theory does not use this paradigm. A future theory might.

---

[403] See Akhiezer (1962), Blaha (2003), Gelfand (2000), Giaquinta (1996) and (1998), Jost (1998), and Sagan (1993),

In the unified theory that we will develop all particle fields including the graviton field are defined as a mapping of a Minkowski space-time y to a "particle" space-time X with the mapping determined as an extremum of a variation of a fundamental field (a type 3 variational problem in the above classification). Our theory could be generalized to include a back-reaction of the particle fields on the fundamental field (a type 4 variational problem in the above classification). We will not discuss this possibility in this book.

D.2 Simple Physical Example – Strings On Springs

In this section we will describe a simple physical example that illustrates a variational problem of type 3 in the Calculus of Variations. We view it as a composition of extrema. (This problem can be addressed using other calculus of Variations techniques.) The approach used in the solution of this problem is similar to the approach used in Two-Tier quantum field theory.

D.2.1 A Strings on Springs Mechanics Problem

Consider a long string or bar that can oscillate (undulate) in a direction perpendicular to its length. Further assume that one end of this bar or string is attached to a spring that cause the entire bar or string to oscillate back and forth in a direction parallel to its long side. This configuration is illustrated in Fig. D.1.

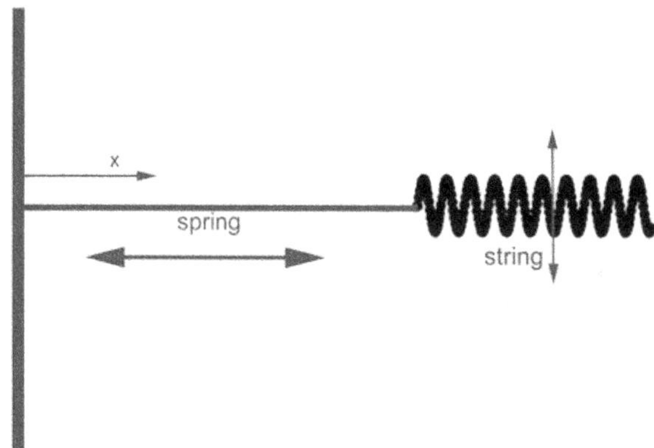

Figure D.1. An oscillating string attached to a spring.

Let x denote the distance to a point on the string when the spring is at equilibrium. If 2π times the frequency of the spring is ω_1, then the location of this point when the spring is oscillating is

$$X(t) = x + A \sin(\omega_1 t + \phi_1) \tag{D.1}$$

where ϕ_1 is a phase, and A is the amplitude of the spring oscillation. Then the vertical displacement of a traveling wave on the *string* can take the form

$$\psi(t) = B \sin(\omega_2 t - k_2(x + A \sin(\omega_1 t + \phi_1)) + \phi_2) \tag{D.2}$$

where B is the amplitude of the string wave, and k_2, ω_2 and ϕ_2 are the parameters of the string wave. These simple mechanical formulae are well known. But they lead to an interesting new application of the ideas of the Calculus of Variations.

Suppose we treat X as an independent variable with X given by eq. (D.1), and with eq. (D.2) written as:

$$\psi(t) = B \sin(\omega_2 t - k_2 X + \phi_2) \tag{D.3}$$

Defining

$$\psi = \psi(X(t), t) \tag{D.4}$$

we can specify the dynamics of the above motion by finding the extrema of

$$I = \int \mathscr{L}_\psi \, dX(t) + \int \mathscr{L}_X \, dt \tag{D.5}$$

where the Lagrangian terms are

$$\mathscr{L}_\psi = \tfrac{1}{2} \{ \mu \, (\partial\psi/\partial t)^2 - Y \, (\partial\psi/\partial X)^2 \} \tag{D.6}$$

with μ and Y being constants, and

$$\mathscr{L}_X = \tfrac{1}{2} \{ m(\partial X/\partial t)^2 - k(X - x)^2 \} \tag{D.7}$$

where m and k are constants, and where x is a parameter. Applying Hamilton's Principle, and performing independent variations of X and ψ yields the Lagrangian equations:

$$\frac{\partial \mathscr{L}_\psi}{\partial \psi} - \frac{\partial}{\partial X} \frac{\partial \mathscr{L}_\psi}{\partial(\partial\psi/\partial X)} - \frac{\partial}{\partial t} \frac{\partial \mathscr{L}_\psi}{\partial(\partial\psi/\partial t)} = 0 \tag{D.8}$$

and

$$\frac{\partial \mathscr{L}_X}{\partial X} - \frac{\partial}{\partial t} \frac{\partial \mathscr{L}_X}{\partial(\partial X/\partial t)} = 0 \tag{D.9}$$

$$\partial\, X \qquad \partial t \quad \partial(\partial X/\partial t)$$

The resulting equations of motion are:

$$\mu\, \partial^2\psi/\partial t^2 - Y\, \partial^2\psi/\partial X^2 = 0 \tag{D.10}$$

and

$$m\, \partial^2 X/\partial t^2 + k(X - x) = 0 \tag{D.11}$$

with the solutions given in eqs. D.1 and D.2.

The procedure that we use to obtain these results may look a bit strange but they illustrate a type 3 problem in the Calculus of Variations involving the composition of extrema—the composition of an extremum that specifies a manifold in a space (possibly including all of space in a $R^n \rightarrow R^n$ mapping) with an extremum determining a function on that manifold. The procedure is described in detail in the next section.

D.3 The Composition of Extrema – A Lagrangian Formulation

In this section we will explore the general case of the composition of extrema for fields. We will discuss the case of a scalar field ϕ that is a function of a vector field X^μ in a D-dimensional space with coordinate variables that we will denote as y^μ. (The discussion for other types of fields is a straightforward extension of this discussion.) Thus

$$\phi = \phi(X) \tag{D.12}$$

and

$$X^\mu = X^\mu(y) \tag{D.13}$$

We assume that the dynamics can be described by a Lagrangian formulation using an extension of Hamilton's principle:

$$I = \int \mathscr{L}\, d^4 y \tag{D.14}$$

with

$$\mathscr{L} = \mathscr{L}(\phi(X),\, \partial\phi/\partial X^\nu,\, X^\mu(y),\, \partial X^\mu(y)/\partial y^\nu,\, y) \tag{D.15}$$

If we perform a standard variation[404] in ϕ for fixed y (and thus fixed X) we find

---

[404] Bogoliubov, N. N., & Shirkov, D. V., Volkoff, G. M. (tr), *Introduction to the Theory of Quantized Fields* (Wiley-Interscience, New York, 1959); Goldstein H., *Classical Mechanics* (Addison-Wesley, Reading, MA 1965).

$$\delta I = \int [\delta\phi\, \partial\mathcal{L}/\partial\phi + \delta(\partial\phi/\partial X^\nu)\, \partial\mathcal{L}/\partial(\partial\phi/\partial X^\nu)]\, d^4y \qquad (D.16)$$

We can rewrite the variation in the derivative of ϕ as

$$\delta(\partial\phi/\partial X^\nu) = \partial(\delta\phi)/\partial X^\nu \qquad (D.17)$$

$$= \partial y^\mu/\partial X^\nu\, \partial(\delta\phi)/\partial y^\mu \qquad (D.18)$$

with an implied summation over repeated indices. After substituting eq. D.18 in eq. D.16, and performing an integration by parts (and discarding the surface term which is assumed to yield zero in the standard fashion) we obtain:

$$\delta I = \int \delta\phi\, \{\partial\mathcal{L}/\partial\phi - \partial/\partial y^\mu\, [\partial\mathcal{L}/\partial(\partial\phi/\partial X^\nu)\, \partial y^\mu/\partial X^\nu\,]\}\, d^4y$$

Since the variation of $\delta\phi$ is arbitrary we conclude

$$\partial\mathcal{L}/\partial\phi - \partial/\partial y^\mu\, [\partial\mathcal{L}/\partial(\partial\phi/\partial X^\nu)\, \partial y^\mu/\partial X^\nu] = 0 \qquad (D.19a)$$

The second term in eq. D.19a shows the effect of the dependence of ϕ on the field X, $\phi = \phi(X)$, rather than directly on the coordinate system y.

Similarly we can perform a variation in X^μ and obtain

$$\partial\mathcal{L}/\partial X^\mu - \partial/\partial y^\nu\, [\partial\mathcal{L}/\partial(\partial X^\mu/\partial y^\nu)] = 0 \qquad (D.19b)$$

The X field defines a "manifold" or, more properly, specifies a transformation from $R^n \to R^n$. If we make standard assumptions about the mapping: that it is continuous and piece-wise invertible, then we can establish the following lemmas:

Lemma 1: *If the transformation $X^\mu = X^\mu(y)$ is a transformation from $R^n \to R^n$ that is of class C' and piece-wise invertible, then*

$$\frac{\partial}{\partial y^\nu}\, \frac{\partial y^\nu}{\partial X^\mu} = -\frac{\partial \ln J}{\partial X^\mu} \qquad (D.20)$$

where

$$J = |\partial(X)/\partial(y)| \qquad (D.21)$$

is the absolute value of the Jacobian of the transformation.

Proof:
Consider two equivalent forms of an integral:

$$I = \int \mathscr{L} J \, d^4y = \int \mathscr{L} \, d^4X$$

where $\mathscr{L}$ is specified as in eq. D.15. Then the first expression for I leads to eq. D.19a which can be written in the form

$$\partial\mathscr{L}/\partial\phi - \partial/\partial X^\mu \left[\partial\mathscr{L}/\partial(\partial\phi/\partial X^\mu)\right] - \partial\mathscr{L}/\partial(\partial\phi/\partial X^\mu)\{\partial[J\partial y^\nu/\partial X^\mu]/\partial y^\nu\} = 0$$

Using the second expression for I above we obtain the following equation by variation in ϕ:

$$\partial\mathscr{L}/\partial\phi - \partial/\partial X^\mu \left[\partial\mathscr{L}/\partial(\partial\phi/\partial X^\mu)\right] = 0$$

Comparing these two expressions and realizing that $\partial[J\partial y^\nu/\partial X^\mu]/\partial y^\nu$ is totally independent of ϕ and its derivatives leads us to conclude

$$\partial[J\partial y^\nu/\partial X^\mu]/\partial y^\nu = 0 \tag{D.22}$$

It is a general relationship for a transformation between X and y based on continuity and piece-wise invertibility. After a few elementary manipulations eq. D.22 can be rewritten in the form of eq. D.20. ∎

Lemma 2: *If the transformation $X^\mu = X^\mu(y)$ is a transformation from $R^n \to R^n$ that is of class C' and piece-wise invertible and $\mathscr{L} = \mathscr{L}(\phi(X), \partial\phi/\partial X^\nu, X^\mu(y), \partial X^\mu(y)/\partial y^\nu, y)$, then*

$$\partial\mathscr{L}/\partial(\partial\phi/\partial X^\nu)\, \partial y^\mu/\partial X^\nu = \partial\mathscr{L}/\partial(\partial\phi/\partial y^\mu) \tag{D.23}$$

Proof:
Let us express $\mathscr{L}$ as a power series in derivatives of ϕ:

$$\mathscr{L} = \sum_{n=0} a_{n\mu_1\mu_2\cdots\mu_n}(\phi(X), X^\mu(y), \partial X^\mu(y)/\partial y^\nu, y) \prod_{j=1}^{n} \partial\phi/\partial X^{\mu_j}$$

which can rewritten using piece-wise invertibility as

$$\mathscr{L} = \sum_{n=0} a_{n\mu_1\mu_2\cdots\mu_n}(\phi(X),\, X^\mu(y),\, \partial X^\mu(y)/\partial y^\nu,\, y) \prod_{j=1}^{n} \partial\phi/\partial y^{\nu_j}\; \partial y^{\nu_j}/\partial X^{\mu_j}$$

Taking the derivative of this equation with respect to $\partial\phi/\partial y^\mu$ immediately yields the result. ■

Eq. D.23 enables us to rewrite eq. D.19a as:

$$\partial\mathscr{L}/\partial\phi \;-\; \partial/\partial y^\mu\,[\partial\mathscr{L}/\partial(\partial\phi/\partial y^\mu)] = 0 \tag{D.24}$$

which is as one would expect.

In order to get a feeling for the effect of eq. D.19a we will look at a simple example where we specify the relation of the X and y variables directly. Then we will look at the composition of extrema where the transformation between X and y is itself determined as an extremum solution.

D.3.1 Example: a hyperplane

We assume eq. D.19b yields the transformation:

$$X^i = ay^i \qquad \text{for } i = 1,2,3$$

$$X^0 = 0$$

Then eq. D.19a becomes

$$\partial\mathscr{L}/\partial\phi \;-\; \partial/\partial y^i\,[\partial\mathscr{L}/\partial(\partial\phi/\partial y^i)] = 0 \tag{D.25}$$

with the time derivative disappearing. Effectively the variation of ϕ on the hyperplane $X^0 = 0$ is determined by the differential equation generated by D.25. On this hyperplane the transformation between the X and y variables is invertible.

D.3.2 Coordinate Transformation Determined as an Extremum Solution

We now develop a formalism that determines a mapping from space onto itself as the solution of an extremum problem and also determines the dynamics of one or more fields as a function of this mapping. To this author's knowledge this area in the Calculus of Variations – the determination of an extremum on a manifold where the manifold itself is determined by an extremum – has not been previously explored. We will also develop a hamiltonian formulation. Then we will proceed to quantize the theory.

D.3.3 Separable Lagrangian Case

Although there are many forms that the composition of extrema could take, one fairly general form that is directly useful in quantum field theory applications is based on a Lagrangian that can be split into two parts which we will call a *separable Lagrangian*:

$$\mathscr{L} = \mathscr{L}_F J + \mathscr{L}_C(X^\mu(y), \partial X^\mu(y)/\partial y^\nu, y) \tag{D.26}$$

where J is defined in eq. D.21, where $\mathscr{L}_F$ contains all the dynamics of the fields and their interactions, and where $\mathscr{L}_C$ defines the coordinate mapping as an extremum solution. The procedure to determine the differential equations that specify the mapping, and the field equations that specify field interactions and evolution, is to vary in the coordinates X^μ and in the fields independently, using Hamilton's Principle. The extrema are to be determined for

$$I = \int \mathscr{L} \, d^4y \tag{D.27}$$

We will begin by considering the case of one scalar field:

$$\mathscr{L}_F = \mathscr{L}_F(\phi(X), \partial\phi/\partial X^\nu) \tag{D.28}$$

and

$$\mathscr{L}_C = \mathscr{L}_C(X^\mu(y), \partial X^\mu(y)/\partial y^\nu, y) \tag{D.29}$$

Eq. D.27 can be written in the form:

$$I = \int \mathscr{L}_F(\phi(X), \partial\phi/\partial X^\nu) \, dX + \int \mathscr{L}_C(X^\mu(y), \partial X^\mu(y)/\partial y^\nu, y) \, d^4y \tag{D.30}$$

using the Jacobian to transform to an integral over dX in the first term. A standard variation of ϕ and the application of Hamilton's Principle yields

$$\partial\mathscr{L}_F/\partial\phi - \partial/\partial X^\mu [\partial\mathscr{L}_F/\partial(\partial\phi/\partial X^\mu)] = 0 \tag{D.31}$$

reflecting the fact that ϕ is a function of X^μ only, with X^μ a function of the y coordinates.

Next we perform a variation of X^μ determining the mapping from $y \rightarrow X$ as an extremum of the integral in eq. D.27. We note the piece-wise invertibility of the coordinate mapping $X^\mu(y)$ allows us to write the Jacobian J as a function of y^μ only. A standard variation of X^μ and the application of Hamilton's Principle yields

$$\partial \mathscr{L}_C / \partial X^\mu - \partial/\partial y^\nu \, [\partial \mathscr{L}_C / \partial(\partial X^\mu/\partial y^\nu)] = 0 \qquad (D.32)$$

D.3.4 Klein-Gordon Example

The Klein-Gordon scalar field theory furnishes us with a simple example of the application of the preceding development. The Lagrangian is

$$\mathscr{L}_F = \tfrac{1}{2} \, [\, (\partial\phi/\partial X^\nu)^2 - m^2\phi^2 \,] \qquad (D.33)$$

From eq. D.31 we obtain the field equation:

$$(\Box + m^2) \, \phi(X) = 0 \qquad (D.34)$$

where

$$\Box = \partial/\partial X^\nu \, \partial/\partial X_\nu$$

$$(D.34a)$$

A fourier representation of the solution of eq. D.34 is:

$$\phi(X) = \int dp \, \delta(p^2 - m^2)\theta(p^0) \, [A(p) \, e^{-ip\cdot X} + A(p)^* \, e^{ip\cdot X}] \qquad (D.35)$$

where $A(k)$ is a function of k and $*$ indicates complex conjugation.

The determination of $X^\mu(y)$ depends on the Lagrangian $\mathscr{L}_C$ and the solutions of eq. D.3A. If we chose

$$\mathscr{L}_C = -\tfrac{1}{2} \, (\partial X^\mu/\partial y^\nu)^2 \qquad (D.36)$$

Then we obtain the equation

$$\Box \, X^\mu = 0 \qquad (D.37)$$

with the solution

$$X^\mu = \int dk \, \delta(k^2)\theta(k^0) \, [a^\mu(k) \, e^{-ik\cdot y} + a^\mu(k)^* \, e^{ik\cdot y}] \qquad (D.38)$$

where $a^\mu(k)$ are complex vector functions of k in general. (We ignore positivity issues for the moment.) Substitution of eq. D.38 in eq. D.35 yields an expression with a form reminiscent of bosonic string expressions.[405] We will take up this point later in subsequent chapters.

D.4 The Composition of Extrema – Hamiltonian Formulation

The previous section established a Lagrangian formulation of dynamics based on the composition of extrema. In this section we will develop an equivalent hamiltonian formulation. We will assume a Minkowskian space-time with X^0 and y^0 playing the role of the time coordinates in the respective coordinate systems.

Initially, we will assume a scalar field ϕ with a Lagrangian of the form in eq. D.15 and define canonical momenta with

$$\Pi_\phi = \partial\mathcal{L}/\partial\dot{\phi} \equiv \partial\mathcal{L}/\partial(\partial\phi/\partial X^\mu)\,\partial y^0/\partial X^\mu \tag{D.39}$$

$$\Pi_X{}^\mu = \partial\mathcal{L}/\partial\dot{X}_\mu \tag{D.40}$$

where

$$\dot{\phi} = \partial\phi/\partial y^0 \equiv \partial\phi/\partial X^\mu\,\partial X^\mu/\partial y^0 \tag{D.41}$$

$$\dot{X}^\mu = \partial X^\mu/\partial y^0 \tag{D.42}$$

Then we define the hamiltonian density as

$$\mathcal{H} = \Pi_\phi\,\dot{\phi} + \Pi_X{}^\mu\,\dot{X}_\mu - \mathcal{L}(\phi(X), \partial\phi/\partial X^\nu, X^\mu(y), \partial X^\mu(y)/\partial y^\nu, y) \tag{18-A.43}$$

and the hamiltonian

$$H = \int \mathcal{H}\,d^3y \tag{D.44}$$

The hamiltonian density has the general form

$$\mathcal{H} = \mathcal{H}(\phi(X), \partial\phi/\partial X^i, \Pi_\phi, X^\mu(y), \partial X^\mu(y)/\partial y^j, \Pi_X{}^\mu, y^\nu) \tag{D.45}$$

for the case of one scalar field where the indices i and j represent space coordinates; time coordinates are assigned index value 0.

---

[405] See for example Polchinski (1998) and Bailin (1994).

If we calculate the differential change in H using eq. D.45 we obtain

$$dH = \int \Big\{ \partial\mathscr{H}/\partial\phi \; d\phi + \partial\mathscr{H}/\partial\Pi_\phi \; d\Pi_\phi - \partial/\partial y^\nu [\partial\mathscr{H}/\partial(\partial\phi/\partial X^j)\partial y^\nu/\partial X^j] d\phi + \\ + \partial\mathscr{H}/\partial X^\mu \; dX^\mu + \partial\mathscr{H}/\partial\Pi_X{}^\mu \; d\Pi_X{}^\mu - \partial/\partial y^j \, [\partial\mathscr{H}/\partial(\partial X^\mu/\partial y^j)] \; dX^\mu \Big\} d^3y \qquad \text{(D.46)}$$

after some partial integrations. (Repeated indices indicate summations. Indices labeled i and j indicate space coordinates. Greek indices include all space-time components of a variable.)

Expressing the differential in H using eq. D.43 we obtain

$$dH = \int dy \; \Big\{ \Pi_\phi \, d\dot{\phi} + \dot{\phi} \, d\Pi_\phi - \partial\mathscr{L}/\partial\phi \; d\phi - \partial\mathscr{L}/\partial(\partial\phi/\partial X^\mu) d(\partial\phi/\partial X^\mu) + \qquad \text{(D.47a)}$$

$$+ \Pi_X{}^\mu \; d\dot{X}^\mu + \dot{X}^\mu \, d\Pi_X{}^\mu - \partial\mathscr{L}/\partial X^\mu \; dX^\mu - \partial\mathscr{L}/\partial(\partial X^\mu/\partial y^j) d(\partial X^\mu/\partial y^j) \Big\}$$

After some manipulations we find

$$dH = \int \Big\{ \; \dot{\phi} \; d\Pi_\phi + \dot{X}_\mu d\Pi_X{}^\mu \; - \partial/\partial y^0 \; \Pi_\phi \, d\phi - \partial/\partial y^0 \; \Pi_X{}^\mu \, dX_\mu \Big\} dy \qquad \text{(D.47b)}$$

using the equations of motion eqs. D.19a and D.19b.

Comparing eqs D.46 and D.47 we obtain Hamilton's equations in the case of the composition of extrema:

$$\dot{\phi} \; = \; \partial\mathscr{H}/\partial\Pi_\phi \qquad \text{(D.48a)}$$

$$\dot{\Pi}_\phi = -\partial\mathscr{H}/\partial\phi + \partial/\partial y^\nu \; [\partial\mathscr{H}/\partial(\partial\phi/\partial X^j) \; \partial y^\nu/\partial X^j] \qquad \text{(D.48b)}$$

$$\dot{X}_\mu \; = \partial\mathscr{H}/\partial\Pi_X{}^\mu \qquad \text{(D.48c)}$$

where

$$\dot{\Pi}_X{}^\mu = -\partial\mathscr{H}/\partial X^\mu + \partial/\partial y^j \; [\partial\mathscr{H}/\partial(\partial X^\mu/\partial y^j)] \qquad \text{(D.49a)}$$

$$\dot{\Pi}_\phi = \partial\Pi_\phi/\partial y^0$$

$$\dot{\Pi}_X{}^\mu = \partial\Pi_X{}^\mu/\partial y^0 \qquad \text{(D.49b)}$$

D.5 Translational Invariance

If the Lagrangian of a field theory has no explicit dependence on the coordinates then one expects translational invariance accompanied by a conservation law for an energy-momentum stress tensor. We will show this is the case for Lagrangians implementing the composition of extrema. We assume a Lagrangian without an explicit dependence on the coordinates y^ν:

$$\mathscr{L} = \mathscr{L}(\phi(X), \partial\phi/\partial X^\nu, X^\mu(y), \partial X^\mu(y)/\partial y^\nu) \tag{D.50}$$

Under an infinitesimal displacement,

$$y'^\nu = y^\nu + \epsilon^\nu \tag{D.51a}$$

$$\delta\phi = \phi(X(y+\epsilon)) - \phi(X(y))$$

$$= \epsilon^\alpha\, \partial\phi/\partial y^\alpha \tag{D.51b}$$

$$\delta X^\mu = \epsilon^\alpha\, \partial X^\mu/\partial y^\alpha \tag{D.51c}$$

$$\delta(\partial\phi/\partial X^\mu) = \epsilon^\alpha\, \partial(\partial\phi/\partial y^\alpha)/\partial X^\mu \tag{D.51d}$$

$$\delta(\partial X^\mu/\partial y^\nu) = \epsilon^\alpha\, \partial(\partial X^\mu/\partial y^\alpha)/\partial y^\nu \tag{D.51e}$$

and the Lagrangian changes by

$$\delta\mathscr{L} = \epsilon^\alpha\, \partial\mathscr{L}/\partial y^\alpha \tag{D.52}$$

The change can also be expressed in terms of the changes in the fields, their derivatives and the mapping X^μ:

$$\delta\mathscr{L} = \partial\mathscr{L}/\partial\phi\, \delta\phi + \partial\mathscr{L}/\partial(\partial\phi/\partial X^\mu)\, \delta(\partial\phi/\partial X^\mu) + \partial\mathscr{L}/\partial X^\mu\, \delta X^\mu +$$
$$+ \partial\mathscr{L}/\partial(\partial X^\mu/\partial y^\nu)\, \delta(\partial X^\mu/\partial y^\nu) \tag{D.53}$$

Combining eqs. D.51, D.52 and D.53 we obtain (after some manipulations):

$$\epsilon^{\nu}\,\partial/\partial y_{\mu}\,\mathscr{T}_{\mu\nu} = 0 \tag{D.54}$$

where

$$\mathscr{T}_{\mu\nu} = -g_{\mu\nu}\mathscr{L} + \partial\mathscr{L}/\partial(\partial\phi/\partial X^{\delta})\,\partial y_{\mu}/\partial X^{\delta}\,\partial\phi/\partial y^{\nu} + \partial\mathscr{L}/\partial(\partial X^{\delta}/\partial y_{\mu})\partial X^{\delta}/\partial y^{\nu} \tag{D.55a}$$

or, alternately using Lemma 2,

$$\mathscr{T}_{\mu\nu} = -g_{\mu\nu}\mathscr{L} + \partial\mathscr{L}/\partial(\partial\phi/\partial y_{\mu})\,\partial\phi/\partial y^{\nu} + \partial\mathscr{L}/\partial(\partial X^{\delta}/\partial y_{\mu})\,\partial X^{\delta}/\partial y^{\nu} \tag{D.55b}$$

Since ϵ^{α} is an arbitrary displacement we obtain the conservation law:

$$\partial/\partial y_{\mu}\,\mathscr{T}_{\mu\nu} = 0 \tag{D.56}$$

Eq. D.56 implies the energy-momentum vector

$$P_{\beta} = \int d^{3}y\,\mathscr{T}_{0\beta} \tag{D.57}$$

is conserved. We note

$$\partial/\partial y^{0}\,P_{\beta} = 0 \tag{D.58}$$

since eq. D.56 and D.57 can be used to obtain the integral of a divergence, which results in zero. The hamiltonian (eqs. D.43-44) is

$$H = P_{0} \tag{D.59}$$

We note for later use that the total energy, H, which is conserved, contains a term that represents the energy in the X^{μ} mapping. Thus energy can be exchanged in principle between the ϕ field sector and the X^{μ} sector.

D.6 Lorentz Invariance and Angular Momentum Conservation

We can also verify Lorentz invariance and obtain the form of the conserved angular momentum by considering the effect of an infinitesimal Lorentz transformation. We will consider the case of a scalar field ϕ.

Under an infinitesimal Lorentz transformation ($\epsilon_{\mu\nu} = -\epsilon_{\nu\mu}$):

$$y'_\mu = y_\mu + \delta y_\mu = y_\mu + \epsilon_{\mu\nu} y^\nu \qquad \text{(D.60a)}$$

$$\delta\phi = \phi(X(y')) - \phi(X(y))$$

$$= \epsilon^{\mu\nu} y_\nu \, \partial\phi/\partial X^a \, \partial X^a/\partial y^\mu \qquad \text{(D.60b)}$$

$$\delta X^\mu = S^\mu{}_a X^a(y') - X^\mu(y) \qquad \text{(D.60c)}$$

$$= \epsilon^\mu{}_a X^a(y) + \partial X^\mu/\partial y^\beta \, \delta y^\beta \qquad \text{(D.60d)}$$

where $S^\mu{}_a$ is the matrix for the Lorentz transformation of a vector. (If X^μ were a gauge field then an additional operator gauge term would have to be added to eq. D.60d.)

The Lagrangian changes by

$$\delta\mathscr{L} = \epsilon^{\mu\nu} y_\nu \partial\mathscr{L}/\partial y^\mu \qquad \text{(D.61)}$$

under the infinitesimal Lorentz transformation. The change in the Lagrangian can also be expressed as:

$$\delta\mathscr{L} = \partial\mathscr{L}/\partial\phi \, \delta\phi + \partial\mathscr{L}/\partial(\partial\phi/\partial X^\mu) \, \delta(\partial\phi/\partial X^\mu) + \partial\mathscr{L}/\partial X^\mu \, \delta X^\mu +$$
$$+ \, \partial\mathscr{L}/\partial(\partial X^\mu/\partial y^\nu) \, \delta(\partial X^\mu/\partial y^\nu) \qquad \text{(D.62)}$$

Combining eqs. D.61 and D.62, and substituting and simplifying terms leads to:

$$\epsilon_{\mu\nu} \, \partial/\partial y^\sigma \mathscr{M}^{\sigma\mu\nu} = 0 \qquad \text{(D.63)}$$

where

$$\mathscr{M}^{\sigma\mu\nu} = (g^{\mu\sigma} y^\nu - g^{\nu\sigma} y^\mu)\mathscr{L} + \partial\mathscr{L}/\partial(\partial\phi/\partial X^a) \, \partial y^\sigma/\partial X^a \, (y^\mu \partial\phi/\partial y_\nu - y^\nu \partial\phi/\partial y_\mu) +$$
$$+ \, \partial\mathscr{L}/\partial(\partial X^\delta/\partial y^\sigma) \, (g^{\delta\nu} X^\mu - g^{\delta\mu} X^\nu + y^\mu \, \partial X^\delta/\partial y_\nu - y^\nu \, \partial X^\delta/\partial y_\mu) \qquad \text{(D.64)}$$

The conserved angular momentum is:

$$M^{\mu\nu} = \int d^3y \, \mathscr{M}^{0\mu\nu} \qquad \text{(D.65)}$$

with

$$\partial M^{\mu\nu}/\partial y^0 = 0 \tag{D.66}$$

The angular momentum density can be written in the familiar form:

$$\mathcal{M}^{\sigma\mu\nu} = y^\mu\,\mathcal{T}^{\sigma\nu} - y^\nu\,\mathcal{T}^{\sigma\mu} + \partial\mathcal{L}\big/\partial(\partial X^\delta/\partial y^\sigma)\,(g^{\delta\nu}X^\mu - g^{\delta\mu}X^\nu) \tag{D.67}$$

taking account of the vector nature of X^μ. The spatial part of $M^{\mu\nu}$ is the angular momentum.

D.7 Internal Symmetries

We will now consider the case of a set of scalar fields ϕ_r in a Lagrangian with an internal symmetry. Under a local transformation

$$\phi_r(X) \rightarrow \phi_r(X) - i\epsilon\lambda_{rs}\,\phi_s(X) \tag{D.68}$$

If the Lagrangian is invariant under this transformation, then

$$\delta\mathcal{L} = 0 = \partial\mathcal{L}\big/\partial\phi_r\,\delta\phi_r + \partial\mathcal{L}\big/\partial(\partial\phi_r/\partial X^\alpha)\,\delta(\partial\phi_r/\partial X^\alpha) \tag{D.69}$$

Using the equation of motion eq. D.19a satisfied by all the components ϕ_r we obtain a conserved current:

$$\mathcal{J}^\nu = -i\,\partial\mathcal{L}\big/\partial(\partial\phi_r/\partial X^\delta)\,\partial y^\nu/\partial X^\delta\,\lambda_{rs}\,\phi_s \tag{D.70}$$

which satisfies

$$\partial\mathcal{J}^\nu/\partial y^\nu = 0 \tag{D.71}$$

The conserved charge is

$$Q = \int d^3y\,\mathcal{J}^0 \tag{D.72}$$

$$\partial Q/\partial y^0 = 0 \tag{D.73}$$

D.8 Separable Lagrangians

We now consider the case of a separable Lagrangian such as in eq. D.26. Adopting the definitions:

$$\phi' = \partial\phi/\partial X^0 \tag{D.74}$$

$$X_\mu' = \partial X_\mu / \partial y^0 \tag{D.75}$$

we define canonical momenta as

$$\pi_\phi = \partial \mathcal{L} / \partial \phi' \equiv \partial \mathcal{L} / \partial(\partial \phi / \partial X^0) \tag{D.76}$$

$$\pi_X{}^\mu = \partial \mathcal{L} / \partial X_\mu' \equiv \partial \mathcal{L} / \partial(\partial X_\mu / \partial y^0) \tag{D.77}$$

We now define the separable hamiltonian density as

$$\mathcal{H}_s = J\pi_\phi \, \phi' + \pi_X{}^\mu X_\mu' - \mathcal{L}_s \tag{D.78}$$

where J is the Jacobian (eq. D.21) and

$$H_s = \int \mathcal{H}_s \, d^3y \tag{D.79}$$

The separable Lagrangian (from eq. D.26) is:

$$\mathcal{L}_s = \mathcal{L}_F(\phi(X), \partial\phi/\partial X^\mu) \, J + \mathcal{L}_C(X^\mu(y), \partial X^\mu(y)/\partial y^\nu, y) \tag{D.80}$$

In the case of one scalar field the separable hamiltonian density has the general form

$$\mathcal{H}_s = \mathcal{H}_s(\phi(X), \pi_\phi, \partial\phi/\partial X^i, X^\mu(y), \pi_X{}^\mu, \partial X^\mu(y)/\partial y^i, y^\nu) \tag{D.81}$$

where the indices i and j indicate spatial components. In particular, the terms in the separable hamiltonian are:

$$\mathcal{H}_s = \mathcal{H}_F J + \mathcal{H}_C \tag{D.82}$$

with

$$\mathcal{H}_F(\phi(X), \pi_\phi, \partial\phi/\partial X^i) = \pi_\phi \, \phi' - \mathcal{L}_F \tag{D.83}$$

$$\mathcal{H}_C(X^\mu(y), \pi_X{}^\mu, \partial X^\mu(y)/\partial y^j, y^\nu) = \pi_X{}^\mu X_\mu' - \mathcal{L}_C \tag{D.84}$$

where J is the absolute value of the Jacobian defined in D.21.

We now define the time integral of H as we did in eq. D.14 when considering the Lagrangian formulation:

$$G \;=\; \int dy^0\, H_s \tag{D.85}$$

Thus G is an integral over all space-time coordinates. Using G we can develop a hamiltonian formulation. First we calculate the differential change in G. Using eqs. D.81-2 and D.85 we obtain

$$
\begin{aligned}
dG = \int \Big\{ & J\, \partial\mathscr{H}_{\mathrm{F}}/\partial\phi\, d\phi + J\, \partial\mathscr{H}_{\mathrm{F}}/\partial\pi_\phi\, d\pi_\phi + \\
& + J\, \partial\mathscr{H}_{\mathrm{F}}/\partial(\partial\phi/\partial X^j)\, d(\partial\phi/\partial X^j) + \partial\mathscr{H}_{\mathrm{C}}/\partial X^\mu\, dX^\mu + \\
& + \partial\mathscr{H}_{\mathrm{C}}/\partial\pi_X{}^\mu\, d\pi_X{}^\mu + \partial\mathscr{H}_{\mathrm{C}}/\partial(\partial X^\mu/\partial y^j)\, d(\partial X^\mu/\partial y^j) \Big\}\, d^4y
\end{aligned}
\tag{D.86}
$$

with summations implied by repeated indices. (Index labels i and j label spatial coordinates only; Greek indices label space-time coordinates.) Rewriting dG as two integrals and performing partial integrations yields:

$$
\begin{aligned}
dG = & \int d^4X \Big\{ \partial\mathscr{H}_{\mathrm{F}}/\partial\phi\, d\phi + \partial\mathscr{H}_{\mathrm{F}}/\partial\pi_\phi\, d\pi_\phi - \partial/\partial X^j[\partial\mathscr{H}_{\mathrm{F}}/\partial(\partial\phi/\partial X^j)]\, d\phi \Big\} + \\
& + \int d^4y \Big\{ \partial\mathscr{H}_{\mathrm{C}}/\partial X^\mu\, dX^\mu + \partial\mathscr{H}_{\mathrm{C}}/\partial\pi_X{}^\mu\, d\pi_X{}^\mu - \partial/\partial y^j[\partial\mathscr{H}_{\mathrm{C}}/\partial(\partial X^\mu/\partial y^j)]\, dX^\mu \Big\}
\end{aligned}
\tag{D.87}
$$

Alternately, expressing the differential in G using eqs. D.82-4 we obtain

$$
\begin{aligned}
dG = & \int d^4X \Big\{ \pi_\phi\, d\phi' + \phi'\, d\pi_\phi - \partial\mathscr{L}_{\mathrm{F}}/\partial\phi\, d\phi - \partial\mathscr{L}_{\mathrm{F}}/\partial(\partial\phi/\partial X^\mu)d(\partial\phi/\partial X^\mu) \Big\} + \\
& + \int d^4y \Big\{ \pi_{X\mu}\, dX^{\mu\prime} + X^{\mu\prime}\, d\pi_{X\mu} - \partial\mathscr{L}_{\mathrm{C}}/\partial X^\mu\, dX^\mu - \partial\mathscr{L}_{\mathrm{C}}/\partial(\partial X^\mu/\partial y^j)d(\partial X^\mu/\partial y^j) \Big\}
\end{aligned}
\tag{D.88}
$$

which becomes

$$dG = \int d^4X \Big\{ -\pi_\phi'\, d\phi + \phi'\, d\pi_\phi \Big\} + \int d^4y \Big\{ -\pi_{X\mu}'\, dX^\mu + X^{\mu\prime}\, d\pi_{X\mu} \Big\} \tag{D.89}$$

using the equations of motion eqs. D.31-2.

Comparing eqs D.87 and D.89 we obtain Hamilton's equations for the case of the composition of extrema for a separable Lagrangian:

$$\phi' = \partial\mathscr{H}_{\mathrm{F}}/\partial\pi_\phi \tag{D.90}$$

$$\pi_\phi{}' = -\partial \mathcal{H}_{\text{F}}/\partial\phi + \partial/\partial X^j \, [\partial \mathcal{H}_{\text{F}}/\partial(\partial\phi/\partial X^j)] \qquad \text{(D.91)}$$

$$X_\mu{}' = \partial \mathcal{H}_{\text{C}}/\partial\pi_X{}^\mu \qquad \text{(D.92)}$$

$$\pi_{X\mu}{}' = -\partial \mathcal{H}_{\text{C}}/\partial X^\mu + \partial/\partial y^j \, [\partial \mathcal{H}_{\text{C}}/\partial(\partial X^\mu/\partial y^j)] \qquad \text{(D.93)}$$

where

$$\pi_\phi{}' = \partial \, \pi_\phi/\partial X^0 \qquad \text{(D.94)}$$

$$\pi_{X\mu}{}' = \partial \, \pi_{X\mu}/\partial X^0 \qquad \text{(D.95)}$$

Notice that $\mathcal{L}_{\text{F}}$, $\mathcal{H}_{\text{F}}$ and π_ϕ have precisely the same form, as a function of X^μ, as one sees in a conventional field theory formalism. Yet X^μ is a mapping/function of the coordinates y. In reality, it can be viewed as a field as we shall see.

D.9 Separable Lagrangians and Translational Invariance

The general rule for conventional Lagrangians is: if a Lagrangian has no explicit dependence on the coordinates then translational invariance follows accompanied by a conservation law for an energy-momentum tensor. We will show that this rule needs modification for separable Lagrangians that implement the composition of extrema.

Consider the Lagrangian:

$$\mathcal{L}_s = J \, \mathcal{L}_{\text{F}}(\phi(X), \partial\phi/\partial X^\mu) + \mathcal{L}_{\text{C}}(X^\mu(y), \partial X^\mu(y)/\partial y^\nu) \qquad \text{(D.96)}$$

in which the X^μ play a dual role as both fields and coordinates. Let us consider a variation in X^μ:

$$X^\mu(y) \rightarrow X^\mu(y) + \delta X^\mu(y) \qquad \text{(D.97)}$$

where $\delta X^\mu(y)$ is an arbitrary function of y that vanishes at the endpoints of the integration region of the integral. The action is:

$$I = \int \mathcal{L}_s d^4 y \qquad \text{(D.98)}$$

We will show that a variation in $X^\mu(y)$ leads to a conserved energy-momentum tensor. But we will use integrals of the Lagrangian density since it provides a simpler derivation of the result. Under the variation of eq. D.97 we find

$$\delta\phi = \phi(X(y) + \delta X^\mu(y)) - \phi(X(y))$$

$$= \delta X^\mu \, \partial\phi/\partial X^\mu \tag{D.99a}$$

$$\delta(\partial\phi/\partial X^\nu) = \delta X^\mu \, \partial(\partial\phi/\partial X^\mu)/\partial X^\nu \tag{D.99b}$$

$$\delta(\partial X^\mu/\partial y^\nu) = \partial(\delta X^\mu)/\partial y^\nu \tag{D.99c}$$

The integral in eq. D.98 changes by

$$\delta I = \int d^4y \, \delta\mathscr{L}_s = \int d^4y \, [\delta(J\mathscr{L}_F) + \delta\mathscr{L}_C] \tag{D.100a}$$

which becomes:

$$\delta I = \int d^4y \, [\delta X^\mu \, \partial(J\mathscr{L}_F)/\partial X^\mu + \partial(\delta X^\mu \partial\mathscr{L}_C / \partial(\partial X^\mu/\partial y^\nu))/\partial y^\nu] \tag{D.100b}$$

due to the equations of motion of X^μ (eq. D.19b) in X^μs role. Since the second term is a total divergence its contribution to δI is zero. Thus we can express eq. D.100b as:

$$\delta I = \int d^4y \, [J \, \delta\mathscr{L}_F + \mathscr{L}_F \, \delta J] \tag{D.101}$$

realizing that the Jacobian J depends on y and thus X:

$$\delta J = \delta X^\mu \, \partial J/\partial X^\mu \tag{D.102}$$

A partial integration gives

$$\mathscr{L}_F \, \delta J = \delta X^\mu \, \partial(J\mathscr{L}_F)/\partial X^\mu - \delta X^\mu J \, \partial\mathscr{L}_F/\partial X^\mu \tag{D.103}$$

Evaluating $\delta\mathscr{L}_F$ we find:

$$\delta\mathscr{L}_F = \partial\mathscr{L}_F/\partial\phi \; \delta\phi + \partial\mathscr{L}_F/\partial(\partial\phi/\partial X^\mu) \; \delta(\partial\phi/\partial X^\mu) \qquad (D.104)$$

which gives

$$\delta\mathscr{L}_F = \delta X^\nu \; \partial/\partial X^\mu \, [\partial\mathscr{L}_F/\partial(\partial\phi/\partial X^\mu) \; \partial\phi/\partial X^\nu \,] \qquad (D.105)$$

using the equations of motion eq. D.31, and using eq. D.99b. Combining eqs. D.100, D.101, D.103 and D.105 we obtain:

$$\int d^4y \, J \, \delta X^\nu \; \partial/\partial X_\mu \, \mathscr{T}_{F\mu\nu} = \int d^4X \, \delta X^\nu \; \partial/\partial X_\mu \, \mathscr{T}_{F\mu\nu} = 0 \qquad (D.106)$$

where

$$\mathscr{T}_{F\mu\nu} = -\, g_{\mu\nu}\,\mathscr{L}_F + \partial\mathscr{L}_F/\partial(\partial\phi/\partial X_\mu) \; \partial\phi/\partial X^\nu \qquad (D.107)$$

after some manipulations. Since δX^ν is an arbitrary function of y the differential conservation law follows:

$$\partial/\partial X_\mu \, \mathscr{T}_{F\mu\nu} = 0 \qquad (D.108)$$

Eq. D.108 implies the energy-momentum vector

$$P_{F\beta} = \int d^3X \, \mathscr{T}_{F0\beta} \qquad (D.109)$$

is conserved:

$$\partial/\partial X^0 \, P_{F\beta} = 0 \qquad (D.110)$$

The hamiltonian density (eq. D.83) is

$$\mathscr{H}_F = \mathscr{T}_{F0\beta} \qquad (D.111)$$

Thus the field energy

$$H_F = P_{F0} = \int d^3X \, \mathscr{T}_{F00} \qquad (D.112)$$

is conserved with respect to the "time" X^0. Later we will see that H_F is trivially conserved in the Coulomb gauge of X_μ. (We will also establish an electromagnetic-like quantum field theory for X_μ with gauge invariance.) In other gauges the conservation of H_F is not trivial.

D.10 Separable Lagrangians and Angular Momentum Conservation

We can also verify Lorentz invariance and obtain the form of the conserved angular momentum for a separable Lagrangian by considering the effect of an infinitesimal Lorentz transformation. We will consider the case of a scalar field ϕ.

Under an infinitesimal Lorentz transformation as specified by eqs. D.60a – D.60d the separable Lagrangian changes by

$$\delta\mathscr{L}_s = \epsilon^{\mu\nu}\, y_\nu\, \partial\mathscr{L}_s/\partial y^\mu \tag{D.113}$$

which can also be expressed as

$$\delta\mathscr{L}_s = \partial\mathscr{L}_s/\partial\phi\,\delta\phi + \partial\mathscr{L}_s/\partial(\partial\phi/\partial X^\mu)\,\delta(\partial\phi/\partial X^\mu) + \partial\mathscr{L}_s/\partial X^\mu\,\delta X^\mu +$$
$$+ \,[\partial\mathscr{L}_s/\partial(\partial X^\mu/\partial y^\nu)]\,\delta(\partial X^\mu/\partial y^\nu) \tag{D.114}$$

Combining eqs. D.113 and D.114 leads to:

$$\epsilon_{\mu\nu}\,\partial/\partial y^\sigma \mathscr{M}_s^{\ \sigma\mu\nu} = 0 \tag{D.115}$$

where

$$\mathscr{M}_s^{\ \sigma\mu\nu} = J\,\mathscr{M}_F^{\ \sigma\mu\nu} + \mathscr{M}_C^{\ \sigma\mu\nu} + \mathscr{M}_M^{\ \sigma\mu\nu} \tag{D.116}$$

$$\mathscr{M}_F^{\ \sigma\mu\nu} = (g^{\mu\sigma}y^\nu - g^{\nu\sigma}y^\mu)\mathscr{L}_F + \partial\mathscr{L}_F/\partial(\partial\phi/\partial y_\sigma)\,(y^\mu\partial\phi/\partial y_\nu - y^\nu\partial\phi/\partial y_\mu) \tag{D.117}$$

$$\mathscr{M}_C^{\ \sigma\mu\nu} = (g^{\mu\sigma}y^\nu - g^{\nu\sigma}y^\mu)\mathscr{L}_C + $$
$$+ \,\partial\mathscr{L}_C/\partial(\partial X^\delta/\partial y^\sigma)(g^{\delta\nu}X^\mu - g^{\delta\mu}X^\nu + y^\mu\,\partial X^\delta/\partial y_\nu - y^\nu\,\partial X^\delta/\partial y_\mu) \tag{D.118}$$

$$\mathscr{M}_M^{\ \sigma\mu\nu} = \mathscr{L}_F\partial J/\partial(\partial X^\delta/\partial y^\sigma)(g^{\delta\nu}X^\mu - g^{\delta\mu}X^\nu + y^\mu\,\partial X^\delta/\partial y_\nu - y^\nu\,\partial X^\delta/\partial y_\mu) \tag{D.119}$$

where the third term originates in the dependence of J on derivatives of X^μ. Eq. D.117 was obtained in part by using the identity:

$$\partial\mathscr{L}/\partial(\partial\phi/\partial y^\sigma) = \partial\mathscr{L}/\partial(\partial\phi/\partial X^\alpha)\, \partial y^\sigma/\partial X^\alpha \qquad \text{(D.120)}$$

where $\mathscr{L}$ and ϕ have the form specified in eq. D.15.

The conserved angular momentum is:

$$M_s^{\mu\nu} = \int dy\, \mathscr{M}_s^{0\mu\nu} \qquad \text{(D.121)}$$

with

$$\partial M_s^{\mu\nu}/\partial y^0 = 0 \qquad \text{(D.122)}$$

D.10.1 Angular Momentum and $\mathscr{L}_F$

An alternate conserved angular momentum can be obtained by considering the "field" part of the Lagrangian $\mathscr{L}_F$ under an infinitesimal Lorentz transformation ($\epsilon_{\mu\nu} = -\epsilon_{\nu\mu}$):

$$X'_\mu = X_\mu + \delta X_\mu \qquad \text{(D.123a)}$$

$$\begin{aligned}\delta\phi &= \phi(X'(y)) - \phi(X(y)) \\ &= \delta X^\mu\, \partial\phi/\partial X^\mu\end{aligned} \qquad \text{(D.123b)}$$

$$\delta X^\mu = S^\mu_{\ a} X^a(y) - X^\mu(y) \qquad \text{(D.123c)}$$

$$= \epsilon^\mu_{\ a} X^a(y) \qquad \text{(D.123d)}$$

where $S^\mu_{\ a}$ is the Lorentz transformation matrix for a vector. (If X^μ is a gauge field then an additional operator gauge term would have to be added to eq. D.123d.)

The Lagrangian changes by

$$\delta\mathscr{L}_F = \epsilon^{\mu\nu} X_\nu\, \partial\mathscr{L}_F/\partial X^\mu \qquad \text{(D.124)}$$

under an infinitesimal Lorentz transformation. The change can also be expressed as:

$$\delta\mathscr{L}_F = \partial\mathscr{L}_F/\partial\phi\, \delta\phi + \partial\mathscr{L}_F/\partial(\partial\phi/\partial X^\mu)\, \delta(\partial\phi/\partial X^\mu) \qquad \text{(D.125)}$$

Combining eqs. D.124 and D.125 leads to:

$$\epsilon_{\mu\nu}\partial/\partial X^{\sigma}\,\mathcal{M}_{FX}{}^{\sigma\mu\nu} = 0 \tag{D.126}$$

where

$$\mathcal{M}_{FX}{}^{\sigma\mu\nu} = (g^{\mu\sigma}X^{\nu} - g^{\nu\sigma}X^{\mu})\mathcal{L}_{F} + \partial\mathcal{L}_{F}/\partial(\partial\phi/\partial X^{\sigma})\,(X^{\mu}\partial\phi/\partial X_{\nu} - X^{\nu}\partial\phi/\partial X_{\mu}) \tag{D.127}$$

The conserved angular momentum associated with the X coordinates is:

$$M_{FX}{}^{\mu\nu} = \int d^{3}X\,\mathcal{M}_{FX}{}^{0\mu\nu} \tag{D.128}$$

with

$$\partial M_{FX}{}^{\mu\nu}/\partial X^{0} = 0 \tag{D.129}$$

The angular momentum density can be written in the familiar form:

$$\mathcal{M}_{FX}{}^{\sigma\mu\nu} = X^{\mu}\,\mathcal{T}_{F}{}^{\sigma\nu} - X^{\nu}\,\mathcal{T}_{F}{}^{\sigma\mu} \tag{D.130}$$

using eq. D.107.

D.11 Separable Lagrangians and Internal Symmetries

We will now consider the case of a set of scalar fields ϕ_{r} in a separable Lagrangian with an internal symmetry under a local transformation

$$\phi_{r}(X) \rightarrow \phi_{r}(X) - i\epsilon\lambda_{rs}\,\phi_{s}(X) \tag{D.131}$$

If the Lagrangian is invariant under this transformation, then

$$\delta\mathcal{L}_{S} \equiv \delta\mathcal{L}_{F} = 0 = \partial\mathcal{L}_{F}/\partial\phi_{r}\,\delta\phi_{r} + \partial\mathcal{L}_{F}/\partial(\partial\phi_{r}/\partial X^{\alpha})\,\delta(\partial\phi_{r}/\partial X^{\alpha}) \tag{D.132}$$

Using the equation of motion eq. D.31, which is satisfied by all components ϕ_{r}, we obtain a conserved current:

$$\mathcal{J}^{\nu} = -i\,\partial\mathcal{L}_{F}/\partial(\partial\phi_{r}/\partial X^{\nu})\,\lambda_{rs}\,\phi_{s} \tag{D.133}$$

satisfying

$$\partial \mathcal{J}^\nu / \partial X^\nu = 0 \qquad\qquad (D.134)$$

The conserved charge is

$$Q = \int d^3X \, \mathcal{J}^0 \qquad\qquad (D.135)$$
$$\partial Q / \partial X^0 = 0 \qquad\qquad (D.136)$$

We note eq. D.71 provides a corresponding conservation law for the y coordinate system.

Appendix E. CQ Mechanics – PseudoQuantum Mechanics

E.1 Why Quantum Theory?

A question that is not often considered in these days is the reason that Nature 'chose' to be quantum rather than based on classical, deterministic mechanics. In our Theory of Everything presented in Blaha (2015c) and subsequent books in 2015 and 2016 we assumed that al Natural phenomena were ultimately based on quantum field theory.

There is a more fundamental assumption that we could posit that leads to quantum field theory and then to quantum mechanics (which is based on quantum field theory.[406]) If we assume the following postulate:

All entities in the universe are composed of discrete particles, that are integer countable, and all interactions can ultimately be reduced to the interactions of these particles.

Then, when we define field theories, they must be quantum field theories – describe particles (quanta) – and thus the field theories must be second quantized. *Particles are integer countable*[407] whether free or in perturbation theory interaction terms. And the interactions of these theories must be based on the exchange of particles although the particles have a 'cloud' of virtual particles surrounding them. The quantum mechanics of the particles constituting atoms (matter) then follows as a consequence of quantum field theory.

Classical mechanics then becomes an approximation to quantum mechanics under certain conditions that turn out to be the common conditions of everyday experience.

In basing the origin of quantum theory on the particulate nature of the entities in the universe we assume that particles exist, and can be defined, in our mostly flat space-time (which itself is generated by amalgamations of graviton particles). We also assume that the particle concept can be extended to unusual space-time coordinates such as non-static space-times. However it became apparent many years ago when accelerating coordinate systems and other

---

[406] Heitler (1954) shows how the Heisenberg Uncertainty Principle is a consequence of quantum field theory using an example from Quantum Electrodynamics. The Uncertainty Principle and the Correspondence Principle lead directly to quantum mechanics. Quantum mechanics is thus a consequence of quantum field theory – not an independent fundamental theory.

[407] Theories with continuous matter have not been shown to exist experimentally.

non-static[408] coordinate systems were considered, that the definition of particles in quantum field theory is problematic.[409]

Sections 2 and 3 describe the correct definition of particles in quantum field theory. The correct definition of bosons furnishes a basis for a better definition of the Higgs Mechanism. The necessity of higher derivative theories of Gravity and the Strong Interactions[410] to obtain explicit color confinement and to reconcile gravity theory with data on the rotation of stars around galactic centers leads to an extension of the definition of particles to have principal value propagators and thus gives *particulate* action-at-a-distance.[411]

A further issue, that emerges in perturbation theory calculations in quantum field theories, leads to the introduction of a vector field as part of each propagator that eliminates the point-like nature of particle interactions in the high energy (short distance) limit in favor of 'fuzzy' interacting particles. This extension of quantum field theory is called Two-Tier quantum field theory.[412] It is required for the Theory of Everything since a conventional renormalization procedure is not known to exist – and is not likely to exist.

The combination of features that we have developed enables us to create a divergence-free[413] Theory of Everything where particles can be defined in any static or non-static coordinate system and where bosons, neglecting interactions, are stable against decay to negative energy states.[414]

The formalism that we present can be applied to quantum and classical dynamics. We define a quantum-classical formalism, that we call *QCMechanics*, that has a fully quantum sector, a classical sector, and an intermediate sector bridging the quantum and classical sectors.

We then proceed to develop the harmonic oscillator theory within this framework. Subsequently we discuss a generalized Feynman path integral formalism, a generalized Schrödinger equation, a generalized Boltzmann equation, the Fokker-Planck equation, a generalized approach to quantum and classical chaos, and to quantum entanglement as well as semi-quantum entanglement. Our formalism applies to both Quantum Field Theory and Quantum Mechanics as well as the path integrals, the Fokker-Planck equation and the Boltzmann equation.

---

[408] A non-static coordinate system mixes space and time coordinates.

[409] S. Blaha, Il Nuovo Cimento **49 A**, 35 (1979). ___, **49 A**, 113 (1979), which appear in appendices A and B and is discussed in the following chapters, describe how to define particles in any coordinate system. The particle definition issued is discussed in papers which they reference.

[410] Unified theory: See Blaha (2016e).

[411] See appendices A and B.

[412] See Blaha (2005a).

[413] There are no divergences in perturbation theory calculations and no need for renormalization programs to remove divergences. Physical quantities do get renormalized by finite amounts.

[414] Negative energy boson states are equivalent to classical fields.

E.2 Boson Particle Formulation

There are two issues confronting the usual approach to the quantization of boson fields that require resolution through a 'new' quantization procedure for boson particles. One problem is the need to quantize boson fields in unconventional coordinate systems such as accelerating coordinate systems and coordinate systems defined for highly curved space-time. The other major problem is the need to quantize boson fields in such a way that bosons of negative energy have a physical interpretation.[415]

In this section we will define a new quantization procedure for bosons that will eliminate both of these problems. In section 3 we will describe the analogous quantization procedure for fermions that will enable us to create well-defined Dirac field particle states in any coordinate system. (A filled Dirac sea of negative energy fermions will exist in this formulation as it does in the conventional formulation.)

In the case of both bosons and fermions we will see that the flat space-time, static coordinate systems that are normally used will remain valid special case approximations to the new formulations of quantum field theory.

Having resolved these problems for quantum field theories of bosons and fermions we will point out in section 5 that 'ordinary' quantum mechanics also has a problem with quantization in unconventional coordinate systems. There are also difficulties in the transition between classical and quantum 'analogues.' For example the transition from the classical Boltzmann equation to a quantum version is uncertain.

Using a framework analogous to our 'new' approach to the quantum field theory of bosons and fermions we will establish a generalization of quantum mechanics that contains both quantum mechanics and classical mechanics, and an intermediate mixed form. This generalization supports a smooth transition between classical mechanics and quantum mechanics. With this generalization we will be able to examine the transition from quantum to classical mechanics in detail without recourse to methods such as expansions in Planck's constant $\hbar$.

E.2.1 Quantization of Boson Fields in Unconventional (Static and Non-Static) Coordinate Systems

The problems associated with the definition of asymptotic particle states in arbitrary coordinate systems have been pointed out by numerous authors.[416] Our 1978 paper (Appendix A) resolves this problem with a consistent procedure for the local definition of asymptotic boson particle states in any coordinate system, which may or my not have a time-like Killing vector.

---

[415] There is no Pauli Exclusion Principle for bosons. Negative energy fermions 'fill' their Dirac negative energy sea due to the limitation of fermions to one fermion per state imposed by the Pauli Exclusion Principle.

[416] See appendix A, which contains our 1978 paper Il Nuovo Cimento **49 A**, 35, for references.

The general procedure is described in section 2 in Appendix A starting with eq. 6. The boson particle interpretation is described in section 4. In this section we wish to bring out salient details of the procedure which we will be relevant for our consideration of the generalization of quantum mechanics that we will consider later.

The first distinctive feature of this form of boson second quantization is the use of two fields to define a second quantized boson theory. The use of *two* fields enables us to define states which correspond to quantum field particles in any coordinate system. Further they give us the scope to define, not only quantum field particle states, but also classical boson field states. States, which are composites of both classical fields and quantum particles, can also be defined.

In our formulation[417] the simplest lagrangian density for a generic massless, scalar Klein-Gordon particle is:

$$\mathcal{L} = \partial\varphi_1/\partial x_\mu \partial\varphi_2/\partial x^\mu \tag{E.2.1}$$

with hamiltonian density

$$\mathcal{H} = \pi_1\,\pi_2 + \partial\varphi_1/\partial x_i \partial\varphi_2/\partial x^i \tag{E.2.2}$$

where i labels spatial coordinates, and $\pi_1 = \partial\varphi_2/\partial t$ and $\pi_2 = \partial\varphi_1/\partial t$. Eqs. E.2.1 and E.2.2 are without a potential or mass term. Eq. 6, and the discussion following it, in appendix A describe the massive boson case.

The fields can be fourier expanded in terms of creation and annihilation operators:

$$\varphi_i(\mathbf{x}, t) = \int d^3k \, [a_i(k)f_k(x) + a_i^\dagger(k)f_k{}^*(x)] \tag{E.2.3}$$

for i = 1, 2 where

$$f_k(x) = e^{-ik\cdot x} /(2\omega_k(2\pi)^3)^{\frac{1}{2}}$$

with $\omega_k = |\mathbf{k}|$ in the massless case and $\omega_k = (|\mathbf{k}|^2 + m^2)^{\frac{1}{2}\,i}$ for a massive boson.

The creation and annihilation operators satisfy the commutation relations:

$$[a_i(k), a_j^\dagger(k')] = (1 - \delta_{ij})\delta^3(\mathbf{k} - \mathbf{k}') \tag{E.2.4}$$
$$[a_i(k), a_j(k')] = 0$$
$$[a_i^\dagger(k), a_j^\dagger(k')] = 0$$

for i, j = 1, 2. The vacuum state |0> satisfies

$$a_1(k)|0> = a_1^\dagger(k)|0> = 0 \tag{E.2.5}$$

---

[417] In earlier books we have called this approach to second quantization *pseudoquantum field theory.*

$$a_2(k)|0> \neq 0 \qquad\qquad a_2^\dagger(k)|0> \neq 0 \qquad\qquad (E.2.6)$$

The dual vacuum state satisfies

$$<0|a_2(k) = <0|a_2^\dagger(k) = 0 \qquad\qquad (E.2.7)$$
$$<0|a_1(k) \neq 0 \qquad\qquad <0|a_1^\dagger(k) \neq 0 \qquad\qquad (E.2.8)$$

Positive energy single particle *ket* states are defined using $a_2^\dagger(k)$ while negative energy ket states are defined using $a_2(k)$. Positive energy single particle *bra* states are defined using $a_1(k)$ while negative energy bra states are defined using $a_1^\dagger(k)$.

E.2.1.1 Transformations to Other (Possibly Non-Static) Coordinate Systems

The preceding discussion applies directly in the rectangular coordinates with which we are familiar. In eqs. 2 – 4, and their discussion, in appendix A we show that the definition of boson field orthonormal sets according to a different definition of positive frequency is related to the definition above in eq. E.2.3 by a local Bogoliubov transformation. The definition of particle states and vacua are different in general. However, as eqs. 15 – 31 (Appendix A) show, we can define the transformation to preserve the invariance of the particle number operator and thus make the theory under a different definition of positive frequency fully unitarily equivalent to the original theory. Thus the particle interpretation of states is preserved.

The general form of the Bogoliubov transformation (eq. 23) is

$$a_i(k, \lambda_1(k), \lambda_2(k)) = B(\lambda_1(k), \lambda_2(k))a_i(k)B^{-1}(\lambda_1(k), \lambda_2(k)) \qquad (23)$$
$$= \exp(i\lambda_1(k))\cosh(\lambda_2(k))a_i(k) + \exp(-i\lambda_1(k))\sinh(\lambda_2(k))a_i^\dagger(k)$$

with $B(\lambda_1(k), \lambda_2(k))$ given by eq. 24 and

$$B^{-1}(\lambda_1(k), \lambda_2(k)) = B^\dagger(\lambda_1(k), \lambda_2(k)) \qquad\qquad (E.2.9)$$

where † indicates hermitean conjugate. The text following eq. 23 provide the definition of bra and ket states, inner products, the energy-momentum tensor, equal-time commutation relations, the Green's functions, and the perturbation theory of the 'new' formalism.

Appendix A shows the general form of transformations between type '1' and type '2' creation and annihilation operators in this excerpt:[418]

---

[418] Excerpt used with the kind permission of Il Nuovo Cimento A.

The equal-time commutation relations, and the self-adjointness of H and φ_2 place six constraints on the constants C_{ij} and $\tilde{C}_{ij}$ in eqs. (15) and (16). After some algebra we find that we are able to express the field operators in the form

$$(40) \qquad \varphi_1(x) = \int d^3k \left[\left(\frac{\cos(\theta_1 - \theta_2)}{\sin\theta_1} A_{1k} + \frac{\sin(\theta_1 - \theta_2)}{\sin\theta_1} A_{2k} \right) f_k(x) + \right.$$
$$\left. + \left(\frac{\cos(\theta_1 - \theta_2)}{\cos\theta_1} A_{1k}^\dagger - \frac{\sin(\theta_1 - \theta_2)}{\cos\theta_1} A_{2k}^\dagger \right) f_k^*(x) \right],$$

$$(41) \qquad \varphi_2(x) = \int d^3k \left[(\cos\theta_2 A_{2k} + \sin\theta_2 A_{1k}) f_k(x) + (\sin\theta_2 A_{2k}^\dagger - \cos\theta_2 A_{1k}^\dagger) f_k^*(x) \right],$$

where θ_1 and θ_2 are arbitrary constants which fix the boundary conditions of the Green's functions. (They are *not* related to the Bogoliubov transformations

Thus PseudoQuantized Field Theory resolves the particle interpretation ambiguities of second quantization in non-static coordinate systems through Bogoliubov rotations.

E.2.1.2 Negative Energy Bosons

Traditional boson second quantization has the problem of the absence of a barrier to the decay of positive energy states to negative energy states since the Pauli Exclusion Principle does not apply to bosons. This problem has been masked ('overcome') by a clever choice of boundary conditions that are embodied in the creation/annihilation momentum space operator conditions:

$$a|0> = 0 \qquad \text{Conventional Approach} \qquad (E.2.10)$$
$$a^\dagger|0> \neq 0$$

In this conventional approach the creation of negative energy boson states is eliminated *ab initio* by these conditions. Yet boson quantum fields still have a conceptual physical cloud hanging over them that spin ½ fields do not. A spin ½ particle cannot transition to negative energy because there is a filled sea of negative energy particles. No additional particles can fall into the sea due to the Pauli Exclusion Principle that forbids two fermions with the same 4-momentum and quantum numbers.

In the case of scalar particles the Pauli Exclusion Principle does not apply and so, *physically*, a *filled* negative energy sea of bosons is not possible and positive energy bosons should be able to transition to negative energy states. This problem was "resolved" by the above definition of boson vacuums to exclude transitions to negative energy. But the rationale for the definition is lacking. Dirac was once asked about this issue many years ago. He said he had a solution to the problem. However he did not present it – presumably in keeping with his well-known taciturn nature. So the issue remained an open question.

In this book and earlier work[419] we showed that a more physically satisfactory method exists for avoiding the negative energy state problem. This method relies on the use of a larger Fock space in which *negative energy states (or partially negative energy states) are interpreted as states containing classical fields or a mix of classical fields and individual boson particles.* This approach resolves the negative energy boson issue and provides a common framework for boson particles and classical boson fields.

The issue of the spontaneous decay of a positive energy boson into a negative energy state still seems to exist. However all known fundamental scalar bosons are Higgs bosons that have a vacuum expectation value and a 'heavy' quantum field part of positive energy that immediately decays into other particles such as a pair of photons. The decay of a positive energy boson to a negative energy state is precluded by a separation of the formalism into separate positive energy and negative energy sectors as shown in section 4.5 below.

E.2.2 Classical Field States for Bosons

Classical c-number boson fields exist in our PseudoQuantum Field Theory. A classical c-number field has the form

$$\Phi(\mathbf{x}, t) = \int d^3k \, [\alpha(k)f_k(x) + \alpha^*(k)f_k^*(x)] \tag{E.2.11}$$

A corresponding classical state is a coherent state with the form

$$| \Phi, \Pi\rangle = C \exp\left\{\int d^3k \, [\alpha(k)a_2^\dagger(k) + \alpha^*(k)a_2(k)]\right\}|0\rangle \tag{E.2.12}$$

and correspondingly for $\Pi(x)$ where C is a normalization constant.

The defining properties of a classical field state are:

$$\varphi_1(x)|\Phi, \Pi\rangle = \Phi(x)|\Phi, \Pi\rangle \tag{E.2.13}$$
$$\pi_1(x)|\Phi, \Pi\rangle = \Pi(x)|\Phi, \Pi\rangle$$

where $\Phi(x)$ and $\Pi(x)$ are sharp on the states and where $\varphi_i(x)$ is given by eq. E.2.3.

Additional details on coherent states, which differ somewhat from conventional coherent states such as those of Kibble[420] and others, can be found in Appendix C.

E.2.3 The Enigma of Higgs Particles and the Higgs Mechanism

Our PseudoQuantum Field Theory is ideally suited for describing Higgs Mechanism phenomena. In our previous work on the Standard Model, and its generalization to The Unified

---

[419] See Appendix 2-A and references therein.
[420] T. W. B. Kibble, Jour. Math. Phys. **2**, 212 (1961).

SuperStandard Model described in a series of books entitled *Physics is Logic,* we showed that the fermion spectrum results from Complex Special Relativity, the gauge interactions result from the Reality group, the fermion generations result from the Generation group, the layers of fermions result from the U(4) Layer group, and from the combination with Complex General Relativity in our Theory of Everything. Higgs particles and the Higgs Mechanism were inserted *ad hoc* to generate particle masses and symmetry breaking effects.

The apparent recent discovery of Higgs particles at CERN seems to solidify the existence of the Higgs sector of the Standard Model and of our Unified SuperStandard Model as described in earlier volumes of *Physics is Logic.*[421]

But whence arises Higgs particles? There does not appear to be a more fundamental cause than the need for particle masses obtained through symmetry breaking. And so the Higgs sector was an expedient mechanism. With our method of avoiding divergences in perturbation theory using Two-Tier quantum field theory the need for the Higgs Mechanism appears to have disappeared with the former need for a symmetry breaking mechanism to generate particle masses. The ElectroWeak sector has no divergences in our approach and thus does not need the renormalization program previously developed that was based on symmetry breaking using the Higgs Mechanism.

In considering the Higgs Mechanism a number of peculiarities appear that diminishes its attractiveness:

6. As remarked above, it is selective in the sense that some gauge fields have associated Higgs particles and utilize the Higgs Mechanism, and some gauge fields do not have associated Higgs particles. In particular, the ElectroWeak gauge fields, the Generation group gauge fields, the Layer group gauge fields, and the complex gravitation fields have associated Higgs particles. The strong interaction (gluon) gauge fields do not.

7. The conventional Higgs potentials have a quadratic mass term of the "wrong" sign plus a quartic interaction term, which together, generate non-zero vacuum expectation values. They obviously accomplish their goal. But the source of these potentials, and why they have their form, is unknown. One suspects a fundamental principle should be operative here.

8. One can imagine creating a Higgs microscope at some super-accelerator. Using this microscope in the presence of a (classical) condensate could enable the Uncertainty Principle to be violated. This possibility, in the case of a microscope using electromagnetic fields, was the source of a heuristic argument for the need to quantize the electromagnetic field.[422]

---

[421] Blaha (2015a) and (2015b).
[422] Heitler (1954) p. 86 provides a good discussion of the need to quantize the electromagnetic field.

9.　The standard formulation of the Higgs Mechanism uses classical fields under the assumption that a path integral formulation justifies their use. While this may be true, the path integral formulation relies on implicit, unstated boundary conditions that obscure the physics of the quantum field theoretic nature of the mechanism. A direct quantum field theoretic study of the Higgs Mechanism is needed and would further elucidate its character. It is possible, and it has been shown in our earlier books, that the apparently "true" mechanism described below reveals a number of important new results in a properly formulated version of the Higgs Mechanism.

E.2.4 "True" Origin of an Acceptable Mass Creation Mechanism

In this section we are using *pseudoquantization*[423] and *pseudoquantum field theory.* It combines both quantum and classical fields within the same framework. In this extended theory vacuum expectation values appear as coherent ground states that are strictly classical in nature.

This section is based on our 1978 paper that appeared in the peer-reviewed journal *Physical Review D.* The paper is reproduced in appendix C for the reader's convenience.

We suggest the reader skim or read the paper before proceeding. The paper also presents a new formulation of Quantum Theory that incorporates both quantum and classical mechanics within one framework that is of interest in its own right. See section 4 for details. Recently, experimenters have been investigating the possibility of macroscopic and other strange quantum phenomena. The new formulation is ideally suited for tracing the transition from a quantum to a classical regime. For example, it is applicable to "large n atoms" where the outermost electrons approach classical behavior with an almost continuous energy spectrum.

E.2.5 Higgs-Like Vacuum Expectation Value Generation of Masses

The Higgs Mechanism is based on the appearance of non-zero, c-number vacuum expectation values for Higgs fields due to potential terms directly appearing in lagrangians.

E.2.5.1 Pseudoquantization of Higgs Particles

We will now consider the pseudoquantization of a scalar particle using two fields in a manner shown earlier. It will become a "Higgs" particle with a non-zero vacuum expectation value.

Using the formalism described earlier we define $\varphi_1(x)$ and $\varphi_2(x)$[424] for a generic boson suppressing any internal symmetry indices for simplicity. We define a "vacuum state" containing a coherent superposition that satisfies

$$\varphi_1(x)|\Phi, \Pi> = \Phi|\Phi, \Pi> \qquad (E.2.14)$$

---

[423] This new formalism was first described in S. Blaha, Phys. Rev. **D17**, 994 (1978).
[424] The subscripts on the fields are not gauge symmetry indices but simply identifiers distinguishing the fields from one another.

where Φ is a constant. Evaluating a fermion interaction term we find a mass term emerges[425]

$$\overline{\psi}(\varphi_1 + \varphi_2)\psi \;\; \rightarrow \;\; \overline{\psi}(\Phi + \varphi_2)\psi \qquad (E.2.15)$$

It can also generate a mass for an interaction with a gauge field of the form

$$A^{\mu}(\varphi_1 + \varphi_2)^2 A_{\mu} \;\; \rightarrow \;\; A^{\mu}(\Phi + \varphi_2)^2 A_{\mu} \qquad (E.2.16)$$

for ElectroWeak and other gauge fields. The φ_2 term leads to the production of Higgs particles in interactions. (The production of Higgs particles that decay into ElectroWeak gauge particles has recently been found at CERN.)

The present formalism provides a clean way to separate the vacuum expectation value of a scalar particle from its quantum field part in contrast to the conventional Higgs Mechanism where one has to separate a Higgs field into parts manually.

To obtain both the vacuum expectation value and the interaction with the quantum part of the pseudoquantum fields we choose to always specify interactions with fermions and gauge fields using $\varphi = \varphi_1 + \varphi_2$ as seen above.

It appears that our formulation of the mass generation mechanism sheds significant light on the reason for the special prominence of inertial frames. Consider massive scalars.[426] Eq. 6 in appendix A describes a massive scalar particle. If the scalar is massive, then the rest frame particle "vacuum" coherent state below yields a non-zero expectation value Φ:

$$|\Phi, \Pi\rangle = C\exp\{[(2\pi)^3 m/2]^{\frac{1}{2}}\Phi[a_2^{\dagger}(0,m) + a_2(0,m)]\}|0\rangle \qquad (E.2.17)$$

where m is a generic mass. (We note that the conventional Higgs Mechanism also has mass terms.) *Thus our pseudoquantum formalism allows us to define coherent "vacuum" states that lead to particle masses and Higgs particles.*

E.2.6 PseudoQuantized Non-Abelian Fields

The previous sections have considered scalar boson field theory. PseudoQuantum Field Theory also applies to non-abelian fields. See appendix B, Blaha (2016e) and earlier papers by the author.[427]

---

[425] When matrix elements with a "vacuum state" are calculated.

[426] Experiments at CERN have apparently discovered a Higgs particle with a 125 GeV/c mass.

[427] S. Blaha, Phys. Rev. **D10**, 4268 (1974); Phys. Rev. **D11**, 2921 (1975).

E.3 Fermion Quantization

Fermion field quantization is problematic in unconventional coordinate systems such as accelerating coordinate systems and coordinate systems defined for highly curved space-time.

In this section we will define a PseudoQuantization procedure for fermions that supports second quantization in non-static and unusual non-rectangular coordinate systems.

Having resolved these problems for both bosons and fermions using PseudoQuantum field theory we will see in section 4 that 'ordinary' quantum mechanics also has a problem with quantization in unconventional coordinate systems. It also has difficulties in the transition between classical and quantum mechanics. For example the transition from the classical Boltzmann equation to a quantum version is uncertain.

Using a framework analogous to our PseudoQuantum field theory formalisms for bosons and fermions we will establish a generalization of quantum mechanics, PseudoQuantization Mechanics, which contains both quantum mechanics and classical mechanics, and intermediate mixed mechanic states. This generalization supports a smooth transition between classical mechanics and quantum mechanics. With this generalization we will be able to examine the transition from quantum to classical mechanics in detail without recourse to methods such as expansions in Planck's constant $\hbar$.

The fermion PseudoQuantization procedure is described in section 3 in Appendix A to which the reader is referred. It is similar to boson PseudoQuantization in that it requires two fermion fields for each fermion particle. Eq. 61 and the following discussion in appendix A show a simple illustrative canonical lagrangian formulation of fermion PseudoQuantization including fourier expansions, equal time commutation relations, and creation and annihilation operators. Eq. 69 shows the general form of fourier expansions while eqs. 77 and 78 show the form restricted by anti-commutation relations and adjointness of the hamiltonian to have rotations of creation and annihilation operators: b_{1k} and b_{2k}, and d_{1k} and d_{2k} in a manner similar to the analogous boson case in eqs. 40 and 41.

Thus boson and fermion PseudoQuantization field theory both have pairs of fields associated with each particle and utilize rotations to implement unitary equivalence between static and non-static coordinate systems.

E.4 PseudoQuantization Mechanics – Joint Quantum-Classical Mechanics Formalism

Having established the need for paired fields for bosons and fermions using PseudoQuantization field theory to achieve unitary equivalence of quantization in both static and non-static coordinate systems we now turn to the case of quantum mechanics. Here again we find there is a problem associated with the transformations between coordinate systems in certain cases. There is also the problem in determining the transition between quantum mechanical entities and their classical equivalents.

These problems are analogous to those described in previous sections for second quantization. Since quantum mechanics is derived from quantum field theory it is reasonable to suspect that the resolution of quantum mechanics problems will ultimately be found in an analogue of PseudoQuantization which we earlier saw resolved quantum field theory problems.

This section will present PseudoQuantization Mechanics,[428] which contains both fully quantum, and fully classical, sectors as well as an intermediate sector that provides a transition between the quantum and classical regimes. With this formalism we can overcome coordinate system issues as well as the challenge of the correspondence between classical and quantum physics.

E.4.1 Coordinate Systems Problems of Quantum Mechanics

Problems exist in coordinate transformations in certain quantum mechanical situations which are 'fixed' through the use of 'recipes' that patch over the difficulties. One example is the change of coordinates in path integrals.[429] Gutzwiller (1990) points out that there is no simple rule for general canonical transformations of coordinates in path integrals. Given the central role of path integrals in quantum theory the difficulties of canonical transformations in path integrals is of concern. Other problems with canonical transformations in quantum mechanics are also discussed in Gutzwiller (1990).

We shall develop the PseudoQuantization formalism with a view towards facilitating canonical transformations in quantum mechanical studies as well as elucidating the transition from the quantum to the classical regimes.

E.4.2 PseudoQuantization Harmonic Oscillator

The harmonic oscillator plays a central role in classical and quantum mechanics due to its appearance in a variety of physical problems. In this section we will describe the PseudoQuantum formulation of the one-dimensional simple harmonic oscillator as a prelude to the general PseudoQuantum description.

Appendix C contains a paper on the PseudoQuantum harmonic oscillator.[430] In this section we will describe it, in part, with some changes, as a formalism that embodies both classical and quantum sectors, and provides a graceful transition between classical and quantum harmonic oscillator dynamics. Thus we will have an example of a new approach to understanding the classical-quantum transition. In later sections we will apply this approach to physical phenomena where the transition from a classical description to a quantum equivalent is problematic.

We begin with (appendix C) two commuting variables x_1 and p_1, which we augment with two new variables x_2 and p_2, defined by

---

[428] Much of this chapter was presented in S. Blaha, Phys. Rev. D17, 994 (1978) which is reprinted in appendix C. See also S. Blaha, Phys. Rev. D10, 4268 (July, 1974) and Phys. Rev. D11, 2921 (1974). See appendix D.

[429] See pp. 202-3 in Gutzwiller (1990).

[430] See its description in appendix C – section II of S. Blaha, Phys Rev D17, 994 (1978). Excerpts used with the kind permission of Physical Review D.

$$x_i = (m\omega/\hbar)^{-\frac{1}{2}} Q_i \qquad\qquad (E.4.1)$$
$$p_i = (m\omega\hbar)^{\frac{1}{2}} P_i$$

for i, j = 1, 2 where

$$P_2 = -i\, d/dQ_1 \qquad\qquad (E.4.2)$$
$$Q_2 = i\, d/dP_1$$

with the commutation relations:

$$[Q_i, P_j] = i(1 - \delta_{ij}) \qquad\qquad (E.4.3)$$

for i, j = 1, 2.

Next we define raising and lowering operators

$$a_i = 2^{-\frac{1}{2}}(Q_i + iP_i) \qquad\qquad (E.4.4)$$
$$a_i^\dagger = 2^{-\frac{1}{2}}(Q_i - iP_i)$$
$$Q_i = (a_i + a_i^\dagger)/\sqrt{2}$$
$$P_i = (a_i - a_i^\dagger)/(\sqrt{2}i)$$

with

$$[a_i, a_j^\dagger] = (1 - \delta_{ij}) \qquad\qquad (E.4.5)$$
$$[a_i, a_j] = 0$$
$$[a_i^\dagger, a_j^\dagger] = 0$$

for i, j = 1, 2.

We now define an alternate set of raising and lowering operators that will use an angle θ to provide a continuous transition from classical to quantum (and vice versa)[431]

$$b_1 = Q_1\cos\theta + iP_2\sin\theta \qquad\qquad (E.4.6)$$
$$b_2 = -Q_2\sin\theta + iP_1\cos\theta$$

$$b_1^\dagger = Q_1\cos\theta - iP_2\sin\theta \qquad\qquad (E.4.7)$$
$$b_2^\dagger = -Q_2\sin\theta - iP_1\cos\theta$$

Their commutation relations are

$$[b_1, b_1^\dagger] = \sin(2\theta) \qquad\qquad (E.4.8a)$$
$$[b_2, b_2^\dagger] = -\sin(2\theta)$$
$$[b_1, b_2^\dagger] = [b_2, b_1^\dagger] = 0$$
$$[b_1, b_2] = [b_1^\dagger, b_2^\dagger] = 0$$

The PseudoQuantum Hamiltonian[432] is

$$\hat{H} = p_1 p_2/m + m\omega^2 x_1 x_2 \qquad\qquad (E.4.8b)$$
$$= \tfrac{1}{2}\omega(\{a_1, a_2^\dagger\} + \{a_2, a_1^\dagger\})$$

---

[431] This definition differs from that appearing in appendix C.
[432] Eqs. 3, 12, 21 in appendix C with ω made explicit.

$$= \omega(P_1 P_2 + Q_1 Q_2)$$

In terms of the original P and Q variables we find

$$Q_1 = (b_1 + b_1^{\dagger})/(2\cos\theta) \tag{E.4.9}$$
$$Q_2 = -(b_2 + b_2^{\dagger})/(2\sin\theta)$$
$$P_1 = -i(b_2 - b_2^{\dagger})/(2\cos\theta)$$
$$P_2 = -i\sin(\theta)(b_1 - b_1^{\dagger})/(2\sin\theta)$$

E.4.2.1 Dirac Metric Operator ζ Transforming From Classical to Quantum Oscillator

At this point we define 'number' states with a_2 and $a_2^{\dagger}$:

$$|n_+, n_-> = (a_2^{\dagger})^{n_+}(a_2)^{n_-}|0,0> \tag{E.4.10}$$

where

$$\hat{H}|n_+, n_-> = \omega(n_+ - n_-)|n_+, n_-> \tag{E.4.11}$$

In view of the commutation relations we wish to transform eq. E.4.10 to

$$|n_+, n_-> = (b_1^{\dagger})^{n_+}(b_2^{\dagger})^{n_-}|0,0> \tag{E.4.12}$$

where the vacuum in eqs. E.4.10 and E.4.12 will be seen to be the same:

$$a_1^{\dagger}|0,0> = a_1|0,0> = 0 \tag{E.4.13}$$
$$b_1|0,0> = b_2|0,0> = 0$$

We define a 'Dirac-like' metric operator ζ. It satisfies

$$\zeta^{-1}a_2^{\dagger}\zeta = b_1^{\dagger} \tag{E.4.14}$$
$$\zeta^{-1}a_2\zeta = b_2^{\dagger}$$

We provisionally define

$$\zeta = \exp(aP_1Q_1 + bP_1^2 + cQ_2P_1 + dP_2P_1 + eQ_1^2 + fQ_2Q_1 + gP_2Q_1) \tag{E.4.15}$$

Eqs. E.4.4 and E.4.14 imply the values of the constants in eq. 4.15 so that

$$\zeta = \exp[(-i\cos\theta\, P_1Q_1 - (\cos\theta)/2\, P_1^2 + i\sin\theta\, Q_2P_1 - \sin\theta\, P_2P_1 - (\cos\theta)/2\, Q_1^2 - \sin\theta\, Q_2Q_1 + i\sin\theta\, P_2Q_1)/\sqrt{2}] \tag{E.4.16}$$
$$= \exp[(\cos\theta\,(a_1^{\dagger 2} - a_1^2)/2 - (\cos\theta)/4\,(a_1 - a_1^{\dagger})^2 + (\sin\theta)/2\,(a_2 + a_2^{\dagger})(a_1 - a_1^{\dagger}) +$$
$$+ (\sin\theta)/2\,(a_2 - a_2^{\dagger})(a_1 - a_1^{\dagger}) - (\cos\theta)/4\,(a_1 + a_1^{\dagger})^2 - (\sin\theta)/2\,(a_2 + a_2^{\dagger})(a_1 + a_1^{\dagger}) +$$
$$+ (\sin\theta)/2\,(a_2 - a_2^{\dagger})(a_1 + a_1^{\dagger}))/\sqrt{2}]$$

$$\zeta^{-1} = \exp[(\cos\theta\,(a_1^{\dagger 2} - a_1^2)/2 - (\cos\theta)/4\,(a_1 - a_1^\dagger)^2 + (\sin\theta)/2\,(a_2 + a_2^\dagger)(a_1 - a_1^\dagger) +$$
$$+ (\sin\theta)/2\,(a_2 - a_2^\dagger)(a_1 - a_1^\dagger) - (\cos\theta)/4\,(a_1 + a_1^\dagger)^2 - (\sin\theta)/2\,(a_2 + a_2^\dagger)(a_1 + a_1^\dagger) +$$
$$+ (\sin\theta)/2\,(a_2 - a_2^\dagger)(a_1 + a_1^\dagger))/\sqrt{2}]$$

(E.4.17)

We note the ground state (vacuum) explicitly satisfies:

$$|0,0> = \zeta^{-1}|0,0>$$

(E.4.18)

by eq. E.4.13.

We also note

$$\zeta^{-1}[a_2, a_1^\dagger]\zeta = 1$$

and

$$\zeta^{-1}[a_1, a_2^\dagger]\zeta = 1$$

imply

$$\zeta^{-1}a_1\zeta = b_1/\sin(2\theta)$$
$$\zeta^{-1}a_1^\dagger\zeta = -b_2/\sin(2\theta)$$

(E.4.19)

using eq. E.4.8.

The transformed Hamiltonian H can be expressed as

$$\hat{H}_\zeta = \zeta^{-1}\hat{H}\zeta = \tfrac{1}{2}\omega(\{b_1, b_1^\dagger\} - \{b_2, b_2^\dagger\})/\sin(2\theta)$$

(E.4.20)

E.4.2.2 Classical, Intermediate, and Quantum Wave Functions

The classical $(n_+, n_-)^{th}$ coordinate space wave function (eq. 28 appendix A) has the form:

$$\Psi_{n_+,n_-}(x_1, p_1, x_2, p_2, \theta) = (n_+!n_-!)^{-\frac{1}{2}}<x_1, p_1, x_2, p_2|(a_2^\dagger)^{n_+}(a_2)^{n_-}|0, 0>$$
$$= (n_+!n_-!)^{-\frac{1}{2}}<x_1, p_1, x_2, p_2|\zeta\zeta^{-1}(a_2^\dagger)^{n_+}\zeta\zeta^{-1}(a_2)^{n_-}\zeta\zeta^{-1}|0, 0>$$
$$= (n_+!n_-!)^{-\frac{1}{2}}<x_1, p_1, x_2, p_2|\zeta^\dagger b_1^{\dagger n_+}b_2^{\dagger n_-}|0,0>$$

(E.4.21)

using the conventional normalization of states with the form $|n> = (n_-!)^{-\frac{1}{2}}a^{\dagger n}|0>$.

Next we note

$$<x_1, p_1, x_2, p_2|\zeta^\dagger = <x_1, p_1, x_2, p_2|$$

(E.4.22)

similarly to eq. E.4.18.

$$b_1^\dagger = Q_1\cos\theta - iP_2\sin\theta \equiv \cos\theta\,Q_1 - \sin\theta\,d/dQ_1 = \cos\theta\,\eta_1 - \sin\theta\,\partial/\partial\eta_1$$

(E.4.23)

$$b_2^\dagger = -Q_2\sin\theta - iP_1\cos\theta \equiv -i(\sin\theta\,d/dP_1 + \cos\theta\,P_1) = -i(\sin\theta\,\partial/\partial\eta_2 + \cos\theta\,\eta_2)$$
$$= \sin\theta\,\partial/\partial\eta_3 - \cos\theta\,\eta_3$$

where $\eta_3 = i\eta_2 = iQ_2$ and $\eta_1 = Q_1$.

Eq. E.4.21 can be expressed as:

$$\Psi_{n_+,n_-}(x_1, p_1, x_2, p_2, \theta) = (n_+!n_-!)^{-\frac{1}{2}}(-1)^{n_-}[\cos\theta\ \eta_1 - \sin\theta\ \partial/\partial\eta_1]^{n_+} \cdot$$
$$\cdot\ [\cos\theta\ \eta_3 - \sin\theta\ \partial/\partial\eta_3]^{n_-}<x_1, p_1, x_2, p_2|0,0> \qquad (E.4.24)$$

The determination of

$$\Psi_{0,0} \equiv\ <x_1, p_1, x_2, p_2|0,0>$$

begins with noting

$$<x_1, p_1, x_2, p_2|b_1|0,0> = 0$$

or

$$(\cos\theta\ \eta_1 + \sin\theta\ \partial/\partial\eta_1)\Psi_{0,0} = 0$$

and

$$<x_1, p_1, x_2, p_2|b_2|0,0> = 0$$

or

$$(\cos\theta\ \eta_3 + \sin\theta\ \partial/\partial\eta_3)\Psi_{0,0} = 0$$

These conditions require

$$\Psi_{0,0} = C\ \exp[-\tfrac{1}{2}\cot\theta(\eta_1^2 + \eta_3^2)] \qquad (E.4.25)$$

where the normalization $C = [m\omega\cot\theta/(i\pi\hbar)]^{\frac{1}{2}}$ is determined by

$$1 = C^2 \int dx_1 dx_2\ \exp[-\tfrac{1}{2}\cot\theta(\eta_1^2 + \eta_3^2)] \qquad (E.4.26)$$

Then eq. E.4.24 becomes

$$\Psi_{n_+,n_-}(x_1, p_1, x_2, p_2, \theta) = (n_+!n_-!)^{-\frac{1}{2}}[m\omega\cot\theta/(i\pi\hbar)]^{\frac{1}{2}}(-1)^{n_-}[\cos\theta\ \eta_1 - \sin\theta\ \partial/\partial\eta_1]^{n_+} \cdot$$
$$\cdot\ [\cos\theta\ \eta_3 - \sin\theta\ \partial/\partial\eta_3]^{n_-}\exp[-\tfrac{1}{2}\cot\theta(\eta_1^2 + \eta_3^2)] \qquad (E.4.27)$$

$$= (n_+!n_-!)^{-\frac{1}{2}}[m\omega\cot\theta/(i\pi\hbar)]^{\frac{1}{2}}(-1)^{n_-}[\cos\theta\ \eta_1 - \sin\theta\ \partial/\partial\eta_1]^{n_+} \cdot$$
$$\cdot\ [i\cos\theta\ \eta_2 + i\sin\theta\ \partial/\partial\eta_2]^{n_-}\exp[-\tfrac{1}{2}\cot\theta(\eta_1^2 - \eta_2^2)]$$

Note that eq. E.4.27 contains a product of Hermite polynomials if $\theta = \pi/4$. It is not surprising that we obtain quantum harmonic oscillator factors in the wave function since, as eq. E.4.8a shows the b operators have conventional quantum oscillator commutation relations for $\theta = \pi/4$ – thus this value of θ corresponds to the quantum case. We note that Hermite polynomials $H_n(\eta)$ are generated by

$$(\eta - \partial/\partial\eta)^n \exp(-\tfrac{1}{2}\eta^2) = \exp(-\tfrac{1}{2}\eta^2)H_n(\eta) \qquad (E.4.28)$$

We can generalize Hermite polynomials for other values of θ with

$$H_n(\eta, \theta) = \exp[+\tfrac{1}{2}\cot\theta\ \eta^2]\ [\cos\theta\ \eta\ -\sin\theta\ \partial/\partial\eta]^n \exp[-\tfrac{1}{2}\cot\theta\ \eta^2] \qquad (E.4.29)$$

Then eq. E.4.27 can be expressed by

$$\Psi_{n+,n-}(x_1, p_1, x_2, p_2, \theta) = (n_+!n_-!)^{-\frac{1}{2}}[m\omega\cot\theta/(i\pi\hbar)]^{\frac{1}{2}}(-1)^{n_-}H_{n+}(\eta_1,\theta)H_{n-}(\eta_3,\theta)\exp[-\tfrac{1}{2}\cot\theta(\eta_1{}^2 + \eta_3{}^2)]$$
$$(E.4.30)$$
$$= (n_+!n_-!)^{-\frac{1}{2}}[m\omega\cot\theta/(i\pi\hbar)]^{\frac{1}{2}}(-1)^{n_-}H_{n+}(\eta_1,\theta)H_{n-}(i\eta_2,\theta)\exp[-\tfrac{1}{2}\cot\theta(\eta_1{}^2 - \eta_2{}^2)]$$
$$= (-1)^{n_-}\Psi_{n+}(\eta_1, \theta)\Psi_{n-}(\eta_3, \theta)$$
$$= (-1)^{n_-}\Psi_{n+}((m\omega)^{\frac{1}{2}}x_1, \theta)\Psi_{n-}(i(m\omega)^{\frac{1}{2}}x_2, \theta)$$

where

$$\Psi_n(\eta, \theta) = = (n!)^{-\frac{1}{2}}[m\omega\cot\theta/(i\pi\hbar)]^{1/4}H_n(\eta,\theta)\ \exp[-\tfrac{1}{2}\cot\theta\eta^2] \qquad (E.4.30a)$$

At $\theta = \pi/4$ the wave function factorizes into a harmonic oscillator wave function times an inverted harmonic oscillator wave function:

$$\Psi_{n+,n-}(x_1, p_1, x_2, p_2, \theta=\pi/4) = (n_+!n_-!)^{-\frac{1}{2}}[m\omega/(i\pi\hbar)]^{\frac{1}{2}}(-1)^{n_-}2^{-(n_+ + n_-)/2}H_{n+}(\eta_1)H_{n-}(\eta_3)\exp[-\tfrac{1}{2}(\eta_1{}^2+\eta_3{}^2)]$$
$$(E.4.30b)$$
$$= (-1)^{n_-}2^{-(n_+ + n_-)/2}\Psi_{n+}((m\omega)^{\frac{1}{2}}x_1)\Psi_{n-}(i(m\omega)^{\frac{1}{2}}x_2)$$

where $H_n(\eta)$ is a Hermite polynomial of degree n.

For $\theta = 0$ we find the b commutation relations (eq. 8a) are zero indicating that the wave function is classical in nature. In this case, simply substituting $\theta = 0$ would cause eq. E.4.30 to 'blow up.' However for certain values of n+ and n– the limit $\theta \to 0$ yields a physically interesting result – a wave function that is a delta function similar to that appearing in eq. 43 in appendix C.

Consider first the case n+ = 1 and n– = 0. Then

$$\Psi_{1,0}(x_1, p_1, x_2, p_2, \theta) = [m\omega\cot\theta/(i\pi\hbar)]^{\frac{1}{2}}[\cos\theta\ \eta_1\ -\sin\theta\ \partial/\partial\eta_1]\exp[-\tfrac{1}{2}\cot\theta(\eta_1{}^2 + \eta_3{}^2)]$$

As $\theta \to 0$, and for $\eta_3 = 0$, we find

$$\Psi_{1,0}(x_1, p_1, x_2, p_2, \theta\to0) \to [m\omega/(i\hbar)]^{\frac{1}{2}}\eta_1(\pi\sin\theta)^{-\frac{1}{2}}\exp[-\tfrac{1}{2}\eta_1{}^2/\sin\theta]$$
$$\to [m\omega/(i\hbar)]^{\frac{1}{2}}\eta_1(2\pi\sin\theta)^{-\frac{1}{2}}\exp[-\eta_1{}^2/(2\sin\theta)]$$
$$\to [m\omega/(i\hbar)]^{\frac{1}{2}}\eta_1\delta(\eta_1) = 0 \qquad (E.4.31)$$

Now consider the case n+ = 0 and n– = 0:

$$\Psi_{0,0}(x_1, p_1, x_2, p_2, \theta) = [m\omega\cot\theta/(i\pi\hbar)]^{\frac{1}{2}}\exp[-\tfrac{1}{2}\cot\theta(\eta_1{}^2 + \eta_3{}^2)]$$

As $\theta \to 0$, and for $\eta_3 = 0$, we find

$$\Psi_{0,0}(x_1, p_1, x_2, p_2, \theta\to 0) \to [m\omega/(i\hbar)]^{\frac{1}{2}}(\pi\sin\theta)^{-\frac{1}{2}}\exp[-\tfrac{1}{2}\cot\theta\,\eta_1^{\,2}/\sin\theta]$$
$$\to [m\omega/(i\hbar)]^{\frac{1}{2}}\eta_1(2\pi\sin\theta)^{-\frac{1}{2}}\exp[-\eta_1^{\,2}/(2\sin\theta)]$$
$$\to [m\omega/(i\hbar)]^{\frac{1}{2}}\delta(\eta_1) = i^{-\frac{1}{2}}\delta(x_1) \neq 0 \qquad\text{(E.4.32)}$$

using

$$\delta(\eta) = \lim_{\varepsilon\to 0} (\pi\varepsilon)^{-\frac{1}{2}}\exp[-\eta^2/\varepsilon] \qquad\text{(E.4.33)}$$

Thus the Gaussian factor combined with the preceding $(2\pi\sin\theta)^{-\frac{1}{2}}$ grows to a delta-function wave function. *Wave functions corresponding to higher values of $n+$ and $n-$ go to zero in the limit $\theta \to 0$. Only $\Psi_{0,0}(x_1, p_1, x_2, p_2, \theta\to 0)$ is non-zero.*

The introduction of the time dependence and a shift of the location of the minimum of the harmonic oscillator potential to x_0 would lead to a wave function such as:

$$\Psi_{0,0}(x_1, p_1, x_2, p_2, \theta\to 0) = i^{-\frac{1}{2}}\delta(x_1 - x_0\sin(\omega t)) \qquad\text{(E.4.34)}$$

A similar behavior may be seen in the case $\eta_1 = 0$. Then we find a wave function with a factor of $\delta(x_2)$.

Lastly, the case of $\theta \to \pi/2$ is of interest. Eq. 4.30 yields

$$\Psi_{n_+,n_-}(x_1, x_2, \theta\to\pi/2) = (n_+!n_-!)^{-\frac{1}{2}}[m\omega\cos\theta/(i\pi\hbar)]^{\frac{1}{2}}[-\partial/\partial\eta_1]^{n_+}[\partial/\partial\eta_3]^{n_-}\exp[-\tfrac{1}{2}\cos\theta(\eta_1^{\,2}+\eta_3^{\,2})]|_{\theta\to\pi/2}$$
$$= 0 \qquad\text{(E.4.35)}$$

Figuratively speaking, the wave function progresses from one non-zero 'classical' wave function at $\theta = 0$, to a quantum mechanical wave function at $\theta = \pi/4$, to a zero value wave function at $\theta = \pi/2$. Thus one might say "The good Lord by giving us a quantum universe put us in a position halfway between nothingness and classical mechanics." By implementing Quantum theory we get Second Quantization of particle fields, and thereby, integer countability of particle numbers – a distinct simplification in Nature.

E.4.2.3 Energy Eigenvalues

From eq. E.4.8b, E.4.10, and E.4.11 we see that eq. E.4.11 for the state

$$|n_+, n_-> = (a_2^\dagger)^{\,n_+}(a_2)^{\,n_-}|0, 0>$$

shows the energy of the wave function (eqs. E.4.27 and E.4.30) to be

$$E_{n+,n-} = (n_+ - n_-)\hbar\omega = [n_+ + \tfrac{1}{2} - (n_- + \tfrac{1}{2})]\hbar\omega \qquad\text{(E.4.36)}$$

Eq. E.4.27 satisfies the PseudoQuantized Schrödinger equation:

$$\hat{H}\Psi_n(x_1, p_1, x_2, p_2, \theta, t) = i\partial\Psi_n(x_1, p_1, x_2, p_2, \theta, t)\partial t \qquad\text{(E.4.37)}$$

E.4.3 Wave Function as a Function of Position and Momentum

We can define the wave function in terms of x_1 and p_1 with a fourier transform:

$$\Psi_{n+,n-}(x_1, p_1, \theta) = \int dx_2 \, e^{-ip_1 x_2} \, \Psi_{n+,n-}(x_1, x_2, \theta) \qquad (E.4.38)$$

$\Psi_{n+,n-}(x_1, p_1, \theta=\pi/4)$, is a wave function for the combined 'normal', and inverted, harmonic oscillators. Thus the full PseudoQuantum theory enables us to define a wave function that is a function of both position and momentum without inconsistency. We discuss this topic in more detail later when we compare it to the Wigner distribution function.

E.4.4 Intermediate Classical-Quantum Wave Functions

For other values of θ in eq. E.4.27 and E.4.30 we obtain wave functions that are intermediate between quantum and classical operator wave functions. Later we will find it of interest to trace the evolution of a classical wave function to a quantum wave function and vice versa in the general case of non-harmonic oscillator dynamics.

It is interesting to note the dependence of the energy level spacing on the angle θ. The transformed energy (eq. E.4.14 implements the transformation to the b operators) has the form:

$$\hat{H}_\zeta = \zeta^{-1}\hat{H}\zeta = \tfrac{1}{2}\omega(\{b_1, b_1^{\dagger}\} - \{b_2, b_2^{\dagger}\})/\sin(2\theta) \qquad (E.4.20)$$

If either n_+ and n_- change by one unit, then the energy changes by

$$\Delta E = \tfrac{1}{2}\omega/\sin(2\theta)$$

For $\theta = \pi/4$ (the quantum case)

$$\Delta E = \tfrac{1}{2}\omega \qquad (E.4.39)$$

For $\theta = \pi/8$ (the quantum approaching classical case)

$$\Delta E = \tfrac{1}{2}\omega/0.383 = 1.307\omega \qquad (E.4.40)$$

As $\theta \to 0$ (the classical case)

$$\Delta E \to \infty \qquad (E.4.41)$$

Thus 'higher' (lower) energy states beyond the $n_+ = 0$ and $n_- = 0$ state are inaccessible energy-wise. This corresponds to our above finding that only the wave function $\Psi_{0,0}$ is non-zero in the classical limit.

E.5 General Formalism for a PseudoQuantized System

The basic procedure of our PseudoQuantization Formalism are described in our paper Phys. Rev **D17**, 994 (1978) reprinted in appendix C.[433] The relevant excerpt is

We shall now briefly outline the procedure for embedding a classical-mechanical system in a quantum system.[6] Consider a classical Hamiltonian system with one degree of freedom, and commuting canonical variables, x_1 and p_1, which have the equations of motion

$$\dot{x}_1 = -i[x_1, \hat{H}] , \tag{1}$$

$$\dot{p}_1 = -i[p_1, \hat{H}] , \tag{2}$$

where defining

$$\hat{H} = -i\left(\frac{\partial H(x_1,p_1)}{\partial p_1}\frac{\partial}{\partial x_1} - \frac{\partial H(x_1,p_1)}{\partial x_1}\frac{\partial}{\partial p_1}\right) \tag{3}$$

allows us to write Hamilton's equations in commutator form. With Sudarshan[6] we define

$$x_2 = i\frac{\partial}{\partial p_1} \tag{4}$$

and

$$p_2 = -i\frac{\partial}{\partial x_1} \tag{5}$$

so that

$$[x_1, x_2] = [p_1, p_2] = 0 , \tag{6}$$

$$[x_1, p_2] = [x_2, p_1] = i , \tag{7}$$

and $\hat{H}$ can now be taken to be the operator

$$\hat{H} = \frac{\partial H(x_1,p_1)}{\partial p_1}p_2 + \frac{\partial H(x_1,p_1)}{\partial x_1}x_2 . \tag{8}$$

---

[433] Excerpt used with the kind permission of Physical Review D.

It is now apparent that we can take the above quantities and equations of motion to describe a quantum mechanical system with two degrees of freedom in the "coordinate" representation where the "coordinates" are (x_1, p_1) and the canonical momenta are $\Pi = (p_2, -x_2)$. As we will see below the linearity of $\hat{H}$ in the momenta is crucial for the maintenance of the classical character of x_1 and p_1, and for the observability of the phase-space trajectory. Since we choose to identify the physical observables with the commutative algebra of the coordinate operators, x_1 and p_1, we are led to impose the superselection condition that the momenta, Π, are unobservable. As a result the Hamiltonian and other generators of canonical transformations, which are all linear in the momenta, are also unobservable. However, in each case there is an associated dynamical quantity which is observable.

The required unobservability of the momenta restricts the form of the interaction between a classical-made-quantum system and an inherently quantum system to

$$H_{\text{int}} = \Phi_1 x_2 + \Phi_2 p_2 + X , \tag{9}$$

where Φ_1, Φ_2, and X are functions of x_1, p_1, and the quantum system variables. The commutation relations of these functions are also constrained[6] by the superselection rule and the commutativity of the classical variables, x_1 and p_1, and their time derivatives. In the next section we will study the simple harmonic oscillator in order to exemplify the quantum-mechanical case described above and also for direct use in the field-theoretic generalizations of subsequent sections.

Based on the above discussion we assume that we start with a conventional Hamiltonian that we express as

$$H = H(x_1, p_1) = \tfrac{1}{2}p_1^2 + V(x_1) \tag{E.5.1}$$

Introducing x_2 and p_2, as in eqs. 4 and 5 above, we can generalize H to a PseudoQuantum Hamiltonian $\hat{H}$:

$$\hat{H} = p_1 p_2 + x_2 \partial V / \partial x_1 \qquad \text{(E.5.2)}$$

where V is a function of x_1.

We can introduce raising and lowering operators a_i and $a_i^{\dagger}$ using the procedure of section E.4. Then we can proceed as in the harmonic oscillator case to calculate wave functions. In the next section we apply this procedure to the Boltzmann equation, which has some similarity to the Schrödinger equation.

E.6 PseudoQuantization of the Boltzmann Equation

The Boltzmann equation is a classical dynamics equation that describes the dynamics of a multi-particle system with interactions. The equivalent quantum formulation is not known. However Wigner and others have proposed possible quantum equivalents that of some of the expected features of the quantum Boltzmann function. In this section we will follow a procedure similar to that of section 4 for the Vlasov approximation. For special cases of the collision term of the Boltzmann equation we will obtain quantum equivalents.

E.6.1 Non-Relativistic Boltzmann Equation

The non-relativistic Boltzmann equation for identical particles of one chemical species is

$$[\mathbf{p}\cdot\nabla/m + F\cdot\partial/\partial\mathbf{p}]f = -\partial f/\partial t + (\partial f/\partial t)_{\text{coll}}$$

where $f(\mathbf{r}, \mathbf{p}, t)$ is Boltzmann's probability density function. It has often been remarked that this Boltzmann equation strongly resembles the Schrödinger equation.

E.6.2 PseudoQuantum Form of the Boltzmann Equation

We can make the case that it even more strongly resembles the PseudoQuantized Schrödinger equation (see eq. E.5.2) by defining

$$-i[\mathbf{p_1}\cdot\mathbf{p_2}/m - \mathbf{x_2}\cdot F]f = -\partial f/\partial t + (\partial f/\partial t)_{\text{coll}} \qquad \text{(E.6.1)}$$

where $\mathbf{F} = \mathbf{F}(x_1, t)$ and

$$\mathbf{p_2} = i\nabla \qquad \text{(E.6.2)}$$
$$x_2 = -i\partial/\partial\mathbf{p_1}$$

Comparing eq. E.6.1 with section E.4, we find we can define

$$\hat{H} = \mathbf{p_1}\cdot\mathbf{p_2}/m + x_2\cdot\partial V/\partial\mathbf{x_1} \qquad \text{(E.6.3)}$$

where

$$\partial V / \partial \mathbf{x}_1 = \mathbf{F}(\mathbf{x}_1, t)$$

Given the close similarity of the Boltzmann equation and the Schrödinger equation it is sensible to treat the solution of the equation as a 'wave function' that initially represents a classical state such as the classical harmonic oscillator wave function that we saw in section E.4. Then we will define the Boltzmann distribution in terms of the wave function solution.

The PseudoQuantized Boltzmann wave equation is

$$\hat{H}\psi = -i\partial\psi/\partial t + i(\partial\psi/\partial t)_{coll} \qquad (E.6.4)$$

where

$$\psi = \psi(\mathbf{x}_1, \mathbf{x}_2, \theta)$$

with θ defined later in specific cases. The value of θ determines whether ψ is classical, quantum, or in an intermediate state.

In a manner somewhat analogous to that of Wigner[434] we define a Boltzmann distribution with

$$f_q(\mathbf{r}_1, \mathbf{p}_1, t, \theta) = \int d^3r_2 \, \psi(\mathbf{r}_1, \mathbf{r}_2, t, \theta)\psi^{\dagger}(\mathbf{r}_1, \mathbf{r}_2, t, \theta) \exp(-2i\mathbf{r}_2{\cdot}\mathbf{p}_1/\hbar) \quad (E.6.5)$$

in three spatial dimensions where we use the suffix 'q' of f_q to signify the PseudoQuantum Boltzmann probability density function $f_q(\mathbf{r}, \mathbf{p}, t)$. The function f_q can be classical, quantum, or intermediate between classical and quantum depending on the value of θ. We will consider examples that illustrate the dependence of f_q on θ. Later we will also see that our definition of f_q eliminates the problems of the Wigner density function.

E.6.3 PseudoQuantum Form of the Vlasov Equation

The collision-less Boltzmann equation is called the Vlasov equation. It is of interest because of the difficulties associated with solving the full Boltzmann equation. Its PseudoQuantized equivalent is

$$\hat{H}\psi = -i\partial\psi/\partial t \qquad (E.6.6)$$

This equation has 'only' the difficulty of its solution for the various forces $\mathbf{F}(\mathbf{x}_1, t)$.

We note that the one-dimensional version of eq. 6.6 where $V = \frac{1}{2} x_1^2$ is solved in section 4.

E.6.4 PseudoQuantum Vlasov Equation Solution for a Three-dimensional Harmonic Oscillator Force with $\theta = \pi/4$

The choice of $\theta = \pi/4$ gives 'quantum' harmonic oscillator solutions consisting of a harmonic oscillator factor and an inverted harmonic oscillator factor.

---

[434] E. P. Wigner, Phys. Rev. **40**, 749 (1932).

The three-dimensional harmonic oscillator PseudoQuantum 'Hamiltonian' equation is

$$(\mathbf{p}_1 \cdot \mathbf{p}_2/m + x_2 \cdot x_1)\psi = -i\partial\psi/\partial t \tag{E.6.7}$$

or

$$(\mathbf{p}_1 \cdot \mathbf{p}_2/(2m') + x_2 \cdot x_1)\,\psi = -i\partial\psi/\partial t \tag{E.6.8}$$

This equation is fully separable[435] for $\theta = \pi/4$ in rectangular coordinates which we label x, y, and z. The solution is a product of one-dimensional PseudoQuantum harmonic oscillator wave function factors of the form of E.4.30b:

$$\psi_{n+,n-}(\mathbf{r}_1, \mathbf{r}_2, t, \pi/4) = \Psi_{nx+,nx-}(x_{1x},p_{1x},x_{2x},p_{2x},t,\theta)\,\Psi_{ny+,ny-}(x_{1y},p_{1y},x_{2y},p_{2y},t,\theta)\,\Psi_{nz+,nz-}(x_{1z},p_{1z},x_{2z},p_{2z},t,\theta) \tag{E.6.9}$$

$$= (-1)^{n_{x-}+n_{y-}+n_{z-}}2^{-(n_{x+}+n_{x-}+n_{y+}+n_{y-}+n_{z+}+n_{z-})/2}\Psi_{nx+}((m\omega)^{\frac{1}{2}}x_1)\Psi_{nx-}(i(m\omega)^{\frac{1}{2}}x_2)\Psi_{ny+}((m\omega)^{\frac{1}{2}}y_1)\cdot$$
$$\cdot\Psi_{ny-}(i(m\omega)^{\frac{1}{2}}y_2)\Psi_{nz+}((m\omega)^{\frac{1}{2}}z_1)\Psi_{nz-}(i(m\omega)^{\frac{1}{2}}z_2)$$

$$= A\Psi_{nx+}((m\omega)^{\frac{1}{2}}x_1)\Psi_{ny+}((m\omega)^{\frac{1}{2}}y_1)\Psi_{nz+}((m\omega)^{\frac{1}{2}}z_1)\Psi_{nx-}(i(m\omega)^{\frac{1}{2}}x_2)\Psi_{ny-}(i(m\omega)^{\frac{1}{2}}y_2)\,\Psi_{nz-}(i(m\omega)^{\frac{1}{2}}z_2)$$
$$= A\Psi_1(\mathbf{r}_1)\Psi_2(\mathbf{r}_2) \tag{E.6.9a}$$

with the time dependence not displayed and where

$$A = (-1)^{n_{x-}+n_{y-}+n_{z-}}2^{-(n_{x+}+n_{x-}+n_{y+}+n_{y-}+n_{z+}+n_{z-})/2} \tag{E.6.9c}$$

using eq. E.4.30b.

The energy, which is constant since the Hamiltonian is not explicitly time dependent, is

$$E_{n+,n-} = (n_+ - n_-)\hbar\omega \tag{E.6.10}$$

with

$$n_+ = n_{x+} + n_{y+} + n_{z+} \tag{E.6.11}$$
$$n_- = n_{x-} + n_{y-} + n_{z-}$$

Following Wigner, a fourier transform for $\theta = \pi/4$ of eq. E.6.9a factors gives a *quantum* Boltzmann density function:

$$f_q(\mathbf{q}, \mathbf{p}_1, t, \pi/4) = \int d^3Q\,\Psi(\mathbf{q} - \mathbf{Q})\Psi^\dagger(\mathbf{q} + \mathbf{Q})\exp(-2i\mathbf{Q}\cdot\mathbf{p}_1/\hbar)$$

$$= \int d^3Q\,\Psi_1(\mathbf{q} - \mathbf{Q})\Psi_1^\dagger(\mathbf{q} + \mathbf{Q})\Psi_2(\mathbf{q} - \mathbf{Q})\Psi_2^\dagger(\mathbf{q} + \mathbf{Q})\exp(-2i\mathbf{Q}\cdot\mathbf{p}_1/\hbar) \tag{E.6.12}$$

where we let $\mathbf{r}_1 = \mathbf{q} - \mathbf{Q}$ and $\mathbf{r}_2 = \mathbf{q} + \mathbf{Q}$.

---

[435] For other values of θ the solutions of the equation do not separate type '1' coordinates from type '2' coordinates. See eq. E.4.30 for the general case.

If we define the fourier transform of a wave function $\Psi(\mathbf{r})$ by

$$\Phi(\mathbf{p}) = (2\pi\hbar)^{-3/2} \int d^3r \ \Psi(\mathbf{r}) \exp(-i\mathbf{r}\cdot\mathbf{p}/\hbar)$$

then

$$f_{qp}(\mathbf{p}_1, t, \pi/4) = \int d^3q \ f_q(\mathbf{q}, \mathbf{p}_1, t, \pi/4) = \int d^3p \ \Phi(\mathbf{p})\Phi^\dagger(\mathbf{p}) \qquad \text{(E.6.13)}$$

yields a projection of the phase space distribution into momentum space. In addition

$$f_{qp}(\mathbf{p}_1, t, \pi/4) = \Psi(\mathbf{q})\Psi^\dagger(\mathbf{q}) \qquad \text{(E.6.14)}$$

yields a projection of the phase space distribution into coordinate space.

Thus $f_q(\mathbf{q}, \mathbf{p}_1, t, \pi/4)$ can be interpreted as the quantum equivalent of the (classical) Boltzmann distribution. *PseudoQuantization gives us 2n variables just as there are 2n variables in phase space.*

E.6.5 PseudoQuantum Vlasov Equation Solution for a Three-dimensional Harmonic Oscillator Force for Arbitrary θ

The general representation of our PseudoQuantized Vlasov equation solution for any value of θ is given by eq. E.4.30. The 3-dimensional Vlasov representation is

$$\psi_{n-,n-}(\mathbf{r}_1, \mathbf{r}_2, t, \theta) = \psi_{n_{x+},n_{x-}}(x_{1x},p_{1x},x_{2x},p_{2x},t,\theta)\psi_{n_{y+},n_{y-}}(x_{1y},p_{1y},x_{2y},p_{2y},t,\theta)\psi_{n_{z+},n_{z-}}(x_{1z},p_{1z},x_{2z},p_{2z},t,\theta)$$

$$\text{(E.6.15)}$$

$$= (-1)^{n_{x-} + n_{y-} + n_{z-}} 2^{-(n_{x+} + n_{x-} + n_{y+} + n_{y-} + n_{z+} + n_{z-})/2}\Psi_{n_{x-}}((m\omega)^{1/2} x_1, t,\theta)\Psi_{n_{x-}}(i(m\omega)^{1/2} x_2, t,\theta) \cdot$$
$$\cdot\Psi_{n_{y-}}((m\omega)^{1/2} y_1,t,\theta)\Psi_{n_{y-}}(i(m\omega)^{1/2} y_2,t,\theta)\Psi_{n_{z-}}((m\omega)^{1/2} z_1, t,\theta)\Psi_{n_{z-}}(i(m\omega)^{1/2} z_2, t,\theta)$$

$$= A\Psi_{n_{x-}}((m\omega)^{1/2} x_1, t, \theta)\Psi_{n_{y-}}((m\omega)^{1/2} y_1, t, \theta)\Psi_{n_{z-}}((m\omega)^{1/2} z_1, t, \theta)\Psi_{n_{x-}}(i(m\omega)^{1/2} x_2, t, \theta) \cdot$$
$$\cdot\Psi_{n_{y-}}(i(m\omega)^{1/2} y_2, t, \theta) \ \Psi_{n_{z-}}(i(m\omega)^{1/2} z_2, t, \theta)$$

$$= A\Psi_1(\mathbf{r}_1, t, \theta)\Psi_2(\mathbf{r}_2, t, \theta)$$

Following similar steps as in the previous section we find

$$f_q(\mathbf{q}, \mathbf{p}_1, t, \theta) = \int d^3Q \ \psi_{n-,n-}(\mathbf{q} - \mathbf{Q}, t, \theta) \ \psi_{n-,n-}^\dagger(\mathbf{q} + \mathbf{Q}, t, \theta) \exp(-2i\mathbf{Q}\cdot\mathbf{p}_1/\hbar)$$

$$= A^2 \int d^3Q \ \Psi_1(\mathbf{q} - \mathbf{Q}, t, \theta)\Psi_1^\dagger(\mathbf{q} + \mathbf{Q}, t, \theta)\Psi_2(\mathbf{q} - \mathbf{Q}, t, \theta)\Psi_2^\dagger(\mathbf{q} + \mathbf{Q}, t, \theta)\exp(-2i\mathbf{Q}\cdot\mathbf{p}_1/\hbar)$$
$$\text{(E.6.16)}$$

where we let $\mathbf{r}_1 = \mathbf{q} - \mathbf{Q}$ and $\mathbf{r}_2 = \mathbf{q} + \mathbf{Q}$.

Following similar steps we can again obtain eqs. E.6.13 and E.6.14 and establish a connection between phase space, and momentum and coordinate space projections.

If we define the fourier transform of a wave function $\Psi(\mathbf{r})$ by

$$\Phi(\mathbf{p}, \theta) = (2\pi\hbar)^{-3/2} \int d^3r \, \Psi(\mathbf{r}, \theta) \exp(-i\mathbf{r}\cdot\mathbf{p}/\hbar)$$

then

$$f_{qp}(\mathbf{p}_1, t, \theta) = \int d^3q \, f_q(\mathbf{q}, \mathbf{p}_1, t, \theta) = \int d^3p \, \Phi(\mathbf{p}, \theta)\Phi^\dagger(\mathbf{p}, \theta) \qquad (E.6.17)$$

yields a projection of the phase space distribution into momentum space. In addition

$$f_{qq}(\mathbf{p}_1, t, \theta) = \Psi(\mathbf{q}, \theta)\Psi^{\dagger}(\mathbf{q}, \theta) \qquad (E.6.18)$$

yields a projection of the phase space distribution into coordinate space.

Thus $f_q(\mathbf{q}, \mathbf{p}_1, t, \theta)$ can be interpreted as the quantum equivalent of the (classical) Boltzmann distribution. PseudoQuantization gives us 2n variables just as there are 2n variables in phase space.

E.6.6 PseudoQuantum Vlasov Equation Solution for a Three-dimensional Harmonic Oscillator Force for $\theta = 0$ – The Classical Case

In the $\theta = 0$ case, which is the classical mechanics limit, we find that eq. E.4.34 gives a precise expression for the Vlasov Boltzmann equation solution of eq. E.6.16. We note that only the 0-0 wave functions are non-zero. In one dimension we have:

$$\psi = \Psi_{0,0}(x_1, p_1, x_2, p_2, \theta\rightarrow0) = i^{-\frac{1}{2}}\delta(x_1 - x_0\sin(\omega t)) \qquad (E.4.34)$$

The 3-dimensional case (eq. E.6.16) gives

$$\begin{aligned}
f_q(\mathbf{q}, \mathbf{p}_1, t, \theta{=}0) &= \int d^3Q \, \delta^3(\mathbf{q} - \mathbf{Q} - \mathbf{x}_0\sin(\omega t))\delta^3(\mathbf{q} + \mathbf{Q} - \mathbf{x}_0\sin(\omega t))\exp(-2i\mathbf{Q}\cdot\mathbf{p}_1/\hbar) \\
&= \delta^3(\mathbf{q} - \mathbf{x}_0\sin(\omega t))
\end{aligned} \qquad (E.6.19)$$

a classical solution specifying the classical harmonic oscillator trajectory. We note that all other solutions (for other values of n_+ and n_-) are 'pushed' to $E = \infty$ according to eq. E.4.41.

We note $f_q(\mathbf{q}, \mathbf{p}_1, t, \theta{=}0)$ is positive definite as a probability should be. The integral

$$\int d^3q \, f_q(\mathbf{q}, \mathbf{p}_1, t, \theta{=}0) = 1$$

shows the sum of the probabilities of the normalized Boltzmann distribution in coordinate space is unity.

We thus have achieve a quantum-classical Boltzmann distribution in phase space in both coordinates and momenta using PseudoQuantization where the number of phase space parameters is 2n = 6 in this case—unlike the case of the Wigner density alternative.

E.6.7 Comparison to the Wigner Density Function

The Wigner density function in n dimensions is defined as:

$$\Psi(p, q) = \int d^n Q \, \psi(q - Q) \, \psi^\dagger(q + Q) \, \exp(-2ipQ/\hbar) \qquad (E.6.20)$$

where $\psi(q)$ of the wave function of the system. The interpretation of $\Psi(p, q)$ as the quantum probability in phase space corresponds to $f(p, q)$ – the classical Boltzmann distribution. It is often interpreted in that manner.

However, several concerns are usually expressed about this interpretation:

1. Although real-valued $\Psi(p, q)$ can have a negative value making a probability interpretation problematic.

2. $\Psi(p, q)$ appears to depend on 2n values. However the wave function $\psi(q)$, upon which it is defined, only depends on n variables. Thus the domains of each function are different and $\Psi(p, q)$ can only be viewed as dependent on n variables.

Wigner attempted to overcome these objections by using the quantum mechanics density matrix $\rho(q, q')$ in an attempt to reflect the usual situation that a quantum system is in a mixed state consisting of a superposition of orthogonal states $\psi_k(q)$ with a probability of $\rho_k \geq 0$ with the $\Sigma \, \rho_k = 1$. The density matrix for this case is defined to be

$$\rho(q, q') = \Sigma \, \rho_k \, \psi_k(q)\psi_k^\dagger(q') \qquad (E.6.21)$$

Using the density matrix the Wigner distribution now is

$$\Psi(p, q) = \int d^n Q \, \rho(q - Q, q + Q) \, \exp(-2ipQ/\hbar) \qquad (E.6.22)$$

The new form of the Wigner distribution is a function of 2n variables and $\Psi(p, q)$ is positive definite. However, the density matrix (eq. E.6.14) has all eigenvalues between 0 and 1 and a trace equal to one. These properties are not shared by every potential Boltzmann probability $f(p, q, t)$. Thus the representation is limited to 'special cases.'

Our form of the *quantum* Boltzmann probability distribution is $f_q(\mathbf{r}_1, \mathbf{p}_1, t, \theta)$ which we have shown overcomes the redundancy of variables in the Wigner quantum generalization of the Boltzmann distribution and gives a sensible result in the classical limit.[436]

---

[436] Our PseudoQuantum equivalent density has the same form as the Wigner density (eq. E.6.21).

E.6.8 PseudoQuantum Form of the BGK Approximation to the Boltzmann Equation

The BKG approximation to the Boltzmann equation is

$$[\mathbf{p}\cdot\nabla/m + \mathbf{F}\cdot\partial/\partial\mathbf{p}]f = -\partial f/\partial t + \upsilon(f_0 - f) \qquad (E.6.23)$$

where f_0 is the local Maxwell distribution $f_0 = f_0\,(\mathbf{r}, \mathbf{p})$ and υ is the molecular collision frequency. This model of the collision term due to Bhatnagar, Gross, and Crook[437] has been a much studied approximation.

Before introducing the PseudoQuantum form of the BKG approximation we use the local Maxwell-Boltzmann distribution to re-express the BKG approximation in the form

$$f_0 = n[m/(2\pi kT)]^{3/2}\exp[-m(p - p_0)^2/(2kT)] \qquad (E.6.24)$$

where n is the particle density (assumed constant at equilibrium), k is Boltzmann's constant, T is the temperature, p_0 is the average momentum, and m is the mass of a particle. Inserting f_0 in eq. E.6.23 and letting

$$f = f_0 g \qquad (E.6.25)$$

we obtain

$$[\mathbf{p}\cdot\nabla/m + \mathbf{F}\cdot\partial/\partial\mathbf{p} - (m/kT)\mathbf{F}\cdot(\mathbf{p} - \mathbf{p_0}) + \upsilon]g = -\partial g/\partial t + \upsilon \qquad (E.6.26)$$

The PseudoQuantized equivalent of the expanded BKG approximation (eq. E.6.26) is

$$[\mathbf{p_1}\cdot\mathbf{p_2}/m - \mathbf{x_2}\cdot\mathbf{F}(\mathbf{x_1}) + (m/kT)\mathbf{F}(\mathbf{x_1})\cdot(\mathbf{p_2} + i\mathbf{p_0}) + i\upsilon]g = -i\partial g/\partial t + i\upsilon \qquad (E.6.27)$$

with the Maxwell-Boltzmann distribution term acting as a 'driving force.'

Given a force F we can proceed to PseudoQuantize using operators that are similar to those of eqs. E.4.1 – E.4.7 but adapted to the force and the Maxwell-Boltzmann distribution 'driving force.' We will not consider BKG examples in this book although a harmonic driving force $\mathbf{F}(\mathbf{x_1}) = -m\omega^2\mathbf{x_1}$ is an interesting case to consider.

E.6.9 Relativistic Boltzmann Equation

The Boltzmann equation is non-relativistic and is appropriate in systems that are at rest or moving at non-relativistic velocities. If a system is traveling at relativistic velocities then the relativistic Boltzmann equation must be used. In this section we first generalize the Boltzmann equation to its special relativistic form by making all terms covariant. Then, when we 'go to' a rest frame, the relativistic equation becomes the non-relativistic Boltzmann equation.

---

[437] P. L. Bhatnagar, E. P. Gross, and M. Crook, Phys. Rev. **94**, 511 (1954).

E.6.9.1 Relativistic Generalization of the Boltzmann Equation

The relativistic form of the non-relativistic Boltzmann equation of section E.6.1 is

$$[p^\mu \nabla_\mu/m + F^\mu \partial/\partial p^\mu]f = (\partial f/\partial t)_{collRelativistic} \qquad (E.6.28)$$

where we use indices to transform vectors into 4-vectors: the momentum, derivative operators and the force become Lorentz 4-vectors. The collision term must now be in a relativistic form.

E.6.9.2 Pseudoquantized Relativistic Boltzmann Equation

The PseudoQuantum form of the Boltzmann equation is discussed earlier:

$$-i[\mathbf{p}_1 \cdot \mathbf{p}_2/m - \mathbf{x}_2 \cdot \mathbf{F}]f = -\partial f/\partial t + (\partial f/\partial t)_{coll} \qquad (E.6.1)$$
$$\mathbf{p}_2 = i\nabla \qquad (E.6.2)$$
$$\mathbf{x}_2 = -i\partial/\partial \mathbf{p}_1$$

We make it relativistic in a manner similar to the approach in subsection E.6.9.1. The result is the relativistic PseudoQuantum Boltzmann equation:

$$[p_1{}^\mu p_{2\mu} - mx_2{}^\mu F_\mu(x_1{}^\alpha)]f = im(\partial f/\partial t)_{collRelativistic} \qquad (E.6.29)$$

where the collision term is relativistic. This formalism, superficially, has two times. However when the wave functions are calculated only one time $x_1{}^0$ is relevant as the calculations in section 4 in the Vlasov approximation suggest.

E.6.10 Quantum and Classical Entropy

The von Neumann entropy for a system described by a density matrix ρ is defined as

$$S = -\,tr[\rho ln\rho] \qquad (E.6.30)$$

Using eigenvectors |n> the density matrix can be expressed as

$$\rho = \sum_i \eta_i |i><i| \qquad (E.6.31)$$

Then ρ can be expressed in the information theory Shannon formulation of entropy:

$$S = -\sum_i \eta_i ln\,\eta_i \qquad (E.6.32)$$

If we use the harmonic oscillator development of section 4 we can express the von Neumann entropy in a form which ranges from classical to quantum as a function of the angle θ. If we define the harmonic oscillator states

$$| n+, n-> = b_1^{\dagger n+}b_2^{\dagger n-}|0,0>$$

as in section 4 where

$$b_1^\dagger = Q_1\cos\theta - iP_2\sin\theta$$
$$b_2^\dagger = -Q_2\sin\theta - iP_1\cos\theta$$

(E.4.23)

then the density matrix is

$$\rho(\theta) = \sum_{n+,n-} |n+, n-><n+, n-| = \sum_{n+,n-} \eta_{n+,n-}\, b_1^\dagger(\theta)^{n+}b_2^\dagger(\theta)^{n-}|0,0><0,0|b_1(\theta)^{n+}b_2(\theta)^{n-}$$

(E.6.33)

In the quantum limit where $\theta = \pi/4$ we see

$$\rho(\pi/4) = \sum_{n+,n-} (\eta_{n+,n-}/2^{n+ + n-})(Q_1 - iP_2)^{n+}(Q_2 + iP_1)^{n-}|0,0><0,0|(Q_1 + iP_2)^{n+}(Q_2 - iP_1)^{n-}$$

(E.6.34)

yielding a quantum density matrix and thus a von Neumann quantum entropy.

In the classical limit where $\theta = 0$ we see

$$\rho(0) = \sum_{n+,n-} (\eta_{n+,n-}/2^{n+ + n-})Q_1^{n+}P_1^{n-}|0,0><0,0|Q_1^{n+}P_1^{n-} = \rho(Q_1, P_1)$$

(E.6.35)

yielding a purely classical function of Q_1 and P_1 as the density matrix. The von Neumann entropy's classical limit in this case is the classical phase space quantity

$$S = -\{\Sigma(\eta_{n+,n-}/2^{n+ + n-})Q_1^{2n+}P_1^{2n-}\}\ln\{\Sigma(\eta_{n+,n-}/2^{n+ + n-})Q_1^{2n+}P_1^{2n-}\}$$
$$= S(Q_1, P_1)$$

(E.6.36)

since Q_1 and P_1 commute.

E.7 PseudoQuantum Path Integral Formulation

The path integral formulation of quantum mechanics (and also of quantum field theory) plays an important role in the understanding of quantum physics. One of its major issues is the transition from a quantum mechanical framework to a classical mechanical framework. We will examine this issue from the point of view of a PseudoQuantum path integral formulation. Earlier we have seen that we can embody both quantum and classical mechanics phenomena within the PseudoQuantum framework and 'rotate' between quantum and classical mechanics solutions.

In order to establish a PseudoQuantum path integral formalism we must first generate a lagrangian from a PseudoQuantum Hamiltonian. In section E.5 we described the general formalism for deriving a PseudoQuantum Hamiltonian from a classical Hamiltonian. We now construct the PseudoQuantum lagrangian. Starting with the equations in section E.5:

$$x_2 = id/dp_1$$
$$p_2 = -id/dx_1$$

$$\hat{H}(x_1, p_1, x_2, p_2) = \partial H(x_1, p_1)/\partial p_1\, p_2 + \partial H(x_1, p_1)/\partial x_1\, x_2$$

we define the velocities[438]

$$x'_1 = \partial\hat{H}(x_1, p_1, x_2, p_2)/\partial p_1 = \partial^2 H(x_1, p_1)/\partial p_1^2\, p_2 + \partial^2 H(x_1, p_1)/\partial x_1 \partial p_1\, x_2 \qquad \text{(E.7.1)}$$

$$x'_2 = \partial\hat{H}(x_1, p_1, x_2, p_2)/\partial p_2 = \partial H(x_1, p_1)/\partial p_1|_{p_2 = p_1} \qquad \text{(E.7.2)}$$

The lagrangian L is constructed in the canonical way using Legendre transformations

$$L = p_1 x'_1 + p_2 x'_2 - \hat{H}(x_1, p_1, x_2, p_2) \qquad \text{(E.7.3)}$$

$$= p_1[\partial^2 H(x_1, p_1)/\partial p_1^2\, p_2 + \partial^2 H(x_1, p_1)/\partial x_1 \partial p_1\, x_2] - \partial H(x_1, p_1)/\partial x_1\, x_2$$

$$= \partial^2 H(x_1, p_1)/\partial p_1^2\, p_1 p_2 + \partial^2 H(x_1, p_1)/\partial x_1 \partial p_1\, p_1 x_2 - \partial H(x_1, p_1)/\partial x_1\, x_2$$
$$\text{(E.7.3a)}$$

where p_1 and p_2 are extracted from eqs. E.7.1 and E.7.2.

We now consider the example of a harmonic oscillator where

$$H = p^2/(2m) + \tfrac{1}{2} m\omega^2 x^2$$
$$\hat{H} = p_1 p_2/m + m\omega^2 x_1 x_2 \qquad \text{(E.4.8b)}$$

Substituting in eq. E.7.3 we find

$$L = \partial^2 H(x_1, p_1)/\partial p_1^2\, p_1 p_2 + \partial^2 H(x_1, p_1)/\partial x_1 \partial p_1\, p_1 x_2 - \partial H(x_1, p_1)/\partial x_1\, x_2$$
$$= p_1 p_2/m - m\omega^2 x_1 x_2$$
$$= m x'_1 x'_2 - m\omega^2 x_1 x_2 \qquad \text{(E.7.4)}$$

with p_1 and p_2 determined, and replaced, as functions of x'_1 and x'_2 by eqs. E.7.1 and E.7.2.

The Lagrange equations of motion determined for $i = 1, 2$ are:

---

[438] The velocity x'_2 is a defined quantity, which is defined in a manner consistent with the definition of x'_1.

$$d/dt \, (\partial L/\partial x'_i) - \partial L/\partial x_i = 0 \qquad (E.7.5)$$

$$mx''_2 + m\omega^2 x_2 = 0 \qquad (E.7.6)$$
$$mx''_1 + m\omega^2 x_1 = 0 \qquad (E.7.7)$$

E.7.1 Feynman Path Integral formulation

The propagator $K(x - y, t)$ for the Feynman path integral formulation has the form:

$$K(x - y, T) = A \lim_{n \to \infty} \underset{-\infty}{\overset{+\infty}{\int\!\!\int\!\!\int\!\!\int}} \dots \int dx_0 dx_1 \dots dx_n \exp[i/\hbar \int_t^{t+T} L(x, v, t_a) \, dt_a] \qquad (E.7.8)$$

where A is a constant, and the integral over the dx's ranges from $-\infty$ to ∞.

E.7.1.1 Conventional Formulation – Free Particle Case

In the one dimensional free particle case the path integral is a product of n infinitesimal paths of time interval ε:

$$K(x - y, T) = \prod_n G_\varepsilon \qquad (E.7.9)$$

where $T = n\delta$. Using ~ to denote proportionality up to a constant we find the fourier transform of an interval of path

$$G_\delta = \int dx \, e^{-ipx} \exp[-ix^2/(2\varepsilon)] \qquad (E.7.10)$$
$$\sim \exp[-p^2/(2\varepsilon)]$$

Then the product of the incremental factors that total to time T give

$$K(p, T) \sim \exp[-iTp^2/2] \qquad (E.7.11)$$

A fourier transformation yields the free particle propagator

$$K(x - y, T) \sim \int dp \, e^{-ip(x-y)} \exp[-iTp^2/2]$$
$$\sim \exp[-i(x-y)^2/T] \qquad (E.7.12)$$

where we normalize the propagator to unity

$$\int dy \, K(x - y, T) = 1 \qquad (E.7.13)$$

E.7.1.2 PeseudoQuantum Formulation – Free Particle Case

We will now develop the PseudoQuantum path integral formalism for the case of a free particle. The form of the path integral now is

$$K(x - y, T) = A \lim_{n \to \infty} \int\int\int\int_{-\infty}^{+\infty} \ldots \int dx_{10}dx_{11}\ldots dx_{1n} \, dx_{20}dx_{21}\ldots dx_{2n} \exp[i/\hbar \int_{t}^{t+T} L(x_1, x_2, x'_1, x'_2, v, t_a) \, dt_a]$$

$$(E.7.14)$$

where A is a constant, and all integrals over the dx's ranges from $-\infty$ to ∞. Note that we use two sets of coordinates and momenta.

Following a similar path to subsection E.7.1.1 we first we determine the fourier transform on the path integral for an infinitesimal time interval ε:

$$G_\varepsilon = \int\int dx_1 dx_2 \, \exp[-ip_1x_1 - ip_2x_2] \, \exp[-imx_1x_2/\varepsilon]$$
$$\sim \exp[-i\varepsilon p_1p_2/m] \tag{E.7.15}$$

Upon combining the intervals to a total time T we obtain the fourier transform of the total path integral

$$K(p, T) \sim \exp[-iTp_1p_2/m] \tag{E.7.16}$$

which yields the spatial path integral

$$K(x_1 - y_1, x_2 - y_2, T) \sim \int d \, p_1 dp_2 \, \exp[ip_1(x_1 - y_1) + ip_2(x_2 - y_2)] \, \exp[-iTp_1p_2/m]$$

$$\sim \exp[-im(x_1 - y_1)(x_2 - y_2)/T] \tag{E.7.17}$$

which we normalize to unity

$$\int dy_1 dy_2 \, K(x_1 - y_1, x_2 - y_2, T) = 1 \tag{E.7.18}$$

with the result

$$K(x_1 - y_1, x_2 - y_2, T) = (m/T)\exp[-im(x_1 - y_1)(x_2 - y_2)/T] \tag{E.7.19}$$

If we now use the path integral on a fre particle wave function we see that it displaces the wave function by the time T:

$$\Psi_0(x_1, x_2, t) = \exp[-ip_1x_1 - ip_2x_2 - iEt] \qquad \text{where } E = p_1p_2/m \tag{E.7.20}$$
$$\Psi(y_1, y_2, t + T) = \int\int dx_1 dx_2 \, K(x_1 - y_1, x_2 - y_2, T)\Psi_0(x_1, x_2, t) \tag{E.7.21}$$
$$= \exp[-ip_1y_1 - ip_2y_2 - iE(t + T)]$$
$$= \Psi_0(y_1, y_2, t + T) \tag{E.7.22}$$

E.7.1.3 Introducing the Rotation Between the Quantum and Classical Cases of the Path Integral

We begin by expressing the coordinates in terms of new 'rotated' coordinates:

$$x_1 = u_1 \cos\theta + u_2 \sin\theta \qquad\qquad (E.7.23)$$
$$x_2 = -u_1 \sin\theta + u_2 \cos\theta$$
$$p_1 = -p_{u1} \sin\theta + p_{u2} \cos\theta$$
$$p_2 = p_{u1} \cos\theta + p_{u2} \sin\theta$$

Then the quantities of interest are now expressed as

$$\Psi_0(x_1, x_2, t) = \exp[-ip_1x_1 - ip_2x_2 - iEt] \qquad\qquad \text{where} \quad E = p_1p_2/m \qquad (E.7.24)$$

$$= \Psi_0(u_1, u_2, t) = \exp\{-i[(p_{u2}u_2 - p_{u1}u_1)\sin(2\theta) + (p_{u2}u_1 + p_{u1}u_2)\cos(2\theta)] - iEt\}$$
$$(E.7.25)$$

where the energy E now having the form

$$E = (-p_{u1} \sin\theta + p_{u2} \cos\theta)(p_{u1} \cos\theta + p_{u2} \sin\theta)/m$$
$$= [(p_{u2}{}^2 - p_{u1}{}^2)\sin(2\theta)]/(2m) + [p_{u2}p_{u1} \cos(2\theta)]/m \qquad (E.7.26)$$

We will now examine the two special cases: $\theta = 0$ corresponding to classical mechanics and $\theta = \pi/4$ corresponding to quantum mechanics.

$\underline{\theta = 0}$

The wave equation in this case is

$$\Psi_0(u_1, u_2, t) = \exp\{-i(p_{u2}u_1 + p_{u1}u_2) - iEt\} \qquad\qquad (E.7.27)$$

with

$$E = p_{u2}p_{u1}/m$$

The wave function has a 'classical' form as we showed in eq. 47 in appendix C.

$\underline{\theta = \pi/4}$

The wave equation in this case is

$$\Psi_0(u_1, u_2, t) = \exp\{-i(p_{u2}u_2 - p_{u1}u_1) - iEt\} \qquad\qquad (E.7.28)$$

with

$$E = (p_{u2}{}^2 - p_{u1}{}^2)/(2m)$$

This wave function has a quantum form with a positive energy part and a negative energy part:

E.7.1.4 Free Path Integral with Rotation Between Classical and Quantum Mechanics

The incremental path integral factor, expressed in terms of u_1 and u_2, and then fourier transformed is:

$$G_\varepsilon(p_{u1}, p_{u2}) = \iint du_1 du_2 \exp\{-i[(p_{u2}u_2 - p_{u1}u_1)\sin(2\theta) + (p_{u2}u_1 + p_{u1}u_2)\cos(2\theta)]\} \cdot$$
$$\cdot \exp\{-im[(u_2{}^2 - u_1{}^2)\sin(2\theta)/2 + u_2 u_1 \cos(2\theta)]\}/\varepsilon\}$$

$$\text{(E.7.29)}$$

$$\sim \exp\{-i\varepsilon[(p_{u2}{}^2 - p_{u1}{}^2)\sin(2\theta)/2 + p_{u1}p_{u2}\cos(2\theta)]/m\}$$

yielding the cumulative product for the time interval T

$$K(p_{u1}, p_{u2}, T) \sim \exp[-iT[(p_{u2}{}^2 - p_{u1}{}^2)\sin(2\theta)/2 + p_{u1}p_{u2}\cos(2\theta)]/m] \qquad \text{(E.7.30)}$$

Upon fourier transforming to coordinate space we find

$$K(u_1 - v_1, u_2 - v_2, T) \sim \int dp_{u1}dp_{u2}\exp\{i[(p_{u2}w_2 - p_{u1}w_1)\sin(2\theta) + (p_{u2}w_1 + p_{u1}w_2)\cos(2\theta)]\}K(p_{u1}, p_{u2}, T)$$

$$\sim \exp[im[(w_2{}^2 - w_1{}^2)\sin(2\theta)/2 + w_2 w_1 \cos(2\theta)]/T] \qquad \text{(E.7.31)}$$

where

$$w_i = u_i - v_i$$

The special cases of interest are:

$\underline{\theta = 0}$

$$K(u_1 - v_1, u_2 - v_2, T) \sim \exp[imw_2 w_1]/T] \qquad \text{(E.7.32)}$$

This gives the *Classical* path integral without use of any limiting or approximation procedure such as one often finds in the literature.

$\underline{\theta = \pi/4}$

This case yields the familiar free particle path integral – but with a part for a positive energy particle and a part for a negative energy particle. The negative energy part can be removed easily yielding the conventional free particle quantum path integral.

$$K(u_1 - v_1, u_2 - v_2, T) \sim \exp[im(w_2{}^2 - w_1{}^2)/(2T)] \qquad \text{(E.7.33)}$$

E.7.2 General PseudoQuantum Formulation

The propagator $K(x - y, t)$ for the conventional Feynman path integral formulation has the form:

$$K(x - y, T) = A \lim_{n \to \infty} \underset{-\infty}{\overset{+\infty}{\iiint}} \dots \int dx_0 dx_1 \dots dx_n \exp[i/\hbar \overset{t+T}{\underset{t}{\int}} L(x, v, t_a)\, dt_a] \qquad \text{(E.7.34)}$$

where A is a constant, and the integral over the dx's ranges from $-\infty$ to ∞.

The PseudoQuantum form of the path integral formalism is based on the PseudoQuantum lagrangian

$$L(x_1, v_1, x_2, v_2) = \partial^2 H(x_1, p_1)/\partial p_1^2\, p_1 p_2 + \partial^2 H(x_1, p_1)/\partial x_1 \partial p_1\, p_1 x_2 - \partial H(x_1, p_1)/\partial x_1\, x_2$$

$$(E.7.35)$$

$$K(x_1 - y_1, x_2 - y_2, T) = A \lim_{n \to \infty} \underset{-\infty}{\overset{+\infty}{\iiiint}} \ldots \int dx_{10}dx_{11}\ldots dx_{1n}\, dx_{20}dx_{21}\ldots dx_{2n}\, \exp[i/\hbar \overset{t+T}{\underset{t}{\int}} L(x_1, v_1, x_2, v_2)\, dt_a]$$

$$(E.7.36)$$

From subsection E.7.1.2, where

$$G_\varepsilon = \iint dx_1 dx_2\, \exp[-ip_1 x_1 - ip_2 x_2]\, \exp[-imx_1 x_2/\varepsilon]$$
$$\sim \exp[-i\varepsilon p_1 p_2/m]$$

$$(E.7.37)$$

we can perform the x_2 integration in G_ε using

$$x'_2 \cong x_2(t + \varepsilon) - x_2(t)\, /\varepsilon$$

$$(E.7.38)$$

if

$$p_2 = mx'_2$$

Then

$$G_\varepsilon = \iint dx_1 dx_2\, \exp[-ip_1 x_1 - ip_2 x_2]\, \exp\{-i\varepsilon[\partial^2 H(x_1, p_1)/\partial p_1^2\, p_1 mx_2/\varepsilon + \partial^2 H(x_1, p_1)/\partial x_1 \partial p_1\, p_1 x_2 - \partial H(x_1, p_1)/\partial x_1\, x_2]]$$

$$= \int dx_1 \exp[-ip_1 x_1]\, \delta(p_2 - (m\partial^2 H(x_1, p_1)/\partial p_1^2\, p_1 + \varepsilon \partial^2 H(x_1, p_1)/\partial x_1 \partial p_1\, p_1 - \varepsilon \partial H(x_1, p_1)/\partial x_1))$$

$$= \int dx_1 \exp[-ip_1 x_1]\, \delta(p_2 - (m\partial^2 H(x_1, p_1)/\partial p_1^2\, p_1 + \varepsilon \partial^2 H(x_1, p_1)/\partial x_1 \partial p_1\, p_1 - \varepsilon \partial H(x_1, p_1)/\partial x_1)) \qquad (E.7.39)$$

E.7.2.1 PseudoQuantum Wave Functions and Schrödinger equation

In the presence of a potential, the path integral formulation leads us to transform it into a Schrödinger equation. The incremental time displacement form of the wave equation is

$$\Psi(y_{1k+1}, y_{2k+1}, t + \varepsilon) = \iint dx_{1k} dx_{2k}\, \exp\{(i\varepsilon/\hbar)[m(x_{1k+1} - x_{1k})(x_{2k+1} - y_{2k})/\varepsilon^2 -$$
$$- x_{2k+1}(\partial H(x_1, p_1)/\partial x_1)|_{x_1 = x_{1k+1}}]\}\Psi_0(x_{1k}, x_{2k}, t) \qquad (E.7.40)$$

$$= \iint dx_{1k} dx_{2k}\, \exp\{(i\varepsilon/\hbar)[m(x_{1k+1} - x_{1k})(x_{2k+1} - y_{2k})/\varepsilon^2 -$$
$$- x_{2k+1}(\partial V(x)/\partial x)|_{x = x_{1k+1}}]\}\Psi_0(x_{1k}, x_{2k}, t)$$

for a potential $V(x)$. In the case of the harmonic oscillator the PseudoQuantum potential term is

$$x_{2k+1}\partial V(x)/\partial x|_{x = x_{1k+1}} = m\omega^2 x_{1k+1} x_{2k+1} \qquad (E.7.41)$$

The PseudoQuantum Schrödinger equation that results is

$$i\hbar\partial\Psi(x_1, x_2, t)/\partial t = (-\hbar^2/m)\partial^2\Psi(x_1, x_2, t)/\partial x_1\partial x_2 + V(x_1, x_2)\Psi(x_1, x_2, t) \qquad (E.7.42)$$

E.7.2.2 Decomposition of PseudoQuantum Schrödinger Equation into Quantum and Classical Parts

The Schrödinger equation can be decomposed into a quantum and a classical part. Starting from eq. E.7.42:

$$i\hbar\partial\Psi(x_1, x_2, t)/\partial t = (-\hbar^2/m)\partial^2\Psi(x_1, x_2, t)/\partial x_1\partial x_2 + V(x_1, x_2)\Psi(x_1, x_2, t) \qquad (E.7.43)$$

we find

$$i\hbar\partial\Psi(u_1, u_2, t, \theta)/\partial t = (-\hbar^2/m)[\sin(2\theta)(\partial^2/\partial u_2^2 - \partial^2/\partial u_1^2)/2 + \cos(2\theta)\partial^2/\partial u_1\partial u_2]\Psi(u_1, u_2, t, \theta) +$$
$$+ V(u_1\cos\theta + u_2\sin\theta, -u_1\sin\theta + u_2\cos\theta)\Psi(u_1, u_2, t, \theta) \qquad (E.7.44)$$

using the relation to u_1 and u_2:

$$x_1 = u_1\cos\theta + u_2\sin\theta$$
$$x_2 = -u_1\sin\theta + u_2\cos\theta$$

Then

$$\Psi(u_1, u_2, t, \theta) = \exp\{-i[(p_{u2}u_2 - p_{u1}u_1)\sin(2\theta) + (p_{u2}u_1 + p_{u1}u_2)\cos(2\theta)] - iEt\} \qquad (E.7.45)$$

and the energy is

$$E = [(p_{u2}^2 - p_{u1}^2)\sin(2\theta)]/(2m) + [p_{u2}p_{u1}\cos(2\theta)]/m \qquad (E.7.46)$$

Again there are two cases of interest:

<u>The Quantum part $\theta = \pi/4$</u>

$$i\hbar\partial\Psi(u_1, u_2, t, \theta = \pi/4)/\partial t = (-\hbar^2/m)[(\partial^2/\partial u_2^2 - \partial^2/\partial u_1^2)/2]\Psi(u_1, u_2, t, \theta = \pi/4) +$$
$$+ V((u_1 + u_2)/\sqrt{2}, (-u_1 + u_2)/\sqrt{2})\Psi(u_1, u_2, t, \theta = \pi/4) \qquad (E.7.47)$$

In the quantum free particle case

$$\Psi(u_1, u_2, t, \theta = \pi/4) = \exp\{-i(p_{u2}u_2 - p_{u1}u_1) - iEt\} \qquad (E.7.48)$$

where

$$E = (p_{u2}^2 - p_{u1}^2)/(2m)$$

The Classical part $\theta = 0$

$$i\hbar\partial\Psi(u_1, u_2, t, \theta = 0)/\partial t = (-\hbar^2/m)\partial^2/\partial u_1\partial u_2 \, \Psi(u_1, u_2, t, \theta = 0) + V(u_1, u_2)\Psi(u_1, u_2, t, \theta = 0)$$

In the classical free particle case

$$\Psi(u_1, u_2, t, \theta = 0) = \exp\{-i(p_{u2}u_1 + p_{u1}u_2) - iEt\} \tag{E.7.49}$$

where

$$E = p_{u2}p_{u1}/m$$

E.7.3 Classical Part of Free Particle PseudoQuantum Feyman Path Propagator

The classical part of the free particle PseudoQuantum path integral is described as follows. First the time incremental factor is

$$G_\varepsilon (p_{u1}, p_{u2}) = \iint du_1 du_2 \exp\{-i[(p_{u2}u_2 - p_{u1}u_1)\sin(2\theta) + (p_{u2}u_1 + p_{u1}u_2)\cos(2\theta)]\} \cdot$$
$$\cdot \exp\{-im[(u_2^2 - u_1^2)\sin(2\theta)/2 + u_2u_1\cos(2\theta)]/\varepsilon\}$$

$$\sim \exp\{-i\varepsilon[(p_{u2}^2 - p_{u1}^2)\sin(2\theta)/2 + p_{u1}p_{u2}\cos(2\theta)]/m\} \tag{E.7.50}$$

The product of the incremental terms is

$$K(p_{u1}, p_{u2}, T) \sim \exp[-iT[(p_{u2}^2 - p_{u1}^2)\sin(2\theta)/2 + p_{u1}p_{u2}\cos(2\theta)]/m] \tag{E.7.51}$$

which yields the *classical* path integral

$$K(u_1 - v_1, u_2 - v_2, T) \sim \int dp_{u1}dp_{u2} \exp\{i[(p_{u2}w_2 - p_{u1}w_1)\sin(2\theta) + (p_{u2}w_1 + p_{u1}w_2)\cos(2\theta)]\}K(p_{u1}, p_{u2}, T)$$
$$\sim \exp[im[(w_2^2 - w_1^2)\sin(2\theta)/2 + w_2w_1\cos(2\theta)]/T] \tag{E.7.52}$$

where

$$w_i = u_i - v_i$$

For $\theta = 0$ the classical path integral steps are:

$$G_\varepsilon (p_{u1}, p_{u2}) \sim \exp\{-i\varepsilon p_{u1}p_{u2}/m\} \tag{E.7.53}$$

$$K(p_{u1}, p_{u2}, T) \sim \exp[-iTp_{u1}p_{u2}/m] \tag{E.7.54}$$

$$K(u_1 - v_1, u_2 - v_2, T) \sim \exp[im(u_1 - v_1)(u_2 - v_2)/T] \tag{E.7.55}$$

This gives us a classical path integral formulation that avoids approximation techniques which have hitherto been used. The above development can be rewritten in terms of the x and p variables.

E.7.4 Fokker-Planck Equation

The Feynman path integral formulation and equations can be transformed into similar forms by letting $i\hbar$ be changed to a positive constant. The Fokker-Planck equation gives the probability density of a particle velocity's time evolution under the impact of forces. This equation is also known as the Smoluchowski equation—named after its originator.

In its formulation a variable x_2 is introduced as the 'response' variable to the 'primary' variable x_1. The Fokker-Planck equation for the probability density is

$$p(x_1', t + \varepsilon) = (1/2\pi i) \int_{-\infty}^{\infty} dx_1 \int_{-i\infty}^{i\infty} dx_2 \, \exp\{\varepsilon[-x_2(x_1' - x_1)/\varepsilon + x_2 D_1(x_1, t) + x_2^2 D_2(x_1, t)]\} p(x_1, t) \quad (E.7.56)$$

The origin of x_2 in our formalism, and in the Fokker-Planck equation, are very different. The Fokker-Planck equation has the lagrangian:

$$L = \int dt \, [x_2 D_1(x_1, t) + x_2^2 D_2(x_1, t) - x_2 \, \partial x_1/\partial t]$$

A comparison of this formulation with the PseudoQuantum path integral formulation shows an apparently remarkable similarity. Thus one might view the Fokker-Planck equation as a precursor of the PseudoQuantum formulation. The papers appearing in the appendices of this book show that the origin of our formalism and the Fokker-Planck formalism are very different.

E.8 The Transition Between Classical and Quantum Chaos

Chaos has become an increasingly important field of activity. While classical chaos has been the more studied aspect of chaos there has been an increasingly larger interest in quantum chaos. In this section we wish to show that the PseudoQuantum formalism appears to be a useful means of relating classical chaos to quantum chaos for many systems. It makes it possible to trace the transition from a classical chaos situation to quantum chaos. It also offers the possibility to determine the quantum analogue of a classical chaos phenomenon.

Given a quantum system it is often difficult to determine whether it has a chaotic regime. Frequently extensive numerical analysis is needed for this determination.

In this section we show that the PseudoQuantum formalism enables us to determine a classical Hamiltonian from a quantum formalism where chaos is known to occur based on extensive numerical investigations.

A well-studied[439,440] Hamiltonian for a quantum theory, known to have chaotic regions, is

$$H = (p_x^2 + p_y^2/2 + x^2y^2 + \beta(x^4 + y^4)/4 \qquad (E.8.1)$$

Creating the equivalent PseudoQuantum hamiltonian we obtain

$$\hat{H} = p_{x1}p_{x2} + p_{y1}p_{y2} + y_1^2x_1x_2 + x_1^2y_1y_2 + \beta(x_1^3x_2 + y_1^3y_2) \qquad (E.8.2)$$

Introducing new variables we can develop a form of eq. E.8.2 that allows us to trace the transition from a quantum theory to a classical theory, which also should have chaotic regimes.

$$x_1 = u_{x1}\cos\theta + u_{x2}\sin\theta \qquad (E.8.3)$$
$$x_2 = -u_{x1}\sin\theta + u_{x2}\cos\theta$$

$$p_{x1} = p_{ux1}\cos\theta + p_{ux2}\sin\theta$$
$$p_{x2} = -p_{ux1}\sin\theta + p_{ux2}\cos\theta$$
$$y_1 = u_{y1}\cos\theta + u_{y2}\sin\theta$$
$$y_2 = -u_{y1}\sin\theta + u_{y2}\cos\theta$$

$$p_{y1} = p_{uy1}\cos\theta + p_{uy2}\sin\theta$$
$$p_{y2} = -p_{uy1}\sin\theta + p_{uy2}\cos\theta$$

Then we obtain the PseudoQuantum Hamiltonian

$$\hat{H}(\theta) = p_{x1}p_{x2} + p_{y1}p_{y2} + y_1^2x_1x_2 + x_1^2y_1y_2 + \beta(x_1^3x_2 + y_1^3y_2)$$

$$= (p_{ux2}^2 - p_{ux1}^2 + p_{uy2}^2 - p_{uy1}^2)\sin(2\theta)/2 + (p_{uy1}p_{uy2} + p_{ux1}p_{ux2})\cos(2\theta) + (u_{y1}\cos\theta + u_{y2}\sin\theta)^2[(u_{x2}^2 - u_{x1}^2)\sin(2\theta)/2 + u_{x1}u_{x2}\cos(2\theta)] + (u_{x1}\cos\theta + u_{x2}\sin\theta)^2[(u_{y2}^2 - u_{y1}^2)\sin(2\theta)/2 + u_{y1}u_{y2}\cos(2\theta)] + \beta\{(u_{x1}\cos\theta + u_{x2}\sin\theta)^2[(u_{x2}^2 - u_{x1}^2)\sin(2\theta)/2 + u_{x1}u_{x2}\cos(2\theta)] + (u_{y1}\cos\theta + u_{y2}\sin\theta)^2[(u_{y2}^2 - u_{y1}^2)\sin(2\theta)/2 + u_{y1}u_{y2}\cos(2\theta)]\} \qquad (E.8.4)$$

---

[439] The Hamiltonian model above has been studied by: Y. Y. Bai, G. Hose, K. Stefański, and H. S. Taylor, Phys. Rev. **A31**, 2821 (1985), R. L. Waterland, J.-M. Yuan, C. C. Martens, R. E. Gillilan, and W. P. Reinhardt, Phys. Rev. Lett. **61**, 2733 (1988, and other papers..

[440] Another much studied model—the 2-Dimensional stadium, Quantum billiard ball model has classical chaotic dynamics, and a quantum approximation that is chaotic: S. W. McDonald and A. N. Kaufman, Phys. Rev. **A37**, 3067 (1988); ________, Phys. Rev. Lett., **42**, 1189 (1979), and other papers.

The classical Hamiltonian that results from this analysis is

$$\hat{H}(0) = (p_{uy1}p_{uy2} + p_{ux1}p_{ux2}) + u_{y1}{}^2 u_{x1} u_{x2} + u_{x1}{}^2 u_{y1} u_{y2} + \beta\{u_{x1}{}^2 u_{x1} u_{x2} + u_{y1}{}^2 u_{y1} u_{y2}\} \qquad (E.8.5)$$

The Quantum Hamiltonian that emerges is

$$\hat{H}(\pi/4) = (p_{ux2}{}^2 - p_{ux1}{}^2 + p_{uy2}{}^2 - p_{uy1}{}^2)/2 + u_{y2}{}^2(u_{x2}{}^2 - u_{x1}{}^2)/2 + u_{x2}{}^2(u_{y2}{}^2 - u_{y1}{}^2)/2 + \beta\{u_{x2}{}^2(u_{x2}{}^2 - u_{x1}{}^2)/2 + u_{y2}{}^2(u_{y2}{}^2 - u_{y1}{}^2)/2\} \qquad (E.8.6)$$

While these Hamiltonians require numerical analysis to understand their chaotic features, they offer the possibilities of comparative studies of quantum and classical chaos.

The study of other models of a similar character would appear to be of importance in elucidating quantum chaos.

E.9 The Transition Between Classical & Quantum Entanglement Dynamics

Quantum Entanglement has become of great importance judging from its increasing number of papers. It offers the possibility of new forms of communication that might be of great value in interstellar communication should Mankind reach the stars.

In this section we will study a prototype example of quantum entanglement with a view towards investigating the transition from quantum entanglement through 'semi-classical' quantum entanglement to a 'classical' limit.

The example that we consider will use a combination of a positive energy single particle state entangled with a negative energy particle state to simulate the more commonly studied case of entangled spins. The particles, in a superposed state, are assumed to separate, and one particle will be measured thus determining the state of the other particle due to entanglement. We define the 'entangled' NOON-type state:

$$\Psi = (|n+ = 1, n- = 0\rangle + |\, n+ = 0, n- = 1\rangle)/\sqrt{2} \qquad (E.9.1)$$

where one pure state $|n+ = 1, n- = 0\rangle$ is a one particle state of positive energy and the other state $|\, n+ = 0, n- = 1\rangle$ has a one negative energy particle.

We define projection operators of the form:[441]

$$\rho(\theta) = ||n+(\theta),n-(\theta)\rangle\langle n+(\theta),n-(\theta)| \qquad (E.9.2)$$

---

[441] We will be using the harmonic oscillator formalism of chapter 4 although the conclusions will be more far reaching.

using an angle θ as we have done previously to specify the quantum-classical content of the projection.

More generally we will define harmonic oscillator states:

$$| n+, n-> = b_1^{\dagger n_+} b_2^{\dagger n_-} |0,0> \qquad (E.9.3)$$

with a density operator

$$\rho(\theta) = \sum_{n+,n-} |n+, n-><n+, n-| = \sum_{n+,n-} b_1^\dagger(\theta)^{n_+} b_2^\dagger(\theta)^{n_-} |0,0><0,0| b_1(\theta)^{n_+} b_2(\theta)^{n_-} \qquad (E.9.4)$$

We will consider the particular projection:

$$P(\theta) = |1(\theta),0><1(\theta),0| \qquad (E.9.5)$$

which we will apply to Ψ:

$$P(\theta)\Psi = |1(\theta), 0>/\sqrt{2} = \sin(2\theta)b_1^\dagger(\theta)|0, 0>/\sqrt{2} = \sin(2\theta)(Q_1\cos\theta - iP_2\sin\theta)|0, 0>/\sqrt{2} \qquad (E.9.6)$$

Using the familiar relations:

$$b_1 = Q_1\cos\theta + iP_2\sin\theta \qquad (E.4.6)$$
$$b_2 = -Q_2\sin\theta + iP_1\cos\theta$$
$$b_1^\dagger = Q_1\cos\theta - iP_2\sin\theta \qquad (E.4.7)$$
$$b_2^\dagger = -Q_2\sin\theta - iP_1\cos\theta$$

with commutation relations

$$[b_1, b_1^\dagger] = \sin(2\theta) \qquad (E.4.8a)$$
$$[b_2, b_2^\dagger] = -\sin(2\theta)$$
$$[b_1, b_2^\dagger] = [b_2, b_1^\dagger] = 0$$
$$[b_1, b_2] = [b_1^\dagger, b_2^\dagger] = 0$$

We find

$$\Psi' = P(\theta)\Psi \qquad (E.9.7)$$

has the following forms for $\theta = \pi/4$ and $\theta = 0$:

$\underline{\theta = \pi/4}$

$$\Psi' = (Q_1 - iP_2)|0, 0>/2 \qquad (E.9.8)$$

gives a quantum state.

$\underline{\theta = 0}$

$$\Psi' = 0 \qquad\qquad (E.9.9)$$

We note that we showed in section 4 that 'classical' states containing particles have infinite energy—This above result, $\Psi' = 0$, reflects the infinite energy of states containing one or more particles. The value of $\Psi' = 0$ leaves the positivity or negativity of the other particle undetermined since entanglement is a purely quantum phenomena.

$\underline{\text{Other values of } \theta}$

$$\Psi' = \sin(2\theta)(Q_1\cos\theta - iP_2\sin\theta)|0,\,0>/\sqrt{2} \qquad\qquad (E.9.10)$$
$$= \sin(2\theta)b_1^{\dagger}(\theta)|0,\,0>/\sqrt{2} = \sin(2\theta)|1(\theta),0>/\sqrt{2}$$

An intermediate result occurs but the other entangled separated particle state's energy is determined to be negative.

E.10 PseudoQuantum Transition Between Quantum and [442]Classical Dynamics

The preceding sections have shown that the use of the PseudoQuantum framework, which contains a purely quantum sector (albeit with both positive and negative energy parts that are separable), a purely classical sector, and an intermediate sector that is partly quantum and partly classical, enables us to

1. Relate the corresponding quantum and classical dynamics of a physical phenomenon.

2. Study the transition from quantum to classical behavior without recourse to approximations or limits such as $\hbar \to 0$.

3. Determine the classical equivalent of a quantum dynamical system.

4. Determine the quantum equivalent of a classical dynamical system.

These advantages appear to be fairly general in nature—as evidenced by the harmonic oscillator case studied in section 4—since the harmonic oscillator plays such a prominent role in many physical situations.

We conclude that the PseudoQuantum formalism, which is not only relevant for quantum-classical mechanics dynamics, but is also of importance in Quantum Field Theories as shown in our earlier sections and in our papers in the appendices. It has also recently been used

---

[442] Gutzwiller (1990) points out the use of harmonic oscillator wave functions in several studies of the quantum-classical connection.

in a new GraviStrong unified theory that relates quark confinement to deviations from Newtonian gravitation at galactic distance scales using a canonical PseudoQuantum formulation of a higher derivative Quantum Field Theory. (See Blaha (2016e).)

Appendix F. Complex Lorentz Group Details

F.1 Transformations Between Coordinate Systems

The measurement of time and space is simple in practice but raises weighty questions when their underlying basis is examined. We shall begin by measuring spatial distances with a ruler, and by measuring time with a clock. Earlier we determined that four dimensions: one space dimension and three space dimensions were required. We now define rectangular coordinate systems with x, y, and z axes as pictured in Fig. F.1 below. We then postulate:

Postulate F.1. Any observer can define a set of time and space coordinates called a coordinate system in which the observer is at rest. One can define a transformation that relates the coordinate systems of two observers traveling at a any constant velocity with respect to each other.

One can always relate the coordinates of two coordinate systems by having an observer in each coordinate system specify the coordinates of objects located at each spatial point, and then creating a map between the coordinates of corresponding spatial locations.

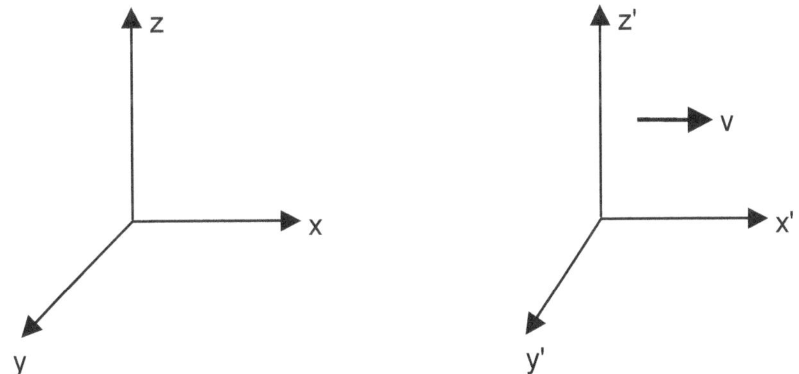

Figure F.1. Depiction of two coordinate systems. The "primed" coordinate system is moving with velocity **v** in the positive x direction with respect to the "unprimed" coordinate system. We choose parallel axes for convenience.

If space is flat the relation between the respective coordinates is linear. (One could reverse the logic of that statement by defining a flat space to be one in which the coordinates of a point in any coordinate system are linearly related to the coordinates of any other coordinate system moving at a constant velocity with respect to it.) Thus we can express the relation between the coordinates in the "unprimed" system to the coordinates in the "primed" system as a transformation between coordinate systems:

$$\mathbf{a'} = A\mathbf{a} + \mathbf{B}t + \mathbf{C}$$
$$t' = Dt + \mathbf{E} \cdot \mathbf{a}$$

(F.1)

where A is a 3×3 matrix, $\mathbf{B}$, $\mathbf{C}$ and $\mathbf{E}$ are 3-vectors, and D is a number (scalar value).

Having restricted the set of transformations between coordinate systems to the form of eq. F.1 we now assert postulates that restrict the form of the transformation to Lorentz transformations and transformations similar to Lorentz transformations.

Postulate F.2. The speed of light, c, is the same in all coordinate systems.

Postulate F.3. The invariant interval or distance $d\tau$ is defined by

$$d\tau^2 = g_{\mu\nu}dx^\mu dx^\nu$$

(F.2)

It is invariant under a change of coordinate systems. The 16 quantities $g_{\mu\nu}$ are known as the metric tensor.[443] The four quantities dx^μ are infinitesimal displacements in space and time.

If we expand eq. F.2 in rectangular coordinates it is equivalent to

$$d\tau^2 = g_{00}dx^0dx^0 + g_{11}dx^1dx^1 + g_{22}dx^2dx^2 + g_{33}dx^3dx^3$$

(F.3)

which equals

$$d\tau^2 = c^2dt^2 - dx^2 - dy^2 - dz^2$$

(F.4)

using the familiar form of the time and rectangular space coordinates.

F.2 The Lorentz Group

The metric tensor $g_{\mu\nu}$ for rectangular coordinates has the matrix form G = diag(1, −1, −1, −1):

---

[443] The repeated indices indicate a summation. In this case from 0 to 3 as shown in eq. 4-A.3

$$G = \begin{bmatrix} 1 & 0 & 0 & 0 \\ 0 & -1 & 0 & 0 \\ 0 & 0 & -1 & 0 \\ 0 & 0 & 0 & -1 \end{bmatrix} \qquad (F.5)$$

The invariant interval under a transformation between rectangular coordinate systems (with "primed" and "unprimed" coordinates) has the form of eq. F.4 for the unprimed coordinates and the same form for the primed coordinates:

$$d\tau^2 = c^2 dt'^2 - dx'^2 - dy'^2 - dz'^2 \qquad (F.6)$$

In matrix form we can define an "unprimed" coordinate column vector with

$$a = \begin{bmatrix} t \\ x \\ y \\ z \end{bmatrix} \qquad (F.7a)$$

and its corresponding "primed" coordinate with

$$a' = \begin{bmatrix} t' \\ x' \\ y' \\ z' \end{bmatrix} \qquad (F.7b)$$

If a and a' are the coordinates of the same point in the respective coordinate systems then, by postulates F.2 and F.3, they are related by a boost Lorentz transformation $\Lambda(\mathbf{v})$ with the form

$$a' = \Lambda(\mathbf{v})a \qquad (F.8)$$

(and possibly a spatial rotation matrix factor), where $\mathbf{v}$ is the relative velocity of the coordinate systems. The form of the transformation eq. F.8, which is called a Lorentz *boost*, is constrained by postulates F.2 and F.3 to be[444]

---

[444] We shall consider only the proper, orthochronous Lorentz group at this point. We assume that the primed and unprimed coordinate systems have parallel axes. So there is no rotation of axes embodied in eq. 15.9.

$$\Lambda(\mathbf{v}) = \begin{bmatrix} \gamma & -\gamma v_x & -\gamma v_y & -\gamma v_z \\ -\gamma v_x & 1 + (\gamma - 1)v_x^2/v^2 & (\gamma - 1)v_x v_y/v^2 & (\gamma - 1)v_x v_z/ \\ -\gamma v_y & (\gamma - 1)v_x v_y/v^2 & 1 + (\gamma - 1)v_y^2/v^2 & (\gamma - 1)v_y v_z/v^2 \\ -\gamma v_z & (\gamma - 1)v_x v_z/v^2 & (\gamma - 1)v_y v_z/v^2 & 1 + (\gamma - 1)v_z^2/v^2 \end{bmatrix} \qquad \text{(F.9)}$$

where $\gamma = (1 - v^2)^{-\frac{1}{2}}$, $\mathbf{v} = (v_x, v_y, v_z)$, $v = |\mathbf{v}|$ and we set $c = 1$ for convenience.[445] The set of all matrices of the form of $\Lambda(\mathbf{v})$, or $\Lambda(\mathbf{v})\mathcal{R}(\boldsymbol{\theta})$ or $\mathcal{R}(\boldsymbol{\theta})\Lambda(\mathbf{v})$ where $\mathcal{R}(\boldsymbol{\theta})$ is a spatial rotation with angle vector $\boldsymbol{\theta}$, for $v < c$ form a matrix representation of the Lorentz group. Elements, $\Lambda(\mathbf{v}, \boldsymbol{\theta})$, of the Lorentz group satisfy the defining relation of the Lorentz group:

$$\Lambda(\mathbf{v}, \boldsymbol{\theta})^T G \Lambda(\mathbf{v}, \boldsymbol{\theta}) = G \qquad \text{(F.10)}$$

where the superscript ^{T} specifies the transpose of the matrix.

The Lorentz group, with which we are familiar, relates the coordinates of an event in two coordinate systems that differ by a a spatial rotation, and a relative velocity whose magnitude is less than the speed of light. The inhomogenous Lorentz group includes coordinate displacements.[446]

The group elements of the homogeneous Lorentz group can be expressed in terms of the generators $\mathbf{K}$ of boosts to coordinate systems moving at a constant velocity $\mathbf{v}$ and the generators $\mathbf{J}$ of purely spatial rotations by

$$\Lambda(\mathbf{v}, \boldsymbol{\theta}) = \exp[i\omega\hat{\mathbf{u}}\cdot\mathbf{K} + i\boldsymbol{\theta}\cdot\mathbf{J}] \qquad \text{(F.11)}$$

where the vector $\boldsymbol{\theta}$ is a 3-vector specifying the rotation angles, and where $\mathbf{v} = \hat{\mathbf{u}}\,\tanh\omega$, $\hat{\mathbf{u}}\cdot\hat{\mathbf{u}} = 1$.

The boost transformation $\Lambda(\mathbf{v}) = \Lambda(\mathbf{v}, \mathbf{0})$ has the form

$$\Lambda(\mathbf{v}) = \exp[i\omega\hat{\mathbf{u}}\cdot\mathbf{K}] \qquad \text{(F.12)}$$

Its matrix form is eq. F.9. The matrix form can be expressed in terms of the unit normalized velocity vector $\mathbf{u} = (u_x, u_y, u_z)$ and ω as

$$\Lambda(\omega, \mathbf{u}) = \Lambda(\mathbf{v}) \qquad \text{(F.13)}$$

---

[445] One can set $c = 1$ by an appropriate choice of time and spatial distance scales. The demonstration that $\Lambda(\mathbf{v})$ has the form given by eq. 15.9 can be found in many textbooks.

[446] See Weinberg (1995) for a discussion of the inhomogeneous Lorentz group.

$$
= \begin{bmatrix}
\cosh(\omega) & -\sinh(\omega)u_x & -\sinh(\omega)u_y & -\sinh(\omega)u_z \\
-\sinh(\omega)u_x & 1 + (\cosh(\omega)-1)u_x^2 & (\cosh(\omega)-1)u_xu_y & (\cosh(\omega)-1)u_xu_z \\
-\sinh(\omega)u_y & (\cosh(\omega)-1)u_xu_y & 1 + (\cosh(\omega)-1)u_y^2 & (\cosh(\omega)-1)u_yu_z \\
-\sinh(\omega)u_z & (\cosh(\omega)-1)u_xu_z & (\cosh(\omega)-1)u_yu_z & 1 + (\cosh(\omega)-1)u_z^2
\end{bmatrix}
$$

where $\Lambda(\omega, \mathbf{u}) = \Lambda(\omega, \mathbf{u}, \theta = 0)$ in the previous notation. This definition of the general form of proper, orthochronous, Lorentz boost matrices $\Lambda(\omega, \mathbf{u})$ will be used in subsequent sections to define faster-than-light boost transformations.

The vector form of a Lorentz boost transformation is

$$
\mathbf{x'} = \mathbf{x} + (\gamma - 1)\mathbf{x \cdot v}\, \mathbf{v}/v^2 - \gamma \mathbf{v}t \tag{F.14}
$$
$$
t' = \gamma(t - \mathbf{v \cdot x}/c^2)
$$

where $\gamma = (1 - \beta^2)^{-\frac{1}{2}}$ with $\beta = v/c = v$ (since we set $c = 1$).

F.3 The Nature of $\Lambda(\omega, \mathbf{u})$ for Complex ω

We now turn to the case of complex ω wich includes superluminal (faster-than-light) Lorentz transformations as well as conventional Lorentz transformations. Since, for any complex value z

$$
\cosh^2(z) - \sinh^2(z) = 1 \tag{F.15}
$$

it follows that for any complex value of ω, $\Lambda(\omega, \mathbf{u})$ is a member of the Lorentz group, and/or of the complex Lorentz group[447] for complex ω:

$$
\Lambda(\omega, \mathbf{u})^{\mathrm{T}}G\Lambda(\omega, \mathbf{u}) = G \tag{F.16}
$$

For certain values of the imaginary part of ω the matrix $\Lambda(\omega, \mathbf{u})$ has a particularly simple form, similar to that of $\Lambda(\omega, \mathbf{u})$ for real ω, but which generates boosts to relative velocities greater than the speed of light. Among these values are:

$$
\omega = \omega_{\pm} = \omega \pm i\pi/2 \tag{F.17}
$$

Later we will see that these alternate choices $\omega_{\pm}$ correspond to specific choices of parity.

---

[447] The complex Lorentz group is defined as the group of all complex transformations that satisfy eq. 15.16.

F.4 Complex Lorentz Group

In the preceding section we saw that the parameter ω can be complex and the boost transformation will still satisfy the Lorentz condition eq. F.10. More generally we can consider complex homogeneous Lorentz transformations $\Lambda(\mathbf{v}, \boldsymbol{\theta})$ which can be represented by eq. F.11 with complex parameters ω, $\hat{\mathbf{u}}$, and $\boldsymbol{\theta}$ where $\hat{\mathbf{u}}$ and $\boldsymbol{\theta}$ are complex 3-vectors. $\boldsymbol{\theta}$ specifies a rotation angle.

In general $\Lambda(\mathbf{v}, \boldsymbol{\theta})$ is then a transformation between coordinate systems that have complex coordinates. One coordinate system is moving at a constant complex velocity with respect to the other. Coordinate systems do not necessarily have parallel spatial axes in general.

Within the complex Lorentz group, denoted L(C),[448] there are subsets of boosts that play important physical roles in the derivation of the form of The Standard Model. In particular we will see that certain classes of boosts generate faster-than-light transformations. These transformations can be further divided into subclasses of "left-handed" and "right-handed" transformations based on the quantum field theories to which they lead. Further within each subclass there are subclasses of transformations that naturally lead to Dirac-like free field equations that can be described as lepton-like and quark-like.

Thus these boosts are a key ingredient to understanding the form of The Standard Model.

F.5 Faster-than-Light Transformations

In this section we will substitute $\omega_\pm$ for ω in $\Lambda(\omega, \mathbf{u})$ and then show that we obtain two sets of possible transformations from sublight reference frames to faster-than-light reference frames. One set of transformations, where $\omega_L = \omega + i\pi/2$, will be called *left-handed superluminal boosts*. They eventually lead to the "left-handed" part of The Standard Model. We denote members of this set, $\Lambda_L(\omega, \mathbf{u})$, with the subscript "L" for left-handed.

The other set of boosts where $\omega_R = \omega - i\pi/2$ will be called *right-handed superluminal boosts*. They eventually lead to a right-handed, unphysical,[449] version of The Standard Model. We denote members of this set of boosts, $\Lambda_R(\omega, \mathbf{u})$, with the subscript "R" for right-handed.

Before considering faster-than-light boosts we note the relation between a real-valued ω in a *conventional* Lorentz boost $\Lambda(\omega, \mathbf{u})$, and the magnitude of the relative velocity v for v < 1, is

$$\mathbf{v} = \hat{\mathbf{u}}\,\tanh\omega \qquad \text{with} \qquad \hat{\mathbf{u}}\boldsymbol{\cdot}\hat{\mathbf{u}} = 1$$

$$\cosh(\omega) = \gamma = (1 - v^2)^{-\frac{1}{2}}$$
$$\sinh(\omega) = v\gamma = \beta\gamma \tag{F.18}$$

---

[448] Streater (2000) points out that the complex Lorentz group is essential to the proof of the CPT theorem.
[449] Currently the case. If a right-handed counterpart to the current Standard Model surfaces at higher energies then the features emerging from right-handed superluminal boosts then become physically important.

where $\beta = v = |\mathbf{v}|$.

F.6 Left-Handed Superluminal Transformations

Left-handed (proper orthochronous) superluminal boost transformations $\Lambda_L(\mathbf{v})$ have the same form as eq. F.9 for ordinary (proper orthochronous) Lorentz boost transformations. However the magnitude of the relative velocity $\mathbf{v}$ is greater than the speed of light. Thus $\gamma = (1 - v^2)^{-\frac{1}{2}}$ is pure imaginary and $\Lambda_L(\mathbf{v})$ is complex.

$$\Lambda_L(\mathbf{v}) = \begin{bmatrix} \gamma & -\gamma v_x & -\gamma v_y & -\gamma v_z \\ -\gamma v_x & 1 + (\gamma - 1)v_x^2/v^2 & (\gamma - 1)v_x v_y/v^2 & (\gamma - 1)v_x v_z/v^2 \\ -\gamma v_y & (\gamma - 1)v_x v_y/v^2 & 1 + (\gamma - 1)v_y^2/v^2 & (\gamma - 1)v_y v_z/v^2 \\ -\gamma v_z & (\gamma - 1)v_x v_z/v^2 & (\gamma - 1)v_y v_z/v^2 & 1 + (\gamma - 1)v_z^2/v^2 \end{bmatrix} \quad (F.19)$$

This transformation raises several issues – the most prominent of which is the interpretation of the imaginary coordinates generated by the transformation. Imaginary coordinates would appear at first glance to be unphysical. However we view the measurement of these quantities operationally: an observer measures distances with "rulers", and time with clocks, which both give real numeric values. Thus an observer *in any coordinate system* will always measure real numbers for time and space distances. However an observer *in another coordinate system* that is related to the first coordinate system by a superluminal transformation will view the coordinates in the first system as complex as eq. F.19 indicates.

The reconciliation of these points of view requires the introduction of a new transformation, called a Reality group transformation, in addition to a superluminal Lorentz transformation for the case of faster than light transformations. Reality group transformations maps the complex coordinates generated by a Lorentz transformation to the real coordinates seen by the observer[450] in the "faster than light" reference frame. We describe the Reality group transformations in detail earlier. We show how they imply the Reality group is $SU(3)\otimes SU(2)\otimes U(1)\otimes SU(2)\otimes U(1)$.

F.6.1 Cosh-Sinh Representation of Left-Handed Superluminal Boosts

We will now develop the representation of left-handed superluminal boost transformations in terms of $\cosh(\omega)$ and $\sinh(\omega)$ for later use in our discussion of tachyons. We find that we must use a complex $\omega_L \equiv \omega + i\pi/2$ to properly describe left-handed superluminal

---

[450] The linearity of the superluminal transformation makes this secondary transformation physically possible.

boosts. The relation between ω_L and v is different from eq. F.18 for the case of left-handed superluminal boosts:

$$\cosh(\omega_L) = i \sinh(\omega) = -\gamma = i\gamma_s$$
$$\sinh(\omega_L) = i \cosh(\omega) = -\beta\gamma = i\beta\gamma_s \tag{F.20}$$

where $\beta = v > 1$, $\boldsymbol{\omega \geq 0}$, and

$$\gamma_s = (\beta^2 - 1)^{-\frac{1}{2}} \tag{F.21}$$

Eq. F.20 implies

$$\sinh(\omega) = \gamma_s \tag{F.22}$$
$$\cosh(\omega) = \beta\gamma_s$$

Upon substituting $\boldsymbol{\omega_L}$ for ω in eq. F.13 we obtain another form for a left-handed superluminal transformation (equivalent to that of eq. F.19):

$$\Lambda_L(\omega, \mathbf{u}) = \Lambda(\omega + i\pi/2, \mathbf{u})$$

$$= \begin{bmatrix} \cosh(\omega_L) & -\sinh(\omega_L)u_x & -\sinh(\omega_L)u_y & -\sinh(\omega_L)u_z \\ -\sinh(\omega_L)u_x & 1+(\cosh(\omega_L)-1)u_x^2 & (\cosh(\omega_L)-1)u_xu_y & (\cosh(\omega_L)-1)u_xu \\ -\sinh(\omega_L)u_y & (\cosh(\omega_L)-1)u_xu_y & 1+(\cosh(\omega_L)-1)u_y^2 & (\cosh(\omega_L)-1)u_yu_z \\ -\sinh(\omega_L)u_z & (\cosh(\omega_L)-1)u_xu_z & (\cosh(\omega_L)-1)u_yu_z & 1+(\cosh(\omega_L)-1)u_z^2 \end{bmatrix}$$

$$= \begin{bmatrix} i\gamma_s & -i\beta\gamma_su_x & -i\beta\gamma_su_y & -i\beta\gamma_su_z \\ -i\beta\gamma_su_x & 1 + (i\gamma_s - 1)u_x^2 & (i\gamma_s - 1)u_xu_y & (i\gamma_s - 1)u_xu_z \\ -i\beta\gamma_su_y & (i\gamma_s - 1)u_xu_y & 1 + (i\gamma_s - 1)u_y^2 & (i\gamma_s - 1)u_yu_z \\ -i\beta\gamma_su_z & (i\gamma_s - 1)u_xu_z & (i\gamma_s - 1)u_yu_z & 1 + (i\gamma_s - 1)u_z^2 \end{bmatrix} = \Lambda_L(\mathbf{v}) \tag{F.23}$$

A simple case that illustrates a left-handed superluminal boost is to assume the relative velocity is in the x direction. Then eq. F.23 becomes

$$\Lambda_L(\omega, \mathbf{u} = (1,0,0)) = \begin{bmatrix} i\gamma_s & -i\beta\gamma_s & 0 & 0 \\ -i\beta\gamma_s & i\gamma_s & 0 & 0 \\ 0 & 0 & 1 & 0 \\ 0 & 0 & 0 & 1 \end{bmatrix} \tag{F.24}$$

implementing the coordinate transformation:

$$X' = \Lambda_L(\omega, \mathbf{u} = (1,0,0))X$$

or

$$t' = i\gamma_s(t - \beta x)$$
$$x' = i\gamma_s(x - \beta t) \qquad\qquad (F.25)$$
$$y' = y$$
$$z' = z$$

The addition rule for the x-component of velocity can be computed for infinitesimal displacements in space and time:

$$v_x' = \Delta x' /\Delta t' = (\Delta x\, \gamma_s - \Delta t\, \beta\gamma_s)/(\Delta t\, \gamma_s - \Delta x\, \beta\gamma_s)$$

$$= (v_x - \beta)/(1 - \beta v_x) \qquad\qquad (F.26)$$

in the limit $\Delta t \to 0$ where the x component of a particle's velocity in the unprimed frame is $v_x = \Delta x/\Delta t$. $\Delta t'$ is determined by

$$\Delta t' = i\Delta t\, \gamma_s(1 - \beta v_x) \qquad\qquad (F.27)$$

Note the velocity of light is the same in the primed and unprimed reference frames. (If $v_x = 1$ then $v_x' = 1$.) *Thus left-handed superluminal transformations preserve the constancy of the speed of light in all reference frames.* (Postulate F.2)

Further note that increasing the value of ω in $\Lambda_L(\omega, \mathbf{u})$ corresponds to decreasing the magnitude of the relative velocity v since

$$v = \cotanh(\omega) \qquad\qquad (F.28)$$

by eq. F.22. Thus when $\omega = 0$ then $v = \infty$, and when $\omega = \infty$ then $v = 1$. This is the reverse of the sublight case: by eq. F.18 $v = \tanh(\omega)$. Thus when $\omega = 0$ then $v = 0$, and when $\omega = \infty$ then $v = \infty$.

F.6.2 General Velocity Transformation Law – Left-Handed Superluminal Boosts

The general velocity transformation law for a particle moving with velocity **v** in the unprimed reference frame and velocity **v'** in the primed reference frame is

$$\mathbf{v'} = [\mathbf{v} + (\gamma - 1)\mathbf{w}\cdot\mathbf{v}\, \mathbf{w}/w^2 - \gamma\mathbf{w}]/[\, \gamma(1 - \mathbf{w}\cdot\mathbf{v})] \qquad\qquad (F.29)$$

where **w** is the relative velocity of the primed reference frame with respect to the unprimed reference frame, and $\gamma = (1 - w^2)^{-\frac{1}{2}}$. Eq. F.29 is obtained by calculating the derivative dx'/dt' using eqs. F.14. The relative velocity **w** can be greater or less than the speed of light. Eq. F.29 implies

$$v'^2 = 1 + (v^2 - 1)(1 - w^2)/(1 - \mathbf{w} \cdot \mathbf{v})^2 \qquad (F.30)$$

The relation of the velocities (eq. F.30) will be used to determine the multiplication rules for subluminal and superluminal Lorentz transformations (next subsection).

F.6.3 Left-Handed Transformations Multiplication Rules

In this subsection we will determine the multiplication rules of left-handed subluminal and superluminal Lorentz boosts. To do this we will consider three reference frames: an "unprimed" frame, a "primed" frame moving with velocity $\mathbf{w}$ with respect to the unprimed frame, and a "double-primed" frame moving with velocity $\mathbf{v}$ with respect to the unprimed frame and velocity $\mathbf{v'}$ with respect to the primed frame. See Fig.F.2.

The velocity $\mathbf{v'}$ is related to $\mathbf{v}$ by eqs. F.29 and F.30. Think of the double-primed coordinate system as attached to a particle. In addition note that the transformation law from the unprimed to the double-primed reference frame can be viewed as the product of consecutive transformations (boosts) from the unprimed to the primed reference frames and then from the primed to the double-primed reference frames.

Thus the transformations have the general form:

$$\Lambda_?(\mathbf{v}) = \Lambda_?(\mathbf{v'})\Lambda_?(\mathbf{w}) \qquad (F.31)$$

where the "?" subscripts indicate subluminal or superluminal transformations (boosts) depending on the magnitude of the relative velocity in the transformation's parentheses.

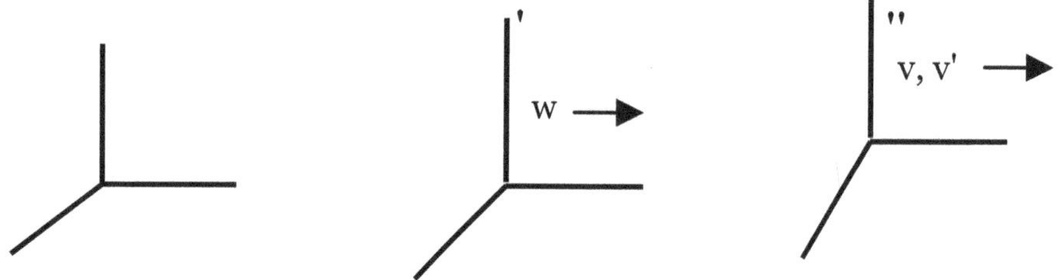

Figure F.2. Three reference frames used to establish transformation multiplication rules.

We now consider the various cases using eq. F.30:

1) If $w > 1$ and $v' > 1$

then eq. F.30 implies $v < 1$ and thus the left $\Lambda_?(\mathbf{v})$ is a subluminal transformation

$$\Lambda(\mathbf{v}) = \Lambda_L(\mathbf{v'})\Lambda_L(\mathbf{w}) \tag{F.32}$$

2) If $w > 1$, $v' < 1$

   then eq. F.30 implies $v > 1$ and thus the left $\Lambda_?(\mathbf{v})$ is a superluminal transformation

$$\Lambda_L(\mathbf{v}) = \Lambda(\mathbf{v'})\Lambda_L(\mathbf{w}) \tag{F.33}$$

3) If $w < 1$, $v' > 1$

   then eq. F.30 implies $v > 1$ and thus the left $\Lambda_?(\mathbf{v})$ is a superluminal transformation

$$\Lambda_L(\mathbf{v}) = \Lambda_L(\mathbf{v'})\Lambda(\mathbf{w}) \tag{F.34}$$

4) If $w < 1$, $v' < 1$

   then eq. F.30 implies $v < 1$ and thus the left $\Lambda_?(\mathbf{v})$ is a Lorentz transformation

$$\Lambda(\mathbf{v}) = \Lambda(\mathbf{v'})\Lambda(\mathbf{w}) \tag{F.35}$$

where, in each above case, the transformation on the left side of the equation may be a boost or a combination of a boost and a spatial rotation. Thus we have obtained the multiplication rules for left-handed subluminal and superluminal Lorentz transformations.

F.6.4 Inverse of Left-Handed Transformations

The inverse of a Lorentz boost is

$$\Lambda^{-1}(\omega, \hat{\mathbf{u}}) = \exp[-i\omega\hat{\mathbf{u}}\cdot\mathbf{K}] \tag{F.36}$$

where $\omega \geq 0$. Thus the inverse is generated by letting $\omega \rightarrow -\omega$. Note that since $v = \tanh\omega$, the effect of $\omega \rightarrow -\omega$ is to let $v \rightarrow -v$. In the case of superluminal left-handed boosts, since

$$\Lambda_L(\omega, \mathbf{u}) = \Lambda(\omega + i\pi/2, \mathbf{u}) = \exp[i(\omega + i\pi/2)\hat{\mathbf{u}}\cdot\mathbf{K}] \tag{F.37}$$

we find the inverse is

$$\Lambda_L^{-1}(\omega, \mathbf{u}) = \Lambda(-(\omega + i\pi/2), \mathbf{u}) = \exp[-i(\omega + i\pi/2)\hat{\mathbf{u}}\cdot\mathbf{K}] \tag{F.38}$$

where $\omega \geq 0$. Since $\Lambda_L^{-1}(\omega, \mathbf{u})$ is not the hermitean conjugate of $\Lambda_L(\omega, \mathbf{u})$, superluminal boosts are not unitary. However unitarity is not required since complex Lorentz group elements satisfy the defining relation of the Lorentz group (eq. F.10).

F.7 Right-Handed Superluminal Transformations

When we transform between reference frames using a *right-handed*[451] superluminal boost the relation between ω and v is different. The variable ω becomes $\omega_R = \omega - i\pi/2$ and

$$\cosh(\omega_R) = -i\sinh(\omega) = \gamma = -i\gamma_s \tag{F.39}$$
$$\sinh(\omega_R) = -i\cosh(\omega) = \beta\gamma = -i\beta\gamma_s \tag{F.40}$$

where $\beta = v > 1$ and $\boldsymbol{\omega \geq 0}$. Note that $\omega = \mathrm{Re}\,\omega_R$

$$\sinh(\omega) = \gamma_s \tag{F.41}$$
$$\cosh(\omega) = \beta\gamma_s \tag{F.42}$$

with

$$\gamma_s = (\beta^2 - 1)^{-\frac{1}{2}} \tag{F.43}$$

Upon substituting ω_R for ω in eq. F.13 we obtain the form of the right-handed superluminal boost:[452]

$$\Lambda_R(\omega, \mathbf{u}) = \Lambda(\omega - i\pi/2, \mathbf{u}) \tag{F.44}$$

$$= \begin{bmatrix} -i\gamma_s & i\beta\gamma_s u_x & i\beta\gamma_s u_y & i\beta\gamma_s u_z \\ i\beta\gamma_s u_x & 1 + (-i\gamma_s - 1)u_x^2 & (-i\gamma_s - 1)u_x u_y & (-i\gamma_s - 1)u_x u_z \\ i\beta\gamma_s u_y & (-i\gamma_s - 1)u_x u_y & 1 + (-i\gamma_s - 1)u_y^2 & (-i\gamma_s - 1)u_y u_z \\ i\beta\gamma_s u_z & (-i\gamma_s - 1)u_x u_z & (-i\gamma_s - 1)u_y u_z & 1 + (-i\gamma_s - 1)u_z^2 \end{bmatrix}$$

A simple case that illustrates right-handed superluminal transformations is to assume a relative velocity in the x direction. Then eq. F.44 becomes

$$\Lambda_R(\omega, \mathbf{u} = (1,0,0)) = \begin{bmatrix} -i\gamma_s & i\beta\gamma_s & 0 & 0 \\ i\beta\gamma_s & -i\gamma_s & 0 & 0 \\ 0 & 0 & 1 & 0 \\ 0 & 0 & 0 & 1 \end{bmatrix} \tag{F.45}$$

implementing the coordinate transformation:

---

[451] We call these transformations right-handed because they lead eventually to an alternate right-handed Standard Model This alternate right-handed Standard Model does not appear to correspond to current experimental reality.

[452] We note the singularities at $\beta = \pm 1$ or $\omega = \pm\infty$. **As a result we have a branch cut in the complex ω-plane consisting of the entire real ω axis. Therefore three left-handed boosts are not equivalent to a right-handed boost but rather appear on a different Riemann sheet.**

$$X' = \Lambda_R(\omega, \mathbf{u})X$$

or

$$t' = -i\gamma_s(t - \beta x)$$
$$x' = -i\gamma_s(x - \beta t)$$
$$y' = y$$
$$z' = z$$

(F.46)

Comparing eq. F.45 with eq. F.24 for a left-handed superluminal boost we see that

$$PT\Lambda_L(\omega, \mathbf{u} = (1,0,0)) = \begin{bmatrix} -i\gamma_s & i\beta\gamma_s & 0 & 0 \\ i\beta\gamma_s & -i\gamma_s & 0 & 0 \\ 0 & 0 & -1 & 0 \\ 0 & 0 & 0 & -1 \end{bmatrix}$$

where P is the parity operator and T is the time reversal operator. If we now apply a spatial rotation $\mathcal{R}$ of π radians around the x axis then we obtain

$$\mathcal{R}PT\Lambda_L(\omega, \mathbf{u} = (1,0,0))\mathcal{R}^{-1} = \begin{bmatrix} -i\gamma_s & i\beta\gamma_s & 0 & 0 \\ i\beta\gamma_s & -i\gamma_s & 0 & 0 \\ 0 & 0 & 1 & 0 \\ 0 & 0 & 0 & 1 \end{bmatrix}$$

(F.47)

$$= \Lambda_R(\omega, \mathbf{u} = (1,0,0))$$

Since P and T commute with spatial rotations we find

$$\Lambda_R(\omega, \mathbf{u} = (1,0,0)) = PT\mathcal{R}\Lambda_L(\omega, \mathbf{u} = (1,0,0))\mathcal{R}^{-1}$$

(F.48)

or, more generally, performing additional spatial rotations:

$$\Lambda_R(\omega, \mathbf{u}) = PT\mathcal{R}_u\mathcal{R}\mathcal{R}_w\Lambda_L(\omega, \mathbf{w})\mathcal{R}_w^{-1}\mathcal{R}^{-1}\mathcal{R}_u^{-1}$$

(F.49)

or,

$$\Lambda_R(\omega, \mathbf{u}) = PT\mathcal{R}_{tot}\Lambda_L(\omega, \mathbf{w})\mathcal{R}_{tot}^{-1}$$

(F.50)

where $\mathbf{u}$ and $\mathbf{w}$ are unit vectors, and $\mathcal{R}_{tot} = \mathcal{R}_u\mathcal{R}\mathcal{R}_w$. Alternately,

$$\Lambda_L(\omega, \mathbf{w}) = PT\mathcal{R}_{tot}^{-1}\Lambda_R(\omega, \mathbf{u})\mathcal{R}_{tot}$$

(F.51)

or

$$\Lambda_L(\omega, \mathbf{w}) = PT\Lambda_R(\omega, \mathbf{u}')$$

(F.52)

for some unit vector $\mathbf{u}'$.

Thus we have shown that PT can be used to relate left-handed and right-handed boosts in a one-to-one fashion. *The appearance of the* parity *operator P takes on great significance when we derive features of the Standard Model. The appearance of left-handed form of The Standard Model stems directly from the implicit parity dependence of the left-handed sector of the superluminal part of the complex Lorentz group.*

For a right-handed boost the addition rule for the x-component of velocity can be computed for infinitesimal displacements in space and time:

$$v_x' = \Delta x' / \Delta t' = (\Delta x\, \gamma_s - \Delta t\, \beta\gamma_s)/(\Delta t\, \gamma_s - \Delta x\, \beta\gamma_s)$$
$$= (v_x - \beta)/(1 - \beta v_x) \tag{F.53}$$

in the limit $\Delta t \to 0$ where the x component of a particle's velocity in the unprimed frame is $v_x = \Delta x/\Delta t$. Note if $v_x = 1$ then $v_x' = 1$. *Thus right-handed superluminal transformations also preserve the constancy of the speed of light in all reference frames.*

F.8 Inhomogeneous Left-Handed Lorentz Group Transformations

The *Left-Handed transformations of the complex Lorentz group* consist of the elements of the real Lorentz group plus left-handed superluminal boost transformations, and combinations of boosts and spatial rotations. Thus the homogeneous left-handed superluminal transformations have the general form:

$$\Lambda_L(\mathbf{v}, \boldsymbol{\theta}) = \exp[i\omega_L \hat{\mathbf{u}}\cdot\mathbf{K} + i\boldsymbol{\theta}\cdot\mathbf{J}] \tag{F.54}$$

where $\omega_L' = \omega + i\pi/2$, $\boldsymbol{\theta}$ is the angular vector, and $\mathbf{J}$ is the angular momentum operator vector. Inhomogeneous left-handed superluminal transformations, which include displacements, can be expressed as

$$\Lambda_L(\mathbf{v}, \boldsymbol{\theta}, \mathbf{d}) = \exp[i\omega_L \hat{\mathbf{u}}\cdot\mathbf{K} + i\boldsymbol{\theta}\cdot\mathbf{J} - i\mathbf{d}\cdot\mathbf{P}] \tag{F.55}$$

where $\mathbf{P}$ is the momentum operator vector and $\mathbf{d}$ is a displacement vector.

We note

$$\det \Lambda_L(\omega, \mathbf{u}) = \pm 1 \tag{F.56}$$

The ordinary Lorentz group is divided into four disjoint subgroups that are often denoted:

$$
\begin{aligned}
L_+^\uparrow: &\quad \det \Lambda(\omega, \mathbf{u}) = +1; \;\; \mathrm{sgn}\, \Lambda(\omega, \mathbf{u})^0{}_0 = +1 \\
L_-^\uparrow: &\quad \det \Lambda(\omega, \mathbf{u}) = -1; \;\; \mathrm{sgn}\, \Lambda(\omega, \mathbf{u})^0{}_0 = +1 \\
L_+^\downarrow: &\quad \det \Lambda(\omega, \mathbf{u}) = +1; \;\; \mathrm{sgn}\, \Lambda(\omega, \mathbf{u})^0{}_0 = -1 \\
L_-^\downarrow: &\quad \det \Lambda(\omega, \mathbf{u}) = -1; \;\; \mathrm{sgn}\, \Lambda(\omega, \mathbf{u})^0{}_0 = -1
\end{aligned}
\tag{F.57}
$$

where sgn $\Lambda(\omega, \mathbf{u})^0{}_0$ is the sign of the 00 component of the $\Lambda(\omega, \mathbf{u})$ matrix. The various subgroups are related by the discrete transformations of parity P and time reversal T:

$$L^{\uparrow}_{+} \xrightarrow{P} L^{\uparrow}_{-}$$

$$L^{\uparrow}_{+} \xrightarrow{PT} L^{\downarrow}_{+}$$

$$L^{\uparrow}_{+} \xrightarrow{T} L^{\downarrow}_{-}$$

The left-handed superluminal transformations are disjoint in a somewhat different way. By eq. F.56 the determinants are ± 1. However the 0-0 matrix element of eq. F.16 gives

$$\Lambda_L{}^0{}_0{}^2 - \Sigma_i\, (\Lambda_L{}^i{}_0)^2 = 1 \tag{F.58}$$

The representation of superluminal boosts shows that each factor in eq. F.58 is imaginary. Thus eq. F.58 implies

$$\Sigma_i\, |\Lambda_L{}^i{}_0|^2 \geq 1 \tag{F.59}$$

$$|\Lambda_L{}^0{}_0| \geq 0 \qquad (\text{not } \geq 1) \tag{F.60}$$

where $||$ indicates absolute value since the quantities in eq. F.58 are squares – not in absolute value. Thus the magnitude of $\Lambda_L{}^0{}_0$ does not have a gap. Therefore left-handed superluminal transformations can be divided into two categories:

$$_LL_{+}: \quad \det \Lambda_L(\omega, \mathbf{u}) = +1 \tag{F.61}$$
$$_LL_{-}: \quad \det \Lambda_L(\omega, \mathbf{u}) = -1$$

as one expects for complex Lorentz group transformations.[453]

Earlier we saw that under a PT transformation a left-handed superluminal transformation becomes a right handed superluminal transformation. Again, as in the left-handed case, the various disjoint pieces are related by the discrete transformations of parity P and time reversal T:

$$_LL_{+} \xrightarrow{P} {}_LL_{-}$$

$$_LL_{+} \xrightarrow{T} {}_LL_{-} \tag{F.62}$$

---

[453] Streater (2000) p. 13.

F.9 Inhomogeneous Right-Handed Extended Lorentz Group

The inhomogeneous right-handed part of the complex Lorentz group[454] consists of the real Lorentz group plus right-handed superluminal transformations plus rotations and displacements that have the form:

$$\Lambda_R(\mathbf{v}, \boldsymbol{\theta}, \mathbf{d}) = \exp[i\omega_R \hat{\mathbf{u}}\cdot\mathbf{K} + i\boldsymbol{\theta}\cdot\mathbf{J} - i\mathbf{d}\cdot\mathbf{P}] \tag{F.63}$$

in general where $\omega_R = \omega - i\pi/2$.

F.10 General Forms of Superluminal Boosts

The group elements of the homogeneous complex Lorentz group L(C) can be expressed in terms of the group generators as

$$\Lambda_C = \exp[i(\omega_r\hat{\mathbf{u}}_r + i\omega_i\hat{\mathbf{u}}_i)\cdot\mathbf{K} + i\boldsymbol{\theta}_c\cdot\mathbf{J}] \tag{F.64}$$

where the vector $\boldsymbol{\theta}_c$ is a complex 3-vector, $\omega_r \geq 0$ and $\omega_i \geq 0$ are real numbers, and $\hat{\mathbf{u}}_r$ and $\hat{\mathbf{u}}_i$ are real normalized 3-vectors such that $\hat{\mathbf{u}}_r\cdot\hat{\mathbf{u}}_r = 1 = \hat{\mathbf{u}}_i\cdot\hat{\mathbf{u}}_i$. The generators of the homogeneous complex Lorentz group are $\mathbf{K}$, and $\mathbf{J}$ just as for the homogeneous real Lorentz group.

We now focus on boosts because they will be crucial in the determination of the equations of motion of various types of spin ½ particles. A boost has the form

$$\Lambda_C(\mathbf{v}_c) = \exp[i\omega\hat{\mathbf{w}}\cdot\mathbf{K}] \tag{F.65}$$

where

$$\omega = (\omega_r^2 - \omega_i^2 + 2i\omega_r\omega_i\,\hat{\mathbf{u}}_r\cdot\hat{\mathbf{u}}_i)^{\frac{1}{2}} \tag{F.66}$$

and

$$\hat{\mathbf{w}} = (\omega_r\hat{\mathbf{u}}_r + i\omega_i\hat{\mathbf{u}}_i)/\omega \tag{F.67}$$

Since $\hat{\mathbf{u}}_r\cdot\hat{\mathbf{u}}_r = 1 = \hat{\mathbf{u}}_i\cdot\hat{\mathbf{u}}_i$ we see

$$\hat{\mathbf{w}}\cdot\hat{\mathbf{w}} = 1 \tag{F.68}$$

The complex relative velocity is

$$\mathbf{v}_c = \hat{\mathbf{w}}\tanh(\omega) \tag{F.69}$$

Having placed boost transformations in the form of eq. F.12 we can take advantage of the form of real proper orthochronous Lorentz boost transformations, eq. F.13, and analytically

---

[454] Since γ_s has branch points at v = ±1 (which corresponds to ω = ±∞ for both the left-handed and right-handed groups) there is a cut along the real ω axis between −∞ and +∞ in the ω complex plane. Therefore, we note, the product of three left-handed Lorentz transformations does not yield a right-handed transformation (as might be supposed from eqs. 2.43 and 2.51) but rather a left-handed transformation on the second sheet. A transformation with ω + 3iπ/2 is not equivalent to a transformation with ω − iπ/2.

continue to complex ω and complex unit vectors $\hat{\mathbf{w}}$ provided eq. F.69 is satisfied. The resulting complex generalization will be the matrix form of proper boosts:

$$\Lambda_C(\mathbf{v_c}) = \exp[i\omega\hat{\mathbf{w}}\cdot\mathbf{K}] \equiv \Lambda_C(\omega, \hat{\mathbf{w}})$$

$$= \begin{bmatrix} \cosh(\omega) & -\sinh(\omega)\hat{w}_x & -\sinh(\omega)\hat{w}_y & -\sinh(\omega)\hat{w}_z \\ -\sinh(\omega)\hat{w}_x & 1+(\cosh(\omega)-1)\hat{w}_x^2 & (\cosh(\omega)-1)\hat{w}_x\hat{w}_y & (\cosh(\omega)-1)\hat{w}_x\hat{w}_z \\ -\sinh(\omega)\hat{w}_y & (\cosh(\omega)-1)\hat{w}_x\hat{w}_y & 1+(\cosh(\omega)-1)\hat{w}_y^2 & (\cosh(\omega)-1)\hat{w}_y\hat{w}_z \\ -\sinh(\omega)\hat{w}_z & (\cosh(\omega)-1)\hat{w}_x\hat{w}_z & (\cosh(\omega)-1)\hat{w}_y\hat{w}_z & 1+(\cosh(\omega)-1)\hat{w}_z^2 \end{bmatrix} \quad \text{(F.70)}$$

Since analytic continuations are unique, the above form for $\Lambda_C(\mathbf{v_c})$ is well-defined and unique. It spans the complete set of proper complex Lorentz boosts.

We now will study six classes of boosts that have the property that they boost from a coordinate system with real time and space coordinates to a coordinate system with either a purely real or purely imaginary time, and real, imaginary or complex spatial coordinates. These boosts produce left-handed lepton-like and "quark-like" free Dirac-like equations. They also produce right-handed lepton-like and "quark-like" free Dirac-like equations. We will discuss these Dirac-like equations in detail later. First we describe the four categories of boosts that have the property that they transform the reference frame of a particle at rest to a reference frame where the energy is either purely real or purely imaginary – the distinguishing feature of these four sets of transformations.

F.10.1 "Lepton-like" Left-Handed Boosts

If we let

$$\hat{\mathbf{u}}_i = \hat{\mathbf{u}}_r \equiv \hat{\mathbf{u}} \tag{F.71}$$

so that the vector $\hat{\mathbf{u}}_i$ is parallel to $\hat{\mathbf{u}}_r$, and let

$$\omega_i = \pi/2 \tag{F.72}$$

then $\Lambda_C(\mathbf{v_c})$ becomes a lepton-like left-handed boost:[455]

$$\Lambda_C = \exp[i(\omega_r + i\,\pi/2)\hat{\mathbf{u}}_r\cdot\mathbf{K}] \tag{F.73}$$

---

[455] We say "lepton-like" because we obtain a lepton-like Dirac-like equation using these boosts later. Similarly for "quark-like.'

F.10.2 "Lepton-like" Right-Handed Boosts

If we let

$$\hat{\mathbf{u}}_i = -\hat{\mathbf{u}}_r \equiv -\hat{\mathbf{u}} \tag{F.74}$$

so that the vector $\hat{\mathbf{u}}_i$ is anti-parallel to $\hat{\mathbf{u}}_r$, and

$$\omega_i = -\pi/2 \tag{F.75}$$

then $\Lambda_C(\mathbf{v}_c)$ becomes a right-handed boost:

$$\Lambda_C = \exp[i(\omega_r - i\,\pi/2)\hat{\mathbf{u}}_r\cdot\mathbf{K}] \tag{F.76}$$

F.10.3 "Quark-like" Left-Handed Boosts

If the real and imaginary relative vectors parts of $\hat{\mathbf{w}}$, namely $\hat{\mathbf{u}}_r$ and $\hat{\mathbf{u}}_i$, are perpendicular, $\hat{\mathbf{u}}_r\cdot\hat{\mathbf{u}}_i = 0$, then by eq. F.66

$$\omega = (\omega_r^2 - \omega_i^2)^{\frac{1}{2}} \tag{F.77}$$

Thus ω is either pure real ($\omega_r \geq \omega_i$) or pure imaginary ($\omega_r < \omega_i$). We choose ω real, and then reset

$$\omega = (\omega_r^2 - \omega_i^2)^{\frac{1}{2}} \rightarrow \omega' = (\omega_r^2 - \omega_i^2)^{\frac{1}{2}} + i\pi/2 = \omega + i\pi/2 \tag{F.78}$$

by adding $i\pi/2$ to the ω factor in eq. F.65 since ω is a free parameter. Then the resulting Lorentz transformation then becomes a "quark-like" left-handed boost:[456]

$$\Lambda_C = \exp[i((\omega_r^2 - \omega_i^2)^{\frac{1}{2}} + i\pi/2)(\omega_r\hat{\mathbf{u}}_r + i\omega_i\hat{\mathbf{u}}_i)\cdot\mathbf{K}/\omega] \tag{F.79}$$

F.10.4 "Quark-like" Right-Handed Boosts

If the real and imaginary relative vectors parts of $\hat{\mathbf{w}}$, namely $\hat{\mathbf{u}}_r$ and $\hat{\mathbf{u}}_i$, are perpendicular, $\hat{\mathbf{u}}_r\cdot\hat{\mathbf{u}}_i = 0$, then by eq. F.66

$$\omega = (\omega_r^2 - \omega_i^2)^{\frac{1}{2}} \tag{F.80}$$

Thus ω again starts out either pure real ($\omega_r \geq \omega_i$) or pure imaginary ($\omega_r < \omega_i$). In this case we also choose ω real, and then reset

$$\omega = (\omega_r^2 - \omega_i^2)^{\frac{1}{2}} \rightarrow \omega' = (\omega_r^2 - \omega_i^2)^{\frac{1}{2}} - i\pi/2 \tag{F.81}$$

---

[456] We say "quark-like" because we will later obtain a quark-like left-handed Dirac-like equation with complex spatial momentum terms using these boosts.

by subtracting $i\pi/2$ from ω in eq. F.65 since ω is a free parameter. The resulting Lorentz boost

$$\Lambda_C = \exp[i((\omega_r{}^2 - \omega_i{}^2)^{\frac{1}{2}} - i\pi/2)(\omega_r\hat{\mathbf{u}}_r + i\omega_i\hat{\mathbf{u}}_i)\cdot\mathbf{K}/\omega] \qquad (F.82)$$

becomes a quark-like right-handed boost.[457]

F.10.5 "Quark-like" Boosts

If the real and imaginary relative vectors parts of $\hat{\mathbf{w}}$, namely $\hat{\mathbf{u}}_r$ and $\hat{\mathbf{u}}_i$, are perpendicular, $\hat{\mathbf{u}}_r\cdot\hat{\mathbf{u}}_i = 0$, then by eq. F.66

$$\omega = (\omega_r{}^2 - \omega_i{}^2)^{\frac{1}{2}} \qquad (F.83)$$

Thus ω again starts out either pure real ($\omega_r \geq \omega_i$) or pure imaginary ($\omega_r < \omega_i$). In this case choose ω_r real and use ω as defined by eq. F.83.

Then the resulting Lorentz boost

$$\Lambda_C = \exp[i(\omega_r{}^2 - \omega_i{}^2)^{\frac{1}{2}}(\omega_r\hat{\mathbf{u}}_r + i\omega_i\hat{\mathbf{u}}_i)\cdot\mathbf{K}/\omega] \qquad (F.84)$$

becomes a quark-like boost without handedness.[458]

F.10.6 Conventional "Dirac" Boosts

If we let

$$\hat{\mathbf{u}}_i = \hat{\mathbf{u}}_r \equiv \hat{\mathbf{u}} \qquad (F.85)$$

so that the vector $\hat{\mathbf{u}}_i$ is parallel to $\hat{\mathbf{u}}_r$, and let

$$\omega_i = 0 \qquad (F.86)$$

then $\Lambda_C(\mathbf{v}_c)$ becomes a Dirac boost:[459]

$$\Lambda = \exp[i\omega_r\hat{\mathbf{u}}_r\cdot\mathbf{K}] \qquad (F.87)$$

This boost can be used to generate the free Dirac equation.

---

[457] We say "quark-like" because we obtain a quark-like right-handed Dirac-like equation with complex spatial momentum terms using these boosts later.

[458] We again say "quark-like" because we obtain a quark-like Dirac equation with complex spatial momentum terms using these boosts later.

[459] We say "Dirac" because we obtain a Dirac equation using this boost later.

Appendix G. Phenomena Beyond the Light Barrier

G.1 Superluminal (Faster-than-Light) Transformations

In this Appendix we will briefly survey some of the very different features of faster-than light physical phenomena. We will frame our discussion in terms of the two simple reference frames depicted in Fig. G.1. The prime frame is moving at a speed v > c (the speed of light) in the positive x direction with respect to the unprimed reference frame.

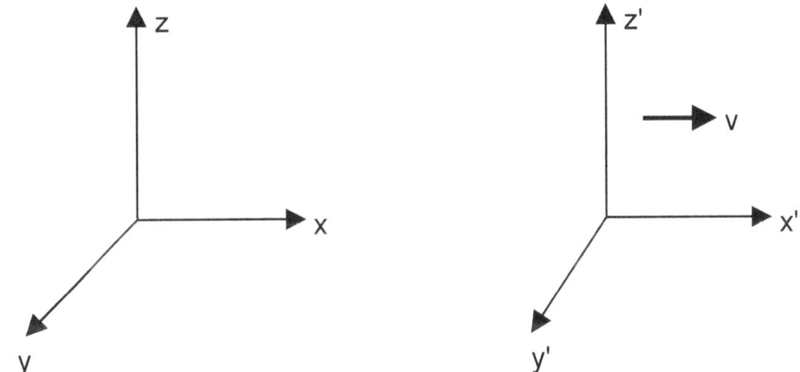

Figure G.1. Two coordinate systems having a relative speed v in the x direction.

As shown later in the text we define a superluminal (faster-than-light) transformation between coordinates in these reference frames with (Eqd. 2.16 and 2.13 are in Blaha (2007b))

$$t' = \gamma_s(t - \beta x/c)$$
$$x' = \gamma_s(x - \beta ct) \tag{2.16}$$
$$y' = iy$$
$$z' = iz$$

where

$$\gamma_s = (\beta^2 - 1)^{-\frac{1}{2}} \tag{2.13}$$

and $\beta = v/c > 1$. The appearance of imaginary values for y' and z' is not a cause for alarm. An observer resident in the prime coordinate system will measure real y and z distances with a ruler. The only purpose of the factors of i is to relate the y and z coordinates to y' and z'. An observer in either coordinate system will view his/her coordinates as real.

The energy and momentum of a tachyon (faster-than-light) particle of mass m traveling at a speed v > c is

$$E = \gamma_s mc^2 \tag{G.1}$$

and

$$\mathbf{p} = m\gamma_s \mathbf{v} \tag{G.2}$$

Note that the tachyon defining condition is satisfied:

$$E^2 - c^2 \mathbf{p}^2 = -m^2 c^4 \tag{G.3}$$

Also note that in the limit $\beta \to \infty$ that

$$E = 0 \tag{G.4}$$

and

$$p = mc \tag{G.5}$$

where $p = |\mathbf{p}|$. Tachyons are always in motion. The minimal momentum of a tachyon is given by eq. G.5. It corresponds to zero energy. It is the tachyon equivalent of Einstein's famous $E = mc^2$.

G.2 Length Dilations and Time Contractions

In ordinary Lorentz transformations a moving ruler will appear to be shorter in the direction of its motion when measured in another reference frame. This phenomenon is called *Lorentz contraction.*

Superluminal Length Dilation/Contraction

In the case of a superluminal transformation we find precisely the opposite effect, *superluminal length dilation*, is a possibility. Consider the case of the transformation of eq. 2.16 above (coresponding to Fig. G.1), which relates the prime reference frame traveling at speed v in the positive x direction to the unprimed reference frame. A ruler perpendicular to the x-axis will have the same length in both reference frames if its endpoints are simultaneously measured – perhaps by photographing it. The y and z equations in eqs. 2.16 specify this fact up to an extraneous factor of i.

If the ruler is at rest in the prime reference frame and parallel to the x' axis, then a simultaneous measurement of its endpoints at the same time t_0 by an observer in the unprimed reference frame (perhaps by photographing it) will reveal both *length contraction and dilation* depending on the value of β. If the length is $L' = x'_2 - x'_1$ in the prime frame and $L = x_2 - x_1$ in the unprimed frame, then the equations:

$$x'_1 = \gamma_s(x_1 - \beta c t_0) \tag{G.7}$$
$$x'_2 = \gamma_s(x_2 - \beta c t_0) \tag{G.8}$$

imply

$$L' = \gamma_s L = (\beta^2 - 1)^{-\frac{1}{2}} L \tag{G.9}$$

Thus we have three cases:

Case 1: $\beta \in \langle 1, \sqrt{2} \rangle$:	$L < L'$	Contraction	(G.10)
Case 2: $\beta = \sqrt{2}$:	$L = L'$	Equality	(G.11)
Case 3: $\beta \in \langle \sqrt{2}, \infty \rangle$:	$L > L'$	Dilation	(G.12)

Superluminal Time Contraction/Dilation

In the case of a superluminal transformation we find *superluminal time contraction* is a possibility. Consider again the case of the transformation of eq. 2.16 above coresponding to Fig. G.1 relating the prime reference frame traveling at speed v in the positive x direction to the unprimed reference frame. Consider the time interval between two events occurring at the same point x'_0 in the prime reference frame. From the viewpoint of an observer in the unprimed frame the events take place at different points x_1 and x_2. If the time interval is $T' = t'_2 - t'_1$ in the prime frame and $T = t_2 - t_1$ in the unprimed frame, then the inverse of eqs. 2.16 give:

$$t_1 = \gamma_s(t'_1 + \beta x'_0/c) \tag{G.13}$$
$$t_2 = \gamma_s(t'_2 + \beta x'_0/c) \tag{G.14}$$

and imply

$$T = \gamma_s T' = (\beta^2 - 1)^{-\frac{1}{2}} T' \tag{G.15}$$

Again we have three cases:

Case 1: $\beta \in \langle 1, \sqrt{2} \rangle$:	$T > T'$	Dilation	(G.16)
Case 2: $\beta = \sqrt{2}$:	$T = T'$	Equality	(G.17)
Case 3: $\beta \in \langle \sqrt{2}, \infty \rangle$:	$T < T'$	Contraction	(G.18)

The time interval in the unprimed frame can be less than, equal to, or greater than the time interval in the frame where the events take place at the same spatial point.

Thus superluminal transformations are more complex than Lorentz transformations with respect to space and time, dilation and contraction.

G.3 Tachyon Fission to More Massive Particles – Reverse Fission

Another way in which faster-than-light phenomena differ from sublight phenomena is particle fission. Normally when a particle or nucleus decays or fissions the masses of the particles produced by the decay are smaller than the mass of the original particle or nucleus. And energy is released. We are familiar with fission as the source of nuclear energy.

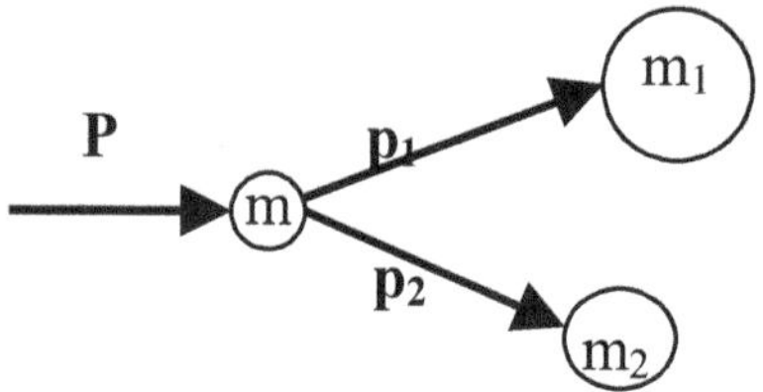

Figure G.2. Two particle decay of a tachyon.

In the case of faster-than-light particles, tachyons, a much different possibility is present: a tachyon can decay into heavier tachyons: *a particle's spatial 3-momentum can be transformed into mass.* We will consider the specific case of a tachyon decaying into two particles to illustrate this possibility. (See Fig. G.2.)

We will assume the initial tachyon has zero energy[460] and thus the tachyons emerging from the decay also have zero energy. The analysis is based on conservation of total energy and momentum.

Momentum conservation implies

$$\mathbf{P} = \mathbf{p_1} + \mathbf{p_2} \tag{G.19}$$

Since all energies are zero

$$(c\mathbf{P})^2 = (c\mathbf{P})^2 = m^2$$
$$(c\mathbf{p_1})^2 = (c\mathbf{p_1})^2 = m_1^2 \tag{G.20}$$
$$(c\mathbf{p_2})^2 = (c\mathbf{p_2})^2 = m_2^2$$

where $P = |\mathbf{P}|$, $p_1 = |\mathbf{p_1}|$, and $p_2 = |\mathbf{p_2}|$. If we now square eq. G.19 and use eqs. G.20 we obtain

$$m^2 = m_1^2 + m_2^2 + 2m_1m_2 \cos\theta \tag{G.21}$$

where θ is the angle between the emerging particles momenta $\mathbf{p_1}$ and $\mathbf{p_2}$.

Eq. G.21 has a number of interesting cases:

Case $\theta = 0$:

$$m = m_1 + m_2 \tag{G.22}$$

---

[460] If a particle has zero energy its velocity is infinite so the case considered is somewhat artificial. However the results would still be approximately true for very large velocities. The simplicity of the kinematics led us to consider this case.

The masses of the outgoing tachyons sum to the mass of the original tachyon.

Case $\theta = \pi/2$:

$$m^2 = m_1{}^2 + m_2{}^2 \qquad (G.23)$$

The masses of each outgoing tachyon is less than the mass of the original tachyon.

Case $\theta = \pi$:

$$m^2 = (m_1 - m_2)^2 \qquad (G.24)$$

In this case either $m_1 > m$ or $m_2 > m$. Thus one of the outgoing tachyons has a greater mass than the original tachyon. Mass is effectively created from the spatial momentum of the particle. This process is the inverse of normal particle decay or fission where the sum of the outgoing masses is always less than the original particle's mass and the difference is mass converted into energy in the form of additional photons via $E = mc^2$.

This last case, where one of the outgoing particles is more massive than the original particle, is not just for $\theta = \pi$. Since

$$\cos\theta = (m^2 - m_1{}^2 - m_2{}^2)/(2m_1 m_2) \qquad (G.25)$$

we see that *the sum of the outgoing tachyon masses is always greater than the original tachyon mass (except when $\theta = 0$)* since

$$\cos\theta = 1 + [m^2 - (m_1 + m_2)^2]/(2m_1 m_2) \leq 1 \qquad (G.26)$$

and thus

$$[m^2 - (m_1 + m_2)^2]/(2m_1 m_2) \leq 0 \qquad (G.27)$$

Note $m = m_1 + m_2$ only if $\theta = 0$.

Since we can transform the above discussion to the case of tachyons with a non-zero energy using an ordinary Lorentz transformation the above discussion in this subsection is general.

We therefore conclude that when a tachyon decays into two tachyons the sum of the masses of the produced tachyons is greater than the mass of the original tachyon except if the angle between the momenta of the produced tachyons is zero. In that case the sum of the masses of the produced tachyon equals the mass of the original tachyon.

*Thus tachyons can engage in **reverse fission** in which **momentum is converted into mass so the outgoing particles have a total mass greater than the incoming particle**.* In the case of "normal" fission part of the mass of a particle can be converted to energy and the sum of the masses of the decay product particles is less than the mass of the original particle.

G.4 Light Chasing Faster-than-Light Particles?

Einstein told a story that he imagined positioning himself in a (Galilean) reference frame moving at the speed of light and seeing electromagnetic waves "frozen" in time so that they were no longer vibrating. This vision inspired him to reconsider the transformation laws between coordinate systems and to derive the theory of Special Relativity. In Special Relativity the speed of light is the same in all reference frames.

In this subsection we will consider a light pulse from the points of view of two reference frames whose relative speed v is greater than the speed of light. We will use the example considered earlier and add a pulse of light traveling in the positive x direction. (See Fig. G.3.)

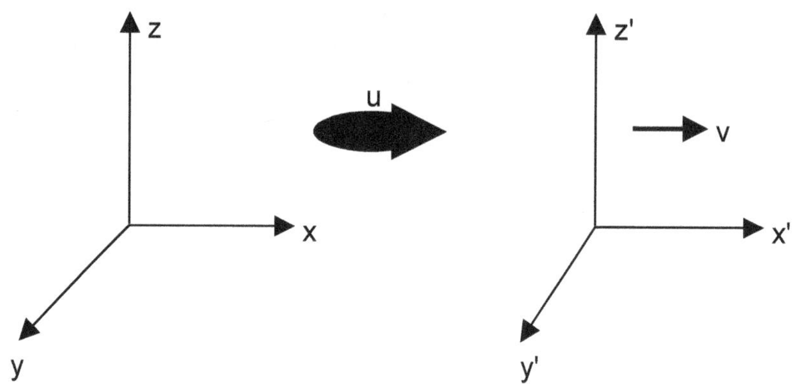

Figure G.3. Two coordinate systems having a relative speed v in the x direction. A pulse of light is displayed as a thick arrow.

The general law for the addition of velocities in a situation such as depicted in Fig. G.3 is well known. If we adapt it to the present example and let u be the speed of the pulse in the unprimed frame (temporarily forgetting it is a light pulse) we find it implies

$$u' = (u - \beta c)/(1 - \beta u/c) \qquad (2.17a)$$

where $\beta = v/c > 1$, and u' is the speed of the pulse in the prime frame. Then if we set u = c we see that u' = c as well. *Thus our superluminal transformations preserve the constancy of the speed of light just like Lorentz transformations.*

As a result the pulse of light will intersect the z' axis eventually. However <u>if</u> superluminal transformations did not preserve the speed of light in all frames the pulse might never reach the z' axis. For example under a Galilean transformation the speed of the pulse would be u' = u – v = c – v and the pulse would actually be falling further and further behind the z' axis.

G.5 Electromagnetic Field of a Charged Tachyon – A Pancake Effect?

The electric field of a charge q at rest in a reference frame is:

$$\mathbf{E} = (q/(4\pi\varepsilon_0)\check{\mathbf{r}}/r^2 \tag{G.28}$$

in spherical coordinates where $\check{\mathbf{r}}$ is a unit vector in the radial direction.

Sublight Charged Particle

The electric and magnetic fields of a charge q moving in the positive x direction with speed v < c are

$$\mathbf{E} = (q/(4\pi\varepsilon_0)\check{\mathbf{r}}(1 - \beta^2)/[r^2(1 - \beta^2\sin^2\theta)^{\frac{3}{2}}] \tag{G.29}$$
$$\mathbf{B} = (q/(4\pi\varepsilon_0)\check{\mathbf{r}}\beta(1 - \beta^2)\sin\theta/[r^2(1 - \beta^2\sin^2\theta)^{\frac{3}{2}}] \tag{G.30}$$

where $\check{\mathbf{r}}$ is the radial unit vector, β = v/c, and θ is measured with respect to the polar axis which is taken to be the x axis. As $\beta \to 1$ the electric and magnetic fields develop a "pancake" form with large field strengths in the directions perpendicular to the direction of motion similar to the transverse fields of electromagnetic quanta. This feature is the basis of the Weizsäcker-Williams method of virtual quanta.

Charged Tachyon

The electric and magnetic fields of a tachyon of charge q moving in the positive x direction with speed v > c are

$$\mathbf{E} = (q/(4\pi\varepsilon_0)\check{\mathbf{r}}(\beta^2 - 1)/[r^2(\beta^2\sin^2\theta - 1)^{\frac{3}{2}}] \tag{G.31}$$
$$\mathbf{B} = (q/(4\pi\varepsilon_0)\check{\mathbf{r}}\beta(\beta^2 - 1)\sin\theta/[r^2(\beta^2\sin^2\theta - 1)^{\frac{3}{2}}] \tag{G.32}$$

where β = v/c > 1, and θ is again measured with respect to the polar axis which is taken to be the x axis. In the case of tachyons there are three cases of interest.

Case $\beta^2\sin^2\theta - 1 < 0$:
The electric and magnetic fields are pure imaginary and are excluded from the forward and backward cones surrounding the x axis defined by $|\sin\theta| < \beta^{-1}$.

Case $\beta^2\sin^2\theta - 1 = 0$:

The electric and magnetic fields are infinite. Thus the field strengths are infinite on a cone at the angle θ with respect to the x-axis. By comparison, a magnetic monopole only has a one-dimensional, singularity line extending from the monopole to infinity.

Case $\beta^2\sin^2\theta - 1 > 0$:

The electric and magnetic fields decrease in strength as $\sin^2\theta$ increases. Thus the region of maximum field strength are the forward and backward cones where $|\sin\theta|$ is greater than but near β^{-1} in value. The pancake picture of the sublight charged particle does not hold for charged tachyons.

Are Tachyonic Cones in the Au-Au Scattering Quark-Gluon Plasma?

Cones have been observed in high energy Au-Au scattering in which a quark-gluon plasma is created. These cones have been attributed to a variety of causes such as hydrodynamically generated Mach cones, and Cherenkov radiation. The possibility exists that tachyonic excitations may transiently exist in the quark-gluon plasma and may, in part, explain the observed cones and dips. The above described cones in the case of a moving charged tachyon are remarkably similar in character. See the CERES collaboration paper arXiv:nucl-ex/0701023, and references therein, for experimental findings.

G.6 Superluminal (Tachyon) Physics is Different

The simple classical examples presented in this appendix demonstrate that superluminal physics has many interesting new features that are worthy of interest. Since tachyons exist in Black Holes, and, perhaps, in other contexts, their study is a worthwhile endeavor.

Appendix H. Superluminal (Faster Than Light) Kinetic Theory and Thermodynamics

This appendix[461] changes the flow of topics of these volumes from gravitation and particle theory to superluminal many particle dynamics. We will progress from superluminal Kinetic theory to Thermodynamics. We will see that there are strong similarities with non-relativistic Thermodynamics.

H.1 Superluminal Kinetic Theory

Assemblages of large numbers of particles embody the Maxwell-Boltzmann distribution. The Boltzmann H theorem is the beginning point for derivations of the non-relativistic Maxwell-Boltzmann distribution. The non-relativistic Maxwell-Boltzmann distribution has the form

$$f(\mathbf{v}, \mathbf{r}) = n(m/(2\pi kT))^{3/2}\exp\{-[m(\mathbf{v} - \mathbf{v}_0)^2/2 + V(r)]/(kT)\} \qquad (H.1)$$

where n is the particle density, T is the temperature, $\mathbf{v}_0$ is the average velocity, m is the particle mass, V(r) is an external conservative force, and k is Boltzmann's constant. In terms of a hamiltonian

$$H(\mathbf{v}, \mathbf{r}) = mv^2/2 + V(r) \qquad (H.2)$$

we can express the Maxwell-Boltzmann distribution as

$$f(\mathbf{v}, \mathbf{r}) = n(m/(2\pi kT))^{3/2}\exp\{-H(\mathbf{v} - \mathbf{v}_0, \mathbf{r})/(kT)\} \qquad (H.3)$$

H.1.1 Relativistic Form of the Maxwell-Boltzmann Distribution

If we assume that we have a container containing a distribution of relativistic (sublight) particles with an average velocity $\mathbf{v}_0 = 0$, and no external force, then the form of eq. H.3 generalizes to the relativistic Maxwell-Boltzmann distribution

$$f_R(\mathbf{v}) = C_R \exp\{-H/(kT)\} \qquad (H.4)$$

---

[461]From Blaha (2012a).

where C_R is a normalization constant and H is the relativistic hamiltonian for a free particle:

$$H = c(m^2c^2 + \mathbf{p}^2)^{½}$$

(H.5)

with $\mathbf{p} = \gamma m\mathbf{v}$ and $\gamma = (1 - v^2/c^2)^{-½}$. C_R is determined by the condition

$$\int d^3 v f_R(\mathbf{v}) = 1$$

(H.6)

H.1.2 Superluminal Form of the Maxwell-Boltzmann Distribution

The superluminal form of Maxwell-Boltzman distribution is based on the form of the mass shell condition for superluminal particles:

$$E^2 - c^2\mathbf{p}^2 = m^2c^4$$

(H.7)

which implies a free hamiltonian

$$H_S = c(\mathbf{p}^2 - m^2c^2)^{½}$$

(H.8)

where

$$\mathbf{p} = \gamma_s m\mathbf{v}$$

(H.9)

and

$$\gamma_s = (v^2/c^2 - 1)^{-½}$$

The seemingly slight difference between eqs. H.8 and H.9, and eq. H.5 causes major differences between superluminal and relativistic kinetic theory and thermodynamics. On the other hand relativistic kinetic theory and thermodynamics are qualitatively similar in many ways with their non-relativistic counterparts.

One major difference is the behavior of kinematic variables near the speed of light:

<u>As $v \rightarrow c$ Below the Speed of Light</u>

$$p \rightarrow \infty$$
$$H \rightarrow \infty$$

<u>As $v \rightarrow c$ From Above the Speed of Light</u>

$$p \rightarrow \infty$$
$$H_S \rightarrow \infty$$

<u>As $v \rightarrow \infty$</u>

$$p \rightarrow mc$$
$$H_S \rightarrow 0$$

Thus as v ranges from c to ∞, H_S decreases monotonically from ∞ to zero and p decreases from ∞ to mc. This behavior contrasts with H in eq. H.5, which increases monotonically with p as v increases from 0 to c. Thus the sublight Maxwell-Boltzmann distribution decreases with v as v increases from 0 to c.

The superluminal Maxwell-Boltzmann distribution *increases* with v as v increases from c to ∞ as we see below. The superluminal Maxwell-Boltzmann distribution decreases with p as p increases from mc to ∞. *As a result the natural physical parametrization of the Maxwell-Boltzmann distribution should be in terms of the momentum rather than the velocity.* Thus Boltzmann's H function which normally is

$$H_B(t) = \int d^3v \; f(\mathbf{v}, t) \log f(\mathbf{v}, t)$$

must be replaced with[462]

$$H_{BS}(t) = \int d^3p \; f_S(\mathbf{p}, t) \log f_S(\mathbf{p}, t)$$

The equilibrium superluminal Maxwell-Boltzman distribution can be derived from $H_S(t)$. It has the same general form as the relativistic distribution

$$f_S(\mathbf{p}) = C_S \exp\{-H_S/(kT)\} \tag{H.10}$$

where C_S is a normalization constant and H_S is the superluminal hamiltonian for a free particle. We now apply the normalization condition[463]

$$n = N/V = \int d^3p f_S(\mathbf{p}) = C_S \int d^3p \, \exp\{-H_S/(kT)\} \tag{H.11}$$

where n is the particle density, N is the number of particles in the system, and V is the volume of the system. We calculate C_S by evaluating the integral:

$$n = 4\pi C_S \int_m^\infty dp \; p^2 \, \exp\{-H_S/(kT)\} \tag{H.12}$$

Letting $x = p/(mc)$ and $\alpha = mc^2/(kT)$ we see eq. H.12 becomes

$$n = 4\pi m^3 c^3 C_S \int_1^\infty dx \; x^2 \, \exp\{-\alpha(x^2 - 1)^{\frac{1}{2}}\} \tag{H.13}$$

Then letting $y^2 = x^2 - 1$ we find

---

[462] Note the additional factor of m^3 in $\int d^3p$ will be absorbed in the normalization (eq. H.11).
[463] We note that using $\int d^3v$ rather than $\int d^3p$ in eq. H.11 would result in a divergence – another reason for our choice of integration parameter.

$$n = 4\pi m^3 c^3 C_S \int_0^\infty dy\; y(y^2 + 1)^{\frac{1}{2}} \exp(-\alpha y)$$

$$= -m^3 c^3 C_S G^{31}_{13}(\alpha^2/4 \mid {}^{\;0}_{-3/2,0,\,\frac{1}{2}}) \tag{H.14}$$

where $G^{31}_{13}(\ldots)$ is Meijer's G-Function.[464] Therefore

$$C_S = -[m^3 c^3 G^{31}_{13}((mc^2/(2kT))^2 \mid {}^{\;0}_{-3/2,0,\,\frac{1}{2}})/n]^{-1} \tag{H.15}$$

The most probable momentum of a particle p_p is the maximum of

$$p_p = \mathrm{Max}\{p^2 \exp[-H_S/(kT)]\}$$

$$= \{(2(kT)^2/c^2)[1 + (1 - m^2 c^4/(kT)^2)^{\frac{1}{2}}]\}^{\frac{1}{2}} \tag{H.16}$$

For large T or small T the maximum is

$$p_p \approx 2kT/c\; > mc$$

The velocity v_p corresponding to the maximum in the momentum is

$$v_p = cp_p/(p_p^2 - m^2 c^2)^{\frac{1}{2}}$$

For large T or small T, the velocity v_p corresponding to the maximum in the momentum is approximately

$$v_p \approx c + \tfrac{1}{2}\, m^2 c^5/(2kT)^2$$

H.2 Superluminal Thermodynamics

Turning now to the thermodynamics of a dilute superluminal gas implied by the superluminal Maxwell-Boltzman distribution we begin by calculating the average energy per particle

$$\varepsilon = C_S \int d^3 p\; H_S \exp[-H_S/(2kT)]/\int d^3 p\; C_S \exp[-H_S/(kT)] \tag{H.17}$$

$$= (C_S/n) \int d^3 p\; H_S \exp[-H_S/(kT)]$$

---

[464] See Gradshteyn (1965) integral 3.389.2 and p. 1068 for the properties of Meijer's G-Function.

$$= (C_S/n)2kT\alpha\, 4\pi m^2 c^3 \int_0^\infty dy\; y^2 (y^2 + 1)^{\frac{1}{2}} \exp(-\alpha y)$$

$$= -(C_S/n)m^3 c^5 G^{31}_{13}(\alpha^2/4 \mid^{-\frac{1}{2}}_{-2,0,\,\frac{1}{2}})$$

$$= mc^2\, G^{31}_{13}((mc^2/(2kT))^2 \mid^{-\frac{1}{2}}_{-2,0,\,\frac{1}{2}})/G^{31}_{13}((mc^2/(2kT))^2 \mid^{0}_{-3/2,0,\,\frac{1}{2}}) \tag{H.18}$$

The Maxwell-Boltzman normalization factor is related to the energy per particle by

$$C_S = -n\varepsilon/(m^3 c^5 G^{31}_{13}(\alpha^2/4 \mid^{-\frac{1}{2}}_{-2,0,\,\frac{1}{2}})) \tag{H.19}$$

Note that C_S is proportional to the energy in contrast to the non-relativistic case where the Maxwell-Boltzman normalization factor $C = (3m/(4\pi\varepsilon))^{3/2}$.

We now calculate the superluminal pressure for the case of a distribution of superluminal particles bouncing on a wall perpendicular to the z-axis. The wall is assumed to be a perfectly reflecting plane. The pressure is the average force per unit area due to the gas of superluminal particles. The number of particles bombarding the wall per second is with $v_z > 0$ is $v_z f_S(\mathbf{p})d^3p$. Thus the pressure is

$$P = \int d^3p\, 2p_z v_z f_S(\mathbf{p}) \tag{H.20}$$

where the particle momentum changes by $2p_z$ due to reflection. Due to spherical symmetry one expects the average values for the various components of $\mathbf{v}$ to be equal. Consequently we can re-express eq. H.20 as

$$P = 1/3 \int d^3p\, 2m\gamma_s v^2 f_S(\mathbf{p}) \tag{H.21}$$
$$= 1/3 \int d^3p\, 2pv f_S(\mathbf{p})$$

Since

$$v = cp/(p^2 - m^2 c^2)^{\frac{1}{2}} \tag{H.22}$$

we see

$$P = 8\pi c/3 \int_m^\infty dp\, p^4 f_S(\mathbf{p})/(p^2 - m^2 c^2)^{\frac{1}{2}}$$

Following steps similar to eqs. H.12 – H.15 leads to

$$P = m^4 c^4 C_S G^{31}_{13}((mc^2/(2kT))^2 \mid^{\frac{1}{2}}_{-2,0,\,\frac{1}{2}}) \tag{H.23}$$

The *equation of state* relating the pressure and energy is

$$P = -(m/c)\{G^{31}_{13}((mc^2/(4kT))^2 \mid^{1/2}_{-2,0,\,1/2})/G^{31}_{13}(\alpha^2/4 \mid^{-1/2}_{-2,0,\,1/2}))\}n\varepsilon \tag{H.24}$$

Substituting for ε we find

$$P = -(nm^2c)\{G^{31}_{13}(\rho \mid^{1/2}_{-2,0,\,1/2})/ G^{31}_{13}(\rho \mid^{0}_{-3/2,0,\,1/2})\} \tag{H.25}$$

where

$$\rho = (mc^2/(2kT))^2 \tag{H.26}$$

Turning now to the consideration of a dilute gas the internal energy of the gas can be defined to be[465]

$$U(t) = N\varepsilon \tag{H.27}$$

We note that the work done by the superluminal gas if its volume increases by dV is PdV. Then the superluminal (and usual) form of the first law of thermodynamics is

$$dQ = dU + PdV \tag{H.28}$$

where Q is the heat absorbed. The heat capacity of the system for constant volume is

$$C_V = (\partial U/\partial T)_V \tag{H.29}$$

The second law of thermodynamics, Boltzmann's H theorem, is based on

$$H = -S/kV \tag{H.30}$$

where H is the negative of the entropy divided k times the volume V. In systems where there are no superluminal particles, the H theorem states that the entropy never decreases for an isolated gas of fixed volume.

We can calculate H for a superluminal system under equilibrium conditions, H_e, from[466]

$$H_e = \int d^3p f_S(\mathbf{p})\ln(f_S(\mathbf{p})) \tag{H.31}$$

$$= \int d^3p f_S(\mathbf{p})[\ln C_S - H_S/(kT)] \tag{H.32}$$

---

[465] The internal energy of a gas of non-interacting non-relativistic particles is $U(t) = 3NkT/2$. In the superluminal case it appears that it is eq. H.18.

[466] We consistently assume that integrals over the momentum $\int d^3p$ are the proper integration (rather than integrations over velocity $\int d^3v$) because, for example, the calculation of the normalization constant eq. H.11 would diverge if the integration were over $\int d^3v$.

$$= n \ln C_S - \int d^3p f_S(\mathbf{p}) H_S/(kT)$$

$$= n \ln C_S - n\varepsilon/(kT) \tag{H.33}$$

by eqs. H.11 and H.17. Therefore

$$S = -kVH_{Se} = -kN \ln C_S + N\varepsilon/T \tag{H.34}$$

Consequently we obtain the superluminal *and* standard non-relativistic result

$$1/T = (\partial S/\partial U)_x \tag{H.35}$$

where x represents all other extensive variables.

H.3 Approximate Calculation of Kinetic and Thermodynamic Quantities

We can obtain more tractable expressions for kinetic and thermodynamic quantities by assuming $\mathbf{p}^2 \gg m^2c^2$ and approximating the hamiltonian (eq. H.8) with

$$H_{Sa} = cp \tag{H.36}$$

The approximate normalization condition is

$$n = N/V = \int d^3p f_{Sa}(\mathbf{p}) = C_{Sa}\int d^3p \exp\{-H_{Sa}/(kT)\} \tag{H.37}$$

where n is the particle density, N is the number of particles in the system, and V is the volume of the system. C_S is determined by

$$n = 4\pi C_{Sa} \int_{mc}^{\infty} dp\, p^2 \exp\{-cp/(kT)\} \tag{H.38}$$

Letting $\alpha = c/(kT)$ we see eq. H.38 becomes

$$n = 4\pi C_{Sa}\, d^2/d\alpha^2 \int_{mc}^{\infty} dp \exp(-\alpha p)$$

$$= 4\pi C_{Sa}\, d^2/d\alpha^2\, [(1/\alpha) \exp(-\alpha mc)] \tag{H.39}$$

Therefore the normalization factor is

$$C_{Sa} = n/\{4\pi \ d^2/d\alpha^2 \ [(1/\alpha)\exp(-\alpha mc)]\} \tag{H.40}$$

The most probable momentum of a particle p_p is the maximum of

$$p_{pa} = \mathrm{Max}\{p^2\exp[-H_{Sa}/(kT)]\}$$
$$= 2kT/c \tag{H.41}$$

The velocity v_{pa} corresponding to the maximum in the momentum is

$$v_{pa} = cp_{pa}/(p_{pa}^{\ 2} - m^2c^2)^{\frac{1}{2}}$$

For large T or small T, the velocity v_{pa} corresponding to the maximum in the momentum is approximately

$$v_{pa} \approx c + \tfrac{1}{2} \ m^2c^5/(2kT)^2$$

Turning now to the thermodynamics implied by the superluminal Maxwell-Boltzman distribution we begin by calculating the average energy per particle

$$\varepsilon_a = \int d^3p \ H_{Sa} \ \exp[-H_{Sa}/(kT)]\Big/\int d^3p \ \exp[-H_{Sa}/(kT)] \tag{H.42}$$

$$= (C_{Sa}/n) \int d^3p \ H_{Sa} \ \exp[-H_{Sa}/(kT)]$$

$$= -(4\pi c C_{Sa}/n) \ d^3/d\alpha^3[(1/\alpha)\exp(-\alpha mc)] \rightarrow 3kT \ \text{for} \ T \gg mc$$

where $\alpha = c/(kT)$.[467]

The Maxwell-Boltzman normalization factor is related to the energy per particle by

$$C_{Sa} = -n\varepsilon_a/\{4\pi c \ d^3/d\alpha^3[(1/\alpha)\exp(-\alpha mc)]\} \tag{H.43}$$

Note that C_{Sa} is proportional to the energy ε_a in contrast to the non-relativistic case where the Maxwell-Boltzman normalization factor $C = (3m/(4\pi\varepsilon))^{3/2}$.

We now calculate the superluminal pressure for the case of a distribution of superluminal particles bouncing on a wall perpendicular to the z-axis. The wall is assumed to be a perfectly reflecting plane. The pressure is the average force per unit area due to the gas of superluminal particles. The number of particles bombarding the wall per second is with $v_z > 0$ is $v_z f_{Sa}(\mathbf{p})d^3p$. Thus the pressure is

---

[467] The Superluminal case differs from the non-relativistic case: $\varepsilon_a = 3kT/2$. An example of $\varepsilon_a = 3kT$ is a crystal with a potential energy of compression. See p. 192 Morse (1964).

$$P_a = \int d^3p \, 2p_z v_z f_{Sa}(\mathbf{p}) \tag{H.44}$$

where the particle momentum changes by $2p_z$ due to reflection. Due to spherical symmetry one expects the average values for the various components of $\mathbf{v}$ to be equal. Consequently we can re-express eq. H.44 as

$$P_a = 1/3 \int d^3p \, 2m\gamma_s v^2 f_{Sa}(\mathbf{p}) \tag{H.45}$$

$$= 1/3 \int d^3p \, 2pv f_{Sa}(\mathbf{p})$$

Since

$$v = cp/(p^2 - m^2c^2)^{\frac{1}{2}} \tag{H.46}$$

we see

$$P_a = 8\pi c/3 \int_{mc}^{\infty} dp \, p^4 f_{Sa}(\mathbf{p})/(p^2 - m^2c^2)^{\frac{1}{2}} \tag{H.47}$$

$$\cong 8\pi c/3 \int_{mc}^{\infty} dp \, p^3 f_{Sa}(\mathbf{p})$$

Evaluating eq. H.47 yields

$$P_a = -(8\pi c/3)\, C_{Sa}\, d^3/d\alpha^3 [(1/\alpha)\exp(-\alpha mc)] \tag{H.48}$$

The *equation of state* relating the pressure and energy is[468]

$$P_a = 2/3 \, n\varepsilon_a \tag{H.49}$$

For $T \gg mc$ we found[469]

$$\varepsilon_a \rightarrow 3kT \tag{H.50}$$

then, contrary to non-relativistic kinetic theory, we find ($T \gg mc$)

$$P_a = 2nkT \tag{H.51}$$

Turning now to the consideration of a dilute gas the internal energy of the gas for $T \gg mc$ is

$$U(t) = N\varepsilon \rightarrow 3NkT \tag{H.52}$$

We note again that the work done by the superluminal gas if its volume increases by dV is PdV. Then the superluminal (and usual) form of the first law of thermodynamics is

---

[468] The same equation of state as non-relativistic kinetic theory. See p. 72 Huang (1965).
[469] Later we will define temperature in terms of the entropy S as $1/T = (\partial S/\partial U)_x$ where x is all other extensive variables.

$$dQ = dU + PdV \tag{H.53}$$

where Q is the heat absorbed. The heat capacity of the system for constant volume is $(T \gg mc)$

$$C_V \rightarrow 3Nk \tag{H.54}$$

The second law of thermodynamics, Boltzmann's H theorem, is based on

$$H = -S/kV \tag{H.55}$$

where H is the negative of the entropy divided k times the volume V. In systems where there are no superluminal particles, the H theorem states that the entropy never decreases for an isolated gas of fixed volume.

We can calculate H_{BS} for a superluminal system under equilibrium conditions, H_{BSea}, from[470]

$$H_{BSea} = \int d^3 p f_{Sa}(\mathbf{p}) \ln(f_{Sa}(\mathbf{p})) \tag{H.56}$$
$$= \int d^3 p f_{Sa}(\mathbf{p})[\ln C_{Sa} - H_{Sa}/(kT)] \tag{H.57}$$
$$= n \ln C_{Sa} - \int d^3 p f_{Sa}(\mathbf{p}) H_{Sa}/(kT)$$
$$= n \ln C_{Sa} - n\varepsilon_a/(kT) \tag{H.58}$$

by eqs. H.11 and H.17. Therefore

$$S_a = -kVH_{BSea} = -kN \ln C_{Sa} + N\varepsilon_a/T \tag{H.59}$$

The superluminal *and* standard non-relativistic result still holds

$$1/T = (\partial S/\partial U)_x \tag{H.60}$$

where x represents all other extensive variables.

H.4 Superluminal Kinetics and Themodynamics Are Similar to the Non-Relativistic Case

In the previous sections we have shown that kinetic theory and the laws of themodynamics are usually similar in the superluminal and non-relativistic cases modulo detail

---

[470] We consistently assume that integrals over the momentum $\int d^3 p$ are the proper integration (rather than integrations over velocity $\int d^3 v$) because, for example, the calculation of the normalization constant eq. H.11 would diverge if the integration were over $\int d^3 v$.

differences in the values of the various quantities due to differences between superluminal kinematics and non-relativistic kinematics.

Appendix I. PseudoQuantization – Embedding Classical Fields in Quantum Field Theories

This refereed paper is S. Blaha, Phys. Rev. **D17**, 994 (1978). Reprinted with the kind permission of Physical Review D.

PHYSICAL REVIEW D   VOLUME 17, NUMBER 4   15 FEBRUARY 1978

Embedding classical fields in quantum field theories

Stephen Blaha*

Physics Department, Syracuse University, Syracuse, New York 13210

(Received 2 August 1976; revised manuscript received 7 November 1977)

We describe a procedure for quantizing a classical field theory which is the field-theoretic analog of Sudarshan's method for embedding a classical-mechanical system in a quantum-mechanical system. The essence of the difference between our quantization procedure and Fock-space quantization lies in the choice of vacuum states. The key to our choice of vacuum is the procedure we outline for constructing Lagrangians which have gradient terms linear in the field variables from classical Lagrangians which have gradient terms which are quadratic in field variables. We apply this procedure to model electrodynamic field theories, Yang-Mills theories, and a vierbein model of gravity. In the case of electrodynamics models we find a formalism with a close similarity to the coherent-soft-photon-state formalism of QED. In addition, photons propagate to $t = +\infty$ via retarded propagators. We also show how to construct a quantum field for action-at-a-distance electrodynamics. In the Yang-Mills case we show that a previously suggested model for quark confinement necessarily has gluons with principal-value propagation which allows the model to be unitary despite the presence of higher-order-derivative field equations. In the vierbein-gravity model we show that our quantization procedure allows us to treat the classical and quantum parts of the metric field in a unified manner. We find a new perturbation scheme for quantum gravity as a result.

I. INTRODUCTION

The relation between classical and quantum systems has been a subject of continuing interest over the years: First, in the original development of quantum mechanics, second, in the study of the classical limit and infrared divergences of quantum-electrodynamic processes,[1,2] and third, in recent attempts to construct strong-interaction models of quark confinement which are for the most part either classical field theory models in search of quantization[3] or quantized gluon models wherein quark confinement is a consequence of infrared behavior.[4,5]

We will describe a new quantization procedure (called pseudoquantization) for field theory which is the analog of Sudarshan's method for embedding a classical-mechanical system in a quantum-mechanical system. It can be used with advantage to either embed a classical field theory in a quantum field theory in such a way as to maintain the classical character of the embedded fields (while studying the interaction between the classical and quantum sectors on essentially the same footing), or to quantize a class of field theories, members of which have been used as models for gravity and as models for the strong interaction with quark confinement.[7-9]

We shall begin (Sec. II) by pseudoquantizing a classical simple harmonic oscillator. This case is of particular importance because of the analogy between the mode amplitudes of a quantum field and the coordinates of a set of simple harmonic oscillators which we will take advantage of in later sections.

In Sec. III we describe the pseudoquantization procedure for field theory. We apply it to electrodynamic models and show that the propagation of photons to $t = +\infty$ is necessarily retarded in this formalism. Further, we display a close analogy between the present formalism and the coherent-soft-photon-state formalism[10] of QED.

In Sec. IV we apply the pseudoquantization procedure to a classical Yang-Mills field. The resulting field theory (with a slight but important modification) has been used as a model for the strong interactions with quark confinement.[7-9] We also apply the pseudoquantization procedure to a vierbein model of gravity and obtain a new perturbation theory for quantum gravity.

In Sec. V we show that principal-value propagators naturally arise in certain sectors of pseudoquantized theories thus verifying an *ad hoc* procedure devised to unitarize a model of quark confinement.[7-9] We also show how to construct a quantum version of action-at-a-distance electrodynamics.

We shall now briefly outline the procedure for embedding a classical-mechanical system in a quantum system.[6] Consider a classical Hamiltonian system with one degree of freedom, and commuting canonical variables, x_1 and p_1, which have the equations of motion

$$\dot{x}_1 = -i[x_1, \hat{H}] , \tag{1}$$

$$\dot{p}_1 = -i[p_1, \hat{H}] , \tag{2}$$

where defining

$$\hat{H} = -i\left(\frac{\partial H(x_1, p_1)}{\partial p_1}\frac{\partial}{\partial x_1} - \frac{\partial H(x_1, p_1)}{\partial x_1}\frac{\partial}{\partial p_1}\right) \tag{3}$$

allows us to write Hamilton's equations in commutator form. With Sudarshan[6] we define

$$x_2 = i\frac{\partial}{\partial p_1} \tag{4}$$

and

$$p_2 = -i\frac{\partial}{\partial x_1} \tag{5}$$

so that

$$[x_1, x_2] = [p_1, p_2] = 0 , \tag{6}$$

$$[x_1, p_2] = [x_2, p_1] = i , \tag{7}$$

and $\hat{H}$ can now be taken to be the operator

$$\hat{H} = \frac{\partial H(x_1, p_1)}{\partial p_1} p_2 + \frac{\partial H(x_1, p_1)}{\partial x_1} x_2 . \tag{8}$$

It is now apparent that we can take the above quantities and equations of motion to describe a quantum mechanical system with two degrees of freedom in the "coordinate" representation where the "coordinates" are (x_1, p_1) and the canonical momenta are $\Pi = (p_2, -x_2)$. As we will see below the linearity of $\hat{H}$ in the momenta is crucial for the maintenance of the classical character of x_1 and p_1, and for the observability of the phase-space trajectory. Since we choose to identify the physical observables with the commutative algebra of the coordinate operators, x_1 and p_1, we are led to impose the superselection condition that the momenta, Π, are unobservable. As a result the Hamiltonian and other generators of canonical transformations, which are all linear in the momenta, are also unobservable. However, in each case there is an associated dynamical quantity which is observable.

The required unobservability of the momenta restricts the form of the interaction between a classical-made-quantum system and an inherently quantum system to

$$H_{\text{int}} = \Phi_1 x_2 + \Phi_2 p_2 + X , \tag{9}$$

where Φ_1, Φ_2, and X are functions of x_1, p_1, and the quantum system variables. The commutation relations of these functions are also constrained[6] by the superselection rule and the commutativity of the classical variables, x_1 and p_1, and their time derivatives. In the next section we will study the simple harmonic oscillator in order to exemplify the quantum-mechanical case described above and also for direct use in the field-theoretic generalizations of subsequent sections.

II. SIMPLE HARMONIC OSCILLATOR

In this section we discuss the embedding of a classical simple harmonic oscillator in a quantum system. We shall see that the space of states for the indefinite-metric classical-made-quantum system is far larger than the set of states of a classical harmonic oscillator. However, there is a subset of coherent states which may be placed in one-to-one correspondence with the classical harmonic-oscillator states. The classical-made-quantum oscillator is necessarily an indefinite-metric quantum theory for the simple physical reason that the classical bound states cannot have quantized energy levels. Indefinite-metric quantum theories normally have severe problems of physical interpretation. The present work raises the possibility of a partial resolution of some of these problems through a reinterpretation of an indefinite-metric quantum system as a system composed of a classical subsystem interacting with an essentially quantum subsystem of positive metric.

The classical simple harmonic oscillator of frequency ω has the Hamiltonian

$$\mathcal{K} = \frac{1}{2m}(p_1^2 + m^2\omega^2 x_1^2) , \tag{10}$$

and the motion is described by

$$x_1 = A\sin(\pi t + \delta) , \tag{11}$$

where A and δ are constants. To embed this classical system in a quantum-mechanical system we introduce the variables x_2 and p_2, and, using Eq. (8), obtain the quantum Hamiltonian

$$\hat{H} = \frac{1}{m}p_1 p_2 + m\omega^2 x_1 x_2 . \tag{12}$$

We eliminate constants by defining (for $i = 1, 2$)

$$x_i = \left(\frac{1}{m\omega}\right)^{1/2} Q_i , \tag{13}$$

$$p_i = (m\omega)^{1/2} P_i , \tag{14}$$

and

$$\hat{H} = H\omega \tag{15}$$

so that

$$H = P_1 P_2 + Q_1 Q_2 . \tag{16}$$

The raising and lowering operators are defined by

$$a_j = \frac{1}{\sqrt{2}}(Q_j + iP_j) , \tag{17}$$

and

$$a_j^\dagger = \frac{1}{\sqrt{2}}(Q_j - iP_j) \tag{18}$$

for $j = 1, 2$. They have the commutation relations

$$[a_i, a_j] = [a_i^\dagger, a_j^\dagger] = 0 , \tag{19}$$

$$[a_i, a_j^\dagger] = 1 - \delta_{ij} \tag{20}$$

or $i, j = 1, 2$. As a result H is seen to have the form

$$H = \tfrac{1}{2}(a_1 a_2^\dagger + a_2 a_1^\dagger + a_1^\dagger a_2 + a_2^\dagger a_1) \,. \tag{21}$$

The number operators are defined by

$$N_1 = a_2 a_1^\dagger \tag{22}$$

and

$$N_2 = a_2^\dagger a_1 \tag{23}$$

and are not Hermitian. However, their sum is Hermitian and we see that

$$H = N_1 + N_2 \,. \tag{24}$$

The number operators have the following commutation relations with the raising and lowering operators:

$$N_i a_j = a_j (N_i + \delta_{ij} - 1) \tag{25}$$

and

$$N_i a_j^\dagger = a_j^\dagger (N_i - \delta_{ij} + 1) \tag{26}$$

for $i, j = 1, 2$.

Up to this point we have maintained a symmetry of the dynamics under the exchange of the subscripts, $1 \longleftrightarrow 2$. Now we must break that symmetry by choosing a vacuum state which is an eigenstate of Q_1 and P_1 or alternately a_1 and $a_1^\dagger$. The commutativity of Q_1 and P_1 permit this. The observability of Q_1 and P_1 for all time requires it. So we define

$$a_1^\dagger |0\rangle = a_1 |0\rangle = 0 \,. \tag{27}$$

As a result $a_2 |0\rangle \neq 0$ and $a_2^\dagger |0\rangle \neq 0$. The eigenstates of the number operators are

$$|n_+, n_-\rangle = (a_2^\dagger)^{n_+} (a_2)^{n_-} |0, 0\rangle \tag{28}$$

and satisfy

$$N_1 |n_+, n_-\rangle = -n_- |n_+, n_-\rangle \,, \tag{29}$$

$$N_2 |n_+, n_-\rangle = n_+ |n_+, n_-\rangle \,, \tag{30}$$

so that

$$H |n_+, n_-\rangle = (n_+ - n_-) |n_+, n_-\rangle \,. \tag{31}$$

The lack of a lower bound to the energy spectrum is in a sense a problem but a necessary one in that it leads to the possibility of bound states with a continuous energy spectrum—a requirement of a faithful representation of the classical oscillator states. There is a subset of coherent states which can be put in a one-to-one relation with the set of classical oscillator states. The defining property of that subset is that its elements are eigenstates of the operators a_1 and $a_1^\dagger$. If we expand an element of that subset in terms of the number eigenstates

$$|z\rangle = \sum_{n_+, n_- = 0}^{\infty} f(z \,|\, n_+, n_-) |n_+, n_-\rangle \tag{32}$$

and use

$$a_1^\dagger |n_+, n_-\rangle = -n_- |n_+, n_- - 1\rangle \,, \tag{33}$$

$$a_1 |n_+, n_-\rangle = n_+ |n_+ - 1, n_-\rangle \tag{34}$$

to evaluate the eigenvalue equations

$$a_1 |z\rangle = iz^* |z\rangle \,, \tag{35}$$

$$a_1^\dagger |z\rangle = -iz |z\rangle \,, \tag{36}$$

we find

$$f(z \,|\, n_+, n_-) = \frac{C (iz^*)^{n_+} (iz)^{n_-}}{n_+! \, n_-!} \,, \tag{37}$$

where C is a constant. As a result

$$|z\rangle = C \exp[i(z a_2 + z^* a_2^\dagger)] |0, 0\rangle \,. \tag{38}$$

We shall call the $|z\rangle$ states coherent states because of their close formal resemblance to the coherent states used in the study of the classical limit of harmonic oscillators, and of quantum electrodynamics[11] (which were eigenstates of the lowering operator but not of the raising operator).

Since $[H, a_1] = -a_1$, and $[H, a_1^\dagger] = a_1^\dagger$, it is clear that the (x_1, p_1) phase-space trajectory is sharp on the set of coherent $|z\rangle$ states. The classical trajectory represented by the state $|z\rangle$ is easily seen to be

$$x_1 = \left(\frac{2}{m\omega}\right)^{1/2} R \sin(\omega t + \delta) \tag{39}$$

and

$$p_1 = (2m\omega)^{1/2} R \cos(\omega t + \delta) \,, \tag{40}$$

where $z = Re^{i\delta}$. The linearity of H in the "momenta", $\Pi = (p_2, -x_2)$, is crucial for the observability of the phase-space trajectory. In fact, the linearity of all generators of canonical transformations in the momenta is necessary if the canonical transformations are not to take states out of the subset of coherent states.

The superselection rule which follows from the unobservability of the momenta, Π, is best approached by a consideration of the momentum-and coordinate-space representations of the coherent states. In the coordinate-space representation we find that Eqs. (35) and (36) give

$$\left[\left(\frac{m\omega}{2}\right)^{1/2} x_1 + i\left(\frac{1}{2m\omega}\right)^{1/2} p_1\right]\langle x_1 p_1 | z\rangle = iz^* \langle x_1 p_1 | z\rangle \tag{41}$$

and

$$\left[\left(\frac{m\omega}{2}\right)^{1/2} x_1 - i\left(\frac{1}{2m\omega}\right)^{1/2} p_1\right]\langle x_1 p_1 | z\rangle = -iz \langle x_1 p_1 | z\rangle \,, \tag{42}$$

so that

$$\langle x_1 p_1 | z \rangle = \sqrt{2}\, \delta\!\left(x_1 - \left(\frac{2}{m\omega}\right)^{1/2} \mathrm{Im}z \right)$$
$$\times\, \delta(p_1 - (2m\omega)^{1/2}\mathrm{Re}z). \tag{43}$$

We have normalized $\langle x_1 p_1 | z \rangle$ so that

$$\langle z' | z \rangle = \int_{-\infty}^{\infty} dx_1 dp_1 \langle z' | x_1 p_1 \rangle \langle x_1 p_1 | z \rangle$$
$$= \delta(\mathrm{Re}z - \mathrm{Re}z')\delta(\mathrm{Im}z - \mathrm{Im}z'). \tag{44}$$

In momentum space Eqs. (35) and (36) lead to the differential equations

$$\left[\left(\frac{m\omega}{2}\right)^{1/2} i\frac{d}{dp_2} + \left(\frac{1}{2m\omega}\right)^{1/2}\frac{d}{dx_2} \right]\langle x_2 p_2 | z \rangle = iz^{*}\langle x_2 p_2 | z \rangle \tag{45}$$

and

$$\left[\left(\frac{m\omega}{2}\right)^{1/2} i\frac{d}{dp_2} - \left(\frac{1}{2m\omega}\right)^{1/2}\frac{d}{dx_2} \right]\langle x_2 p_2 | z \rangle = -iz\langle x_2 p_2 | z \rangle. \tag{46}$$

They are easily integrated to give

$$\langle x_2 p_2 | z \rangle = \frac{1}{\sqrt{2}\,\pi}\exp\!\left[-ip_2\left(\frac{2}{m\omega}\right)^{1/2}\mathrm{Im}z \right.$$
$$\left. + ix_2(2m\omega)^{1/2}\mathrm{Re}z \right] \tag{47}$$

with the normalization condition

$$\langle z' | z \rangle = \int_{-\infty}^{\infty} dx_2 dp_2 \langle z' | x_2 p_2 \rangle \langle x_2 p_2 | z \rangle$$
$$= \delta(\mathrm{Re}z - \mathrm{Re}z')\delta(\mathrm{Im}z - \mathrm{Im}z'). \tag{48}$$

The transformation function between the two representations is

$$\langle x_1 p_1 | x_2 p_2 \rangle = \frac{1}{2\pi}\exp(+ip_2 x_1 - ip_1 x_2), \tag{49}$$

so that

$$\langle x_1 p_1 | z \rangle = \int_{-\infty}^{\infty} dx_2 dp_2 \langle x_1 p_1 | x_2 p_2 \rangle \langle x_2 p_2 | z \rangle. \tag{50}$$

Each coherent state, $|z\rangle$, is a superselection sector in itself. There is no measurable dynamical variable $F = F(a_1, a_1^{\dagger})$ which connects different states:

$$\langle z' | F(a_1, a_1^{\dagger}) | z \rangle = F(iz^{*}, -iz)\delta^2(z - z'). \tag{51}$$

This reflects the lack of a superposition principle in classical mechanics.

The operator formalism for coherent states is incomplete in that we have not defined an inner product. To remedy this deficiency we define the vacuum dual to $|0,0\rangle$ to satisfy

$$\langle 0,0 | a_2 = \langle 0,0 | a_2^{\dagger} = 0 \tag{52}$$

with $\langle 0,0 | 0,0 \rangle = 1$. The dual state corresponding to the physical state, z, we define to be

$$\langle z | = \langle 0,0 | \delta(ia_1 + z^{*})\delta(ia_1^{\dagger} - z)$$
$$\equiv \langle 0,0 | \int_{-\infty}^{\infty} \frac{d\alpha\, d\beta}{(2\pi)^2}\exp[i\alpha\,(\mathrm{Im}z - 2^{-1/2}Q_1)$$
$$+ i\beta(\mathrm{Re}z - 2^{-1/2}P_1)] \tag{53}$$

so that Eqs. (48) and (51) follow if we choose $C = 1$.

Sometimes the dynamical state of a classical system is incompletely known and one only has a set of probabilities that the system is at a particular phase-space point at $t = 0$. If we let $P(z)$ be the probability that the system is at a phase-space point corresponding to z (as defined above), then using the properties

$$P(z) \geq 0, \quad \int d^2z\, P(z) = 1 \tag{54}$$

one sees that a density operator

$$\rho\delta^2(0) = \int d^2z\, |z\rangle P(z)\langle z | \tag{55}$$

may be defined which satisfies

$$\mathrm{Tr}\rho = 1 \tag{56}$$

and

$$\langle z' | \rho | z' \rangle \equiv \lim_{z'' \to z'} \langle z'' | \rho | z' \rangle = P(z'). \tag{57}$$

The mean value of an observable $A = A(a_1, a_1^{\dagger})$ is given by

$$\langle A \rangle = \mathrm{Tr}\rho A = \int d^2z\, A(iz^{*}, -iz)P(z), \tag{58}$$

and one can develop a formalism similar to the density-matrix formalism of quantum mechanics.

We now turn to a closer investigation of the relation of the pseudoquantum mechanics discussed above and true quantum-mechanical systems. We shall be particularly interested in the relation of the coherent states described above and the coherent states of a quantum-mechanical harmonic oscillator—to which they bear such a remarkable resemblance. We shall see that the pseudoquantum oscillator system is equivalent to an indefinite-metric quantum system composed of a harmonic oscillator (thus the connection to the coherent-state quantum oscillator formalism) and an "inverted" oscillator to be described below.

Let us define the following rotated raising and lowering operators in terms of the operators defined in Eqs. (17) and (18):

$$b_1 = a_1 \cos\theta + a_2 \sin\theta, \tag{59}$$

$$b_2 = -a_1 \sin\theta + a_2 \cos\theta. \tag{60}$$

Their commutation relations are

$$[b_1, b_1^\dagger] = \sin(2\theta) , \qquad (61)$$

$$[b_2, b_2^\dagger] = -\sin(2\theta) , \qquad (62)$$

$$[b_2, b_1^\dagger] = [b_1, b_2^\dagger] = \cos(2\theta) \qquad (63)$$

with all other commutators equal to zero. The Hamiltonian of Eq. (21) becomes

$$H = \tfrac{1}{2}(\{b_1, b_1^\dagger\} - \{b_2, b_2^\dagger\}) \sin(2\theta)$$
$$+ \tfrac{1}{2}(\{a_1, a_2^\dagger\} + \{a_2, a_1^\dagger\}) \cos(2\theta) , \qquad (64)$$

where $\{u, v\} = uv + vu$.

Now θ is an arbitrary angle and it is obvious that choosing $\theta = 0$ gives the commutation relations and Hamiltonian studied above. However, the choice $\theta = \pi/4$ results in a new form for H and the commutation relations, which can be interpreted as a harmonic oscillator (the b_1 and $b_1^\dagger$ sector) and an "inverted" harmonic oscillator (the b_2 and $b_2^\dagger$ sector) where the commutator and b_2 terms in the Hamiltonian have the wrong sign. The commutativity of the oscillator raising and lowering operators with the inverted oscillator raising and lowering operators leads to a simple factorization of the coherent states which lays bare the basic of the close similarity of form for our coherent states and the coherent states of a quantum oscillator[10]:

$$|z\rangle = \frac{1}{\sqrt{2\pi}} \exp\left[\frac{i}{\sqrt{2}}(zb_1 + z^*b_1^\dagger)\right]$$

$$\times \exp\left[\frac{i}{\sqrt{2}}(zb_2 + z^*b_2^\dagger)\right]|0, 0\rangle , \qquad (65)$$

while the coherent state of Ref. 11 has the form

$$|\alpha\rangle = \exp(\alpha b^\dagger - \alpha^* b)|0\rangle , \qquad (66)$$

where α is a complex numer and $[b, b^\dagger] = 1$. It should be remembered that our choice of vacuum state such that $a_1|0, 0\rangle = a_1^\dagger|0, 0\rangle = 0$ obviates a simple direct relationship.

Since we have uncovered an interesting relation between a classical-made-quantum system and a "quantum" system of indefinite metric the possibility of reinterpreting indefinite-metric quantum systems as systems containing classical subsystems naturally arises.

III. EMBEDDING OF CLASSICAL FIELDS

In this section we shall discuss the embedding of a classical field theory in a quantum field theory. We shall study the embedding in detail for a scalar field and then describe the features of a classical-made-quantum electrodynamics which we shall call pseudoquantum electrodynamics for the sake of brevity.

Consider a classical field, $\phi_1(x)$, with canonically conjugate momentum, $\pi_1(x)$, and Hamiltonian equations of motion

$$\frac{d}{dt}\phi_1(x) = \frac{\delta\hat{H}}{\delta\pi_1(x)} , \qquad (67)$$

$$\frac{d}{dt}\pi_1(x) = \frac{-\delta H}{\delta\phi_1(x)} , \qquad (68)$$

where $\hat{H}$ is the Hamiltonian. We wish to define a "quantum" Hamiltonian, H, which allows us to rewrite Eqs. (67) and (68) in commutator form:

$$\frac{d}{dt}\phi_1(x) = i[H, \phi_1(x)] , \qquad (69)$$

$$\frac{d}{dt}\pi_1(x) = i[H, \pi_1(x)] . \qquad (70)$$

Equations (69) and (70) are satisfied if

$$H = \int d^3x \left[\frac{\delta H}{\delta\pi_1(x)}\frac{1}{i}\frac{\delta}{\delta\phi_1(x)}\right.$$
$$\left. - \frac{\delta H}{\delta\phi_1(x)}\frac{1}{i}\frac{\delta}{\delta\pi_1(x)}\right] . \qquad (71)$$

We now formally define

$$\phi_2(x) = i\frac{\delta}{\delta\pi_1(x)} \qquad (72)$$

and

$$\pi_2(x) = -i\frac{\delta}{\delta\phi_1(x)} , \qquad (73)$$

so that

$$H = \int d^3x \left[\frac{\delta\hat{H}}{\delta\pi_1(x)}\pi_2(x)\right.$$
$$\left. + \frac{\delta\hat{H}}{\delta\phi_1(x)}\phi_2(x)\right] . \qquad (74)$$

The fields satisfy the equal-time commutation relations

$$[\phi_i(x), \pi_j(y)] = i(1 - \delta_{ij})\delta^3(\vec{x} - \vec{y}) , \qquad (75)$$

$$[\phi_i(x), \phi_j(y)] = 0 , \qquad (76)$$

$$[\pi_i(x), \pi_j(y)] = 0 , \qquad (77)$$

where δ_{ij} is the Kronecker δ.

We note that the linearity of H in ϕ_2 and π_2 is necessary to maintain the classical character of ϕ_1 and π_1. This is best seen by an examination of Eqs. (69) and (70) and the corresponding Hamiltonian equations for ϕ_2 and π_2. (Other generators of canonical transformations are also linear in π_2 and ϕ_2.)

$\phi_2(x)$ and $\pi_2(x)$ will not be observables on the set of physical states, so that $\phi_1(x)$ and $\pi_1(x)$ will both be sharp on the set of physical states and satisfy superselection rules.

If we wish to couple the classical field to a truly quantum system and maintain the classical nature of the field then certain restrictions exist on the form of the total Hamiltonian H_{tot} and on the commutation relations of the various terms occurring in it. First, the coupling must satisfy the requirement that H_{tot} is linear in $\phi_2(x)$ and $\pi_2(x)$. If we denote the quantum fields by ψ and write the general form of the Hamiltonian as

$$H_{\text{tot}} = H + H_Q(\psi) + H_{\text{int}} , \tag{78}$$

where H is given by Eq. (74), $H_Q(\psi)$ depends only on the quantum fields, ψ, and

$$H_{\text{int}} = \int d^3x [\bar{A}(\phi_1, \pi_1, \psi)\phi_2(x) \\ + \bar{B}(\phi_1, \pi_1, \psi)\pi_2(x) \\ + \bar{C}(\phi_1, \pi_1, \psi)] , \tag{79}$$

then we can rearrange the Hamiltonian so that

$$H_{\text{tot}} = \int d^3x [A(\phi_1, \pi_1, \psi)\phi_2(x) \\ + B(\phi_1, \pi_1, \psi)\pi_2(x) \\ + C(\phi_1, \pi_1, \psi)] , \tag{80}$$

where

$$A = \frac{\delta \hat{H}}{\delta \phi_1(x)} + \bar{A} , \tag{81}$$

$$B = \frac{\delta \hat{H}}{\delta \pi_1(x)} + \bar{B} , \tag{82}$$

and

$$C = \bar{C} + \mathcal{H}_Q \tag{83}$$

with $H_Q = \int d^3x\, \mathcal{H}_Q$. An examination of the equations of motion of $\phi_1(x)$, $\pi_1(x)$, and ψ,

$$\frac{d}{dt}\phi_1 = B(\phi_1, \pi_1, \psi) , \tag{84}$$

$$\frac{d}{dt}\pi_1 = A(\phi_1, \pi_1, \psi) , \tag{85}$$

$$\frac{d}{dt}\psi = i[H_{\text{tot}}, \psi] , \tag{86}$$

and the second time derivatives of ϕ_1 and π_1, such as

$$\frac{d^2}{dt^2}\phi_1(x) = i[H, B] \\ = \int d^3y \left(-A\frac{\delta B}{\delta \pi_1(y)} + B\frac{\delta B}{\delta \phi_1(y)} + i\phi_2(y)[A, B] \right. \\ \left. + i\pi_2(y)[B(y), B(x)] + i[C, B] \right) , \tag{87}$$

leads us to require the equal-time commutation

relations

$$[A(x), A(y)] = [A(x), B(y)] = [B(x), B(y)] = 0 , \tag{88}$$

where $A(x) = A(\phi_1(x), \pi_1(x), \psi(x))$, etc., so that $\phi_1(x)$ and $\pi_1(x)$ are independent of ϕ_2 and π_2 and hence observable for all time. An examination of higher time derivatives of ϕ_1 and π_1 lead to further restrictions on the equal-time commutation relations of A, B, and C. Examples are

$$[A, [C, B]] = 0 , \tag{89}$$

$$[B, [C, B]] = 0 , \tag{90}$$

$$[A, [C, [C, [C, B]]]] = 0 , \tag{91}$$

etc. A sufficient condition for satisfying all relations of this class consists of having equal-time commutation relations with the form

$$[A, C] = F_1(A, B, \phi_1, \pi_1) \tag{92}$$

and

$$[B, C] = F_2(A, B, \phi_1, \pi_1) . \tag{93}$$

Finally, we note that another obvious requirement [cf. Eqs. (84) and (85)] for the observability of ϕ_1 and π_1 is that A and B depend only on an (equal-time) commutative subset of the quantum field variables, ψ.

The above restrictions on the equal-time commutation relations have a direct interpretation in terms of Feynman diagrams for quantum corrections to the classical field behavior. For example, consider the interaction of the classical field sector with a scalar quantum field, ψ, expressed in the interaction

$$H_{\text{int}} = g\phi_2(x)\psi^2(x). \tag{94}$$

If $H_Q(\psi)$ is the conventional free Klein-Gordon Hamiltonian, then we find that Eq. (92) is not satisfied so that the Green's function for the classical ϕ_1 field receives quantum corrections from vacuum polarization loops of ψ particles and thus loses its classical character.

We now define a Lagrangian appropriate to our pseudoquantum field theory and then verify the reasonableness of our definition, and the pseudoquantization procedure described above, by studying the equivalent path-integral formulation. The Lagrangian corresponding to the pseudoquantum Hamiltonian, H, is

$$L = \int d^3x(\pi_1\dot{\phi}_2 + \pi_2\dot{\phi}_1) - H , \tag{95}$$

where $L = L(\phi_1, \dot{\phi}_1, \phi_2, \dot{\phi}_2)$ and

$$\pi_1 = \frac{\delta L}{\delta \dot{\phi}_2} , \tag{96}$$

$$\pi_2 = \frac{\delta L}{\delta \dot{\phi}_1} \, . \tag{97}$$

The vacuum-vacuum transition amplitude for the field theory corresponding to the H_{tot} of Eq. (78) will be shown to be

$$W = \int \prod_x d\phi_1(x) d\phi_2(x) d\pi_1(x) d\pi_2(x) d\psi(x) \exp(iS) \, , \tag{98}$$

where $S = \int dt \, L_{tot}$ up to external source terms. We begin by considering the vacuum-vacuum transition amplitude corresponding to H_Q,

$$W_Q = \int \prod_x d\psi(x) \exp(iS_Q) \, , \tag{99}$$

where ϕ_1 has the character of an external source. We can now introduce the classical behavior of the ϕ_1 field through functional δ functions

$$\int \prod_x d\psi(x) d\phi_1(x) d\pi_1(x) \delta(B(\phi_1,\pi_1,\psi) - \dot{\phi}_1)$$
$$\times \delta(A(\phi_1,\pi_1,\psi) + \dot{\pi}_1) e^{iS_Q} \, , \tag{100}$$

which can be put in the form

$$\int \prod_x d\phi_1(x) d\pi_1(x) d\phi_2(x) d\pi_2(x)$$
$$\times \exp\left\{ i \int d^4x [(\dot{\phi}_1 - B)\pi_2 - (\dot{\pi}_1 + A)\phi_2] + iS_Q \right\} \, . \tag{101}$$

After performing a partial integration on the $\dot{\pi}_1 \phi_2$ term and discarding a surface term we see that the definition of L in Eq. (95) is correct and that the vacuum-vacuum transition amplitude is indeed given by Eq. (98).

The restrictions on the commutation relations of the various terms in the H_{tot} [expressed in Eqs. (88)–(93)] translate into the requirement that the "quantum completion"[11] of the ϕ_2 field does not take place, i.e., that all N-point functions of the ϕ_2 field are zero:

$$\frac{\delta^n W}{\delta J_2(x_1) \delta J_2(x_2) \cdots \delta J_2(x_n)} = 0 \, , \tag{102}$$

where J_2 is an external source coupled to ϕ_2.

We now discuss the embedding of a free classical Klein-Gordon field in a quantum field theory. The Lagrangian density is

$$\mathcal{L} = \frac{\partial \phi_1}{\partial x^\mu} \frac{\partial \phi_2}{\partial x_\mu} - m^2 \phi_1 \phi_2 \, . \tag{103}$$

From which one obtains the Euler-Lagrange equations (for $i = 1, 2$)

$$(\Box + m^2)\phi_i(x) = 0 \, . \tag{104}$$

The canonical momenta are (note that π_2 is conjugate to ϕ_1, etc.)

$$\Pi_i = \dot{\phi}_i \tag{105}$$

for $i = 1, 2$ with the equal-time commutation relations given by Eqs. (75)–(77). We expand the fields in Fourier integrals:

$$\phi_1(\vec{x}, t) = \int d^3k [a_1(k) f_k(x) + a_1^\dagger f_k^*(x)] \tag{106}$$

and

$$\phi_2(\vec{x}, t) = \int d^3k [a_2(k) f_k(x) + a_2^\dagger(k) f_k^*(x)] \, , \tag{107}$$

where

$$f_k(x) = (2\pi)^{-3/2} (2\omega_k)^{-1/2} e^{-ik \cdot x} \tag{108}$$

with $\omega_k = (\vec{k}^2 + m^2)^{1/2}$. The Fourier component operators satisfy the commutation relations

$$[a_i(k), a_j^\dagger(k')] = (1 - \delta_{ij}) \delta^3(\vec{k} - \vec{k}') \tag{109}$$

and

$$[a_i(k), a_j(k')] = [a_i^\dagger(k), a_j^\dagger(k')] = 0 \tag{110}$$

for $i, j = 1, 2$.

In terms of the Fourier coefficients

$$H = \int d^3x (\dot{\phi}_1 \dot{\phi}_2 + \vec{\nabla}\phi_1 \cdot \vec{\nabla}\phi_2 + m^2 \phi_1 \phi_2) \tag{111}$$

becomes

$$H = \int d^3k \, \omega_k [\{a_1(k), a_2^\dagger(k)\}$$
$$+ \{a_2(k), a_1^\dagger(k)\}] \, . \tag{112}$$

The analogy between the mode amplitudes of the fields and the raising and lowering operators of the simple harmonic oscillator has been previously remarked. We can therefore use the considerations of Sec. II to establish the spectrum of physical states. The defining properties of a physical state are that $\phi_1(x)$ and $\pi_1(x)$ are sharp on it for all time:

$$\phi_1(x)|\Phi, \Pi\rangle = \Phi(x)|\Phi, \Pi\rangle \tag{113}$$

and

$$\pi_1(x)|\Phi, \Pi\rangle = \Pi(x)|\Phi, \Pi\rangle \, , \tag{114}$$

where $\Phi(x)$ and $\Pi(x)$ are c-number functions of x:

$$\Phi(x) = \int d^3k [\alpha(k) f_k(x) + \alpha^*(k) f_k^*(x)] \tag{115}$$

and

$$\Pi(x) = -i \int d^3k \, \omega_k [\alpha(k) f_k(x) - \alpha^*(k) f_k^*(x)] \tag{116}$$

with $\alpha(k)$ a c-number function of k.

As a result we are led to define a set of physical states, $|\alpha\rangle$, which are in one-to-one correspondence with the classical solutions of the Klein-Gordon equation and satisfy

$$a_1(k)|\alpha\rangle = \alpha(k)|\alpha\rangle, \tag{117}$$

$$a_1^\dagger(k)|\alpha\rangle = \alpha^*(k)|\alpha\rangle. \tag{118}$$

In analogy with the states of the simple harmonic oscillator (Sec. II) we further define

$$|\alpha\rangle = C \exp\left\{ \int d^3k'[\alpha(k')a_2^\dagger(k') \right.$$

$$\left. - \alpha^*(k')a_2(k')]\right\}|0\rangle, \tag{119}$$

where the vacuum state, $|0\rangle$, satisfies

$$a_1(k)|0\rangle = a_1^\dagger(k)|0\rangle = 0. \tag{120}$$

The physical states, $|\alpha\rangle$, lie in a space which is the infinite tensor product of single-mode spaces. While ϕ_1 and π_1 are sharp for all time on the subset of physical states, we see that ϕ_2 and π_2 are not and, in fact, when applied to a physical state map it into an unphysical state. The superselection rules are embodied in

$$\langle \alpha' | \mathcal{O} | \alpha \rangle = \mathcal{O}_\alpha \delta^2(\alpha - \alpha'), \tag{121}$$

where $\mathcal{O}$ is the operator corresponding to any observable, $\mathcal{O}_\alpha$ is its eigenvalue for the state $|\alpha\rangle$, and $\delta^2(\alpha - \alpha')$ is a functional δ function in the real and imaginary parts of $\alpha - \alpha'$. The functional δ functions have their origin in the definition of the dual set of physical states. We define the dual vacuum state $\langle 0 |$ by

$$\langle 0 | a_2(k) = 0 \tag{122a}$$

and

$$\langle 0 | a_2^\dagger(k) = 0 \tag{122b}$$

for all k with $\langle 0 | 0 \rangle = 1$. The dual state corresponding to $\alpha(k)$ we define by

$$\langle \alpha | = \langle 0 | \prod_k \delta(\alpha(k) - a_1(k))\delta(\alpha^*(k) - a_1^\dagger(k))$$

$$\equiv \langle 0 | \delta(\alpha - a_1)\delta(\alpha^* - a_1^\dagger), \tag{123}$$

so that

$$\langle \alpha' | \alpha \rangle = \delta^2(\alpha' - \alpha) \tag{124}$$

if $C = 1$.

We have now established a procedure for embedding a classical field in a quantum field theory. Given a Lagrangian, L, for a classical field theory describing a field $\phi_1(x)$, the Lagrangian density for the pseudoquantum field theory, $\mathcal{L}_{PQ}$ is

$$\mathcal{L}_{PQ}(\phi_1, \dot\phi_1, \phi_2, \dot\phi_2) = \frac{\delta L}{\delta\phi_1(x)} \phi_2(x)$$

$$+ \frac{\delta L}{\delta\dot\phi_1(x)} \pi_2(x) \tag{125}$$

up to a divergence with

$$\pi_2(x) = \frac{\delta}{\delta\dot\phi_1(x)} \int d^3x\, \mathcal{L}_{PQ}. \tag{126}$$

In the case of a classical electromagnetic field interacting with a quantum electron field, one pseudoquantum model, which describes some electromagnetic processes, has the Lagrangian

$$\mathcal{L} = -\tfrac{1}{2} F_{\mu\nu}^1 F_{\mu\nu}^2 + \overline{\psi}(i\not{\nabla} - e\not{A}_1 - m_0)\psi, \tag{127}$$

where $A_\mu^1(x)$ is the classical electromagnetic field, ψ is the electron field, $A_\mu^2(x)$ is the unobservable auxiliary field, and $F_{\mu\nu}^i = \partial_\nu A_\mu^i - \partial_\nu A_\nu^i$ for $i = 1, 2$. Although our interpretation of the free electromagnetic part of the Lagrangian, $-\tfrac{1}{2}F_{\mu\nu}^1 F_{\mu\nu}^2$, is new, the actual form of this term appeared some time ago in a generalization of electrodynamics by Mie,[12] and was recently used in an Abelian prototype model for quark confinement.[8] The equations of motion are

$$\partial^\mu F_{\mu\nu}^1 = 0, \tag{128}$$

$$\partial^\mu F_{\mu\nu}^2 + eJ_\nu = 0, \tag{129}$$

and

$$(i\not{\nabla} - e\not{A}^1 - m)\psi = 0. \tag{130}$$

The canonical momentum which is conjugate to A_μ^1 is

$$\Pi_\mu^2 = F_{0\mu}^2 \tag{131}$$

and that conjugate to A_μ^2 is

$$\Pi_\mu^1 = F_{0\mu}^1. \tag{132}$$

We take A_μ^1 and Π_μ^1 to be classical fields which are observable for all time. A_μ^2 and Π_μ^2 are not observable. Note that $\mathcal{L}$ is invariant under the independent gauge transformations

$$A_\mu^1 \rightarrow A_\mu^1 + \partial_\mu \Lambda^1(x) \tag{133}$$

and

$$A_\mu^2 \rightarrow A_\mu^2 + \partial_\mu \Lambda^2(x). \tag{134}$$

Since $\Pi_0^1 = \Pi_0^2 = 0$, it is apparent that A_0^1 and A_0^2 are c numbers. If we chose the Coulomb gauge for A_μ^1,

$$\vec\nabla \cdot \vec{A}^1 = 0, \tag{135}$$

and for A_μ^2,

$$\vec\nabla \cdot \vec{A}^2 = 0, \tag{136}$$

then we can establish the equal-time commutation relations

$$[\Pi_i^a(\vec{x},t), A_j^b(\vec{y},t)] = i(1 - \delta_{ab})$$

$$\times \int \frac{d^3k}{(2\pi)^3} e^{i\vec{k}\cdot(\vec{x}-\vec{y})} \left(\delta_{ij} - \frac{k_i k_j}{|\vec{k}|^2}\right)$$

$$= i(1 - \delta_{ab})\delta_{ij}^{tr}(\vec{x} - \vec{y}) \qquad (137)$$

for $a, b = 1, 2$ and $i, j = 1, 2, 3$.

This pseudoquantum field theory describes the dynamics of quantum electron fields interacting with a free, classical electromagnetic field. A typical perturbation theory matrix element would have the form

$$\langle \mathcal{G}', 0 | T(\bar{\psi}(x)J^{\mu_1}(x_1)A_{\mu_1}^1(x_1)J^{\mu_2}(x_2)A_{\mu_2}^1(x_2)\cdots J^{\mu_n}(x_n)A_{\mu_n}^1(x_n)\psi(y)) | \mathcal{G}, 0 \rangle, \qquad (138)$$

where $|\mathcal{G}, 0\rangle$ is the tensor product of an electron vacuum state and an electromagnetic state corresponding to the classical field $\mathcal{G}_\mu(z)$. Because $A_\mu^1(x)$ is sharp on this state, the matrix element becomes

$$\langle 0 | T(\bar{\psi}(x)J^{\mu_1}(x_1)\cdots J^{\mu_n}(x_n)\psi(y)) | 0 \rangle \mathcal{G}_{\mu_1}(x_1)\mathcal{G}_{\mu_2}(x_2)\cdots \mathcal{G}_{\mu_n}(x_n) \qquad (139)$$

modulo a functional δ function in $\mathcal{G}' - \mathcal{G}$. Thus this model is equivalent to a quantized electron field interacting with an external electromagnetic field.

Another possibility for a model electrodynamics is realized by letting the interaction term in Eq. (127) above be replaced with

$$L_{int} = - e\bar{\psi}A_2\psi. \qquad (140)$$

Because the equivalent of the equal-time commutation relation, Eq. (92), is not true in this model, the A_μ^1 field loses its purely classical character due to quantum corrections. However, this model may be of value for the study of the modification of the A_μ^1 field resulting from the emission of many soft photons by a current.

Since vacuum polarization effects modify the electromagnetic field in this case we define in-field eigenstates (in the transverse gauge) by

$$\vec{A}_{in}^1 | \mathcal{G} \rangle_{in} = \vec{\mathcal{G}}_{in} | \mathcal{G} \rangle_{in}, \qquad (141)$$

where

$$| \mathcal{G} \rangle_{in} = \exp\left[\int d^3k \sum_{\lambda=1}^2 (\alpha(k,\lambda)a_2^\dagger(k,\lambda)\right.$$

$$\left. - \alpha^*(k,\lambda)a_2(k,\lambda)) \right] | 0 \rangle \qquad (142)$$

and

$$\vec{\mathcal{G}}_{in} = \int d^3k \sum_{\lambda=1}^2 \vec{\epsilon}(k,\lambda)[\alpha(k,\lambda)f_k(x)$$

$$+ \alpha^*(k,\lambda)f_k^*(x)] \qquad (143)$$

with

$$\vec{A}_{in}^i = \int d^3k \sum_{\lambda=1}^2 \vec{\epsilon}(k,\lambda)[a_i(k,\lambda)f_k(x)$$

$$+ a_i^\dagger(k,\lambda)f_k^*(x)] \qquad (144)$$

for $i = 1, 2$. The vacuum state is defined by

$$a_1(k,\lambda)|0\rangle = a_1^\dagger(k,\lambda)|0\rangle = 0$$

for all k, λ. The interacting field, $\vec{A}^1$, is apparently not sharp on $|\mathcal{G}\rangle_{in}$ but is sharp on

$$| \mathcal{G} \rangle = U^{-1}(t, -\infty) | \mathcal{G} \rangle_{in}, \qquad (145)$$

where

$$U(t, -\infty) = T\left(\exp\left[- i \int_\infty^t d^4x\, H_{int}(A_{in}^2, \psi_{in}) \right] \right) \qquad (146)$$

because

$$\vec{A}^1(\vec{x}, t) = U^{-1}(t, -\infty)\vec{A}_{in}^1(\vec{x}, t)U(t, -\infty). \qquad (147)$$

With these preliminaries completed, the study of physical processes within the framework of these models is now possible, although we shall not pursue it in this report.

Before turning to a discussion of non-Abelian gauge field theories, it is worth noting that the choice of vacuum state we have made necessitates a redefinition of normal-ordering. By normal-ordering a Lagrangian term we shall mean that the observable fields (to which we have consistently appended the superscript or subscript one) are to be placed to the right, and unobservable fields, labeled by two, are to be placed to the left. Thus Wick's theorem (with our definition of normal-ordering) becomes in the case of two fields

$$T(\phi_{1\,in}(x_1)\phi_{2\,in}(x_2)) = \,: \phi_{1\,in}(x_1)\phi_{2\,in}(x_2):$$

$$+ \langle 0 | T(\phi_{1\,in}(x_1)\phi_{2\,in}(x_2)) | 0 \rangle$$

$$= \phi_{2\,in}(x_2)\phi_{1\,in}(x_1)$$

$$+ \theta(x_{10} - x_{20})[\phi_{1\,in}(x_1), \phi_{2\,in}(x_2)].$$
$$(148)$$

Note that the Green's function

$$G(x_1, x_2) = \langle 0 | T(\phi_{1\,in}(x_1)\phi_{2\,in}(x_2)) | 0 \rangle \qquad (149)$$

is necessarily retarded. From this we can conclude that the models of electrodynamics, which we have considered, naturally embody the observed retarded nature of classical electrodynamics. Another way of stating this result is: If classical electrodynamics is to have a pseudoquantum formulation, its Green's functions are necessarily retarded. The origin of the asymmetry is the definition of the vacuums (which is equivalent to a specification of boundary conditions). Just as in classical electrodynamics retarded propagation is implemented by a choice of boundary conditions which do not require a commitment to any specific cosmological model.

Finally we would like to note that the Lagrangian obtained from adding L_{int} of Eq. (140) to the Lagrangian of Eq. (127) is equivalent to the usual Lagrangian of electrodynamics plus a term describing a massless Abelian gauge field with the wrong sign. (This is seen by defining new fields equal to the sum and difference of A_μ^1 and A_μ^2.) This field theory may be quantized following the procedure we have outlined. A_μ^1 loses its classical character due to quantum corrections.

IV. NON-ABELIAN GAUGE THEORIES

In this section we shall describe the procedure for embedding a classical non-Abelian Yang-Mills field in a quantum field theory. Then we will discuss a vierbein formulation of quantum gravity which could have been interpreted as a pseudoquantum field theory for a classical metric field if it were not for one term in the Lagrangian which makes it a truly quantum field theory. Nevertheless we suggest a new canonical quantization procedure based on our pseudoquantum approach.

Consider a classical Yang-Mills field, $A_\mu^1 = A_\mu^1 \cdot T$, where the jth component of T is a matrix representing a generator of a non-Abelian group G in the defining representation with commutation relations

$$[T_j, T_k] = it_{jkl} T_l. \tag{150}$$

We can define a pseudoquantum field theory, wherein the classical character of A_μ^1 is maintained, which has the Lagrangian density

$$\mathcal{L} = \tfrac{1}{2} \underline{F}_{\mu\nu}^1 \cdot \underline{F}^{2\mu\nu} - \tfrac{1}{2} \underline{F}^{2\mu\nu} \cdot (\partial_\mu \underline{A}_\nu^1 - \partial_\nu \underline{A}_\mu^1 + g \underline{A}_\mu^1 \times \underline{A}_\nu^1)$$
$$- \tfrac{1}{2} \underline{F}^{1\mu\nu} \cdot (\partial_\mu \underline{A}_\nu^2 - \partial_\nu \underline{A}_\mu^2 + g \underline{A}_\mu^1 \times \underline{A}_\nu^2 - g \underline{A}_\nu^1 \times \underline{A}_\mu^2)$$
$$+ \bar{\psi}(i\slashed{\nabla} + g\slashed{A}^1 - m)\psi, \tag{151}$$

where ψ is a fermion field. The theory is invariant under the local gauge transformation, $S \in G$,

$$\psi' = S^{-1}\psi, \tag{152}$$

$$A_\mu^{1'} = S^{-1}A_\mu^1 S + \frac{i}{g}S^{-1}\partial_\mu S, \tag{153}$$

$$F_{\mu\nu}^{1'} = S^{-1}F_{\mu\nu}^1 S, \tag{154}$$

$$A_\mu^{2'} = S^{-1}A_\mu^2 S, \tag{155}$$

$$F_{\mu\nu}^{2'} = S^{-1}F_{\mu\nu}^2 S. \tag{156}$$

Except for one important term this Lagrangian with its attendant gauge invariance properties has been suggested as a possible model for the quark-confining strong interaction.[8] Since the omitted term has a masslike character $\Lambda^2 A_\mu^2 \cdot A^{2\mu}$, where Λ has the dimensions of a mass, it is clear that the strong-interaction model's ultraviolet behavior approaches that of the present pseudoquantum theory if the same quantization procedure is followed in both cases. We shall discuss this question further in the next section and show that the *ad hoc* procedure followed in Ref. 8 leads to the same result as the quantization procedure developed in this report.

The Euler-Lagrange equations of motion which are obtained from $\mathcal{L}$ in the canonical manner are

$$\underline{F}_{\mu\nu}^1 = \partial_\mu \underline{A}_\nu^1 - \partial_\nu \underline{A}_\mu^1 + g \underline{A}_\mu^1 \times \underline{A}_\nu^1, \tag{157}$$

$$\underline{F}_{\mu\nu}^2 = \partial_\mu \underline{A}_\nu^2 - \partial_\nu \underline{A}_\mu^2 + g \underline{A}_\mu^1 \times \underline{A}_\nu^2 - g \underline{A}_\nu^1 \times \underline{A}_\mu^2, \tag{158}$$

$$(\partial_\mu + g\underline{A}_\mu^1 \times)\underline{F}^{1\mu\nu} = 0, \tag{159}$$

$$(\partial_\mu + g\underline{A}_\mu^1 \times)\underline{F}^{2\mu\nu} + g\underline{A}_\mu^2 \times \underline{F}^{1\mu\nu} + g\underline{J}^\nu = 0, \tag{160}$$

$$(i\slashed{\nabla} + g\slashed{A}^1 - m)\psi = 0, \tag{161}$$

with the conservation law

$$(\partial_\nu + g\underline{A}_\nu^1 \times)\underline{J}^\nu = 0. \tag{162}$$

The canonical momentum which is conjugate to $\underline{A}_j^1$ is

$$\underline{\Pi}_j^2 = \underline{F}_{0j}^2 \tag{163}$$

and the canonical momentum conjugate to $\underline{A}_j^2$ is

$$\underline{\Pi}_j^1 = \underline{F}_{0j}^1 \tag{164}$$

for $j = 1, 2, 3$. The canonical momentum corresponding to the fields A_0^i is zero for $i = 1, 2$. The existence of equations of constraint among the Euler-Lagrange equations implies that not all field components are independent, so that we must isolate the independent components prior to defining the canonical equal-time commutation relations.

Following Ref. 8 we choose to work in the Coulomb gauge, $\nabla_i A_i^1 = 0$, and define the field variables

$$A_i^2 = A_i^{2T} + A_i^{2L}, \tag{165}$$

$$\underline{\Pi}_i^a = \underline{\Pi}_i^{aT} + \underline{\Pi}_i^{aL}, \tag{166}$$

where

$$\nabla_i \cdot \underline{A}_i^{2T} = \nabla_i \cdot \underline{\Pi}_i^{aT} = 0 \tag{167}$$

and $a = 1, 2$. Then the nonzero equal-time commutation relations are

$$[\Pi_{ip}^{aT}(x), A_{jq}^{bT}(y)] = i\delta_{pq}(1 - \delta_{ab})\delta_{ij}^{tr}(\vec{x} - \vec{y}), \tag{168}$$

where p and q are internal-symmetry indices, $a, b = 1, 2$, and $i, j = 1, 2, 3$.

While the classical character of A_μ^1 can be maintained with our choice of $\mathcal{L}$, this theory has features due to its non-Abelian nature which make it less trivial and therefore more interesting than the corresponding Abelian theory discussed in the last section. If we follow a procedure similar to that in the Abelian case [Eq. (127)] and introduce a set of states appropriate to the quadratic part of the Lagrangian, then the cubic and quartic Yang-Mills terms in the interaction part of the Lagrangian will act to transform $A_{\text{in}\,\mu}^1$ eigenstates into eigenstates of the interacting field A_μ^1. This is, of course, necessary for the classical Yang-Mills equations of motion to be satisfied. Our formalism, thus, offers a perturbative method for calculating solutions of the classical Yang-Mills equations. In addition, it gives an interesting interpretation to the short-distance behavior of the quark-confining field theory of Ref. 8. At short distances the gluon field A_μ^1 effectively decouples from the quark sector and becomes, in effect, a free field. This type of short-distance behavior is certainly not at odds with the seemingly simple behavior observed in hadron processes at high energy. Therefore, it is possible that pseudoquantum field theory may be relevant to the short-distance behavior of hadron interaction. Certainly, it is interesting that elementary fermions fall into two similar groups: those which appear to be individually observable (leptons) and those which are not individually observable (quarks).

We now turn to a consideration of a vierbein model of gravity which has certain close similarities to the pseudoquantum field theories we have been studying. In Weyl's formulation[13] of the Einstein-Cartan theory of gravity a vierbein field, $l^{\mu a}(x)$, is introduced which is the "square root" of the metric tensor

$$g^{\mu\nu} = \eta_{ab} l^{\mu a} l^{\nu b}, \tag{169}$$

where η_{ab} is the constant metric tensor of special relativity, where Roman indices transform as vectors under the $SL(2,C)$ group of local Lorentz transformations, and where Greek indices transform as vectors under general coordinate transformations. It is useful to introduce the constant Dirac matrices, γ_a and $4S_{ab} = i[\gamma_a, \gamma_b]$. Under an $SL(2,C)$ transformation,

$$S = \exp[iC^{ab}(x)S_{ab}], \tag{170}$$

a spinor, $\psi(x)$, becomes

$$\psi' = S\psi. \tag{171}$$

The local nature of the transformation requires the introduction of a gauge field

$$B_\mu^{ab} = -B_\mu^{ba} \tag{172}$$

which transforms inhomogeneously,

$$B_\mu \to SB_\mu S^{-1} - \frac{i}{g} S\partial_\mu S^{-1}, \tag{173}$$

so that a Lorentz transformation gauge-covariant derivative can be defined

$$\nabla_\mu \psi = (\partial_\mu + igB_\mu)\psi, \tag{174}$$

where $B_\mu = B_\mu^{ab} S_{ab}$ and $g = 12\pi G$ where G is Newton's constant. Under a gauge transformation we have

$$l^\mu = l^{\mu a}\gamma_a - Sl^\mu S^{-1}, \tag{175}$$

so that the gauge-covariant derivative of l^μ is defined to be

$$\nabla_\nu l^\mu = (\partial_\nu + igB_\nu \times) l^\mu, \tag{176}$$

where $B_\nu \times l^\mu = [B_\nu, l^\mu]$. The commutator

$$igB_{\mu\nu} = [\partial_\mu + igB_\mu, \partial_\nu + igB_\nu] \tag{177}$$

transforms homogeneously under a gauge transformation

$$B_{\mu\nu} \to SB_{\mu\nu}S^{-1}, \tag{178}$$

and as a second-rank tensor under general coordinate transformations. With these field quantities we are able to construct a Lagrangian $\mathcal{L}_{\text{Weyl}}$ which reduces to the Einstein Lagrangian for gravity when no matter is present,[13]

$$\mathcal{L} = \mathcal{L}_{\text{Weyl}} + \mathcal{L}_{\text{matter}}, \tag{179}$$

where

$$\mathcal{L}_{\text{Weyl}} = \frac{i}{8l} \operatorname{Tr} l^\mu l^\nu B_{\mu\nu} \tag{180}$$

and where, for example, we might let

$$l\mathcal{L}_{\text{matter}} = \bar{\psi}(il^\mu \nabla_\mu + m)\psi \tag{181}$$

with $l = \det(l^{\mu a})$.

We observe that the terms containing derivatives in $\mathcal{L}_{\text{Weyl}}$ are linear in the field B_μ—a suggestive feature in view of our previous discussion. However, the quadratic term in B_μ eliminates the possibility of regarding $\mathcal{L}_{\text{Weyl}}$ as a pseudoquantum field theory for a classical field $l^{\mu a}$. But, regardless of this consideration, the fact that $l^{\mu a}$ is necessarily classical in part leads us to consider quantizing vierbein gravity in a manner which is based on the pseudoquantization procedure described above. Remembering that a successful perturbation theory requires the perturbation to be around known solutions we introduce a quadratic Lagrangian term via

$$\mathcal{L} = \mathcal{L}_0 + (\mathcal{L} - \mathcal{L}_0) = \mathcal{L}_0 + \mathcal{L}_{\text{int}}, \tag{182}$$

where

$$\mathcal{L}_0 = -\tfrac{1}{4} i \operatorname{Tr}(B'_{\mu a} l^\mu \gamma^a + ig [B_a, B_b] \gamma^a \gamma^b) \tag{183}$$

and

$$B'_{\mu a} = \partial_\mu B_a - \partial_a B_\mu . \qquad (184)$$

Our plan is to follow the pseudoquantization procedure for the "free" part of the Lagrangian $\mathcal{L}_0$. Therefore we will (i) choose a particular coordinate system (harmonic coordinates) and a particular gauge, the "Lorentz" gauge, $\partial^\mu B_\mu = 0$, (ii) establish equal-time commutation relations, (iii) define a set of eigenstates of $l^{\mu a}$, and (iv) proceed to calculate quantum corrections in perturbation theory.

The equations of motion for the "free" Lagrangian $\mathcal{L}_0$ are

$$\partial_\mu B_b^{ab} - \partial_b B_\mu^{ab} = 0 \qquad (185)$$

and

$$\partial_\mu(l^{\mu a}\eta^{\nu b} - l^{\nu a}\eta^{\mu b}) + 2g\,(\eta^{\nu a}B_c^{cb} - \eta^{\nu b}B_c^{ca}$$
$$-\eta^{ac}B_c^{\nu b} + \eta^{bc}B_c^{\nu a}) = 0 . \qquad (186)$$

We work in the gravitational equivalent of the Lorentz gauge of electrodynamics,

$$\partial^\mu B_\mu^{ab} = 0 , \qquad (187)$$

and choose harmonic coordinates

$$\partial_\mu l^{\mu a} = \tfrac{1}{2}\,\partial^a \eta_{\sigma\,\tau}\,l^{\sigma\,\tau} . \qquad (188)$$

The Green's function associated with Eq. (185) is

$$G_{\alpha ef,\,\rho\sigma}(x,y) = -\tfrac{1}{2} \int \frac{d^4k}{k^2}\, e^{-ik\cdot(x-y)} g_{\alpha ef,\,\rho\sigma}(k) , \qquad (189)$$

where

$$g_{\alpha ef,\,\rho\sigma}(k) = k_e\left(\eta_{\alpha\rho}\eta_{f\sigma} + \eta_{\alpha\sigma}\eta_{f\rho} - \eta_{\alpha f}\eta_{\rho\sigma} - \frac{k_\alpha k_\rho \eta_{f\sigma} + k_\alpha k_\sigma \eta_{f\rho}}{k^2}\right)$$
$$-k_f\left(\eta_{\alpha\rho}\eta_{e\sigma} + \eta_{\alpha\sigma}\eta_{e\rho} - \eta_{\alpha e}\eta_{\rho\sigma} - \frac{k_\alpha k_\rho \eta_{e\sigma} + k_\alpha k_\sigma \eta_{e\rho}}{k^2}\right). \qquad (190)$$

In order to relate the above Green's function to a time-ordered product of the quantum fields it is first necessary to introduce a set of coherent states, $|L\rangle$, which are eigenstates of $l^{\mu a}$:

$$l^{\mu a}(x)\,|L\rangle = L^{\mu a}(x)\,|L\rangle , \qquad (191)$$

where $L^{\mu a}(x)$ is a c-number function of x. In particular, we define $|\eta\rangle$ to satisfy

$$l^{\mu a}|\eta\rangle = \eta^{\mu a}|\eta\rangle , \qquad (192)$$

where $\eta^{\mu a}$ is the constant Lorentz metric tensor of special relativity. Given a state $|L\rangle$ we define the field

$$l_L^{\mu a} = l^{\mu a} - L^{\mu a} . \qquad (193)$$

This field corresponds to the quantum part of $l^{\mu a}$ and when applied to the purely classical state $|L\rangle$ has the eigenvalue zero.

We now make the identification

$$iG_{\alpha ef,\,\rho\sigma}(x,y) = \langle L\,|\,T(B_{\alpha ef}(x), l_{L\rho\sigma}(y))\,|L\rangle . \qquad (194)$$

If we desire to calculate quantum corrections to $l_{\rho\sigma} = \eta_{\rho\sigma}$ we choose $|L\rangle = |\eta\rangle$. (It should be noted that $G_{\alpha ef,\,\rho\sigma}$ is independent of the choice of $|L\rangle$ as we have defined it.) Because $l_{L\rho\sigma}(y)$ is sharp on $|L\rangle$ we find that the right side of Eq. (194) becomes

$$iG_{\alpha ef,\,\rho\sigma}(x,y) = \theta(y_0 - x_0)[l_{\rho\sigma}(y), B_{\alpha ef}(x)] \qquad (195)$$

up to a functional δ function. From the form of $\mathcal{L}_0$ we see that the commutator is not zero. It is fully determined by an equal-time commutation

relation of $l_{\rho\sigma}$ and $B_{\alpha ef}$ (which by the way is the only nonzero equal-time commutator if the canonical procedure is followed), the equations of motion, and the requirement that it be zero at space-like distances. The "retarded" form of $G_{\alpha ef,\,\rho\sigma}$ fixes the integration contour around poles in Eq. (192). The other nonzero Green's function in the free Lagrangian model specified by $\mathcal{L}_0$ is

$$iH^{\mu\nu,\,\rho\sigma}(x,y) = \langle L\,|\,T(l_L^{\mu\nu}(x), l_L^{\rho\sigma}(y))\,|L\rangle . \qquad (196)$$

It is nonzero owing to the presence of the $[B_\mu, B_\nu]$ term in $\mathcal{L}_0$. We shall show in the next section that it is a principal-value propagator rather than a Feynman propagator. In coordinate space this results in $H^{\mu\nu,\,\rho\sigma}$ being the sum of the advanced and retarded propagators. As a result our model is equivalent to an action-at-a-distance theory in some sectors.

The classical part of $l_{\mu\alpha}$ is the solution of the classical linearized field equations with appropriate matter sources. The linearized field equations are derived from a Lagrangian consisting of $\mathcal{L}_0$ plus matter terms. (Note that the form of $\mathcal{L}_0$ is obtained by substituting $l_{\mu a} = \eta_{\mu a} + h_{\mu a}$ in $\mathcal{L}_{\text{Weyl}}$, expanding, and keeping quadratic terms.) Thus the class of possible background metrics is restricted.

A simplification occurs in perturbation theory when the classical part of $l_{\mu\alpha}$ is $\eta_{\mu\alpha}$. In this case $(\mathcal{L}_{\text{Weyl}} - \mathcal{L}_0)|\eta\rangle = 0$ when $\mathcal{L}_0$ and $\mathcal{L}_{\text{Weyl}}$ are expressed in terms of asymptotic fields.

V. PRINCIPAL-VALUE PROPAGATORS AND ACTION AT A DISTANCE

In this section we shall show that certain propagators, in field theories where the pseudoquantization procedure has been followed, are principal-value propagators (i.e., the sum of the advanced and retarded Green's functions in coordinate space) rather than Feynman propagators. We also describe a quantum field theory for action-at-a-distance electrodynamics which completes the program initiated by Schwarzschild, Tetrode, and Fokker.[14]

To illustrate the origin of the principal-value propagator we return to the scalar field model of Eq. (103) which described a classical field, $\phi_1(x)$. We introduce an interaction term

$$L_{int} = - \int d^3z \, \tfrac{1}{2} \lambda^2 [\phi_2(z)]^2 \tag{197}$$

(where λ is a constant), which destroys the purely classical nature of ϕ_1. Suppose we consider the Green's function

$$i\tilde{G}(x,y) = \langle 0 | T(\phi_1(x)\phi_1(y)) | 0 \rangle , \tag{198}$$

which would be zero if L_{int} were not present. In terms of in-fields we have

$$i\tilde{G}(x,y) = \left\langle 0 \left| T\left(\phi_{1in}(x)\phi_{1in}(v) \exp\left(i \int dt \, L_{int} \right) \right) \right| 0 \right\rangle , \tag{199}$$

where the vacuum states, $|0\rangle$ and $\langle 0|$, are defined as in Eqs. (120) and (122). From the definition of the vacuum we find (dropping "in" labels)

$$i\tilde{G}(x,y) = \frac{-i\lambda^2}{2} \int d^4z \langle 0 | T(\phi_1(x)\phi_1(y)\phi_2^2(z)) | 0 \rangle , \tag{200}$$

which becomes

$$i\tilde{G}(x,y) = \frac{-i\lambda^2}{2} \epsilon(x_0 - y_0) \frac{\partial}{\partial m^2} \Delta(x-y) \tag{201}$$

with

$$\Delta(x-y) = -i \int \frac{d^4k}{(2\pi)^3} \delta(k^2 - m^2)\epsilon(k_0)e^{-ik\cdot(x-y)} . \tag{202}$$

Using

$$\tfrac{1}{2} \epsilon(x_0 - v_0)\Delta(x-v) = \int \frac{d^4k}{(2\pi)^4} \, P \, \frac{1}{k^2 - m^2}$$

$$\times e^{-ik\cdot(x-y)} , \tag{203}$$

we see that

$$\tilde{G}(x,y) = -\lambda^2 \int \frac{d^4k}{(2\pi)^4} \, P \, \frac{1}{(k^2 - m^2)^2} e^{-ik\cdot(x-y)} , \tag{204}$$

where

$$P \, \frac{1}{(k^2 - m^2)^2} \equiv \frac{1}{2} \left[\frac{1}{(k^2 - m^2 + i\epsilon)^2} + \frac{1}{(k^2 - m^2 - i\epsilon)^2} \right]. \tag{205}$$

The form of $\tilde{G}$ is consistent with the equations of motion:

$$(\Box + m^2)\phi_1 + \lambda^2 \phi_2 = 0 , \tag{206}$$

$$(\Box + m^2)\phi_2 = \delta^4(x-y) . \tag{207}$$

The appearance of the principal-value dipole propagator rather than the Feynman dipole propagator in Eq. (204) is useful because it eliminates certain unitarity problems associated with indefinite-metric fields. However, depending on the model under consideration, it could lead to difficulties with causality. To illustrate the manner in which unitarity problems are resolved, consider the interaction of the ϕ_1 dipole field with a scalar quantum field ψ with

$$L'_{int} = g\phi_1(x)[\psi(x)]^2 . \tag{208}$$

Suppose we consider the subset of in and out states containing arbitrary numbers of ψ particles but no ϕ_1 or ϕ_2 particles. These states have positive metric. If one could systematically exclude indefinite-metric ϕ_1 and ϕ_2 particles from physical states one would avoid negative probabilities and other problems. But the sum over states in a unitarity sum would normally include states with ϕ_1 particles if the ϕ_1 field had Feynman propagators. In the case of principal-value propagators, no intermediate states with ϕ_1 particles occur, since the pole term is not present. The interaction mediated by the ϕ_1 field is a form of action at a distance and ϕ_1 is properly described by the phrase adjunct field, coined by Feynman and Wheeler.[14] A more detailed discussion of the unitarity question is given in Refs. 7 and 8. In those articles a dipole gluon model for quark confinement was proposed which introduced principal-value propagators in an *ad hoc* manner to resolve unitarity problems. It was pointed out that causality problems did not necessarily exist in those models because the non-Abelian dipole gluons were confined for the same reason as the quarks so that— at the worst— there would be unobservable causality violations at distances of the order of hadron dimensions.

The pseudoquantization procedure may be used to construct a quantum field-theoretic version of action-at-a-distance electrodynamics. Consider the Lagrangian

$$\mathcal{L} = -\tfrac{1}{2} F^{\mu\nu}(\partial_\nu A_\mu - \partial_\mu A_\nu) + \tfrac{1}{4} F^{\mu\nu}F_{\mu\nu}$$

$$+ \bar{\psi}(i\slashed{\partial} - e\slashed{A} - m_0)\psi . \tag{209}$$

We define the momentum

$$\Pi_\mu = \frac{\delta \mathcal{L}}{\delta \dot{A}^\mu} = F_{0\mu}. \tag{210}$$

Going to the transverse gauge as in Sec. IV, we define the equal-time commutation relation

$$[\Pi_i(\vec{x}, t), A_j(\vec{y}, t)] = i\delta^{tr}_{ij}(\vec{x} - \vec{y}). \tag{211}$$

Suppose we neglect interaction terms in $\mathcal{L}$ for the moment and choose $F_{\mu\nu}$ to be an observable classical field (as it is up to quantum corrections which we neglect) and A_μ to be unobservable (as it is because it is not gauge invariant). Then we follow our pseudoquantization procedure for

$$\mathcal{L}_0 = -\tfrac{1}{2} F^{\mu\nu}(\partial_\nu A_\mu - \partial_\mu A_\nu) + \tfrac{1}{4} F^{\mu\nu} F_{\mu\nu}. \tag{212}$$

In particular, we define a vacuum such that

$$F_{\mu\nu}|0\rangle = 0, \quad A_\mu|0\rangle \neq 0, \tag{213}$$

while

$$\langle 0|A_\mu = 0, \quad \langle 0|F_{\mu\nu} \neq 0. \tag{214}$$

Then

$$iG_{\mu\nu}(x, y) = \langle 0| T(A_\mu(x) A_\nu(y)) |0\rangle \tag{215}$$

would be zero were it not for $F_{\mu\nu} F^{\mu\nu}$ in $\mathcal{L}_0$. In terms of appropriate in-fields it becomes

$$2iG_{\mu\nu}(x, y) = \int d^4z \, (\theta(x_0 - y_0)\theta(y_0 - z_0)$$
$$+ \theta(y_0 - x_0)\theta(x_0 - z_0))$$
$$\times [A_{\mu\,in}(x), F_{\alpha\beta\,in}(z)][A_{\mu\,in}(y), F^{\alpha\beta}_{in}(z)]. \tag{216}$$

Note that we are treating $F_{\mu\nu} F^{\mu\nu}$ in $\mathcal{L}_0$ as an interaction term. The structure of $G_{\mu\nu}(x, y)$ is the same as that of Eq. (200) so we can conclude that

$$G_{\mu\nu}(x, y) = -g_{\mu\nu} \int \frac{d^4k}{(2\pi)^4} \, P \, \frac{1}{k^2} \, e^{-ik\cdot(x-y)} \tag{217}$$

in the Feynman gauge. Thus the action-at-a-distance interaction follows from the pseudoquantization of electrodynamics. The classical character of $F_{\mu\nu}$ is lost owing to quantum corrections resulting from the presence of $J_\mu A^\mu$ in the Lagrangian.

The example we have just studied has a certain parallel in the vierbein model of gravitation studied in the last section. The forms of the Lagrangian and commutation relations are similar. As a result it is clear that

$$D^{\mu\nu,\lambda\sigma}(x, v) \equiv \left\langle L \left| T\left(l^{\mu\nu}_{L\,in}(x)l^{\lambda\sigma}_{L\,in}(v) \int d^4z \, \bar{\mathcal{L}}_{int}(z)\right) \right| L \right\rangle \tag{218}$$

with

$$\bar{\mathcal{L}}_{int} = \tfrac{1}{4} g \operatorname{Tr}[B_{\mu\,in}, B_{\nu\,in}|\gamma^\mu\gamma^\nu \tag{219}$$

is a principal-value propagator. Therefore we have constructed an action-at-a-distance version of quantum gravity. Our motivation was to take account of the classical part of $l^{\mu a}$ in a way which did not divorce it from the quantum part to which it is intimately related.

VI. CONCLUSION

We have seen that an alternative to Fock-space quantization exists for a class of field theories which have Lagrangian gradient terms which are linear in field variables. A method was also proposed for constructing Lagrangians of that type from classical Lagrangians with gradient terms which are quadratic in field variables. To some extent this process has a parallel in the passage from Klein-Gordon field Lagrangians which are quadratic in derivatives to Dirac field Lagrangians which are linear in derivatives.

The quantization procedure we have outlined is canonical so far as the fields are concerned. We do, however, make a choice of vacuum states which differs from the usual choice. As a result we have found free propagators which were either retarded, or half-advanced and half-retarded. The choice of vacuum state does not in itself preclude the appearance of Feynman propagators. If one has a good reason to modify the canonical commutation relations then it is possible to obtain Feynman propagators.[15] The procedure we have outlined has, therefore, a greater generality than the particular class of models studied in the present work. It can enable one to embed a classical field theory in a quantum field theory in such a way as to maintain its classical character. It can also be applied to study classical field theories which obtain quantum corrections. Finally it can be applied in order to obtain a fully second-quantized field theory (cf. Ref. 15).

ACKNOWLEDGMENT

This work was supported in part by the U.S. Energy Research and Development Administration.

---

*Present address: Physics Department, Williams College, Williamstown, Mass. 01267.

[1]D. R. Yennie, S. C. Frautschi, and H. Suura, Ann. Phys. (N.Y.) 13, 379 (1961).

[2]R. J. Glauber, Phys. Rev. 131, 2766 (1963).

[3]W. A. Bardeen, M. S. Chanowitz, S. D. Drell, M. Weinstein, and T.-M. Yan, Phys. Rev. D 11, 1094 (1975).

J. M. Cornwall and G. Tiktopoulos, Phys. Rev. D 13, 3370 (1976).

S. Blaha, Phys. Lett. 56B, 373 (1975).

E. C. G. Sudarshan, Center for Particle Theory report Univ. of Texas—Austin, 1976 (unpublished).

S. Blaha, Phys. Rev. D 10, 4268 (1974).

S. Blaha, Phys. Rev. D 11, 2921 (1975).

S. Blaha, Lett. Nuovo Cimento 18, 60 (1977).

[0]Cf. Ref. 2; T. W. B. Kibble, J. Math. Phys. 9, 315 (1968); Phys. Rev. 173, 1527 (1968); 174, 1882 (1968); 175, 1624 (1968);

[1]A. Salam, lecture at Center for Theoretical Studies, Miami, Florida, 1973 (unpublished).

[2]G. Mie, Ann. Phys. (Leipzig) 37, 511 (1912); 39, 1 (1912); 40, 1 (1913); H. Weyl, *Space, Time, Matter* (Dover, N.Y. 1952).

[13]H. Weyl, Z. Phys. 56, 330 (1929); T. W. B. Kibble, J. Math. Phys. 2, 212 (1961); J. Schwinger, Phys. Rev. 130, 1253 (1963); C. J. Isham, A. Salam, and J. Strathdee, Lett. Nuovo Cimento 5, 969 (1972); F. W. Hehl, P. von der Heyde, G. D. Kerlick, and J. Nester, Rev. Mod. Phys. 48, 393 (1976); and references therein.

[14]K. Schwarzschild, Göttinger Nachrichten 128, 132 (1903); H. Tetrode, Z. Phys. 10, 317 (1922); A. D. Fokker, *ibid*. 58, 386 (1929); J. Wheeler and R. P. Feynman, Rev. Mod. Phys. 17, 157 (1945); 21, 425 (1949).

[15]S. Blaha (unpublished).

Appendix J. PseudoQuantum – Second Quantized Non-Abelian Field Theory for Hadrons with Quark Confinement …

This refereed paper is S. Blaha, Phys. Rev. **D11**, 2921 (1975). Reprinted with the kind permission of Physical Review D.

PHYSICAL REVIEW D   VOLUME 11, NUMBER 10   15 MAY 1975

Second-quantized non-Abelian field theory for hadrons with quark confinement and scaling deep-inelastic structure functions*

Stephen Blaha

Laboratory of Nuclear Studies, Cornell University, Ithaca, New York 14853

(Received 30 December 1974)

A four-dimensional second-quantized field theory with quarks bound by "colored" non-Abelian gluons is described which has the following properties: (1) the only physical particles are color singlets composed solely of quarks, (2) the deep-inelastic structure functions have Bjorken scaling, (3) gluon loops and Faddeev-Popov ghost loops are identically zero in any gauge, (4) Regge trajectories are apparently linear on a Chew-Frautschi plot, and (5) constituent motion within hadrons can be nonrelativistic.

I. INTRODUCTION

After a period of some skepticism the possibility that hadronic interactions might be understood within the framework of quantum field theory is again being seriously considered.[1] This is partly the result of the psychological climate created by the apparently successful unification of weak and electromagnetic interactions in a renormalizable field theory and partly the result of a greater appreciation of the variety of phenomena which can occur in field theories.

In this article we shall describe a field-theoretic model of hadron binding which has two major features: (1) Hadrons only occur as quark-antiquark or three-quark bound states, and (2) quarks behave as quasifree particles within hadrons. We assume that the suggestions of an internal symmetry called color[2] are correct and that the strong interaction consists of the exchange of colored Yang-Mills gluons. The nature of the interaction allows only color singlet states to occur in the gauge-invariant physical particle spectrum and consequently the first feature will be realized by choosing the color group to be SU(3). Since the (Schwinger) mechanism which produces this result is an infrared phenomenon, the second feature is not precluded and the model is essentially free in the ultraviolet region of the quark sector.

Our model is a non-Abelian version of a recently investigated Abelian field theory which had quark confinement and scaling electroproduction structure functions.[3] In that theory the free propagator of the massless gluon field embodying the quark-quark interaction was proportional to

$$\lambda^2/k^4, \tag{1}$$

where λ is a constant with the dimensions of mass and k is the gluon four-momentum. As a result the Schwinger mechanism[4] manifestly occurred, and it was shown that any charged particle was

totally screened by vacuum polarization effects. In addition, explicit calculations of the deep-inelastic electroproduction structure functions in perturbation theory were in agreement with Bjorken scaling with corrections of $O(q^{-4})$, where q is the virtual photon four-momentum. These features of the Abelian model will also be shown to be true in the non-Abelian version. In addition, we shall argue that the quarks can be nonrelativistic within hadrons and that the spectrum of states has linearly rising Regge trajectories.

In spite of these salutary properties an interaction of the form of Eq. (1) could be questioned because of well-known[5] indefinite-metric difficulties which result in the violation of unitarity. While an optimist may hope that the nonappearance of colored gluons in asymptotic (color singlet) states might eliminate unitarity problems it is almost certain that the approximation techniques which will necessarily be used to find the bound states will lead to the occurrence of negative-metric states. Whether these states are "real" or artifacts of the approximation will not be clear. In view of this we suggested[3] that the gluon propagator be taken in principal value rather than as a Feynman propagator:

$$P\frac{\lambda^2}{k^4} \equiv \frac{\lambda^2}{2}\left[\frac{1}{(k^2+i\epsilon)^2} + \frac{1}{(k^2-i\epsilon)^2}\right]. \tag{2}$$

As a result unitarity is maintained order by order in perturbation theory. Gluons do not appear in asymptotic states. All components of the vector-gluon propagator are "Coulombized" and the gluon field reduced to the embodiment of a direct quark interaction. There are a number of other decided advantages to principal-value propagators in the present context: (1) no color singlet states composed solely of gluons, (2) the elimination of substantial infrared divergences, (3) the suppression of corrections to Bjorken scaling in the electroproduction structure functions by a factor of q^2

vis-à-vis the corresponding Feynman-propagator result which sets the stage for precocious scaling, and (4) the elimination of closed loops of vector gluons and thus the elimination of Faddeev-Popov ghost loops.

In Sec. II we give a brief recapitulation of the Abelian model. In Sec. III we describe the canonical properties of the non-Abelian model. In Sec. IV we describe the qualitative features of the model and describe an approximation technique which appears to be naturally adopted to "solving" the theory. We shall restrict our discussion to the color binding interaction and defer the introduction of other interactions to a later work. The properties of the bound states in the non-Abelian model are currently under study and will be the subject of the next report.

II. ABELIAN MODEL

The possibility that the physical particle spectrum of a field theory consisted only of neutral states and did not include states of charged fields was first investigated in massless two-dimensional quantum electrodynamics.[6] In that case the absence of the "electron" from the gauge-invariant physical particle spectrum was directly related to the acquisition of a mass by the photon via the Schwinger mechanism. The Schwinger mechanism was manifest in the lowest-order contribution to the vacuum polarization (Fig. 1), and, taking account of the dimensionality of the coupling constant, $e \sim$ mass, could almost be considered a consequence of dimensional analysis. These vacuum polarization effects led to the total screening of the "electronic" charge, and, as a result, the "electron" was removed from the gauge-invariant physical particle spectrum. Our Abelian and non-Abelian models will display a similar pattern of events.

The Lagrangian of the Abelian model contains two gluon fields, $A_\mu^1(x)$ and $A_\mu^2(x)$, and the quark field $\psi(x)$:

$$\mathcal{L} = -\tfrac{1}{2} F_{\mu\nu}^1 F_{\mu\nu}^2 - \tfrac{1}{2}\lambda^2 A_\mu^2 A_\mu^2 + \overline{\psi}(i\,\nabla\!\!\!\!/ - g A\!\!\!/^1 - m)\psi ,$$

$$(3)$$

where for typographic convenience we denote the inner product of four vectors, $a \cdot b = a_\mu b_\mu = a_0 b_0 - \vec{a} \cdot \vec{b}$ throughout, λ is a constant with the dimensions of mass, g is dimensionless, and $F_{\mu\nu}^i$

$$= \partial_\nu A_\mu^i - \partial_\mu A_\nu^i.$$

Following the canonical procedure we find the equations of motion;

$$\partial_\mu F_{\mu\nu}^1 + \lambda^2 A_\nu^2 = 0 , \tag{4}$$

$$\partial_\mu F_{\mu\nu}^2 + g J_\nu = 0 , \tag{5}$$

$$(i\nabla\!\!\!\!/ - g A\!\!\!/^1 - m)\psi = 0 , \tag{6}$$

and nonzero equal-time commutation relations [in the Coulomb gauge $\vec{\nabla} \cdot \vec{A}^1 = 0$; note $\partial_\mu A_\mu^2 = 0$ by Eq. (4)]

$$\left[F_{0i}^1(x), A_j^2(y) \right] = i\Delta_{ij}^{tt}(x-y) , \tag{7}$$

$$\left[F_{0i}^2(x), A_j^1(y) \right] = i\Delta_{ij}^{tt}(x-y) , \tag{8}$$

with $i, j = 1, 2, 3$ and

$$\Delta_{ij}^{tt}(x-y) = \int \frac{d^3k}{(2\pi)^3} e^{i\vec{k}\cdot(\vec{x}-\vec{y})} \left(\delta_{ij} - \frac{k_i k_j}{|\vec{k}|^2} \right). \tag{9}$$

It is clear from the equations of motion, Eqs. (4) and (5), that A_μ^2 may be eliminated to obtain

$$\Box \partial_\mu F_{\mu\nu}^1 + g\lambda^2 J_\nu = 0 . \tag{10}$$

The form of the quark-gluon interaction and Eq. (10) show that only the Green's function of A_μ^1 is relevant to quark-quark scattering. The perturbation theory rules of QED may be used if the photon propagator is replaced with the gluon propagator for A_μ^1:

$$iG_{\mu\nu}^{11}(k) = \frac{i\lambda^2(g_{\mu\nu} - \chi k_\mu k_\nu /k^2)}{k^4} , \tag{11}$$

where χ is constant, determined by the gauge choice.

In Ref. 3 we showed that choosing $G_{\mu\nu}^{11}$ to be a principal-value propagator allowed us to develop a perturbation theory which was unitary order by order:

$$G_{\mu\nu}^{11}(k^2) \equiv \tfrac{1}{2}\left[G_{\mu\nu}^{11}(k^2 + i\epsilon) + G_{\mu\nu}^{11}(k^2 - i\epsilon) \right]. \tag{12}$$

In addition, the equivalent of the Nambu representation of a Feynman diagram was given and some features of the perturbation theory discussed. Of particular interest was a calculation of the deep-inelastic electroproduction structure functions which scaled in the Bjorken limit. Leading corrections to scaling were of $O(q^{-4})$ as $q^2 \to \infty$ with q being the virtual photon four-momentum, and were given by the diagrams of Fig. 2(b), 2(c), and 2(d). This is to be contrasted with the logarithmic deviations from scaling found in pseudoscalar or vector meson models previously studied.[7]

The Schwinger mechanism manifestly occurred in low orders of perturbation theory. As a result quarks (and all charged objects) are removed from the gauge-invariant spectrum of physical

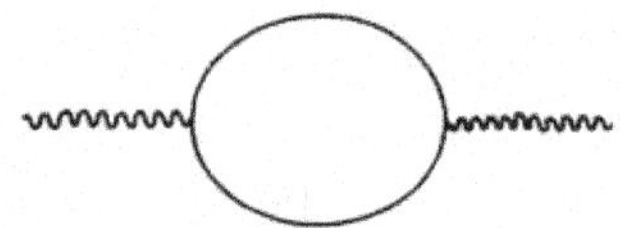

FIG. 1. A vacuum polarization diagram.

states. The total screening of charge can be seen from the following argument.[3] Consider a spatially bounded system of charge density ρ. The total charge is

$$Q = \int d^3x\, \rho(x) \tag{13}$$

$$= \frac{-1}{g\lambda^2} \int d^3x\, \Box \nabla^2 A_0^1 \tag{14}$$

using the equations of motion in the Coulomb gauge. By Gauss's law

$$Q = \frac{-1}{g\lambda^2} \int d\vec{S} \cdot \vec{\nabla}\, \Box A_0^1 . \tag{15}$$

From the definition of a Green's function, we have

$$A_0^1(x) = \int d^4y\, G_{00}^{11}(x - y)\rho(y) \tag{16}$$

in the Coulomb gauge. If, for simplicity, we choose ρ to describe a static point quark charge and use the free gluon propagator [Eq. (11)], then $Q \neq 0$. However, if we take account of the effect of vacuum polarization processes (the Schwinger mechanism) we find $A_0^1(x)$ is a monotonically decreasing function of $|\vec{x}|$ for large $|\vec{x}|$ and consequently $Q = 0$ in the limit where the integration surface is taken to infinity in Eq. (15). Thus the spectrum of physical states does not include states of nonzero charge. In the next section we shall show that the proof of quark confinement is essentially the same in the non-Abelian model.

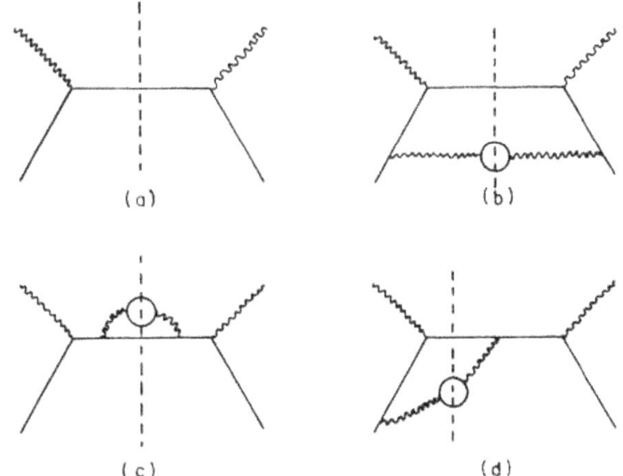

FIG. 2. Lowest-order diagrams contributing to the inelastic electroproduction structure functions. The dashed lines indicate the only contributions to the electroproduction structure functions of the absorptive part of the forward virtual Compton scattering diagram. External "wiggly" lines represent photons while internal "wiggly" lines represent gluons.

III. NON-ABELIAN MODEL

The non-Abelian model for the color sector of hadronic interactions is a direct generalization of the model of the last section.[8] There are two colored Yang-Mills fields, $A_{\mu a}^1(x)$ and $A_{\mu a}^2(x)$, which when regarded as vectors in the adjoint representation of the color group are denoted $\underline{A}_\mu^1$ and $\underline{A}_\mu^2$. The Lagrangian is

$$\mathcal{L} = \tfrac{1}{2}\underline{F}_{\mu\nu}^1 \cdot \underline{F}_{\mu\nu}^2 - \tfrac{1}{2}\underline{F}_{\mu\nu}^2 \cdot (\partial_\mu \underline{A}_\nu^1 - \partial_\nu \underline{A}_\mu^1 + g\underline{A}_\mu^1 \times \underline{A}_\nu^1)$$
$$- \tfrac{1}{2}\underline{F}_{\mu\nu}^1 \cdot (\partial_\mu \underline{A}_\nu^2 - \partial_\nu \underline{A}_\mu^2 + g\underline{A}_\mu^1 \times \underline{A}_\nu^2 - g\underline{A}_\nu^1 \times \underline{A}_\mu^2)$$
$$- \tfrac{1}{2}\lambda^2 \underline{A}_\mu^2 \cdot \underline{A}_\mu^2 + \bar\psi(i\nabla + gA^1 - m)\psi \tag{17}$$

$$= \mathcal{L}_0 + \bar\psi(i\nabla + gA^1 - m)\psi . \tag{18}$$

with ψ being the quark field.

It is invariant under the local gauge transformation

$$\psi' = S^{-1}\psi , \tag{19}$$

$$A_\mu^{1\prime} = S^{-1}A_\mu^1 S + \frac{i}{g} S^{-1}\partial_\mu S , \tag{20}$$

$$A_\mu^{2\prime} = S^{-1}A_\mu^2 S , \tag{21}$$

$$F_{\mu\nu}^{1\prime} = S^{-1}F_{\mu\nu}^1 S , \tag{22}$$

$$F_{\mu\nu}^{2\prime} = S^{-1}F_{\mu\nu}^2 S , \tag{23}$$

where S is an element in the gauge group G [which is color SU(3) in our case], and A_μ^1 is a matrix in the defining representation of G formed from

$$A_\mu^1 = \underline{A}_\mu^1 \cdot \underline{T} . \tag{24}$$

T_a is a matrix in the defining representation of G satisfying

$$[T_a, T_b] = i f_{abc} T_c , \tag{25}$$

and $\underline{T}$ is a vector formed from such matrices. We note that the homogeneity of the gauge transformation of A_μ^2 allows a mass term to occur in $\mathcal{L}$ without breaking the gauge symmetry. We shall see that the natural gauge-fixing term to add to the Lagrangian has the form

$$- \frac{1}{\beta} \partial_\mu \underline{A}_\mu^1 \cdot \partial_\nu \underline{A}_\nu^2 . \tag{26}$$

The Euler-Lagrange equations of motion are obtained in the canonical manner:

$$(\partial_\mu + g\underline{A}_\mu^1 \times)\underline{F}_{\mu\nu}^1 - \lambda^2 \underline{A}_\nu^2 = 0 , \tag{27}$$

$$(\partial_\mu + g\underline{A}_\mu^1 \times)\underline{F}_{\mu\nu}^2 + g\underline{A}_\mu^2 \times \underline{F}_{\mu\nu}^1 + g\underline{J}_\nu = 0 , \tag{28}$$

$$\underline{F}_{\mu\nu}^1 = \partial_\mu \underline{A}_\nu^1 - \partial_\nu \underline{A}_\mu^1 + g\underline{A}_\mu^1 \times \underline{A}_\nu^1 , \tag{29}$$

$$\underline{F}_{\mu\nu}^2 = \partial_\mu \underline{A}_\nu^2 - \partial_\nu \underline{A}_\mu^2 + g\underline{A}_\mu^1 \times \underline{A}_\nu^2 - g\underline{A}_\nu^1 \times \underline{A}_\mu^2 , \tag{30}$$

$$(i\nabla + gA^1 - m)\psi = 0 . \tag{31}$$

The antisymmetry of $\underline{F}_{\mu\nu}^1$ and $\underline{F}_{\mu\nu}^2$ leads to two conservation laws,

$$\partial_\nu (g\underline{A}_\mu^1 \times \underline{F}_{\mu\nu}^1 - \lambda^2 \underline{A}_\mu^2) = 0 , \tag{32}$$

$$\partial_\nu (\underline{A}_\mu^1 \times \underline{F}_{\mu\nu}^2 + \underline{A}_\mu^2 \times \underline{F}_{\mu\nu}^1 + \underline{J}_\nu) = 0 , \tag{33}$$

which can be reexpressed as

$$(\partial_\nu + g\underline{A}_\mu^1 \times)\underline{A}_\nu^2 = 0 \tag{34}$$

and

$$(\partial_\nu + g\underline{A}_\mu^1 \times)\underline{J}_\nu = 0 \tag{35}$$

using the equations of motion. The first of these relations acts in effect as a gauge-fixing term for A_μ^2 if a gauge is chosen for A_μ^1. The second relation has the familiar form of current-conservation equations in conventional Yang-Mills theories.

We turn now to the derivation of the perturbation-theory rules in the gluon sector. We consider the vacuum-vacuum transition amplitude in the presence of external sources[9]:

$$W(\underline{J}_\mu^1, \underline{J}_\mu^2) = \int \prod_x dA_\mu^1 dA_\nu^2 \exp\left[i \int d^4x \left(\mathcal{L}_0 - \frac{1}{\beta} \partial_\mu \underline{A}_\mu^1 \cdot \partial_\nu \underline{A}_\nu^2 + \underline{A}_\mu^1 \cdot \underline{J}_\mu^1 + \underline{A}_\mu^2 \cdot \underline{J}_\mu^2 \right) \right] . \tag{36}$$

After some functional translations we find

$$W(\underline{J}_\mu^1, \underline{J}_\nu^2) = \exp\left\{ -i \int d^4x \, d^4y \left[\underline{J}_\mu^1(x) \cdot G_{\mu\nu}^{12}(x-y) \cdot \underline{J}_\nu^2(y) + \tfrac{1}{2}\underline{J}_\mu^1(x) \cdot G_{\mu\nu}^{11}(x-y) \cdot \underline{J}_\nu^1(y) \right] \right\} , \tag{37}$$

where we have dropped an irrelevant factor independent of $\underline{J}_\mu^1$ and $\underline{J}_\mu^2$ on the right-hand side, and

$$G_{\mu\nu ab}^{12}(x) = -\delta_{ab} \int \frac{d^4k\, e^{-ik\cdot x}}{(2\pi)^4 k^2} \left[g_{\mu\nu} + (\beta - 1)\frac{k_\mu k_\nu}{k^2} \right] \tag{38}$$

and

$$G_{\mu\nu ab}^{11}(x) = \frac{\lambda^2 \delta_{ab}}{(2\pi)^4} \int \frac{d^4k\, e^{-ik\cdot x}}{k^4} \left[g_{\mu\nu} + (\beta^2 - 1)\frac{k_\mu k_\nu}{k^2} \right] , \tag{39}$$

with a and b labeling color indices. The free propagators corresponding to the time-ordered products are

$$\langle TA_{\mu a}^1(x) A_{\nu b}^1(y) \rangle = i G_{\mu\nu ab}^{11}(x-y) \tag{40}$$

and

$$\langle TA_{\mu a}^1(x) A_{\nu b}^2(y) \rangle = i G_{\mu\nu ab}^{12}(x-y). \tag{41}$$

The somewhat unusual Green's functions of Eqs. (40) and (41) have their origin in the canonical equal-time commutation relations which we shall now find.

From Eqs. (27)–(30) we obtain the equations of motion

$$\partial_0 \underline{A}_k^1 = \underline{F}_{0k}^1 + \partial_k \underline{A}_0^1 + g\underline{A}_k^1 \times \underline{A}_0^1 , \tag{42}$$

$$\partial_0 \underline{A}_k^2 = \underline{F}_{0k}^2 + \partial_k \underline{A}_0^2 + g\underline{A}_k^1 \times \underline{A}_0^2 - g\underline{A}_0^2 \times \underline{A}_k^1 , \tag{43}$$

$$\partial_0 \underline{F}_{0k}^1 = (\partial_i + g\underline{A}_i \times)\underline{F}_{ik}^1 - g\underline{A}_0^1 \times \underline{F}_{0k}^1 + \lambda^2 \underline{A}_k^2 , \tag{44}$$

$$\partial_0 \underline{F}_{0k}^2 = (\partial_i + g\underline{A}_i^1 \times)\underline{F}_{ik}^2 - g\underline{A}_0^1 \times \underline{F}_{0k}^2 - g\underline{A}_\mu^2 \times \underline{F}_{\mu k}^1 - g\underline{J}_k . \tag{45}$$

and equations of constraint

$$\underline{F}_{ik}^1 = \partial_i \underline{A}_k^1 - \partial_k \underline{A}_i^1 + g\underline{A}_i^1 \times \underline{A}_k^1 , \tag{46}$$

$$\underline{F}_{ik}^2 = \partial_i \underline{A}_k^2 - \partial_k \underline{A}_i^2 + g\underline{A}_i^1 \times \underline{A}_k^2 - g\underline{A}_k^1 \times \underline{A}_i^2 . \tag{47}$$

$$(\partial_i + g\underline{A}_i^1 \times)\underline{F}_{i0}^1 + \lambda^2 \underline{A}_0^2 = 0 , \tag{48}$$

$$(\partial_i + g\underline{A}_i^1 \times)\underline{F}_{i0}^2 + g\underline{A}_i^2 \times \underline{F}_{i0}^1 - g\underline{J}_0 = 0 . \tag{49}$$

The Lagrangian indicates that the canonical momenta are

$$\underline{\Pi}_j^1 = \underline{F}_{0j}^2 \tag{50}$$

and

$$\underline{\Pi}_j^2 = \underline{F}_{0j}^1 . \tag{51}$$

for $j = 1, 2, 3$ with $\underline{\Pi}_j^i$ conjugate to $\underline{A}_j^i$. and $\underline{A}_0^i$ having no conjugate momentum for $i = 1, 2$. However, the equations of constraint indicate that not all components are independent. We now find the independent components. Let us define

$$\underline{F}_{0i}^a = \underline{F}_{0i}^{aT} + \underline{F}_{0i}^{aL} \tag{52}$$

and

$$\underline{F}_{0i}^{aL} = \partial_i \underline{\varphi}^a , \tag{53}$$

where

$$\partial_i \underline{F}_{0i}^{aT} = 0 . \tag{54}$$

Then Eq. (48) gives

$$(\partial_i + g\underline{A}_i^1 \times)\partial_i \underline{\varphi}^1 - \lambda^2 \underline{A}_0^2 = -g\underline{A}_i^1 \times \underline{F}_{i0}^{1T} \tag{55}$$

and Eq. (49) gives

$$(\partial_i + g\underline{A}_i^1 \times)\partial_i \underline{\varphi}^2 + g\underline{A}_i^2 \times \partial_i \underline{\varphi}^1$$
$$= g\underline{A}_i^1 \times \underline{F}_{i0}^{2T} + g\underline{A}_i^2 \times \underline{F}_{0i}^{1T} - g\underline{J}_0 . \tag{56}$$

Rewriting Eqs. (42) and (43) after taking the divergence with respect to spatial components gives

$$(\partial_0 + g\underline{A}_0^1 \times)\partial_k \underline{A}_k^1 = (\partial_k + g\underline{A}_k^1 \times)\partial_k \underline{A}_0^1 + \partial_k \partial_k \underline{\varphi}^1 \tag{57}$$

and

$$(\partial_0 + g\underline{A}_0^1 \times)\partial_k \underline{A}_k^2 + g\underline{A}_0^2 \times \partial_k \underline{A}_k^1$$
$$= \partial_k \partial_k \underline{A}_0^2 + g\underline{A}_k^2 \times \partial_k \underline{A}_0^1 + g\underline{A}_k^1 \times \partial_k \underline{A}_0^2 + \partial_k \partial_k \underline{\varphi}^2 \tag{58}$$

If we choose the Coulomb gauge, $\vec{\nabla}\cdot\underline{A}^1 = 0$, then

$$\partial_k\partial_k\underline{A}_0^1 + g\underline{A}_k^1\times\partial_k\underline{A}_0^1 + \partial_k\partial_k\underline{\varphi}^1 = 0 \tag{59}$$

and

$$(\partial_0 + g\underline{A}_0^1\times)\partial_k\underline{A}_k^2 - \partial_k\partial_k\underline{A}_0^2 - g\underline{A}_k^2\times\partial_k\underline{A}_0^1 - g\underline{A}_k^1\times\partial_k\underline{A}_0^2$$
$$-\partial_k\partial_k\underline{\varphi}^2 = 0 , \tag{60}$$

thus determining $\underline{A}_0^1$ and $\underline{A}_0^2$. Suppose we now define

$$\vec{\underline{A}}^2 = \vec{\underline{A}}^{2T} + \vec{\underline{A}}^{2L} , \tag{61}$$

$$\vec{\underline{A}}^{2L} = \vec{\nabla}\underline{\varphi}^3 , \tag{62}$$

with

$$\vec{\nabla}\cdot\vec{\underline{A}}^{2T} = 0 \tag{63}$$

Taking the divergence of Eq. (44) leads to our final equation for dependent variables

$$\lambda^2\partial_k\partial_k\underline{\varphi}^3 = \partial_0\partial_k\partial_k\underline{\varphi}^1 + g\,\partial_k(\underline{A}_\mu^1\times\underline{F}_{\mu k}^1) . \tag{64}$$

The independent dynamical variables are thus seen to be F_{0i}^{1T}, F_{0i}^{2T}, A_i^{1T}, and A_i^{2T}. Their equal-time commutation relations are

$$[F_{0ia}^{1T}(x), A_{jb}^2(y)] = i\delta_{ab}\Delta_{ij}^{tt}(x-y) , \tag{65}$$

$$[F_{0ia}^{2T}(x), A_{jb}^1(y)] = i\delta_{ab}\Delta_{ij}^{tt}(x-y) . \tag{66}$$

with $i, j = 1, 2, 3$, Δ_{ij}^{tt} given by Eq. (9), and a and b are color indices. All other commutators of the forms $[A^1, A^1]$, $[A^2, A^2]$, $[F^1, F^1]$, $[F^2, F^2]$, $[F^1, F^2]$ are zero.

We return to our development of perturbation-theory rules. The cubic and quartic gluon vertices of our model are given by (see Fig. 3)

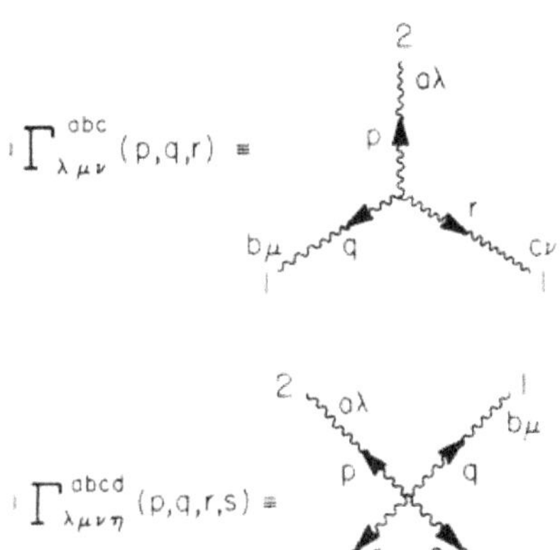

FIG. 3. Cubic and quartic vertices which are given in Eqs. (67) and (68). They introduce $1/r$ potentials in the model and may have an important effect in the baryon spectrum. The numbers 1 and 2 indicate fields $\underline{A}_\mu^1$ and $\underline{A}_\mu^2$, respectively, while p, q, r, and s are momenta, and a, b, c, and d are color indices.

$$i\,\Gamma_{\lambda\mu\nu}^{abc}(p,q,r) = f^{abc}[g_{\lambda\mu}(r_\mu - p_\mu) + g_{\mu\lambda}(p_\nu - q_\nu)$$
$$+ g_{\mu\nu}(q_\lambda - r_\lambda)] . \tag{67}$$

with $p + q + r = 0$, and

$$i\,\Gamma_{\lambda\mu\nu\eta}^{abcd}(p,q,r,s) = -i\,f^{abf}f^{cdf}(g_{\lambda\nu}g_{\mu\eta} - g_{\lambda\eta}g_{\mu\nu})$$
$$- i\,f^{acf}f^{bdf}(g_{\lambda\eta}g_{\mu\nu} - g_{\lambda\nu}g_{\mu\eta})$$
$$- i\,f^{adf}f^{bcf}(g_{\lambda\nu}g_{\mu\eta} - g_{\lambda\mu}g_{\nu\eta}) . \tag{68}$$

with $p + q + r + s = 0$.

The Faddeev-Popov ghost loops will not be relevant to our line of development so we omit their discussion. The necessity for their introduction[10] is closely related to the requirement of unitarity in Yang-Mills theories. In the present model unitarity will be necessarily violated irrespective of the ghost loops if the Green's functions [Eqs. (38) and (39)] pole ambiguities are resolved by using Feynman's $i\epsilon$ procedure. To avoid unitarity violation we have suggested an alternative procedure where the Green's function singularities are taken in principal value.

$$G_{\mu\nu ab}^{kL}(k^2) = \tfrac{1}{2}[G_{\mu\nu ab}^{kL}(k^2 + i\epsilon) + G_{\mu\nu ab}^{kL}(k^2 - i\epsilon)] , \tag{69}$$

in momentum space (cf. the Appendix). This choice has the advantage stated in the Introduction. The effects are the same as in the Abelian model[3] and may be summarized as: (1) Only states composed solely of quarks contribute to unitarity sums, (2) gluons do not appear in asymptotic states, (3) unitarity is achieved but at the price of possible advanced effects whose range is limited to hadronic dimensions and thus apparently unobservable, and (4) nonscaling corrections to Bjorken scaling in the deep-inelastic electroproduction structure functions are suppressed by a factor of q^2 vis-à-vis the corresponding result using Feynman propagators with q being the virtual photon four-momentum.

A novel feature of the use of principal-value propagators in non-Abelian models is the elimination of closed loops composed solely of gluons. If we consider a subdiagram consisting of a gluon loop with p lines, then Eq. (51) of Ref. 3 gives the Feynman parameter representation

$$I = \int_{-\infty}^{\infty}\prod_{j=1}^{p}\alpha_j\,d\alpha_j\,\frac{\epsilon(\alpha_1\alpha_2\cdots\alpha_p C)}{C^2}\,N\mathrm{e}^{iD/C} , \tag{70}$$

where C is a polynomial consisting of Feynman parameters only, while D contains scalar products of external momenta, N symbolizes appropriate numerator factors, and $\epsilon(\alpha) = \pm 1$ if $\alpha \gtrless 0$. Since N can be written as a sum of terms each of which is homogeneous in the Feynman parameters, we can take N to be homogeneous without loss of generality. Then scaling all parameters with u, assuming

$$N(u\alpha_1, u\alpha_2, \ldots, u\alpha_p) = u^r N(\alpha_1, \alpha_2, \ldots, \alpha_p), \quad (71)$$

with r an integer, and using

$$\int_0^\infty \frac{du}{u} \, \delta\left(1 - \frac{|\alpha_1 + \alpha_2 + \cdots + \alpha_p|}{u} \right) = 1 \quad (72)$$

we find

$$I = \Gamma(r + 2p - 2L)$$
$$\times \int_{-\infty}^\infty \frac{\prod_{j=1}^p \alpha_j \, d\alpha_j \epsilon (\alpha_1 \alpha_2 \cdots \alpha_p C) N \delta\left(1 - \left|\sum_k \alpha_k\right| \right)}{C^2(-iD/C)^{r+2p-2L}}, \quad (73)$$

with $L = $ number of loops $= 1$. Suppose we let $\alpha_j \to -\alpha_j$ for all j in I. Then we find $I = -I$ or

$$I = 0. \quad (74)$$

Thus any closed loop containing only principal-value propagators is zero. Since Faddeev-Popov ghosts appear only in closed loops and consistency[11] requires we use principal-value propagators for them if we use such propagators for gluons, we see that ghosts do not appear in our model. Physically we can understand this result if we remember that ghost loops were introduced to cure problems arising from contributions to unitarity sums of "opened" gluon loops.[10] In our model "opened" loops do not contribute to unitarity sums in any case so the raison d'être for ghosts is lacking.

We now derive the Ward-Takahashi-Slavnov identities using functional methods. Since we take our gluon propagators in principal value it might appear that our use of functional techniques is unjustified. We shall take the view that the functional representation of the vacuum-vacuum transition amplitude embodies the combinatorics of perturbation theory and acts as a generating function for identities, such as the Ward-Takahashi-Slavnov identities. Thus, questions of convergence of functional integrals are irrelevant—the important question is whether identities are valid in perturbation theory.

We define $W(J)$, the vacuum-vacuum transition amplitude, by

$$W(J) = \int \prod_x dA_\mu^1 \, dA_\mu^2 \, d\psi \, d\bar{\psi} \exp\left(i \int \bar{\mathcal{L}} \, dx \right), \quad (75)$$

with

$$\bar{\mathcal{L}} = \mathcal{L} - \frac{1}{\beta} \partial_\mu \underline{A}_\mu^1 \cdot \partial_\nu \underline{A}_\nu^2 + \underline{A}_\mu^1 \cdot \underline{J}_\mu^1 + \underline{A}_\mu^2 \cdot \underline{J}_\mu^2 + \bar{\psi}\eta + \bar{\eta}\psi ,$$

with $\mathcal{L}$ given by Eq. (17). Under the infinitesimal gauge variation

$$\underline{A}_\mu^1 \to \underline{A}_\mu^1 - (\partial_\mu + g\underline{A}_\mu^1 \times) \underline{\theta} , \quad (76)$$

$$\underline{A}_\mu^2 \to \underline{A}_\mu^2 - g\underline{A}_\mu^2 \times \underline{\theta} , \quad (77)$$

$$\psi \to \psi - ig\,\theta\psi , \quad (78)$$

$$\bar{\psi} \to \bar{\psi} + ig\,\bar{\psi}\theta , \quad (79)$$

with $\theta = \underline{T} \cdot \underline{\theta}$, $\mathcal{L}$ is invariant but the remaining terms in $\bar{\mathcal{L}}$ lead to

$$\delta\bar{\mathcal{L}} = \frac{1}{\beta} \left[(\partial_\mu + g\underline{A}_\mu^1 \times) \partial_\nu \partial_\mu \underline{A}_\nu^2 + g\underline{A}_\nu^2 \times \partial_\mu \partial_\nu \underline{A}_\mu^1 \right] \cdot \underline{\theta}$$
$$- (\partial_\mu + g\underline{A}_\mu^1 \times) J_\mu^1 \cdot \underline{\theta} - g\underline{J}_\mu^2 \times \underline{A}_\mu^2 \cdot \underline{\theta}$$
$$+ ig\,\bar{\psi}\theta\eta - ig\,\bar{\eta}\theta\psi . \quad (80)$$

Since a transformation of the integration variables does not change the value of the functional integral, the variation of W with respect to θ can be taken to be zero and our equivalent of the Ward-Takahashi-Slavnov identity is

$$\left\{ \frac{1}{\beta} \left[D_\nu\left(\frac{\delta}{i\delta \underline{J}_\alpha^1} \right) \partial_\nu \partial_\mu \frac{\delta}{i\delta \underline{J}_\mu^2} + g \frac{\delta}{i\delta \underline{J}_\nu^2} \times \partial_\nu \partial_\mu \frac{\delta}{i\delta \underline{J}_\mu^1} \right] + D_\mu\left(\frac{\delta}{i\delta \underline{J}_\alpha^1} \right) J_\mu^1 - g\underline{J}_\mu^2 \times \frac{\delta}{i\delta \underline{J}_\mu^2} + g\,\underline{T}\eta \frac{\delta}{\delta\eta} - g\,\bar{\eta}\underline{T} \frac{\delta}{\delta\bar{\eta}} \right\} W = 0 , \quad (81)$$

with

$$D_\mu\left(\frac{\delta}{i\delta \underline{J}_\alpha^1} \right) = \partial_\mu + g \frac{\delta}{i\delta \underline{J}_\mu^1} \times . \quad (82)$$

In order to investigate the structure of the gluon propagators we shall obtain the proper vertex identity equivalent to Eq. (81). We focus on the novelties of the gluon sector and neglect the quark field terms in $\mathcal{L}$ and Eq. (81). Let us define

$$W(J) = e^{iZ(J)} , \quad (83)$$

$$\underline{B}_\mu^i = -\frac{\delta Z(J)}{\delta \underline{J}_\mu^i} , \quad i = 1, 2 \quad (84)$$

$$\Gamma(B) = Z(J) + \int d^4x (\underline{J}_\mu^1 \cdot \underline{B}_\mu^1 + \underline{J}_\mu^2 \cdot \underline{B}_\mu^2) , \quad (85)$$

where $\Gamma(B)$ is the generating functional of proper vertices. An immediate consequence is

$$\underline{J}^i_\mu = \frac{\delta\Gamma}{\delta \underline{B}^i_\mu}\,, \quad i=1,2 \tag{86}$$

and as a result Eq. (81) can be rewritten in the form

$$\frac{1}{\beta}\left[\Box\partial_\mu \underline{B}^2_\mu - g\underline{B}^1_\nu\times\partial_\nu\partial_\mu \underline{B}^2_\mu - g\underline{B}^2_\nu\times\partial_\mu\partial_\nu \underline{B}^1_\mu + g\frac{\delta}{i\delta \underline{J}^1_\nu}\times\partial_\nu\partial_\mu \underline{B}^2_\mu + g\frac{\delta}{i\delta \underline{J}^2_\nu}\times\partial_\nu\partial_\mu \underline{B}^1_\mu\right] - \partial_\mu\frac{\delta\Gamma}{\delta \underline{B}^1_\mu} + \underline{B}^1_\mu\times\frac{\delta\Gamma}{\delta \underline{B}^1_\mu} + \underline{B}^2_\mu\times\frac{\delta\Gamma}{\delta \underline{B}^2_\mu} = 0.$$

$$\tag{87}$$

If we apply $\delta/\delta\underline{B}^1_\alpha$ to Eq. (87) and set $\underline{B}^i_\mu = 0$ afterwards, we find

$$-\partial_\mu\frac{\delta^2\Gamma}{\delta \underline{B}^1_\alpha\delta \underline{B}^1_\mu}\Bigg|_{B^1=B^2=0} = 0. \tag{88}$$

The second-order functional derivative of Γ is the inverse of the full propagator $G^{11'}_{\mu\nu ab}$ and Eq. (88) implies that the proper part of $(G^{11'}_{\mu\nu ab})^{-1}$ is purely transverse. We note that the "free" propagator (Eq. 39) contribution to $(G^{11'}_{\mu\nu ab})^{-1}$ is not one-particle irreducible and thus not constrained by Eq. (88). Therefore we find the general form

$$G^{11'}_{\mu\nu ab}(k) = \delta_{ab}\left(g_{\mu\nu} - \frac{k_\mu k_\nu}{k^2}\right)G^{11}(k^2) + \delta_{ab}\beta^2\lambda^2\frac{k_\mu k_\nu}{k^6}\,,$$

$$\tag{89}$$

so that the longitudinal part of the full propagator is not renormalized.

The longitudinal part of the full propagator $G^{12'}_{\mu\nu ab}(k)$ is also not renormalized. This may be seen by applying $\delta/\delta\underline{B}^2_\mu$ to Eq. (87) and setting $\underline{B}^i_\mu = 0$ afterwards:

$$\frac{1}{\beta}\Box\partial_\mu\delta^4(x-y) - \partial_\nu\frac{\delta^2\Gamma}{\delta \underline{B}^2_\mu\delta \underline{B}^1_\nu}\Bigg|_{B^1=B^2=0} = 0. \tag{90}$$

This implies

$$(G^{12'}_{\mu\nu ab})^{-1} = \frac{\delta_{ab}(g_{\mu\nu} - k_\mu k_\nu/k^2)}{G^{12}} - \frac{k_\mu k_\nu\delta_{ab}}{\beta} \tag{91}$$

or

$$G^{12'}_{\mu\nu ab}(k) = \delta_{ab}\left(g_{\mu\nu} - \frac{k_\mu k_\nu}{k^2}\right)G^{12}(k) - \beta\delta_{ab}\frac{k_\mu k_\nu}{k^4}. \tag{92}$$

Having now developed the general form of the propagators we now will define the gluon vacuum polarization tensors.

$$\Pi^{11}_{\mu\nu ab}(k) = [G^{11'}_{\mu\nu ab}(k)]^{-1} - [G^{11}_{\mu\nu ab}(k)]^{-1}, \tag{93}$$

$$\Pi^{12}_{\mu\nu ab}(k) = [G^{12'}_{\mu\nu ab}(k)]^{-1} - [G^{12}_{\mu\nu ab}(k)]^{-1}, \tag{94}$$

which are transverse by our previous discussion:

$$k_\mu\Pi^{11}_{\mu\nu ab} = k_\mu\Pi^{12}_{\mu\nu ab} = 0. \tag{95}$$

Rather than write the Schwinger-Dyson equations for our polarization tensors we have given a diagrammatic representation in Fig. 4.

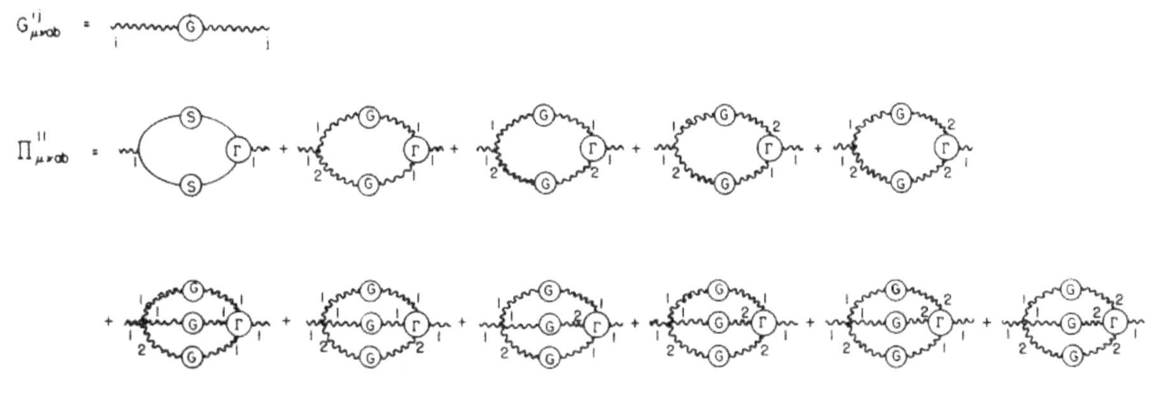

FIG. 4. Diagrammatic representation of the Schwinger-Dyson equation for the proper gluon self-energy, $\Pi^{11}_{\mu\nu ab}$. The numbers at the end of a gluon line specify whether $\underline{A}^1_\mu$ or $\underline{A}^2_\mu$ correspond to that end. The quark propagator is denoted S while Γ denotes the appropriate proper (one-particle irreducible) vertex function. A similar diagrammatic expression can be written for $\Pi^{12}_{\mu\nu ab}$.

IV. OBSERVATIONS

The Schwinger mechanism forces quark confinement to bound color singlet states in a manner which is identical to the Abelian case as described in Sec. II. In order to demonstrate that only color singlets exist in the gauge-invariant physical particle spectrum it is sufficient to show

$$\underline{Q}\,\psi_{\rm phys} = 0 \,, \tag{96}$$

where

$$\underline{Q} = \int d^3x\, \underline{J}_0(x) \tag{97}$$

for any physical state $\psi_{\rm phys}$ corresponding to a spatially localized distribution of quarks. We consider a single static quark located at the origin and choose to work in the Coulomb gauge ($\vec{\nabla}\cdot\vec{A}^1 = 0$). Then the time components of the equations of motion [Eqs. (27) and (28)] lead to (at large distance)

$$\Box\nabla^2 \underline{A}_0^1 = g\lambda^2 \underline{J}_0 \tag{98}$$

if we take into account the elimination of gluons' degrees of freedom through the choice of principal-value propagators and their consequent inability to act as sources. We may now repeat the arguments of Eqs. (13)–(16) for the Abelian case after noticing the occurrence of the Schwinger mechanism in the non-Abelian case which can be verified in low orders of perturbation theory for $\Pi_{\mu\nu ab}^{11}$. Thus the expectation value of the charge in the one-quark state is zero. Since the one-quark state is a charge eigenstate, we find Eq. (96) to be true in this case and more generally through the additivity of the charge operator. Thus only color singlet bound states of quarks are physical.[12]

While the infrared behavior of the theory leads to quark confinement, the ultraviolet behavior allows the quarks to appear quasifree. This is particularly noticeable when we take $\lambda^2 = 0$ in our Lagrangian and examine the corresponding perturbation theory. Taking $\lambda^2 = 0$ is equivalent to examining the short-distance behavior of the theory. The only diagrams which exist in this limit are given in Fig. 5. The quark sector of the theory is free. The only nontree structures are one-quark-loop diagrams for the scattering of gluons associated with A_μ^2 (which of course can only be generated by a hypothetical external source). (As a point of comparison we have shown in Fig. 6 the additional diagrams which would occur in the even that Feynman propagators were used—these diagrams necessarily involve gluon loops which principal-value propagators force to be zero.) The vital role of the $\lambda^2 A_\mu^2 A_\mu^2$ term in the Lagrangian in generating the interacting theory and the fact that λ^2 has the dimensions of (mass)^2 allow a natural approximation procedure in this model. This is perhaps best seen within the context of deep-inelastic electroproduction. Just as in the Abelian case we find that the structure functions scale with leading corrections of $O(q^{-1})$, where q equals the virtual photon four-momentum. We can establish a parton picture of scattering wherein the photon is absorbed on one of the quasifree nucleon constituents [as in Fig. 2(a)] if $|q^2| \gg g^2\lambda^2$. Then leading corrections to such a picture [e.g., the diagrams of Fig. 2(b)–2(d)] will be suppressed by $(g^2\lambda^2/q^2)^2$. Thus the dimensional nature of the effective coupling constant allows a particularly simple picture to exist of the region of large spacelike virtual photon mass and the parton picture emerges as a natural approximation.

The k^{-4} form of the quark interaction also appears to have decidedly good features as far as the bound-state structure is concerned. Ignoring the numerator tensor (which does not affect our conclusions), we find the Fourier transform of the gluon propagator,

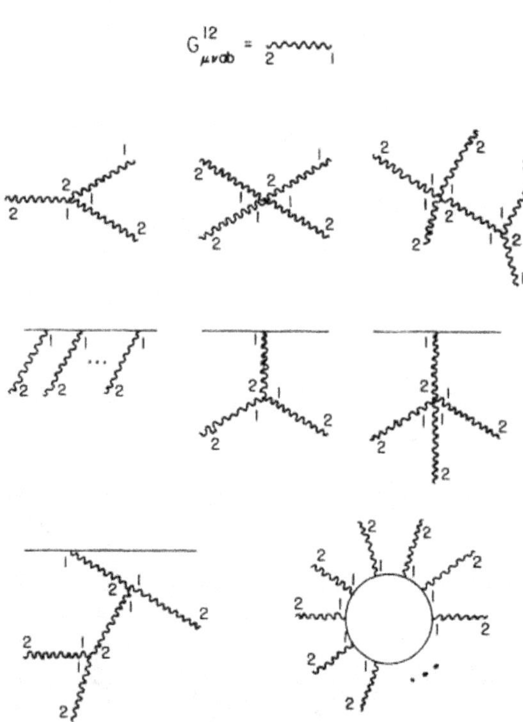

FIG. 5. Some examples of the surviving diagrams in the $\lambda^2 = 0$ limit of the non-Abelian model with principal-value gluon propagators. Except for the class of one-fermion-loop diagrams only tree diagrams exist in this limit. Note that there are no four (or more) external quark line diagrams and no two (or more) external $\underline{A}_\mu^1$ gluon "external" lines.

$$G(k) = \mathrm{P}\frac{1}{k^4},\tag{99}$$

to be proportional to

$$\bar{G}(x) = \theta(x^2).\tag{100}$$

Since $\bar{G}$ has a smooth finite limit as $x^2 \to 0$, the short-distance limit, arguments can be made[13] that low-mass bound states can occur in this model. In addition, Dalitz[14] has pointed out that the linearity of trajectories on the Chew-Frautschi plot would follow from a flat-bottomed, smooth interaction—a criterion which is met by Eq. (100). It is interesting to note that had we used a Feynman propagator rather than principal value, then $\bar{G}$ would have been $\ln(x^2)$ and thus the general criterion just stated would not have been met. This would appear to be another point in favor of our choice of principal-value propagators.

Another property which is desirable in the bound-state solutions is nonrelativistic motion of the bound-state constituents.[15] Again an interaction of the form of Eq. (99) appears to realize this feature—even in the strong-binding limit. To see this we shall first take account of the Schwinger mechanism and in the spirit of Hartree-Fock theory modify the quark interaction to

$$G'(k) = \mathrm{P}\frac{1}{(k^2 - \mu^2)^2}.\tag{101}$$

If we now take Eq. (101) to be the Green's function for the effective gluon field and calculate the "Coulomb potential" of a static, point quark source located at the origin we find

$$\varphi(r) = \frac{\varphi_0}{\mu}\, e^{-\mu r},\tag{102}$$

where φ_0 is a constant independent of μ. In the limit $\mu \to 0$ we find

$$\varphi(r) \cong \varphi_0\left(\frac{1}{\mu} - r + \cdots\right).\tag{103}$$

The first two terms of Eq. (103) correspond to choosing Eq. (99) rather than Eq. (101) as the gluon Green's function (in the limit $\mu \to 0$). Equation (102) includes vacuum polarization effects which damp the interaction at large distances. Thus Eq. (102) imperfectly reflects the possibility that a quark-antiquark pair can separate and induce another quark-antiquark pair to be created from the vacuum so that two color singlet mesons will result (presuming it is energetically favored). At shorter distances Eq. (102) appears to be a reasonable approximation. This exponential potential was studied within the framework of the Schrödinger equation in the strong-binding limit (φ_0/μ large) by Greenberg.[16] He showed

that the average momentum of the bound constituent in the s state satisfied

$$\frac{p}{m} \sim \left(\frac{\mu}{m}\right)^{1/3}\tag{104}$$

with m being the quark mass. Thus for μ/m small the quark motion is self-consistently nonrelativistic.

In conclusion, we have shown that a four-dimensional, Lorentz-invariant second-quantized field theory of hadron binding is possible with scaling electroproduction structure functions, only zero-triality physical particle states, and, apparently, linearly rising Regge trajectories and nonrelativistic constituents. A detailed study of the bound states is now in progress.

ACKNOWLEDGMENT

I am grateful to the members of the Newman Laboratory for interesting conversations.

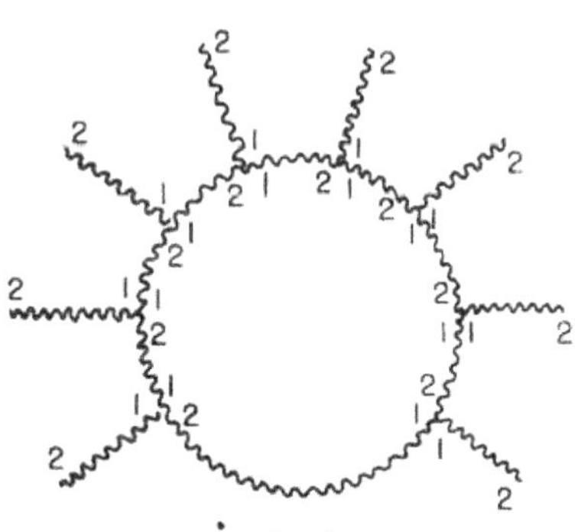

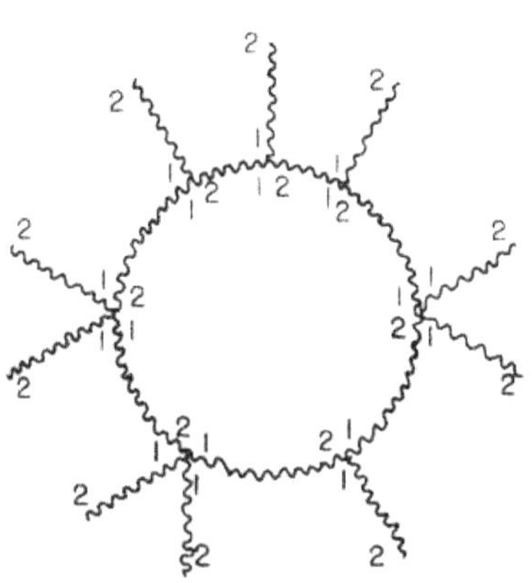

FIG. 6. Some additional diagrams which occur in the $\lambda^2 = 0$ limit of the non-Abelian model if Feynman gluon propagators are used. In addition, there will be Faddeev-Popov ghost-loop diagrams depending on the choice of gauge.

APPENDIX

In Ref. 3 semiclassical arguments based on Dirac's theory of constraints were given to introduce the use of principal-value propagators. We will now describe a second-quantized realization of those arguments for the case of a scalar Klein-Gordon field $\varphi(x)$ with the Lagrangian

$$\mathcal{L} = \tfrac{1}{2}(\partial_\mu \varphi)^2 - \tfrac{1}{2}m^2\varphi^2 . \tag{A1}$$

The generalization to vector gluons is immediate. The canonical equal-time commutation relations are

$$[\varphi, \varphi] = [\dot{\varphi}, \dot{\varphi}] = 0 , \tag{A2}$$

$$[\dot{\varphi}(\vec{x}, t), \varphi(\vec{y}, t)] = -i\delta^3(\vec{x} - \vec{y}) . \tag{A3}$$

If we expand $\varphi(x)$ in plane waves,

$$\varphi(\vec{x}, t) = \sum_{\vec{k}} (A_{\vec{k}}\, e^{-ik \cdot x} + A_{\vec{k}}^\dagger e^{ik \cdot x}) , \tag{A4}$$

then the q-number Fourier components $A_{\vec{k}}$ must satisfy

$$[A_{\vec{k}}, A_{\vec{k}'}] = [A_{\vec{k}}^\dagger, A_{\vec{k}'}^\dagger] = 0 , \tag{A5}$$

$$[A_{\vec{k}}, A_{\vec{k}'}^\dagger] = \delta^3(\vec{k}' - \vec{k}) \tag{A6}$$

for consistency with Eqs. (A2) and (A3). Now the time-ordered product satisfies

$$T(\varphi(x)\varphi(y)) = \epsilon(x_0 - y_0)[\varphi(x), \varphi(y)]$$
$$+ \{\varphi(x), \varphi(y)\} , \tag{A7}$$

with $\epsilon(x_0) = \pm 1$ for $x_0 \gtrless 0$ and $\{A, B\} = AB + BA$. The first term on the right-hand side is a c number completely determined by Eqs. (A5) and (A6). If the second q-number expression were zero, then we would obtain a principal-value propagator from Eq. (A7):

$$T(\varphi(x)\varphi(y)) = i \int \frac{d^4k}{(2\pi)^4} e^{-ik \cdot (x-y)} \, \mathrm{P}\, \frac{1}{(k^2 - m^2)} . \tag{A8}$$

We therefore require

$$\{\varphi(x), \varphi(y)\} = 0 , \tag{A9}$$

with the consequence

$$\{A_{\vec{k}}, A_{\vec{k}'}\} = \{A_{\vec{k}}^\dagger, A_{\vec{k}'}^\dagger\}$$
$$= \{A_{\vec{k}}, A_{\vec{k}'}^\dagger\}$$
$$= 0 . \tag{A10}$$

Equations (A5), (A6), and (A10) imply

$$A_{\vec{k}}, A_{\vec{k}'} = A_{\vec{k}}^\dagger A_{\vec{k}'}^\dagger = 0 , \tag{A11}$$

$$A_{\vec{k}} A_{\vec{k}'}^\dagger = \tfrac{1}{2}\delta^3(\vec{k} - \vec{k}') , \tag{A12}$$

$$A_{\vec{k}}^\dagger A_{\vec{k}'} = -\tfrac{1}{2}\delta^3(\vec{k} - \vec{k}') \tag{A13}$$

for all $\vec{k}$ and $\vec{k}'$. Thus quadratic terms in A and $A^\dagger$ are reduced to c numbers. It should further be noted that the multiplication rule is not associative.[17] In fact, the multiplication rules of the A and $A^\dagger$ operators in Eqs. (A11)–(A13) are realized by taking multiplication to be

$$UV = \tfrac{1}{2}[U, V] \tag{A14}$$

for U, V being any $A_{\vec{k}}$ or $A_{\vec{k}'}^\dagger$. If we take an analogy to Lie-algebra theory seriously, where the adjoint representation of an algebra has a multiplication rule defined by commutators

$$\tilde{U} * \tilde{V} = [\tilde{U}, \tilde{V}] \tag{A15}$$

then we could call Eqs. (A11)–(A13) the adjoint representation of the Fourier components of φ.

The c-number nature of AA, $A^\dagger A^\dagger$, or $AA^\dagger$ can be understood physically in the following manner. Since the φ field has principal-value propagators it is not associated with a particle but is merely the embodiment of an interaction between other objects (which we have suppressed in our Lagrangian). Consequently an emission of a φ field quantum must be directly correlated with a subsequent absorption—it cannot propagate into empty space. The c-number nature of $AA^\dagger$ reflects this correlation between emission and absorption.

Finally, it should be noted that the existence of a vacuum is inconsistent with Eqs. (A11)–(A13).

---

*Work supported in part by the National Science Foundation.

[1] K. Johnson, Phys. Rev. D **6**, 1101 (1972); C. M. Bender, J. E. Mandula, and G. S. Guralnik, Phys. Rev. Lett. **32**, 1467 (1974); A. Chodos et al., Phys. Rev. D **9**, 3471 (1974); W. A. Bardeen et al., ibid. **11**, 1094 (1975); M. Creutz, ibid. **10**, 1749 (1974); P. Vinciarelli, Nuovo Cimento Lett. **4**, 905 (1972); R. Dashen, B. Hasslacher, and A. Neveu, Phys. Rev. D **10**, 4114 (1974); **10**, 4130 (1974); **10**, 4138 (1974).

[2] Y. Nambu, in *Preludes in Theoretical Physics*, edited by A. de-Shalit, H. Feshbach, and L. Van Hove (North-Holland, Amsterdam, 1966), p. 133; H. J. Lipkin, Phys. Lett. **45B**, 267 (1973).

[3] S. Blaha, Phys. Rev. D **10**, 4268 (1974).

[4] J. Schwinger, Phys. Rev. **128**, 2425 (1962).

[5] A. Pais and G. Uhlenbeck, Phys. Rev. **79**, 145 (1950); J. Kiskis, Phys. Rev. D **11**, 2178 (1975).

[6] A. Casher, J. Kogut, and L. Susskind, Phys. Rev. D **10**, 732 (1974); J. Lowenstein and J. Swieca, Ann. Phys. (N.Y.) **68**, 172 (1971).

[7] R. Jackiw and G. Preparata, Phys. Rev. Lett. **22**, 975 (1969); S. Adler and W. Tung, ibid. **22**, 978 (1969); S. Blaha, Phys. Rev. D **3**, 510 (1971).

[8]The Lagrangian of Eq. (17) was first written by D. Sinclair as a generalization of the Abelian model of Ref. 3. An alternative non-Abelian model for quark confinement has been suggested by S. K. Kauffmann [Nucl. Phys. B87, 133 (1975)]. I am grateful to Dr. Kauffmann for sending me a copy of his paper prior to publication.

[9]E. Abers and B. W. Lee [Phys. Rep. 9C, 1 (1973)] provide a useful review of conventional Yang-Mills theories.

[10] R. P. Feynman, Acta Phys. Pol. 24, 697 (1963).

[11]B. W. Lee and J. Zinn-Justin [Phys. Rev. D 5, 3121 (1972)] point out that the $i\epsilon$ prescription in their Eq. (2.8) for the ghost loop is dictated by unitarity considerations.

[12]This does not preclude color-singlet states of the gluons from playing a role in the theory. They are not particles but can be exchanged between color-singlet quark states in scattering events. On naive dimensional grounds they should be most important in forward scattering. This leads to the possibility that the Pomeron might possibly be interpreted as a "two-gluon bound state". In the case of wide-angle scattering the predominant mechanism for large momentum transfer would appear to be constituent interchange due to the strong damping effects of k^{-4} propagators on momentum transfer.

[13]M. Böhm, H. Joos, and M. Krammer, in Recent Developments in Mathematical Physics, proceedings of the XII Schladming Conference (Acta Phys. Austriaca Suppl. XI), edited by P. Urban (Springer, New York, 1970), p. 3.

[14]R. H. Dalitz, a paper presented at the Topical Conference on Meson Spectroscopy, Philadelphia, 1968 (unpublished).

[15]H. J. Lipkin, Phys. Rep. 8C, 175 (1973).

[16]O. W. Greenberg, Phys. Rev. 147, 1077 (1966).

[17]Nonassociative field operators have been previously used by M. Günaydin and F. Gürsey, Phys. Rev. D 9, 3387 (1974).

Appendix K. Towards a Field Theory of Hadron Binding

This refereed paper is Phys. Rev. **D10**, 4268 (1974). Reprinted with the kind permission of Physical Review D.

Landau-Ginzburg theory, but $\rho \sim i(\varphi^* \dot\varphi - \dot\varphi^* \varphi) = 0$ in the Higgs theory.

[9]M. Kalb and P. Ramond, Phys. Rev. D **9**, 2273 (1974). See also E. Cremmer and J. Scherk, Nucl. Phys. **B72**, 117 (1974).

[10]L. N. Chang and F. Mansouri, Phys. Rev. D **5**, 2235 (1972); Goto, Ref. 7; G. Goddard, J. Goldstone,

C. Rebbi, and C. B. Thorn, Nucl. Phys. **B56**, 109 (1973).

[11]A clearcut answer to this problem seems to be lacking. See, however, Nielsen and Olesen, Ref. 1; G. 't Hooft, CERN Report No. TH-1873-CERN, 1974 (unpublished); Y. Nambu, Ref. 2.

PHYSICAL REVIEW D   VOLUME 10, NUMBER 12   15 DECEMBER 1974

Towards a field theory of hadron binding*

Stephen Blaha

Laboratory of Nuclear Studies, Cornell University, Ithaca, New York 14850

(Received 17 July 1974)

A field-theoretic model for hadron binding is described in which free quarks are totally screened. Quarks interact via a dipole vector-gluon field. A second-quantization procedure for the gluon field, which reduces the field to an embodiment of a direct particle interaction, eliminates unitarity problems. A detailed description of perturbation-theory rules is given. In contrast to the results of the pseudoscalar-meson and massive-vector-meson models (without cutoff), scaling occurs in the electroproduction structure functions. Another possible model having some resemblance to the relativistic harmonic-oscillator quark model of Feynman, Kislinger, and Ravndal is also described. It is unitary and has scaling structure functions.

I. INTRODUCTION

The current understanding of hadronic structure allows two apparently contradictory statements to be made: The constituents of the hadron appear to be loosely bound, quasifree particles. The constituents of the hadron are not produced and do not occur outside of hadrons. Several attempts have been made to resolve this paradoxical situation. They may be divided into two categories: "conventional" field-theoretic approaches,[1,2] and *ad hoc* approaches which postulate manifestly non-field-theoretic structures for confinement, e.g., the "bag" model.[3] In the first approach Casher, Kogut, and Susskind[1] and Wilson[2] showed that quarks could be totally screened and not observed. However, a four-dimensional, Lorentz-invariant field-theoretic model of hadron binding with its attendant conceptual and computational advantages appears to be lacking. We shall discuss a possible candidate, the dipole gluon model, in detail. In addition, another possibility is briefly described in Appendix B which bears some comparison with the quark model of Feynman *et al.*[4] The dipole gluon model has two major qualitative features in common with the bag model[3] and the two-dimensional quantum-electrodynamic model[1]: (1) The dipole gluon field has no independent degrees of freedom; neither does a bag or the two-dimensional electromagnetic field. (2) The "Coulomb"

potential between quarks is proportional to the distance between them in all three models. In a sense the bag model may be regarded as a phenomenological approximation to the dipole model, and the dipole model as a generalization of the two-dimensional model to four dimensions.

In Sec. II we describe a quantization procedure which avoids the introduction of indefinite-metric in or out states and thus leads to a unitary S matrix. In Sec. III we describe the properties of the "free" gluon Lagrangian model. In Sec. IV we describe the perturbation-theory rules of the dipole model. Section V contains a discussion of unitarity, causality, quark confinement, and scaling properties of the electroproduction structure functions. For simplicity we shall ignore all but the dipole quark interaction and do not introduce internal quark quantum numbers.

II. SECOND-QUANTIZATION PROCEDURE FOR THE GLUON FIELD

We shall not quantize the gluon field in the conventional manner for three reasons: (1) to be consistent with experiment where no such particle has been identified, (2) to avoid unitarity problems in the S matrix, and (3) to avoid infrared problems in perturbation theory. We attribute no dynamical degrees of freedom to the gluon field. Instead we regard the field as the embodiment of a direct quark-quark interaction. The gluon field can thus be removed from the Lagrangian in favor of a non-local current-current interaction. However, it will be of no small technical advantage to keep the gluon field in the Lagrangian. In order to do this we shall second-quantize the field following the normal prescription and then, instead of introducing a Fock space for free gluons, reduce q-number expressions in the gluon field to c-number expressions via suitable operator boundary conditions.

To illustrate this procedure we consider a scalar boson field, ϕ, with Lagrangian L. We second-quantize the in field, ϕ_{in}, with equal-time commutators

$$[\phi_{in}(x), \phi_{in}(y)] = 0 , \tag{1}$$

$$[\Pi_{in}(x), \Pi_{in}(y)] = 0 , \tag{2}$$

$$[\phi_{in}(x), \Pi_{in}(y)] = -i\delta^3(\vec{x} - \vec{y}) , \tag{3}$$

where

$$\Pi_{in}(x) = \frac{\delta L}{\delta \dot{\phi}_{in}(x)} \tag{4}$$

and L_F is the free Lagrangian part of L. The usual operator expressions and identities are established. In particular the formal expansion of the S matrix in terms of time-ordered products of in fields can be made. (We are using only in fields for convenience—our remarks apply to out fields also.)

The unequal-time commutator, $[\phi_{in}(x), \phi_{in}(y)]$, is a c-number expression which is completely determined if we require that it be consistent with the equations of motion, that it be consistent with Eqs. (1)–(3) in the limit of equal times, and that it vanish at spacelike distances. Consequently all terms with an even number of factors of $\phi(x)$ reduce to sums of c numbers times products of anticommutators $\{\phi(x), \phi(y)\}$. Terms with an odd number of factors have one factor, $\phi(x)$, times sums of c numbers times products of anticommutators. At this point we could introduce a Fock space of states to complete the reduction of q-number expressions to c number expressions. For reasons stated above we do not. In analogy to Dirac's theory[5] of Hamiltonian constraints we impose operator boundary conditions which complete the specification of the dynamics of the system. We choose

$$\Pi_{in}(x) \approx 0 \approx \phi_{in}(x) \tag{5}$$

for all x, where $\approx$ means weakly equal in the sense of Dirac, i.e., evaluate all commutators before imposing the constraints. This eliminates ϕ's degrees of freedom. The free Hamiltonian,

$$H_F = \int \Pi\dot{\phi} - L_F , \tag{6}$$

is now arbitrary to the extent that H_F may be replaced by

$$H_T = H_F + \int A\phi_{in} + \int b\Pi_{in} , \tag{7}$$

where A and b will be completely determined by requiring Eq. (5) be true for all time:

$$[\Pi_{in}, H_T] \approx 0 \tag{8}$$

and

$$[\phi_{in}, H_T] \approx 0 . \tag{9}$$

Thus

$$A = -\frac{\delta H_F}{\delta \phi_{in}} \tag{10}$$

and

$$b = -\frac{\delta H_F}{\delta \Pi_{in}} . \tag{11}$$

To see the effects of this procedure more concretely let

$$L_F = \int (\tfrac{1}{2}\partial_\mu \phi \partial^\mu \phi - \tfrac{1}{2}m^2\phi^2) ; \tag{12}$$

then (suppressing the subscript "in" for notational convenience)

$$i\Delta(x - y) = [\phi(x), \phi(y)] \tag{13}$$

$$= \int \frac{d^4k}{(2\pi)^3} \epsilon(k_0)\delta(k^2 - m^2)e^{-ik \cdot (x-y)} \tag{14}$$

and the time-ordered product becomes

$$i\tilde{\Delta}(x - y) \equiv T(\phi(x)\phi(y)) = \tfrac{1}{2}i\epsilon(x_0 - y_0)\Delta(x - y) , \tag{15}$$

with $\epsilon(x) = \pm 1$ for $x \gtrless 0$. More generally, for even N

$$T(\phi(1)\phi(2)\cdots\phi(N))$$

$$= \sum_{permutations} i^{N/2}\tilde{\Delta}(x_1 - x_2)\tilde{\Delta}(x_3 - x_4)\cdots\tilde{\Delta}(x_{N-1} - x_N) , \tag{16}$$

where $\phi(i) = \phi(x_i)$. The natural correspondence to the Wick expansion

$$\langle 0| T(\psi(x_1)\cdots\psi(x_N))|0\rangle$$

$$= \sum_{permutations} i^{N/2}\Delta_F(x_1 - x_2)\cdots\Delta_F(x_{N-1} - x_N) \tag{17}$$

(where Δ_F is the Feynman propagator corresponding to the field ψ) allows us to use conventional perturbation-theory rules, except that diagrams with incoming or outgoing ϕ lines do not contribute to the S matrix and the Feynman propagator

$$\Delta_F(k) = \frac{1}{k^2 - m^2 + i\epsilon} \tag{18}$$

is to be replaced with

$$\bar{\Delta}(k) = P\,\frac{1}{k^2 - m^2 + i\epsilon} = \frac{1}{2}\left(\frac{1}{k^2 - m^2 + i\epsilon} + \frac{1}{k^2 - m^2 - i\epsilon}\right) \tag{19}$$

for internal ϕ lines. In configuration space the Green's function corresponding to Eq. (19) is half the sum of the advanced and retarded Green's functions.

If we follow the above procedure in second-quantizing the electromagnetic field the resulting model quantum electrodynamics corresponds to the classical action-at-a-distance electrodynamics of Schwarzschild, Tetrode, and Fokker.[6] The fact that photon production does not occur in the model QED corresponds to the absence of radiation reaction in the classical theory. In Sec. V this will be shown to be the key to maintaining the unitarity of the S matrix in the dipole gluon model.

III. THE DIPOLE GLUON MODEL

We now consider a model[4] for hadron binding which has several major qualitative features in agreement with experimental results: large-transverse-momenta damping, scaling electroproduction structure functions, and complete screening of free quarks. The Lagrangian is

$$\mathcal{L} = -\tfrac{1}{2}F^1_{\mu\nu}F^{2\mu\nu} - \tfrac{1}{2}\lambda^2 A^2_\mu A^{2\mu} + \bar{\psi}(i\,\overset{\leftrightarrow}{\nabla} - gA^1 - m)\psi \,, \tag{20}$$

where A^1_μ and A^2_μ are massless gluon fields, $F^i_{\mu\nu} = \partial_\nu A^i_\mu - \partial_\mu A^i_\nu$ for $i = 1, 2$, ψ is the quark field, and g is a dimensionless and λ a dimensional coupling constant. The equations of motion are

$$\partial^\mu F^1_{\mu\nu} + \lambda^2 A^2_\nu = 0 \,, \tag{21}$$

$$\partial^\mu F^2_{\mu\nu} + gJ_\nu = 0 \,, \tag{22}$$

$$(i\,\overset{\leftrightarrow}{\nabla} - gA^1 - m)\psi = 0 \,, \tag{23}$$

with J_μ the quark current. Equation (21) implies $\partial^\mu A^2_\mu = 0$ while A^1_μ is a gauge-invariant field. As a result we have

$$\Box\,\partial^\mu F^1_{\mu\nu} + g\lambda^2 J_\nu = 0 \,. \tag{24}$$

We now consider the "free" gluon case whose Lagrangian is the first two terms on the right-hand side of Eq. (20). The canonical momentum conjugate to A^1_μ is

$$\Pi^1_\mu = F^2_{0\mu} \tag{25}$$

and that conjugate to A^2_μ is

$$\Pi^2_\mu = F^1_{0\mu} \,. \tag{26}$$

Since $\Pi^1_0 = \Pi^2_0 = 0$ we find that A^1_0 and A^2_0 are c numbers and thus $\vec{\nabla}\cdot\vec{A}^2$ is also a c number with the possible exception of the zero-frequency mode. If we choose the Coulomb gauge for A^1_μ

$$\vec{\nabla}\cdot\vec{A}^1 = 0 \tag{27}$$

then we obtain the equal-time commutation relations

$$[\Pi^a_i(x), A^b_j(y)] = i\delta^{ab}\int \frac{d^3k}{(2\pi)^3}\, e^{i\vec{k}\cdot(\vec{x}-\vec{y})}\left(\delta_{ij} - \frac{k_i k_j}{|\vec{k}|^2}\right) \tag{28}$$

for $i, j = 1, 2, 3$, in analogy to similar expressions in quantum electrodynamics. All other equal-time commutators are zero. We can define an electric field, $\vec{E}$, and magnetic field, $\vec{B}$, by

$$\vec{E} = -\vec{\nabla}A^{10} - \frac{\partial}{\partial t}\vec{A}^1 \tag{29}$$

and

$$\vec{B} = \vec{\nabla}\times\vec{A}^1 \,, \tag{30}$$

which imply

$$\vec{\nabla}\times\vec{E} = \frac{\partial\vec{B}}{\partial t} \tag{31}$$

and

$$\vec{\nabla}\cdot\vec{B} = 0 \,. \tag{32}$$

In the Coulomb gauge Eq. (24) can be restated as

$$\Box\,\vec{\nabla}\cdot\vec{E} = g\lambda^2 J^0 \,, \tag{33}$$

$$\Box\left(\vec{\nabla}\times\vec{B} - \frac{\partial\vec{E}}{\partial t}\right) = g\lambda^2\vec{J} \,, \tag{34}$$

where J^μ is the quark current. Equations (29) and (33) give our analog to the differential equation for the instantaneous Coulomb potential of QED,

$$\Box\nabla^2 A^{10} = -g\lambda^2 J^0 \,, \tag{35}$$

while the equivalent vector-potential differential equation is

$$\Box(\Box\vec{A}^1 + \vec{\nabla}\dot{A}^{10}) = g\lambda^2\vec{J} \,. \tag{36}$$

The free gluon unequal-time commutators may be determined from Eqs. (21), (22), (28), (35), and (36) (with the current, of course, set to zero):

$$i\Delta_{ij}^{11}(x-y) \equiv \left[A_i^1(x), A_j^1(y)\right] = -i\lambda^2(\delta_{ij} - \nabla_i\nabla_j\nabla^{-2})\left[\frac{\partial}{\partial\mu^2}\Delta(x-y,\mu)\right]_{\mu=0} \; , \tag{37}$$

$$i\Delta_{ij}^{12}(x-y) \equiv \left[A_i^1(x), A_j^2(y)\right] = i(\delta_{ij} - \nabla_i\nabla_j\nabla^{-2})\Delta(x-y,0) \; , \tag{38}$$

$$i\Delta_{ij}^{22}(x-y) \equiv \left[A_i^2(x), A_j^2(y)\right] = 0 \; , \tag{39}$$

with $i, j = 1, 2, 3$ and

$$i\Delta(x-y,\mu) = \int \frac{d^4k}{(2\pi)^3}\epsilon(k_0)\delta(k^2-\mu^2)e^{-ik\cdot(x-y)} \; . \tag{40}$$

The commutators, Δ^{11} and Δ^{12}, are zero at space-like separations due to the form of Δ. They are consistent with the equal-time commutation relations in that limit and they are also consistent with the equations of motion due to the identity

$$\Box\left[\frac{\partial}{\partial\mu^2}\Delta(x-y,\mu)\right]_{\mu=0} = -\Delta(x-y,0) \; . \tag{41}$$

Assuming that we have established all operator expressions we are now in a position to apply operator boundary conditions to the gluon field. The key quantities so far as the perturbation theory we will consider in the next section is concerned are the time-ordered propagators of the gluon field

$$T(A_i^a(x)A_j^b(y)) \equiv \tfrac{1}{2}\epsilon(x_0-y_0)\left[A_i^a(x), A_j^b(y)\right] \tag{42}$$

$$= \tfrac{1}{2}i\epsilon(x_0-y_0)\Delta_{ij}^{ab}(x-y) \; , \tag{43}$$

where we have suppressed the "in" subscript on the field operator. We can take advantage of the gauge invariance of A_μ^1 to express $T(A_\mu^1 A_\nu^1)$ in the Feynman gauge,

$$T(A_\mu^1(x)A_\nu^1(y)) = i\lambda^2 g_{\mu\nu}\int\frac{d^4k}{(2\pi)^4}\left(P\frac{1}{k^4}\right)e^{-ik\cdot(x-y)} \tag{44}$$

with

$$P\frac{1}{k^4} \equiv \frac{1}{2}\left[\frac{1}{(k^2+i\epsilon)^2} + \frac{1}{(k^2-i\epsilon)^2}\right] \; . \tag{45}$$

In coordinate space

$$T(A_\mu^1(x)A_\nu^1(y)) = i g_{\mu\nu}\lambda^2\theta((x-y)^2)/16\pi \; . \tag{46}$$

The equations of motion of the dipole model display a close analogy to those of quantum electrodynamics. The main difference (with important physical consequences) is the increased degree of the differential equation for A_μ^1 vis-à-vis the corresponding QED equations. The result is a dipole propagator rather than a monopole propagator in momentum space. One could have second-quantized the dipole field in a manner which leads to dipole Feynman propagators. In that case the S matrix would not be unitary in perturbation theory.

IV. PERTURBATION THEORY

The rules for forming the integral corresponding to a Feynman diagram in the dipole model are identical with those of quantum electrodynamics[7] except that we use

$$iB_{\mu\nu}(q) = ig_{\mu\nu}\lambda^2 P\frac{1}{q^4} \tag{47}$$

rather than the Feynman photon propagator

$$iD_{F\mu\nu}(q) = -\frac{ig_{\mu\nu}}{q^2+i\epsilon} \; . \tag{48}$$

The choice of a principal-value propagator has substantial effects in perturbation theory. For example we shall show that consistency with unitarity requires no diagrams with in or out gluon lines contribute to the S matrix. In addition, there are novelties in the type of divergences in diagrams and the analytic structure of the S matrix. It also appears that conclusions based on summing only a finite number of graphs contributing to an S-matrix element may be misleading. This follows from the fact (to be shown in Sec. V) that free quarks do not exist upon summation of all orders of perturbation theory [Eq. (65)], though this is not seen in a summation to any finite order. The physical states are neutral bound states and thus it appears that the best methods of exploring the physics embodied in this model will involve Bethe-Salpeter equations[8] or eikonal summations. They are currently under study.

We now describe the modifications necessary to compute diagrams in perturbation theory. The propagator of Eq. (47) may be exponentiated through the use of the identity

$$P\frac{1}{k^4} = -\tfrac{1}{2}\int_{-\infty}^{\infty}d\alpha\,\alpha\,\epsilon(\alpha)\exp(i\alpha k^2) \; , \tag{49}$$

where $\epsilon(\alpha) = \pm 1$ for $\alpha \gtrless 0$. Since Feynman parameters are not necessarily positive the following identity will be useful in evaluating loop integrations:

$$\int d^4k\exp[iC(\alpha)k^2] = i\pi^2\epsilon(C)/C^2 \; . \tag{50}$$

As a result the Feynman parameter representation of a diagram will have the form

$$I = \int_{-\infty}^{\infty} \prod_{j=1}^{p} \alpha_j \, d\alpha_j \int_{0}^{\infty} \frac{d\beta_1 \cdots d\beta_q}{C^2} \, \epsilon(\alpha_1 \alpha_2 \cdots \alpha_p C) N e^{iD/C} \ . \tag{51}$$

where α_i corresponds to an internal gluon line and β_i to an internal fermion line, N symbolizes numerator terms, and C and D are determinantal functions.[9] If we had given the gluons dipole Feynman propagators we would have obtained

$$\int_{0}^{\infty} \frac{\prod_{j=1}^{p} \alpha_j \, d\alpha_j \, d\beta_1 d\beta_2 \cdots d\beta_q \, N \exp(iD/C)}{C^2} \ , \tag{52}$$

in comparison to Eq. (51). If we now scale all Feynman parameters with u and use the identity

$$\int_{0}^{\infty} \frac{du}{u} \, \delta\left(1 - \frac{|\alpha_1 + \alpha_2 + \cdots + \alpha_p + \beta_1 + \beta_2 + \cdots + \beta_q|}{u}\right) = 1 \tag{53}$$

(where $|\ |$ indicates absolute value) to introduce an integration over u in Eq. (51), we obtain

$$I = \Gamma(q + 2p - 2l) \int_{-\infty}^{\infty} \prod_{j=1}^{p} \alpha_j \, d\alpha_j \int_{0}^{\infty} \frac{d\beta_1 \cdots d\beta_q \, \epsilon(\alpha_1 \cdots \alpha_p C) \tilde{N}}{C^2 (-iD/C)^{q+2p-2l}} \, \delta\left(1 - \left|\sum_i \alpha_i + \sum_j \beta_j\right|\right) \tag{54}$$

where l = the number of loop integrations in the original diagram and $\tilde{N}$ is obtained from N. An example of this procedure is given in Appendix A. As an alternative to the above method one can introduce light-cone coordinates and evaluate pole terms by contour integrations with Eq. (45) specifying the location of the poles relative to the contour.

The divergences occurring in this model are somewhat novel. As one would expect, with a dipole propagator the ultraviolet divergences are restricted to some lower-order diagrams and are logarithmic in nature (see Fig. 1). The dipole propagator, because it is in principal value, does not induce infrared divergences in loop integrations. However, a third type of divergence, which may be called a light-cone divergence,[10] does occur and is connected with a divergence in a loop integration, $\int d^4k$, associated with the region where $k^2 = k_0^2 - k_3^2 - \vec{k}_\perp^2 \approx 0$ and $k_0, k_3 \to \infty$. In the Feynman parameter representation of Eq. (54) the divergence will appear at the $\pm\infty$ limits of Feynman parameter integrals. The worst divergence is quadratic and associated with one-loop diagrams with one internal gluon line (Fig. 2). These divergences can be managed through the use of Pauli-Villars regularization. Some diagrams containing light-cone divergences are given in Fig. 2. It should be noted that they are necessarily one-loop diagrams. We can demonstrate this by an examination of the overall degree of light-cone divergence

of a graph in the representation of Eq. (54). Let us scale all Feynman parameters in Eq. (54) with Λ and determine the leading behavior as $\Lambda \to \infty$. We find, for l = number of loops > 1,

$$\tilde{N} \sim \Lambda^0 \ , \tag{55}$$

$$C \sim \Lambda^l \ , \tag{56}$$

$$D \sim \Lambda^{l+1} \ , \tag{57}$$

and as a result

$$I \sim \Lambda^{2p+q-1-2l-(q+2p-2l)} = \Lambda^{-1} \ , \tag{58}$$

or convergence of the integral as a whole. However, for one-loop diagrams $(l = 1)$

$$C \sim \Lambda^0 \tag{59}$$

and consequently

$$\tilde{N} \sim \Lambda^q \ , \tag{60}$$

$$I \sim \Lambda^{3-2p} \ . \tag{61}$$

For example, the diagram of Fig. 2(a) has $p = 1$ and diverges quadratically.

Light-cone divergences stem directly from the use of principal-value propagators for the gluon. As such they reflect the nontrivial nature of Wick rotation in this model and they lead to divergences in the vertex renormalization constant, the wave-function renormalization constant, and the quark self-mass which prevent this model from being a superrenormalizable theory of the conventional variety.

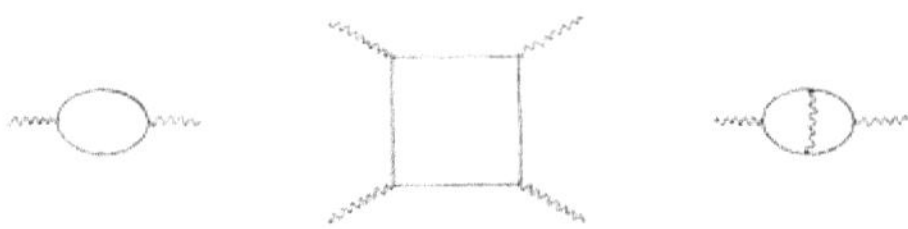

FIG. 1. Some ultraviolet-divergent diagrams.

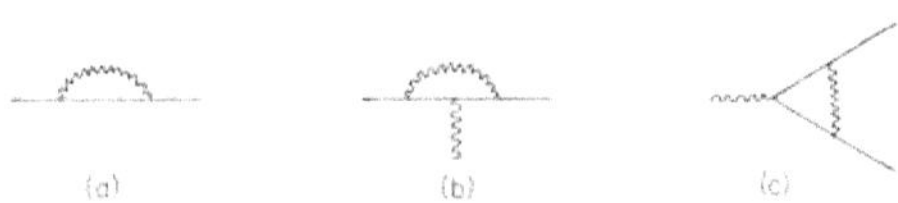

FIG. 2. Some light-cone divergent diagrams.

The Schwinger-Dyson equations for dipole electrodynamics are quite similar to those of quantum electrodynamics, with the exception of the gluon Green's functions, which we now discuss. The proper gluon self-energy, $\Pi_{\mu\nu}(q)$, which couples only to the A_μ^1 channel due to the form of our Lagrangian, satisfies

$$\Pi_{\mu\nu}(q) = i \int \frac{d^4k}{(2\pi)^4} \, \mathrm{Tr}\, \gamma_\mu S_F'(k) \Gamma_\nu(k, k+q) S_F'(k+q) \tag{62}$$

$$= (q_\mu q_\nu - g_{\mu\nu} q^2) \Pi(q^2) \, , \tag{63}$$

where S_F' is the quark propagator and Γ_ν is the proper vertex function (see Fig. 3). The gluon self-energy is related to the complete gluon propagator, $B_{\mu\nu}'$, by

$$iB_{\mu\nu}' = iB_{\mu\nu} + ig^2 iB_{\mu\lambda}\Pi^{\lambda\sigma} iB_{\sigma\nu}' \, , \tag{64}$$

with $B_{\mu\nu}$ given by Eq. (47). Using Eq. (63) we find

$$B_{\mu\nu}'(q) = \frac{\lambda^2 g_{\mu\nu}}{q^4 + g^2\lambda^2 q^2 \Pi(q)} - \frac{q_\mu q_\nu g^2 \lambda^4 \Pi(q)}{q^8 + g^2\lambda^2 q^6 \Pi(q)} \, . \tag{65}$$

The Green's function for the gluon field, A_μ^2, which is zero within the context of the free gluon Lagrangian [cf. Eq. (39)], is nonzero in the interacting theory due to vacuum-polarization effects. It is related to $B_{\mu\nu}'$ by

$$B_{\mu\nu}^2(q) = \frac{(q^4/\lambda^2)B_{\mu\nu}'(q) - g_{\mu\nu}}{\lambda^2} \tag{66}$$

$$= \frac{-g^2\Pi(q)g_{\mu\nu}}{q^2 + g^2\lambda^2\Pi(q)} \tag{67}$$

up to terms proportional to $q_\mu q_\nu$. Equation (65) will play an instrumental role in the demonstration of free-quark screening in the next section.

V. SOME GENERAL PROPERTIES

In this section we will first consider the screening mechanism for quarks and then discuss unitarity, causality, and scaling properties of the lowest-order contributions to the electroproduction structure functions.

In Ref. 1 attention was drawn to the screening of free quarks due to vacuum polarization. Our mechanism is a variation of the Schwinger mechanism[11] but differs from it in an important respect: It is manifest in low order and thus not a matter of conjecture—an important consideration for insoluble field theories. Even in lowest order (Fig. 1), where $\Pi(q)$ is a constant up to a logarithmic term, we find manifest screening.

Let us consider a system containing free quarks in some bounded region. We choose to work in the Coulomb gauge. Because of Eq. (35) the total charge is proportional to

$$Q \propto \int d^3x \, \nabla^2 \Box A_0^1 \tag{68}$$

$$= \int d\vec{S} \cdot \vec{\nabla} \Box A_0^1 \, . \tag{69}$$

However, an examination of Eq. (65) shows that important vacuum-polarization effects occur at large distances. The potential corresponding to a static free quark located at the origin is

$$A_0^1 = \frac{-\lambda^2 g |\vec{x}|}{8\pi} \tag{70}$$

(if we ignore vacuum-polarization effects) and a finite contribution to Q would result if substituted in Eq. (69). At large distances A_0^1 is substantially modified from the expression in Eq. (70). From Eq. (65) we see that the large-distance behavior of A_0^1 is controlled by

$$\frac{1}{g^2 q^2 \Pi(q)} \, , \tag{71}$$

and since $\Pi(q)$ is a constant up to logarithms in lowest order and not proportional to a positive power of q^2 in any finite order of perturbation theory we find A_0^1 to be proportional to at most an inverse power of $|\vec{x}|$ at large distances. Substituting an inverse power of $|\vec{x}|$ for A_0^1 in Eq. (69) and letting the surface of integration go to infinity shows $Q = 0$. Thus isolated free quarks do not exist in this model. Only neutral bound states occur.

We have chosen a propagator for the gluon which allows the S matrix to be unitarity. Our gluons are dipole ghosts, and, having indefinite metric, they would normally destroy the unitarity of the S matrix. But the quantization procedure eliminates their appearance in in or out states and their principal-value propagator precludes states containing gluons from contributing to the absorptive part of any Feynman diagram.[12] This is required if the S matrix is to be unitary. But as a result the S matrix is not analytic. The nonanalyticity is closely associated with advanced noncausal effects. Our procedure forces unitarity to

FIG. 3. Representation of the Schwinger-Dyson equation for the gluon self-energy.

be valid at the expense of noncausality. Tradeoffs of this type have recently been discussed by Coleman.[13] We return to the question of causality later.

We have verified that unitarity is maintained in perturbation theory by an explicit calculation of the lowest-order quark self-energy [Fig. 2(a)], which is

$$\Sigma(q) = \frac{-\lambda^2 g^2}{8\pi^2} \, P \sum_i \frac{c_i \Lambda_i^2 \ln(\Lambda_i^2/q^2)}{q^2(q^2 - \Lambda_i^2)} \left(\frac{\Lambda_i^2}{q^2} \, \rlap{/}q - 2m \right) , \tag{72}$$

where $c_1 = 1$, $\Lambda_1 = m$, the regulator identities $\sum_i c_i = \sum_i c_i \Lambda_i^2 = 0$ hold, and P signifies $\Sigma(q^2 + i\epsilon) = \Sigma(q^2 - i\epsilon)$ as is demonstrated in detail in Appendix A. The fact that the singularities in Eq. (72) occur in principal value implies Σ has no absorptive part. This is to be contrasted with the corresponding quantity in QED which has an absorptive part reflecting the physically allowed decay of an off-shell electron into an electron and a photon. No similar possibility exists in our model.

We now will show that the absorptive part (in the physical region) of a Feynman diagram with one internal gluon line only receives contributions from intermediate states (obtained by appropriately "cutting" internal lines in all possible ways) which do not contain the gluon. The generalization to diagrams with many gluon lines is immediate. First we note that a principal-value propagator may be decomposed:

$$P \, \frac{1}{k^2 - m^2} = \frac{1}{k^2 - m^2 + i\epsilon} + i\pi\delta(k^2 - m^2) . \tag{73}$$

For the sake of simplicity we shall write the integral corresponding to our hypothetical diagram as

$$I = \int d^4k \, \tilde{I} \, P \, \frac{1}{k^4} \tag{74}$$

$$= \frac{\partial}{\partial\mu^2} \int d^4k \, \tilde{I} \, P \, \frac{1}{k^2 - \mu^2} \bigg|_{\mu^2 = 0} , \tag{75}$$

where I and $\tilde{I}$ have indices and momenta appropriate to the diagram in question and the limits we have introduced engender no infrared difficulties due to the choice of a principal-value propagator. Substituting Eq. (73) into Eq. (75) we can decompose I into three Feynman integrals (actually their derivative with respect to mass, etc.),

$$I = \frac{\partial}{\partial\mu^2} \int d^4k \left[\frac{\tilde{I}}{k^2 - \mu^2 + i\epsilon} + i\pi\theta(k_0)\delta(k^2 - \mu^2) + i\pi\theta(-k_0)\delta(k^2 - \mu^2) \right]_{\mu^2 = 0} , \tag{76}$$

in each of which only Feynman propagators are used. The last two terms correspond to opening up the loop containing the gluon. Their Feynman diagrams have in and out gluon lines of momentum k, which is summed over. Let us now restrict ourselves to the physical region[14] of our diagram so that we can take the absorptive part of I in the following way:

$$\mathrm{Abs}(I) = i\pi \frac{\partial}{\partial\mu^2} \int d^4k [-\tilde{I}'\theta(k_0)\delta(k^2 - \mu^2) + \tilde{I}'\theta(k_0)\delta(k^2 - \mu^2) + \tilde{I}'\theta(-k_0)\delta(k^2 - \mu^2)] + R$$

$$= R . \tag{77}$$

The term in square brackets contains all contributions from intermediate states containing a gluon, while R contains the remainder of the absorptive part. The first two terms cancel, while the third term is zero in the physical region. Thus we have shown that the absorptive part receives no contributions from states containing the gluon. Consequently only states containing quarks contribute to unitarity sums for absorptive parts and diagrams containing external gluons do not contribute to the S matrix.

The principal-value gluon propagator has introduced noncausal effects into our model in the sense that the corresponding configuration-space Green's function is half-advanced and half-retarded. However, because we have maintained the commutativity of field operators at spacelike distances the principle of microscopic causality is not violated. Although advanced effects are not observed in everyday life they do not lead to internal inconsistency or paradoxes.[6] On the microscopic level, for example, within the confines of a hadron, advanced effects are not necessarily ruled out on physical grounds. From the earlier discussion of vacuum-polarization effects it is clear that noncausalities must be limited to very short distances. It thus appears that the only significant question involving causality is whether the nonanalyticity of the S matrix for low-order quark-quark scattering will be reflected in the scattering amplitudes of bound states in a manner which is in substantial disagreement with our understanding of S-matrix analyticity for physical particle scattering. The answer to this question is not known.

As an application of the dipole model we shall

study the deep-inelastic electroproduction structure functions in low-order perturbation theory. Previous calculations[15] of the structure functions in pseudoscalar-meson or neutral-vector-meson field-theoretic models (without transverse-momentum cutoffs) contained logarithmic deviations from scaling in apparent conflict with experimental results. The dipole model has strong transverse-momentum damping and as a result one obtains scaling structure functions—in fact, the only asymptotically leading contribution appears to be the diagram of Fig. 4(a). Higher-order diagrams do not scale by powers of q^2, the photon mass squared. For example the diagrams of lowest order in q^2 [Figs. 4(b)–4(d)] contributing to νW_2 are of $O(q^{-4})$. Thus the dipole model establishes a parton picture of the deep-inelastic structure functions since quarks appear to be pointlike particles in the scaling region. The choice of a principal-value propagator for the gluon has the effect of suppressing corrections to the scaling part of νW_2 which would have been of $O(q^{-2})$, such as the contribution of the diagram of Fig. 5. The absorptive part of that diagram is zero due to principal-value gluon propagator. Thus the precocious nature of scaling could be connected with the properties of the principal-value gluon propagator in electroproduction. On the other hand, the principal-value propagator will not play such an important role (at least in low order) in suppressing nonscaling contributions to the absorptive part of the amplitude associated with $e^+ e^- \rightarrow$ hadrons. Thus low-order calculations are suggestive so far as scaling phenomena are concerned.

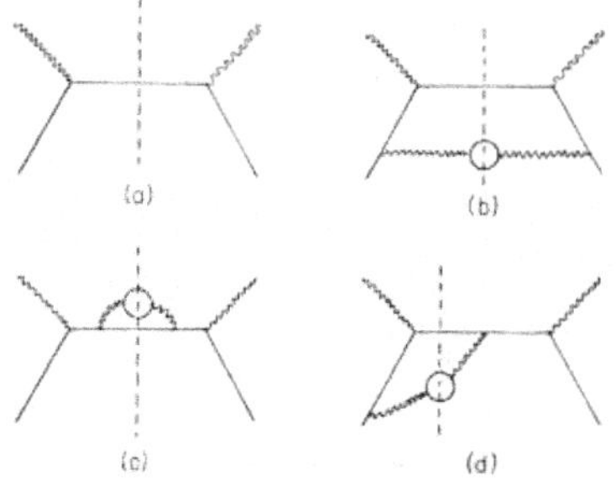

FIG. 4. Lowest-order diagrams contributing to the inelastic electroproduction structure functions. The dashed lines indicate the only contributions to the electroproduction structure functions of the absorptive part of the forward virtual Compton scattering diagram. External "wiggly" lines represent photons, while internal "wiggly" lines represent dipole gluons.

VI. CONCLUSION

The dipole electrodynamics model which we have discussed in the preceding sections is a prototype for a field theory of hadron binding. It has a number of desirable qualitative features such as quark confinement and scaling electroproduction structure functions. The physical content of this model is in the bound-states sector. This sector is currently under study using Bethe-Salpeter and eikonal techniques.

In a more realistic version of this model charge will be replaced by color in such a way that only zero-triality states are physical. The fields A_μ^1 and A_μ^2 will then become Yang-Mills fields. In that case the use of principal-value propagators appears to substantially simplify the model since closed loops of Yang-Mills fields are necessarily zero.[16]

ACKNOWLEDGMENT

I am grateful to the members of the Newman Laboratory of Nuclear Studies for helpful discussions and particularly to Dr. J. Kogut, Dr. D. K. Sinclair, and Dr. L. Susskind.

APPENDIX A

As an example of the modifications in perturbation-theory calculations resulting from principal-value propagators we evaluate the self-energy contribution of Fig. 2(a) and verify Eq. (72):

$$\Sigma(q) = \frac{i g^2 \lambda^2}{(2\pi)^4} \int \frac{d^4 k}{(q+k)^2 - m^2 + i\epsilon} \left(\mathrm{P} \frac{1}{k^4} \right) \gamma_\nu (\slashed{q} + \slashed{k} + m) \gamma^\nu$$

$$(A1)$$

in the Feynman gauge. Feynman parameters can be introduced, and using Eqs. (49) and (50) we obtain

$$\Sigma(q) = \frac{i g^2 \lambda^2}{32\pi^2} \int_{-\infty}^{\infty} d\alpha \, \epsilon(\alpha)\alpha \int_0^{\infty} \frac{d\beta}{C^2} \left(\frac{-2\alpha}{C} \slashed{q} + 4m \right)$$

$$\times \epsilon(C) e^{iD/C}, \quad (A2)$$

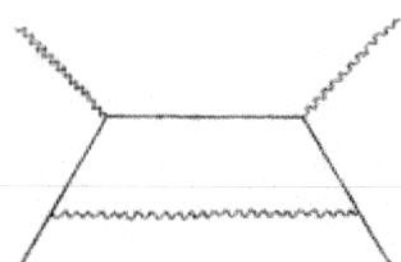

FIG. 5. A forward virtual Compton scattering diagram *not* contributing to the electroproduction structure functions.

where

$$C = \alpha + \beta \tag{A3}$$

and

$$D = \alpha\beta q^2 - \beta C m^2 \ . \tag{A4}$$

Scaling α and β and using Eq. (53) converts $\Sigma(q)$ to the form

$$\Sigma(q) = \frac{-g^2\lambda^2}{32\pi^2} \int_{-\infty}^{\infty} d\alpha \ \epsilon(\alpha) \alpha \int_{-\infty}^{\infty} d\beta \ \frac{\delta(1 - \alpha - \beta)(-2\alpha q\!\!\!/ + 4m)}{\alpha\beta q^2 - \beta m^2} \ . \tag{A6}$$

This may be shown by letting $\alpha \to -\alpha$ and $\beta \to -\beta$ in the term in question. Equation (A6) has divergences at $\alpha = \pm\infty$ after the β integration. These may be handled by introducing Pauli-Villars regulators of mass Λ_i satisfying

$$\sum_i c_i = 0 \ , \tag{A7}$$

$$\sum_i c_i \Lambda_i^2 = 0 \ , \tag{A8}$$

$$c_1 = 1 \ , \tag{A9}$$

$$\Lambda_1 = m \ . \tag{A10}$$

Equation (A6) then becomes

$$\Sigma(q) = \frac{-g^2\lambda^2}{32\pi^2} \sum_i c_i \int_{-\infty}^{\infty} \frac{d\alpha \ \alpha \ \epsilon(\alpha)(-2\alpha q\!\!\!/ + 4m)}{(1 - \alpha)\{\alpha[q^2 + i\epsilon(\alpha)\delta] - \Lambda_i^2\}} \ . \tag{A11}$$

which may be shown to give Eq. (72) by elementary integrations. Apparent singularities in the denominator of the integrand of Eq. (A11) do not lead to difficulties if we take account of the $i\epsilon$'s which we have suppressed. The $i\epsilon(\alpha)\delta$ term (δ is infinitesimal) shows $\Sigma(q)$ to be in principal value $[\Sigma(q^2 + i\delta) = \Sigma(q^2 - i\delta)]$. It originates in the exponentiation of the principal-value propagator using Eq. (49).

APPENDIX B

We will briefly describe another possible model for hadron binding. Like the dipole model it is a member of a class of null-metric gluon theories with multipole Green's functions. The physical motivation for considering this model is a gross similarity to a quark model of Feynman *et al.*[4] which posited a relativistic harmonic oscillator potential, $x^\mu x_\mu$, between quarks and obtained quite successful agreement with experiment. If we neglect factors due to its vectorial nature (and also vacuum polarization effects) the interaction between quarks in our model is $x^\mu x_\mu \theta(x_\mu x^\mu)$ (note that $x_\mu x^\mu = x_0^2 - \vec{x}^2$). The $\theta(x^\mu x_\mu)$ factor, which is necessary for unitarity to be maintained, seems

$$\Sigma(q) = \frac{-g^2\lambda^2}{32\pi^2} \int_{-\infty}^{\infty} d\alpha \ \alpha \epsilon(\alpha) \int_0^{\infty} \frac{d\beta}{D} \left[-2\alpha\epsilon(C)q\!\!\!/ + 4m \right]$$
$$\times \delta(1 - |\alpha + \beta|) \ . \tag{A5}$$

Of the two "points" contributing to the integral, $\alpha + \beta = \pm 1$, the contribution of the term $\alpha + \beta = -1$ can be included in the other term by an extension of the β integration domain:

to imply that only "timelike" excitations are physical, and as a result the analysis of Feynman *et al.* cannot be directly appropriated for our use.

We shall introduce three vector-gluon fields, $A_\mu^i(x)$ ($i = 1, 2, 3$), of which only one will interact directly with the prototype spin-$\frac{1}{2}$ quark field, $\psi(x)$:

$$\mathcal{L} = -\tfrac{1}{2}F_{\mu\nu}^1 F^{3\mu\nu} + \tfrac{1}{4}F_{\mu\nu}^2 F^{2\mu\nu} - \lambda^2 A_\mu^2 A^{3\mu}$$
$$+ \bar\psi(i\nabla\!\!\!\!/ - gA\!\!\!/^1 - m)\psi \ , \tag{B1}$$

with $F_{\mu\nu}^i = \partial_\nu A_\mu^i - \partial_\mu A_\nu^i$, and λ and g coupling constants. Following the canonical procedure we obtain the equations of motion

$$\partial^\mu F_{\mu\nu}^1 + \lambda^2 A_\nu^2 = 0 \ , \tag{B2}$$

$$\partial^\mu F_{\mu\nu}^2 - \lambda^2 A_\nu^3 = 0 \ , \tag{B3}$$

$$\partial^\mu F_{\mu\nu}^3 + gJ_\nu = 0 \ , \tag{B4}$$

$$(i\nabla\!\!\!\!/ - gA\!\!\!/^1 - m)\psi = 0 \ , \tag{B5}$$

with J^ν the quark current. The equations of motion reveal the Lagrangian to be invariant under local gauge transformations of A_μ^1 and ψ while

$$\partial^\mu A_\mu^2 = \partial^\mu A_\mu^3 = 0 \ . \tag{B6}$$

Furthermore, Eqs. (B2) and (B3) imply

$$\Box F_{\mu\nu}^1 - \lambda^2 F_{\mu\nu}^2 = 0 \ , \tag{B7}$$

$$\Box F_{\mu\nu}^2 + \lambda^2 F_{\mu\nu}^3 = 0 \ , \tag{B8}$$

and as a result

$$\Box^2 \partial^\mu F_{\mu\nu}^1 = g\lambda^4 J_\nu \tag{B9}$$

irrespective of the gauge choice for A_μ^1.

Following the conventional procedure we find that the canonical equal-time commutation relations in the radiation gauge ($\vec\nabla \cdot \vec{A}^1 = 0$) are

$$[F_{0i}^a(x), A_j^b(y)]$$
$$= ih^{ab} \int \frac{d^3k}{(2\pi)^3} e^{i\vec{k}\cdot(\vec{x}-\vec{y})} \left(\delta_{ij} - \frac{k_i k_j}{|\vec{k}|^2}\right) \ . \tag{B10}$$

with $h^{13} = h^{31} = -h^{22} = 1$ and all other $h^{ab} = 0$. We can now choose to quantize the theory as described in the text. Again we may use the perturbation-theory rules of QED if the photon propagator is replaced with the gluon propagator in the following manner:

$$iD_{F\mu\nu} \to iG_{\mu\nu} = i\lambda^4 g_{\mu\nu} \, \mathrm{P} \, \frac{1}{k^6} \qquad (B11)$$

in the Feynman gauge, with

$$\mathrm{P} \, \frac{1}{k^6} = \frac{1}{2} \, \frac{1}{(k^2 + i\epsilon)^3} + \frac{1}{2} \, \frac{1}{(k^2 - i\epsilon)^3} \, . \qquad (B12)$$

In coordinate space the gluon propagator is

$$T(A_\mu^1(x)A_\nu^1(y)) = i\lambda^4 g_{\mu\nu} \int d^4k \left(\mathrm{P} \, \frac{1}{k^6} \right) e^{-ik \cdot (x-y)} \qquad (B13)$$

$$= \frac{i}{64\pi} \, \lambda^4 g_{\mu\nu} (x - y)^2 \theta((x - y)^2) \qquad (B14)$$

in the Feynman gauge, which suggests a relationship between our model and that of Feynman, Kislinger, and Ravndal[4] as stated previously.

The discussions of unitarity, causality, and quark confinement given in the text apply to this model with only superficial changes. The light-cone divergences encountered in the dipole model are not so extreme here. For example, the overall degree of light-cone divergence for one-loop diagrams is $3-3p$ [where p is the number of internal gluon lines; cf. Eq. (61)] and thus the lowest-order quark self-energy (Fig. 2) is only logarithmically divergent,

$$\Sigma(q) = \frac{\lambda^4 g^2}{16\pi^2} \, \mathrm{P} \, \frac{1}{q^4} \left\{ \not{q} \left[\ln\left(\frac{\Lambda^2}{m^2}\right) - \frac{2q^2 m^2}{(q^2 - m^2)^2} + \frac{3q^4 m^2 - q^6}{(q^2 - m^2)^3} \ln\left(\frac{q^2}{m^2}\right) \right] + 2m \left[\frac{q^2(q^2 + m^2)}{(q^2 - m^2)^2} - \frac{2q^4 m^2}{(q^2 - m^2)^3} \ln\left(\frac{q^2}{m^2}\right) \right] \right\} , \qquad (B15)$$

where q is the quark four-momentum, m the quark mass, Λ^2 is a regulator mass, and P signifies that all singularities are to be taken in principal value.

Finally, we would like to note again that the deep-inelastic structure functions scale in this model with leading nonscaling corrections of $O(1/q^6)$, where q is the virtual-photon four-momentum. These corrections come from the diagrams of Fig. 4.

---

*Work supported in part by the National Science Foundation.

[1] A. Casher, J. Kogut, and L. Susskind, Phys. Rev. Lett. 31, 792 (1973).

[2] K. Wilson, Phys. Rev. D 10, 2445 (1974).

[3] A. Chodos, R. L. Jaffe, K. Johnson, C. B. Thorn, and V. F. Weisskopf, Phys. Rev. D 9, 3471 (1974).

[4] R. Feynman, M. Kislinger, and F. Ravndal, Phys. Rev. D 3, 2706 (1971).

[5] P. A. M. Dirac, Lectures on Quantum Mechanics (Yeshiva Univ., New York, 1964).

[6] K. Schwarzschild, Göttinger Nachrichten 128, 132 (1903); H. Tetrode, Z. Phys. 10, 317 (1922); A. D. Fokker, ibid. 58, 386 (1929); Physica (The Hague) 9, 33 (1929); 12, 145 (1932); J. Wheeler and R. P. Feynman, Rev. Mod. Phys. 17, 157 (1945); 21, 425 (1949).

[7] J. D. Bjorken and S. D. Drell, Relativistic Quantum Fields (McGraw-Hill, New York, 1965), p. 382.

[8] E. Salpeter and H. Bethe, Phys. Rev. 84, 1232 (1951).

[9] R. J. Eden, P. V. Landshoff, D. I. Olive, and J. C. Polkinghorne, The Analytic S-Matrix (Cambridge Univ. Press, Cambridge, 1966), p. 34.

[10] Suggested by Dr. D. K. Sinclair.

[11] J. Schwinger, Phys. Rev. 128, 2425 (1962).

[12] After this work was completed Dr. K. Subbarao pointed out that Professor E. C. G. Sudarshan has considered the use of action-at-a-distance interactions to remedy unitarity problems in indefinite-metric theories: E. C. G. Sudarshan, Fields and Quanta 2, 175 (1972).

[13] S. Coleman, in Subnuclear Phenomena, edited by A. Zichichi (Academic, New York, 1970), Part A, p. 282.

[14] R. J. Eden, P. V. Landshoff, D. I. Olive, and J. C. Polkinghorne, The Analytic S-Matrix (Ref. 9), p. 112.

[15] R. Jackiw and G. Preparata, Phys. Rev. Lett. 22, 975 (1969); S. Adler and W. Tung, ibid. 22, 978 (1969); S. Blaha, Phys. Rev. D 3, 510 (1971).

[16] This may be proved using the representation of Eq. (54) with $q = 0$ (only principal-value propagators) and $l = 1$. If we let $\alpha_i \to -\alpha_i$ for $i = 1, 2, \ldots, p$ then we can show that $I = -I$ and thus $I = 0$. $\tilde{N}$ is arbitrary except that it can be written as a sum of terms which are homogeneous in the Feynman parameters. This condition can always be satisfied in perturbation theory.

Appendix L. Modified Gravitation Extract From Blaha (2016e)

Earlier we determined the dynamics of the Strong Interaction sector of this unified Gravity-Strong Interaction theory that constitutes a subsector of our Theory of Everything. In this section we will develop the gravitation sector of this unified subsector.[471] We will see that the theory has higher derivative dynamic equations but unlike the Strong Interaction sector, which yields color (quark-gluon) confinement, the gravitation dynamic equations do not have confinement – but do yield a modified form of gravity at intermediate distances of the order of the average galactic radius. The modification of gravity implied by our theory is consistent with the need for Dark Matter described in our Theory of Everything books (Blaha (2015a) and (2016c)).[472] It is also consistent with a MOND theory[473] with the addition of sixth order derivatives in the Gravitation sector lagrangian in a manner consistent with the higher order terms appearing in the Strong Interaction sector of the unified theory. Thus our approach to MOND does not require a major change in Mechanics, quantum theory, or General Relativity (modulo higher order derivatives). The consistency between the need for higher order derivatives in both the Strong Interaction and Gravitation sectors is encouraging.

The development in this appendix is a generalization with higher order derivative terms of the unified theory presented in chapter 6 of Blaha (2016d).

L.1 Gravitation Sector Dynamic Equations

Our gravity sector has two metric fields, $g_{\mu\nu}$ and $g^2_{\mu\nu}$ derived from the unified formalism described earlier. Some of the relevant gravitation equations found in sections 22.2 and 22.3 are:

$$H^{\sigma}{}_{\nu\mu} = \Gamma^{\sigma}{}_{\nu\mu} + \Gamma^{2\sigma}{}_{\nu\mu}$$

---

[471] There are other higher derivative theories – some with two metrics, and some with a metric plus vector plus scalar field formulation. The present work is based on a unification of Strong and Gravity sectors and a totally different formalism. Some significant references are: M. Milgrom, Phys. Rev. **D80**, 123536 (2009); C. Skordis et al, Phys. Rev. Lett. **96**, 011301 (2006); R. H. Sanders, Astrophysical Journal **480**, 492 (1997); and references therein; J. D. Bekenstein, Phys. Rev. **D70**, 083509 (2004) and references therein; J-P. Bruneton, Phys. Rev. **D76**, 124012 (2007) with a higher derivative gravity and metric, vector, scalar fields. See also references within these articles.

[472] I. Ferreras et al, Phys. Rev. Lett., **96**, 011301 (2006) shows the need for both Dark Matter and MOND based on studies of astrophysical data.

[473] Our gravity theory has aspects that are virtually identical to the MOND theories described in A. Balakin et al, Phys. Rev. **D70**, 064027 (2004); H-S Zhao et al, Phys. Rev. **D82**, 103001 (2010); and references therein. However our approach is very different.

$$H^{\beta}{}_{\sigma\nu\mu} = \partial_{\mu}H^{\beta}{}_{\sigma\nu} - \partial_{\nu}H^{\beta}{}_{\sigma\mu} + H^{\gamma}{}_{\nu\sigma}H^{\beta}{}_{\gamma\mu} - H^{\gamma}{}_{\mu\sigma}H^{\beta}{}_{\gamma\nu}$$

$$H_{\sigma\mu} = H^{\beta}{}_{\sigma\beta\mu}$$

$$H = g^{\sigma\mu}H_{\sigma\mu}$$

$$\mathcal{H} = R'^{1} + R'^{2}$$

We use the gravitational sector lagrangian:[474]

$$\mathcal{L}_{G} = \sqrt{g}[MD_{\nu}R'^{1}{}_{G\sigma\mu}D^{\nu}R'^{2}{}_{G}{}^{\sigma\mu} + aR'^{1}{}_{G\sigma\mu}R'^{2}{}_{G}{}^{\sigma\mu} + bg^{\sigma\mu}(R'^{1}{}_{G}{}^{\beta}{}_{\sigma\beta\mu} + R'^{2}{}_{G}{}^{\beta}{}_{\sigma\beta\mu}) + cg^{\sigma\mu}g^{2}{}_{\sigma\mu} + eg^{2\sigma\mu}g^{2}{}_{\sigma\mu}]$$

$$\mathcal{L}_{G} = \sqrt{g}[MD_{\nu}R'^{1}{}_{G\sigma\mu}D^{\nu}R'^{2}{}_{G}{}^{\sigma\mu} + aR'^{1}{}_{G\sigma\mu}R'^{2}{}_{G}{}^{\sigma\mu} + bH + cg^{\sigma\mu}g^{2}{}_{\sigma\mu} + eg^{2\sigma\mu}g^{2}{}_{\sigma\mu}]$$

where

$$D_{\nu}V_{\mu} = (\partial_{\nu} + iF_{\nu})V_{\mu} - H^{\sigma}{}_{\nu\mu}V_{\sigma}$$
$$= [g^{\sigma}{}_{\mu}\partial_{\nu} + ig^{\sigma}{}_{\mu}F_{\nu} - H^{\sigma}{}_{\nu\mu}]V_{\sigma}$$
$$= [g^{\sigma}{}_{\mu}\partial_{\nu} + iD^{\sigma}{}_{\mu\nu}]V_{\sigma}$$

where M, a, b, c, and d are constants from earlier with[475]

$$a = 1/(2f) = 0.906$$

and

$$b = (8\pi G)^{-1}$$

where G is Newton's gravitational constant.

The lagrangian dynamic equations that result are difficult. We consequently will examine the weak gravitation limiting case where we can approximate the two metrics with

$$g_{\mu\nu} \cong \eta_{\mu\nu} + h_{\mu\nu} \qquad (L.1)$$
$$g^{2}{}_{\mu\nu} \cong \eta_{\mu\nu} + h^{2}{}_{\mu\nu} \qquad (L.2)$$

where

$$|h_{\mu\nu}| \ll 1 \qquad (L.3)$$
$$|h^{2}{}_{\mu\nu}| \ll 1$$

Using the relations

$$\partial_{\mu}h^{\mu}{}_{\nu} = \tfrac{1}{2}\partial_{\nu}h^{\mu}{}_{\mu} \qquad (L.4)$$

---

[474] We note the constant a, that appears in chapters 23 and 24, and this appendix, is NOT the Charmonium constant a with one exception..

[475] This value of a is obained in the Charmonium calculation.

$$\partial_\mu h^{2\mu}{}_\nu = \tfrac{1}{2}\partial_\nu h^{2\mu}{}_\mu \tag{L.5}$$

and neglecting higher order terms in $h_{\mu\nu}$ and $h^2{}_{\mu\nu}$ we find

$$R'^1{}_{\mu\nu} = \tfrac{1}{2}[\Box h_{\mu\nu} - \partial_\mu\partial_\lambda h^\lambda{}_\nu - \partial_\nu\partial_\lambda h^\lambda{}_\mu + \partial_\nu\partial_\mu h^\lambda{}_\lambda]$$
$$\cong \tfrac{1}{2}\Box h_{\mu\nu}$$
$$R'^1 \cong \tfrac{1}{2}\Box h_\mu{}^\mu \tag{L.6}$$

$$R'^2{}_{\mu\nu} = \tfrac{1}{2}[\Box h^2{}_{\mu\nu} - \partial_\mu\partial_\lambda h^{2\lambda}{}_\nu - \partial_\nu\partial_\lambda h^{2\lambda}{}_\mu + \partial_\nu\partial_\mu h^{2\lambda}{}_\lambda]$$
$$\cong \tfrac{1}{2}\Box h^2{}_{\mu\nu}$$
$$R'^2 \cong \tfrac{1}{2}\Box h^2{}_\mu{}^\mu \tag{L.7}$$

with

$$D_\nu = \partial_\nu$$

upon neglecting higher order terms.

Substituting we find the *effective quadratic* part of the lagrangian (in $h_{\mu\nu}$ and $h^2{}_{\mu\nu}$) is

$$\mathcal{L}_G = \sqrt{g}[M\partial_\nu R'^1{}_{G\sigma\mu}\partial^\nu R'^2{}_G{}^{\sigma\mu} + a\Box h_{\sigma\mu}\Box h^{2\sigma\mu}/4 + \tfrac{1}{2}b(\partial_\alpha h^{\sigma\mu}\partial^\alpha h_{\sigma\mu} + \partial_\alpha h^{2\sigma\mu}\partial^\alpha h^2{}_{\sigma\mu}) + c(4 + \eta^{\sigma\mu}h^2{}_{\sigma\mu} +$$
$$+ h^{\sigma\mu}\eta_{\sigma\mu} + h^{\sigma\mu}h^2{}_{\sigma\mu}) + e(2\eta^{\sigma\mu}h^2{}_{\sigma\mu} + h^{2\sigma\mu}h^2{}_{\sigma\mu}) + 1/4(h_{\mu\nu} + h^2{}_{\mu\nu})T^{\mu\nu}] \tag{L.8}$$

Using partial integrations, we find the standard technique for determining the equations of motion from a lagrangian for independent variations with respect to $h_{\mu\nu}$ and $h^2{}_{\mu\nu}$ yields

$$-M\Box^3 h^{2\mu\nu}/4 + a\Box^2 h^{2\mu\nu}/4 + \tfrac{1}{2}b\Box h^{\mu\nu} + c(\eta^{\mu\nu} + h^{2\mu\nu}) + 1/4 T^{\mu\nu} = 0 \tag{L.9}$$

$$-M\Box^3 h^{\mu\nu}/4 + a\Box^2 h^{\mu\nu}/4 + \tfrac{1}{2}b\Box h^{2\mu\nu} + c(\eta^{\mu\nu} + h^{\mu\nu}) + 2e(\eta^{\mu\nu} + h^{2\mu\nu}) + 1/4 T^{\mu\nu} = 0 \tag{L.10}$$

The term $c\eta^{\mu\nu}$ can be viewed as part of the total energy-momentum tensor $T'^{\mu\nu}$:

$$T'^{\mu\nu} = T^{\mu\nu} + 2c\eta^{\mu\nu} \tag{L.10a}$$

It plays a role similar to the Cosmological Constant. Subtracting the equations we find

$$M(\Box^3 h^{\mu\nu}/4 - \Box^3 h^{2\mu\nu}/4) + a\Box^2 h^{2\mu\nu}/4 - a\Box^2 h^{\mu\nu}/4 + \tfrac{1}{2}b\Box h^{\mu\nu} - \tfrac{1}{2}b\Box h^{2\mu\nu} + c(h^{2\mu\nu} - h^{\mu\nu}) - 2e(\eta^{\mu\nu} + h^{2\mu\nu}) = 0 \tag{L.11}$$

and thus

$$[-M\Box^3 + a\Box^2 - \tfrac{1}{2}b\Box + (4c - 8e)]h^{2\mu\nu} = 8e\eta^{\mu\nu} - M\Box^3 h^{\mu\nu} + a\Box^2 h^{\mu\nu} - \tfrac{1}{2}b\Box h^{\mu\nu} + 4c h^{\mu\nu} \tag{L.12}$$

Therefore we determine the metric equation for $h^{2\mu\nu}$

$$h^{2\mu\nu} = [-M\Box^3 + a\Box^2 - \tfrac{1}{2}b\Box + (4c - 8e)]^{-1}[8e\eta^{\mu\nu} - M\Box^3 h^{\mu\nu} + a\Box^2 h^{\mu\nu} - \tfrac{1}{2}b\Box h^{\mu\nu} + 4ch^{\mu\nu}] \quad (L.13)$$

Substituting in eq. L.9 we obtain

$$[-M\Box^3/4 + a\Box^2/4 + c][-M\Box^3 + a\Box^2 - \tfrac{1}{2}b\Box + (4c - 8e)]^{-1}[8e\eta^{\mu\nu} - M\Box^3 + a\Box^2 h^{\mu\nu} - \tfrac{1}{2}b\Box h^{\mu\nu} + 4ch^{\mu\nu}] +$$
$$+ \tfrac{1}{2}b\Box h^{\mu\nu} + 1/4 T'^{\mu\nu} = 0$$

$$(L.14)$$

We now redefine the energy-momentum tensor with the result eq. L.14 becomes

$$\{[-M\Box^3/4 + a\Box^2/4 + c][-M\Box^3 + a\Box^2 - \tfrac{1}{2}b\Box + (4c - 8e)]^{-1}[-M\Box^3 + a\Box^2 - \tfrac{1}{2}b\Box + 4c] +$$
$$+ \tfrac{1}{2}b\Box\}h^{\mu\nu} = -1/4 T'^{\mu\nu}$$

$$(L.15)$$

L.2 Gravity Potential

Assuming that we are dealing with non-relativistic matter we can calculate the gravity potential contribution from eq. L.15

$$V_{G1}(\mathbf{x}) = - \int d^3k \, \exp(i\mathbf{k}\cdot\mathbf{x})V_{G1}(\mathbf{k})/(2\pi)^3 \ldots\ldots \qquad (L.16)$$
where

$$V_{G1}(\mathbf{k}) = \{[Mk^6/4 + ak^4/4 + c][Mk^6 + ak^4 + \tfrac{1}{2}bk^2 + (4c - 8e)]^{-1}[Mk^6 + ak^4 + \tfrac{1}{2}bk^2 + 4c] -$$
$$- \tfrac{1}{2}bk^2\}^{-1}$$

$$= [Mk^6 + ak^4 + \tfrac{1}{2}bk^2 + (4c - 8e)]/\{[Mk^6/4 + ak^4/4 - \tfrac{1}{2}bk^2 + c][Mk^6 + ak^4 + \tfrac{1}{2}bk^2 + 4c] +$$
$$+ 4ebk^2\} \qquad (L.17)$$

Similarly eq. L.13 implies the other contribution to the total gravity potential is

$$V_{G2}(\mathbf{x}) = - \int d^3k \, \exp(i\mathbf{k}\cdot\mathbf{x})V_{G2}(\mathbf{k})/(2\pi)^3 \ldots\ldots \qquad (L.18)$$
where

$$V_{G2}(\mathbf{k}) = \{[Mk^6 + ak^4 + \tfrac{1}{2}bk^2 + 4c + 8e]^{-1}[Mk^6 + a\,k^4 + \tfrac{1}{2}bk^2 + 4c - 8e]\}\,V_{G1}(\mathbf{k}) \qquad (L.19)$$

The total gravity potential energy is thus

$$V^{tot}{}_G(\mathbf{x}) = V_{G1}(\mathbf{x}) + V_{G2}(\mathbf{x}) \qquad (L.20)$$

L.3 Solution for the Case of a Massless Graviton

Eqs. L.16-L.17 can generate a massless graviton, which sems a requirement based on cosmological considerations, and also generate a pair of massive gravitons of very low mass. The massive gravitons generate a MOND-like potential that might explicate the anomalous gravitation effects seen at distances of the order of galactic dimensions.

If we set the 'Cosmological Constants' $c = e = 0$, then eq. L.17 becomes

$$V_{G1}(\mathbf{k}) = 4/\{Mk^2[k^4 + ak^2/M - 2b/M]\} \tag{L.21}$$

The denominator of eq. L.21 can be factored into the form

$$V_{G1}(\mathbf{k}) = 4/\{Mk^2[k^2 + \tfrac{1}{2}m_{SI}^2 + \tfrac{1}{2}(m_{SI}^4 + 8bm_{SI}^2/a)^{1/2}][k^2 + \tfrac{1}{2}m_{SI}^2 - \tfrac{1}{2}(m_{SI}^4 + 8bm_{SI}^2/a)^{1/2}]\} \tag{L.22}$$

where

$$m_{SI} = (a/M)^{1/2}$$

Assuming $m_{SI}^2 \ll 8b/a$, or $a^2/8 < 1/8 \ll bM = M(8\pi G)^{-1}$, which is reasonable since a is approximately one, we see

$$V_{G1}(\mathbf{k}) \cong 4/\{Mk^2[k^2 + (2bm_{SI}^2/a)^{1/2}][k^2 - (2bm_{SI}^2/a)^{1/2}]\} \tag{L.23}$$
$$= [2/(ba)]\{-2/k^2 + 1/[k^2 + (2bm_{SI}^2/a)^{1/2}] + 1/[k^2 - (2bm_{SI}^2/a)^{1/2}]\}$$

up to negligible terms.

From eq. L.19 we see $V_{G2}(\mathbf{k}) = V_{G1}(\mathbf{k})$ if $c = e = 0$. Thus the total gravitational potential by eq. L.8 (in momentum space) is

$$V^{tot}_{Gt}(\mathbf{k}) = \tfrac{1}{2}(V_{G1}(\mathbf{k}) + V_{G2}(\mathbf{k})) \tag{L.24}$$
$$= [2/(ba)]\{-2/k^2 + 1/[k^2 + (2bm_{SI}^2/a)^{1/2}] + 1/[k^2 - (2bm_{SI}^2/a)^{1/2}]\}$$
$$\cong 8\pi G\{-2/k^2 + 1/[k^2 + (2bm_{SI}^2/a)^{1/2}] + 1/[k^2 - (2bm_{SI}^2/a)^{1/2}]\}$$

and the coordinate space potential is[476]

$$V^{tot}_{G}(\mathbf{r}) = -G/r + a_1 Ge^{-m_G r}/r + a_2 G\cos(m_G r)/r \tag{L.25}$$

where

$$m_G^2 = (2bm_{SI}^2/a)^{1/2} = (2b/M)^{1/2} \tag{L.25a}$$

$$a_1 = \tfrac{1}{2}$$
$$a_2 = \tfrac{1}{2}$$

and where m_G is an extremely small mass.

---

[476] Since the theory has higher order drivatives that could lead to unitarity problems Feynman propagators must be taken in Principal order. Since potentials are a part of Feynman propagators the potentials real value must be used.

Note that the real part of the expansion of the third term to third order yields

$$V^{tot}_G(\mathbf{r}) \sim -G/r + a_1 Gm_G^3 r^2 - a_2 Gm_G^2 r + constants \qquad (L.26)$$

with the resultant force

$$\mathbf{F} = \nabla V^{tot}_G(\mathbf{r}) \sim Gr/r^3 + 2a_1 Gm_G^3 \mathbf{r} - a_2 Gm_G^2 \mathbf{r}/r + \ldots \qquad (L.27)$$

This result is to be compared to the MOND force of A. Balakin et al, Phys. Rev. **D70**, 064027 (2004):

$$F = -\lambda Gm[M/r^2 - |\Pi_c| r/c^2] \qquad (L.28)$$

and the vector form suggested by H-S Zhao et al, Phys. Rev. **D82**, 103001 (2010):

$$\partial\Phi/\partial\mathbf{r} = Gm\mathbf{r}/r^3 + (Gm)^{\frac{1}{2}}\mathbf{r}/r^2 \qquad (L.29)$$

The resemblance to MOND estimates is clearly striking.

We chose the value of M such that the gravity potential at distances of the order of a 100,000 light years (the radius of the Andromeda galaxy) is increased by the Yukawa like terms due to the small value of m_G.[477] In addition the masses in the Yukawa-like parts of the Strong Interaction potential are extremely small but their effects are shielded so we find the effective Strong Interaction potential.

Thus our unified theory resolves the MOND gravity problem with a correct choice of M and maintains the Strong Interaction potential found for Charmonium. A modification of Newton's law (F = ma) becomes unnecessary.

---

[477] The third term in eq. 6.25 is an oscillating Yukawa term that, because of the smallness of m_G, is slowly varying towards the end of a galaxy and thus could be well within observational error bounds. It appears that the real part of the third term is the contribution to the total gravitational potential using Principal value propagators.

REFERENCES

Akhiezer, N. I., Frink, A. H. (tr), 1962, *The Calculus of Variations* (Blaisdell Publishing, New York, 1962).

Bjorken, J. D., Drell, S. D., 1964, *Relativistic Quantum Mechanics* (McGraw-Hill, New York, 1965).

Bjorken, J. D., Drell, S. D., 1965, *Relativistic Quantum Fields* (McGraw-Hill, New York, 1965).

Blaha, S., 1998, *Cosmos and Consciousness* (Pingree-Hill Publishing, Auburn, NH, 1998).

________, 2002, *A Finite Unified Quantum Field Theory of the Elementary Particle Standard Model and Quantum Gravity Based on New Quantum Dimensions™ & a New Paradigm in the Calculus of Variations* (Pingree-Hill Publishing, Auburn, NH, 2002).

________, 2003, *A Finite Unified Quantum Field Theory of the Elementary Particle Standard Model and Quantum Gravity Based on New Quantum Dimensions™ and a New Paradigm in the Calculus of Variations* (Pingree-Hill Publishing, Auburn, NH, 2003).

________, 2004, *Quantum Big Bang Cosmology: Complex Space-time General Relativity, Quantum Coordinates™Dodecahedral Universe, Inflation, and New Spin 0, ½, 1 & 2 Tachyons & Imagyons* (Pingree-Hill Publishing, Auburn, NH, 2004).

________, 2005a, *Quantum Theory of the Third Kind: A New Type of Divergence-free Quantum Field Theory Supporting a Unified Standard Model of Elementary Particles and Quantum Gravity based on a New Method in the Calculus of Variations* (Pingree-Hill Publishing, Auburn, NH, 2005).

________, 2005b, *The Metatheory of Physics Theories, and the Theory of Everything as a Quantum Computer Language* (Pingree-Hill Publishing, Auburn, NH, 2005).

________, 2005c, *The Equivalence of Elementary Particle Theories and Computer Languages: Quantum Computers, Turing Machines, Standard Model, Superstring Theory, and a Proof that Gödel's Theorem Implies Nature Must Be Quantum* (Pingree-Hill Publishing, Auburn, NH, 2005).

______, 2006a, *The Foundation of the Forces of Nature* (Pingree-Hill Publishing, Auburn, NH, 2006).

______, 2006b, *A Derivation of ElectroWeak Theory based on an Extension of Special Relativity; Black Hole Tachyons; & Tachyons of Any Spin.* (Pingree-Hill Publishing, Auburn, NH, 2006).

______, 2007a, *Physics Beyond the Light Barrier: The Source of Parity Violation, Tachyons, and A Derivation of Standard Model Features* (Pingree-Hill Publishing, Auburn, NH, 2007).

______, 2007b, *The Origin of the Standard Model: The Genesis of Four Quark and Lepton Species, Parity Violation, the ElectroWeak Sector, Color SU(3), Three Visible Generations of Fermions, and One Generation of Dark Matter with Dark Energy* (Pingree-Hill Publishing, Auburn, NH, 2007).

______, 2008a, *A Direct Derivation of the Form of the Standard Model From GL(16) (Pingree-Hill Publishing, Auburn, NH, 2008).*

______, 2008b, *A Complete Derivation of the Form of the Standard Model With a New Method to Generate Particle Masses Second Edition* (Pingree-Hill Publishing, Auburn, NH, 2008)

______, 2009, *The Algebra of Thought & Reality: The Mathematical Basis for Plato's Theory of Ideas, and Reality Extended to Include A Priori Observers and Space-Time Second Edition* (Pingree-Hill Publishing, Auburn, NH, 2009).

______, 2010a, *Operator Metaphysics: A New Metaphysics Based on a New Operator Logic and a New Quantum Operator Logic that Lead to a Mathematical Basis for Plato's Theory of Ideas and Reality* (Pingree-Hill Publishing, Auburn, NH, 2010).

______, 2010b, *The Standard Model's Form Derived from Operator Logic, Superluminal Transformations and GL(16)* (Pingree-Hill Publishing, Auburn, NH, 2010).

______, 2010c, *SuperCivilizations: Civilizations as Superorganisms* (McMann-Fisher Publishing, Auburn, NH, 2010).

______, 2011a, *21st Century Natural Philosophy Of Ultimate Physical Reality* (McMann-Fisher Publishing, Auburn, NH, 2011).

______, 2011b, *All the Universe! Faster Than Light Tachyon Quark Starships & Particle Accelerators with the LHC as a Prototype Starship Drive Scientific Edition* (Pingree-Hill Publishing, Auburn, NH, 2011).

________, 2011c, *From Asynchronous Logic to The Standard Model to Superflight to the Stars* (Blaha Research, Auburn, NH, 2011).

________, 2012a, *From Asynchronous Logic to The Standard Model to Superflight to the Stars volume 2: Superluminal CP and CPT, U(4) Complex General Relativity and The Standard Model, Complex Vierbein General Relativity, Kinetic Theory, Thermodynamics* (Blaha Research, Auburn, NH, 2012).

________, 2012b, *Standard Model Symmetries, And Four And Sixteen Dimension Complex Relativity; The Origin Of Higgs Mass Terms* (Blaha Reasearch, Auburn, NH, 2012).

________, 2013a, *Multi-Stage Space Guns, Micro-Pulse Nuclear Rockets, and Faster-Than-Light Quark-Gluon Ion Drive Starships* (Blaha Research, Auburn, NH, 2013).

________, 2013b, *The Bridge to Dark Matter; A New Sister Universe; Dark Energy; Inflatons; Quantum Big Bang; Superluminal Physics; An Extended Standard Model Based on Geometry* (Blaha Reasearch, Auburn, NH, 2013).

________, 2014a, *Universes and Megaverses: From a New Standard Model to a Physical Megaverse; The Big Bang; Our Sister Universe's Wormhole; Origin of the Cosmological Constant, Spatial Asymmetry of the Universe, and its Web of Galaxies; A Baryonic Field between Universes and Particles; Megaverse Extended Wheeler-DeWitt Equation* (Blaha Reasearch, Auburn, NH, 2014).

________, 2014b, *All the Megaverse! Starships Exploring the Endless Universes of the Cosmos Using the Baryonic Force* (Blaha Research, Auburn, NH, 2014).

________, 2014c, *All the Megaverse! II Between Megaverse Universes: Quantum Entanglement Explained by the Megaverse Coherent Baryonic Radiation Devices – PHASERs Neutron Star Megaverse Slingshot Dynamics Spiritual and UFO Events, and the Megaverse Microscopic Entry into the Megaverse* (Blaha Research, Auburn, NH, 2014).

________, 2015a, *PHYSICS IS LOGIC PAINTED ON THE VOID: Origin of Bare Masses and The Standard Model in Logic, U(4) Origin of the Generations, Normal and Dark Baryonic Forces, Dark Matter, Dark Energy, The Big Bang, Complex General Relativity, A Megaverse of Universe Particles* (Blaha Research, Auburn, NH, 2015).

________, 2015b, *PHYSICS IS LOGIC Part II: The Theory of Everything, The Megaverse Theory of Everything, U(4)⊗U(4) Grand Unified Theory (GUT), Inertial Mass = Gravitational Mass, Unified Extended Standard Model and a New Complex General Relativity with Higgs Particles, Generation Group Higgs Particles* (Blaha Research, Auburn, NH, 2015).

______, 2015c, *The Origin of Higgs ("God") Particles and the Higgs Mechanism: Physics is Logic III, Beyond Higgs – A Revamped Theory With a Local Arrow of Time, The Theory of Everything Enhanced, Why Inertial Frames are Special, Universes of the Mind* (Blaha Research, Auburn, NH, 2015).

______, 2015d, *The Origin of the Eight Coupling Constants of The Theory of Everything: U(8) Grand Unified Theory of Everything (GUTE), S^8 Coupling Constant Symmetry, Space-Time Dependent Coupling Constants, Big Bang Vacuum Coupling Constants, Physics is Logic IV* (Blaha Research, Auburn, NH, 2015).

______, 2016a, *New Types of Dark Matter, Big Bang Equipartition, and A New U(4) Symmetry in the Theory of Everything: Equipartition Principle for Fermions, Matter is 83.33% Dark, Penetrating the Veil of the Big Bang, Explicit QFT Quark Confinement and Charmonium, Physics is Logic V* (Blaha Research, Auburn, NH, 2016).

______, 2016b, *The Periodic Table of the 192 Quarks and Leptons in The Theory of Everything: The U(4) Layer Group, Physics is Logic VI* (Blaha Research, Auburn, NH, 2016).

______, 2016c, *New Boson Quantum Field Theory, Dark Matter Dynamics, Dark Matter Fermion Layer Mixing, Genesis of Higgs Particles, New Layer Higgs Masses, Higgs Coupling Constants, Non-Abelian Higgs Gauge Fields, Physics is Logic VII* (Blaha Research, Auburn, NH, 2016).

______, 2016d, *Unification of the Strong Interactions and Gravitation: Quark Confinement Linked to Modified Short-Distance Gravity; Physics is Logic VIII* (Blaha Research, Auburn, NH, 2016).

______, 2016e, *MoND: Unification of the Strong Interactions and Gravitation II, Quark Confinement Linked to Large-Scale Gravity, Physics is Logic IX* (Blaha Research, Auburn, NH, 2016).

______, 2016f, *CQ Mechanics: A Unification of Quantum & Classical Mechanics, Quantum/Semi-Classical Entanglement, Quantum/Classical Path Integrals, Quantum/Classical Chaos* (Blaha Research, Auburn, NH, 2016).

______, 2016g, *GEMS: Unified Gravity, ElectroMagnetic and Strong Interactions: Manifest Quark Confinement, A Solution for the Proton Spin Puzzle, Modified Gravity on the Galactic Scale* (Pingree Hill Publishing, Auburn, NH, 2016).

______, 2016h, *Unification of the Seven Boson Interactions based on the Riemann-Christoffel Curvature Tensor* (Pingree Hill Publishing, Auburn, NH, 2016).

______, 2017a, *Unification of the Eleven Boson Interactions based on 'Rotations of Interactions'* (Pingree Hill Publishing, Auburn, NH, 2017).

______, 2017b, *The Origin of Fermions and Bosons, and Their Unification* (Pingree Hill Publishing, Auburn, NH, 2017).

______, 2017c, *Megaverse: The Universe of Universes* (Pingree Hill Publishing, Auburn, NH, 2017).

______, 2017d, *SuperSymmetry and the Unified SuperStandard Model* (Pingree Hill Publishing, Auburn, NH, 2017).

Eddington, A. S., 1952, *The Mathematical Theory of Relativity* (Cambridge University Press, Cambridge, U.K., 1952).

Fant, Karl M., 2005, *Logically Determined Design: Clockless System Design With NULL Convention Logic* (John Wiley and Sons, Hoboken, NJ, 2005).

Feinberg, G. and Shapiro, R., 1980, *Life Beyond Earth: The Intelligent Earthlings Guide to Life in the Universe* (William Morrow and Company, New York, 1980).

Gelfand, I. M., Fomin, S. V., Silverman, R. A. (tr), 2000, *Calculus of Variations* (Dover Publications, Mineola, NY, 2000).

Giaquinta, M., Modica, G., Souchek, J., 1998, *Cartesian Coordinates in the Calculus of Variations* Volumes I and II (Springer-Verlag, New York, 1998).

Giaquinta, M., Hildebrandt, S., 1996, *Calculus of Variations* Volumes I and II (Springer-Verlag, New York, 1996).

Gradshteyn, I. S. and Ryzhik, I. M., 1965, *Table of Integrals, Series, and Products* (Academic Press, New York, 1965).

Heitler, W., 1954, *The Quantum Theory of Radiation* (Claendon Press, Oxford, UK, 1954).

Huang, Kerson, 1992, *Quarks, Leptons & Gauge Fields 2nd Edition* (World Scientific Publishing Company, Singapore, 1992).

Jost, J., Li-Jost, X., 1998, *Calculus of Variations* (Cambridge University Press, New York, 1998).

Kaku, Michio, 1993, *Quantum Field Theory*, (Oxford University Press, New York, 1993).

Kirk, G. S. and Raven, J. E., 1962, *The Presocratic Philosophers* (Cambridge University Press, New York, 1962).

Landau, L. D. and Lifshitz, E. M., 1987, *Fluid Mechanics 2nd Edition*, (Pergamon Press, Elmsford, NY, 1987).

Misner, C. W., Thorne, K. S., and Wheeler, J. A., 1973, *Gravitation* (W. H. Freeman, New York, 1973).

Rescher, N., 1967, *The Philosophy of Leibniz* (Prentice-Hall, Englewood Cliffs, NJ, 1967).

Sagan, H., 1993, *Introduction to the Calculus of Variations* (Dover Publications, Mineola, NY, 1993).

Sakurai, J. J., 1964, *Invariance Principles and Elementary Particles* (Princeton University Press, Princeton, NJ, 1964).

Streater, R. F. and Wightman, A. S., 2000, *PCT, Spin, Statistics, and All That* (Princeton University Press, Princeton, NJ 2000).

Weinberg, S., 1972, *Gravitation and Cosmology* (John Wiley and Sons, New York, 1972).

Weinberg, S., 1995, *The Quantum Theory of Fields Volume I* (Cambridge University Press, New York, 1995).

Weinberg, S., 2000, *The Quantum Theory of Fields Volume III Supersymmetry* (Cambridge University Press, New York, 2000).

Weyl, H., 1950, *Space, Time, Matter* (Dover, New York, 1950).

Weyl, H., (Tr. S. Pollard et al), 1987, *The Continuum* (Dover Publications, New York, 1987).

INDEX

About the Author

Stephen Blaha is a well known Physicist and Man of Letters with interests in Science, Society and civilization, the Arts, and Technology. He had an Alfred P. Sloan Foundation scholarship in college. He received his Ph.D. in Physics from Rockefeller University. He has served on the faculties of several major universities. He was also a Member of the Technical Staff at Bell Laboratories, a manager at the Boston Globe Newspaper, a Director at Wang Laboratories, and President of Blaha Software Inc and of Janus Associates Inc. (NH).

Among other achievements he was a co-discoverer of the "r potential" for heavy quark binding developing the first (and still the only demonstrable) non-abelian gauge theory with an "r" potential; first suggested the existence of topological structures in superfluid He-3; first proposed Yang-Mills theories would appear in condensed matter phenomena with non-scalar order parameters; first developed a grammar-based formalism for quantum computers and applied it to elementary particle theories; first developed a new form of quantum field theory without divergences (thus solving a major 60 year old problem that enabled a unified theory of the Standard Model and Quantum Gravity without divergences to be developed); first developed a formulation of complex General Relativity based on analytic continuation from real space-time; first developed a generalized non-homogeneous Robertson-Walker metric that enabled a quantum theory of the Big Bang to be developed without singularities at t = 0; first generalized Cauchy's theorem and Gauss' theorem to complex, curved multi-dimensional spaces; received Honorable Mention in the Gravity Research Foundation Essay Competition in 1978; first developed a physically acceptable theory of faster-than-light particles; first derived a composition of extrema method in the Calculus of Variations; first quantitatively suggested that inflationary periods in the history of the universe were not needed; first proved Gödel's Theorem implies Nature must be quantum; provided a new alternative to the Higgs Mechanism, and Higgs particles, to generate masses; first showed how to resolve logical paradoxes including Gödel's Undecidability Theorem by developing Operator Logic and Quantum Operator Logic; first developed a quantitative harmonic oscillator-like model of the life cycle, and interactions, of civilizations; first showed how equations describing superorganisms also apply to civilizations. A recent book shows his theory applies successfully to the past 14 years of history and to *new* archaeological data on Andean and Mayan civilizations as well as Early Anatolian and Egyptian civilizations.

He first developed an axiomatic derivation of the forms of The Standard Model from geometry – space-time properties – The Extended Standard Model. It has a Dark Matter sector that approximates the ElectroWeak sector with Dark doublets and Dark gauge interactions. It also uses quantum coordinates to remove infinities that crop up in most interacting quantum field theories and additionally to remove the infinities that appear in the Big Bang and generate an inflationary growth of the universe. The Extended Standard Model has an ultra-high energy

GUT (Grand Unified Theory) limit with a U(4)⊗U(4) symmetry; and can be united with gravitation to form a Theory of Everything. (See *Physics is Logic Part II*.)

Blaha has had a major impact on a succession of elementary particle theories: his Ph.D. thesis (1970), and papers, showed that quantum field theory calculations to all orders in ladder approximations could not give scaling deep inelastic electron-nucleon scattering. He later showed the eigenvalue equation for the fine structure constant α in Johnson-Baker-Willey QED had a zero at $\alpha = 1$ not 1/137 by solving the Schwinger-Dyson equations to all orders in an approximation that agreed with exact results to 4th order in α thus ending interest in this theory. In 1979 at Prof. Ken Johnson's (MIT) suggestion he calculated the proton-neutron mass difference in the MIT bag model and found the result had the wrong sign reducing interest in the bag model. These results all appear in Physical Review papers. In the 2000's he repeatedly pointed out the shortcomings of SuperString theory and showed that The Standard Model's form could be derived from space-time geometry by an extension of Lorentz transformations to faster than light transformations. This deeper space-time basis greatly increases the possibility that it is part of THE fundamental theory.Recently, Blaha showed that the Weak interactions differed significantly from the Strong, electromagnetic and gravitation interactions in important respects while these interactions had similar features, and suggested that ElectroWeak theory, which is essentially a glued union of the Weak interactions and Electromagnetism, possibly modulo unknown Higgs particle features, be replaced by a unified theory of the other interactions combined with a stand-alone Weak interaction theory. Blaha also showed that, if Charmonium calculations are taken seriously, the Strong interaction coupling constant is only a factor of five larger than the electromagnetic coupling constant, and thus Strong interaction perturbation theory would make sense and yield physically meaningful results.

In graduate school (1965-71) he wrote substantial papers in elementary particles and group theory: The Inelastic E- P Structure Functions in a Gluon Model. Phys. Lett. B40:501-502,1972; Deep-Inelastic E-P Structure Functions In A Ladder Model With Spin 1/2 Nucleons, Phys.Rev. D3:510-523,1971; Continuum Contributions To The Pion Radius, Phys. Rev. 178:2167-2169,1969; Character Analysis of U(N) and SU(N), J. Math. Phys. <u>10</u>, 2156 (1969); and The Calculation of the Irreducible Characters of the Symmetric Group in Terms of the Compound Characters, (Published as Blaha's Lemma in D. E. Knuth's book: *The Art of Computer Programming Vols. 1 – 4*).

In the early 1980's Blaha was also a pioneer in the development of UNIX for financial, scientific and Internet applications: benchmarked UNIX versions showing that block size was critical for UNIX performance, developing financial modeling software, starting database benchmarking comparison studies, developing Internet-like UNIX networking (1982) and developing a hybrid shell programming technique (1982) that was a precursor to the PERL programming language. He was also the manager of the AT&T ten-year future products development database. His work helped lead to commercial UNIX on computers such as Sun Micros, IBM AIX minis, and Apple computers.

In the 1980's he pioneered the development of PC Desktop Publishing on laser printers. and was nominated for three "Awards for Technical Excellence" in 1987 by PC Magazine for PC software products that he designed and developed.

Recently he has developed a theory of Megaverses – actual universes of which our universe is one – with quantum particle-like properties based on the Wheeler-DeWitt equation of Quantum Gravity. He has developed a theory of a baryonic force, which had been conjectured many years ago, and estimated the strength of the force based on discrepancies in measurements of the gravitational constant G. This force, operative in 15-dimensinal space, can be used to escape from our universe in "uniships" which are the equivalent of the faster-than-light starships proposed in the author's earlier books. Thus travel to other universes, as well as to other stars is possible.

Blaha also considered the complexified Wheeler-DeWitt equation and showed that its limitation to real-valued coordinates and metrics generated a Cosmological Constant in the Einstein equations.

The author has also recently written a series of books on the serious problems of the United States and their solution as well as a book on the decline of Mankind that will follow from current social and genetic trends in Mankind.

In the past twelve years Dr. Blaha has written over 40 books on a wide range of topics. Some recent major works are: *From Asynchronous Logic to The Standard Model to Superflight to the Stars, All the Universe!, SuperCivilizations: Civilizations as Superorganisms, America's Future: an Islamic Surge, ISIS, al Qaeda, World Epidemics, Ukraine, Russia-China Pact, US Leadership Crisis,The Rises and Falls of Man – Destiny – 3000 AD: New Support for a Superorganism MACRO-THEORY of CIVILIZATIONS From CURRENT WORLD TRENDS and NEW Peruvian, Pre-Mayan, Mayan, Anatolian, and Early Egyptian Data, with a Projection to 3000 AD,* and *Mankind in Decline: Genetic Disasters, Human-Animal Hybrids, Overpopulation, Pollution, Global Warming, Food and Water Shortages, Desertification, Poverty, Rising Violence, Genocide, Epidemics, Wars, Leadership Failure.*

He has taught approximately 4,000 students in undergraduate, graduate, and postgraduate corporate education courses primarily in major universities, and large companies and government agencies.

The above paragraphs summarize much of his work over the past fifty years. This work is fully documented. He continues to engage in research and writing at Blaha Research.